D0903188

Geological Society of America
Memoir 185

# *Historical Perspective of Early Twentieth Century Carboniferous Paleobotany in North America*

*In Memory of William Culp Darrah*

Edited by

Paul C. Lyons
U.S. Geological Survey
MS 956, National Center
Reston, Virginia 22092

Elsie Darrah Morey
Morey Paleobotanical Laboratory
1729 Christopher Lane
Norristown, Pennsylvania 19403

and

Robert H. Wagner
Jardín Botánico de Córdoba
Avenida de Linneo, s/n
14004 Córdoba, Spain

**1995**

Published by The Geological Society of America, Inc.
3300 Penrose Place, P.O. Box 9140, Boulder, Colorado 80301

Printed in U.S.A.

GSA Books Science Editor Richard A. Hoppin

**Library of Congress Cataloging-in-Publication Data**

Historical perspective of early twentieth century Carboniferous
paleobotany in North America : in memory of William Culp Darrah /
edited by Paul C. Lyons, Elsie Darrah Morey, and Robert H. Wagner.
p. cm. — (Memoir / Geological Society of America ; 185)
Includes bibliographical references and index.
ISBN 0-8137-1185-1
1. Paleobotany—Carboniferous. 2. Plants, Fossil—North America.
3. Paleobotany—North America—History—20th century.
4. Paleobotanists—Europe—Biography. 5. Paleobotanists—United
States—Biography. I. Darrah, William Culp, 1909– . II. Lyons,
Paul C. III. Morey, Elsie Darrah. IV. Wagner, Robert Herman.
V. Series: Memoir (Geological Society of America) ; 185.
QE919.H58 1995
560'.1727—dc20 95-15891
CIP

10 9 8 7 6 5 4 3 2 1

# Contents

## PORTRAITS OF SELECTED NORTH AMERICAN PALEOBOTANISTS



## THE AMATEUR COLLECTORS

## SPECIAL PAPERS REFLECTING W. C. DARRAH'S INTEREST OR INFLUENCE

# *Foreword*

The William Culp Darrah Memorial Volume is a tribute to William (Bill) Culp Darrah (1909–1989), American paleobotanist, who was a pioneer in studies of late Carboniferous floras of North America. He was the focal point for paleobotanical correlations of the European and North American upper Carboniferous (Pennsylvanian) sequences. As a twentieth-century Renaissance man with a gift for historical studies—as indicated by his classic book, *Powell of the Colorado* (1951, Princeton University Press)—Bill Darrah kept extensive records and correspondence, beginning in the late 1920s, of his contacts with the leading paleobotanists of the early twentieth century. These files—The William Culp Darrah Collection, including a wealth of correspondence and photographs organized by his daughter, Elsie Darrah Morey—form a fundamental source for this volume. They are a part of the rich legacy of North American and European paleobotany. It is our desire in this volume to share this legacy with the paleobotanical community worldwide. Darrah's historical reservoir of paleobotanical information and his influence on the paleobotanical development of two of the editors (E. D. Morey and P. C. Lyons) provided the impetus for this memoir. The third editor, R. H. Wagner, shares with Darrah a common mentor—W. J. Jongmans, one of the foremost twentieth-century authorities on Carboniferous stratigraphic paleobotany. As an undergraduate, Wagner was also influenced by Darrah's book, *Principles of paleobotany* (1939, Chronica Botanica), which was the textbook Wagner used at the University of Leiden, The Netherlands.

An underlying theme in the W. C. Darrah volume is the influence that Europeans had on the development of Carboniferous paleobotany in North America. Leo Lesquereux, the father of North American paleobotany, was of European birth. W. J. Jongmans and Paul Bertrand came to the United States in 1933 to study firsthand the Mississippian and Pennsylvanian floral succession from a European perspective. Walther Gothan's student, Hans Bode, also focused on biostratigraphic problems between the European and North American floral successions. The development of American Carboniferous paleobotany was further enhanced by some North Americans—notably W. A. Bell, W. C. Darrah, and H. N. Andrews, Jr.—going to Europe during the early twentieth century to study and examine floras in European museum collections.

A strictly U.S. tradition was started in the 1890s with the paleobotanical research of David White. A similar tradition was started earlier in Canada by J. W. Dawson and was carried on by W. A. Bell and E. L. Zodrow, Bell's successor. White's influence permeated both European and North American paleobotany. As noted in the various portraits in this volume, White's activity planted the seeds of research for the next generation of North American Carboniferous paleobotanists, notably E. H. Sellards, Reinhardt Thiessen, M. K. Elias, A. C. Noé, C. A. Arnold, C. B. Read, W. C. Darrah, J. M. Schopf, L. R. Wilson, and H. N. Andrews, Jr. They, in turn, passed along the White heritage to the next generation of paleobotanists—A. T. Cross, R. M. Kosanke, W. N. Stewart, and S. H. Mamay, to name a few.

Because Darrah's influence also permeated both North American and European Carboniferous paleobotany, we have included in this volume biographical portraits of some of the leading figures in Carboniferous paleobotany of the early twentieth century who influenced Darrah's thinking. Darrah had personal contact with virtually all of the major European and American paleobotanists of that period. Those having the most significant effect on his development as a paleobotanist were David White (1862–1935), Paul Bertrand (1879–1944), and W. J. Jongmans (1878–1957), who were all pioneers in Carboniferous paleobotany. Other influences—E. H. Sellards (1875–1961), M. K. Elias (1889–1982), A. C. Noé (1873–1939), C. R. Florin (1894–1965), M. C. Stopes (1880–1958), F. D. Reed (1894–1988), Walther Gothan (1879–1954), W. A. Bell (1889–1969), and Reinhardt Thiessen (1867–1938)—are also featured in this memoir. In addition to these central figures, Darrah's North American paleobotanical contemporaries—C. A. Arnold (1901–1977), H. N. Andrews, Jr. (1910———), J. M. Schopf (1911–1978), L. R. Wilson (1906———), and C. B. Read (1907–1979)—are profiled in order to show the birth of new paleobotanical ideas and discoveries in early-twentieth-century Carboniferous paleobotany.

In order to help the reader follow the text, a modern-day interpretation of the correlations of the European and North American upper Carboniferous is shown in Figure 1. The common elements of this correlation scheme and the early-twentieth-century upper Carboniferous schemes that are highlighted in Chapters 25 and 26 are obvious.

The contributions of selected amateur collectors in the pre-1950 period are also covered by portraits in this memoir. These include F. O. Thompson (1883–1953), George Langford, Sr. (1876–1964), and J. E. Jones (1883–1957). Thompson was a major influence, as both benefactor and collector, on Darrah's development into one of the leading paleobotanists of his time. Because of this relationship, Darrah's 1969 monograph, *A critical review of the Upper Pennsylvanian floras of eastern United States with notes on the Mazon Creek flora of Illinois,* on compressional and impressional floras of the eastern United States was dedicated to Thompson.

All these portraits, together with a portrait of William C. Darrah (1909–1989), comprise about half of the W. C. Darrah volume. The remaining part of the volume consists of chapters that focus on topics of central concern to Darrah's paleobotanical studies. These include Pennsylvanian coal-ball studies, late Carboniferous floral zonation schemes, the Stephanian problem in North America, museum collections of Pennsylvanian (late Carboniferous) plant fossils in the United States, and Darrah's experience collecting in the western European coalfields and his examination of classic European museum collections of Carboniferous plant fossils. In addition, a chapter on Carboniferous roof-shale floras, of which the senior author is R. A. Gastaldo, one of Darrah's students at Gettysburg College, is included to demonstrate Darrah's continued influence as an educator and promoter of paleobotany. Finally, the chapter by Cross and Kosanke is a detailed history of the roots of Carboniferous palynology in North America.

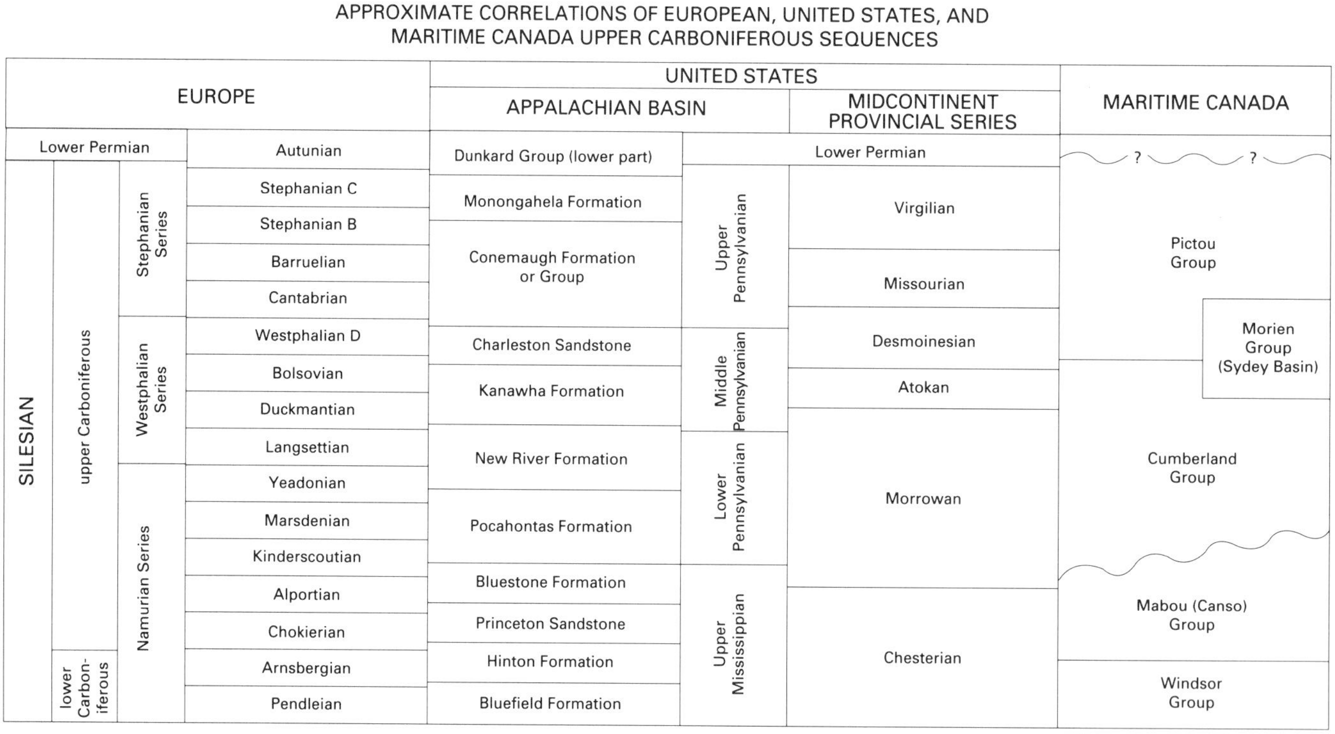

APPROXIMATE CORRELATIONS OF EUROPEAN, UNITED STATES, AND MARITIME CANADA UPPER CARBONIFEROUS SEQUENCES

| EUROPE | | | | UNITED STATES: APPALACHIAN BASIN | | UNITED STATES: MIDCONTINENT PROVINCIAL SERIES | MARITIME CANADA |
|---|---|---|---|---|---|---|---|
| Lower Permian | | | Autunian | Dunkard Group (lower part) | Lower Permian | | ? ? |
| SILESIAN | upper Carboniferous | Stephanian Series | Stephanian C | Monongahela Formation | Upper Pennsylvanian | Virgilian | Pictou Group |
| | | | Stephanian B | | | | |
| | | | Barruelian | Conemaugh Formation or Group | | Missourian | |
| | | | Cantabrian | | | | |
| | | Westphalian Series | Westphalian D | Charleston Sandstone | Middle Pennsylvanian | Desmoinesian | Morien Group (Sydey Basin) |
| | | | Bolsovian | Kanawha Formation | | Atokan | |
| | | | Duckmantian | | | | |
| | | Namurian Series | Langsettian | New River Formation | Lower Pennsylvanian | Morrowan | Cumberland Group |
| | | | Yeadonian | | | | |
| | | | Marsdenian | Pocahontas Formation | | | |
| | | | Kinderscoutian | | | | |
| | | | Alportian | Bluestone Formation | Upper Mississippian | | Mabou (Canso) Group |
| | | | Chokierian | Princeton Sandstone | | Chesterian | |
| | lower Carbon-iferous | | Arnsbergian | Hinton Formation | | | Windsor Group |
| | | | Pendleian | Bluefield Formation | | | |

Figure 1. Approximate correlations of upper Carboniferous sequences in Europe, the United States, and Maritime Canada. Not to scale. *Sources of information:* Bell (1944), Cridland et al. (1963), Englund et al. (1979), Wagner and Winkler Prins (1985, 1991), Zodrow and Cleal (1985), Ryan et al. (1991), Gibling et al. (1992), and Lyons et al. (1993). The Riversdale Group of Bell (1944) has been abandoned by Ryan et al. [(1991)] and is now considered partly equivalent to the Cumberland Group.

Until this multiauthored volume was written, Darrah's influence on the fabric and development of North American paleobotany was not known to most paleobotanists. Darrah wrote two of the earliest textbooks on paleobotany (*Textbook of paleobotany*, 1939, D. Appleton-Century; *Principles of paleobotany*, 1939, Chronica Botanica). Paleobotany was in its infancy in North America when he began his paleobotanical studies in 1926, and Darrah's legacy is shrouded in controversy that is explained in the Darrah portrait in this volume.

Since embarking on this venture, we received much encouragement and support from the paleobotanical community. S. H. Mamay was always ready to give sage advice and help. We thank T. N. Taylor and R. M. Kosanke for graciously volunteering to do portraits of two paleobotanists (F. D. Reed and L. R. Wilson, respectively) who made notable contributions to North American paleobotany and palynology. T. L. Phillips, H. W. Pfefferkorn, and A. H. V. Smith gave helpful suggestions. J. H. F. Kerp provided a review and translation of the Walther Gothan portrait.

Aureal T. Cross (professor emeritus, Michigan State University)—who knew Darrah from his youthful days in the late 1930s—provided much encouragement, advice, and information on both European and North American paleobotanists. He also graciously provided photographs of European and North American paleobotanists, both past and present, including those of C. A. Arnold, W. A. Bell, Hans Bode, Robert Crookall, W. C. Darrah, Walther Gothan, W. J. Jongmans, Richard Kräusel, H. H. Thomas, and L. R. Wilson. Cross also helped with the revision of a few chapters that needed further consideration. For all these contributions, in addition to coauthorship of four chapters, we are greatly indebted. We give him our warm and sincere thanks.

Many other individuals offered or supplied photographs of paleobotanists or plant fossils highlighted in this volume. T. Delevoryas provided photograph of T. M. Harris and John Walton. H. N. Andrews, Jr., graciously provided photographs of J. M. Schopf and C. R. Florin. S. H. Mamay was the source of some of the photographs of C. B. Read. E. C. Beaumont is also acknowledged for a photograph of C. B. Read. S. T. Stebbins of the National Park Service (Grand Canyon) provided photographs of David White doing fieldwork in the 1920s in the Grand Canyon. J. Tuck, also of the National Park Service (Grand Canyon), took photographs of David White's grave site at the Grand Canyon. Linwood Thiessen provided many photographs of his father, Reinhardt Thiessen. Alan Graham and E. L. Zodrow are thanked for each providing a photograph of C. A. Arnold. Harry Stopes-Roe provided a photograph of his mother, Marie Stopes. T. M. Stout provided photographs of M. K. Elias. Ruth Thompson Grimes generously allowed the use of a photograph of F. O. Thompson (her father) and A. A. Stoyanow. A rare photograph of Reinhardt Thiessen with his coworkers—H. J. O'Donnell and George Sprunk—was provided by H. J. O'Donnell, Thiessen's coworker in the 1930s. Olha Oleksyshyn Holowecky provided a photograph of her father, John Oleksyshyn, which is reproduced here with the permission of The Geological Society of America. William H. Gillespie graciously provided a photograph of *Danaeites emersonii* Lesq., which is reproduced here with the permission of the West Virginia Geological and Economic Survey. George Massey supplied a photograph of *Triletis reinschii*. Jack A. Simon provided one of Robert M. Kosanke. Margreet Jongmans and Jacinto Talens kindly provided photographs of W. J. Jongmans.

Nora Tamberg of the U.S. Geological Survey (Reston, Virginia) provided a translation of some German correspondence. Also, Frieda Zierhoffer and Emily H. Zukovich translated some letters written in German.

Access to the U.S. Geological Survey's (Reston, Virginia) photographic archives was provided by L. V. Thompson and W. R. Reckert. Thompson made many superb copies of original photographs supplied by many contributors to this volume. Parts of a group photograph of scientists attending the 1933 International Geological Congress in Washington, D.C., which was copied by Thompson, are shown in the David White and Paul Bertrand profiles in this memoir. R. L. Hadden of the U.S. Geological Survey Library (Reston, Virginia) found the original group photograph and its key in the U.S. Geological Library (Reston Virginia). We are greatly indebted to the late William C. Darrah, whose photographic collection provided many unpublished photographs used in this volume as well as a photograph of *Lescuropteris moorii* Lesq., which is reproduced here with the permission of the West Virginia Geological and Economic Survey.

We are indebted to the following organizations for photographs used in this volume and/or permission to publish or republish photographs of historical importance: The Bergius Foundation, Stockholm, Sweden; Cornell University Press, Ithaca, New York; Elsevier Science Publishers, Amsterdam, The Netherlands; The Field Museum of Natural History, Chicago, Illinois; Museum of the Grand Canyon National Park, Grand Canyon, Arizona; The Museum of Comparative Zoology Library, Harvard University, Cambridge, Massachusetts; Hunt Institute for Botanical Documentation, Carnegie Mellon University, Pittsburgh, Pennsylvania; Illinois State Geological Survey, Champaign, Illinois; Library/Archives, Mount Holyoke College, South Hadley, Massachusetts; National Academy of Sciences, Washington, D.C.; The State Museum of Pennsylvania, Harrisburg, Pennsylvania; Queens University Archives, Queens University, Kingston, Canada; Richard Rush Studio, Inc., Chicago, Illinois; Smithsonian Institution Press, Washington, D.C.; Texas Memorial Museum, The University of Texas at Austin, Austin, Texas; The Archives of the

Paleobotanical Collection, University College of Cape Breton, Sydney, Nova Scotia, Canada; The University of Chicago Library, Chicago, Illinois; West Virginia Geological and Economic Survey, Morgantown, West Virginia; The Geological Society of America, Boulder, Colorado; the Wyoming Historical and Geological Society, Wilkes Barre, Pennsylvania; the U.S. Bureau of Mines, Washington, D.C.; and the American Journal of Science, New Haven, Connecticut.

We are sincerely grateful to the numerous reviewers of manuscripts for this volume, whose timely and helpful reviews kept this volume on track. They include: H. N. Andrews, Jr., H. P. Banks, E. S. Belt, A. R. Cameron, W. G. Chaloner, C. J. Cleal, J. C. Crelling, A. T. Cross, T. Delevoryas, W. A. DiMichele, J. T. Dutro, Jr., C. Eble, W. H. Gillespie, P. A. Hacquebard, M. D. Henderson, F. M. Hueber, J. H. F. Kerp, R. M. Kosanke, J. P. Laveine, R. L. Leary, A D. Lesnikowska, R. J. Litwin, S. H. Mamay, G. Mapes, L. C. Matten, C. M. Nelson, H. W. Pfefferkorn, T. L. Phillips, A. Raymond, Gar W. Rothwell, E. I. Robbins, S. P. Schweinfurth, the late N. F. Sohl, T. N. Taylor, T. K. Treadwell, H. W. J. van Amerom, P. H. von Bitter, C. Wnuk, and E. L. Zodrow. We extend special thanks to C. Wnuk, who did a number of timely reviews.

We extend our sincerest thanks to the contributors to the W. C. Darrah volume, who did a conscientious and thorough job with their contributions, which took a considerable amount of detailed historical research. Without them, this volume would not have been possible.

Connie Gilbert of the U.S. Geological Survey typed many of the manuscripts in this volume. We extend to her our sincere thanks for her skills and infinite patience.

We also owe a debt of gratitude to various curators of special collections and curatorial organizations. J. Fraser Cocks III, Curator of Special Collections in the Knight Library, University of Oregon, provided data on the R. W. Chaney papers. We thank H. W. J. van Amerom for access to correspondence in the W. J. Jongmans Collection at the Geological Bureau in Heerlen, The Netherlands. The Center of the History of Science, Stockholm, Sweden, provided copies of correspondence from W. C. Darrah to C. R. Florin and T. G. Halle. The National Museum of Natural History, Paris, France, provided numerous documents on Paul Bertrand. The U.S. Geological Survey Library, Reston, Virginia, rendered bibliographic data on the publications of C. R. Florin and R. Thiessen. The U.S. Geological Survey, Reston, Virginia, provided access to the personnel files of D. White and R. Thiessen. Roxanne Elias Podhaisky provided information about her father, M. K. Elias. The Geological Survey of Canada; the local History Department, Elgin County, Ontario; Queen's University, National Archives of Canada; and the Royal Ontario Museum rendered data on the life and work of W. A. Bell. Mount Holyoke College Library provided bibliographic data on F. D. Reed. Meredith A. Lane, McGregor Herbarium, University of Kansas, provided information about the specimens of E. H. Sellards.

For errors of fact or interpretation, we take full responsibility. We tried our best to give the most reliable information that was available, but after a half century or more since some of these pioneers in North American paleobotany had died, information on their personal and professional lives was sometimes scarce or nonexistent. We also thank Richard B. Rush and the Pennsylvania State Museum for permission to use the photograph appearing on the cover of this memoir. This photograph is a reconstruction, prepared under Darrah's leadership, of a late Carboniferous swamp.

Figure 2. Left to right: Robert H. Wagner, Elsie D. Morey, and Paul C. Lyons at the former personal library of the late William C. Darrah, Gettysburg, Pennsylvania, October 20, 1994.

Last, we are eternally grateful to the late William Culp Darrah, who inspired this volume. His paleobotanical insight and breadth of understanding were constant sources of inspiration to the first two editors of this volume, who consider this historical undertaking a labor of love.

Paul C. Lyons
Elsie Darrah Morey
Robert H. Wagner
Editors

## REFERENCES CITED

Bell, W. A., 1944, Carboniferous rocks and fossil floras of northern Nova Scotia: Geological Survey of Canada Memoir 238, 276 p., 79 pls.

Cridland, A. A., Morris, J. E., and Baxter, R. W., 1963, The Pennsylvanian plants of Kansas and their stratigraphic significance: Palaeontographica, Abt. B, Band 112, p. 58–92.

Englund, K. J., Arndt, H. H., and Henry, T. W., eds., 1979, Proposed Pennsylvanian System stratotype, Virginia and West Virginia: Falls Church, Virginia, American Geological Institute, Selected Guidebook Series 1, 136 p.

Gibling, M. R., Calder, J. H., and Naylor, R. D., 1992, Carboniferous coal basins of Nova Scotia: Geological Association of Canada and Mineralogical Association of Canada Joint Annual Meeting, Wolfville, Nova Scotia, Field Excursion C-1 Guidebook, 84 p.

Lyons, P. C., Congdon, R. D., and Webster, J. D., 1993, Mid-continent correlation of the Fire Clay tonstein using Cl and Y/Th data: Geological Society of America Abstracts with Programs, v. 25, no. 6, p. A-76.

Ryan, R. J., Boehner, R. C., and Calder, J. H., 1991, Lithostratigraphic revisions of the upper Carboniferous to lower Permian strata in the Cumberland Basin, Nova Scotia and the regional implications for the Maritimes Basin in Atlantic Canada: Bulletin of Canadian Petroleum Geology, v. 39, p. 289–314.

Wagner, R. H., and Winkler Prins, C. F., 1985, The Cantabrian and Barruelian stratotypes—A summary of basin development and biostratigraphic information, *in* Lemos de Sousa, M. J., and Wagner, R. H., eds., Papers on the Carboniferous of the Iberian Peninsula: Anais Faculdade de Ciencias, Universidade do Porto, supplement to v. 64 (for 1983), p. 359–410.

Wagner, R. H., and Winkler Prins, C. F., 1991, Major subdivisions of the Carboniferous System: Congrès international de stratigraphie et de géologie du Carbonifère, 11th, Beijing, 1987: Compte Rendu, Volume 1, p. 213–245.

Zodrow, E. L., and Cleal, C. J., 1985, Phyto- and chronostratigraphical correlations between the late Pennsylvanian Morien Group (Sydney, Nova Scotia) and the Silesian Pennant Measures (south Wales): Canadian Journal of Earth Sciences, v. 22, p. 1465–1473.

Printed in U.S.A.

# *Photographs of Contributors to the William Culp Darrah Memorial Volume*

Figure 1. S. R. Ash, 1990, in his laboratory at Weber State University, Ogden, Utah. Photograph by Jeff Kida.

Figure 2. W. G. Chaloner, 1976, when he was professor of botany, Birkbeck College, University of London.

Figure 3. A. T. Cross, 1984, at Succor Creek, Oregon, with the Miocene Succor Creek type locality in the background. Photograph taken by Kyle Walden.

Figure 5. R. A. Gastaldo, October 1988, at the detrital peat beach, Tandjung Bayor, Mahakam River Delta, Kalimantan, Indonesia.

Figure 4. W. A. DiMichele, May 1989, Jack County, Texas, collecting in the Upper Pennsylvanian–Lower Permian transition. Photograph courtesy of Kenneth Craddock, Denton, Texas.

Figure 6. P. G. Gensel, 1992, during a field trip to the Gaspé Peninsula, Quebec.

Figure 8. M. D. Henderson, 1992, at the Burpee Museum of Natural History, Rockford, Illinois.

Figure 7. W. Hartung and students, 1990, at the Westfälische Wilhelms-Universität, Münster, Germany.

Figure 9. R. M. Kosanke, 1989, in Denver, Colorado, when he received the Geological Society of America's Gilbert H. Cady Award.

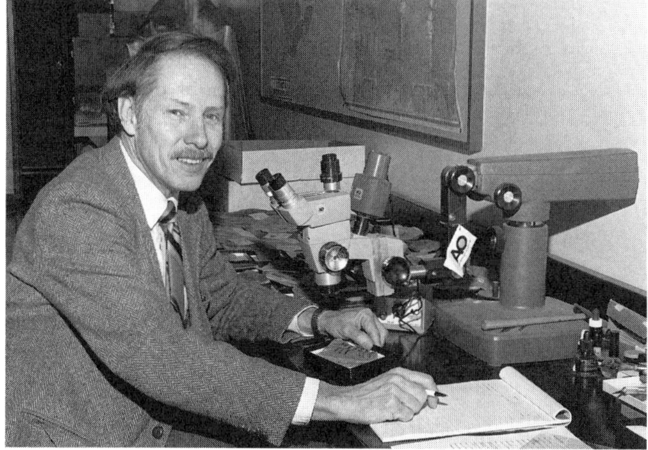

Figure 10. R. L. Leary, April 28, 1992, the Illinois State Museum, Springfield, Illinois. Photograph taken by Marlin Roos and used courtesy of the Illinois State Museum.

Figure 11. J. P. Laveine, September 1989, looking at part of a seed-fern plant, Guangzhou area, South China.

Figure 12, A. D. Lesnikowska, 1992, holding a coal ball from the Pittsburg-Midway Mineral mine (Mineral/Fleming coal bed), near West Mineral, Kansas.

Figure 13. P. C. Lyons, April 1992, at his office in the U.S. Geological Survey (Reston, Virginia), with a large slab of roof shale containing *Neuropteris ovata* from the Upper Kittanning horizon in Grantsville, Maryland.

Figure 15. E. D. Morey, March 1992, taken by P. R. Morey in the Moreys' library at Norristown, Pennsylvania.

Figure 14. S. H. Mamay, 1992, at his office in the Smithsonian Institution, Washington, D.C.

Figure 16. H. W. Pfefferkorn, September 1993, in his laboratory at Hayden Hall, University of Pennsylvania. Photograph taken by M. F. Kleinlerer.

Figure 17. T. L. Phillips, July 1992, in his laboratory at the University of Illinois, Urbana, Illinois.

Figure 18. R. Remy (*left*) and P. Gensel (*right*) in Germany, 1992, during an outing near Münster, Germany.

Figure 19. W. Remy, 1990, in the garden, Münster, Germany.

Figure 20. R. A. Scott, 1991, outside the Morrison Museum, Morrison, Colorado.

Figure 21. T. N. Taylor, 1992, at his office, Ohio State University, Columbus, Ohio.

Figure 23. H. W. J. van Amerom, 1993, in his office at the Geologisches Landesamt Nordrhein-Westfalen, Krefeld, Germany.

Figure 22. M. Teichmüller, 1993, when she received the Gilbert H. Cady Award. Photograph republished with the permission of the Geological Society of America.

Figure 24. R. H. Wagner, 1993, in an opencast coal mine at Puertollano, Province Ciudad Real, Spain.

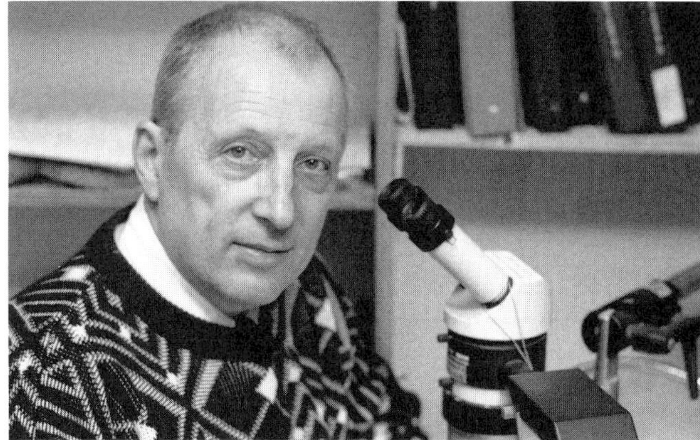

Figure 25. E. L. Zodrow, March 30, 1992, in his laboratory at the University College of Cape Breton, Sydney, Nova Scotia, Canada.

Geological Society of America
Memoir 185
1995

# William Culp Darrah (1909–1989): A portrait

**Elsie Darrah Morey**
*Morey Paleobotanical Laboratory, 1729 Christopher Lane, Norristown, Pennsylvania 19403*
**Paul C. Lyons**
*U.S. Geological Survey, MS 956, National Center, Reston, Virginia 22092*

## ABSTRACT

**William C. Darrah was, above all, a paleobotanist by both inclination and training. His life and career exemplified an enthusiastic search for knowledge, which was fostered by an unflagging interest in the world around him. His creativity crossed over into several diverse fields, including the history of geology and photography. He will be long remembered for his startling paleobotanical discoveries in the 1930s, his 1969 monographic work on late Carboniferous floras, and his role as an educator and promoter of paleobotany.**

## THE EARLY YEARS

William (Bill) Culp Darrah was born in Reading, Pennsylvania, on January 12, 1909. He was the eldest of three sons born to Dorothy Culp (Darrah) and William Henry Darrah. Bill's father was a draftsman with the Reading Crane and Hoist Works and later a mechanical engineer with what is now the Aluminum Company of America, Pittsburgh, Pennsylvania, which later became ALCOA.

For as long as he could remember, Bill Darrah was always very inquisitive about his surroundings. At the age of five, he was fascinated with some postage stamps that his mother gave him one afternoon. His father encouraged the boy's evident interest and took him to the local "old-fashioned" stamp store to meet the stamp dealer; from a mysterious barrel filled with bags of loose stamps of all sizes and shapes from all over the world, Bill made his first purchase. After Bill entered elementary school, he traded stamps with his stamp-collector friends and built his own collection. Trips to the stamp store soon became a weekly event. He bought stamps with money from his allowance, and in a short time the dealer took a shine to this eager young philatelist. Besides helping Bill choose among the "five cents each" and the medium-sized issues for 25 cents, the dealer would steer the boy to excellent higher-denomination U.S. and foreign stamps.

Bill became an avid collector of stamps from all over the world. His interest in stamps continued for the rest of his life, although his active purchasing subsided after 1980. The attraction went beyond hobby acquisition. Systematic treatment and study of his stamp collection, a style later to be applied to his systematic paleobotany, became his source of knowledge about the world, the stamps themselves, and postal history. By 1978, dealers in stamps considered Darrah an authority on the subject and sent him stamps to confirm authenticity.

Young Bill's first memorable encounter with minerals occurred at about the age of eight or nine while he was on an exploratory trip to the "forbidden territory" in the family attic. With delight, he found a handful of specimens—amethyst, malachite, apatite, chalcedony, and sphalerite—and thoroughly enjoyed playing with them. Only later did he casually ask his father where the minerals had come from. "I collected them while studying geology in high school," replied the senior Darrah. "If you want some minerals, why don't you find your own!" he added as encouragement.

Soon Bill was busy collecting his own specimens—rocks, minerals, fossils, arrowheads, and others—around Bethlehem, Pennsylvania, where the family was living at the time. The treasure grew first to fill a dresser drawer, then a wooden box, and then a cabinet. The demonstration of genuine interest in collecting minerals and fossils led Bill's father to introduce his 11-year-old son to his friend, George Currier, a mineralogist from Michigan who was also a collector of stamps, minerals, and fossils. In later life, Bill recalled the first specimen that Currier gave him to handle, "a small meteorite perhaps four

Morey, E. D., and Lyons, P. C., 1995, William Culp Darrah (1909–1989): A portrait, *in* Lyons, P. C., Morey, E. D., and Wagner, R. H., eds., Historical Perspective of Early Twentieth Century Carboniferous Paleobotany in North America (W. C. Darrah volume): Boulder, Colorado, Geological Society of America Memoir 185.

inches in diameter. The weight of the specimen was a complete surprise." During the weekly mineralogy lessons with Currier, Bill learned to identify minerals group by group according to Dana's System of Mineralogy.

Through Bill's membership in the Boy Scouts of America, he found early outlets and directions for the development of his special interests, talents, and abilities. Already with a notably independent temperament and at first determined to resist the urging of his father as well as all invitations, he "somehow in a moment of enthusiasm or weakness, or both" took the pledge in 1921 and joined Troop 12 in Bethlehem, Pennsylvania. Right away, he demonstrated his strong-willed determination to look after himself when it was time for summer camp. His food preferences were well defined—a "picky eater with eccentric tastes," as he put it, a habit that continued throughout his life. But despite his mother's concern for the possibility of his slow starvation, he emerged victorious from the experience by "out-resisting" the camp director, the camp doctor, and the cook.

The Scoutmaster of Troop 12 was Thomas Z. Phillips, a Welshman who as a youth had worked in the anthracite mines in Pennsylvania. Phillips encouraged young Darrah at every opportunity. As Bill passed the required Boy Scouts tests and obtained the rank of Boy Scout First Class, he made it clear that he wished to earn the Mining Merit Badge, the badge requiring the most knowledge of mineralogy and geology. On June 16, 1923, "Mr. William Darrah appeared before me [Dr. Fretz, Lehigh University] this day and satisfied the requirements for the Mining Merit Badge. He seems to have a special aptitude and a real interest in geology." Young Darrah was proud of that badge. He had worked hard, and it was the only Mining Merit Badge awarded in Bethlehem.

In 1923, the Darrah family moved to Pittsburgh, Pennsylvania, where Bill promptly joined Boy Scout Troop 9, headed by Scoutmaster Edward M. Porter. There Bill truly excelled, as indicated by 61 merit badges, 79 medals, the Hornaday Award for significant work in conservation, attainment of the rank of Eagle Scout in 1925 (Fig. 1), and appointment as Assistant Scoutmaster in 1929.

During the period 1924 to 1927, young Bill came face to face with a health problem that would plague him for the rest of his life. Although his health was average as a small child, by age 13 or 14 Bill began to experience a regular succession of skin eruptions due to allergic dermatitis; he also experienced sore throats, chest colds, chronic tonsillitis, and its resultant systemic infection, which all left the boy with periods of extreme weakness and subject to common infections. At the age of 15, a severe case of double-lobar pneumonia nearly killed him and forced him to remain indoors and miss school for an entire year. The damage done to his lungs was irreversible, and the reduction in lung capacity would only get worse as he grew older.

Not one to suffer from boredom—"not with all that's so interesting in the world"—Bill indulged himself during the periods of confinement by reading extensively in his father's library and, later, the public library. The books included classics, biographies, Indian history, mineralogy, plant-life studies, and on and on. Indeed, years later Darrah was heard to reflect and marvel at his good fortune. Almost every subject he encountered interested him. By the age of 14 or so, he had discovered that he had an unusual ability to read with "lightning" speed with high levels of comprehension and that he also had a "photographic" memory. Both assets were to serve him well throughout his life. He recognized and acknowledged that "because of those talents [he] had always scattered [his] talents to accommodate many interests—without hesitation or guilt."[1] Early in life he had learned to accept and appreciate what life had to offer.

Figure 1. W. C. Darrah, Eagle Scout, 1925. The William Culp Darrah Collection.

A newly created scholarship from the Harmon Foundation loan fund was granted to an Eagle Scout in recognition of outstanding character and academic potential. Bill was awarded the first Harmon scholarship in 1927 upon graduation from Pittsburgh's Peabody High School. The award letter from William E. Harmon was a constant reminder of how the young man should act and excel throughout his career.

## THE UNIVERSITY OF PITTSBURGH YEARS

Darrah entered the University of Pittsburgh in 1927, and by the time he graduated in 1931 with a B.S. degree in geology and a minor in botany, the groundwork had been laid for a lifetime of intellectual pursuits and achievements. For a philosophy course, he attempted to trace the genealogy of the philosophic views of Darwinism to certain American scientists. It was this experience that led Darrah to John Wesley Powell (1834–1902), whose thoughts and activities fascinated the young scholar: "The revelation of science is this: Every generation of life is a step in progress to a higher and fuller life; science has discovered hope."[2]

Darrah's geology professor, Henry Leighton, took him and his classmates on a field trip detour to Pithole, the ruins of an oil boom town that flourished in the 1860s around Oil City, Pennsylvania. There Bill's interest in the origins of the petroleum industry was stirred. In several courses under Rowell Johnson, professor of oil and gas engineering, Darrah was impressed by Johnson's emphasis on taking detailed notes during field-trip studies. This discipline suited Darrah's own proclivities and was to serve him well as he expanded his investigations into paleobotany, which he began in the summer of 1927. Indeed, integration of data became an important aspect of his approach to studies for the rest of his life.

It is also of special note that during these years at the University of Pittsburgh, young Darrah met a fellow student, Helen Hilsman, who was majoring in zoology (with hopes of going into medicine). On December 28, 1934, she became his wife and very capable assistant.

Continuing with Otto E. Jennings, professor of botany, and a taxonomist, as his adviser, Darrah began graduate studies in 1931 as a University of Pittsburgh Fellow at the Carnegie Museum (Fig. 2). There he studied Baron de Bayet's collection of Belgium fossils, which includes specimens from Mazon Creek, Illinois, purchased about 1890.

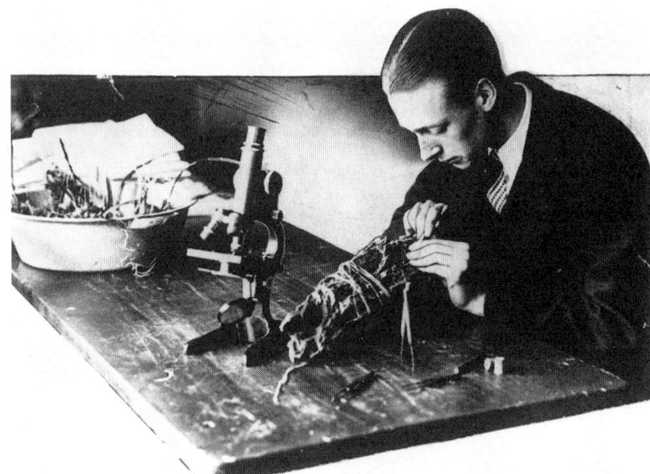

Figure 2. W. C. Darrah, Carnegie Museum, University of Pittsburgh, Pittsburgh, Pennsylvania, 1931. The William Culp Darrah Collection.

Jenning's and Darrah's work was reported by the eminent paleobotanist David White of the U.S. Geological Survey in the Report of the Committee on Paleobotany (National Research Council, 1931, p. 9) in this fashion: "The Carboniferous flora in the vicinity of Pittsburgh, Pennsylvania, has been under investigation by Dr. O. E. Jennings (Carnegie Museum, Pittsburgh) and one of his students, William Darrah, who has collected hundreds of specimens of this flora in that locality. As yet their studies have not progressed far enough for publication." In the Report of the Committee on Paleobotany (National Research Council, 1932, p. 13) more details are given: "The study of the floras of the Conemaugh and Monongahela Formations of western Pennsylvania has been continued by William C. Darrah (Carnegie Museum, Pittsburgh, Pa.). Some results will appear in a thesis on the fossil flora of the Mahoning shales. . . . Darrah reports the assemblage of 30,000 specimens of fossil plants at the Carnegie Museum half of which are Paleozoic, including the Baron de Bayet collection of 10,000 specimens from Continental Europe. . . ."

Before the Report of the Committee on Paleobotany appeared in 1933 (National Research Council, 1933), David White wrote Darrah, "Will you note carefully what I say about your discoveries of Permian species in the Conemaugh and Monongahela. Have I quoted you correctly? Further will it be quite agreeable to you for me to mention your discoveries in a paper that I preparing for presentation at the 16th International Geological Congress which meets in Washington [1933] the 22nd of July?"[3] Then again, "In view of the fact that the question may stimulate debate as to the occurrence of *Callipteris* or *Walchia* in the Monongahela or Conemaugh, as you have reported, I am writing to ask if you will immediately send me specimens of one or both genera with mention of their horizon (I do not care about the locality); or if you are coming to attend the International Geological Congress, will you bring some specimens with you for exhibition?"[4]

When the 1933 Report of the Committee on Paleobotany appeared, it continued this account (p. 10). "Further studies of the floras of the Conemaugh, Monongahela, and Allegheny of western Pennsylvania and northern West Virginia are in progress by W. C. Darrah. He reports *Callipteris* and *Walchia* in the Conemaugh Formation. 'Recent paleobotanical investigations near Pittsburgh' and 'The Studies of Paleozoic Pteridosperms' by him appeared in Proc. Penn. Acad. Sci., vol. 6, 1932. Darrah announced the occurrence of a Radstockian flora in western Pennsylvania." (See W. C. Darrah's complete bibliography, this chapter.)

During 1932 and 1933, Darrah made extensive collections as he attempted to correlate the floras of the Allegheny

Formation with related floras of Missouri and Illinois. During this period, E. W. Berry wrote Darrah, "Your report of *Walchia* and *Callipteris* from the Conemaugh is of much interest to me. I am personally convinced that the base of the Permian should be at the base of the Pittsburgh Reds or Round Knob. I would appreciate your opinion on this from the standpoint of your discoveries in the Conemaugh."[5].

Interpretation of data is always subject to discussion. As Darrah attempted to define his own views, he wrote to David White in early 1934, and White quickly responded: "As to the correlations of the Allegheny, the Radstockian flora as published by Kidston, and zone 'D' of the Westphalian, I am not sure that I will be able fully to agree that the entire Allegheny is as old as the Radstock. The greater part of the Allegheny may be as old as Westphalian 'D' but can hardly be bracketed opposite Westphalian 'C'."[6] And later White responded to another of Darrah's letters,

> After having seen considerable material from the Radstock Series of Great Britain, I cannot possibly view it as young as the Conemaugh. Nor am I able to conclude that beds found in any part of Europe to contain the distinctive Permian types collected at Cassville, West Va., would be assigned by any living paleobotanist to a stage lower than Permian. *Rarity* of the Permian types due perhaps to local climatic conditions, should not outweigh the significance of their determined *presence*.
>
> I agree with you that Middle Kittanning is probably younger than Mazon Creek, and consistent with such a view, I can hardly regard Mazon Creek as Westphalian 'C.' . . .
>
> You are presenting fairly good evidence in support of the argument that in America the Permian may begin near the middle of the Conemaugh. Such a location of the Permian boundary agrees with the general pre-Permian orogeny of America and the great diastrophic unconformity preceding the Permian in the southwest. As I have pointed out in my International Congress paper, the post-Pennsylvanian orogeny with which I provisionally align the exclusion of the seas from the northern part of the Appalachian Trough in middle Conemaugh time, appears to mark the great climatic changes introductory to the establishment of cosmopolitan Permian types in eastern America.[7]

Darrah attended the 16th International Geological Congress (July 22–29, 1933), and White introduced him to both W. J. Jongmans and P. Bertrand. White had previously requested[8] that the young scientist serve as field guide for the two great European paleobotanists during their extensive and separate private field trips through the coalfields of the American Carboniferous. Darrah hosted both Bertrand and Jongmans on separate trips to the Anthracite Basin and in Pittsburgh, where the distinguished visitors viewed collections at Carnegie as well as specimens in Darrah's home.

Darrah's dealings with the European "greats" went very well, and he was quite satisfied with the entire experience. Following their departures and using the knowledge that he had gained from his association with the two men, Darrah was able to write a comparison between the floras of the Appalachian Trough and those in Europe. In 1933, Darrah and Bertrand coauthored a paper entitled "Observations sur les flores houillères de Pennsylvanie (régions de Wilkes-Barre et de Pittsburgh)."

The following was reported in the Report of the Committee on Paleobotany of 1934 (National Research Council, 1934, p. 11): "Jointly with W. J. Jongmans, of Heerlen, Holland, Darrah has prepared an annotated list, with the occurrence, range, and synonymy of the Pteridosperms recorded from North America" (see Darrah, 1937g).

As has often been the case within the scientific disciplines, paleobotany has not been without its own controversies and misunderstandings, both personal and professional. Young Darrah would soon find himself "right in the thick of it."

## THE HARVARD YEARS

### *Curating, researching, lecturing, and writing*

Encouraged by his activities and accomplishments while at the Carnegie Museum and following up on a suggestion from O. E. Jennings, Darrah anticipated the challenges that would be put before him by Oakes Ames (director of the Harvard Botanical Museum) as Darrah moved to Harvard's Botanical Museum (Fig. 3) in mid-1934. And, right away, he faced the packed collection of Leo Lesquereux's fossils as well as 36 boxes of plant fossils from Mazon Creek that had been collected by Frederick O. Thompson (see Chapter 22, this volume) and donated to the Botanical Museum in 1933. Darrah set to work sorting and identifying the fossil specimens. As he noted for himself,

> I have been at the Botanical Museum about two weeks, and I feel sure that Professor Ames has a contract up his sleeve. There is no place I'd rather be than here at Harvard—collections, libraries, equipment, and *MEN*. It will probably only be a one year contract, but that's all we need!
>
> I've finished Lesquereux's unpacked collections, so next to Thompson's, and then the Boston Society's. Ames asked me today to do a biography of Leo. He wants it in two weeks. I might be able to do it by August 10 for Helen's birthday. God what a chance for a youngster!

The immediate task of providing a biography of Leo Lesquereux, the pioneering American Carboniferous paleobotanist, was Darrah's introduction to the academic writing mill—"write fast, write often—and he would find he had much to learn about the process.

It was not long until Darrah was to make himself known to others in paleobotany at home and abroad. Soon after his arrival at Harvard, he received a letter from White dated July 16, 1934. After acknowledging the benefits of the young man's new situation, White went on to ask for some information.

> Miss Stadnichinko, who is working on the regional carbonization of the Lower Kittanning coal beds between the Allegheny River and the Allegheny Front, has asked me about the characteristics of the flora

Figure 3. W. C. Darrah, the Botanical Museum, 1934. Harvard University, Cambridge, Massachusetts. The William Culp Darrah Collection.

accompanying the Lower Kittanning coal. Two kinds of information she needs: First, she would like to know the character and composition of the flora of the Lower Kittanning coal in Central Pennsylvania; second, she would like to have local floras at a number of points for presentation as evidence in the matter of the origin of the coal.

[In Europe they] still cling to the idea that the different stages of carbonization of the Lower Kittanning or any other single coal bed, as sampled by the different localities, are due to the differences in the vegetation—to some sort of variation in the ingredient plant material. . . .

My own observation is that so far as concerns the chemistry of the plants or the plant products, from which the Lower Kittanning coal was formed, there is no notable difference between the mines along the Allegheny River or along the Allegheny Front or in the intervening area. Rather generally, the fossil plant collecting is not particularly good in the roof of the Lower Kittanning coal due to conditions of burial of the coal bed which could not have been at any time far above tide level, though it was of course above tide level while the coal was being formed.

The Lower Kittanning flora as I recall it is in its botanical composition very similar to other Allegheny floras. It has Lycopods, Calamaria, Pteridosperms, and true ferns, most prominent among which are *Pecopteris*.

I see no chance for notable chemical differences in my original plant debris between Kittanning and Windber [towns in Pennsylvania].

If you have evidence to the contrary, I should be as greatly interested as Miss Stadnichinko to know it.[9]

Following his response to White's letter, Darrah received an acknowledgment from Miss Stadnichenko, in which she wrote, "It was gratifying to read your conclusion that the Lower Kittanning flora 'is everywhere homogeneous and presents the same facies. From its northernmost to southernmost occurrence it is the same; likewise, from the Ohio boundary to the Allegheny Front,' as this confirms my microscopic studies and serves as an additional proof that the rank of the coal cannot be explained by the differences in the original flora contributing to the coal bed."[10]

Encouraged by Charles Schubert,[11] professor emeritus of Yale University's Peabody Museum, Oakes Ames (at his own expense, as Darrah later learned) sent Darrah (Figs. 4 and 5) to represent Harvard at the 1935 international congresses in The Netherlands: The Sixth International Botanical Congress and the Second International Carboniferous Congress. While in Europe, Darrah, accompanied by his wife, traveled and continued his studies with Jongmans in The Netherlands and Belgium and with Bertrand in France, where he made quite an impression. (See Chapters 2 and 7, this volume.) Some years later, after the death of P. Bertrand in 1944, T. S. Mahabale (1954, p. 449) recorded the appearance of the young American: "Acknowledging his [P. Bertrand's] efforts and repute, the Faculty of Sciences of Lille founded for him an Academical chair in Paleobotany in 1927, a chair he made famous, and round father gathered brilliant students: G. Livet, G. Dubois, G. Mathieu, P. Deleau, W. C. Darrah, P. Corsin. . . .

At Harvard, Darrah immediately encountered the more mundane but persistent realities of finding moneys to finance the research being done at the Botanical Museum. As he again noted for himself, "I knew on Monday that *Leo Lesquereux* [Darrah, 1934a] would be out today—right on the nose. #1 copy to Helen . . . Ames said he is well pleased with Leo. He has assured me (on July 23 already) that the means will be found to keep me here."

So, early in his career, Darrah was learning to appreciate the generous financial support lavished by interested parties on the ever-growing paleobotanical collections. Oakes Ames lamented to F. (Fred) O. Thompson (Thompson Trust, Des Moines, Iowa), "I fear we can never look to the University for much aid in the way of allowances because there are so many departments in crying need of assistance . . .[f]or anything substantial, we must, as in the past, look to those who feel a direct interest in what we are attempting to do."[12]

This was by no means the first time the two had exchanged letters concerning young Darrah. Indeed, the same F. O. Thompson in many ways had made it possible for Bill to be at Harvard. As Ames had described in an earlier letter,

Figure 4. Left to right: W. C. Darrah, Helen H. Darrah, and Dr. D. H. Linder at the Sixth International Botanical Congress, Amsterdam, The Netherlands, September 2–7, 1935 (see Chapter 2, this volume). D. H. Linder was professor of mycology, Harvard University, Cambridge, Massachusetts. The William Culp Darrah Collection.

Figure 5. W. C. Darrah at Faulquemont Mine, France, September 27, 1935. Photograph taken by H. H. Darrah. The William Culp Darrah Collection.

"You have made two boys very happy! When Darrah came to me this afternoon with your letter and check, the smile on his face and satisfaction oozing from his soul would have been to you, had you been with me, a magnificent dividend."[13]

In the Thompson-Darrah association, the largesse did not stop with the monetary assistance. Over the years, the two men became close friends. They collected fossils together during the years 1937 to 1940. The younger man learned to rely on Fred's knowledge and persistence in collecting fossil specimens, all the while employing his own methods of keeping meticulous records. This tradition of having such a "benefactor" is now gone forever.

In 1936, Darrah applied for and received funding from the Marsh Fund of the National Academy of Sciences to help support the expenses of the ever-growing paleobotanical collections of the Botanical Museum. With moneys from the Thompson Fund, he was able to employ John Herron to make regular collections in the Illinois basin. In the late 1930s, those employees well known to us today included H. P. Banks, W. Benninghoff, C. E. Cross, E. S. Barghoorn, C. L. Fenton, and R. E. Schultes.

John Burbank, who later became a medical doctor, was one of many who cut, polished, and made the peels of the coal-ball materials. Peels were made in enormous numbers, and A. Traverse, while a Ph.D. student at Harvard in the late 1940s, was employed for an entire summer sorting and labeling them. Darrah's peel slides were stored at Harvard in ice cream cartons. Serge Mamay remembers that Elso Barghoorn, in referring to them jokingly, said something like, "How many quarts of *Cordaianthus* slides would you like to see?" which, according to Mamay, "is a great testimony to Bill's energetic approach."

As the collection grew, Darrah began an extensive program of fossil exchange with other institutions. Exchanges of the Mazon Creek nodules were made with the British Museum and other foreign museums. Specimens in exchange with the British Museum came from Kerguelen Island (near the Antarctic continent), India, New Zealand, Dalmatia (Yugoslavia), Greenland, and Grimell Land (on Ellesmere Island, Canada). Other specimens came from France, Germany, The Netherlands, Sweden, Manchuria, Union of South Africa, Brazil, Japan, Cuba, and Argentina. The exchanges from Cuba and Argentina were so small that Darrah hesitated to call them "exchanges."

As Darrah's expertise in paleobotany developed and expanded, he found himself under constant pressure to earn the "all-powerful" Ph.D. degree. Having started under his adviser Professor Jennings at the University of Pittsburgh, Darrah was still working on his dissertation. Certain reservations led him to seek the advice of his mentor and friend, Oakes Ames, who, in fact, had no such degree. In one response, Ames confessed, "I wish I could give you sound advice pertaining to the embellishment of a Ph.D. degree. I fear I cannot do so because I am prejudiced. Many good men and true have ascended the latter of distinction without a degree, having had hawk's eyes for opportunity." Then, he continued,

Is not the Ph.D. degree, in nine cases out of ten, the *open sesame* won by mediocrity from reluctant committees of scholars? I know of one distinguished Ph.D. candidate who . . . was a borderline case in the minds of his disappointed examiners. . . . It is palpably true that this man went up the ladder . . . because of a type of ability that is not a requisite of the Ph.D. degree.

I say in all frankness: If you have a creditable thesis, and feel that you can bluff your way through an oral examination, get the Ph.D. degree by all means. If you possess character, inborn ability and a love of your task, you can live it down. It will not curtail your prospects of success. . . .[14]

Feeling very confident in his own abilities and self-worth and the likelihood that with hard work he could accomplish whatever he set out to do, Darrah was reassured and encouraged by the words of this man whom he respected so highly.[15] There was so much more "out there" of knowledge and experience than the Ph.D. could embrace, and he decided not to pursue it. "My strongest ambition was to do what I wanted to do . . . to excel in the accomplishment, research, writing . . . history—whatever," he explained many years later. As Ames had cautioned him, without the "embellishment" Darrah was to experience both the rewards and the pitfalls.

Essentially confined to the work of the Botanical Museum for his first two years as a research assistant, Darrah's activities and responsibilities were then expanded to include a faculty position as instructor in elementary biology as well as private tutor. Later, as research curator of paleobotany (appointed, 1936), his attention was directed toward paleobotany for graduate students. But in the beginning, Darrah apparently had some doubts about the task of lecturing to students, and again he consulted his mentor, Oakes Ames:

My Dear Darrah: By all means meet Redfield [A. C. Redfield, professor of biology, Harvard University] more than half way. A dose of elementary botany will do you more good than almost any other form of admirable torture you can subject yourself to in this rapidly advancing world. I feel sure that the discipline centered in preparation will do you more good than almost anything else while drawing attention to the fact that you are struggling onward for a place on the university map.

At present on this map you are floundering about in the region marked "unexplored," "unknown." The opportunity now presented to you will, I am sure, remove you from the "unknown," and may place you in the "explored" regions with definite benefits to your future. I do not know how large the class is in Biology D. When I had it, it was too big, and the effort at lecturing was an ordeal. Somehow I feel that you will succeed. . . . If you succeed, you will soon learn what I mean by the term "explored" used above. Boy! Can you make it without a Ph.D? Can you join that distinguished group of men of which I am so proud to be one, men who owe nothing to a rotting thesis and who won through without an explanation? . . .[16]

It was already evident to Darrah that his teaching and lecturing style had to be his own, not a copy of the "greats" he had encountered at Pittsburgh and Harvard. In all seriousness, he approached this new challenge to impart what he had learned and assist those who needed his help.

On the side, Ames confided to Thompson his own reservations: "I am worrying now about the impression our young friend Darrah will make in his elementary course. I have faith in him. But there is always present the danger that youth may slip. And naturally I want my faith to be substantiated."[17] So, once more, Ames offered some guiding thoughts to the young academician now honing his powers of communication. Darrah already had in mind the advisory, "If you want to be understood, do not be afraid to make your points in a language that others not yet so especially trained can comprehend." He had heard that before. But Ames went a little further on the subject. "I am delighted to learn from your letter of January 17, 1938, of the burst of intelligence characteristic of your closing lectures in Biology D. But why should there be 'mutterings' here and there about how unscientific and elementary are the subjects covered by human interests . . . when they attempt to tell the world that science is wholly a matter of sex-hormones, and culture but an empty dream. . . ."[18]

Again, Ames comments on an examination paper for Biology D: "I think you have framed your questions very cleverly, because I am sure I should have failed the course in royal purple had I had to write the answers. . . . The eighth question would have thrown me completely on my beam end. By the way, is there a verb to 'to diagram'? Furthermore, the term 'list' as it is used in your paper rather disturbs my peace of mind. . . . Look it up in the Oxford Dictionary some day. Then cross it out [of] your vocabulary."[19]

Ames's confidence was well founded. Darrah's classes in elementary biology and, later, paleobotany (for graduate students, primarily) were a great success, with a delivery both learned and comprehensible. In later years, Darrah's lectures and writings were notable for their clarity of explanations and breadth of thought-provoking illustrations.

Harvard was a valuable repository of knowledge in more than just paleobotany, and Darrah indulged his curiosity in other areas related to the history of the development of science and technology. His enormous capacity for work and his developing experience were reflected in his attitude toward performance on the job, at Harvard, as well as later in life. He gave complete loyalty and cooperation to the task (employer) at hand and worked as long as it took to do his absolute best. From another point of view, of course, he was very conscious of the fact that he was, in the end, working for himself, his self-satisfaction and pride in performance. So, when Elmer D. Merrill, director of the Arnold Arboretum, came to Darrah's office at the request of the young man's supervisors and "suggested" that all other academic pursuits be dropped, that instead he become *the* world authority on paleobotany—as he had been hired to do—Darrah was dismayed.

This onus weighed heavily on the young scientist as he tried to discern what course his career should take. For clearly he was establishing himself in this academic community.

It was vitally important to share one's discoveries, observations, conclusions, and new knowledge with the community of one's peers. "Publish or perish" was more than an easy slogan; this was serious business and essential for success for any young scientist. This imperative was emphasized at Harvard, as Darrah had been learning from contacts with his superiors and colleagues. And Ames was right there to guide the young man facing the constant pressure and generating the papers.

It seems to me that you have adopted a trip-hammer style [of writing] that leaves the quasi-layman with his tongue hanging out and his mental machinery pretty well over-heated. To the paleobotanist your presentation may be as clear as nujol, but to the general reader there are places where the literary fog seems a bit thick. . . . After you have given close attention to a subject and are clear in your own mind regarding the critical phases of it, the natural thing to do is to set it up in crisp sentences, but to my mind coming to the task of interpreting what you mean, perplexity intensifies as explanatory sentences fail to appear. . . .[20]

Again, the absurdity of excesses that tend to arise was noted by Ames. "This idea of publishing a paper whenever a cell turns over may be good science in some provincial districts. . . ," he continued to caution, "but I am interested in turning the brain cells over by making thoughtful people think."[21]

Still working on his communication skills a few years later, Darrah was chided by R. C. Moore, who, besides discussing the facts in his manuscript on Kansas coal-ball floras, pointed out to the "prolific worker in paleobotany" that in his eagerness to get something into publication, he did so "at the expense of quality. In this paper you have been very careless. . . . There are many ambiguities and—forbid it from Harvard—examples of impossible English. . . ."[22]

As research assistant and, later, curator at the Botanical Museum, Darrah had made a major contribution to assure the continuity of the museum, having sorted and classified as well as greatly enlarged its collection. The Harvard years were full of challenges and opportunities and had been marked by Darrah's energetic and systematic personal quest for knowledge and experience. He came away from his close association with Oakes Ames with an understanding of the influence and a readiness to act upon a "properly fulfilled moral obligation [to pass along one's knowledge] on the course of academic and biological history."[23]

By 1942, Darrah had distinguished himself as one of the leading authorities in the field of paleobotany. To his credit were over 50 scientific papers, two paleobotany texts (1939a, b), and a textbook on botany (1942a) (see W. C. Darrah's complete bibliography, this chapter). His stratigraphic approach led to correlations between eastern U.S. and western European Carboniferous floras (1934–1937) and between Asian and European late Paleozoic floras (1937). Close and systematic studies of collected specimens revealed the presence of Permian-type plants (*Taeniopteris, Walchia*) in the Appalachians (1935a, 1936a). Darrah wrote to P. C. Lyons,

Will wonders never cease! By sheer coincidence I received in the mails today a *Walchia* from the lower Conemaugh of Ohio. . . . While the aspect is somewhat more compact than your specimen with shorter leaves, I believe they represent a single species. . . . "How sweet it is!" When I found *Walchia* in the Upper Conemaugh (Clarksburg) it was ridiculed. Florin, when he visited the U.S.A. in 1938, confirmed the specimens (then only two) and the Dunkard (single) example. I never dreamed it would turn up beneath the Ames limestones. I think the Conemaugh will be a great problem.[24]

The existence of Cambrian spores (1937) and the female gametophytes of fossil *Selaginella* (1938) were not at first accepted because of the startling nature of these discoveries. However, Darrah's seminal work with coal balls in Iowa during the 1939–1941 period, which was enhanced by his own improved technique of preparing peels (1936d; Fig. 6), gained him widespread recognition as a paleobotanist.

### *Textbook of Paleobotany (1939a)*

Ralph W. Wetmore, professor of biology at Harvard, perhaps observing the abilities of this lecturer in Biology D, encouraged Darrah to write a textbook on paleobotany and publish it through D. Appleton-Century Company. As Wetmore happened to be the associate editor of botany there, he would be able to guide the young man and critique his efforts.

Confirming what he already suspected, Darrah received an encouraging letter from C. A. Arnold in response to Darrah's having sent him a preview of his proposed textbook. At that time, it seemed, "there [was] no convenient text available to American students. . . . Your project has entirely changed that." Moreover, Arnold continued, speaking of his own efforts (Arnold, 1947), "I do not intend to deal with matters of correlation [plant groups and geology] . . . that is entirely outside my field."[25] And this was just what Darrah had in mind—an introduction to paleobotany with an approach largely

Figure 6. W. C. Darrah demonstrating his peel technique. Botanical Museum, Harvard University, Cambridge, Massachusetts, 1937. The William Culp Darrah Collection.

botanical but with some attention to historical geology. The stage was set; opportunity and criticism were the protagonists.

The agreement between W. C. Darrah and D. Appleton-Century Company, to publish *Textbook of paleobotany* was signed on December 31, 1936. As the book neared completion, Wetmore insisted on reviewing and editing the manuscript, checking references, quotations, and permissions granted by the numerous authors. In turn, when the manuscript with its criticisms was returned to him, Darrah then checked each point and stamped each comment to indicate some action had been taken.

Obtaining the permission of those authors whose works would provide original data or interpretations was critical to the process of producing a textbook that attempted to integrate diverse sources of information. With this understanding, and following numerous exchanges with the would-be textbook author, Appleton-Century notified Bill that the book would go to press November 2, 1938.[26] Then F. S. Pease wrote personally to acknowledge Darrah's "sending us the material on permissions for quotations in your book. We shall file it with the contract for the book. I think it will be unnecessary for you to sent us all the personal letters which you have received granting permissions, since you say you will have them at our disposal if by any chance they should be needed."[27].

As he had in letters to other authors, Darrah requested permission from R. W. Chaney to quote from his work. He wrote,

> I have recently submitted to the D. Appleton-Century Company a manuscript for a textbook in paleobotany. Now it has been decided to limit the book in several ways to keep the printed matter well within 500 pages.
>
> In the chapter on Cenozoic Floras I would like to quote from your contribution to *Essays in Geobotany in Honor of William Albert Setchell* ([Chaney] 1936) page 59 as follows: "The plants listed indicate a climate . . . in the Andes of Venezuela near Merida." I intend to omit the remark in parentheses because I believe this would not in any way modify the generalization.
>
> I would like also to use the enumerations (p. 58 & 68) in part, i.e. some of the genera would be omitted.
>
> Will you please grant permission to use the quoted portion as above indicated.
>
> With best personal wishes, . . .
>
> P.S.: Naturally credit and citation will be given.[28]

Chaney's letter was quick to follow: "I see no reason why the phrase in parentheses on page 69 referred to in your letter should not be omitted in your quotation. It is apparent that not all of the names listed on pages 58 and 68 need be included. I should recommend however the inclusion of the genera of the following families (abbreviated): . . . The genera in these families include the more commonly represented plants in the Western Tertiary."[29]

At the outset, Darrah had requested D. Appleton-Century to send complimentary copies of his book (1939a) to many authors, including E. W. Berry, R. W. Chaney, A. J. Eames, R. Florin, T. G. Halle, M. Hirmer, and W. J. Jongmans (Chapter 5, this volume), to mention only a few. Berry sent his thanks, with the addition, "I think you have accomplished a difficult undertaking in a rather satisfactory manner."[30]

Soon Darrah received a letter from Jongmans:

> Mr. Appleton sent to me a copy of your textbook. I rapidly looked through it. The first impression was good, only some figures are not well reproduced but this is not your fault. . . . I shall write to you my opinion. There is one thing in your Principles [Darrah, 1939b]. You mention a "Carboniferous" flora from Persia with fourteen European species. Which flora do you mean? Another thing, the Tete flora is a mistake. Zeiller's assistant had given to him the wrong box which contained plants from the Southern France. Tete has *Glossopteris*. Zeiller himself corrected this mistake.[31]

Chaney had already written to Darrah:

> I wish to thank you for a copy of your textbook. In so far as I can see, this is an excellent piece of work. I shall write you more fully later when I have had more opportunity to study it. I have been rather surprised at your methods of using material from my paper, The Succession and Distribution of Cenozoic Floras around the Pacific Basin . . . [Chaney, 1936]. I believe that is a mistake to include material as you have done without quotation marks. I know it is not a common practice among textbook writers. Ideas, of course, will be used by the textbook writer, but in my opinion they should be synthesized and expressed in his own words, unless quotation marks are used. I have not read critically other pages of yours in comparison with the original writings referred to. I am not, therefore, in a position to know whether the book as a whole has followed the practice above [noted].[32]

Thus alerted, Chaney began to compare the quotes in Darrah's textbook with the original sources to see if Darrah had properly credited materials taken from them. In short order, Chaney wrote to Joseph A. Brandt, manager of Princeton University Press: "At the suggestion of our editor, Samuel T. Farquhar, I am writing to call your attention to the use by William C. Darrah of material from Knowlton's 'Plants of the Past' ([1927] Princeton University Press) without quotation marks. Enclosed are excerpts from Darrah's 'Textbook of Paleobotany' with which I suggest that you make comparison of the text in Chapter 13 of Knowlton. . . . I am writing to my close associate, Dr. Erling Dorf, of your institution, suggesting that he discuss this matter with you."[33]

Chaney had written to Dorf of Princeton University:

> I am writing the manager of your press, Joseph A. Brandt, regarding the copying of portions of Knowlton's book, "Plants of the Past," in Darrah's "Textbook of Paleobotany." It is my hope pressure will be brought to bear on the publishers, Appleton-Century, so that this book will be withdrawn. . . . What I am wondering is whether he has copied his whole book from other people. If you have any ideas on this subject I wish you would let me know at once. I wish further, if you are interested, that you would speak to Mr. Brandt regarding Darrah's use without quotation marks of Knowlton's Pliocene Chapter. . . .[34]

Chaney continued to pursue his suspicions. To A. J. Eames, he wrote, "I shall be interested to know whether you have given W. C. Darrah permission to use without quotation marks material in your book, 'Morphology of Vascular Plants,' McGraw-Hill Book Company, first edition, 1936."[35]

Gathering some momentum, Chaney next contacted Winifred Goldring:

I have been having some fun with Darrah's textbook. Apparently the whole of it is copied. He has copied from Eames, Textbook of Botany and from Berry [1930], The Past Climate of the North Polar Region, Smithsonian Miscellaneous Collections. . . . If Berry were more agreeable I would write to him direct, but he is such a crab that I shall not do so. If you feel in the mood, I wish you would write to him not mentioning me and bring to his attention the following. . . . Reading this paragraph, it sounded a little Berryesque and I looked it up. At this point you are probably thinking that RWC is getting WG to pull his chestnuts out of the fire for him. I guess that is correct, so do not feel the least necessity for getting in touch with Berry. He is such a hard hitter that I should like to have him know about this, and have him join the attack, but I believe the book will have to be withdrawn even without his support. The chances are Darrah has included many other excerpts from Berry without quotes. I shall be glad when I have something more important to do than throw mud at this poor innocent child. On the other hand I have a sufficiently mean disposition to greatly enjoy the tone of your letter in which you tell him what you think of him.[36]

Dorf wasted no time in responding to Chaney's alarm:

Your interesting letter regarding our strange colleague arrived just as I was leaving for a short jaunt to Washington. There was a letter of the same vein from Winifred Goldring in the same mail. Since returning on Saturday night I have spent all my spare time doing detective work in the library and in my office. Here's my report to date. . . . I have been informed by the University Press here that Mr. Halsley, their Editorial Manager, has written a strong letter to Appleton-Century asking that the book be withdrawn. I shall continue to do what I can to find other books from which material has been copied until you tell me to let up. I suspect there are plenty of other passages which are not at all original.[37]

Soon, other recipients began to reply to Chaney's letters. Winifred Goldring wrote him, "I have just had another letter from Prof. Berry. He apparently has done more meditating. He writes, 'I agree 100% that Darrah ought to be spanked, but do not think I should be the instrument. I don't have any appetite for that sort of thing at the present time.' "[38]

A. J. Eames wrote Chaney, "I am sorry to be so slow in replying to your letter. The reason for the delay is partly that I wish to get better acquainted with Darrah's book. . . . No request was made to me (and no suggestion reached me) that my textbook would be quoted or used in any way in Darrah's book. I have been more irritated by the way in which my material has been used than by its inclusion in the book. . . .[39]

The complaints about the book reached the desk of D. H. Ferrin at Appleton-Century, and soon Darrah would learn of his book's torturous path among its critics. In May, Ferrin wrote to him, "I shall bring with me correspondence from two or three people who are claiming and, I believe without foundation, that you quoted them in your Paleobotany without their permission."[40]

Right after his meeting with Ferrin, Darrah wrote to Chaney:

This morning Mr. D. H. Ferrin, Vice President of the Appleton-Century Company, came to Boston to see me concerning what seems to be unethical use—if not plagiarism—of publications by you, in collaboration with Clements, and by the late Dr. Knowlton. May I first say that my transgression has been made not in maliciousness but through carelessness—however in good faith. When I engaged a typist to type my lecture notes, or rather narrative, these were set up in essay form. I did not realize that they followed so closely or in the same phraseology the papers cited above. When I wrote to you for permission to quote and you willingly granted my request, I really assumed that I was quoting only that part requested in the letter. Under this unfortunate circumstance I beg to make rectification.[41]

The letter-writing campaign continued. "One thing I wish you would do for me," Chaney wrote to Dorf, "is to find out the present attitude of Princeton Press toward Darrah's plagiarism. My friends at Harvard are rallying to the defense of the boy wonder, largely, I think, to protect the fair name of Harvard. My own attitude is that Darrah should be definitely squelched or punished, or both. Our University Press does not now seem inclined to press the matter of withdrawal of the book. This, I think, is the only fair solution. Can you determine what Princeton is doing and let me know at once."[42]

From Princeton, Dorf replied, "The Darrah situation is deplorable. From my recent conversation with Mr. Halsey of the Princeton Press, I doubt if the press here will do anything legally, though they are still adamant. Darrah wrote to Halsey and explained in detail how each alleged instance of plagiarism was actually taken from some other publication and not from Knowlton. Halsey asked me to check this, and I found in every instance Darrah was just plain lying, and could not have gotten Knowlton's exact words anywhere except from Knowlton."[43]

The publishers entered the fray directly as letters continued to arrive at Appleton-Century. D. H. Ferrin wrote to S. T. Farquhar, manager of University of California Press, "May I this morning simply acknowledge your letter of August 3, in which you suggest the wording of an errata slip to be inserted in each copy of Darrah's *Paleobotany?* Personally, I believe that the insertion of such a slip should meet the approval of Mr. Darrah, but naturally I must have his formal acceptance of the wording."[44] Appleton-Century contacted Darrah about the errata slip, at which time Darrah advised Chaney, "I have just received a letter from D. Appleton-Century Company enclosing a copy of Mr. Farquhar's letter suggesting an insertion giving full source citation for my use of your published work in *Essays in Geobotany* [Chaney, 1936]. The wording and content which he suggests are just and under the circumstances, agreeable to me."[45]

Some months after Chaney's initial alarm, Eames wrote to him, "This summer I have for the first time noted what I had not expected—that Darrah quoted extensively and directly from my book. Before noting this I have thought that there were merely minor quotations and incorrect statements attributed to me."[46]

Chaney was quick to respond. "I am interested to learn that Darrah has plagiarized so fully from your [Eames] book. I believe it wholly proper for you to urge your publishers to take the matter up with Appleton-Century. It is perhaps too late to achieve the result which I felt proper: that is, the withdrawal of the Darrah book. Although Darrah has copied extensively from a book by Knowlton published by Princeton Press, this organization has made only ineffective protests."[47]

Some time after the event, Wetmore acknowledged to Chaney,

> I have thought a great deal about our discussion of the Darrah textbook and still continue to be sick at heart over the whole thing. I have also had two long conferences with Darrah. There is no question but what he has provided himself with a long road to recovery from the effect which this is bound to have upon him and his work. In a sense, I suppose I am in a measure responsible for the situation: certainly I assume the responsibility. There is no doubt in my mind but that the requests which you, Eames and the Princeton Press have made are absolutely necessary. Despite Darrah's seeming innocence in the matter—and he still maintains that the story which he presented to me in June is the case—it certainly implies a type of neglect that could not be tolerated.

He then went on to explain himself.

> I suppose I might have recognized certain sentences or phrases had I been looking for them while reading the manuscript. Instead, reading for content, with no expectation of finding anything of this sort, did not result in my picking up what I should have.
>
> It may be a naive attitude to hold, but ordinarily I do not look for trouble until there is some suspicion that I may find it. Certainly, in this instance, I was not expecting it. Several times I did check on Darrah to see that he had, theoretically at least, obtained permission for all his quotations. In that I thought I was checking the picture in its entirety, but obviously I was not.[48]

Continuing to detect undercurrents involving a charge of plagiarism in his textbook, Darrah wrote to Ferrin at Appleton-Century:

> Recently I have heard statements to the effect that the complaints against my book have still been coming in, notably from Professor A. J. Eames. I assure you that I have this merely upon rumor and perhaps I have been misinformed. I am wondering if you would be good enough to have copies made of any letters in your files referring to Chaney, Berry and Eames, which have not already been sent to me, and also . . . of the various permissions to quote in my textbook which I sent to you at the time when the book went to press. . . .
>
> I had been under the impression that all of these complaints had been answered to the best of my ability and that the situation had subsided, but apparently this has not been true. I have heard that there has been some resentment over the fact that I published the smaller book in Holland [Darrah, 1939b] at the same time as the illustrated textbook. . . .
>
> I shall be out of town on field work for several weeks and I would appreciate it if you would send copies of this correspondence, which I believe is essential information for me, to Professor Oakes Ames . . . at the Botanical Museum.
>
> I regret that is necessary to make this request to you but I feel if there has been any other material which is not known to me I should have access to it.[49]

Two months later, D. H. Ferrin wrote Darrah:

> Upon the receipt of your letter of September 25th, I forwarded to Mr. Thompson of McGraw-Hill a copy of your suggested acknowledgement to Dr. Eames. Mr. Thompson has just telephoned me that so far as he is concerned it is acceptable, but of course, he will have to pass it on to Dr. Eames for his approval.
>
> As an interesting sidelight, Mr. Thompson intimated that only very recently an author of one of McGraw-Hill's publications drew rather heavily on one of Dr. Eames' books without satisfactory acknowledgment. Your case coming on top of that one probably made Dr. Eames more irritable than he otherwise would have been.[50]

Thompson's "sidelight intimation" to Ferrin seemed to be confirmed by the letter Thompson received at McGraw-Hill from Eames, in which Darrah's belated suggested acknowledgment was deemed not acceptable.

> In the first place I do not agree with you that this is wholly a matter of giving credit. The book will surely compete with mine, at least as a reference text where courses in Paleobotany are given. Eighty pages in my text are "straight" paleobotany and provide an introduction to the field and are a general discussion of those fossil groups which fall within the limits of the material covered by my book. Until my book appeared there was no book covering this material in simple and condensed form.
>
> It is true that a considerable part of the material taken by Darrah from my book is factual and this type of material *could* have been obtained by Darrah from other sources, as I obtained it.
>
> The situation is not at all that outlined in Mr. Ferrin's letter to you wherein he states that "in running down some cases of so-called plagiarism it has been discovered that Dr. Darrah and other authors of other books went to the same sources."
>
> The portion of the unwarranted use, however, which couldn't have been "obtained from sources" and for which I ask fullest recognition is that which represents my personal opinions and conclusions, namely, the general discussions, summaries, and conclusions on pages 103, 409, 410 of Darrah's textbook.
>
> I ask that the errata slip be prepared so that acknowledgments to Professor Chaney, Knowlton and myself stand as *separate paragraphs so spaced that they stand clearly as three acknowledgments.*[51]

Eames then wrote to Darrah. "I am indeed most sorry that this 'book trouble' has arisen; it is as unpleasant for me, I assure you, as it is for you. I had not looked in your book for errors or quotations from my book until a student brought the two books and pointed out that paragraph after paragraph was essentially identical. I was then, naturally, compelled to check the extent to which you had made use of my book without acknowledgment." The letter continues, "I have sent to the McGraw-Hill Company a first draft of the paragraph concerning quotations from my book which I should like to see inserted in the errata slip. I trust that the paragraph will be acceptable to you and your publisher."[52]

In June 1940, a still somewhat perplexed Bill Darrah wrote to Oakes Ames:

I am sending you as much as I have of the correspondence on the Dutch book [Darrah, 1939b] and the Eames and Chaney trouble and I think these are self-explanatory. I am sending the original of Chaney's "permission" and note that he does not give me permission, although I did not discover this until months later. I showed this letter to Mr. Ferrin but he assured me that it was considered as legal permission. I am very anxious to have this letter returned, but I though it would be best if you saw it. I am also including an original by Eames which is not a bad letter at all. I am also including a copy of Eames' letter to Mr. Thompson of the McGraw-Hill Company. This one is also self-explanatory. I have a strong suspicion that Eames' chief irritation centers around the fact that the second volume (proposed) of his plant morphology series has either to be completely rewritten or else he finds that much of its cogency is gone. This may be a very unfair criticism but I believe that it enters into the problem.

Professor Eames' [1936] chapter on the Lepidodendrids (Chapter XV) is a direct compilation from Hirmer's Handbuch [1927], commencing with Pages 180 and running through the generic diagnoses over to about page 240. Of course, there is no possibility of construing his work as plagiarism and you will see that in some respects I follow Hirmer closer than he does. Notice my general recapitulation of the living lycopods on the first page of my Chapter VI, although I checked with Alston of the British Museum before I entered the numbers of species involved. Simultaneously Eames must have had before him Scott [1908–1909], *Studies in Fossil Botany,* Volume 1, because so far as I am aware, this is the only accessible book which gives the dimensions of the various structures outside of Hirmer.

One thing has irritated me more than anything else, and I believe is the strongest ground I stand on: This is the fact that for great numbers of fossil plants, the only description known is that of the original worker who did all the investigating and all of the publishing, no one else has ever seen the material. It is utterly impossible to write on fossil cycads without using the work of Wieland to the exclusion of everybody else. One cannot write on the fossil conifers of Carboniferous age unless he takes all of his information from Florin. I have always thought that this was one of the greatest strengths of science, that one searches after truth and makes his results and opinions known so that they may be incorporated into the work of others. As long as I live and I am able to carry on my work, I shall devote a large part of my energies to boiling down the many small contributions of numerous investigators, synthesizing them and making them useful to students and to other persons who are interested. What good are 50,000 papers a year if only the cytogeneticist and the plant physiologist have access to them? Perhaps it is a frame of mind that has cast me in this mold and perhaps it is some attitude that I do not understand, but most the individual papers, even the best ones, in themselves, amount to very little if they cannot be related to large problems that have broad interests and relationships.

It just occurs to me that while I had my book in preparation and Professor Eames' [1936] text came along, Dr. Wetmore said he was very disappointed that Eames had paid so little attention to the fossils with the exception of one or two groups, and that he had grouped all of them as a few closing sections in the book rather than to spread it among the various groups present.

If I find any additional letters here which may be of interest I shall send them to you. I cannot say at this moment just how many more there are. For your own interest I would like to show you the comments that Professor Wetmore made on my book as he was checking the manuscript. You will see from the painstaking detailed notes what a wonderful help he had been to me. That is why I feel that he has "let me down" when apparently he does not defend the book against Eames. . . .[53]

Appleton-Century Company published a full run of Darrah's *Textbook of paleobotany* (1939a). By early 1940, the errata sheet (Fig. 7) that credited the works of Eames, Chaney, and Knowlton was printed for insertion into the textbook. We were unable to discover the exact date of the printed errata because Appleton-Century is not an extant publisher and was bought by Prentice-Hall. However, the errata sheet was included in each book sold after the time of publication of the errata and even inserted into some previously sold copies, such as the one now at the library of the U.S. Geological Survey in Reston, Virginia.

This difficult learning experience for the gifted young scientist was certainly valuable. Indeed, later in this same thoughtful letter to Ames, Darrah averred, "I might say that at first I was very irritated over some of the criticisms, but as I went along I realized how genuine and how worthwhile the great majority of them were." And so, early in his long career, a message of caution had hit its mark: "This experience led me 'to bend over backwards' in every way to give all credit due," reflected Darrah in later years, especially when advising his own students and young colleagues.

## THE WORLD WAR II YEARS AND POSTWAR CORRESPONDENCE

### *Research*

During World War II, paleobotanical research continued, but at a much slower pace. Contacts with colleagues outside the United States were markedly reduced. Indeed, it would be some years after the war's end before exchanges were to approach their former levels, as much restoration of damaged lies, collections, libraries, and laboratories lay ahead.

In 1942, making his own contribution to the war effort, Darrah went on "indefinite leave" status but maintained his position at Harvard. He joined the Raytheon Manufacturing Company in Waltham, Massachusetts, an organization whose wartime technological efforts were significant, and stayed there until 1951. At first he worked as an engineer (mechanical and materials), but he soon moved into research and development—officially, "assistant head, Magnetron Research and Development Laboratories." His early training in geology and minerals found a new focus in raw materials. Most of his work is probably still restricted militarily, but it can be said that Darrah was able to address a number of research problems. One project related to quality assurance in handling pure metals and resulted in his receiving a patent for his efforts.[54] He also worked on a technique involving X-ray photography, a method to demonstrate structure in position within the surrounding matrix (Darrah, 1952a).

Throughout the nine years he spent with Raytheon, Darrah maintained a close connection with Harvard and never neglected his personal interests in paleontology and geology. In

## ERRATA

Through a regrettable error much of the material appearing in Chapter XIX was not properly credited to the sources from which it was taken. Due acknowledgment is now given to Ralph W. Chaney: The Succession and Distribution of Cenozoic Floras Around the Northern Pacific Basin, appearing in Essays in Geobotany in Honor of William Albert Setchell, University of California Press, Berkeley, 1936, for much of the material appearing on pages 352-359; and to Frederic E. Clements and Ralph W. Chaney: Environment and Life in the Great Plains, Supplementary Publication No. 24 of the Carnegie Institution of Washington, for much of the material appearing on pages 360-361.

Similar acknowledgment is hereby made to F. H. Knowlton, Plants of the Past, Princeton University Press, Princeton, 1927, for material appearing on pages 366-367, and again on pages 342-344 in Chapter XVII.

Due acknowledgment is hereby made to Arthur J. Eames for the following material: a considerable portion of the text of Chapter VI, notably pages 82-92 and 100-103; pages 409 and 410; from Arthur J. Eames' "Morphology of Vascular Plants: Lower Groups," The McGraw-Hill Book Company, New York, N. Y., 1936. The title of this book should have been added to the bibliography for Chapter VI. Pages 103, 409, 410 which deal chiefly with the morphology of the plant body represent largely the conclusions and opinions of Professor Eames and full credit is due him therefor.

Future editions will give proper citations.

Figure 7. Errata sheet for Darrah's Textbook of paleobotany, 1939, published by D. Appleton-Century, New York, London. Permission to reproduce errata page in its entirety was given by Glenn E. Bell, Permissions Editor, Prentice-Hall, New Jersey.

1946, he formally resigned from his position at Harvard and recommended his former student, Elso S. Barghoorn (1915–1984), for the job. Darrah also served as consultant for a number of petroleum companies during those nine years.

### *Postwar correspondence*

Correspondence with colleagues continued, as before the war, primarily within the United States. However, following the end of hostilities in Europe, Darrah soon began to receive requests for reprints of professional articles and copies of technical books. With so many laboratories and libraries having been destroyed, his colleagues were eager to rebuild their collections and go on with their paleobotanical work.

Darrah had been concerned throughout the war for his friends in Europe, and afterwards the letters were most unsettling. He did his best to respond to them all and to help wherever he could. A request for Darrah's papers came from Dr. B. Kozo-Poljansky, USSR: "It's extremely necessary for my scientific research work. The war deprived me of my library. . . ."[55]

From Professor Stanislaw Stopa, College of Mining in Krakow, Poland—through his cousin, Walter F. Stopa, in Chicago—came a request for sources of information on obtaining equipment, books, and other materials for rebuilding institutions of learning in the wake of the German invasion.[56]

Emily Dix, in Cambridge, England, wrote,

> I wonder what has happened to Renier, Bertrand [Chapter 7, this volume], Gothan [Chapter 6, this volume], and Jongmans [Chapter 5, this volume], the four great lords of paleobotany in Europe. Renier escaped to the south of France but I wonder where he is now. We had a happy time together with him at Liège [see Chapter 2, this volume]. . . . Sad to relate our department was bombed in London in May, 1941. I managed to retrieve quite a lot of my fossil plants although the fireman's hose and the heat promoted the growth of much fungus in some of them. . . ."[57]

Many letters that were directed to Darrah's last known address at Harvard continued to arrive. From friends with whom he had been able to correspond during the war, Darrah was brought up to date with what was happening. He was relieved to learn that destruction was not complete. S. Leclercq wrote from Liège, Belgium, "Happily my laboratory has not been damaged during the war. Since the war is finished . . . I'm still interested with [structurally] preserved plants from Carboniferous horizons. . . ."[58]

From Heerlen, Darrah's friend, Jongmans, wrote of the busy work of the geologists in the mining districts, a letter punctuated by the sad news of the death of Jongman's son, Rudi.

It was necessary to bring a number of young geologists in security, as the Germans intended to transport them to Germany. So the [directors] of the Mines opened the possibility to study here the geology of the Netherlands and especially of the mining district. I had more than thirty geologists in my institute. . . . Fortunately, we could save them all and they did much good work. Besides this, I personally have given the opportunity to work here to about ten other geologists who dived here [from] the Germans. These young men had no permission so this was an illegal work. This work too brought a great success. My son Rudi was one of those illegal geologists. . . .[59]

In a later letter, Jongmans continued,

I myself have been working on my flora of Turkey and the volume on ferns and pteridosperms of the Fossilium Catalogus. . . .

Perhaps you can ask those of our colleagues in the States who published papers on Carboniferous or plants in general to send to me their papers as soon as this will be possible. Many thanks in advance. It is a pity we could not yet have a third Carboniferous Congress, but it may be that we find a possibility to arrange such a congress in one of the next years. . . . In Europe as well as in the United States, it was to be "business as usual." Still locating his colleagues, Jongmans wrote Darrah, "As far as today I got no news about Hirmer. Kräusel wrote to me that Hirmer is said to be in Munich, but his address is unknown.[60]

During the late 1940s, Darrah conducted in-depth studies on *Lepidodendropsis* and lepidodendroid embryos (Darrah, 1949a, b). As he wrote Jongmans, "Some time during the next few weeks my paper on the embryos of *Lepidocarpon* will be issued. I am very anxious to have your comments because I have murdered several generic concepts. We have in recent years disregarded the whole tradition of descriptive paleobotany. This has come about because so many pure geologists and morphological botanists have published extensively. There is a middle road somewhere which I hope to find in the course of my work."[61]

Jongmans replied: "I have seen your paper on paleozoic lycopodiaceous embryos. This is very interesting. But I have one serious remark. You compare with *Gymnostrobus*. It may be that you are right in some respect. However, most of the figures published under this name are [not] fructifications but compressed *Stigmaria* stems. This is the case with Arber's and most of the Bureau's figures. This fact has been discovered many years ago by Kidston [1914].[62]

## POWELL OF THE COLORADO (1951)

Darrah was never one to drop anything that held his interest. Having first encountered the thoughts of John Wesley Powell during undergraduate days at the University of Pittsburgh, Darrah was still fascinated by the former soldier, scientist, and second director of the U.S. Geological Survey. In 1940, Darrah began to assemble material—letters, papers, photographs—with the thought of some day writing about this figure he so admired and who, ironically, later would prove to be his personal rescue from health problems. His research into Powell's personality and career continued while Darrah was at Raytheon. He retraced many of Powell's footsteps (Ohio, Illinois, the West), read his works, and read the books that he read. He wrote six articles during the years 1947 to 1949 on Powell's expeditions in the West.

During those years, Darrah was plagued by severe corneal ulcers and had to endure several operations on his eye, procedures that resulted in an eight-month ordeal of confinement to a darkened room. The illness led to a very frightening and depressing state for Darrah, who feared blindness in both eyes and an uncertain future in which he still had to support his family, by then including two daughters, Barbara and Elsie. With the help and encouragement of his wife, Helen, he began what turned out to be the perfect "therapy"—namely, writing, or rather dictating to a "very kind secretary," his manuscript on Powell, who had intrigued him some two decades earlier. Darrah's research and painstaking verification of data became the basis of a classic biography—*Powell of the Colorado* (1951). This valuable contribution to the history of geology was a "labor of love" and a satisfying experience for Darrah.

Years later, Darrah was to have a memorable personal experience associated with his earlier tribute to Powell. Serge Mamay supplied an anecdote about one of Darrah's visits to the National Museum (Washington, D.C.) in the late 1960s and a luncheon date in the Cosmos Club, which Powell had cofounded. As the luncheon group (including Ellis Yochelson and Arthur Watt) looked in the Club Library for *Powell of the Colorado*—which Mamay comments is "fascinating reading and high testimony to Darrah's enormous talent for research and writing"—they discovered it missing, much to the chagrin of the club manager. Later the manager appeared with the library copy, which had been placed in the club's vault for safekeeping along with Powell memorabilia! Darrah graciously handled the awkward situation and signed the copy. All were pleased—especially Darrah.

Darrah's photographs of Powell and other memorabilia were to be on display for the U.S. Geological Survey's Powell Centennial Exhibits of 1969 and the Smithsonian Institution's Powell Exhibit entitled, "The Indomitable Major John Wesley Powell, Scientific Explorer of the American West." As Darrah prepared the materials with Nellie C. Carrico, U.S. Geological Survey, for the exhibit, "Nell" sent copies of "new" materials about Powell to him. In a letter to Nell, a delighted Bill wrote, "It is utterly amazing how many new things concerning Major Powell still turn up. The photographs of the Wisconsin farm and the prints of the oil painting are especially appreciated. Yet the surprising thing to me is that they do not materially change our general opinion of Powell or his work; although it may sound conceited, I think what was accomplished for *Powell of the Colorado* was a pretty fair appraisal of the man and his influence. Of course today I would do things differently, but then there have been 20 years of reflections to modify old opinion."[63]

On still another occasion, Darrah recalled his experience in August 1969, during the John Wesley Powell Centennial celebration in Arizona. He was invited for the dedication of the plaque at the new Powell Memorial at Powell Point, South Rim of the Grand Canyon. In his keynote speech, Edwin D. McKee of the U.S. Geological Survey, U.S. Department of the Interior, Denver, Colorado, provided a review of Major Powell's career. Governor Jack Williams of Arizona then presented Darrah with a numbered set of Powell silver medallions issued by the states of Arizona, Utah, and Wyoming (Fig. 8). After being introduced by Senator Barry Goldwater, Darrah offered brief remarks on Powell's contribution to the development of the West, closing with, "My only comment on this happy occasion is my sincere pleasure in seeing Major Powell come into his own, after receiving so many years of neglect and inattention by the American people he so ably served."

Figure 8. W. C. Darrah holding the set of silver Powell medallion coins presented to him by Governor Williams of Arizona, 1969. The William Culp Darrah Collection.

## THE GETTYSBURG YEARS

As fate would have it, Darrah's partial loss of vision had closed one door of opportunity but opened another. In 1951, he was in a position to pursue areas of interest previously overshadowed by pressing demands upon his time and talents. His experiences in the nonacademic world at Raytheon had taught him that he would no longer be comfortable "cloistered" in academe. At the same time, to remain at Raytheon would mean administrative responsibilities only, as any work in the plants would put him at physical risk now that he was blind in one eye. And purely administrative work had little appeal. Therefore, he decided he would become self-employed and follow his own "instincts."

After seriously considering the proximity of Gettysburg, Pennsylvania, to many libraries and museums in nearby Maryland, Pennsylvania, and the District of Columbia, Darrah decided in 1951 to move his family to a small farm on the outskirts of Gettysburg. Here he could resume his research and writing activities and still be relatively close to professional resources. Although the specialized research was curtailed significantly, he continued to read, write, and make other paleobotanical contributions. Before long, he began to diversify his efforts more freely into other activities and academic disciplines. Settling into the community, he became involved with the local library, the Adams County Historical Society, and, briefly, the local PTA (Parent Teachers Association) and youth council.

By 1953, Darrah was an active participant in the affairs of nearby Gettysburg College. First recruited as a lecturer in contemporary civilization, and then as a lecturer in literary foundations, he was instrumental in the development of the College's general education program. Soon he "answered the call" of Dr. Earl Bowen, head of the Biology Department, and joined the faculty as an associate professor in biology in 1954. By 1963, Darrah, now a full professor of biology, was distinguishing himself by his encouragement of students, friends, and colleagues in their own scholarly pursuits, giving unselfishly of his time and experience and gently guiding and stimulating all those who asked. In 1969, Darrah published a monograph to summarize his years of work on the Carboniferous floras of the eastern United States and their correlation with the floras of Europe. Before his retirement, the local Alpha Phi Omega chapter awarded a special recognition: "APO extends hearty appreciation to William C. Darrah for having rendered outstanding service to the student body. Enacted this 4th day of May 1974." Nor did the expressions of appreciation stop here. Following his retirement in 1974, the College awarded him the honorary degree of Doctor of Humane Letters on June 6, 1977, in recognition of his exceptional contributions to the development of the College and the Gettysburg community (Figs. 9 and 10).

In the Gettysburg setting, Darrah was able to complete his earlier research on Pithole, Pennsylvania. The history of this

Figure 9. Left to right: Dr. C. Arnold Hansen, president of Gettysburg College, presenting the honorary degree of Doctor of Humane Letters to W. C. Darrah, June 5, 1977. The William Culp Darrah Collection.

Figure 10. W. C. Darrah, 1977. Photograph taken by Mason P. Smith. The William Culp Darrah Collection.

petroleum boom town is described in still another book, *Pithole, the vanished city* (1972c).

Among the vast store of materials accumulated during Darrah's research efforts on Powell and on Pithole was a large collection of old photographs. In 1944 Darrah began collecting stereographs (also known as stereoviews), which contain two photographs of an area taken from slightly different angles. When looking at the images through a stereoscope, the observer sees a single image in three dimensions. Darrah was first interested in stereographs while he was writing *Powell of the Colorado,* seeing them as a source of documentation of the American West as it looked in the 1870s. First he collected stereoviews pertaining to the Powell Surveys (1869 and 1871). These stereographs were published by the Powell Survey because all equipment, cameras, and chemicals needed to make the photographs were purchased with Powell Survey money. However, Powell retained the title to these stereographs. How Major Powell engineered this arrangement, Darrah did not know. The U.S. Geological Survey was created in 1879, and Powell became its second director (1881–1894). These stereographs were distributed among Congressmen and friends in order to win votes and favors (Waldsmith, 1978a, b).

Darrah's active period for collecting the majority of the stereographs for his collection was the late 1950s to early 1960s, when he received 3,000 to 4,000 stereographs a month. From these stereographs, he selected what he wanted in order to get a complete range of work that had been done, among them the stereographs that illustrated American technology. Darrah realized that his selective purchases had passed up stereographs that illustrated the first 10 years of stereophotography. Thus, to round out his collection, he saved stereographs of early European classics and early American stereoviews just for their representation. He sold the stereographs he did not wish to keep in the collection in lots of 1,000 cards at a time. By 1977, he had amassed a collection of 110,000 cards (Waldsmith, 1978a, b), having purchased them from antique dealers over the years. His reference collection was one of the most comprehensive in the world.

Darrah began his collection of the carte de visite in 1954. These are small images, usually portraits, mounted on cards measuring about 5.3 × 10 cm (2.5 × 4 in). This format revolutionized the profession of photography. Darrah's purpose was to document the existence of as many nineteenth-century photographers as possible. Thus, he recorded the work of about 21,000 photographers, which represents nearly every aspect of the carte de visite subjects. Darrah's collection contains about 57,000 images. Today, this collection is housed in the Rare Book and Special Collections Department of Pattee Library at The Pennsylvania State University, University Park, Pennsylvania.

Darrah became fascinated by the photographs in their own right, the art form, the early methods of preparation, and the clues that they provided about "social science" and produced three authoritative books: *Stereo views—A history of stereographs in America and Their Collections* (1964), *The world of stereographs* (1977), and *Cartes de visite in nineteenth-century photography* (1981). Seminal writings (1990) for a fourth authoritative work on the history of nineteenth-century women photographers were left unfinished by Darrah, to be completed by others with similar interests. Indeed, until the last minute, he was scheduled to be the keynote speaker at the "Women in Photography" conference held at Bryn Mawr, Pennsylvania, in June 1989.

For two of his literary accomplishments in the history of photography (1977, 1981), Darrah twice received the Benjamin Award, presented by the Association of American Publishers for creativity and innovation in publishing. That Darrah received his honor from commercial publishers, for books he published himself, made the award especially meaningful. He was named the First Fellow of the National Stereoscopic Association in 1983 (Treadwell, 1989).

And still Darrah continued to follow the paths of his own choosing. Responding to the attraction of his first intellectual love, paleobotany, he collaborated with Dr. Horace G. Richards from 1971 to 1975 to prepare a revised catalog for the preservation of the sadly deteriorating paleobotanical collection of the Academy of Natural Sciences in Philadelphia, Pennsylvania (see Chapter 4, this volume).

Also during the early 1970s, Darrah was hired by Richard Rush of the Richard Rush Studio of Chicago, Illinois, to act as the technical consultant and director of the studio's construction of the walk-in Carboniferous forest for what was then called William Penn Memorial Museum (now the State Museum of Pennsylvania) in Harrisburg, Pennsylvania. Darrah was responsible for the accuracy of each of the models being constructed as well as for the special relationships within the exhibit. Under this meticulous care, the *Lepidodendron, Psaronius,* and *Medullosa* faced the public in 1977 (see the cover photograph of this volume). Richard Rush later wrote, "I so enjoyed working with [Bill] as he was not only a renowned scientist but also a 'regular guy' who helped us get things done properly and authentically."[64]

This little-known part of Darrah's talents is the life-size Carboniferous swamp reconstruction, complete with lycopod trees, calamiteans, tree ferns, seed ferns, and even reptiles and insects. A. Traverse remarked (personal communication, 1990) that it was the best late Carboniferous reconstruction that he had ever seen.

The First I. C. White Memorial Symposium, called "The Age of the Dunkard," was held in 1972 in Morgantown, West Virginia. Feeling able to participate in the field trip (Darrah, 1972b) and to attend the conference (Fig. 11), Darrah presented a small part of his lifelong work on the Dunkard in a paper entitled, "Historical aspects of the Permian flora of Fontaine and White," which was published in 1975 in the I. C. White symposium proceedings (1975b). His paleobotanical observations and interpretations of the mid-1930s could now be sustained by additional evidence (Darrah and Barlow, 1972). The discovery of the *Taeniopteris,* foliage that might belong to the cycads, in the

Figure 11. The participants in the First I. C. White Memorial Symposium held at West Virginia University, Morgantown, West Virginia, September 25–29, 1972. Front row, left to right: A. T. Cross, H. Bode, R. B. Erwin, J. Clendening, W. C. Darrah. Back row, left to right: R. Lund, P. Rich, E. L. Yochelson, G. L. Wilde, I. G. Sohn, N. Hotten, H. Falke, T. P. Wilson, W. H. Gillespie, and J. A. Barlow. The William Culp Darrah Collection.

Conemaugh also supported Darrah's position on the early occurrence of cycads (Darrah, 1935a). White had so many years before written, "With reference to the Appalachian Trough, I am saying that it probably was generally relatively stable and moist for *Callipteris, Taeniopteris* and *Walchia,* though they were already in America in upland and probably slightly drier areas of redbeds deposition. They either were masked and localized in the midst of the lush vegetation, largely composed of derivatives from the Stephanian flora, or they were invaders transgressing the area of present Dunkard deposits during climatically favorable intervals."[65]

Following the conference, Darrah wrote to Paul Lyons, "The Dunkard flora requires careful revision. I've been at it nearly 25 years, but I am painfully slow. I am afraid I cannot do any more collecting in the field—unless it be so located that physical exertion would be minimal."[66]

And still not finished with this topic, Darrah (Fig. 12) prepared yet another paper on the enigmatic Dunkard plants to present at the 28th International Geological Congress in Washington, D.C., in August 1989. This time, he was pleased to note, his enthusiasm was partly that of an "old timer" who had attended the last International Congress that was held in the United States *over 50 years before* (the 16th, in 1933). With Paul Lyons, Darrah had just completed a manuscript entitled, "Earliest conifers of North America: Upland and/or paleoclimatic indicators?" (Lyons and Darrah, 1989).

Figure 12. W. C. Darrah in his library at his home in Gettysburg, Pennsylvania, 1975. The William Culp Darrah Collection.

For as long as he lived, Bill Darrah was in active pursuit of a breadth of knowledge: knowledge for himself, which he enjoyed, and knowledge for passing on to others, which was his "obligation." "I know I shall die leaving behind me a long trail of uncompleted projects," he admitted. His search for knowledge and excellence came to a close only on the day he died, May 21, 1989.

His career was marked throughout by a willingness to share his learning and experience and to stimulate and guide others, much as he himself had benefited so memorably from the guidance of such people as Currier, Phillips, Porter, Johnson, Jennings, White, Jongmans, Bertrand, Thompson, and Ames. Additional information on his life and work is to be found in Morey (1989a, b), Morey and Lyons (1990), and Lyons and Morey (1991a, b, 1994).

In a speech of appreciation and recognition at the 1987 Kuhnert Photography Conference, with Darrah in the audience, Heinz Henisch stated that Darrah was "an academician who allows his mind to roam, to settle in many places and to convey his findings to us, not only with scholarly distinction but with great charm."[67]

## ACKNOWLEDGMENTS

The authors thank Richard Rush of the Richard Rush Studios of Chicago, Illinois, for details of the project on the Carboniferous swamp reconstruction at the State Museum of Pennsylvania in Harrisburg, Pennsylvania. Thanks and acknowledgment are also extended to Donald Hoff, retired curator of earth sciences, State Museum of Pennsylvania, Harrisburg. He also provided information about the walk-in Carboniferous reconstruction at the State Museum of Pennsylvania.

Special thanks to Albert Traverse of The Pennsylvania State University for speaking candidly about the Darrah plagiarism charge, for giving his insights into the matter, and for putting the charge into proper perspective. Appreciation is also extended to J. Fraser Cocks III, curator of special collections in the Knight Library, University of Oregon, for providing copies of letters from the Ralph Works Chaney Papers, a gesture of invaluable assistance to E. D. Morey in her efforts to put to rest this umbrage that had persisted for so many years. Thanks are also given to Daniel Axelrod, Teresa Barghoorn, Jane Gray, and T. K. Treadwell for their help in answering questions about the Darrah plagiarism charge. To K. Collins, special thanks for obtaining information about the Benjamin Award. The authors thank Glenn E. Bell, permissions editor, Prentice-Hall, New Jersey, for his time in searching for the data on publication of the errata sheet and for permission to reproduce the errata sheet in its entirety. Lastly, the authors acknowledge H. W. Pfefferkorn and T. K. Treadwell for their insightful reviews of this manuscript.

## ENDNOTES

1. Notes in W. C. Darrah Collection, Morey Paleobotanical Laboratory.

2. W. C. Darrah, 1951, *Powell of the Colorado* (Princeton, New Jersey: Princeton University Press), p. 356.

3. Letter from David White to W. C. Darrah, June 26, 1933. W. C. Darrah Collection.

4. Letter from David White to W. C. Darrah, July 14, 1933. W. C. Darrah Collection.

5. Letter from E. W. Berry to W. C. Darrah, July 14, 1933. W. C. Darrah Collection.

6. Letter from David White to W. C. Darrah, January 27, 1934. W. C. Darrah Collection.

7. Letter from David White to W. C. Darrah, May 1, 1934. W. C. Darrah Collection.

8. Letter from David White to W. C. Darrah, July 10, 1933. W. C. Darrah Collection.

9. Letter from David White to W. C. Darrah, July 16, 1934. W. C. Darrah Collection.

10. Letter from Taisia Stadnichinko to W. C. Darrah, August 3, 1934. W. C. Darrah Collection.

11. Letter from Charles Schuchert to W. C. Darrah, June 27, 1935. W. C. Darrah Collection.

12. Letter from Oakes Ames to F. O. Thompson, January 30, 1937. W. C. Darrah Collection.

13. Letter from Oakes Ames to F. O. Thompson, November 27, 1936. W. C. Darrah Collection.

14. Letter from Oakes Ames to W. C. Darrah, April 18, 1937. W. C. Darrah Collection.

15. For academic recognition among hi peers, W. C. Darrah received these honors: Alpha Phi Omega (1928); Phi Sigma (Biology) (1931); Sigma Xi (Science) (1935); Gamma Gamma Gamma (Science) (1937); Fellow, American Academy of Sciences: Botany (1938) and Geology (1945).

16. Letter from Oakes Ames to W. C. Darrah, March 25, 1937. W. C. Darrah Collection.

17. Letter from Oakes Ames to F. O. Thompson, September 26, 1937. W. C. Darrah Collection.

18. Letter from Oakes Ames to W. C. Darrah, January 24, 1938. W. C. Darrah Collection.

19. Letter from Oakes Ames to W. C. Darrah, February 14, 1938. W. C. Darrah Collection.

20. Letter from Oakes Ames to W. C. Darrah, March, 14, 1937. W. C. Darrah Collection.

21. Letter from Oakes Ames to W. C. Darrah, January 24, 1938. W. C. Darrah Collection.

22. Letter from R. C. Moore, state geologist and director of the State Geological Survey of Kansas, to W. C. Darrah, November 11, 1942. W. C. Darrah Collection.

23. Letter from Oakes Ames to W. C. Darrah, March 24, 1938. W. C. Darrah Collection.

24. Letter from W. C. Darrah to P. C. Lyons, February 27, 1984.

25. Letter from C. A. Arnold to W. C. Darrah, March 6, 1937. W. C. Darrah Collection.

26. Letter from Ruth Colton, Assistant to F. S. Pease, to W. C. Darrah, November 2, 1938. W. C. Darrah Collection.

27. Letter from F. S. Pease to W. C. Darrah, February 25, 1939. W. C. Darrah Collection.

28. Letter from W. C. Darrah to R. W. Chaney, January 30, 1939: Ralph Works Chaney Papers (Ax 482), Special Collections, Knight Library, University of Oregon, Eugene, Oregon.

29. Letter from R. W. Chaney to W. C. Darrah, February 6, 1939. W. C. Darrah Collection.

30. Letter from E. W. Berry to W. C. Darrah, April 15, 1939. W. C. Darrah Collection.

31. Letter from W. J. Jongmans to W. C. Darrah, April 28, 1939. W. C. Darrah Collection.

32. Letter from R. W. Chaney to W. C. Darrah, April 5, 1939: Ralph Works Chaney Papers (Ax 482), Special Collections, Knight Library, University of Oregon, Eugene, Oregon.

33. Letter from R. W. Chaney to J. A. Brandt, April 18, 1939: Ralph Works Chaney Papers (Ax 482), Special Collections, Knight Library, University of Oregon, Eugene, Oregon.

34. Letter from R. W. Chaney to Erling Dorf, April 17, 1939: Ralph Works Chaney Papers (Ax 482), Special Collections, Knight Library, University of Oregon, Eugene, Oregon.

35. Letter from R. W. Chaney to A J. Eames, April 20, 1939: Ralph Works Chaney Papers (AX 482), Special Collections, Knight Library, University of Oregon, Eugene, Oregon.

36. Letter from R. W. Chaney to W. Goldring, April 21, 1939: Ralph Works Chaney Papers (AX 487), Special Collections, Knight Library, University of Oregon, Eugene, Oregon.

37. Letter from Erling Dorf to R. W. Chaney, April 27, 1939: Ralph Works Chaney Papers (Ax 482), Special Collections, Knight Library, University of Oregon, Eugene, Oregon.

38. Letter from Winifred Goldring to R. W. Chaney, May 3, 1939: Ralph Works Chaney Papers (Ax 482), Special Collections, Knight Library, University of Oregon, Eugene, Oregon.

39. Letter from A. J. Eames to R. W. Chaney, May 15, 1939: Ralph Works Chaney Papers (Ax 482), Special Collections, Knight Library, University of Oregon, Eugene, Oregon.

40. Letter from D. H. Ferrin to W. C. Darrah, May 16, 1939. W. C. Darrah Collection.

41. Letter from W. C. Darrah to R. W. Chaney, May 18, 1939: Ralph Works Chaney Papers (Ax 482), Special Collections, Knight Library, University of Oregon, Eugene, Oregon.

42. Letter from R. W. Chaney to Erling Dorf, July 27, 1939: Ralph Works Chaney Papers (Ax 482), Special Collections, Knight Library, University of Oregon, Eugene, Oregon.

43. Letter from Erling Dorf to R. W. Chaney, August 4, 1939: Ralph Works Chaney Papers (Ax 482), Special Collections, Knight Library, University of Oregon, Eugene, Oregon.

44. Letter from D. H. Ferrin to S. T. Farquhar, August 8, 1939: Ralph Works Chaney Papers (Ax 482), Special Collections, Knight Library, University of Oregon, Eugene, Oregon.

45. Letter from W. C. Darrah to R. W. Chaney, August 9, 1939: Ralph Works Chaney Papers (Ax 482), Special Collections, Knight Library, University of Oregon, Eugene, Oregon.

46. Letter from A. J. Eames to R. W. Chaney, August 17, 1939: Ralph Works Chaney Papers (Ax 482), Special Collections, Knight Library, University of Oregon, Eugene, Oregon.

47. Letter from R. W. Chaney to A. J. Eames, August 24, 1939: Ralph Works Chaney Papers (Ax 482), Special Collections, Knight Library, University of Oregon, Eugene, Oregon.

48. Letter from R. H. Wetmore to R. W. Chaney, September 28, 1939: Ralph Works Chaney Papers (Ax 482), Special Collections, Knight Library, University of Oregon, Eugene, Oregon.

49. Letter from W. C. Darrah to D. H. Ferrin, July 1, 1940. W. C. Darrah Collection.

50. Letter from D. H. Ferrin to W. C. Darrah, September 27, 1939. W. C. Darrah Collection.

51. Letter from A. J. Eames to J. S. Thompson, McGraw-Hill Publishing Company, September 18, 1939. W. C. Darrah Collection.

52. Letter from A. J. Eames to W. C. Darrah, September 25, 1939. W. C. Darrah Collection.

53. Letter from W. C. Darrah to Oakes Ames, June 29, 1940. W. C. Darrah Collection.

54. U.S. Patent 2,470,341, issued May 17, 1949, entitled *Fluorescent Method of Flow Detection*. W. C. Darrah Collection.

55. A form letter from B. Kozo-Poljansky, Sc.D., Moscow. W. C. Darrah Collection.

56. Letter from Walter F. Stopa for Professor Stanislaw Stopa, Krakow, Poland, to W. C. Darrah, October 5, 1947. W. C. Darrah Collection.

57. Letter from Emily Dix to W. C. Darrah, May 4, 1944. W. C. Darrah Collection.

58. Letter from Suzanne Leclercq to W. C. Darrah, May 20, 1949. W. C. Darrah Collection.

59. Letter from W. J. Jongmans to W. C. Darrah, October 1, 1945. W. C. Darrah Collection.

60. Letter from W. J. Jongmans to W. C. Darrah, March 23, 1947. W. C. Darrah Collection.

61. Letter from W. C. Darrah to W. J. Jongmans, June 22, 1949. W. C. Darrah Collection.

62. Letter from W. J. Jongmans to W. C. Darrah, December 1, 1949. W. C. Darrah Collection.

63. Letter from W. C. Darrah to N. C. Carrico, February 5, 1970. W. C. Darrah Collection.

64. Letter from R. B. Rush to E. D. Morey, August 6, 1991. W. C. Darrah Collection.

65. Letter from David White to W. C. Darrah, June 26, 1933. W. C. Darrah Collection.

66. Letter from W. C. Darrah to P. C. Lyons, August 2, 1978.

67. January 17, 1987, speech by Heinz K. Henisch at the Kuhnert Photography Conference sponsored by the State Museum of Pennsylvania, Harrisburg, Pennsylvania. W. C. Darrah Collection.

## REFERENCES CITED


Arnold, C. A., 1947, An introduction to paleobotany: New York, McGraw-Hill Book Company, p. 432.

Berry, E. W., 1930, The past climate of the North Polar region: Smithsonian Institution Miscellaneous Collections, v. 82, 29 p.

Chaney, R. W., 1936, The succession and distribution of Cenozoic floras around the northern Pacific Basin, *in* Essays in geobotany in honor of William Albert Setchell: Berkeley, California, University of California Press, p. 55–58.

Darrah, W. C. See Bibliography of William Culp Darrah that follows.

Eames, A. C., 1936, Morphology of vascular plants, lower groups (Philophytales to Filicales): New York, McGraw Hill, 411 p.

Hirmer, M., 1927, Handbuch der palaobotanik. Volume 1: Thallophyta, Bryophyta, Pteridophyta: Munich and Berlin, Druck und verlag von R. Oldenbourg, 708 p.

Kidston, R., 1914, On the fossil flora of Staffordshire Coalfields, Part III: Royal Society of Edinburgh Transactions, v. 50, p. 148–150.

Knowlton, F. H., 1927, Plants of the past, a popular account of fossil plants: Princeton, New Jersey, Princeton University Press, 275 p.

Lyons, P. C., and Darrah, W. C., 1989, Earliest conifers of North America: Upland and/or paleoclimatic indicators?: Palaios, v. 4, p. 480–486.

Lyons, P. C., and Morey, E. D., 1991a, A tribute to an American paleobotanist: William Culp Darrah (1909–1989): Congrès international de la stratigraphie et geologie du Carbonifère et Permien, 12th, (Buenos Aires), Abstracts, p. 57–58.

Lyons, P. C., and Morey, E. D., 1991b, Memorial to William Culp Darrah (1909–1989): Torrey Botanical Club Bulletin, v. 118, p. 195–200.

Lyons, P. C., and Morey, E. D., 1993, A tribute to an American paleobotanist: William Culp Darrah (1909–1989): Congrès international de la stratigraphie et géologie du Carbonifère et Permien, 12th, (Buenos Aires): Compte Rendu, v. 2, p. 117–126.

Lyons, P. C., and Morey, E. D., 1994, Birbal Sahni—His North American paleobotanical connection with William Culp Darrah: Geophytology (in press).

Mahabale, T. S., 1954, Two French savants: Charles-Eugène Bertrand, the botanist and Paul Bertrand, the paleobotanist: Bulletin du Museum, 22 series, p. 444–453.

Morey, E. D., 1989a, Memorial to William Culp Darrah (1909–1989): Pennsylvania Geology, v. 20, p. 2.

Morey, E. D., 1989b, Memorial [to] William Culp Darrah (1909–1989): Paleontology, v. 64, p. 2896.

Morey, E. D., and Lyons, P. C., 1990, Dedication and publications of William Culp Darrah (1909–1989), *in* Dolph, G. E., ed., Bibliography of American paleobotany for 1989: Botanical Section of the Botanical Society of America, p. i–ii, 1–3.

National Research Council, Division of Geology and Geography, 1931, Report of the Committee on Paleobotany, April 25, 1931: Washington, D.C., 13 p.

National Research Council, Division of Geology and Geography, 1932, Report of the Committee on Paleobotany, April 23, 1932: Washington, D.C., 20 p.

National Research Council, Division of Geology and Geography, 1933, Report of the Committee on Paleobotany, April 22, 1933: Washington, D.C., 16 p.

National Research Council, Division of Geology and Geography, 1934, Report of the Committee on Paleobotany, April 28, 1934: Washington, D.C., 17 p. and Exhibit A, 9 p.

Scott, D. H., 1908–1909, Studies in fossil botany (second edition), 2 vols.: London, A. C. Black, 683 p.

Treadwell, T. K., 1989, Dr. William Culp Darrah: Stereo World, v. 16, p. 17.

Waldsmith, J., 1978a, Interview: William Culp Darrah: Stereo World, v. 4, p. 13, 16.

Waldsmith, J., 1978b, Interview: William Culp Darrah: Stereo World, v. 5, p. 7.


## BIBLIOGRAPHY OF WILLIAM CULP DARRAH


Darrah, W. C., 1932a, Evidence of Permian floras in the Appalachian region [abs.]: The Biologist, March, p. 38–39.

Darrah, W. C., 1932b, Recent paleobotanic investigations near Pittsburgh, Pennsylvania: Proceedings of the Pennsylvania Academy of Science, v. 6, p. 110–114.

Darrah, W. C., 1932c, The status of Paleozoic pteridosperms: Proceedings of the Pennsylvania Academy of Science, v. 6, p. 115–119.

Darrah, W. C., and Bertrand, P., 1933a, Observations sur les flores houillères de Pennsylvanie (régions de Wilkes-Barre et de Pittsburgh) [abs.]: Académie des Sciences (Paris), Comptes rendus, v. 197, p. 1451–1452.

Darrah, W. C., and Bertrand, P., 1933b, Observations sur les flores houillères de Pennsylvanie: Annales de la Société géologique du Nord (Lille), v. 58, p. 211–224.

Darrah, W. C., 1934a, Leo Lesquereux (1806–89): Harvard University Botanical Museum Leaflets, v. 2, p. 113–119.

Darrah, W. C., 1934b, Stephanian in America [abs.]: Geological Society of America, Proceedings for 1933, p. 451.

Darrah, W. C., 1935a, Permian elements in the fossil flora of the Appalachian Province. I: *Taeniopteris:* Harvard University Botanical Museum Leaflets, v. 3, p. 137–148.

Darrah, W. C., 1935b, American Carboniferous floras: Harvard University Botanical Museum Leaflets, v. 3, p. 1–18.

Darrah, W. C., 1935c, Some Late Carboniferous correlations in the Appalachian province [abs.]: Geological Society of America, Proceedings for 1934, p. 442–443.

Darrah, W. C., 1935d, Recent studies of American pteridosperms [abs.]: Proceedings, International Botanical Congress, 6th, Amsterdam, The Netherlands, Volume 2, p. 234–235.

Darrah, W. C., 1936a, Permian elements in the fossil flora of the Appalachian Province. II: *Walchia:* Harvard University Botanical Museum Leaflets, v. 4, p. 9–19.

Darrah, W. C., 1936b, Antarctic fossil plants: Science, v. 83, p. 390–391.

Darrah, W. C., 1936c, A new *Macrostachya* from the Carboniferous of Illinois: Harvard University Botanical Museum Leaflets, v. 4, p. 52–63.

Darrah, W. C., 1936d, The peel method in paleobotany: Harvard University Botanical Museum Leaflets, v. 4, p. 69–83.

Darrah, W. C., 1936e, The Harvard collections of fossil plants: Harvard University Alumni Bulletin, v. 39, p. 383–385.

Darrah, W. C., 1936f, Sur la présence d'équivalents des Terrain stéphaniens dans l'Amerique du Nord: Annales de la Société géologique du Nord, v. 61, p. 187–197.

Darrah, W. C., 1937a, A review of "The Cerro Cuadrado Petrified Forest": American Journal of Science, v. 33, p. 480.

Darrah, W. C., 1937b, *Codonotheca* and *Crossotheca*—Polliniferous structures of pteridosperms: Harvard University Botanical Museum Leaflets, v. 4, p. 153–172.

Darrah, W. C., 1937c, Some floral relations between the Late Paleozoic of Asia and North America: Problems of Paleontology, Moscow, v. 2–3, p. 195–205.

Darrah, W. C., 1937d, Spores of Cambrian plants: Science, v. 86, p. 154–155.

Darrah, W. C., 1937e, New approach to the study of spores in coal [abs.]: Geological Society of America, Proceedings for 1936, p. 402.

Darrah, W. C., 1937f, American Carboniferous floras: Congrès pour l'avancement des études de stratigraphie du Carbonifère, 2d, Heerlen, The Netherlands, 1935: Compte Rendu, v. 1, p. 109–129.

Darrah, W. C., 1937g, Recent studies of American pteridosperms: Congrès pour l'avancement des études de stratigraphie carbonifère, 2d, Heerlen, The Netherlands, 1935: Compte Rendu, v. 1, p. 131–137.

Darrah, W. C., 1937h, *Oligocarpia* and the antiquity of the Gleicheniaceae [abs.]: American Journal of Botany, v. 24, p. 743.

Darrah, W. C., 1938a, Fossil plants and evolution: Evolution, v. 4, p. 5–6.

Darrah, W. C., 1938b, The occurrence of the genus *Tingia* in Texas: Harvard University Botanical Museum Leaflets, v. 5, p. 173–188.

Darrah, W. C., 1938c, Technical contributions to the study of archeological materials: American Antiquity, v. 3, p. 269–270.

Darrah, W. C., 1938d, A remarkable fossil *Selaginella* with preserved female gametophytes: Harvard University Botanical Museum Leaflets, v. 6, p. 113–136.

Darrah, W. C., 1938e, A new transfer method for studying fossil plants: Harvard University Botanical Museum Leaflets, v. 7, p. 35–36.

Darrah, W. C., 1938f, The embryo of *Cordaites* [abs.]: Botanical Society of America Abstracts, p. 2.

Darrah, W. C., 1938g, A new fossil gleicheniaceous fern from Illinois: Harvard University Botanical Museum Leaflets, v. 5, p. 145–160.

Darrah, W. C., 1939a, Textbook of paleobotany: New York, D. Appleton-Century, 441 p.

Darrah, W. C., 1939b, Principles of paleobotany: Leiden, Holland, Chronica Botanica, 239 p.

Darrah, W. C., 1939c, Devonian floras and beginnings of leafy glories: Pan-American Geologist, v 77, p. 29–36.

Darrah, W. C., 1939d, The fossil flora of Iowa coal balls. I: Discovery and occurrence: Harvard University Botanical Museum Leaflets, v. 7, p. 125–136.

Darrah, W. C., 1939e, The fossil flora of Iowa coal balls. II: The fructification of *Botryopteris:* Harvard University Botanical Museum Leaflets, v. 7, p. 157–168.

Darrah, W. C., 1940a, The fossil flora of Iowa coal balls. III: *Cordaianthus:* Harvard University Botanical Museum Leaflets, v. 8, p. 1–20.

Darrah, W. C., 1940b, The four great problems of paleobotany: Scientia, ser. 4, v. 68, p. 14–20 (suppl. p. 10–14, French).

Darrah, W. C., 1940c, Supposed fossil orchids: American Orchid Society Bulletin, v. 9, p. 149–150.

Darrah, W. C., 1940d, The position of the Nematophytales: Chronica Botanica, v. 6, p. 52–53.

Darrah, W. C., 1941a, The fossil floras of Iowa coal balls. IV: *Lepidocarpon:* Harvard University Botanical Museum Leaflets, v. 9, p. 85–100.

Darrah, W. C., 1941b, Studies of American coal balls: American Journal of Science, v. 239, p. 33–53.

Darrah, W. C., 1941c, The coenopterid ferns in American coal balls: American Midland Naturalist, v. 25, p. 233–269.

Darrah, W. C., 1941d, Notas sobre la historia de la paleobotanica sudamericana: De Lilloa, Revista de Botánica del Instituto Miguel Lillo (Tucumán, Argentina), v. 6, p. 213–239.

Darrah, W. C., 1941e, Utilitarian aspects of paleobotany: Chronica Botanica, v. 6, p. 342–344.

Darrah, W. C., 1941f, Fossil embryos in Iowa coal balls: Chronica Botanica, v. 6, p. 388–389.

Darrah, W. C., 1941g, Changing views of petrifaction: Pan-American Geologist, v. 76, p. 13–26.

Darrah, W. C., 1941h, Observations on the vegetable constituents of coal: Economic Geology, v. 36, p. 589–611.

Darrah, W. C., 1941i, La paleobotanica Sudamerica: Lilloana, v. 6, p. 312–329.

Darrah, W. C., 1942a, An introduction to the plant sciences: New York, John Wiley and Sons, 332 p.

Darrah, W. C., 1942b, Lesquereux collections and type specimens, *in* Sarton, G., ed., Lesquereux (1806–89): ISIS, v. 34, p. 104–106.

Darrah, W. C., 1945a, Paleobotanical work in Latin America, *in* Verdoorn, F., ed., Plants and plant science in Latin America: Waltham, Massachusetts, Chronica Botanica, v. 16, p. 181–183.

Darrah, W. C., 1945b, A brief account of the geology of South America, *in* Verdoorn, F., ed., Plants and plant science in Latin America: Waltham, Massachusetts, Chronica Botanica, v. 16, p. 318–322.

Darrah, W. C., 1945c, A geological sketch of Central America and the Antilles, *in* Verdoorn, F., ed., Plants and plant sciences in Latin America: Waltham, Massachusetts, Chronica Botanica, v. 16, p. 153–156.

Darrah, W. C, 1947a, Recent paleobotanical literature: Chronica Botanica, v. 9, p. 84–88.

Darrah, W. C., 1947b, Major Powell prepares for a second expedition: Utah Historical Quarterly, v. 15, p. 149–153.

Darrah, W. C., 1947c, Biographical sketches and original documents of the first Powell expedition of 1869: Utah Historical Quarterly, v. 15, p. 9–148.

Darrah, W. C., 1948–1949a, John F. Steward (1841–1915): Utah Historical Quarterly, v. 16–17, p. 175–179.

Darrah, W. C., 1948–1949b, Journal of John F. Steward: Utah Historical Quarterly, v. 16–17, p. 181–251.

Darrah, W. C., 1948–1949c, Beaman, Fennemore, Hillers, Dellenbaugh, Johnson and Hattan: Utah Historical Quarterly, v. 16–17, p. 491–503.

Darrah, W. C., 1948–1949d, Three letters by Andrew Hall: Utah Historical Quarterly, v. 16–17, p. 505–508.

Darrah, W. C., 1949a, Paleozoic lepidodendroid embryos: Paleobotanical Notices, v. 2, p. 1–39.

Darrah, W. C., 1949b, Notes on *Lepidodendropsis:* Paleobotanical Notices, v. 1, p. 1–16.

Darrah, W. C., 1951, Powell of the Colorado: Princeton, New Jersey, Princeton University Press, 426 p. (reprinted, 1969).

Darrah, W. C., 1952a, The materials and methods of paleobotany: The Palaeobotanist, v. 1, Birbal Sahni Memorial Volume, p. 145–153.

Darrah, W. C., 1952b, A new cordaicarp from the Pennsylvanian of Iowa: Paleobotanical Notices, v. 5, p. 1–16.

Darrah, W. C., 1955, The materials and methods of paleobotany: Botaniska Zhurnal Akademia Nauk SSSR, v. 40, p. 861–863.

Darrah, W. C., 1960a, Powell of the Colorado: Utah Historical Quarterly, v. 28, p. 223–231.

Darrah, W. C., 1960b, Principles of paleobotany (second edition): New York, Ronald Press, 295 p.

Darrah, W. C., 1964, Stereo views: A history of stereographs in America and their collection: Gettysburg, Pennsylvania, Times and News Publishing, 255 p.

Darrah, W. C., 1966, The structure of *Cardiocarpus florini* (Darrah), a Pennsylvanian cordaite seed from Iowa [abs.]: Proceedings of the Pennsylvania Academy of Science, v. 40, p. 12.

Darrah, W. C., 1967a, The structure of *Cardiocarpus florini* (Darrah), a Pennsylvanian cordaite seed from Iowa: Proceedings of the Pennsylvania Academy of Science, v. 40, p. 80–86.

Darrah, W. C., 1967b, *Rhodea* in the Mississippian (Pocono) floras of the Appalachian region: Proceedings of the Pennsylvanian Academy of Science, v. 41, p. 71–73.

Darrah, W. C., 1968, The pteridosperm genus *Lescuropteris*—Characteristics, distribution, and significance [abs.]: American Journal of Botany, v. 55, p. 725.

Darrah, W. C., 1969a, The age of the highest coals of the Southern Anthracite field [abs.]: Proceedings of the Pennsylvania Academy of Science, v. 43, p. 14.

Darrah, W. C., 1969b, John Wesley Powell and an understanding of the West: Utah Historical Quarterly, v. 37, p. 146–151.

Darrah, W. C., 1969c, John Wesley Powell—His western explorations: Geotimes, v. 14, p. 13–15.

Darrah, W. C., 1969d, A critical review of the Upper Pennsylvanian floras of eastern United States, with notes on the Mazon Creek flora of Illinois: Gettysburg, Pennsylvania, privately printed, 220 p., 80 pls.

Darrah, W. C., 1971, Paleobotany of the Pocono and Pottsville on Interstate 80, Pine Township, Clearfield County, *in* Edmunds, W. E., and Berg, T. M., eds., Geology and mineral resources of the southern half of the Penfield 15-minute quadrangle, Pennsylvania: Pennsylvania Geological Survey Atlas 74, p. 167–174.

Darrah, W. C., 1972a, Historical aspects of the Permian flora of Fontaine and White [abs.], *in* Barlow, J. A., ed., The Age of the Dunkard, Symposium Abstracts and References Papers, I. C. White Memorial Symposium: Morgantown, West Virginia Geological and Economic Survey, p. 1–4.

Darrah, W. C., 1972b, "Brown's Bridge" or "Brown's Mill" (Worley), West Virginia—The classical locality of Fontaine and White's *Callipteris, in* Arkle, T. Jr., ed., I. C. White Memorial Symposium Field Trip (September 27–29, 1972): Morgantown, West Virginia Geological and Economic Survey, p. 16–17.

Darrah, W. C., 1972c, Pithole the vanished city: A story of the early days of the petroleum industry: Gettysburg, Pennsylvania, privately printed, 252 p.

Darrah, W. C., and Barlow, J. A., 1972, Fossil plants of the Waynesburg coal and Cassville Shale at Mount Morris, Pennsylvania, *in* Arkle, T. Jr., ed., I. C. White Memorial Symposium Field Trip (September 27–29, 1972): Morgantown, West Virginia Geological and Economic Survey, p. 14–16.

Darrah, W. C., 1973a, Appendix, *in* Edmunds, W. E., and Berg, T. M., Geology and mineral resources of the southern half of the Penfield 15-minute quadrangle: Pennsylvania Geological Survey Atlas, 184 p.

Darrah, W. C., 1973b, Engineering at Gettysburg College: Gettysburg College History Series, v. 2, 36 p.

Darrah, W. C., 1974, Engineering at Gettysburg College: Gettysburg College Bulletin, v. 65, 24 p.

Darrah, W. C., 1975a, Honor, expectation, and satisfaction: Gettysburg College Bulletin, v. 64, p. 9–11.

Darrah, W. C., 1975b, Historical aspects of the Permian flora of Fontaine and White, *in* Barlow, J. A., ed., The Age of the Dunkard: Proceedings, I. C. White Memorial Symposium, 1st: Morgantown, West Virginia Geological and Economic Survey, p. 81–101.

Darrah, W. C., 1976, The ecology of Big Round Top: Gettysburg, Pennsylvania, Eastern National Park and Monument Association in cooperation with Gettysburg National Military Park, p. 2–15.

Darrah, W. C., 1977, The world of stereographs: Gettysburg, Pennsylvania, W. C. Darrah Publisher, 246 p.

Darrah, W. C., 1981, Cartes de visites in nineteenth century photography: Gettysburg, Pennsylvania, privately printed, 221 p.

Darrah, W. C., 1985, Intentions and techniques—An exhibition of photographs from the Lehigh University collection and nineteenth century Pennsylvania photography as a folk art from the William C. Darrah collection: Bethlehem, Pennsylvania, Lehigh University Art Galleries, p. 26–27.

Darrah, W. C., 1990 (posthumous), Nineteenth-century women photographers, *in* Collins, K., Shadow and substance: Essays on the history of photography: Bloomfield Hills, Michigan, The Amorphous Institute Press, p. 89–103.

Darrah, W. C., and Barghoorn, E. S. Jr., 1938, *Horneophyton,* a necessary change of name for *Hornea:* Harvard University Botanical Museum Leaflets, v. 6, p. 142–144.

Darrah, W. C., and Barlow, J. A., 1972, Fossil plants of the Waynesburg coal and Cassville Shale at Mount Morris, Pennsylvania, *in* Arkle, T. Jr., ed., I. C. White Memorial Symposium Field Trip (September 27–29, 1972): Morgantown, West Virginia Geological and Economic Survey, p. 14–16.

Darrah, W. C., and Bertrand, P., 1933a, Observations sur les flores houillères de Pennsylvanie (régions de Wilkes-Barre et de Pittsburgh) [abs.]: Académie des Sciences (Paris), Comptes rendus, v. 197, p. 1451–1452.

Darrah, W. C., and Bertrand, P., 1933b, Observations sur les flores houillères de Pennsylvanie: Annales de la Société géologique du Nord (Lille), v. 58, p. 211–224.

Darrah, W. C., and Lyons, P. C., 1977, Floral evidence for Upper Pennsylvanian in the Narragansett basin, southeastern New England [abs.]: Geological Society of America Abstracts with Programs, v. 3, p. 297.

Darrah, W. C., and Lyons, P. C., 1978, A late Middle Pennsylvanian flora of the Narragansett basin, Massachusetts: Geological Society of America Bulletin, v. 89, p. 433–438.

Darrah, W. C., and Lyons, P. C., 1979, Fossil floras of Westphalian D and early Stephanian age from the Narragansett basin, Massachusetts and Rhode Island, *in* Cameron, B., ed., Carboniferous basins of southeastern New England, Guidebook for Field Trip no. 5, May 1979, International Congress of Carboniferous Stratigraphy and Geology, 9th, Washington, D.C., and Urbana-Champaign, Illinois: Falls Church, Virginia, American Geological Institute, p. 81–89.

Darrah, W. G., and Lyons, P. C., 1987, Paleoecological significance of walchian conifers in Westphalian (Late Carboniferous) horizons of North America and Europe [abs.]: Congrès international de stratigraphie et de géologie du Carbonifère, 11th (Beijing, China): Abstracts of Papers, v. 1, p. 146–147.

Darrah, W. C., and Lyons, P. C., 1989a, Earliest conifers of North America—Upland and/or paleoclimatic indicators?: Palaios, v. 4, p. 480–486.

Darrah, W. C., and Lyons, P. C., 1989b, Paleoenvironmental and paleoecological significance of walchian conifers in Westphalian (Late Carboniferous) horizons of North America: Congrès international de stratigraphie et de géologie du Carbonifère, 11th (Beijing, China), 1987: Compte Rendu, v. 3, 251–261, pl. 1.

Darrah, W. C., and Russack, R., 1977, An album of stereographs or, Our country victorious and now a happy home—From the collections of William Culp Darrah and Richard Russack: Garden City, New Jersey, Doubleday, 109 p.

Jongmans, W. J., and Gothan, W., with additional notes by W. C. Darrah, 1937, Comparison of the floral succession in the Carboniferous of West Virginia with Europe. Congrès pour l'avancement des études de stratigraphie du Carbonifère, 2d, Heerlen, The Netherlands, 1935: Compte Rendu, v. 1, 393–415.

Jongmans, W. J., Gothan, W., and Darrah, W. C., 1937, Beiträge zur Kenntnis der Flora der Pocono-Schichten aus Pennsylvanien und Virginia: Congrès pour l'avancement des études de stratigraphie du Carbonifère, 2d, Heerlen, The Netherlands, 1935: Compte Rendu, v. 1, p. 423–444.

Treadwell, T. K., 1994 (with W. C. Darrah, posthumous), Stereographers of the world: Bryan, Texas, National Stereoscopic Association, 2 vols., 831 p.

Manuscript Accepted by the Society July 16, 1994

Geological Society of America
Memoir 185
1995

# *William C. Darrah's European experience in 1935: Paleobotanical connections and stratigraphic controversies*

**Elsie Darrah Morey**
*Morey Paleobotanical Laboratory, 1729 Christopher Lane, Norristown, Pennsylvania 19403*

## ABSTRACT

**William C. Darrah attended both the Sixth International Botanical Congress and the Second International Carboniferous Congress in The Netherlands in 1935. Darrah exchanged ideas with the paleobotanists attending both conferences, and afterwards he was hosted in The Netherlands by W. J. Jongmans and in France by Paul Bertrand. Darrah collected plant megafossils in the coalfields of western Europe and also examined museum collections in The Netherlands and France. The entire experience made a lasting impression on him and set the stage for his interpretations of the stratigraphic paleobotany of the upper Carboniferous rocks of the eastern United States.**

## DARRAH'S EUROPEAN CONNECTIONS, 1933

In 1933, while William C. Darrah was engaged in graduate studies at the Carnegie Museum (Pittsburgh, Pennsylvania), he received a letter from David White (geologist and paleobotanist, U.S. Geological Survey) requesting that he serve as field guide for Paul Bertrand prior to the 16th International Geological Congress in Washington, D.C. (1933) and host W. J. Jongmans following the completion of the Congress. White wrote,

> Very likely Dr. W. J. Jongmans, paleobotanist of the Geological Survey of The Netherlands, will have written to you of his plans to visit several of our American coal fields immediately after the conclusion of the 16th International Geological Congress, about the last week in July.
>
> Dr. Jongmans wishes to see the type sections of our Carboniferous formations and to have opportunity to collect fossil plants from them.
>
> He will, of course, go to Pottsville and Pittston, and probably Halberstadt, Unger, or someone else will assist him with advice, at least in the Anthracite region.
>
> In western Pennsylvania he will of course wish to see the type sections of the Allegheny, Conemaugh, Monongahela, and Dunkard. Probably he can best approach the Dunkard from the south at Morgantown [West Virginia], where I think Mr. Oscar Haught will go about with him through a considerable region of northern West Virginia and eastern Ohio.
>
> But in any event, Jongmans will want to visit you and your laboratory and to learn more of your work. Probably, too, you can tell him where he will find mines whose dumps give fine chance for fossil plant collecting or outcrops which he may dig into with hope of success. His financial support is so slender that he is obliged to plan his itinerary very economically. Any advice or any sort of a lift that you can give him in these matters he will surely very much appreciate.
>
> Unless Jongmans proceeds from Pittsburgh to Chicago, I think he very likely will drop down to Morgantown or Wheeling and from there get material from the Monongahela especially and find unexampled opportunities to collect within a relatively short radius samples showing the progressive carbonization from Zanesville, Ohio, to Westernport, Md. Probably he will be loath to believe in the progressive carbonization of a single coal bed until he has actually collected coal from its westward outcrop along the B. & O. between Zanesville and St. Clairsville, and again in the Georges Creek Basin.
>
> From this region of the coal field he will proceed southward to the Kanawha and thence to the Pocahontas field, from which he will perhaps decide to go to Chicago and see something of the Illinois basin before proceeding to Kansas, Oklahoma, and Texas, all of which he is ambitious to do, though his financial backing is, I fear, woefully inadequate. Any help he can get in automobile transportation, letters of introduction, and advices as to whom to see, and what points to visit will be greatly appreciated by him and by me also.[1]

A lasting friendship with each of the two European paleobotanists ensued. From these first contacts Bertrand died in 1944, and Jongmans died in 1957. Their lives are described in Chapters 7 and 5, this volume.

Morey, E. D., 1995, William C. Darrah's European experience in 1935: Paleobotanical connections and stratigraphic controversies, *in* Lyons, P. C., Morey, E. D., and Wagner, R. H., eds., Historical Perspective of Early Twentieth Century Carboniferous Paleobotany in North America (W. C. Darrah volume): Boulder, Colorado, Geological Society of America Memoir 185.

At this same time, White had asked for a "gentleman's agreement" (Phillips et al., 1973, p. 23) with both paleobotanists, requesting that following their collecting expeditions in the Appalachian basin, they would not publish any new species from their American collections without first consulting him. Apparently at this time White was preparing a paper on the Pottsville floras of the Illinois basin. Jongmans ignored this agreement completely. Upon his return to The Netherlands, he collaborated with W. Gothan and published an "unfortunate" paper on the American Carboniferous in which they proposed about 10 new species (1934). They maintained that there was "no Stephanian in the United States east of the Mid-Continent. . . ."[2] White was furious. As he wrote to Charles Schuchert, director of Yale University's Peabody Museum of Natural History, "So far as I know [neither Jongmans, Gothan, or Bertrand has] given succinct stratigraphic or paleobotanical definitions of Stephanian. . . . In the same paper he [Jongmans] makes some remarkable and highly conflicting statements. . . . Some of their identifications are, in my judgement, down-right faulty. Others are suggestive and helpful."[3]

White was not alone in his reactions. C. O. Dunbar[4] and Schuchert[5] wrote of their objections in their letters to Darrah. Indeed, as it happened, most American paleobotanists were already irritated by the regularly expressed opinion of the European paleobotanists that Americans had accomplished nothing in Carboniferous paleobotany. It was understandable, therefore, that the Americans who were to attend the 1935 congresses (W. C. Darrah, C. A. Arnold, and R. C. Moore) felt that they would be fighting for their professional integrity and knowledge of stratigraphy in their own research areas.

William (Bill) Darrah was soon to enter into this scientific maelstrom. Unknown to him at the time, Schuchert encouraged Oakes Ames, director of the Botanical Museum of Harvard University, to find the cash to send Darrah to Heerlen so he could present the American position to Jongmans. This assured, he then wrote Bill Darrah and charged him "to tell the paleobotanists what you think about Dr. White's stratigraphy and paleobotany. You will do it in good Darrah style, modest but positive. You have the facts well in hand and will have to stand up against the doubting Gothan—Jongmans will be friendly enough."[6]

## THE CONGRESSES IN THE NETHERLANDS, 1935

Bill Darrah, age 26, was sent from the Botanical Museum of Harvard University in 1935 by Oakes Ames to attend the Sixth International Botanical Congress (Amsterdam) and the Second International Carboniferous Congress (Heerlen). Darrah was to present invited papers on American Carboniferous floras. Darrah, accompanied by his wife, Helen, set out for Europe by ship. Once in Europe, he received additional financial assistance. Thanks to the efforts of Professor Paul Bertrand, Darrah received a grant from the University of Lille so that he could collect in the French coal basins as well as study the museum collections at the University of Lille. In addition, the Commission for Relief in Belgium (Belgian-American) Educational Foundation awarded him a small grant for study and travel in Europe ($75.00, or 2,214 francs). The rest of this chapter contains information from Bill Darrah's field notebooks and lecture notebooks and from a diary kept by Darrah and his wife throughout their European travels in the fall of 1935.

The first several days of the Darrahs' stay in Europe were connected with the Sixth International Botanical Congress, held in Amsterdam (Fig. 1). In the Darrah diary, Darrah wrote on September 1 that after breakfast "I headed to Headquarters to register for the Congress. There I met Drs. Seaver, Sinnott, Weston, and Linder. The paleobotanists got together in a corner with others in attendance. They were Walton, Sahni, Harris, Thomas, Bertrand, Renier, Arnold, Jongmans, Koopmans, Florin, Halle, Gothan and Pia" (Fig. 2). Most of them were mainly speakers on the later program.

The plenary meeting began on September 2, as did the Paleobotanical Section meeting. That evening everyone attended a reception held by The Netherlands Government. Darrah introduced E. D. Merrill, director of the Arnold Arboretum of Harvard University and administrator of the botanical collections of Harvard University, to Paul Bertrand and W. J. Jongmans ("who were at it already!") while they were there. Merrill gave permission for Darrah to accept Bertrand's offer to come to France to study the fossil collections.

On September 3, Darrah presented his paper entitled "Recent studies of American pteridosperms," which was later published in the Compte Rendu (Darrah, 1937b). The diary continued, "The program began with H. H. Thomas, followed by Halle, Harris, Hirmer, Darrah, Arnold, Sahni and ended with Walton" (Fig. 3). "Thomas [1938] was the best on the program but he had plenty of opposition to his paper entitled 'Pteridosperm evolution and the angiosperms.' " In summing up the conference, Darrah wrote, "Like all speeches, some are excellent; others are not."

At the end of the evening of September 5, Jongmans, T. G. Halle, Gothan, R. G. Koopmans, and Darrah went to a café for coffee. The discussion soon became paleobotanical. The Darrahs learned that some paleobotanists did not agree with Bertrand: "It is because he quotes 'authority' when he speaks and because he splits species. As a result of his splitting of the species, he changes the aspect of the fossil record from each horizon and some of the common abundant forms are therefore absent because they have become 3 or 4 other species." After this lengthy discussion, it started to rain, and so then the group heard a coffee story on Gothan, which Helen Darrah recorded. "It seems Gothan [Fig. 4] likes very strong coffee and decided to have some at Maastricht [The Netherlands]. Well the coffee was too weak to suit him so he complained. Just as he got service it started to rain. There was a leak in the awning of the roof so that Gothan had to sit with his umbrella open so his coffee would not become dilute."

The Botanical Congress was summarized by attendee

Figure 1. Paleobotany Section, Sixth International Botanical Congress, Amsterdam, September 2–7, 1935. Front row, left to right: Miss Berridge, R. Kräusel, R. G. Koopmans, H. H. Thomas, J. Pia, Mrs. A. C. Seward, A. Renier, A. C. Seward, T. G. Halle, W. J. Jongmans, F. E. Weiss, B. Sahni, J. Walton, P. Bertrand, W. Gothan, M. Benson, W. L. Jepson, Mrs. H. H. Thomas. Back row, left to right: W. Reichardt, C. A. Arnold, G. Erdtman, H. Gerth, S. Leclercq, T. Reinhold, R. Florin, M. Hirmer, T. M. Harris, W. Zimmermann, F. Knoll, Mrs. W. C. Darrah, W. C. Darrah, H. V. Krick, E. Hoffmann, Mrs. P. Ledoux, F. Stockmans, W. N. Edwards. William Culp Darrah Collection.

Chester Arnold in 1936, under the heading "Paleobotany at the International Botanical Congresses":

The Sixth International Botanical Congress, which convened in Amsterdam during September, 1935, was the second to have a separate paleobotanical session. The first was in Cambridge, England (1930).

Several paleobotanical papers were given at the Ithaca, New York, Congress in 1926 but only in conjunction with the joint session on morphology, histology, and paleobotany. Among the participants were R. Florin, G. R. Wieland, A. C. Noé, R. Chodat, and R. B. Thompson. Most of the discussions dealt with the morphological aspects of fossil plants and their bearing on evolution and phylogeny. Of special interest was Professor Noé's demonstration of the then recently discovered coal balls from Illinois, reportedly the first to be discovered in North America.

Figure 2. From left to right: T. M. Harris, C. A. Arnold, T. G. Halle, W. C. Darrah, D. N. Wadia, H. Krick (Bartoo), W. N. Edwards, R. Florin. Photograph by B. Sahni during the Sixth International Botanical Congress, Amsterdam, September 2–7, 1935. William Culp Darrah Collection.

At the Cambridge Congress, where D. H. Scott presided over the Paleobotanical Section, the papers were grouped according to their bearings on the major problems of paleobotany. These groups included the antiquity and early evolution of the angiosperms, the earliest known terrestrial floras, the position of the pteridosperms in the plant kingdom, the stratigraphic relations of Permian and Carboniferous plants, and the relation of late Paleozoic to early Mesozoic floras. Of special significance was the account by H. H. Thomas on the Caytoniales and T. G. Halle's discussion of the morphology of Whittleseya. Also, the actual attachment of the seed to the foliage of Lyginopteris was reported by W. J. Jongmans for the first time.

The Paleobotanical Section of the Amsterdam Congress, under the chairmanship of T. G. Halle, was organized in much the same manner. The relation between the Permian and the older Paleozoic floras was given special attention, with stress on the Permo-Carboniferous floras of eastern and northern Asia.

The general effect of the Cambridge and Amsterdam Congresses has been to increase acquaintances among paleobotanists of different nationalities, and to promote a general understanding of paleobotanical problems the world over and the relation of these problems to those of one's own country. The scope of the discussions in future congresses would be considerably broadened if more American paleobotanists were to participate.[7]

On September 8, the Darrahs arrived in Heerlen. Dr. Jongmans had made all the arrangements for the congress. They went by taxi to the Retraitehuis, which turned out to be a

Figure 3. From left to right: W. C. Darrah, C. A. Arnold, B. Sahni. Photograph taken by H. H. Darrah during the Sixth International Botanical Congress, Amsterdam, September 2–7, 1935. William Culp Darrah Collection.

Figure 4. From left to right: W. J. Jongmans, A. Renier, and W. Gothan. Photograph taken by W. C. Darrah during the Second Carboniferous Congress, Heerlen, September 9–12, 1935. William Culp Darrah Collection.

Catholic nunnery, the only place large enough to accommodate everyone in attendance.

The Second Carboniferous Congress (Fig. 5), which was organized by Jongmans, opened the next day. The Paleobotanical Session began with work on the lower Carboniferous rocks. Darrah carefully listened to papers presented by those in attendance and made an observation of those he felt were the best in the field at this time. In order of capability were A. Renier, Gothan, Bertrand, and Jongmans—whom Darrah called "the big four of paleobotany of the Paleozoic." All of these men are about the same age, he noted.

In subsequent discussions, it soon became evident that Jongmans did not know of the recent Mississippian-Devonian controversies of Chadwick (1933a, b; 1935a, b). "Moore believes, as I do, that Jongmans does not know the American stratigraphy. Especially the flora of the Pocono and the Price is in a tangle. Some species are Pennsylvanian (Namurian)," Darrah wrote in his lecture notes taken during the symposium.

> Jongmans found the whole series from the Namurian to Westphalian C in West Virginia [in 1933] (Jongmans, 1937a, b, c). On that I agree. Moore says I should challenge or at least ask Jongmans about the fact that we [Americans] have never found what he has found in the same horizons. It is true that Jongmans has a *Lepidodendropsis* and a *Triphyllopteris* flora but from where? Jongmans has absolutely no evidence on Chadwick's proposition. Arnold summarized the work of Chadwick and also cited the objections of David White. Arnold in discussing Chadwick's work gave no new data. Arnold said the *Triphyllopteris* flora occurs only at Pottsville . . . to the south.

The first speaker at the September 11 paleobotanical session, Darrah presented his paper "American Carboniferous floras," which was later published in the Compte Rendu (Darrah, 1937a). The paper included new data on stratigraphic paleobotany and the floras succession and made a convincing case for a perfectly good Stephanian in the eastern United States. Afterward, R. C. Moore, P. Pruvost, Bertrand, and E. Dix were in agreement. Moore was the next speaker, followed by Jongmans. At this time, Jongmans conceded the existence of both Westphalian D and the Stephanian in North America. "We are floored! We came 3000 miles for a fight and it fails to materialize. Oh well, we won!" Darrah noted with mixed feelings. Then, Dix reported that she had found the same thing in Wales (see Dix, 1934, 1937). Halle was next, speaking on China, "and he agrees on our *Lescuropteris* and *Taeniopteris* zones and our ideas of *Gigantopteris*."

In his monograph (1969), Darrah explained that Jongmans and Bertrand held conflicting opinions on the identity of about 10 Carboniferous plant species, though for the most part they were in agreement. The main problem in their conflicting views is the diversity of forms of *Neuropteris ovata* and *N. scheuchzeri,* especially those found in post-Conemaugh (i.e., Monongahela and Dunkard) strata. There is great vertical range of these forms in the United States but ostensibly a very short range in Europe. This fact is poorly understood and is a source of controversy (see Chapters 25 and 26).

Darrah wrote in his lecture notes, "I am fully concerned now about the significance of the zones of *Neuropteris scheuchzeri* and [*N.*] *ovata.* It is not as Jongmans says but as Bertrand says." In his field notebook, Darrah stated that "Bertrand says *N. scheuchzeri* is rare here and occurs so seldom that it is an excellent guide species. This is an extremely important observation."

On September 12, the paleobotanical arguments dealt with the characterization of the Westphalian D and the Stephanian. Bertrand described his own work but failed to convince most people because of his "splitting" and species determinations. Early in the day, Darrah had learned that there was a "conspiracy" to give Bertrand his Westphalian D so that the Stephanian might be defined away. There was no problem in units older than the Stephanian.

After listening to Bertrand's presentation, Darrah noted:

> We never find *P. dentata-plumosa* higher than mid-Allegheny nor [*Sphenophyllum*] *oblongifolium* nor *P. bredovi* with Westphalian D or with [*Sphenophyllum*] *majus* and [*S.*] *emarginatum.* The same is true for *S.* [*Sphenopteris*] *quadridactylites,* which is limited to the lower Allegheny, with the possible exception of two examples from the middle Allegheny of West Virginia and a doubtful specimen from the upper Allegheny of Pottsville. Neither do we find *Odontopteris reichi-minor* in Westphalian D. I must check these problems in the field with Bertrand. I need especially to check the St. Etienne flora. It is very necessary to see what he [Bertrand] calls *Alethopteris grandini* and *Mixoneura neuropteroides.*

Then Darrah noted, "Renier wants to accept the American succession. Gothan, using Moore's argument, wanted to wait until the next congress. Bertrand backs us and Renier will use the Darrah succession. Thus no trouble until the Stephanian. The section decided to scrap it until the next congress because the American succession will be the World Standard!"

The evening was spent with "many new friends. We danced some or rather, Bill danced the closing number with Helen. . . . But the real dancing treat of the evening was performed by [Walther] Gothan and [Emily] Dix. To the orchestra, reinforced with clapping hands, they danced nobly. Gothan must have been a gay blade in his younger day. Gothan turns out to be swell. He was well on his way to bachelorhood and then up and married late. No children. He plays the flute, piano and violin. It was great fun."

## THE COLLECTING TRIPS TO THE COAL MINES IN THE NETHERLANDS AND BELGIUM

The tours of the coal mines began on September 13, 1935. The trips allowed for collecting specimens in the Westphalian A (Finefrau-Nebenbank, Wilhelmina mine, and Oranje-Nassau)

Figure 5. The Second Carboniferous Congress, Heerlen, September 9–12, 1935. Not everyone in the group photo could be identified. Those that could be identified are located by their position relative to the left end of the row. First row: J. Walton (2); J. Pia (3); P. Bertrand (5); W. Gothan (7); E. Dix (8); W. J. Jongmans (9); A. C. Seward (10); A. Renier (11); M. Hirmer (12); P. Pruvost (13); R. C. Moore (14); Mrs. H. H. Thomas (16); Mrs. W. J. Jongmans (17); Mrs. W. C. Darrah (18). Second row: W. Reichardt (2); F. Stockmans (5); R. G. Koopmans (8); H. H. Thomas (16); C. A. Arnold (17); M. K. Elias (18); F. Sans (19); W. C. Darrah (20). Third row: B. Sahni (2); W. N. Edwards (3); L. Emberger (4); T. M. Harris (6); R. Kräusel (8); T. G. Halle (9); R. Florin (10); K. Patteisky (13). Fourth row: Manasse (1); I. Patac (2); R. Jongmans (4).

and in the Mine Emma, Aegir horizon, Westphalian B. On the fourth trip, September 16, *Lonchopteris* was finally found in the Westphalian B, at Mine Maurits, The Netherlands.

Did we find *Lonchopteris!* It was the happy hunting grounds of all British and American paleobotanists. We gathered six boxes full of specimens—two were exclusively *Lonchopteris*. Edwards said it was positively criminal the way we left them to weather. Before we met Renier to go down the mine [in Belgium] we carried out to completion our organization of the "Society for the Preservation of *Lonchopteris* unincorporated" [Fig. 6].

We donned miners garb from trousers, to shoes, to chemise and two or three hats so that the miners helmets would fit. After photos [Figs. 7 and 8], we descended the Wérister colliery by 640 m. We slopped through water and finally saw some marvellous "coal" balls [and] roof coal "balls." One can scarcely imagine what a mass of them can number. They occur literally by the thousands and every one contains plant material. Renier calls the roof coal ball "mal de mer" du charbonnage. We collected a box full of coal balls and one of impressions. So up again to the surface to wash and shower in *warm* but black water. Then, by invitation of the colliery owners, it was to dinner at the engineer's club.

The diary continued, "Up until this time Bill has been having trouble to get enough to eat due to his normal eating problems (see Chapter 1, this volume). The food has been classified as 'cuisine bourgoise' or it's positively poisonous. Thus the meal at Wérister is described. What a meal! What a meal! And did we eat!

Port in the lounge, first rate (seconds)
Then we filed into the salle à manger for

Thin soup, consommée with swell vegetables (seconds)
Moselle (seconds)
Champignons in Pattie shells (seconds)
Beer (seconds)
Beef-steak & boiled potatoes (seconds)
Peas (thirds)
Claret (seconds)

Fromages (seconds)
Fruit and excellent coffee (all one could eat and then some)
Cognac (Alas, no seconds; Helen says, 'Thank God'.)"

Looking back over the day's activities, Bill Darrah observed, "We did some very enthusiastic and buoyant collecting, until I almost had my left little finger guillotined by the

Figure 6. *Lonchopteris rugosa* Brongniart collected by W. C. Darrah in 1935 from Mine Maurits (Westphalian B), Heerlen, Holland. The specimen was purchased by P. C. Lyons in 1974 from the Museum of Comparative Zoology Museum Store, Harvard University. Photography courtesy of P. C. Lyons. Scale in centimeters.

Figure 7. W. C. Darrah (left) and H. H. Darrah (right) in a photograph taken at the Wérister Colliery, Belgium, September 17, 1935. William Culp Darrah Collection.

Figure 8. From left to right: C. A. Arnold, P. Bertrand, E. Dix, A. Renier, and W. C. Darrah. Photograph taken by H. H. Darrah at the Wérister Colliery, Belgium, September 17, 1935. William Culp Darrah Collection.

chief engineer of the Wérister Mine who was helping us remove the specimens."

## EXAMINATION OF THE COLLECTIONS IN THE NETHERLANDS

On September 18, Darrah began studying the collections at the Geologisch Bureau voor het Nederlandsche Mijngebied in Heerlen, which Jongmans admits is "the best American collection in all of Europe." The collections were "remarkably representative of the whole world" though by no means comprehensive. Darrah found "43 mistakes or very doubtful" determinations of the American specimens. He made small cards for most of these and indicated them to Jongmans.

Also at that time Darrah noted in his diary, "There is still some doubt concerning the joint papers on the American Carboniferous [Darrah] with Jongmans and Gothan." Darrah insisted there would be no inclusion of Westphalian D or Stephanian. "So far Jongmans agrees. He will also send a complete manuscript and illustrations for our opinion."

Continuing with his notes on September 19, Darrah recorded, "Jongmans and I continued studying through the collections, species for species and telling me what I wish or what he thinks I should wish for the Harvard Collection. He is also picking out numbered specimens for us, so that we have a complete and comparative set in America. We got through *Pecopteris* and *Sphenopteris* in the morning and *Alethopteris* and *Mariopteris* during the afternoon."

The work was completed on September 20, as Darrah studied some *Sigillaria* and *Neuropteris.* He brought the day to a close studying *Sphenophyllum.* The selection of specimens, when packed, filled three boxes. They represented sets from the Austrian Alps, Sumatra, Saar (Germany), and Czechoslovakia. Among the fossils are wonderful fruiting specimens of *Sphenopteris hoeninghausii* with attached seeds.

On September 21, agreement was reached concerning Darrah's two joint papers with Jongmans and Gothan. In his diary Darrah wrote, "The manuscripts will be sent later. Despite all the warnings of Edwards, Dix, Harris and Arnold we cannot believe Jongmans would doublecross us and publish without our agreement and still use the Darrah name. Of course, it is possible but there is too much at stake."

Darrah stood between two lines of fire, at home and abroad. "The joint authorship abroad can do nothing but good, but in America things lined up differently. Jongmans had encountered difficulty with the [U.S. Geological Survey] and The Geological Society of America who turned down his manuscript for publication. At home they might call me a traitor. If their work is published in full—erroneously—it is tragic. If I can repair some damage by joint authorship, it's worth the try. Their paper [Jongmans and Gothan, 1934] was lousy and they tentatively agree on all of our corrections."

Schuchert wrote Darrah,

> I want to thank you for your surprising letter. So Jongmans admitted in public that his and Gothan's paper was "hasty and unfortunate." That is what the Geological Society of America thought of their effort, which was sent to the Society for publication and was refused.

When Dunbar read your letter, he said I am glad to know that our recommendation to the Geological Society of America was the correct one. We have known all along that we have much in Pennsylvania and Kansas that correlates with all the Stephanian of Europe. . . .

You did well to bring to the Europeans a lot of literature of which they know little or nothing. That is where the whole trouble lies; their correlations are based on the European sequences, and what the American should have. It is fine for Gothan to say in public that the world standard for the Upper Carboniferous and Permian will have to be built on the sequences in the United States."[8]

From his Netherlands studies and touring the mines in Limburg with Jongmans (Fig. 9), Darrah learned two approaches to his investigations: to study the floral succession from the basal Carboniferous to the Permian and to study the limnic and paralic facies floras of the uppermost Carboniferous. Jongmans taught him not to trust descriptions and to publish illustrations.

## DARRAH'S INTERACTION WITH BERTRAND IN FRANCE

With their stay in The Netherlands completed, the Darrahs moved on to France. Darrah began his studies in France under the direction of Professor Paul Bertrand of the University of Lille. From Bertrand he learned the concept of geographic races of fossil species and the genetic implications of such phenomena. This knowledge would be put to good use as Darrah's career advanced and his writings proliferated.

Darrah accompanied Bertrand to his laboratory to view the photographic apparatus, most of which was of Bertrand's own design. The reproductions are simply marvelous. Darrah was impressed with the complicated flora of the Westphalian D from his examination of the laboratory specimens. Aside from the dubious determinations, it was possible to recognize the definite Stephanian species. The whole aspect is different from the flora of the Rive-de-Gier and of typical Westphalian C.

Figure 9. From left to right: W. J. Jongmans, W. C. Darrah, E. Dix, and R. Kräusel. Photograph taken at the Wérister Colliery, Belgium, September 17, 1935. William Culp Darrah Collection.

During the evening, the Bertrands and Darrahs looked over bound copies of "L'Illustration" and two long manuscripts (monographs) that belonged to Paul Bertrand's father, C. E. Bertrand (see Chapter 7, this volume), who submitted them in 1877 for a prize that was never awarded. One paper dealt with the anatomy of *Selaginella;* the other concerned the comparative anatomy of the integumented seeds of angiosperms and gymnosperms. Bertrand the elder was so furious over the decision that he would not receive a prize that he never edited the works for publication. There were more than 150 marvelous plates drawn with camera lucida. C. E. Bertrand died during World War I at Lille at the age of 66—"still a young man," noted the 26-year-old Bill Darrah.

In late September, Darrah demonstrated his peel technique (Darrah, 1936) to Paul Corsin (1904–1983). Corsin later became Paul Bertrand's successor at the University of Lille. Corsin then used the peel method and managed to obtain a good peel from an Autun specimen. Specimens of Westphalian C and D were studied throughout the day. The next stop was St. Arold (Saar-Lorraine Basin) to meet M. Huchet, the director of the mine. In his diary Darrah wrote, "Near Faulquemont is a new mine. It is simply a model of modern equipment [We] collected many specimens of Westphalian D." Darrah and his follow collectors are shown in Figure 10.

In early October, the Darrahs traveled to the School of Mines (Ecole des Mines) at St. Étienne and saw the plant collections made by Paul Bertrand during and after World War I. All the specimens are "Stephanian.

"The decision was made and the chance taken to collect at Grand Croix. No one has been able to collect much since 1900. Bertrand has collected there many times and has been successful in obtaining only two specimens. It rained all morning so the time is ripe for collecting," Darrah wrote in his diary. They located a field that had just been plowed over the "outcrop," which was buried by some meters. It was the precise locality worked by Brongniart and Grand'Eury. Bertrand

Figure 10. From left to right: W. C. Darrah, P. Bertrand, T. G. Halle, and R. Florin. Photograph taken by H. H. Darrah at Faulquemont, France, September 27, 1935. William Culp Darrah Collection.

obtained three specimens, and Helen and Bill gathered twenty-eight. Specimens collected were seeds, wood, fructifications, leaves, and petioles. "The guardian angels of plenty have been with the Darrahs again. It was a grand experience and the stuff is priceless."

Darrah returned to the School of Mines with Bertrand. There they studied the types of Zeiller and a few of the specimens of Grand'Eury. Darrah wrote,

I showed Bertrand in his own backyard!—*Neuropteris* type *ovata* in the upper Stephanian. Zeiller described *N. stipulata* which is a true *Mixoneura.* Bertrand was amazed, although he did know of the supposed *N. heterophylla* of Zeiller from the Stephanian which look the same. Neither Gothan or Jongmans in their many visits to the school have noted these specimens!

Bertrand admits this is the true *ovata* and is the highest Westphalian "D"—This is coal #8. In coal #10 *Mixoneura ovata* declines and we have Stephanian—Rive-de-Gier. Exactly as in America.

This is the Mason flora [of Pennsylvania—*not* the Mazon Creek flora of Illinois]! After we have argued all these years!

During the period 1927 to 1929, Darrah had tabulated some 16,000 specimens from the Mason shales in Allegheny County, Pennsylvania (Darrah, 1969). The Mason shale is in the basal part of the Conemaugh Group between the Mahoning coal bed and the Brush Creek Limestone Member.

Darrah then went to Paris to the Museum of Natural History to view the specimens of Grand'Eury and Renault. Together, Darrah and Bertrand studied the type specimens of Brongniart. Darrah wrote in his diary, "After 4 hours of heckling and blathering we met A. Loubière [director of the Comparative Anatomy Laboratory of Fossil Plants at the Musée d'Histoire Naturelle, Paris, France]. At first his behavior was [not polite] and Bertrand became angry. Professor [Bertrand] changed his tactics and began elaborating upon our importance, our connections, and our reputations. Loubière could not be outdone, so he most obligingly began to show us slides and tomorrow he will show us the specimens of Brongniart." In summary, Darrah noted that he had seen marvelous material at the museum, but the variety was small.

Darrah recorded, "We met Dr. Czeczott who informed us that (without her knowing it!) we have the 'lost' Unger types at Harvard. Czeczott and many other European workers have hunted them for four years not realizing the types are housed in the Harvard's Paleobotanical Collection. What luck for us. We still find that Harvard's collection is better, richer and more representative than in continental Europe! It has no equal."

The diary entry for October 4 reads: "The Bertrands must leave tonight for St. Germain and we gave them parting greetings. Bertrand shed a few honest tears and said it was one of his happiest moments when we arrived in Lille. Mme. Bertrand kissed Helen tenderly. They have been really swell to us—there's no other words for it." A great bond had been formed between the Bertrands and the Darrahs.

## THE DARRAHS RETURN HOME

The week-long return trip across the North Atlantic presented the Darrahs with an alarming problem. As Mrs. Darrah wrote in their diary,

On October 15th, Bill could not sleep; his chest was very congested. At midnight we called the ship's doctor who gave Bill a thorough examination and said it was asthma and gave him sleeping powder but it did not help much. The doctor came at dawn with more pills and a compress. After the compress Bill could sleep and slept to noon. Bill was put to sleep that night with an injection of morphine by the doctor.

The next day, Bill ate lunch at the insistence of the steward. After two hours of wheezing and whistling we called the doctor for help. We made a discovery. The doctor put Bill on his back, bared his chest and stuffed the woolen blanket around his neck in order to take his body temperature. Bill damned near suffocated to death. It's the lint of the woolens causing the asthma. So Bill dressed and "tottered" to the "A" deck, where he recovered "in a fashion" in about an hour and was able to eat supper in the drawing room. Immediately upon returning to the stateroom at 10:30 P.M., however, the wheezing recommenced. So, carrying his covers and fully dressed, Bill returned to the drawing room where he spent the night sitting in a chair—*but without the asthma.* The only interruption was a couple about to attempt a relatively complicated biological phenomenon in a public place." He was awakened at 7 A.M. by the priests preparing for Mass. Before [he] could pick up his linens and leave, the first mutterings were under way. Bill heard 3 full masses and finally was purged enough to find Helen to go to breakfast, . . . feeling fairly swell but with some wheezing.

The doctor was willing to change the Darrahs' cabin, but they were ready to take a chance on sleeping in it after all of the "frowsy" blankets had been covered in sheets. As it turned out, Bill was able to sleep; the covered blankets "did the trick."

## THE AMERICAN REACTION

Following the completion of the Second Carboniferous Congress, Jongmans wished to publish the proceedings quickly, though, as it happened, the three-volume set was not published until 1937.

In the meantime, there was great puzzlement among the American paleobotanists concerning the delayed appearance of their own accounts of the Amsterdam and Heerlen Congresses. What was Jongmans doing with their manuscripts? C. A. Arnold wrote to Darrah, "R. W. Brown says that he is asking you as well as myself to submit an account of the Amsterdam and Heerlen Congresses for inclusion in the annual report [National Research Council, Committee on Paleobotany]. What I want to suggest is that if you haven't done so, to send in a rather full account of the discussions and conclusions arrived at. . . . I appear to have lost the program."[9] Again, on March 18, 1936, Arnold told Darrah, "Enclosed is a copy of the account . . . if he was counting on prompt publication of the proceedings . . . one would hear something by this time."[10]

And just what was Jongmans doing with the Pocono flora papers? "I think you are quite wise in refusing to have anything to do with his publication of the Pocono material. There would be nothing in it for you and you would just be the butt of all the adverse criticism that would result. And that is exactly what he [Jongmans] wants. He is just like lots of the rest of them; he thinks nothing that amounts to anything has ever been done in this country and that he can do anything with American material he wants to and expects to get away with it."[11] It was assumed by some that Darrah would know what was going on if any American would know. He had established a special rapport with the Europeans, or so it seemed.

Certain aspects of the Americans' position seemed to have achieved a new credibility after the Heerlen experience. Paul Bertrand wrote in 1935 and 1936, asking for information on stratigraphy of the Westphanian and Stephanian in the United States. Bertrand was true to the gentleman's agreement with David White; he did not break it. Rather, an exchange of information ensued. He wrote to Darrah, "You ask me for new informations. Of course I am always willing to send to you any advice or information you may think useful for the progression of your search. . . . I offer you my best compliments for the photo enclosed in your letter of May 25. It enabled me to make a close and accurate comparison with French specimens. *There is not the slightest doubt,* that you got there in the Monongahela Formation two guide specimens."[12]

In another letter, Bertrand continued,

> We did present at Heerlen last year before the Congress upon floral characteristics of Stephanian as distinguished from Westphalian. Possibly, Prof. Gothan and Jongmans are not [pleased]. But what can satisfy them? We already made a detailed statement of all available facts. Either they admit a subdivision of the coal formation in two parts, or they prefer to treat it as a whole. That's purely a matter of taste and opinion. It would be possible, of course, to also put Namurian coal and Permian coal into Westphalian series. I fear much the point is they refuse to admit Stephanian series because the series was a long time ignored in Germany. All I can urge on yourself is always to emphasize the real importance and value of Stephanian floras."[13]

Nevertheless, all was not smooth sailing. Darrah's publications were criticized by M. K. Elias in his letter of 1935. Elias was well known among American paleobotanists for his outspoken critiques and comments.

> You may consider me too old-fashioned and conservative, but I do not think it was good idea to attempt to "stimulate debate among colleagues" in Europe by publishing a couple of geographically poorly and generally carelessly written abstracts, which are too long to be considered and excused as mere abstracts. [See Darrah and Bertrand, 1933, 1934.
>
> If you succeeded, as you claim, in making some corrections in regard to plant distribution and stratigraphic relations of some Appalachian geologic sections—they are worth to be published with substantial and carefully written proof. I do not see at all, however, why you should become impatient and start to venture in difficult and responsible problems of intercontinental correlations, for which task you can be hardly prepared. The chief duty of the American stratigraphic paleobotanists, as it seems to me, is the correlation within our own provinces, and you impress me as being not prepared, as yet, to attempt even this.
>
> If your paper of "50 printed pages" which you intend to publish is not to be accompanied by illustrations (and they should be good) of the form-names which you list in your text—I would question the value and usefulness of it. There is so much confusion in the literature in regard to the meaning of many form-names used by paleobotanists that it became absolutely necessary to make it clear, just what an author means under the used names, especially when applying European terms to the American plants. Nobody knows as yet just what conception of genera and species you hold. Your joint paper with Bertrand, indeed, allows one to assume that your conception of species listed is that of Bertrand, as adopted by him primarily for the European forms. This however is only an assumption. Furthermore, we are justified in being careful in accepting preliminary identifications, especially those made by even accepted authorities of European plants, when they hastily identify the American forms with the European species. Some of the superficial resemblances may be entirely misleading. For instance some identifications and correlations just published by Jongmans are of questionable value. His identification of *Callipteris* in Garnett flora is particularly objectionable. Thru personal communication with him I established that he identified for *Callipteris* a beautifully preserved form, which neither I nor D. White ever considered to be a *Callipteris,* but which is considered by both of us a new species. I [Elias] am describing it as *Dichophyllum mooreii* (Elias, 1937); Moore and Elias, 1937; Moore et al., 1936). the plant is related to *Mauerites* (from Russia), *Baieria,* and *Psygmophyllum,* but not to the typical *Callipteris.* It is very unfortunate that Jongmans based his misidentification of *Callipteris* on incomplete duplicates of this new form, which I gave him for representative collection from Garnett.
>
> Now there may be some excuse for the foreign paleobotanists in making hasty superficial identifications and conclusions (for instance about absence of the Stephanian, equivalents in the Appalachian region) in regard of the American paleontology and geology but there is hardly any excuse for us here, when we hurry and publish immature and questionable identifications and correlations, because this does not help the stratigraphy but only adds complication to already complicated and somewhat confused situation.

It seems to be quite commendable that you do not introduce any new names in your articles, but the fact that you avoid to mention discovery of any new forms in the old and new collections at your study impresses me as indication of superficial examination, even if you did discover some fruits which you intend to describe.[14]

The question of Darrah's relationship to Jongmans and his publications on American Carboniferous floras continued. In 1940, in a letter to Charles B. Read (see Read, 1955) of the U.S. Geological Survey, Darrah fully explained his role in the controversy and his efforts to put the matter to rest.

Professor Ames has taken me to task for certain supposed violations of unpublished work of David White. He has told me that you replied to his letter concerning this matter and that you have clarified the situation considerably—especially that the charge refers to Jongmans and not to me. However, several facts relating to my association with Jongmans should be noted and I am writing to you to give a brief account of my part in the controversy.

In 1933 Jongmans visited this country and collected fossil plants and made observations at numerous localities. I accompanied him in the field in western Pennsylvania, West Virginia, and eastern Ohio. He stayed in my home, studied my collections, and we became intimate friends.

Upon his return to The Netherlands he published, in collaboration with Gothan, an unfortunate paper on the American Carboniferous in which there were proposed some eight or ten new species [Jongmans and Gothan, 1934]. These were published in the Jaarverslag over 1933 (1934), Geologisch Bureau voor het Nederlandsche Mijngebied.

In 1935 I attended the Second Carboniferous Congress at Heerlen, Holland, which Jongmans organized and during the meetings was invited to collaborate on a corrected and elaborated paper on the Carboniferous of eastern North America. At my insistence the project was split into two and for the Upper Carboniferous my opinions are carried as dissenting footnotes. For the Pocono plants I collaborated to make sure that the species of Lesquereux, Meek, and Dawson were included. I told Professor R. C. Moore of this and *he saw no reason why harm should come* of my participation. I have underlined these words because I believe they are the exact words of Moore but I am not fully certain and cannot quote them.

My wife and I remained a week longer in Heerlen studying Jongmans' collections, making corrections in his determinations, and picking duplicates of his European material for the Botanical Museum [Harvard University]. The two papers—one without my name with the title—were published in the proceedings of the Congress [International Carboniferous Congress, Heerlen, 1935] [Jongmans et al., 1937; Darrah, 1937b] in 1937 (issued generally 1938).

There are several facts to which I call your attention, in order to dispel any misunderstanding which might have developed.

1. I did not know that Jongmans had transgressed the work of David White. I did not know that Jongmans collected his own material. Whatever assurances he may have given regarding publication were never divulged to me.

2. I did not know that David White was working on Pocono plants. His last extensive and taxonomic treatment of the lower Carboniferous floras (including the Pottsville) was published about thirty years ago.

3. I did not know that Jongmans had previously invited you to collaborate and that you refused to join them. Jongmans may have told others at the Congress of your refusal, but neither he, nor Moore, nor Arnold (the other Americans present) told me. . . .

The only inkling of this issue which I understood was that the Geological Society of America refused to publish a manuscript by Jongmans and Gothan. Jongmans told me it was because of his erroneous opinion of a hiatus in the late [Upper] Pennsylvanian and the supposed absence of Stephanian equivalents in America. Since these two opinions were corrected and abandoned in the published paper (and in the manuscript, of course), and I had no reason to doubt his word, I willingly joined in helping to correct some of the errors.[15]

The work on stratigraphy continued to be corrected. Seeking further information, Professor M. G. Chaney wrote to Darrah,

In preparing a report regarding classification of Carboniferous rocks in North America, we note a contradiction in the literature regarding correlation of portions of our Pennsylvanian section with that of Europe. On page 104 of your "Principles of Paleobotany" [1939] you placed the upper Pottsville beds with the Westphalian C. This agrees with David White's chart [1936] in the Report of the 1933 International Geological Congress.

We note on the other hand that W. J. Jongmans very definitely classifies the Kanawha with the Westphalian B of Europe, Second Heerlen Congress, 1935 [Jongmans, 1937b].

It would be of much assistance to our Subcommittee [American Association of Petroleum Geologists] if you would give us a discussion of this apparent difference of opinion. . . .16

There is no necessary contradiction in these two age determinations because most of the Kanawha Formation is today considered Westphanian B (Gillespie and Pfefferkorn, 1979), and the upper Pottsville is probably, in part, Westphalian C. This simply implies that the Kanawha Formation is mainly equivalent to the Middle Pottsville in the type area (see Lyons et al., 1985; Chapter 25, this volume).

## CLOSING THOUGHTS

Three decades after the 1935 Carboniferous Congress, Darrah's interpretations of the stratigraphic paleobotany of the late Carboniferous floras of the eastern United States were finally published (1969). This monographic work, which has been well received in the paleobotanical community, gives the details of Darrah's collections and interpretations of the age of the floras in the various coal basins of the eastern United States, a task that he began in 1926. Darrah's European experience was the principal impetus to this fruitful conclusion of a career-long activity.

## ACKNOWLEDGMENTS

I wrote this chapter from information obtained from W. C. Darrah's field notebooks and lecture notebooks and the diary that recorded the Darrahs' trip to Europe. My father often talked about his experiences during his European stay during the fall of 1935. During those times I took notes about the trip. Darrah had hoped to write a book about his recollections but sadly that was one of his uncompleted projects at the time of his death in 1989. I hope that in a small way the chapter reflects the spirit of paleobotany. I wish to thank Paul C. Lyons for the use of the photograph of *Lonchopteris rugosa*. Appreciation is extended to Alicia Lesnikowska for sending me the reference for Moore et al., 1936. T. L. Phillips and W. G. Chaloner are thanked for their thoughtful reviews and suggestions. This chapter is dedicated to my parents, William C. Darrah and Helen H. Darrah.

## ENDNOTES (All of the following are in the W. C. Darrah Collection.)

1. Letter from David White to W. C. Darrah, July 10, 1933.
2. Letter from Charles Schuchert to W. C. Darrah, June 11, 1935.
3. Letter from David White to Charles Schuchert, January 28, 1935.
4. Letter from C. O. Dunbar to W. C. Darrah, June 20, 1935.
5. Letter from Charles Schuchert to W. C. Darrah, June 27, 1935.
6. Letter from Charles Schuchert to W. C. Darrah, June 27, 1935.
7. C. A. Arnold submitted this report to W. C. Darrah for his review on March 18, 1936. Darrah added his remarks about the Sixth International Botanical Congress and Second Carboniferous Congress. The abridged version of this report and the report of W. C. Darrah appear in National Research Council, 1936, p. 1–2.
8. Letter from Charles Schuchert to W. C. Darrah, November 6, 1935.
9. Letter from C. A. Arnold to W. C. Darrah, March 11, 1936.
10. Letter from C. A. Arnold to W. C. Darrah, March 18, 1936.
11. Letter from C. A. Arnold to W. C. Darrah, May 11, 1936.
12. Letter from Paul Bertrand to W. C. Darrah, June 6, 1935.
13. Letter from Paul Bertrand to W. C. Darrah, July 28, 1936.
14. Letter from M. K. Elias to W. C. Darrah, June 22, 1935.
15. Letter from W. C. Darrah to C. B. Read, August 5, 1940.
16. Letter from M. G. Chaney to W. C. Darrah, December 3, 1943.

## REFERENCES CITED

Chadwick, G. H., 1933a, Hamilton red beds in eastern New York: Science, v. 77, p. 86–87.

Chadwick, G. H., 1933b, Great Catskill delta—and revision of Late Devonian succession: Pan-American Geology, v. 60, p. 91–107.

Chadwick, G. H., 1935a, What is "Pocono"?: American Journal of Science, v. 29, p. 133–143.

Chadwick, G. H., 1935b, Faunal differentiation in the Upper Devonian: Geological Society of America Bulletin, v. 46, p. 305–342.

Darrah, W. C., 1936, The peel method in paleobotany: Botanical Museum, Harvard University Leaflets, v. 4, p. 69–83.

Darrah, W. C., 1937a, American Carboniferous floras: Congrès de stratigraphie et de Géologie du Carbonifère, 2d, Heerlen, The Netherlands, 1935: Compte Rendu, v. 1, p. 109–129.

Darrah, W. C., 1937b, Recent studies of American Pteridosperms: Congrès de stratigraphie et de géologie du Carbonifère, 2d, Heerlen, The Netherlands, 1935: Compte Rendu, v. 1, p. 131–137.

Darrah, W. C., 1939, Principles of paleobotany: Leiden, The Netherlands, Chronica Botanica, 239 p.

Darrah, W. C., 1969, A critical review of the Upper Pennsylvanian flora of eastern United States with notes on the Mazon Creek flora of Illinois: Gettysburg, Pennsylvania, privately printed, 220 p., 80 pls.

Darrah, W. C., and Bertrand, P., 1933, Observations sur les flores houillères de Pennsylvanie (région de Wilkes-Barre et du Pittsburgh): Académie des Sciences, Paris, 2d: Comptes rendus, v. 197, p. 1451–1454.

Darrah, W. C., and Bertrand, P., 1934, Observations sur les flores houillères de Pennsylvanie: Société Géologique du Nord Annales (Lille, France), v. 58, p. 211–224.

Dix, E., 1934, The sequence of floras in the Upper Carboniferous, with special reference to South Wales: Royal Society of Edinburgh Transactions, v. 57, p. 789–838.

Dix, E., 1937, The succession of fossil plants in the South Wales Coalfield with special reference to the existence of the Stephanian: Congrès de stratigraphie et de géologie du Carbonifère, 2d, Heerlen, The Netherlands, 1935: Compte Rendu, v. 1, p. 159–184.

Elias, M. K., 1937, Elements of the Stephanian flora in the mid-continent of North America: Congrès de stratigraphie et de géologie du Carbonifère, 2d, Heerlen, The Netherlands, 1935: Compte Rendu, v. 1, p. 203–212.

Gillespie, W. H., and Pfefferkorn, H. W., 1979, Distribution of commonly occurring plant megafossils in the proposed Pennsylvanian System Stratotype, *in* Englund, K. J., Arndt, H. H., and Henry, T. W., eds., Proposed Pennsylvanian System stratotype: 9th International Congress of Carboniferous Stratigraphy and Geology, 9th, Field Trip no. 1, A.G.I. Selected Guidebook Series, no. 1: Falls Church, Virginia, American Geological Institute, p. 87–96.

Jongmans, W. J., 1937a, Contributions to a comparison between the Carboniferous floras of the United States and of western Europe: Congrès de stratigraphie et de géologie du Carbonifère, 2d, Heerlen, The Netherlands, 1935: Compte Rendu, v. 1, p. 363–387.

Jongmans, W. J., 1937b, Comparison of the floral succession in the Carboniferous of West Virginia with Europe: Congrès de stratigraphie et de géologie du Carbonifère, 2d, Heerlen, The Netherlands, 1935: Compte Rendu, v. 1, p. 393-415.

Jongmans, W. J., 1937c, Some remarks on *Neuropteris ovata* in the American Carboniferous: Congrès de stratigraphie et de géologie du Carbonifère, 2d, Heerlen, The Netherlands, 1935: Compte Rendu, v. 1, p. 417–422.

Jongmans, W. J., and Gothan, W., 1934, Florenfolge und vergleichende Stratigraphie des Karbons der östlichen Staaten Nord-Amerikas, Vergleich mit West-Europa: Jaarverslag Geologisch Bureau, Heerlen (1933), p. 17–44.

Jongmans, W. J., Gothan, W., and Darrah, W. C., 1937, Beiträge zur Kenntnis der Flora der Pocono-Schichten aus Pennsylvania und Virginia: Congrès de stratigraphie et de géologie du Carbonifère, 2d, Heerlen, The Netherlands, 1935: Compte Rendu, v. 1, p. 423–444.

Lyons, P. C., Meissner, C. R., Jr., Barwood, H. L., and Adinolfi, F. G., 1985, North American and European megafloral correlations with the upper part of the Pottsville Formation of the Warrior Coal field, Alabama, U.S.A.: Congrès de stratigraphie et de géologie du Carbonifère, 10th, Madrid, 1983: Compte Rendu, v. 2, p. 203–245.

Moore, R. C., and Elias, M. K., 1937, Paleontologic evidence bearing on correlations of Late Paleozoic rocks of Europe and North America: Congrès de stratigraphie et de géologie du Carbonifère, 2d, Heerlen, The Netherlands, 1935: Compte Rendu, v. 2, p. 677–681.

Moore, R. C., Elias, M. K., and Newell, N. D., 1936, A "Permian" flora from the Pennsylvanian rocks of Kansas: Journal of Geology, v. 44, p. 1–33.

National Research Council, 1936, Division of Geology and Geography, Report of the Committee on Paleobotany, May 2, 1936: Division of Geology and Geography, Washington, D.C., 22 p.

Phillips, T. L., Pfefferkorn, H. W., and Peppers, R. A., 1973, Development of paleobotany in the Illinois Basin: Illinois State Geological Survey Circular 480, 86 p.

Read, C. B., 1955, Floras of the Pocono Formation and Price Sandstone in parts of Pennsylvania, Maryland, West Virginia, and Virginia: U.S. Geological Survey Professional Paper 263, 32 p., 20 pls.

Thomas, H. H., 1938, Pteriodsperm evolution and the angiospermae: Congrès de stratigraphie et de géologie du Carbonifère, 2d, Heerlen, The Netherlands, 1935: Compte Rendu, v. 3, p. 1311–1321.

White, D., 1936 (posthumous), Some features of the early Permian floras of America, *in* Report of the 16th Session, United States of America, International Congress, 16th, Washington, D.C., 1933, Volume 1: Washington, D.C., Committee of the International Congress, p. 679–690.

MANUSCRIPT ACCEPTED BY THE SOCIETY JULY 6, 1994

Printed in U.S.A.

Geological Society of America
Memoir 185
1995

# *Correspondence and plant-fossil exchanges between William C. Darrah and European paleobotanists (1932–1951)*

**Elsie Darrah Morey**
*Morey Paleobotanical Laboratory, 1729 Christopher Lane, Norristown, Pennsylvania 19403*
**Paul C. Lyons**
*U.S. Geological Survey, MS 956, National Center, Reston, Virginia 22092*

## ABSTRACT

**William C. Darrah was an avid correspondent. He genuinely enjoyed the exchange of ideas with paleobotanists throughout the world. In 1932, while still a student at the University of Pittsburgh, Darrah began what became a 20-year history of sharing ideas, reprints of publications, and specimens with European paleobotanists. The W. C. Darrah correspondence files—from which this historical account is mainly taken (supplemented by his correspondence in European files)—contain a record of the details of the paleobotanical pursuits of many of the early-twentieth-century European paleobotanists, especially P. Bertrand, W. J. Jongmans, and R. Florin, with whom he had the most extensive correspondence and contacts. The correspondence also contains a record of the shipments and receipts of the plant-fossil exchanges between Darrah and the European paleobotanists.**

## INTRODUCTION

During the period 1932 to 1951, William (Bill) C. Darrah exchanged letters with the leading European paleobotanists. He always enjoyed the thoughtful interplay of ideas about plant biostratigraphy and morphology as they related to Carboniferous and other specimens under discussion and offered, in exchange for European specimens, specimens from the Appalachian and Illinois basins. These letters and personal interactions with the European paleobotanists played a profound role in Darrah's development as a leading American paleobotanist. Details can be found in Darrah (1932a, b, 1934, 1935a, b, 1936a, b, c, d, 1937, 1939a, b, c, 1940, 1941, 1949a, b, 1960); Darrah and Bertrand (1933a, b); and Jongmans and others (1937).

## THE CORRESPONDING EUROPEAN PALEOBOTANISTS

Bill Darrah corresponded with the leading European paleobotanists whose names, affiliations, and countries are noted below. Excerpts from their letters comprise the basis of this historical account. Personal sketches of many of these paleobotanists are in Andrews (1980).

Paul Bertrand, University of Lille, Lille, France
Isabel M. P. Browne, University College, London, England
Robert Crookall, Geological Survey and British Museum of Natural History, London, England
Hanna Czeczott, Warsaw, Poland
Emily Dix, Bedford College for Women, University of London, London, England
Wilfred M. Edwards, Department of Geology, British Museum of Natural History, London, England
Rudolf Florin, Naturhistoriska Riksmuseum, Stockholm, Sweden
Walther Gothan, Preussische Geologische Landesanstalt, Berlin, Germany
Thore G. Halle, Naturhistoriska Riksmuseum, Stockholm, Sweden
Tom M. Harris, The University of Reading, Reading, England
Max Hirmer, München (Munich), Germany

Morey, E. D., and Lyons, P. C., 1995, Correspondence and plant-fossil exchanges between William C. Darrah and European paleobotanists (1932–1951), *in* Lyons, P. C., Morey, E. D., and Wagner, R. H., eds., Historical Perspective of Early Twentieth Century Carboniferous Paleobotany in North America (W. C. Darrah volume): Boulder, Colorado, Geological Society of America Memoir 185.

Wilhelmus J. Jongmans, Geologisch Bureau voor het Nederlandsche Mijngebied, Heerlen, The Netherlands
Richard Kräusel, Senckenberg Museum, Frankfurt-am-Main, Germany
Suzanne S. Leclercq, Laboratoire de Paleontologie, University of Liège, Liège. Belgium
Armand Renier, Service Géologique de Belgique, Bruxelles, Belgium
Albert C. Seward, Botany School, Cambridge University, Cambridge, England
John Walton, The University of Glasgow, Glasgow, Scotland

## THE CORRESPONDENCE

While pursuing graduate studies at the Carnegie Museum in Pittsburgh, Pennsylvania, Bill Darrah was introduced by David White to Paul Bertrand and W. J. Jongmans and also met Armand Renier at the time of the 16th International Geological Congress in Washington, D.C., July 1933. These men were the "Big Three" of European paleobotany, as Darrah noted in his European journal (see Chapter 2, this volume). Bill served as host to **Paul Bertrand** in 1933 before the Congress, when Bertrand studied the American Carboniferous from the Anthracite region of eastern Pennsylvania to the bituminous region of western Pennsylvania, West Virginia, and Ohio (see Chapter 1, this volume). After guiding Bertrand on the field trips, Darrah remained with Bertrand as they attended the Congress in Washington, D.C. Thereafter, their acquaintance was reinforced by correspondence, enriched at the Sixth International Botanical Congress (Amsterdam) and the Second Carboniferous Congress (Heerlen) in 1935, and expanded, following the two congresses, by collecting trips in the coalfields of France under the guidance of Bertrand (Fig. 1). Thus, working together professionally, Darrah and Bertrand also became good friends as they shared ideas and corresponded to exchange paleobotanical thoughts on specimens or data. Darrah's association with his mentor was cut short by Bertrand's death in 1944. In recognition of Bertrand's influence on his biostratigraphic studies, Darrah dedicated his *Principles of Paleobotany* (second edition, 1960) to him.

From Lille, Bertrand wrote, "Your letter of 3 April [1934] interested me immensely. Your observations on *Mixoneura* agree with those just published by Sze and Gothan on the *Mixoneura* of China; they have found in the Stephanian of China species which they find are different from *Neuropteris ovata* Hoffmann and also from *Mixoneura neuropteroides* Göppert." Bertrand continued:

> I can only encourage you to continue your research on this important question and to publish your results. For me, I have been very happy to have collaborated with you to study the very rapid deposition of Pittsburgh and Wilkes Barre. Thanks to you, I am convinced that the Allegheny Formation is definitely equivalent to the Upper Flambants of Sarrebruck [Westphalian D] and at the same time this proves the continuity of the zone of *Neuropteris ovata* in America. . . .
>
> Now that I have seen the equivalents, I am satisfied and I will return immediately to my study of the coal flora here [the Saar and Lorraine]. . . . I would encourage you to head the study [in America]. I am convinced that you have the knowledge and capabilities. This study of course should be done in America.[1]

About a year later, Bertrand wrote again, "I see you have found *Neuropteris scheuchzeri* associated with *Callipteris conferta* in the Upper Dunkard; is it not indeed an important discovery. Granted you are right, it would go far to show that some species *N. scheuchzeri, N. ovata* and perhaps some *Mariopteris* have a considerable vertical range. The stratigraphic table you sent me, with the distribution of the species, is highly interesting. It seems to prove the very existence of the Stephanian in America. Probably further research will continue to strengthen more and more my first impression."[2]

In June 1935, Bertrand again corresponded with Darrah and revealed that, "I made last month for the Geology Society of North France, a short review of the paper upon American Carboniferous floras by Jongmans and Gothan. I take [the] liberty of submitting [to you for your opinions] the last pages of my paper, in which I discuss freely both questions of Westphalian D and of the presence of the Stephanian coal measures in Pennsylvania and Kansas." He then went on to request, "I would be [grateful if you could] write your opinion upon these questions and upon the lists of Pennsylvanian species. Of course, it was not my purpose to give too many facts about your interesting discoveries. I wished only to make it clear the species which I consider strongly pointing or even unmistakenly towards a Stephanian age."[3]

With both men anticipating the upcoming conferences in Amsterdam and Heerlen, Bertrand wrote, "I hope you will be able to bring some peels with you. I am very interested in your new [peel] technique [Darrah, 1936c] because we are intending to start research upon carbonized material with the help of such film peels. You must always be very careful with hydrofluoric acid whose steam is very dangerous for the lungs. As to your stay in France, I have somewhat altered my mind. . . . Fieldwork will be most profitable to you. I am in quite good relations with Mr. A. Renier and Mr. Van Straelen. . . . Discussing all the matters with my colleague, Professor Pierre Pruvost, we have thought it advisable for you to visit our French coal measures especially the Saar Lorraine basin where you will see large displays of *Mixoneura ovata* and St. Étienne Basin where you can see the typical Stephanian flora. Neither in Belgium [nor] the Netherlands do they have the slightest trace of the *Mixoneura* zone or the Stephanian flora."[4]

After Darrah's European experience (see Chapter 2, this volume) Bertrand wrote: "All the things we were able to see together either in the field or in the collections turned out to be a great benefit for me. So I hope in some years to meet you again and exchange with you new profitable remarks and

Figure 1. Paul Bertrand in Heerlen, The Netherlands, after the Second Carboniferous Congress, September 1935. Photograph taken by W. C. Darrah. The William Culp Darrah Collection.

ideas. I thank you for your peel method as applied to silicified specimens."[5]

The exchanges continued. Several months later Bertrand again wrote: "As for your paper about some American relatives of the Westphanian D and lower Stephanian, I would be glad to publish it in our *Annales de la Société Géologique du Nord*. . . . I will try to send you some appreciation upon the value of paralic and limnic horizons as stratigraphic limits and [I] will discuss especially the question with Pierre Pruvost."[6]

Letters continued to follow for several years. Bertrand acknowledged Darrah's interpretations: " . . . with your demonstration that in the coal bearing floras of the United States, the Upper Pennsylvanian you can recognize irrefutably the equivalents of the Stephanian in Central France."[7] (For further reference, see Bertrand 1928, 1933, 1935, 1937; Pruvost and Bertrand, 1932; Bertrand and Corsin, 1933; Bertrand and Pruvost, 1937.)

**Isabel M. P. Browne** wrote Darrah several times from London, England, and her letters dealt with article reprints and data interpretations. She wrote, "In the paper on *Walchia* in the Appalachian Province [Darrah, 1936b] that you were kind enough to send me, you mention . . . among the plants associated with *Walchia, Zygopteris erosa.* Can you tell me where I can find a specific diagnosis or better still, a description of this species?"[8]

Some months later in 1936, she wrote again.

> I have read with interest your paper on *Macrostachya thompsonii* [Darrah, 1936a]. One of the most interesting points to me is that you speak of the bracts as being mucronate with 1–3 teeth. Are these teeth at all deep—or lobe like, or is each one a minute micro? The existence of this micro suggests—in fact it implies—that some at least of the bracts were preserved in their full length and the top end of your second figure gives me the impression that, apart from the extreme apex where the internodes have not elongated appreciably, the bracts, at least as preserved, do not extend upwards for more than 1–2 internodes. [W.] Hartung, in his 1933 paper, makes it a characteristic of the form genus *Macrostachya* that the bracts are long and extend over several internodes. . . . As you do not mention [Hartung's] paper . . . it may have escaped your notice. If you have not got Hartung's paper at hand, I will gladly lend it to you. . . . Hartung has some very thoughtful generalizations about the origin of heterospory.
>
> I must confess that I have regarded *Macrostachya* as a receptacle for large cones the structure of which (especially the position of the sporangiophores) was incompletely known. Now that you have been able to establish that in *M. thompsonii* the sporangiophores are inserted as in *Calamostachys,* I feel [tempted] to think of it as belonging to that genus—After all, some *Calamostachys,* such as *C. solmsii* Weiss reach the respectable length of 150 mm.[9]

On another occasion she wrote, "I am much interested in your photograph of a cone of *Tingia taeniata.* Unfortunately, I have not all my botanical papers with me . . . but it may be a new species of yours. So far as my memory—without books to consult—goes the leaves on the peduncle are rather different from the ordinary *Tingia* leaf—I shall be much interested to hear your views. . . . I think that if *Palaeopteridium* be the vegetative shoots of *Discinites* then the view that the Articulatales are primarily large leaved is strengthened. It will be interesting to see what light your cone of *Tingia taeniata* throws on this question."[10]

Browne noted in yet another letter, "To hear of thousands of slides available for study is enough to make one's mouth water—the petrified *Macrostachya* will indeed be a useful pointer about *Macrostachya* cones' morphology."[11] For other references, see Browne (1925, 1927, 1935a, b).

While Darrah was engaged in graduate studies at the University of Pittsburgh, he received letters from **Robert Crookall** (Fig. 2) in London, who wrote, "I am pleased to hear that

Figure 2. Robert Crookall outside the British Museum of Natural History, London, England. Photograph taken by A. T. Cross in 1951. Courtesy of A. T. Cross.

Figure 3. Hanna Czeczott. Courtesy of and permission from the Hunt Institute for Botanical Documentation, Carnegie Mellon University, Pittsburgh, Pennsylvania.

you have recognized a U.S. flora equivalent to our Radstockian [Westphalian D], and hope that you will eventually be sending me a copy of your account."[12]

Some time later Crookall added, "*Asterotheca miltoni* is of little zonal [significance] with us, on account of its great range. Your Lower Conemaugh flora is indeed strikingly Radstockian in type. May I refer you to my short paper on the Correlation of the British and French Upper Coal Measures [Crookall, 1931]. . . . I regard *[Pecopteris] feminaeformis* as a very characteristic species (unknown in our Radstockian)."[13]

**Hanna Czeczott** (Fig. 3), whom Darrah met at the European Conferences in 1935, wrote him from Warsaw, Poland: "If it really will be proved that several type specimens of Unger of beech, for which I looked in vain in the European institutions, are in Harvard University, it will be worthwhile to pay another visit to the United States, for the first time I was in North America [was] in 1915. Another purpose for another trip to the United States would be to collect specimens of living trees and shrubs to compare with the fossil Tertiary specimens of Europe. As could be expected, the scope of my interest in paleobotany has [become] much enlarged. . . . We have found lastly in Poland several localities containing extremely numerous fossils (Miocene and Pliocene), and I intend to sacrifice in [the] future no less time to the study of the fossil plants than to the living ones. I wish to follow in this respect the line which—as it seem has been taken by Professor [R. W.] Chaney: to study the living vegetation in such localities, which are at present in similar climatic conditions to those in which existed the Miocene and Pliocene vegetation in Middle Europe."[14]

As requested, Darrah sent Czeczott the notes on the Unger types at Harvard on February 21, 1939. For further information see Czeczott (1934).

**Emily Dix** (Fig. 4) first met Darrah during the Carboniferous Congress in Heerlen in 1935. She was a contemporary colleague also struggling for professional recognition. Dix ("Dixie") wrote Darrah from London: "Holland, where I had the good fortune to make your acquaintance, seems a long way off now. How did you eventually decide to deal with the Jongmans-Gothan question? . . . My specimens from Belgium and Holland arrived safely and are looking splendid, labeled in white trays. . . . Ye Gods what a blessing we spent just the one

Figure 4. Emily Dix at the Wérister Colliery, Belgium. Photograph taken by W. C. Darrah, September 19, 1935. The William Culp Darrah Collection.

day in Belgium but it was a nipper. . . . I wonder if you could tell me where I could obtain copies of these beautiful restorations of Carboniferous plants which Dr. Jongmans exhibited in his museum. Do they belong to the Field Museum? If you could assist me in this matter, I should be very grateful."[15]

Two months later Dix again wrote, "Dr. Hawkes, the head of this department has great pleasure in exchanging Kidston's monographs on fossil plants for Lesquereux's Coal Flora and his monograph on the Cretaceous Tertiary Flora . . . and if you could also send a few separate papers on Carboniferous and Permian plants I should be very grateful. You will be interested to hear that *Lonchopteris rugosa* has turned up in abundance in an old brick pit in the Bristol Coalfield through the energetic field work of [Professor] Trueman's students. Perhaps you will have the good luck to find it in some isolated part of the U.S.A.—I mean in numbers!!! [see Chapter 2, this volume]. I am very interested in your peels—one day I must experiment with some coals."[16]

In 1944, Dix wrote Darrah of her survival when London was bombed during World War II: "Bedford College was evacuated to Cambridge at the beginning of the war and new plans are well advanced for our return to London." She then continued, writing of her care of 11 students who were placed with her in a house and explaining that she did the housekeeping chores and did the honors of cooking for all invited guests. "I finished off the supper. You can believe it or not, the result was quite good. Gothan, Jongmans, Renier and Bertrand would be much surprised. Even my friends in the Geological Survey here couldn't believe that Dixie could be interested in housekeeping. . . . Naturally, we are not looking forward to returning to London where we must work in a temporary and ill-equipped building. However, we must smile with thumbs up and get on with the work. To return to paleobotany: Guthörl appears to be very [busy] on the species *Sphenopteris damesi* Stur as a zone fossil in the Saar Basin. . . . I can't find the description of this species. Could you send me the reference to Stur where it is figured and described."[17]

Darrah's correspondence with Emily Dix ceased in 1944. He learned later that year in correspondence from Birbal Sahni that Dix was suffering from a mental illness. For further reference see Dix (1934a, b, 1935, 1937) and Dix and Trueman (1937).

In his position at Harvard's Botanical Museum, Darrah was able to exchange fossil specimens with many other museums. **Wilfred N. Edwards** (Fig. 5), from London's British Museum of Natural History, wrote on a number of occasions. "I have been selecting some duplicates to send to you. The first box is now being packed and I enclose a list of the specimens. It is rather a miscellaneous lot, but some of them may be of use to you. We can still send you some more material. . . . P.S. I am much occupied at home just now with a baby son, about 2 months old!—our first, though we have been married 14 years!" Edwards added proudly.[18]

Later he wrote again, "Your box of seventy specimens of fossil plants has arrived safely, and we are glad indeed to have this selection from you. Many of the specimens, especially those named by Lesquereux [1879–1884] will be useful to have in this country."[19]

Continuing the thought of an earlier letter, Edwards wrote, "The second box of duplicate fossil plants is now practically ready for dispatch and I enclose a list of forty-two specimens in it. I have included a few impressions from the Eocene of Grinnell Land [Greenland]. They are poor and are of interest chiefly because they are from the most northerly locality at which fossil plants have ever been found. Some American expeditions have, I know, been to this region, and it is just possible that you already have material from there. If so, I should very much like to know about it, because I have recently been revising Heer's [1870] work on the original collections made by the 'Discovery' expedition in the 70's. A British expedition went there last year and actually used the coal of this locality for fuel, but the blighters did not trouble to collect a single fossil. I was very amused in reading the report of their expedition in our Geographical Journal to see some rather scornful remarks about Captain [R. E.] Peary, and how he sailed up and down those coasts without caring much about scientific observations."[20]

Figure 5. Sitting, left to right: T. M. Harris, C. A. Arnold, T. G. Halle, and W. C. Darrah; standing, left to right: D. N. Wadia and W. N. Edwards. Photograph taken at the Sixth International Botanical Congress, Amsterdam, September 2–7, 1935. The William Culp Darrah Collection.

Edwards continued to write Darrah, as he remarked, "In looking through the proofs of a paper by Miss Dix, my colleague, Dr. H. H. Thomas, and I had a great deal of trouble over a reference to a paper of yours entitled 'American Carboniferous Floras' [see Darrah, 1937]. . . . It turns out to be a summary of the paper which you were to read at the Heerlen Congress and which is marked as issued from the Botanical Museum of Harvard. Technically, I suppose, it has not been published at all . . . and it is very difficult to refer to it correctly in a bibliography. I have not got a copy of this myself, and shall be very glad if you can let me have one . . . [if] it is to be printed in the Proceedings of the Congress, [or] will it be identical or have you modified or altered it in any way?"[21]

Edwards wrote later, "Thank you very much for the [three] copies of your article on American Carboniferous floras. . . . We will send you another batch of duplicates for exchange, and will include the *Williamsonia* flower and some other Jurassic plants. . . . Harris was in here yesterday, and tells me that he is still waiting anxiously for an exciting box of Triassic plants from America. I hope he will get it soon, because he declines to take on any other important job until he has tackled this Triassic flora, and we are anxious to set him on some of our undescribed stuff, which is in his line."[22]

Edwards wrote again two months later, "We have packed two more boxes of duplicates for you, the first of which contains a fairly representative series of plants from the Middle Estuarine beds (Jurassic) of Yorkshire. The second contains, in addition to the *Williamsonia* for which you asked, a miscellaneous batch of specimens."[23]

**Rudolf Florin** (Figs. 6 and 7) first wrote Darrah from Stockholm's Naturhistoriska Riksmuseum following the Congresses held in 1935 (see Chapter 2, this volume). I [would] appreciate [hearing] from you soon about *Cordaianthus* as well as about the machines you use for grinding slides."[24]

Darrah responded to Florin on January 7, 1936. He referred to the grinding machines and their prohibitive costs. He then went on to mention that he had with Bertrand collected 29 silicified nodules at Combe Rigolle ("Comberigault") near Grand Croix.

Florin replied almost immediately: "Thank you also for your kindness in asking Professor [I. W.] Bailey to send separates of his interesting anatomical papers as well as the engineering department of your university to give us information about its grinding machines. It is extremely generous of you to offer me so kindly to borrow your whole material of *Cordaianthus,* sections, transfers and carbonized material. . . . I can only say, that this material will, I am certain, be of the greatest interest in connection with my studies on the Cordaitales. I have got together quite a rich material from European museums and a year or two ago we started making slides of leaves, male and female cones. I have also a rather rich material from the neighborhood of Saint Étienne at my disposal. Only the female cones are rare. It is my firm decision to finish the Paleozoic conifers as well as the Cordaitales before the [Botanical] Congress in 1940, when it is intended to continue the discussion on the phylogeny of these groups." Florin continued: "You mentioned that you were able to collect some cordaitean material at Grand Croix. Could you possibly tell me at which particular spot there would still be something of this valuable material to be found? I am anxious to know if Professor Bertrand perhaps found so much of this material that he will possibly publish a description of the cordaitalean remains. In that case my task would be somewhat complicated. When I paid a visit to him in 1930 he apparently had no such intention." Florin went on to note that he had been chosen as general secretary of the next botanical congress and thus the time involved in planning for the congress would slow down his scientific activities. "But on the other hand," he emphasized, " . . . this situation now forces me to intensify my wok on the Paleozoic conifers and *Cordaites* so as to be able to finish it in time. I can already tell you that the revision of the walchias will be so radical as to upset nearly all previous determinations."[25]

Soon afterward Florin wrote again: "It is my intention to

Figure 6. W. C. Darrah (left) and R. Florin (right). Photograph taken by H. H. Darrah at the Wérister Colliery, Belgium, 1935. The William Culp Darrah Collection.

publish papers on the Grand Croix material from time to time, but until 1940 I shall not have time to deal with any other groups than the Cordaitales. Your [Darrah's] material probably contains just the same material as I already have at my disposal from Comberigault. . . . But as you possess a rather rich material from Grand Croix I venture to propose that we keep in touch with one another only to avoid parallel work on the same subject."[26]

The letter exchanges with Florin continued. Florin wrote to say that the eight *Walchia* specimens borrowed from the Harvard Botanical Museum were being returned. "However, in order to make my monographs as complete as possible," Florin advised, "I intend now to visit certain museums in Central Europe and in the middle of April leave for the United States. I hope to be allowed to pay you a visit and, besides, I am planning to visit Washington (D.C.), Lawrence (Kansas), Urbana (Ill.), Chicago, etc."[27]

Darrah responded to Florin's letter on March 26, 1937, to say that he was pleased Florin was planning to visit the United States and that he had informed Professors [E. C.] Jeffrey and [R.] Wetmore of Florin's plans. Florin was invited to speak at the Biological Colloquium, to be held at Harvard University, Cambridge, Massachusetts.

Florin responded immediately and thanked Darrah for the invitation to speak at the Biological Colloquium. He then suggested that he would be able to speak on the structure of the pollen-grains in the Cordaitales. Later, upon his arrival at Harvard, Florin expressed his wish to see all the Cordaitales before any material would be sent to Stockholm. He said that the specimens had not yet arrived, but when they did, he would write Darrah again.

Florin wrote, "Some few days ago, I received in good condition the box with fossil plants you were so kind as to promise to send. It contained two specimens of *Walchia* and

Figure 7. Rudolf Florin at the age of 50 (1944). Courtesy of the Bergius Foundation, Stockholm, Sweden.

one of *Ullmannia* belonging to your Museum . . . in addition to 15 specimens of Carboniferous plants from Mazon Creek in Illinois. In exchange for the latter specimens Professor Halle will send you duplicates of Upper Paleozoic plants from North China."[28]

Somewhat later, Florin had more to write to Darrah. "On behalf of the Museum, I acknowledge with many thanks the receipt of 23 slides of *Cordaites—Cordaianthus* from the Late Pennsylvanian of Iowa. We are glad to possess some slides of this material for comparison with our own. . . . I congratulate you to have [obtained] such a rich coal-ball material from Iowa and hope that you will be able to make an important contribution to our knowledge of the flora of the Late Pennsylvanian from the structural point of view. I have mentioned to Professor Halle, the recorder of the Paleobotanical section of the Congress, that you would like to present some of your work on this material."[29]

Darrah again wrote Florin on June 28, 1939. In his letter, Darrah acknowledged the return of the Harvard specimens that he had loaned. He went on to explain that F. O. Thompson was a benefactor to the Harvard Botanical Museum and that he had requested that Darrah name the *Cordaianthus* specimen *C. shuleri* in honor of Henry Shuler, who owned the collieries from which the coal-ball material came. Darrah had his publisher, D. Appleton-Century, send a complimentary copy of his book, *Textbook of Paleobotany* (Darrah, 1939a), to Florin. In response, Florin complained that Darrah had failed to cite much of the European literature. Darrah replied that D. Appleton-Century was not expecting many sales of the textbook in Europe and that he had thus cited only about 75% of the German works that he had consulted. He then went on to admit that he might have overestimated the value of the works of Kräusel, Weyland, Wieland, Thomas, and Sahni.

Florin replied, "I understand quite well that you wish to describe the *Cordaianthus* from Iowa yourself, and I am very grateful to you for the material of these coal balls, which will enable me to discuss also your species in my work. I think I mentioned in a former letter to you that I intend to discuss the morphology of *Cordaianthus* already in the general part (Part 6) of my monograph on the Upper Carboniferous and Lower Permian conifers [Florin, 1938–1945]. . . . I shall be glad to receive your description of *Cordaianthus shuleri* before it is published. Perhaps I may be able to make comments on it which will be of use to you."[30]

Darrah did send his *Cordaianthus* paper to Florin for his comments. Florin responded quickly, "You have certainly got a very interesting material of *Cordaianthus,* both female and male strobili being present. I do not suppose you mean to describe them as belonging to one and the same species of *Cordaianthus.* It is decidedly better to keep them apart as Renault did . . . but judging from my experience with the European material of female and male inflorescences of *Cordaites,* your description of the American material seems to be all right. As a matter of fact, your results appear in the main to confirm my own very nicely. But are the strobili really borne spirally? Otherwise they are, as you know, borne in two rows in *Cordaianthus,* whereas the rather similar female strobili of *Lebachia* (Coniferales) are spirally arranged. The numbers of ovules in each strobilus corresponds to that exhibited by Renault's material from Grand Croix. . . . If the fertile appendages ('bracts') of the American species are, as you say, tetragonal in transverse section, they are probably less flattened than those of the Grand Croix material. . . . I hope you will be able to illustrate fully the American material by means of first class microphotographs. It no doubt deserves this, and the photo will facilitate further discussions. *Cordaianthus* will, of course, be dealt with in Part 6 of my monograph on the Upper Carboniferous and the Lower Permian conifers. (Parts 1–4 have appeared, Part 5 will soon be ready in print.)"[31]

Darrah wrote to Florin on April 4, 1940. He was pleased that the Harvard material was in agreement with Florin's observations; Darrah stated his misgivings but he had decided to publish both sexes under the same species name. He admitted that many paleobotanists would disapprove but went on to say, how can we doubt the validity of both these cones on the same type of plant? The letter went on to describe the discovery of cordaitean embryos, some of which were pear-shaped. The mature embryos were dicotyledonous and had quite large cotyledons that fill the whole seed.

Letters continued to be exchanged between the two men for some time, with requests for reprints and material exchange their chief concerns. For further references, see Florin (1936, 1939, 1951, 1954).

**Walther Gothan** (see Chapter 6 and Chapter 2, this volume) wrote Darrah on two occasions following their meeting at both the Sixth International Botanical Congress and the Second Carboniferous Congress in 1935 in The Netherlands. In one of these letters Gothan wrote: "In reference to your communication about the Stephanian of the United States: this can hardly be covered in a letter. I think it is best, that you state your position, based on the circumstances [in America] and by referencing Jongmans-Gothan research, and thus improve our understanding of it. Noé told me earlier about the difficulties associated with *Neuropteris scheuchzeri* and *[N.] decipiens.* These topics can also be treated with greater precision in North America. . . . I have received a manuscript today from Jongmans about the Pocono flora and *N. ovata.* I was under the impression that the three of us were going to collaborate. I would like to hear your opinion on this before I remind Jongmans about it."[32]

**Thore G. Halle** (Fig. 8) wrote to Darrah from the Naturhistoriska Riksmuseum in Stockholm, Sweden, to say that he had seen that Darrah was working on Paleozoic plants. "I am particularly interested in your discovery of seeds in [connection] with *Lescuropteris moorii* and should like to ask you to read what I say about *Emplectopteris triangularis* [Halle, 1931] in my 'Palaeozoic plants from Central Shansi . . .' [Halle, 1937b] and 'on the seeds of the pteridosperm *Emplectopteris triangularis,*" he wrote; previously he had sent Darrah his memoir of Paleozoic plants from China. "I should not be surprised if *Emplectopteris* turns out to be really identical with *Lescuropteris*; according to the illustrations the venation is different and I could not get any specimens of *Lescuropteris,* so I had to make a new genus, but the resemblance is most striking."[33]

The following year, Halle answered Darrah's request for a reprint on *Emplectopteris.* "It seems to me that in light of the new facts the seed of *Emplectopteris* was very like that of *Lescuropteris.* The important thing, of course, is whether *Lescuropteris* had anastomosing venation; if it has, it seems to me very likely that the two genera are identical, in which case *Emplectopteris* will of course have to be dropped."[34]

As Halle's work on *Codonotheca* and *Whittleseya* continued, he again wrote Darrah to obtain further material. He asked for material that would be suitable for maceration. In exchange for the *Whittleseya* material he offered his publication on Whittleseyinae (1933). For further references, see Halle (1910, 1929, 1937a).

**Tom Harris** (Fig. 9) wrote Darrah from Reading, England, "I understand there is an undescribed collection of fossil plants from the [Middle] Keuper of Virginia in the Harvard University Museum. I am writing to ask whether the Museum

Figure 8. Thore G. Halle, Courtesy of Hunt Institute for Botanical Documentation, Carnegie Mellon University, Pittsburgh, Pennsylvania.

Figure 9. Tom Harris in photograph taken at the International Botanical Congress in Montreal in 1959. Photograph taken by and courtesy of Theodore Delevoryas.

would consider lending me this collection for examination and description. . . . I have just finished the description of a large collection of Uppermost Keuper plants from Greenland for the Danish Government. The Virginia material seems to be rather similar in preservation and I believe that similar techniques would give results of very great interest, since virtually nothing is known about the plants of the Mid-Keuper. Should the Museum be generous enough to let me undertake this work, I would of course return all the specimens and the microscope slides I make from them."[35]

Another letter from Harris followed soon thereafter, as he replied to one from Darrah. "Thanks for your letter. I would be very glad to borrow the 7 boxes full of Triassic plants in spite of the difficulties. . . . I am sorry the locality is shut down. However, it is possible it may be opened up again some day. With regard to the other collections in the National Museum, I can understand that they may not want to lend them. . . . Would you [send them to me, the Harvard specimens, straight away?] I can't work at them for a month or two, but I would greatly like to examine them."[36]

Darrah answered Harris on December 10, 1935, and Harris was quick to respond. "While I am a bit disappointed not to see the Triassic stuff till later, it will not make any very serious difference to my work. . . . I hope you get good stuff out of the Byrd coal. I find Greenland coal dissolves nicely if treated for a few weeks with concentrated $HNO_3$ with some $KC10_3$ added after a time often alkali (repeat the acid if necessary). Probably this is the same method you are using."[37]

Writing again from Reading University some months later, Harris continued in the same context. "I expect you have now cleared up after your museum's show for the Harvard University celebrations [Harvard University's 300th anniversary]. I would be grateful if you would send the Virginia Triassic plants as soon as you can, as I have finished my Greenland work and until proofs arrive I am without a proper job. I think the arrangement was that your museum is to pay for the [shipping] here and we are to pay the carriage back. . . . I was to publish if possible in an American journal."[38] For further reference, see Harris (1931, 1932).

In Darrah's written exchanges with **Max Hirmer,** whom he had first met during the Second Carboniferous Congress in Heerlen in 1935, the focus was limited to specific technical concerns of mutual interest. "As the famous Philippines Deeps are together with the Philippines themselves in American possession," Hirmer wrote, "[I hope you] will be able to give us material of the sediment (about 40 grams and not too sandy)."[39]

Darrah was able to send the desired samples, and soon

thereafter Hirmer thanked him for "[his] kindness to provide my friend and collaborator with the Philippines Deep sediment." Hirmer then added, "concerning the paper of Dr. J. Lutz, I have provided a copy for you."[40]

Hirmer's letter of November 22, 1940, was the beginning of a series of exchanges between the two men concerning the publication of Darrah's American coal-ball flora. "I am very glad and honored that you will give your . . . interesting work on American coal balls . . . to my journal *Palaeontographica,* [and] we will reprint it *immediately,*" he wrote. "Surely the equipment will be of the same good quality as in the case of the work of your compatriot Prof. G. R. Wieland we published. . . . Concerning the security of your manuscript-sending I have asked the American General Consulate here and the General Consul [advises] you to apply to the State Department Washington for sending your manuscript with the Diplomatic post-pouch directly to the American General Consultate Munich with the request to inform me after the arrival. Certainly this will be the surest way. If the State Department Washington will not do so, I propose to send the manuscript via *Lisbon/Portugal* to my friend Dr. R. Florin."[41]

Directly from the Department of State, Darrah received this advisory: "You state you are planning to submit a scientific manuscript to Professor Dr. M. Hirmer, for publication in Germany, and asking if it can be sent through diplomatic pouch. I regret very much that under the specific and stringent regulations which, under existing conditions, the Department of State has been compelled to adopt governing the transmission of non-official mail by the diplomatic pouch, this document does not fall within the categories of such mail which may be so transmitted."[42]

World War II had now accelerated, and the United States became heavily involved on December 7, 1941, the date of the Japanese bombing of Pearl Harbor, Hawaii. It was following the war that Darrah wrote an inquiring letter to **W. J. Jongmans:** see Chapter 5, this volume) "Do you happen to know if Hirmer has continued the publication of *Palaeontographica* since the cessation of the hostilities? I know, of course, that he published Florin's monograph on the conifers. Before the war, I had made preliminary arrangements with him to publish my coal ball flora, but I have made no effort as yet to resume correspondence."[43]

Jongmans replied, "As far as today I got no news about Hirmer. Kräusel wrote to me that Hirmer is said to be in Munich, but his address is unknown. I do not intend to write to him before he has made the beginning. The Germans, at least many of them, are very peculiar in this respect. They seem to be of the opinion that we must come first. They have entirely forgotten that it was the [German] Führer who started the war and that it was the German army which occupied so many countries and that it was Germans who terrorized our people and murdered so numerous innocent men. But I will stop politics."[44]

As he did for Bertrand, Darrah hosted Jongmans's field trip to the Appalachian Basin in 1933 (see Chapters 1 and 26, this volume). Jongmans's travels in the United States that year included visits to Pennsylvania, West Virginia, Ohio, Illinois, and Kansas (see Chapter 26, this volume). He wished to see the American Carboniferous from the Mississippian to the Permian.

The letters between the two paleobotanists began in 1933. Jongmans wrote, "As far as I have seen material of the Lower Dunkard, . . . this is certainly not Permian but Stephanian. The Mazon Creek is Upper [Westphalian] so that you see that we perfectly agree to the correlation of these beds. . . . I will be very thankful for any Pottsville material you can get for me and if you can still get some additional material I will be glad to have it. . . . I have seen very much during my whole trip but unfortunately the time is always too short, so that I have to leave the U.S. without having seen all I intended and hoped."[45]

Darrah responded, "During your stay in Pittsburgh, you helped me much—certainly my outlook has been broadened by your experiences. I have since looked for and found *Leaia* in the Lower Conemaugh at the Mason Shale horizon which is at the Westphalian-Stephanian division, but I must add that *Estheria ortoni* is found in the same beds! Thanking you for your many favors and your instructions. . . . "[46]

Jongmans wrote again, "Yesterday I saw the paper by Bertrand and yourself. I did not know that Bertrand visited you. I had no time to read the paper and will do so as soon as possible. But I have seen that the determinations are typical for Bertrand so that many of them seem to be very [disputable]. It is always a danger, when someone makes up his own opinion without paying attention to the opinion of his colleagues."[47]

And the letters from Jongmans continued: "I will be glad to receive as many specimens from as many localities as possible," he wrote. "Whether they are Pottsville or not, is not the most important thing. Of course for the direct comparison between our coal fields and the U.S. we need Pottsville material but this comparison is not the only interest. If you can provide me with material from the Anthracite basins and from Alabama I will be thankful. . . . In the end of August Gothan comes here to look with me over the American material. Now understand that it is useful to have as much as possible."[48]

About the same time, Darrah wrote to Jongmans, "I am sorry that your specimens have not yet been sent to you. I have had your specimens ready since December but Dr. Thiessen's [see Chapter 11, this volume] material is not even ready yet. . . . Your fine series of papers were transmitted to me and I deeply appreciate them, I shall send to you copies of all my papers as I write them. I want to ask you for specimens *Neuropteris schlehani, Neuropteris tenuifolia, Neuropteris gigantea,* and *Lonchopteris rugosa.* The last is not as useful to me as the others, but I need authentic specimens for comparison."[49]

Jongmans replied,

> I was sorry not to have this Waynesburg and Pittsburgh material here at the time of Dr. Gothan's visit. As far as I had material from Pennsylvania all this belongs to the [Westphalian] C and it is [not] Stepha-

nian, although some very Stephanian types or rather plants occurring in both, are represented. But I did not possess material from the highest parts. The lists published by Bertrand and yourself are, according to Gothan and myself, no proof of the existence of [Stephanian]. They all belong to [Westphalian] C. The introduction by Bertrand of [Westphalian] D is entirely useless and has been caused only by the fact, that Bertrand paid too much attention to some forms without considering the type of the flora. Therefore, it would have been of much value to me if I had received the material in time. That the real Permian over the Waynesburg coal is without doubt, but the whole below this belongs to the [Westphalian]. The same for West Virginia. In parts mapped as Dunkard, and even considered as high in the series we found a Westphalian C flora. As far as now I do not know a Stephanian flora from that State. There must be a hiatus between this and the Permian [see Chapter 26, this volume] which overlies it without visible unconformity. In Illinois too there is no [Stephanian]. In Kansas, however, there may be [Stephanian] over the Lansing but the literature [Sellards, 1900] is much too [confused] to [know] for sure. However, it is rather certain that there are [Stephanian] beds in that State. The most interesting thing I found has been in the Pocahontas field. Here I found the complete sequence from the base of the Namurian. In the New River, Kanawha and Allegheny series of [West] Virginia I found the equivalents of the [Westphalian] A, B and C. So that we can make up a complete succession and correlation. In the lower beds in Illinois, I found the equivalents of the [Westphalian] A and B. Do you still intend to attend the International Botanical Congress 1935 at Amsterdam? After the congress we will have a second Heerlen meeting for a further discussion on the correlation of the Carboniferous. I hope you will be able to come and to read a paper on the American correlations.[50] (See Chapter 2, this volume.)

Jongmans soon thereafter wrote a letter of complaint that the specimens shipped had not arrived. After explaining what specimens he had received, he went on to ask for more specimens.

Darrah's reply was immediate. "I am surprised and disappointed upon receipt of your letter dated November 12th. Before I left Pittsburgh in June to come to Harvard, your specimens with a small selection from the Monongahela and Dunkard (which were added) were packed by Dr. Thiessen, Mr. Sprunk, and myself. You should have had all this material long ago. I cannot understand, but I am writing at once to Dr. Thiessen. If you had Monongahela and Lower Dunkard plants for inspection, I am sure that we have Lower Stephanian you will agree. . . . The determinations are mine not Bertrand's. There is no hiatus between the Permian and lower beds, but a local unconformity in the Lower Conemaugh does occur. I am planning to attend the Carboniferous Congress [Heerlen, 1935] and would be very anxious to present some of my observations. An exchange of ideas and a study of your collections would be of inestimable value to me."[51]

Four days later Darrah had more to say: "The Lower Conemaugh is certainly Westphalian, but from my material (which I have given to Harvard University where I am located) the Upper Monongahela and Upper Conemaugh fall within the limits of the Stephanian. So does the Lower Dunkard. The Upper Dunkard with *Callipteris* is Permian as you agree. . . . All of these are characteristic of the Upper Conemaugh and Monongahela. Contrary evidence includes *Neuropteris scheuchzeri,* which is rather common even in the Dunkard. However, *Pecopteris miltoni* and *Neuropteris ovata* do *not* occur despite White's, Sellard's and Lesquereux's citations. This flora of pecopterids and *[Sphenophyllum] oblongifolium* occurs at Pottsville, in Washington and Greene Counties all in Pennsylvania, also in Kansas as you have observed. But it also does occur in West Virginia and Ohio. I have seen it myself. I have never seen Stephanian plants from Illinois or Indiana although I expect [they] are in Illinois."[52]

Jongman's reply came quickly:

Certainly most of the plants which you mention in your letter of November 27 occur in the [Stephanian], but there is no plant in your list which is found in the [Stephanian] only and not in the higher beds of the [Westphalian]. Of course, I make an exception for *Lescuropteris* which is a somewhat isolated type. Isolate types are no direct proofs and occur in every coal basin. The lists published by Bertrand and yourself do not contain a typical [Stephanian] flora and as long as I have not seen one, I do not accept the presence of [Stephanian] in Pennsylvania. Although I do not entirely deny the possibility of its presence. If it is present, it seems to be very much localized and probably not well developed and thin. As to [West] Virginia all the beds which I have seen are not [Stephanian] even not the so-called Dunkard beds. As there is Permian in [West] Virginia, according to the Permian flora, you must accept a hiatus between it and the underlying [Westphalian] C. This agrees with the observations on tectonics and with the orogenic movements in this part of the United States. It is entirely the same as we have it in West Europe where in the paralic basins the [Stephanian] is absent. It is found in the limnic basins (Saar-basin, Central France). The highest beds I have seen in Illinois still always belong to the [Westphalian] C also they may, partly, belong to high beds of this series. The highest beds are in Kansas. Unfortunately, I could not make collections in beds higher than the Lansing [see Chapter 26, this volume]. The collections I have seen are very high up in the [Westphalian] C, and I expect that Stephanian is really present in the younger formations of that state. But the literature in this state is so confused and obscure that it is impossible to make up your opinion. . . . The whole of my American collections has been looked over by Gothan and myself and a paper on the results is in the press in my annual report [Jongmans and Gothan, 1934]. Another paper, with geological and tectonical remarks by Dr. van Waterschoot van der Gracht (whose name you certainly know), knows the United States very well, is presented to the Geological Society of America for their annual meeting.[53] (See Chapter 2, this volume).

Darrah responded, "I will present as you suggest, two papers at the coming Congress (in Heerlen). . . . I am carefully considering your suggestions about the supposed Stephanian here—particularly records of *[Neuropteris] scheuchzeri, [Neuropteris] ovata, Pecopteris miltoni,* etc. I look forward to seeing your collections next summer. I am in no haste to arrive at any conclusion of my own."[54]

And still the letters continued: "I wish to thank you for the most interesting paper published by you and Dr. Gothan," Darrah wrote. "You have presented some ideas which have never been adequately studied before. Certainly now we know the correlation of our Pottsville floras. You have also recognized the significance of the *Neuropteris scheuchzeri* range—something which I have failed to recognize. However, as you will agree, we need more collections from new and old localities. . . . I feel, in spite of your most weighty arguments, that we have an equivalent of what is called Stephanian at present. It is, however, very thin."[55]

Soon after the Congress, Darrah wrote, "I wish to express my thanks for the wonderful experiences and treatment we received in Heerlen during and after the Congress. We made excellent collections, received good ideas, and had a most enjoyable time."[56]

Barely a month later, Darrah had more to report to Jongmans: "I have written to Dr. Bradford Willard concerning Pocono plants on [November] 1st, and have his reply dated [November] 6th. To quote: 'I have yours of the first inquiring about David White's identifications of Mississippian plants. Unfortunately, I have no list of names to supply you. I shipped considerable material to Dr. White and also carried other specimens down to Washington. We went over them together and he identified a number of genera and species at the time, but I did not prepare any list. . . . Consequently, we have only his oral statement as to these plants, which he pronounced [however, as unquestionably] the Lower Mississippian. They were from the Pocono. . . . I told you that the Harvard Museum has a *Lepidodendropsis* flora from Mauch Chunk and a few specimens from Pottsville Gap. The Mauch Chunk Flora is extremely important and is in direct contradiction to [C. A.] Arnold and [G. H.] Chadwick's position. Dr. [G. H.] Ashley is correct as to the Carboniferous nature of the Pocono and Mauch Chunk."[57]

Jongmans and Gothan (1934) published an "unfortunate paper," according to Darrah, that had been turned down by the U.S. Geological Survey and Geological Society of America. To help ameliorate some of the ill will, Darrah agreed to be coauthor in an attempt to resolve some of the problems of the paper (Jongmans et al., 1937). Darrah's letters are inquiries about this manuscript. (See also Chapter 2, this volume).

Addressing the question still not yet resolved, Darrah wrote to Jongmans, "Perhaps it will be better that you do not include me as a co-author in your paper with Doctor Gothan. Since I have not had an opportunity to see the manuscript, nor the plates. Since I do not feel at all familiar with your work, it would be best. Of course I should like to have you send me the manuscript and then include my name. . . . I have finished cataloguing our collections from Holland and find them most interesting. I am rather puzzled about Namurian. It would seem to be a transition flora which is not very distinct in itself. At Epen the species are similar to those in Westphalian A. If the Gulpen flora is typical—Lower Namurian is very different. The time-duration of Namurian does not seem to be comparable to Westphalian—but more like the Stephanian. Will you please criticize this opinion and show me where I am wrong."[58]

Jongmans responded: "A third paper will be sent to you in a few days. This third paper is that on the comparison between the Carboniferous floras of the United States and of western Europe. This is the most difficult one. It is necessary to make clear in that how I understand this comparison. I hope I have now found a good way out of the trouble."[59] (See Chapter 2, this volume)

Soon afterwards, Jongmans wrote again: "With this same mail I sent to you the paper on correlation between the United States and Europe. You will see that much has been changed compared with the first [reaction to] this article. I hope you can agree with the principles of this paper and especially with the solution I give to get out of the Stephanian trouble. I shall be glad if you read this paper carefully and if you return it to me with your remarks. You can add an appendix to the paper, if you like so, and present your opinion on the subject in this way."[60]

And Jongmans had still more to say on the matter: "As to the manuscripts on the Pocono and the West Virginia floras I prefer that you collaborate as much as possible. The best method is that you sign the paper with us as co-author, and that, where you think this to be necessary or useful, you give your personal remarks and notes. As far as we can agree with your remarks, the text can be altered in your sense. In case there is difference of opinion, your remarks can be given as footnotes under your name, the same method which is followed in committee where there are special opinions or remarks of the minority. As to the third paper, which is now signed by me alone, if you agree with my opinions, I propose to you to sign the paper *with me.* In this case, of course, I shall be glad if you send your remarks or additions. I think that this will be the best way to avoid remarks from a certain group of workers."[61]

During Jongmans's travels in the United States in 1933, he visited M. K. Elias in Kansas (see Chapter 26, this volume). At that time, Elias gave Jongmans two fragmented specimens from Garnett, Kansas. Elias named the specimen *Dichophyllum moorei* in 1936. Upon his return to The Netherlands, Jongmans published a paper with W. Gothan in 1934 in which he identified the specimens as *Callipteris flabellifera* (Weiss). A discussion of the correct identification of this specimen resulted between both Elias and Darrah (see Chapter 2, this volume) and Jongmans and Darrah.

Jongmans continued to try to correct his errors. "I do not remember whether I have written to you about p. 19 where you say that I believe *Callipteris* to be common in Kansas. You correctly say that the plant in question was *Dichophyllum moorei* Elias [see Moore et al. 1936; Elias, 1937)]. However, if you compare the original figures by Weiss and those by Elias you shall agree with me that, at least practically, they are the same. I agree with Elias that it is better to put this plant in a separate genus and not to *Callipteris.* However, this is stratigraphically the same, as according to Weiss, the plant he calls *Callipteris flabellifera* is a plant of the German Rotligend, so that the stratigraphical conclusion was right, and this is the only thing I have said, according to our conception of this part of the Carboniferous as we had it at that day. My present opinion is that it is possible that Elias is right, although it is still rather difficult to explain the situation. But the fact remains that I never said that *Callipteris* was common in Kansas, for the simple reason that I did not know enough of the details. I only said that *Callipteris* (i.e., "Rotliegend" type) was found in these beds."[62] (See Chapter 2, this volume.)

Jongmans asked soon thereafter, "On p. 5 of your paper on American Carboniferous floras [Darrah, 1937] you mention some limestones. Where are they? In Missouri? Please let me know. Also what is the Ames invasion? Is this marine?"[63]

Darrah's explanation followed: "All of the limestones mentioned in my paper are found in the type sections in Pennsylvania. The Ames limestone (Ames invasion of D. White and of C. Schuchert) is Middle Conemaugh. This is our best marine horizon in the Pennsylvanian. It occurs in Pennsylvania, Ohio, West Virginia and Maryland. It is supposed to occur in Illinois, but I cannot tell from the literature. . . . I wish to keep the lists of species as I sent them. If you wish to append an editorial remark as a footnote, I give permission for you to do so. Although you have advanced good reasons for the use of them *'Westphalian E'* [see Chapter 26, this volume]—and I see your arguments, for the time being I will use 'Stephanian.' After your papers are published validly that is another matter. For a young man just beginning really serious work, I should publish my papers as I gave them (errors included). The same remark holds for *Gigantopteris* (on p. 12) as a 'Permian' species. I did not know of your Sumatra paper until my visit to Heerlen. . . . I deeply appreciate the carefully prepared criticisms and suggestions which you have made. I think that I should give what may be called the 'American' opinion. If it turns out that it is in error, so much the better, because I can later correct my own work where some else would not. Your West Virginia manuscript was returned to you more than a month ago. The Westphalian 'E' paper was returned on [September] 4, and one set of fossils on [September] 2."[64]

Darrah wrote Jongmans about the mutual papers that he was to coauthor. The letters between them continued as to their disagreement of where species occur and to which horizon they belong. Jongmans finally wrote, "I am very glad to see that, practically spoken, we fully agree. The remarks have been put into the manuscript as footnotes under your name. The Pocono paper will be published as a joint paper with your name as co-editor [Jongmans et al., 1937]." Jongmans, Gothan, and Darrah were able to reach an agreement about which species occur in the Pocono formation. The 1937 paper delivered by Jongmans at the Heerlen Congress (1935), at Darrah's insistence, contains the works of Dawson, Meek, and Lesquereux.

Jongmans continues in his letters to argue his stratigraphic point of view: "In your remarks to the paper on the comparison between the Carboniferous floras of America and Europe you write about my Westphalian E: To be sure this is a synonymous term for what is called Stephanian. In my opinion, it is not necessary to introduce a new name. It is not right that my Westphalian E should be synonymous with the 'Stephanian'. Certainly part, if not all of the 'Lower Stephanian' (Rive de Gier with *Neuropteris ovata,* etc.) belong to the Westphalian D [see Chapter 26, this volume]. Middle and Upper 'Stephanian' may be synonymous of my Westphalian E. They contain [a] number of specialized species, by which these [F]rench floras are characterized as a *special* facies, which is not found in other European youngest Upper Carboniferous beds, nor in the American, which now are mapped 'Stephanian', without possessing the character of the [F]rench facies. How these youngest beds are called in future is relatively equal to me. But I am of the opinion that, calling the youngest beds (without *Neuropteris ovata,* etc.) 'Westphalian E,' we get a much more clean and regular transition into the Rotliegend than we do in introducing 'Stephanian'. The more as the Rive de Gier flora with *Neuropteris ovata,* etc. *cannot* be separated from the Westphalian D, as it is developed in Britain, in Saarbrücken and in other basins. So we would be obliged to diminish the vertical extension of the 'Stephanian' at any rate and to restrict this name to that what is called Middle and Upper Stephanian now."[65]

Jongmans's letters continued with complaints about fossil specimens not arriving. Moreover, he added, "I shall be still much more glad, if you could find time to answer my letters and to give some [explanation] of the fact that you are so taciturn."[66]

Darrah's response was immediate, and Jongmans wrote a reply: "The American papers, which I prepared for our proceedings are still with Gothan in Berlin. He is very occupied and therefore it takes a long time before he returns them. It is rather disagreeable but I cannot help it. As soon as he returns them they will be sent to you for examination and criticism. You can make every addition, every remark you like. It is my intention to get an agreement. I shall be very glad to get [your] opinion as freely as possible especially as to the conception of the Stephanian which is developed in these papers. I hope Gothan does not keep them too long."[67]

Jongmans wrote again to Darrah, "Why are you so silent? In response Darrah wrote, "First let me explain the reasons for my long silence. Two years ago, I was suddenly made the research curator here in the Museum and had our quarters extended to fill four rooms and storage space. This necessitated a great deal of effort in mere mechanical handling of our many specimens. About a month later I was placed in charge of the elementary lectures and instruction in botany at both Harvard and Radcliffe (a girl's college nearby). I had to drop my research to review my knowledge of physiology and organic chemistry. . . . We have no secretary or stenographer in the museum and I must do all work 'long hand'."[68]. Bill then elaborated in the letter to explain the specimens which were ready for exchange.

Jongmans answered promptly: "I am glad to hear the reason of your long silence lies with your busy life. I was afraid there was some misunderstanding like [the] one exists with [C. B.] Read (for the latter misunderstanding I still always ignore the reason) and I am happy to learn from your letter that this is not the case."[69]

Some time later, Darrah wrote: "I wish to acknowledge receipt of your wonderful paper on the fossil plants of Asia. You have made available such material from Russia, Siberia, and China, which is very difficult to obtain from other sources. Recently in the American Journal of Science February 1940, Dr. Charles Read has contributed a paper on a Carboniferous flora from Oregon [Read and Merriam, 1940], on the west coast of America. He points out that this plant assemblage is Arcto-Carboniferous and entirely European in aspect. He has missed the point, because he does not know the succession of floras of Asia, else he would have noted the works of Gothan and Sze [1933]. It is a *Neuropteris schlehani* flora. There are no illustrations nor critical diagnosis in Read's paper."[70]

Five and a half years later after World War II, Jongmans wrote, "Next year [1946] I intend to retire from my post as a director to the Geologische Bureau. In August of that year I will have been working here for forty years and at the first of May for twenty-five years as a director of the institute. I find that date is the right one to retire and to leave my place to a younger geologist. But it is not my intention to take a rest. The Netherlands Mines will give to me the possibility to prepare my large flora of the Carboniferous and the general stratigraphy. I also intend to write a general flora of the Carboniferous of the world and [will] try to get at a vertical and a horizontal parallelization of the different regions and to treat especially the question of the variations under climatic and geographical conditions."[71]

Reporting on his own experiences during the recent years, Darrah wrote, "I was somewhat amused and taken back by the news that I had died two years ago. Though my health was not of the best, I assure you I am still alive. During the war I have been engaged in industrial war work. . . . Most of my work was indirectly concerned with radar. It was of a metallurgical and chemical nature. Some of the problems I worked on were particle size of metals, especially powdered zirconium, ultraviolet examination (fluorescence) of metals and more recently the writing of technical manuals and reports."[72]

Upon the retirement of Jongmans from the Geologische Bureau in 1946, Darrah sent his heartfelt congratulations. Thereafter, he wrote to Dr. S. van der Heide in Schaesberg, The Netherlands, who was in charge of the retirement festivities, "I am honored by your request for my participation in the jubilee celebration of [Professor] Jongmans' distinguished services to the science of Paleobotany. I have for many years been a close friend and colleague of [Professor] Jongmans and I shall do everything in my power to help to fulfill his desire to visit [again] the United States of America. My home will be open to [Professor] Jongmans during his visit to this part of the country. He may use it as a headquarters for local excursions into Rhode Island Coal Basin [Narragansett basin] and to visit the collections of Harvard University. I will accompany him on his visits in this neighborhood. I have enclosed a card with my signature to be included in the album which you plan to present to him. I am very happy that you should use my name among those who form a committee of honor in recognition of Professor Jongmans' well-known works and researches. With every good wish that the jubilee celebration fulfills the purpose which you have planned."[73]

Two months later, "Many thanks for your congratulations and all your contributions to the success of my jubilee," wrote Jongmans. "It has been a beautiful and interesting day and I was very glad with the numerous contributions especially those by my colleagues in other parts of the world. . . . The Netherlands Coal mines have given as a jubilee gift the task to describe the Carboniferous flora and stratigraphy of our coalfields in the widest sense and in comparison with the conditions in other basins. . . . I intend to go to Turkey for the continuation of my work in the Anatolian coalfields, which I started in 1938."[74]

More letters from Jongmans followed. "We are very pleased to learn that you have returned safely to your home following your interesting experiences in Turkey," wrote Darrah. "I have also completed a small paper on *Lepidodendropsis*. Since your visit to America some years ago this interesting type has been found in Pennsylvania, Illinois, Texas, Indiana, Virginia, Kentucky and of course in the Maritime provinces of eastern Canada. You have done a real service in pointing out the importance of this characteristic group."[75]

During February 1947, Jongmans sent a letter of introduction for the new director of the Geologische Bureau, Dr. A. A. Thiadens, who was one of Jongmans's leading collaborators in the geological investigations that were carried out during World War II in The Netherlands Mining District.

Keeping up to date with what was going on in the field, Jongmans wrote Darrah, "Dr. Verdoorn wrote to me that you intended to publish a second edition of your *Principles of Paleobotany* and also on you plan to write a book on coal. . . . I am much interested in what you write about *Lepidodendropsis*. You will remember that I also described specimens from Russia and from Egypt in the publications of the bureau in Heerlen. During my recent visit to Egypt, I found that the plant bed with *Lepidodendropsis* is well recognizable in every borehole in which this special horizon is present. This horizon will probably prove to be valuable guide-horizon at least in that region."[76]

Darrah wrote back:

> It is nice to hear that you have opportunities to travel in many parts of the world, to study additional Carboniferous deposits. Although each of the countries you mention has interesting possibilities, I hope that you will visit the Brazilian coalfields. I have not had personal opportunity to examine very much material from Brazil but what small collections have passed through my hands convince me that there are many Arcto-Carboniferous forms present in the upper portion of the South American Upper Carboniferous. Of course, there are many constituents in the flora which are characteristic of the southern hemisphere. At one time, the United States Geological Survey (David White and C. B. Read) had obtained modest collections from Brazil (see Chapters 10 and 19, this volume), but these have never been investigated and I am quite sure never will be.

You express an interest in knowing what kind of work I have been engaged in during the war. In December 1942, I accepted a position with a manufacturer of vacuum tubes to be used in radar installations. My first problems concerned the use of finely divided metals with a grain size running down from 0.5 microns. We are attempting to disperse such powders in silica gels. After doing this work, which was of a physical chemical nature, I also did some research on the solubility of metallic oxides in glass, but the problem on which I was engaged was solved in a very different way and the work was terminated before we had reached any conclusion. At this stage during the war, America had reached its peak of military production and in those days of expansion and coordination of effort, I was placed in the administration of a department which controlled the quality of raw materials, such as ceramics, pure metal, and similar materials. The chief work was to train adequately a large number of women and girls who, for the most part, had no college training in physics, chemistry, or mathematics. In American high schools, most students take one year of science to learn to use the microscope. These simple skills were of great value to build upon, and we did some very unexpected things with people having very little technical skill and training. Since the war, I have remained with industry and although my duties have changed somewhat, I am not permitted to divulge the nature of my work. So you can see this is a far cry from paleobotany. Nevertheless, I have found the opportunity to develop many interesting techniques applicable to the study of coals, various types of petrifications, and above all, in the preparation of materials for elaborate investigations. I have done considerable work on fossils from Kentucky, Colorado and Iowa during the past five years.[77]

In his own efforts, Jongmans needed some assistance and wrote Darrah, "You know that I am working on a paper on the trees especially Paleozoic and Mesozoic times. For this purpose I should like to have good [photographs] of the fossilized stems as they have been found in the Chalcedony Park, Arizona; the Napa Valley, California; Yellowstone National Park. . . . Is it possible to provide me with some good photos of the [trunks] and of reconstructions of the whole plants."[78]

Darrah responded quickly: "I have made an effort to locate sources of information which you seek. Most of what I tell you will, I am afraid, be disappointing. No material from Chalcedony Park, the Napa Valley or Yellowstone, is available at Harvard, and you will have to request photographs from the institutions where the originals are housed. The Geological Survey in Washington contains Knowlton's specimens and also many of the more recent collections. They can provide you with some good photographs of the figures to which you refer. . . . Dr. Henry Andrews at the Missouri Botanical Garden has made a number of very excellent collections in Yellowstone and Arizona. I am quite certain that he would be happy to assist you in obtaining reproductions of his choice specimens. For California material, you will have to communicate directly with Dr. Ralph Chaney. I am sure that you know all of the men who are in charge of these collections and I am sure that they will cooperate with your work. As for reconstructions of American Cycadophytes, I can give you several drawings which I would like to have returned at some future time. However, you will do better by obtaining from Dr. Wieland photographs of the trunks."[79]

Months later, Jongmans had more to report to Darrah on his recent activities. He wrote, "Since the beginning of October, we have been traveling in Spain and we are now in Portugal. Next Sunday we leave for French Morocco, where I intend to look over the Carboniferous. There is fine [Stephanian] section on the southern part and after having seen that we go to Ouijda, in the coalfield. They have asked me to describe the fossil flora of the coalfield. In Spain, I visited a number of coal districts, especially in the northern part, but also in the south. I made several collections and some interesting and surprising discoveries. In Portugal there is not so much, only some smaller coalfields but with interesting floras."[80]

Still reflecting on the fascinations of his travels, Jongmans wrote Darrah several months later: "In Spain I hope to have done good work. Paleobotany in Spain is an almost undiscovered field. No paper with good figures exists, the principal thing they have are lists by Mallada. . . . The inspector general of mines asked me to go there and look after the stratigraphy. I was quite astonished to find a typical Pocono-flora, a small difference. . . . In the Sahara, we discovered a new coalfield, which may be of some importance as coal in North Africa is rare. It is a remarkable thing to work in the desert collecting fossil plants, the rests of a [luxurious] vegetation. . . . However, I collected many duplicates as I suppose that other paleobotanists may be interested in those desert-Carboniferous plants. The floras belong to the [Westphalian] C, D and perhaps even [Westphalian] E or the rest of the [Stephanian]."[81]

Darrah answered, "Under separate cover I have sent two copies of a paper on *Lepidodendropsis.* The illustrations which include the type of specimens of *Stigmaria minuta* and others identified by Dawson and Meek are obviously identical with various forms which you described some years ago. After deliberating for a long time, I have united all of these forms under *Lepidodendropsis corrugatum* for reasons presented in the article. It is entirely possible that you will find it impossible to agree with what I have done but I believe it helps to clarify some of the confused older American literature and to place your species in perspective."[82]

Jongman's response was almost immediate: "With much interest I read that you [wrote] a paper on *Lepidodendropsis,* and I wonder how you can unite the different species described under one single name. I agree with you that nomenclature is a very difficult thing especially when you refigure such 'species' which have been published in earlier days without sufficient descriptions and with figures which are so bad that it is not possible to decide what they are." Jongmans then went on, as he referred to his work on the (Mesozoic) reconstructions for his upcoming book (Jongmans, 1949). He stated that he had received photographs from several American museums as well as individuals: C. A. Arnold, H. N. Andrews, R. W. Chaney, W. Goldring, and G. R. Wieland. He said he now intended to publish a similar work in English. "It will be published by a [D]utch publisher with the assistance of Dr. Verdoorn. Now the condition is that it must be written in real [E]nglish so that it must be looked over very critically by an American or [E]nglish colleague. I should be very glad if you could help me in this question and be the colleague who corrects my book and makes it possible that it is understood by an [E]nglish speaking public. In that case, I will write it in my own [E]nglish language and send it to you. You would oblige me very much by helping me in this matter."[83]

Darrah's response was immediate: "I am honored by your invitation to assist in editing your manuscript on fossil plants to prepare it for an English edition. I have always admired facility of Europeans to write and speak in many languages but I sympathize with you in finding it necessary to have someone make certain that the literary style conforms to the best usage. I know how difficult it is for me to write in French or German even though I read both languages rather easily. Such an editing will require a very considerable time but I shall do all I can to help. Perhaps it would be well if you would consult your publisher or Dr. Verdoorn [Fig. 10] to see whether they were fully satisfied with my participation. It may be that they would suggest someone better qualified than I. . . . Present plans are for me to complete a revision of my Principles so that Dr. Verdoorn has the copy by the first of November. He hopes to issue the book on April 1, 1950. I mention this because it would be most convenient for me if you would send your chapters in small parcels. This would not only make it easier for me to do the work over a period of time, but it would give you a chance to examine whatever changes I might suggest to make sure that I have not destroyed your meaning. I have among my papers the sketches and photographs I promised to you. I shall locate them and send them."[84]

Some time later, Jongmans wrote, "I have seen that a second edition of your Principles is in the press (Chronica Botanica). Is this edition already [completed] and has it been printed? I will be glad if you can inform me about this."[85]

Jongmans's letters continued to ask for reprints and for Darrah to replace reprints missing from Jongmans's collection and presumably lost. For further references, see Jongmans (1937a, b, 1949); Jongmans and Gothan (1937a, b).

The correspondence between Darrah and **Richard Kräusel** (Figs. 11, 12), whom he met in 1935, lasted from 1937 to

Figure 10. Joop Verdoorn (left) and Frans Verdoorn (right) holding the hands of their grandson, Dana Grootenboer. Frans Verdoorn was managing editor of Chronica Botanica Company and secretary of the International Biohistorical Commission of Waltham, Massachusetts. Verdoorn was also director of the Biohistorical Institution of the University of Utrecht, The Netherlands. He established and edited Annals of Bryologici and Chronica Botanica. Verdoorn worked with many scientists and assisted in publishing their works, including those of W. C. Darrah and W. J. Jongmans. Photograph was taken at their home on April 22, 1973. The William Culp Darrah Collection.

Figure 11. Left to right: R. Kräusel, W. J. Jongmans (with pipe), and F. Stockmans. Photograph taken on a field trip during the Second Carboniferous Congress, Heerlen, The Netherlands, 1935. Photograph taken by H. H. Darrah. The William Culp Darrah Collection.

Figure 12. Richard Kräusel in his study in Frankfurt, Germany. Photograph taken by A. T. Cross in 1951. Courtesy of A. T. Cross.

Figure 13. Suzanne S. Leclercq. Courtesy of and permission from the Hunt Institute for Botanical Documentation, Carnegie Mellon University, Pittsburgh, Pennsylvania.

1949. The letters are discussions of interpretation of Devonian plants. "From your textbook," wrote Kräusel from the Senckenberg Museum in Frankfurt, Germany, in an early letter, "I note that all of this agrees with what I have found out. In the future there will be more explanations about the Devonian flora that will be confirmed besides *Callixyon* from the Rhineland. Many thanks for the stamps, as I am an avid collector, especially from other countries."[86]

Kräusel wrote later that he was concerned that the separates had yet to arrive. "Hopefully, soon the mail will be normal again and deliveries will not take so long," he commented. Then he wrote, "What you have found out about *Pseudosporochnus* is very interesting. Now you are asking me about my opinions concerning *Archaeosigillaria. Archaeosigillaria* has aroused my interest."[87]

**Suzanne S. Leclercq** (Fig. 13), from the Laboratoire de Paléontologie at the University of Liège, Belgium, also attended the 1935 Congress in Heerlen, where Darrah met her. They were both responsible for their respective museum collections, and their correspondence after the meetings reflected their mutual interests. Then, a number of years later, Darrah wrote, "I am delighted to receive a letter from you and to know that you have returned to your post at the University. I hope your laboratory survived serious damage during the war and that your activities have returned to normal. Neither of the two publications mentioned in your letter of April 14 reached me and I would be grateful to have copies of them. Under separate cover I am sending a small parcel of a half dozen or so reprints which appeared between 1940 and the present. Would you be interested in receiving a small box of 25 microscopic slides of typical coal ball plants mostly seed ferns and *Cordaites?* Perhaps you would be willing to send in exchange either some small fragments of Belgium coal balls or a few peels which you have prepared from Belgium coal balls."[88]

Leclercq replied, "My [research is] on Middle Upper Devonian material. I'm still interested [in the structure of] preserved plants from Carboniferous horizons. That is why I agree with your kind proposition to send me a small box of 25 selected microscopic slides, and to send in exchange coal balls from Belgium and a few peels which have been prepared [from] Belgium coal balls."[89]

Leclercq wrote again: "I have received a small grant to go to your country, where I intend to stay from the second fortnight of April until the end of June and afterwards to reach Stockholm for the Botanical Congress. I shall land [in] New York [on] April 11th, where I have to stay about two weeks, then Boston two weeks, Ann Arbor one week, Chicago on week, Urbana and St. Louis 2 weeks-and-a-half. I have also to go to Washington, but I should prefer to meet you and Mrs. Darrah. Please let me know if it is possible with the plan I give you and when? I received safely the parcel of 25 microscope slides. Thank you so much for them, and also for your offer to send me a general assortment of typical American fossil plants. I was very busy until now, but I hope to prepare for you during this month a small collection of peels from Belgium coal-balls and especially of *Sphenophyllum* as you asked me. Unhappily, *Lyginopteris* are not frequently found."[90]

Darrah answered Leclercq right away: "I was delighted to have your letter of January 10 informing us that you contemplate a visit to the United States. We are most anxious to see you again and can arrange our plans to suit whatever schedule you must maintain. . . . I know of your major interests and I am sure you will find many things of curiosity and importance both in the University and in my private collection. Please let me know near the time when you will arrive so that we can be at your command."[91]

In confirmation of Darrah's offers of hospitality and assistance, Leclercq responded, "I shall be glad to see you and Mrs. Darrah again. I have prepared a small collection of peels that will be sent in a fortnight. I'm still very busy now. When I reach New York I shall write you the exact time of my arrival at Harvard to make arrangements to meet one another. I'm curious to examine your own collection which is important and of quality."[92] Thereafter followed her promised note: "I write you a few words to let you know that I landed [in] the U.S. a few days ago. At the same time I write to Professor [Elso] Barghoorn. Let me know if Friday or Saturday 29th or 30th of April will be convenient [for] you."[93]

Leclercq wrote again to Darrah as she traveled across the country: "In Albany, Ithaca and Ann Arbor I examined Devonian plants, some of them very well preserved but generally very fragmentary. I think the young [Harlan] Banks is a good worker; and [Professor C. A.] Arnold is now busy with fine Eocene material found in Oregon. In the Field Museum [Chicago, Illinois] I saw selected fossils collected by the old Mister [George] Langford [see Chapter 23, this volume]. As soon as I shall be back, I [will] send you photos of the *Rhacophyton zygopteroides* and a collection of peels of coal ball plants from Belgium. And thank you for the set of microscope slides you continue to send."[94]

Some months later, Darrah again heard from Leclercq:

Since I flew back from Stockholm, scientific visitors came to Liège keeping me busy. Mrs. Sahni [stayed] a week at Liège to see the equipment of my laboratory and order useful instruments for the Institute of Palaeobotany at Lucknow [India].

Since the war, I got opportunities to provide my laboratory with new machines and all necessities for coal balls and Devonian material as optical microscopes and photography [equipment]. Doctor [Sergius] Mamay from St. Louis [stayed] a week too, looking at the collection of the Belgium coal ball slides and peels. I [enjoyed] his visit for he is a clever and critical [minded] young man. He was interested by a peculiar coal ball full of *Stauropteris oldhamia* rachises (more than 160 rachises in the largest slide). When I began the study of that material I hoped the stem should be found and its structure definitely established; unhappily none of the structure which could have been a stem has ever been found in connection with any of the numerous first order rachises. At that time the negative result disappointed me so much, that the research was abandoned. Now after observing the material again I think that the wonderful bushy organization of the plant probably required a description. [Shortly] before I [went] to the U.S.A. I received a drill core with petrified structure coming from Campine's coal basin of Belgium and found in the Westphalian B. As you know, the coal balls of Bouxharmont I formerly studied were coming from Westphalian A, a lower horizon. . . .

I went with Professor [W. N.] Stewart to the Mazon Creek formation not far from Coal City. . . . When I was in Albany, Miss Winifred Goldring and Professor [H.] Banks took me to Gilboa Quarry. . . .

If [I] compare with the VI Congress the last one [shows] some changes of interest, mainly in application of paleobotany to industrious purposes. For instance: Tertiary flora and lignites are worked more than before, as also the spore determination in coal. . . . Some botanists like Professor [W.] Zimmerman of Tübingen gave interminable accounts on nebulous phyllo-genetic theories, not always in right conformity with the facts; but more in number came the botanists to our meetings. On the whole, every one was glad to meet each other once again and to note [that] there were more paleobotanists than before although some eminent old ones were definitely missing.[95]

In yet another letter, Leclercq notified Darrah, "At last the Bouxharmont coal balls I promised you are ready."[96] And the specimens were indeed shipped right away. Darrah responded two months later: "Forgive my tardy acknowledgment of the receipt of the parcel of coal balls. The box arrived in New York City during the Christmas holiday. . . . I have made a few exploratory peels and found good megaspores of *Bothrodendron* and petioles of *Botryopteris*. The nodules should afford considerable interesting material. . . . [We] have selected another series of preparations among which are several excellent *Cordaianthus* much better than the usual condition."[97] For further references, see Leclercq (1925a, b, 1935, 1936, 1950).

Bill Darrah met **Armand Renier** at the 16th International Congress in Washington, D.C., in 1933. Dr. Renier, a paleobotanist/geologist with the Service Géologique de Belgique in Bruxelles, assisted the young scientist to receive a grant in 1935 from the Belgium Relief Educational Foundation for his travels to attend the congresses in Amsterdam and Heerlen (see Chapter 2, this volume). Renier and his friend P. Bertrand hosted Darrah at the Belgian Wérister mine. Thereafter, through early letters, a friendship between the two men developed. Renier published papers on Wérister mine stratigraphy and specimens, and in one letter Renier wrote Darrah to thank him for his paper on Iowa coal balls and pointed out a mistaken impression concerning the Wérister mine: "As in England (Shore: Littleborough) they are in Belgium (Wérister and other places) two kinds of coal balls: the coal balls which occur in the coal seam contain only plants and the roof-balls [nodules] which are found in the shale above the seam and where, with much [maceration] relicts of plants, marine shells, [and] fish are found. Everywhere coal seams containing coal balls have a roof with a marine fauna. It was the way to discover coal balls in Belgium and also in Spain: First look for marine roofs, second search for coal balls, whose occurrence is more rare. About the confusion between coal balls and roof balls, I mention a paper of the late Professor [Henry] Danville on the coal balls of Yorkshire. . . . I have pointed out the mistake in a short paper. . . . I hope to send you a copy of this pamphlet and also of the note about the [methodical] search for coal balls. The roof balls are ever regular in shape [which] is never the case for coal balls."[98] For further reference, see Renier (1907, 1930).

**Albert C. Seward** (Fig. 14), from the Botany School of Cambridge University, England, also attended the Sixth International Botanical Congress in Amsterdam in 1935. Darrah met him there, and in the years that followed their brief letters primarily concerned exchanges of reprints and technical matters.

To **H. Hamshaw Thomas** (Cambridge University, Cambridge, England) (Fig. 15), whom he had also met at the Botanical Congress in 1935, Darrah wrote: "It was a pleasure to hear from you again and to receive some of your recent papers. . . . The passing of Sahni has dumbfounded us. When he was here a year ago, he appeared to be in excellent health. It seems a shame that his Institute of Paleobotany had barely begun. I hope that from the large following he had developed among younger Indian botanists a suitable successor will bring the Institution to a high place in science."[99]

Figure 14. A. C. Seward. Courtesy of and permission from the Hunt Institute for Botanical Documentation, Carnegie Mellon University, Pittsburgh, Pennsylvania.

Figure 15. H. H. Thomas outside Cambridge University, Cambridge, England. Photograph taken by A. T. Cross in 1951. Courtesy of A. T. Cross.

**John Walton** (University of Glasgow, Scotland) (Fig. 16) and Darrah (Fig. 17) met in Amsterdam at the International Botanical Congress in 1935. In one of Walton's letters, he expressed his concern about Darrah's new peel technique [Darrah, 1936c], a modification of Walton's own technique (Walton, 1928): The solution might be explosive. Darrah's response was quick to follow: "I received your card with the suggestion that cellulose acetate might prove safer and keep colorless in our work on fossil plants, much more satisfactory than the nitrate peels. When I first worked with your techniques some nine or tens years ago, I was using acetate solutions and found them successful, but I could not obtain standardized materials and most of the substrates we had were of the rayon type. I long ago gave up the commercial nitrocelluloses in favor of standardized commercial preparations. One of the chemical houses in St. Louis has sold under the trade name 'Parlodian' a good biological solution. . . . I have been trying various maceration techniques on coal balls.. . . . By maceration with a simple acid solution of very

Figure 16. John Walton at the International Botanical Congress in Montreal in 1959. Photograph taken by and courtesy of Theodore Delevoryas.

Figure 17. W. C. Darrah (facing camera). This photograph taken at the Sixth International Botanical Congress, Amsterdam, The Netherlands, September 2–7, 1935, by H. H. Darrah shows Darrah as he looked at the time when most of the correspondence with the Europeans was written. The William Culp Darrah Collection.

low concentration under gentle conditions, we have recovered cellular elements from *Mesoxylon,* for instance. . . . I do not understand the full significance of this indestructability but have found this property to be of such widespread occurrence that it must have some meaning. The pyritized coal balls which I have been studying are strikingly different from those obtained at Shore, Dulsgate and in Holland and Belgium. With the exception of spores the tissue in the European coal balls are black and fall to pieces when macerated."[100]

## CLOSING THOUGHTS

As revealed in this historical account, Bill Darrah's "obligation" to help others succeed, even as he pursued his own particular interests, was always foremost in his dealings with his friends and colleagues. He generously exchanged ideas and gave away or exchanged fossil-plant specimens with his European counterparts and thus promoted Euramerican paleobotany and biostratigraphy.

## ACKNOWLEDGMENTS

The William C. Darrah Collection in Paleobotany is made up of letters, photographs, field notebooks, and notebooks of information related to the field of paleobotany. Acknowledgment is given to the late William C. Darrah for his penchant for saving such memorabilia. Sadly, he did not save carbon copies of the letters he wrote to his colleagues. As he explained so simply to me (Elsie Darrah Morey, his daughter), "I know what I wrote to them." And with his prodigious memory, we can be certain that he did.

Many thanks are extended to Britta Lundblad for sending E. D. Morey photocopies of the letters of W. C. Darrah to R. Florin. The original letters from W. C. Darrah to Florin and T. G. Halle are housed in the Center of the History of Science under the auspices of the Swedish Academy of Sciences, Stockholm, Sweden.

We thank H. W. J. van Amerom for the photocopies of the W. C. Darrah letters to W. J. Jongmans, which were in the files at the Geological Bureau, Rijks Geologische Dienst, Heerlen, The Netherlands. We also thank Frieda Zierhoffer and Emily Zukovich for translating the German letters written by R. Kräusel and Nora Tamberg of the U.S. Geological Survey Library for the excellent translation of the letter written by W. Gothan.

Theodore Delevoryas is thanked for his kindness in sending photographs of Tom Harris and John Walton that are reproduced in this chapter, as is A. T. Cross for the privilege of using his photographs of R. Crookall, R. Kräusel, and H. H. Thomas. The Hunt Institute for Botanical Documentation of the Carnegie Mellon University, Pittsburgh, Pennsylvania, is thanked for permission to use the photographs of Hanna Czeczott, Thore G. Halle, and A. C. Seward. The Bergius Foundation (Stockholm) is thanked for permission to use the photograph of R. Florin.

Theodore Delevoryas and H. W. J. van Amerom are thanked for their reviews of this manuscript.

## ENDNOTES

1. Letter from P. Bertrand to W. C. Darrah, May 1, 1934. W. C. Darrah Collection.
2. Letter from P. Bertrand to W. C. Darrah, April 12, 1935. W. C. Darrah Collection.
3. Letter from P. Bertrand to W. C. Darrah, June 6, 1935. W. C. Darrah Collection.
4. Letter from P. Bertrand to W. C. Darrah, August 5, 1935. W. C. Darrah Collection.
5. Letter from P. Bertrand to W. C. Darrah, October 7, 1935. W. C. Darrah Collection.
6. Letter from P. Bertrand to W. C. Darrah, July 28, 1936. W. C. Darrah Collection.
7. Letter from P. Bertrand to W. C. Darrah, October 8, 1936. W. C. Darrah Collection.
8. Letter from I. M. P. Browne to W. C. Darrah, April 19, 1936. W. C. Darrah Collection.
9. Letter from I. M. P. Browne to W. C. Darrah, October 3, 1936. W. C. Darrah Collection.
10. Letter from I. M. P. Browne to W. C. Darrah, October 5, 1938. W. C. Darrah Collection.
11. Letter from I. M. P. Browne to W. C. Darrah, July 13, 1938. W. C. Darrah Collection.
12. Letter from R. Crookall to W. C. Darrah, July 14, 1932. W. C. Darrah Collection.
13. Letter from R. Crookall to W. C. Darrah, March 23, 1933. W. C. Darrah Collection.
14. Letter from H. Czeczott to W. C. Darrah, January 26, 1939. W. C. Darrah Collection.
15. Letter from E. Dix to W. C. Darrah, February 24, 1936. W. C. Darrah Collection.
16. Letter from E. Dix to W. C. Darrah, April 30, 1936. W. C. Darrah Collection.
17. Letter from E. Dix to W. C. Darrah, May 4, 1944. W. C. Darrah Collection.
18. Letter from W. N. Edwards to W. C. Darrah, May 22, 1936. W. C. Darrah Collection.
19. Letter from W. N. Edwards to W. C. Darrah, July 1, 1936. W. C. Darrah Collection.
20. Letter from W. N. Edwards to W. C. Darrah, July 9, 1936. W. C. Darrah Collection.
21. Letter from W. N. Edwards to W. C. Darrah, October 12, 1936. W. C. Darrah Collection.
22. Letter from W. N. Edwards to W. C. Darrah, November 20, 1936. W. C. Darrah Collection.
23. Letter from W. N. Edwards to W. C. Darrah, January 18, 1937. W. C. Darrah Collection.
24. Letter from R. Florin to W. C. Darrah, September 25, 1935. W. C. Darrah Collection.
25. Letter from R. Florin to W. C. Darrah, November 30, 1935. W. C. Darrah Collection.
26. Letter from R. Florin to W. C. Darrah, January 22, 1936. W. C. Darrah Collection.
27. Letter from R. Florin to W. C. Darrah, March 12, 1937. W. C. Darrah Collection.
28. Letter from R. Florin to W. C. Darrah, November 18, 1937. W. C. Darrah Collection.
29. Letter from R. Florin to W. C. Darrah, May 30, 1939. W. C. Darrah Collection.
30. Letter from R. Florin to W. C. Darrah, July 13, 1939. W. C. Darrah Collection.
31. Letter from R. Florin to W. C. Darrah, March 2, 1940. W. C. Darrah Collection.
32. Letter from W. Gothan to W. C. Darrah, April 14, 1936. W. C. Darrah Collection.
33. Letter from T. G. Halle to W. C. Darrah, July 17, 1932. W. C. Darrah Collection.
34. Letter from T. G. Halle to W. C. Darrah, March 20, 1933. W. C. Darrah Collection.
35. Letter from T. M. Harris to W. C. Darrah, September 20, 1935. W. C. Darrah Collection.
36. Letter from T. M. Harris to W. C. Darrah, November 29, 1935. W. C. Darrah Collection.
37. Letter from T. M. Harris to W. C. Darrah, February 20, 1936. W. C. Darrah Collection.
38. Letter from T. M. Harris to W. C. Darrah, November 4, 1936. W. C. Darrah Collection.
39. Letter from M. Hirmer to W. C. Darrah, December 15, 1937. W. C. Darrah Collection.
40. Letter from M. Hirmer to W. C. Darrah, March 25, 1938. W. C. Darrah Collection.
41. Letter from M. Hirmer to W. C. Darrah, November 22, 1940. W. C. Darrah Collection.
42. Letter from H. A. Havens, acting chief, Division of Foreign Service Administration, to W. C. Darrah, December 18, 1940. W. C. Darrah Collection.
43. Letter from W. C. Darrah to W. J. Jongmans, January 15, 1947. Archives, Rijks Geologische Dienst, Office Heerlen, P.O. Box 126, 6400 AC Heerlen, The Netherlands.
44. Letter from W. J. Jongmans to W. C. Darrah, March 23, 1947. W. C. Darrah Collection.
45. Letter from W. J. Jongmans to W. C. Darrah, September 14, 1933. W. C. Darrah Collection.
46. Letter from W. C. Darrah to W. J. Jongmans, January 1, 1934. Archives, Rijks Geologische Dienst, Office Heerlen, P.O. Box 126, 6400 AC Heerlen, The Netherlands.
47. Letter from W. J. Jongmans to W. C. Darrah, March 24, 1934. W. C. Darrah Collection.
48. Letter from W. J. Jongmans to W. C. Darrah, April 16, 1934. W. C. Darrah Collection.
49. Letter from W. C. Darrah to W. J. Jongmans, April 2, 1934. Archives, Rijks Geologische Dienst, Office Heerlen, P.O. Box 126, 6400 AC Heerlen, The Netherlands.
50. Letter from W. J. Jongmans to W. C. Darrah, November 12, 1934. W. C. Darrah Collection.
51. Letter from W. C. Darrah to W. J. Jongmans, November 23, 1934. W. C. Darrah Collection.
52. Letter from W. C. Darrah to W. J. Jongmans, November 27, 1934. Archives, Rijks Geologische Dienst, Office Heerlen, P.O. Box 126, 6400 AC Heerlen, The Netherlands.
53. Letter from W. J. Jongmans to W. C. Darrah, December 10, 1934. W. C. Darrah Collection.
54. Letter from W. C. Darrah to W. J. Jongmans, January 31, 1935. Archives, Rijks Geologische Dienst, Office Heerlen, P.O. Box 126, 6400 AC Heerlen, The Netherlands.
55. Letter from W. C. Darrah to W. J. Jongmans, May 24, 1935. Archives, Rijks Geologische Dienst, Office Heerlen, P.O. Box 126, 6400 AC Heerlen, The Netherlands.
56. Letter from W. C. Darrah to W. J. Jongmans, October 9, 1935. Archives, Rijks Geologische Dienst, Office Heerlen, P.O. Box 126, 6400 AC Heerlen, The Netherlands.
57. Letter from W. C. Darrah to W. J. Jongmans, November 15, 1935. Archives, Rijks Geologische Dienst, Office Heerlen, P.O. Box 126, 6400 AC Heerlen, The Netherlands.
58. Letter from W. C. Darrah to W. J. Jongmans, March 2, 1936.

Archives, Rijks Geologische Dienst, Office Heerlen, P.O. Box 126, 6400 AC Heerlen, The Netherlands.

59. Letter from W. J. Jongmans to W. C. Darrah, May 29, 1936. W. C. Darrah Collection.

60. Letter from W. J. Jongmans to W. C. Darrah, June 10, 1936. W. C. Darrah Collection.

61. Letter from W. J. Jongmans to W. C. Darrah, June 23, 1936. W. C. Darrah Collection.

62. Letter from W. J. Jongmans to W. C. Darrah, July 11, 1936. W. C. Darrah Collection.

63. Letter from W. J. Jongmans to W. C. Darrah, August 27, 1936. W. C. Darrah Collection.

64. Letter from W. C. Darrah to W. J. Jongmans, September 9, 1936. Archives, Rijks Geologische Dienst, Office Heerlen, P.O. Box 126, 6400 AC Heerlen, The Netherlands.

65. Letter from W. J. Jongmans to W. C. Darrah, January 4, 1937. W. C. Darrah Collection.

66. Letter from W. J. Jongmans to W. C. Darrah, April 4, 1938. W. C. Darrah Collection.

67. Letter from W. C. Darrah to W. J. Jongmans, May 19, 1938. Archives, Rijks Geologische Dienst, Office Heerlen, P.O. Box 126, 6400 AC Heerlen, The Netherlands.

68. Letter from W. C. Darrah to W. J. Jongmans, March 31, 1939. Archives, Rijks Geologische Dienst, Office Heerlen, P.O. Box 126, 6400 AC Heerlen, The Netherlands.

69. Letter from W. J. Jongmans to W. C. Darrah, April 11, 1939. W. C. Darrah Collection.

70. Letter from W. C. Darrah to W. J. Jongmans, February 22, 1940. Archives, Rijks Geologische Dienst, Office Heerlen, P.O. Box 126, 6400 AC Heerlen, The Netherlands.

71. Letter from W. J. Jongmans to W. C. Darrah, October 1, 1945. W. C. Darrah Collection.

72. Letter from W. C. Darrah to W. J. Jongmans, November 14, 1945. Archives, Rijks Geologische Dienst, Office Heerlen, P.O. Box 126, 6400 AC Heerlen, The Netherlands.

73. Letter from W. C. Darrah to Dr. S. Van der Heide, March 7, 1946. W. C. Darrah Collection.

74. Letter from W. J. Jongmans to W. C. Darrah, May 20, 1946. W. C. Darrah Collection.

75. Letter from W. C. Darrah to W. J. Jongmans, January 15, 1947. Archives, Rijks Geologische Dienst, Office Heerlen, P.O. Box 126, 6400 AC Heerlen, The Netherlands.

76. Letter from W. J. Jongmans to W. C. Darrah, March 23, 1947. W. C. Darrah Collection.

77. Letter from W. C. Darrah to W. J. Jongmans, April 9, 1947. Archives, Rijks Geologische Dienst, Office Heerlen, P.O. Box 126, 6400 AC Heerlen, The Netherlands.

78. Letter from W. J. Jongmans to W. C. Darrah, July 22, 1947. W. C. Darrah Collection.

79. Letter from W. C. Darrah to W. J. Jongmans, August 15, 1947. Archives, Rijks Geologische Dienst, Office Heerlen, P.O. Box 126, 6400 AC Heerlen, The Netherlands.

80. Letter from W. J. Jongmans to W. C. Darrah, January 12, 1948. W. C. Darrah Collection.

81. Letter from W. J. Jongmans to W. C. Darrah, April 20, 1948. W. C. Darrah Collection.

82. Letter from W. C. Darrah to W. J. Jongmans, May 26, 1949. Archives, Rijks Geologische Dienst, Office Heerlen, P.O. Box 126, 6400 AC Heerlen, The Netherlands.

83. Letter from W. J. Jongmans to W. C. Darrah, June 15, 1949. W. C. Darrah Collection.

84. Letter from W. C. Darrah to W. J. Jongmans, June 22, 1949. Archives, Rijks Geologische Dienst, Office Heerlen, P.O. Box 126, 6400 AC Heerlen, The Netherlands.

85. Letter from W. J. Jongmans to W. C. Darrah, June 1, 1950. W. C. Darrah Collection.

86. Letter from R. Kräusel to W. C. Darrah, September 3, 1937. W. C. Darrah Collection.

87. Letter from R. Kräusel to W. C. Darrah, March 9, 1940. W. C. Darrah Collection.

88. Letter from W. C. Darrah to S. Leclercq, April 28, 1949. W. C. Darrah Collection.

89. Letter from S. Leclercq to W. C. Darrah, May 20, 1949. W. C. Darrah Collection.

90. Letter from S. Leclercq to W. C. Darrah, January 10, 1950. W. C. Darrah Collection.

91. Letter from W. C. Darrah to S. Leclercq, January 18, 1950. W. C. Darrah Collection.

92. Letter from S. Leclercq to W. C. Darrah, February 25, 1950. W. C. Darrah Collection.

93. Letter from S. Leclercq to W. C. Darrah, April 17, 1950. W. C. Darrah Collection.

94. Letter from S. Leclercq to W. C. Darrah, May 13, 1950. W. C. Darrah Collection.

95. Letter from S. Leclercq to W. C. Darrah, September 8, 1950. W. C. Darrah Collection.

96. Letter from S. Leclercq to W. C. Darrah, November 21, 1950. W. C. Darrah Collection.

97. Letter from W. C. Darrah to S. Leclercq, January 16, 1951. W. C. Darrah Collection.

98. Letter from A. Renier to W. C. Darrah, July 20, 1939. W. C. Darrah Collection.

99. Letter from W. C. Darrah to H. H. Thomas, April 25, 1949. W. C. Darrah Collection.

100. Letter from W. C. Darrah to John Walton, March 14, 1940. W. C. Darrah Collection.

## REFERENCES CITED


Andrews, H. N., 1980, The fossil hunters, in search of ancient plants: Ithaca, New York, and London, England, Cornell University Press, 421 p.

Bertrand, P., 1928, Stratigraphie du Westphalien et du Stéphanien dans les différents bassins houillers Français: Congrès pour l'avancement des études de stratigraphie carbonifère, 1st, Heerlen, The Netherlands, 1927: Compte Rendu, p. 93–101.

Bertrand, P., 1933, Les flores houillères d'Amérique d'après les travaux de M. David White: Annales de la Société Géologique du Nord, v. 58, p. 231–254.

Bertrand, P., 1935, Nouvelles corrélations stratigraphiques entre le Carbonifère des Etats-Unis et celui de l'Europe occidentale d'après MM. Jongmans et Gothan: Annales de la Société Géologique du Nord, v. 60, p. 25–38.

Bertrand, P., 1937, Tableaux des flores successives du Westphalien supérieur et du Stéphanien: Congrès pour l'avancement des études de stratigraphie du carbonifère, 2d, Heerlen, The Netherlands, 1935: Compte Rendu, v. 1, p. 67–79.

Bertrand, P., and Corsin, P., 1933, La flore houillère de la Sarre et de la Lorraine: Annales de la Société Géologique du Nord, v. 57, p. 193–206.

Bertrand, P., and Pruvost, P., 1937, La question du Westphalian et du Stéphanien en France: Congrès pour l'avancement des études de stratigraphie du carbonifère, 2d, Heerlen, The Netherlands, 1935: Compte Rendu, v. 1, p. 81–83.

Browne, I. M. P., 1925, Notes on the *Calamostachys* Type in the Renault and Ruche Collections: Annals of Botany, v. 39, p. 315–358.

Browne, I. M. P., 1927, A new theory of the morphology of the Calamarian cone: Annals of Botany, v. 41, p. 301.

Browne, I. M. P., 1935a, Some views on the morphology and phylogeny of the leafy vascular sporophyte: Botanical Reviews, v. 1, p. 383–407.

Browne, I. M. P., 1935b, Some views on the morphology and phylogeny of the leafy vascular sporophyte (continued): Botanical Reviews, v. 1, p. 427–447.

Crookall, R., 1931, The correlation of the British and French Upper Coal Measures, Summary Program for 1930: Memoirs of the Geological Survey, 1931, Part III, 62 p.

Czeczott, H., 1934, What is *Fagus Feroniae* Unger: Acta Societatis Botanicorum Poloniae, v. 11, p. 109–116.

Darrah, W. C., 1932a, Evidence of Permian floras in the Appalachian region: The Biologist, March, p. 38–39.

Darrah, W. C., 1932b, Recent paleobotanic investigations near Pittsburgh, Pennsylvania: Proceedings of the Pennsylvania Academy of Science, v. 6, p. 110–114.

Darrah, W. C., 1934, Stephanian in America: Geological Society of America Proceedings for 1933, p. 451.

Darrah, W. C., 1935a, Permian elements in the fossil flora of the Appalachian Province. I: *Taeniopteris:* Harvard University Botanical Museum Leaflets, v. 3, p. 137–148.

Darrah, W. C., 1935b, Some late Carboniferous correlations in the Appalachian Province: Geological Society of America Proceedings for 1934, p. 442–443.

Darrah, W. C., 1936a, A new *Macrostachya* from the Carboniferous of Illinois: Harvard University Botanical Museum Leaflets, v. 4, p. 52–63.

Darrah, W. C., 1936b, Permian elements in the fossil flora of the Appalachian Province. II: *Walchia:* Harvard University Botanical Museum Leaflets, v. 4, p. 9–19.

Darrah, W. C., 1936c, The peel method in paleobotany: Harvard University Botanical Museum Leaflets, v. 4, p. 69–83.

Darrah, W. C., 1936d, Sur la présence d'équivalents des terrains stéphaniens dans l'Amérique du Nord: Annales de la Société Géologique du Nord, v. 61, p. 187–197.

Darrah, W. C., 1937, American Carboniferous floras: Congrès pour l'avancement des études de stratigraphie du carbonifère, 2d, Heerlen, The Netherlands, 1935: Compte Rendu, v. 1, p. 109–129.

Darrah, W. C., 1939a, Textbook of paleobotany: New York and London, D. Appleton-Century, 441 p.

Darrah, W. C., 1939b, Principles of paleobotany: Leiden, Holland, Chronica Botanica Company, 239 p.

Darrah, W. C., 1939c, The fossil flora of Iowa coal balls. II: The fructification of *Botryopteris:* Harvard University Botanical Museum Leaflets, v. 7, p. 157–168.

Darrah, W. C., 1940, The fossil flora of Iowa coal balls. III: *Cordaianthus:* Harvard University Botanical Museum Leaflets, v. 8, p. 1–20.

Darrah, W. C., 1941, Studies of American coal balls: American Journal of Science, v. 239, p. 33–53.

Darrah, W. C., 1949a, Notes on *Lepidodendropsis:* Paleobotanical Notices, v. 1, p. 1–16.

Darrah, W. C., 1949b, Paleozoic Lepidodendroid embryos: Paleobotanical Notices, v. 2, p. 1–39.

Darrah, W. C., 1960, Principles of paleobotany (second ed.): New York, Ronald Press, 295 p.

Darrah, W. C., and Bertrand, P., 1933a, Observations sur les flores houillères de Pennsylvanie (regions de Wilkes-Barre et de Pittsburgh): Academie des Sciences, Paris, Comptes rendus, v. 197, p. 1451–1454.

Darrah, W. C., and Bertrand, P., 1933b, Observations sur les flores houillères de Pennsylvanie: Annales Société Géologique du Nord, v. 58, p. 211–224.

Dix, E., 1934a, The sequence of floras in the Upper Carboniferous, with special reference to South Wales: Royal Society of Edinburgh Transactions, v. 57, p. 789–838.

Dix, E., 1934b, The succession of fossil plants in the Millstone Grit and the lower portion of the Coal Measures of the South Wales Coalfield (near Swansea) and a comparison with that of other areas: Palaeontographica B, v. 78, p. 185–202.

Dix, E., 1935, Note on the flora of the highest "Coal Measures" of Warwickshire: Geological Magazine, v. 62, p. 555–557.

Dix, E., 1937, The succession of fossil plants in the South Wales Coalfield with special reference to the existence of the Stephanian: Congrès pour l'avancement des études de stratigraphie du carbonifère, 2d, Heerlen, The Netherlands, 1935: Compte Rendu, v. 1, 159–184.

Dix, E., and Trueman, A. E., 1937, The value of non-marine lamellibranchs for the correlations of the Upper Carboniferous: Congrès pour l'avancement des études de stratigraphie du carbonifère, 2d, Heerlen, The Netherlands, 1935: Compte Rendu, v. 1, p. 185–201.

Elias, M. K., 1936, Character and significance of the Late Paleozoic Flora at Garnett, *in* Moore, R. C., Elias, M. K., and Newell, N. D., A "Permian" flora from the Pennsylvanian rocks of Kansas: Journal of Geology, v. 44, p. 9–23.

Elias, M. K., 1937, Elements of the Stephanian flora in the Mid-Continent of North America: Congrès pour l'avancement des études de stratigraphie du carbonifère, 2d, Heerlen, The Netherlands, 1935: Compte Rendu, v. 1, p. 205–212.

Florin, R., 1936, On the structure of the pollen-grains in the Cordaitales: Svensk Botanisk Tidskrift, v. 30, p. 624–651.

Florin, R., 1938–1945, Die Koniferen des Oberkarbons und des Unteren Perms, Parts I–VIII: Palaeontographica B, v. 85, p. 1–730.

Florin, R., 1939, The morphology of the female fructifications in cordaites and conifers of Paleozoic age: Preliminary note: Botaniska Notiser, v. 36, p. 547–565.

Florin, R., 1951, Evolution in Cordaites and Conifers: Acta Horti Bergiani, v. 15, p. 285–388.

Florin, R., 1954, The female reproductive organs of Conifers and Taxads: Biological Reviews, v. 29, p. 367–389.

Gothan, W., and Sze, H. C., 1933, Uber *Mixoneura* und ihr Vorkommen in China: Nanking, Memoir National Research Institute of Geology (Akademia Sinica), v. 13, p. 41–57, pl. 5.

Halle, T. G., 1910, A gymnosperm with Cordaitean-like leaves from the Rhaetic beds of Scania: Arkiv för Botanik, v. 9, p. 1–6, pl. 1.

Halle, T. G., 1929, Some seed-bearing pteridosperms from the Permian of China: Kungliga Svenska Vetenskapsakademiens, Handlingar, v. 6, p. 3–24, pl. 6.

Halle, T. G., 1931, On the seeds of the pteridosperms *Emplectopteris triangularis:* Bulletin of the Geological Society of China, v. 11, p. 301–303.

Halle, T. G., 1933, The structure of certain fossil seed-bearing organs believed to belong to pteridosperms: Kungliga Svenska Vetenskapsakademiens, Handlingar, v. 12, p. 1–98.

Halle, T. G., 1937a, The position and arrangement of the spore-producing members of the Paleozoic pteridosperms: Congrès pour l'avancement des études de stratigraphie du carbonifère, 2d, Heerlen, The Netherlands, 1935: Compte Rendu, v. 1, p. 227–235.

Halle, T. G., 1937b, The relation between the late Paleozoic floras of Eastern and Northern Asia: Congrès pour l'avancement des études de stratigraphie du carbonifère, 2d, Heerlen, The Netherlands, 1935: Compte Rendu, v. 1, p. 237–245.

Harris, T. M., 1931, Rhaetic floras: Biological Reviews, v. 6, p. 133–162.

Harris, T., 1932, Fossil flora of Scoresby Sound East Greenland, Part 3: Meddeleiser om Grønland, v. 85, p. 1–133.

Hartung, W., 1933, Die Sporenverhältnisse der Calamariaceen: Arbeiten aus dem Institut für Paläobotanik und Petrographie der Brennstein, Band 3, Heft 1: Geologische Landensanstalt, Berlin, v. 4, p. 95–149, pls. 8–11.

Heer, O., 1870, Contributions to the fossil flora of north Greenland: Royal Society (London), Philosophical Transactions, v. 159, p. 445–488.

Jongmans, W. J., 1937a, Synchronismus und stratigraphie: Congrès pour l'avancement de études de stratigraphie du carbonifère, 2d, Heerlen, The Netherlands, 1935: Compte Rendu, v. 1, p. 327–344.

Jongmans, W. J., 1937b, Contributions to a comparison between the Carboniferous floras of the United States and of Western Europe: Congrès pour l'avancement des études de stratigraphie du carbonifère, 2d, Heerlen, The Netherlands, 1935: Compte Rendu, v. 1, p. 364–387.

Jongmans, W. J., 1949, Het wisselend aspect van het bos in de oudere geologische formaties, *in* Boerhaveo, W., ed., Hout in alle tijden: Amsterdam, Beekman Deventer, 164 p.

Jongmans, W. J., and Gothan, W., 1934, Florenfolge und vergleichende stratigrafie des Karbons der östlichen Staaten Nord-Amerikas; Vergleich mit

West-Europa: Heerlen, Jaarverslag Geologisch Bureau (1933), p. 17–44.

Jongmans, W. J., and Gothan, W., 1937a, Betrachtungen über die Ergebnisse des Zweiten Kongresses für Karbon Stratigraphie: Congrès pour l'avancement des études de stratigraphie du carbonifère, 2d, Heerlen, The Netherlands, 1935: Compte Rendu, v. 1, p. 1–40.

Jongmans, W. J., and Gothan, W., with some additional notes by W. C. Darrah, 1937b, Comparison of the floral succession in the Carboniferous of West Virginia with Europe: Congrès pour l'avancement des études de stratigraphie du carbonifère, 2d, Heerlen, The Netherlands, 1935: Compte Rendu, v. 1, p. 393–415.

Jongmans, W. J., Gothan, W., and Darrah, W. C., 1937, Beiträge zur Kenntnis der Flora der Pocono-Schichten aus Pennsylvanien und Virginia: Congrès pour l'avancement des études de stratigraphie du carbonifère, 2d, Heerlen, The Netherlands, 1935: Compte Rendu, v. 1, p. 423–444.

Leclercq, S., 1925a, Les Végétaux à structure Conservée du Charbonnage de Wérister: Bulletin de la Société Géologique de Belgique, v. 34, p. 31-32.

Leclercq, S., 1925b, Introduction á l'Étude Anatomique des Végétaux Houillers de Belgique: Les Coal-balls de la couche Bouxharmont des Charbonnages de Wérister: Mémoires de la Société Géologique de Belqique, p. 1–79, 49 pls.

Leclercq, S., 1935, Coal-balls de la couche Saurue Synonyme de Bouxharmont: Bulletins de la Société royale des Sciences de Liège, no. 4-5, p. 189–194.

Leclercq, S., 1936, Nouveau "nids" de coal-balls dans la couche Bouxharmont (Charbonnages de Wérister, Bassin de Liége): Annales of the Société Géologique de Belgique, t. 59, p. 166–172.

Leclercq, S., 1950, Note préliminaire sur le *Rhacophyton zygopteroides* nov. sp. Leclercq: Bulletin de l'Académie royale de Belgique (Classe des Sciences), v. 36, p. 77–91.

Lesquereux, L., 1879–1884, Description of the coal flora of the Pennsylvanian and of the Carboniferous formations in Pennsylvania and throughout the United States: 2nd Geological Survey of Pennsylvania, 997 p. (Atlas, 1879; Volumes 1 and 2, 1880; and Volume 3, 1884).

Moore, R. C., Elias, M. K., and Newell, N. D., 1936, A "Permian" flora from the Pennsylvanian rocks of Kansas: Journal of Geology, v. 44, p. 1–31.

Pruvost, M. P., and Bertrand, P., 1932, Quelsques résultats des récentes explorations géologiques du Bassin Houiller du Nord de la France: Revue de l'Industrie Minérale, no. 282, p. 365–379.

Read, C. B., and Merriam, C. W., 1940, A Pennsylvanian flora from central Oregon: American Journal of Science, v. 238, p. 107–111.

Renier, A., 1907, Les nodules à goniatites ne constituent pas une objection réelle à la théorie de la formation autochtone des couches de houille: Annales de la Société Scientifique de Bruxelles, v. 31, p. 169–174.

Renier, A., 1930, La stratigraphie du terrain houiller de la Belgique: Musée Royal d'Histoire Naturelle de Belgique Mémoire 44, 101 p.

Sellards, E. H., 1900, Note on the Permian flora of Kansas: Kansas University Quarterly, v. 9, p. 63–64.

Walton, J., 1928, A method of preparing sections of fossil plants contained in coal balls or in other types of petrification: Nature (London), v. 72, p. 571.

MANUSCRIPT ACCEPTED BY THE SOCIETY, JULY 6, 1994

Printed in U.S.A.

Geological Society of America
Memoir 185
1995

# *History of significant collections of Carboniferous (Mississippian and Pennsylvanian) plant fossils in museums of the United States*

**Elsie Darrah Morey**
*Morey Paleobotanical Laboratory, 1729 Christopher Lane, Norristown, Pennsylvania 19403*
**Richard L. Leary**
*Illinois State Museum, Research and Collections Center, 1101 East Ash Street, Springfield, Illinois 62703*

## ABSTRACT

**This chapter discusses nineteenth- and early-twentieth-century Carboniferous (Mississippian and Pennsylvanian) collections of plant fossils in the United States. There are two extensive pre-1950 collections of Carboniferous plant fossils in the United States: at the U.S. National Museum of Natural History and at Harvard University. Both of these contain many type specimens. Other less prominent historical collections, which include type or figured specimens of fossil plants, are also included in this overview on Carboniferous paleobotanical collections in the United States. The special significance of the early museum collections is that they include the research collections that contain the type specimens on which later determinations are based. In other words, the type specimens are the permanent records for taxonomic purposes. Thus, this report provides useful information for research paleobotanists who need to determine the locations of type and other specimens for taxonomic and other paleobotanical purposes.**

## INTRODUCTION

Without the efforts of early geologists and amateur collectors, many organized collections of Carboniferous (Mississippian and Pennsylvanian) plant fossils would not exist. These collections were crucial for the development of paleobotany in the United States. The achievements of many noteworthy paleobotanists, past and present, resulted from their research in these collections. Collections of historical significance, which contain considerable numbers of North American Carboniferous plant-fossil specimens that were mainly collected prior to 1950, are discussed in this chapter.

Few paleobotanical papers were written before 1850 in the United States. Steinhauer's (1818) report of a fossil flora in coal strata was one of the first. Indeed, from 1824 to the 1850s, paleontologic interests in the United States centered on invertebrate and vertebrate fossils. Beginning in the 1820s, many states established geologic surveys to determine their economically important natural resources. The plant and animal fossils collected during these early surveys served principally to identify the geologic age of the mineral and rock specimens. The report of Edward Hitchcock's (1833) state-sponsored survey of Massachusetts included illustrations of Carboniferous plant fossils collected in Massachusetts and Rhode Island (see Lyons, 1993). A part of Hitchcock's collections is in the Pratt Museum at Amherst College, Amherst, Massachusetts.

H. D. Rogers led the First Geological Survey of the Commonwealth of Pennsylvania from 1836 to 1842. The purpose of this survey was to map the boundaries of the state of Pennsylvania and to collect mineral and fossil specimens to form a collection representative of Pennsylvania. The fate of much of the fossil specimens collected by this first survey is unknown (Darrah, 1969). When Rogers was not paid by the Pennsylvania survey, he claimed all the fossil plants that he had collected at his own expense. Rogers's collections were eventually donated to Harvard University in 1892 and 1912. The Second Geological Survey of Pennsylvania (1874–1888), un-

Morey, E. D., and Leary, R. L., 1995, History of significant collections of Carboniferous (Mississippian and Pennsylvanian) plant fossils in museums of the United States, *in* Lyons, P. C., Morey, E. D., and Wagner, R. H., eds., Historical Perspective of Early Twentieth Century Carboniferous Paleobotany in North America (W. C. Darrah volume): Boulder, Colorado, Geological Society of America Memoir 185.

der the direction of J. Peter Lesley (1819–1903), was required by law to deposit its collections at the Academy of Natural Sciences in Philadelphia, until such time as a state museum could be established in Harrisburg, the state capital (Merrill, 1920). As is the case with all fossil collections, the specimens were of little use until identifications could be made.

Leo Lesquereux did the pioneering monographic work (1879–1884) on Pennsylvanian (late Carboniferous) floras of the United States (Darrah, 1934, 1969). According to Darrah (1934), the first systematic studies of Carboniferous fossil plants in the United States were published in 1854 and 1858 by Lesquereux. Lesquereux (1806–1889)—the father of paleobotany in the United States—examined, named, described, and labeled nine-tenths of the Paleozoic fossil plants then known from the United States (Hayden, 1901). The majority of the fossil plants described in Lesquereux's "Coal flora" (1879–1884) came from the private collections of R. D. Lacoe, an amateur collector, under whose patronage Lesquereux worked from 1878 until his death in 1889. In 1892, Lacoe donated most of his collection of fossil plants to the U.S. National Museum (USNM) in Washington, D.C.

The early museum collections contain the type specimens on which later taxonomic determinations are based. Thus, they are the permanent records for taxonomic purposes.

The order used in describing each of the collections follows that of Darrah (1969). The collections of Pennsylvanian plants are discussed in their order of importance by the number of types or figured specimens that each collection contains. The U.S. National Museum of Natural History and Harvard University Paleobotanical Collection hold the largest number of figured specimens of Carboniferous plant fossils from the United States.

## NATIONAL MUSEUM OF NATURAL HISTORY COLLECTIONS, WASHINGTON, D.C.

### *E. D. Morey and R. L. Leary*

The Smithsonian Institution was founded in 1846 as an establishment for "increase and diffusion among men," according to the donor's bequest. Joseph Henry, the Smithsonian's first secretary, directed the Institution's resources to support original research and the publication of its results.

The growth of government and other collections at the Smithsonian's "National Museum" soon resulted in a lack of adequate storage and exhibit space. The great mass of materials given to the Smithsonian at the close of the Centennial Exhibition of 1876 could neither be accommodated in the original Smithsonian building nor stored in the nearby Armory. Congress appropriated money to construct a United States National Museum in 1878 and the structure, now known as the Arts and Industries Building, was completed in 1881 adjacent to the Smithsonian's "Castle," the old red-brick building still standing near the Mall in Washington, D.C. This space was soon outgrown, and the present U.S. National Museum of Natural History was built across the Mall early in the twentieth century to allow for further storage and display of the ever-increasing volume of specimens. Since the early twentieth century and until the late 1980s, the Paleontology and Stratigraphy Branch of the U.S. Geological Survey (USGS) was housed in the new building, and its collections and staff formed an integral part of the U.S. National Museum of Natural History.

The USGS was founded on March 3, 1879, for the purpose of ". . . the classification of the public lands, and examination of the geological structure, mineral resources, and products of the national domain" (20th *U.S. Statutes at Large,* 394). The establishing law for the USGS required that "all collections of rocks, minerals, soils, fossils, and objects of natural history, Archaeology, and Ethnology, made by the . . . Geological Survey . . . when no longer needed for investigations in progress shall be deposited in the [Smithsonian's] National Museum" (20th *U.S. Statues at Large,* 394).

Clarence King, as the first director of the USGS, helped to found the agency and shaped its staff, methods, standards, and products. When King resigned in 1881, John Wesley Powell, the philosophical father of the USGS, succeeded him as director. Under King, paleontology principally served the USGS's mandated work in economic geology. Powell expanded the program to include topographic mapping and basic geology, including paleontology, at the expense of the economic studies. On July 1, 1881, Powell hired Lester F. Ward (1841–1913; Fig. 1) as a "Paleo-Botanist." By 1890, William M. Fontaine, John S. Newberry, David White, and Frank H. Knowlton were also hired to serve as paleobotanists in the USGS (Nelson and Yochelson, 1980).

In the Annual Reports of the National Museum for 1892, Ward noted, "No gift of greater importance to the department of fossil plants has ever been made than that by Mr. R. D. Lacoe of Pittston, Pennsylvania, under the terms of which his great collection of fossil plants is to be permanently deposited in the National Museum. The value of this collection . . . is far greater than that of the entire amount of the collections in the department prior to the date of its gift."[1]

Lacoe (1825–1901; see White, 1903; Fig. 2) had little formal education, but whatever his pursuits, he worked with great attention to detail. As a well-established businessman in Pittston, Pennsylvania, Lacoe owned extensive acreage and coal, knitting, and paper-box companies—to mention only a few of his enterprises.

In 1865, Lacoe's relatively poor health led to his retirement from business. He began to study geology and how it related to the coal-mining industry. In the process of collecting specimens, Lacoe accumulated a very large number of fossil plants and animals. He arranged the collection as follows: the "Coal Flora," "Fossil Insects of the Carboniferous," and "Fossils of the Paleozoic limestone beds." As Lacoe's collection continued to grow, he decided to give the specimens to

Figure 1. Lester F. Ward (1841–1913) and his secretary, who became Mrs. C. D. Walcott. Ward was the paleobotanical curator of the U.S. National Museum and accepted the gift of fossils from R. D. Lacoe in 1892. From the William Culp Darrah Paleobotanical Collection.

the National Museum, now the U.S. National Museum of Natural History (USNM). The gift was delivered in 1892, and David White (see Chapter 10, this volume) organized the collection and updated the taxonomy. The Lacoe Collection and White's contributions from his own extensive collections form a large part of the Carboniferous plant-fossil collections now at the USNM.

The specimens in the collection are primarily from the Northern Anthracite coalfield of eastern Pennsylvania. Lacoe considered himself an amateur collector, but he did write two papers (Lacoe, 1883, 1884). He engaged Leo Lesquereux from 1875 to 1889 to examine the specimens that Lacoe had amassed in the United States as well as the majority of those that he had obtained from Europe. Lesquereux (1887) marveled that "Mr. R. D. Lacoe of Pittston has procured from almost all the localities where coal is worked in the United States an immense amount of specimens far beyond any seen even in the largest museums of Europe."[2] Even after Lacoe gave much of his collection to the USNM, he continued to collect fossils with the assistance of other collectors. He also purchased private collections, many of which contained type materials. The USNM held only 102 types of Paleozoic plants prior to receiving Lacoe's gift. The 10,000 specimens donated by Lacoe included 575 type or figured specimens. Lacoe was appropriately honored in the publications of Lesquereux, who named eight species for him. Later, David White (1893) named *Sphenopteris lacoei* in honor of Lacoe. Charles B. Read (1946) established the fossil plant genus *Lacoea.*

Read (see Mamay et al., this volume), White's successor as curator of the Paleobotanical Collections at the USNM, also collected extensively. His specimens, too, became part of the USNM collection. Read wrote Darrah, "At one time, I liked best to stay in the office and work. Now I'd rather get out in the field, and generally am out for four or five months each year. . . . My feeling about field work is that while it may slow up production of new papers now, I will benefit by it eventually."[3]

Following Read in the early 1950s, James M. Schopf (see Cross et al., this volume) spent one year at the USNM before he became chief of the USGS Coal Section at Ohio State University, Columbus, Ohio. At the USNM, Schopf made plans for a coal laboratory at Ohio State.

Sergius H. Mamay succeeded Schopf as curator. Mamay, while working with the USGS (S. H. Mamay, personal communication, 1993), was transferred to the USNM to curate the USGS's paleobotanical collections. Through Mamay's efforts in 1983, all the Paleozoic and Mesozoic plant collections of the USGS were accessioned to be holdings of the USNM. The Cenozoic plant collections were turned over from the USGS to the USNM in 1988. The Paleozoic paleobotanists currently employed by the USNM are Francis M. Hueber and William A. DiMichele. Other paleobotanists specializing in Mesozoic and Cenozoic floras who have worked at the USNM include Lester F. Ward, Frank Hall Knowlton, Roland W. Brown, Jack A. Wolfe, and Scott Wing. Brown and Wolfe were USGS employees.

Over the years, able assistants have maintained the USNM-USGS collections of fossil plants. Jack Murata (see Mamay et al., this volume) was Read's assistant during the 1930s and did most of the thin sections for Read's New Albany Shale paleobotanical papers. Arthur D. Watt became

Figure 2. Ralph Dupuy Lacoe (1825–1901) donated his extensive fossil collections to the U.S. National Museum in 1892. Photograph reproduced from Ralph Dupuy Lacoe Memorial (Hayden, 1901). Photographic copy taken by L. V. Thompson, U. S. Geological Survey. Photograph republished courtesy of and with permission from the Wyoming Historical and Geological Society, Wilkes-Barre, Pennsylvania.

Mamay's assistant in 1956 and remained in that position until he retired in 1981. Watt organized the type collections. Watt was a talented field assistant and an important source of historical information about the collections and paleobotanists who worked at the USNM. Watt produced the "Bibliography of American paleobotany" between 1962 and 1979. In that regard, "Art probably produced a greater service to American paleobotany than any other single person."[4] James P. Ferrigno became Hueber's assistant in 1962. At that time, Ferrigno took over the development of the type collection at the USNM. He made it into a well-curated and -documented collection. Ferrigno took many excellent photographs and his photographs (including some for William C. Darrah) appear in many publications that resulted from studies of the USNM Plant Collections (S. H. Mamay, personal communication, July 22, 1993).

An important collection at the Smithsonian Institution is the McLuckie Collection. John and Lucy McLuckie were long-time residents of Coal City, Illinois, and close neighbors of the strip mines of the Mazon Creek region. John McLuckie (Fig. 3) began collecting fossiliferous nodules about 1930 while operating machinery for the Northern Illinois Coal Company. Lucy soon joined her husband and began collecting "out of self defense." Later, their three children often accompanied John and Lucy into the strip mines. Over the years, the McLuckies assembled a large collection of fossils from the area. The basement of their home became a "museum," the walls lined with cabinets filled with Pennsylvanian plant-fossil specimens. Eugene Richardson of the Field Museum (Chicago) called it one of the finest private collections in the country. The fame of the McLuckie Collection spread, and people from all over the world came to view the outstanding specimens. Leading scientists, scholars, and amateurs were given tours of the McLuckie's "museum."

The McLuckies and their children often encountered George Langford and his son in the strip mines (see Chapter 23, this volume). Over the years, a friendship developed and the families shared specimens and experiences. The Mc-

Figure 3. John McLuckie at his home in Coal City, Illinois. The photograph was taken in 1936 by Frederick O. Thompson. From the William Culp Darrah Paleobotanical Collection.

Luckies were friendly, generous people who encouraged many amateur collectors.

John McLuckie died in 1963 and Lucy McLuckie in 1982, leaving the future of their fantastic collection uncertain. Collectors and dealers wanted to buy the collection for resale, spreading decades of accumulation around the world. Museums, not wishing to see the valuable collection dispersed, sought to acquire it intact. Financially, of course, museums and universities cannot compete with dealers. Finally, after years of disagreement among the heirs, the collection was donated in April 1988 to the USNM in Washington, D.C. The Smithsonian Institution received about 2,500 excellent specimens. Most of the nodules are no doubt from the various pits around Coal City, although the exact locations are not recorded. The collection consists primarily of plant fossils; some animal fossils are also included. The collection will be available to scholars. Some specimens are on display in the USNM.

The USNM Paleobotanical Collection is enormous. By 1991, it included more than 20,600 fossil-plant types included in the collection (S. H. Mamay, personal communication, 1992). The USGS Fossil Plant Locality Register includes 11,000 numbers. USNM's Fossil-Plant Catalogue now contains more than 51,430 references (S. H. Mamay, personal communication, 1992).

Watt's (1970) *Catalog of the illustrated Paleozoic plant specimens in the U.S. National Museum of Natural History* indicates that for the 110 papers that were written on Paleozoic plants from 1907 to 1964, most of the specimens are in the USNM collection. These papers illustrate more than 700 specimens from the Carboniferous period alone, bearing witness to the size of the USNM Collection and the materials that the USNM maintains (see also Watt, 1974).

Figure 4. Louis Agassiz (1807–1873). Agassiz was the founder of the Museum of Comparative Zoology, Harvard University, Cambridge, Massachusetts. Photograph republished courtesy of and with permission from the Museum of Comparative Zoology Library, Harvard University.

## HARVARD UNIVERSITY'S PALEOBOTANICAL COLLECTIONS, CAMBRIDGE, MASSACHUSETTS

***E. D. Morey***

Louis Agassiz (Fig. 4), an eminent Swiss paleontologist, emigrated to the United States in 1847 and later helped to establish the Museum of Comparative Zoology at Harvard University, Cambridge, Massachusetts. After settling in at Harvard, Agassiz soon persuaded his colleague, Leo Lesquereux, to join him in the United States (Darrah, 1934). Harvard at that time had a museum with broad-based collections that included plant-fossil collections (Knoll and Barghorn, 1984). Lesquereux described the fossil plants and made plans for building a comprehensive collection of fossil floras of the United States (Darrah, 1936). The Harvard University Botanical Museum was established in 1858 and the Museum of Comparative Zoology the next year. Under the direction of Agassiz, the Museum of Comparative Zoology broadened its holdings to include the study of paleontology in all its branches.

Agassiz engaged Lesquereux to work on Harvard's plant-fossil collection housed in the Museum of Comparative Zoology from 1865 to 1871. In addition to describing all the fossil plants in the museum's existing collections, Lesquereux described his own collections as well as donated collections. During this period, Lesquereux, while maintaining his home in Columbus, Ohio, spent several months of each year working on the paleobotanical collection at Harvard University. Lesquereux's first paper on the collection appeared in 1867, and he subsequently donated his personal collections to the Museum in 1868.

The Commonwealth of Pennsylvania began its First Geological Survey (1836–1842) under the direction of Henry D. Rogers (1808–1866; Fig. 5). Rogers's report, published in 1858, included a section in which Lesquereux described and illustrated the fossil plants that he had studied from the collec-

Figure 5. Henry Darwin Rogers (1808–1866). First director of the Pennsylvania Geological Survey (1846–1852) and state geologist of New Jersey (1835–1840). Photograph republished courtesy of and with permission from the Smithsonian Institution.

Figure 6. Leo Lesquereux (1806–1889), father of paleobotany in the United States. Lesquereux identified most of the early fossil specimens from the early collections of Lacoe. Photograph republished courtesy of and with permission from the National Academy of Sciences, Washington, D.C.

tions belonging to the Geological Survey of Pennsylvania (Darrah, 1969).

Pennsylvania had no legal claim to the portion of this early collection that Rogers had amassed at his own expense. In 1863, H. D. Rogers gave to his brother, William B. Rogers, his part of the collection for the museum at a new institution of higher learning—Massachusetts Institute of Technology in Cambridge. In 1882, this collection was transferred to the Boston Society of Natural History (Darrah, 1969), and it was later given to the Museum of Comparative Zoology of Harvard University in 1892 and 1912 (Darrah, 1969). Some of the Society's collections are today at the Pratt Museum, Amherst College (P. C. Lyons, written communication, March 1, 1994).

Lesquereux (Fig. 6) was hired (1874–1888) by the Second Geological Survey of Pennsylvania to study the paleobotanical specimens that had been collected over a twenty-year period. The significance of these fossils is demonstrated in Lesquereux's monograph on Carboniferous floras in the United States (Atlas, 1879; Volumes 1 and 2, 1880; Volume 3, 1884), which is collectively called the "Coal flora" (1879–1884) and is considered the foundation of American Carboniferous paleobotany (Darrah, 1969). About 60% of the specimens discussed in these four volumes are preserved in Harvard University's Paleobotanical Collections (HUPC; Darrah, 1969).

Robert T. Jackson (1853–1948), curator HUPC, obtained from Lacoe in 1899 (Darrah, 1936) two notebooks that Lesquereux had compiled (1854–1858), which became the catalog to the specimens at Harvard. From the two notebooks, Jackson identified 87 of Lesquereux's type specimens (Darrah, 1969). Lacoe gave the remainder of Lesquereux's notebooks to the USNM minus the catalog of the Harvard University specimens.

The HUPC contains material formerly kept at the Museum of Comparative Zoology and also the extensive collections of the Boston Society of Natural History that were donated to Harvard in 1914. Jackson moved the Society's collections from their former sites in 1892 to the Harvard University Botanical Museum.

In 1934, William C. Darrah (see Chapter 1, this volume) reopened the long-dormant paleobotanical collection, which was nominally under the supervision of Oakes Ames, then director of the Botanical Museum at Harvard. In his position as research assistant, Darrah produced a brief biography of Leo Lesquereux (Darrah, 1934). In it, he described enthusiastically the Harvard Botanical Collection: "In 1868, the collection contained 2,500 specimens belonging to some 500 species. By 1885, the time of the final donations from Lesquereux, it had grown to 10,000 specimens of 2,000 rarities. This entire priceless assortment has passed through the hands of Lesquereux. Within the past fifty years by purchase and donation the collection has been tripled in specimens and doubled in species—truly a remarkable collection unequalled in America" (Darrah, 1934).

David White wrote Darrah in 1934:

> Until informed by you, I had no idea that the material in the line of Paleozoic plants was so large at Harvard. I have formerly supposed that the earlier collections by Lesquereux had been lost, or broken, or stolen long ago. I think that Mr. Lacoe who funded Lesquereux in the latter's last years, carrying him in effect as a semi-pensioner, shared my belief that the earlier collections were lost. Lacoe paid him a round sum which was almost a gratuity for the types now in the Lacoe Collection, besides paying him salary for work, the very last of which was of little or no value, I fear. We had supposed that the types described in the H. D. Rogers, 1858, had either been sent back to Scotland or were totally lost.[5]

C. B. Read, taking issue with Darrah's claim about the importance of Harvard's collection, wrote to Darrah:

> I really do not think that the collection of fossil plants at Harvard has such a unique status as you suggest. It cannot contain all the originals and fundamental floras accumulated in the first thirty years of this country's paleobotanical research for the reason that there are types of these floras in a number of museums both in America and outside of America. I have known that Harvard had a number of "Coal Flora" types. However, we have roughly speaking, fifty [100 would be closer] of the specimens figured in the Atlas. . . . You would seem to indicate by some of your statements that the Harvard Collection is the most outstanding collection of fossil plants in America and suggest that it contains more than 30,000 specimens. The National Museum Collection, which has been largely assembled by the U.S. Geological Survey, contains several times that much material and a number of types of [L.] Lesquereux, [F. H.] Knowlton, [L. F.] Ward, [A.] Hollick, [D.] White, [E. W.] Berry, most of them in fact, and some others. Certainly it must at least share the honors with the Harvard Collection in being the actual basis of study of all American fossil floras. . . . The point that I wish to make is that no museum in the world can boast of possessing all or anywhere near all of the original and fundamental floras accumulated in the first thirty years of paleobotanical research in America, and further I wish to question the statement that your collection is unequalled in America.[6]

Darrah received yet another criticism, this time from R. W. Brown: "Your sketch of 'Lesqereux' was on my desk. It is well written, although I think you lay it on a little thick when you say without qualification that Lesquereux's collection at the Botanical Museum is . . . 'the actual basis for the study of all American fossil floras.' . . . Of course you are working for Harvard now, so we will let is pass this time!"[7]

Darrah continued to build the Harvard Paleobotanical Collection. Frederick O. Thompson (see Chapter 22, this volume) collected Mazon Creek nodules from Coal City, Illinois, and shipped them to Harvard University. Over a period of five years, Thompson sent 12,000 nodules, of which 4,000 were accessioned into the collection (Darrah, 1969).

As Darrah identified the specimens, he compared them to Lesquereux's compressional fossil-plant specimens. Darrah found that the relatively abundant forms represented in the F. O. Thompson Collection reflected the same aspect recognized by Lesquereux in his 1866 report published by the Illinois State Geological Survey (Darrah, 1969).

Darrah (see Chapter 2, this volume) collected with W. J. Jongmans in the Netherlands from the following coal mines: Willem-Sophie, Wilhelmina, Oranje Nassau, Emma, and Maurits. Charles Schuchert wrote Darrah "I am delighted that you collected for Harvard University thirty boxes of fossil plants."[8] The European collection of fossils is an important part of the Carboniferous flora represented at Harvard.

After Thompson's discovery of coal balls in Iowa, he shipped literally tons of them to Harvard (see Chapter 27, this volume). From the coal-ball materials, Darrah and his students—primarily Ralph Witter, John Burbank, and Elso Barghoorn—made peels in great quantity, principally for microscope slides for exchange with other universities at home and abroad. Darrah hoped this program would match the exchange of fossil specimens that he had established with other universities. Unfortunately, Darrah left Harvard on indefinite leave of absence during World War II (see Chapter 1, this volume). Today the microscope coal-ball slides, prepared by and under Darrah's supervision, are used in paleobotanical teaching at Harvard (Knoll and Barghoorn, 1984).

Barghoorn became Darrah's successor in 1946 as curator of the paleobotanical collections at Harvard, and he continued to build up the collection, particularly by the addition of Mesozoic and Tertiary plant fossils. Through donations, Barghoorn acquired collections of Cretaceous and Tertiary fossil woods from Frank Hankins, Triassic fossils from Arizona from Lyman Daugherty, and an extensive collection of Middle Devonian plants from Raymond Baschnagel (Knoll and Barghoorn, 1984). Barghoorn and his students collected large numbers of fossil plants from the Eocene lignite at Brandon, Vermont, during many trips from 1947 to 1977 (Knoll and Barghoorn, 1984).

Important early collections at Harvard include the Lomax Collection and the Jeffrey Collection. The Lomax Collection is made up of 300 thin sections of coal balls showing Carboniferous plant-fossil structures. It is significant that the thin sections were cut by Lomax (Fig. 7) from the surface next to those illustrated in D. H. Scott's *Studies in fossil botany* (1900).

The E. C. Jeffrey Collection originally was utilized in Jeffrey's work on the origin of coal (Jeffrey, 1925). This col-

Figure 7. Joseph R. Lomax. Lomax's father was the preparator of thin sections for W. C. Williamson and D. H. Scott in the 1890s. Harvard's Paleobotanical Collection has a Lomax Collection of microscope slides. The Lomax Collection of coal balls, purchased by W. C. Darrah from Lomax's son in 1955, was purchased from Darrah by E. D. Morey in 1976. This photograph of Lomax was taken about 1946. From the William Culp Darrah Paleobotanical Collection.

lection was expanded by Barghoorn and now contains new samples of coal, thin sections, and macerations from most of the coal basins of the world (Knoll and Barghoorn, 1984). Other early paleobotanical acquisitions at Harvard University are the William Mather Collection (1828–1833), the H. D. Rogers and W. B. Rogers Collection (1838–1858), and the H. G. Bronn Collection (1828–1860).

Recent Carboniferous collections at Harvard University from Pennsylvanian floras of New England are from John Oleksyshyn, P. C. Lyons, C. A. Kaye, E. S. Grew, and C. E. Grant. H. L. Barwood donated a Pennsylvanian plant-fossil collection from the Warrior Basin of Alabama (Knoll and Barghoorn, 1984; see Lyons and others, 1985). Other collections at Harvard include an extensive Precambrian collection (Knoll and Barghoorn, 1984).

The Harvard Collection now contains over 800 type specimens and about 575 figured specimens. The total collection numbers over 63,000 specimens (R. Wilcox, personal communication, 1991), making the Harvard Collection the second-largest collection of plant fossils in the United States, after that of the USNM. The early collections and rare fossil-plant material from Europe, particularly the presence of many specimens from Lesquereux's "Coal Flora," qualify this collection for the title of America's "type" paleobotanical collection, according to Knoll and Barghoorn (1984).

## THE UNIVERSITY OF KANSAS PALEOBOTANICAL COLLECTION, LAWRENCE, KANSAS

***E. D. Morey***

The R. L. McGregor Herbarium of the University of Kansas contains the university's paleobotanical collection. The total number of fossil-plant specimens in the collection is 22,354, including 181 types. The importance of this collection lies mainly in the coal-ball plant fossils; there are 2,087 specimens, including 56 types. With the coal-ball materials are "peel books" that include a peel of every cut surface of every cut coal-ball specimen. Peels from 926 coal balls total 4,637, of which there are 447 type peels. Microscope slides total 1,676, of which there are 536 type slides. A second portion of the collection is the fossil-plant compressions. There are 2,063, including 87 types. The collection contains 218 specimens from the Hamilton Quarry, Kansas, and there are 301 nodules in the collection from the Mazon Creek area, Illinois (M. A. Lane, personal communication, 1993).

The Cretaceous is represented by 298 specimens from the Dakota Formation, including one type. The Cretaceous collections contain 331 specimens. Fossil seeds from the Tertiary are represented by 227 specimens, including 5 types. Miscellaneous collections add another 299 specimens.

The specimens collected by E. H. Sellards (see Chapter 12, this volume) were collected at the time when he was affiliated with the Kansas Geological Survey near the turn of the twentieth century. Sellards had collected extensively in the Permian of northeastern Kansas. The specimens are valuable because Sellards designated hypotypes of a previously described taxon each time he extended the range geographically or stratigraphically.

The University of Kansas Collection also contains the photographic negatives of the plates for R. W. Baxter's publications. There is a paleobotanical reprint collection with about 2,700 references. An interactive database was prepared by Meredith A. Lane and Alicia Lesnikowska (who recently curated this collection) in 1993, so that a search for certain taxon in the collection, a locality, the name of the collector, or accession number can quickly produce the desired information.

## ILLINOIS STATE MUSEUM PALEOBOTANICAL COLLECTIONS, SPRINGFIELD, ILLINOIS

***R. L. Leary***

Richard L. Leary is the curator of the Illinois State Museum paleobotanical collections. The fossil collections of the Illinois State Museum were assembled by the first Illinois State Geological Survey (1851–1875). Many of the plant fossils were collected by A. H. Worthen, who was director of the Illinois Survey from 1858 to 1875, founder of the Illinois State Museum, and its first director (1877–1881). Other specimens were donated to Worthen or to the state collection by associates of Worthen.

Amateur collectors have always played an important role in adding to the State Museum's collections. For example, several collectors, primarily Joseph Even, S. S. Strong, T. P. Winslow, and M. J. Hall, gave specimens of Mazon Creek fossiliferous nodules to Worthen.

In the 1800s, there was no established scientific community in Illinois. However, there was a major center for scientific and geological research in nearby New Harmony, Indiana, beginning in the late 1830s. As a result, as collections of the Illinois State Geological Survey were acquired, they were taken to New Harmony for study. In 1854, the Illinois legislature decided that the collections should be housed in Springfield. The collections were loaded on boats, taken down the Wabash and Ohio Rivers, up the Mississippi River, and finally by train back to Springfield.

Even then the collections were not "home free." For many years the geology cabinet was moved from one inappropriate and inadequate site to another. The collections were at one time or another displayed in the Armory, a Masonic hall, and the State Senate and State Supreme Court chambers in the State Capitol. Specimens were stored in the basement of the Capitol during legislative sessions. As a result of the many moves and improper storage, many specimens were lost and others separated from their labels. Finally, in 1923, the museum achieved an adequate and permanent place in the new Centennial Building in Springfield.

The paleobotanical collections of the Illinois State Museum currently contain 99 type specimens and about 600 figured specimens. Most of these types were described by Leo Lesquereux in the late 1800s and redescribed by Raymond Janssen in 1940. Janssen also described several new genera and species. The majority of the figured specimens were illustrated by George Langford in his volumes on the Wilmington Coal Flora (1958, 1963).

Lesquereux examined the collections of the State "Cabinet" of Worthen, and others, and published descriptions and illustrations (Lesquereux, 1866, 1870). Some of these specimens were also figured and described in Lesquereux's Atlas (1879) and in his two-volume text (1880). Approximately 100 of Worthen's specimens were described as new species (Lesquereux, 1866, 1870). Although a majority of these type specimens were deposited in the State Museum's collections, some were returned to their owners and are now lost. The State Museum has 59 of Lesquereux's type specimens.

The 1930s brought renewed interest in the Illinois State Museum paleobotanical collections. Perhaps the major impetus for this was the acquisition of the Langford Collection of fossiliferous nodules from the world-famous Mazon Creek region. Altogether, George Langford Sr. (see Chapter 23, this volume) and his son, George Jr., split nearly 250,000 nodules, taking away about 10% from strip-mine waste piles and discarding the rest. Of the 25,000 specimens they kept 459, including 16 animal-bearing nodules originally selected for the Illinois State Museum. Altogether the State Museum acquired about 1,670 nodules, in most cases both part and counterpart, from the Langfords.

Raymond Janssen came to the Illinois State Museum in the late 1930s and spent several years studying the collections. Fifty-nine out of Lesquereux's 109 new species were located by Janssen. Janssen redescribed and revised the identification of these specimens in his book *Some fossil plant types of Illinois* (Janssen, 1940). In addition to reviewing and updating the work of Lesquereux, Janssen also described 11 new species and two new genera on the basis of Mazon Creek specimens in the Museum's collections. In 1939, the State Museum published Janssen's popular book, *Leaves and stems from fossil forests.*

Another hiatus in the study of the collections followed, but in the 1960s and 1970s, several notable paleobotanists visited the Illinois State Museum to examine the collections. Maxine Abbott was the first of these when, in 1961, she visited the Museum while studying *Lepidostrobus.* William C. Darrah followed in 1967 when he examined specimens while working on his monograph on compression floras (Darrah, 1969).

Darrah and Abbott were followed in 1974 by H. W. Pfefferkorn, a German paleobotanist who was trained under W. Remy (see Chapter 6, this volume). He worked several years at the Illinois State Geological Survey. In addition to examining existing collections, Pfefferkorn worked with Richard Leary, the Museum's curator of geology, to collect and study a basal Pennsylvanian clastic (nonswamp) flora in Brown County, Illinois. Together Leary and Pfefferkorn added a new element to the Museum's collections and began a study of "upland" floras that has continued for the last 20 years.

The Illinois State Museum paleobotanical collections continue to grow as donations and current research add specimens. Although research has concentrated on Early Pennsylvanian floras, Leary (1985, 1986) described three new genera of Mississippian fossil algae from Jersey County, Illinois. As a result of Leary's research, nearly 10,000 fossil plants were collected in western Illinois and are part of the research (nonaccessioned) collections of the Museum (see Leary, 1976).

Each year, paleobotanists visit the Illinois State Museum

to utilize the collections. In 1988, the geological collections were moved from the Museum into a Research and Collections Center. As new cabinets are acquired and specimens unpacked, accessibility will be improved and use will be increased.

## THE PALEOBOTANICAL COLLECTIONS AT THE STATE MUSEUM OF PENNSYLVANIA, HARRISBURG, PENNSYLVANIA

***E. D. Morey***

The Second Geological Survey of Pennsylvania (1874–1888) was one of the most ambitious state surveys ever attempted. J. P. Lesley (1819–1903; Fig. 8) was appointed state geologist. The central office, located in Harrisburg, Pennsylvania, had a staff of three. In 1874, a small museum was organized to care for the collections of rocks, minerals, and fossils obtained from the field parties made up of about 13 people. The largest portions of the collections of the State Museum consist of the specimens from the First and the Second Geological Surveys.

U. S. NATIONAL MUSEUM BULLETIN 103 PL. 10

J. PETER LESLEY

STATE GEOLOGIST OF PENNSYLVANIA, 1874 1903.

Figure 8. J. Peter Lesley (1819–1903), second director of the Pennsylvania Geological Survey (1874–1903). Photograph republished courtesy of and with permission from the Smithsonian Institution.

The museum continued to develop and later was named the William Penn Memorial Museum. In 1976, it was renamed the State Museum of Pennsylvania. The Museum preserves a small collection of fossil plants collected by Leo Lesquereux, who was employed to continue his work on the "Coal flora" (1879–1884). The collection includes specimens numbered 11074 thru 11623, which are described and illustrated in Lesquereux's "Coal flora" and which originally belonged to the Pennsylvania Second State Geological Survey; 99 of these specimens are missing. There are no type specimens housed in this collection; the type specimens are in the USNM Collection.

The James Collection of fossil plants was amassed by I. E. James of Pittston, Pennsylvania, between 1890 and 1910. The majority of the specimens are from the Northern Anthracite coalfield of eastern Pennsylvania. The James Collection of fossil plants was purchased by the State Museum in two lots—in 1909 and in 1916. The collection comprises 288 specimens, of which none are type specimens. David White identified the fossil plants in this collection, which are of exceptional quality; many of them are on display at the Museum. Eighty-seven localities are represented in the collection, including Pittston, Luzerne County, Pennsylvania; the Mineral Spring Colliery, Coalbrook Slope, Parsons, Pennsylvania; and Lehigh Valley Coal Company, Pleasant Valley, Pennsylvania.

At the time of purchase, no catalog was found for the James Collection, and many of the original labels were separated from specimens. In 1968, William C. Darrah, Professor of Biology at Gettysburg College, was commissioned to restore the James Collection. Darrah found that James was methodical and meticulous in labeling his specimens. Using the labels and numbers provided by James, Darrah authenticated almost the entire collection. From this, Darrah prepared a new catalog that contains six indexes to this collection.

The State Museum has continued to expand its collections. John Oleksyshyn gave 424 specimens to the Museum that represent a portion of his collections of the floras from the Anthracite region (see Oleksyshyn, 1982). The William Klose Collection of the floras from the Anthracite region is made up of 7,085 specimens. There are no type specimens in the Oleksyshyn or the Klose Collections at the State Museum; two of Oleksyshyn's holotypes (Oleksyshyn, 1976) are at Harvard University (HUPC). William Klose has worked extensively on these collections as did A. J. Miklausen, a contemporary of W. C. Darrah.

## THE CLAUDE WESTON UNGER COLLECTION READING PUBLIC MUSEUM AND ART GALLERY, READING, PENNSYLVANIA

***E. D. Morey***

Claude W. Unger (1882–1945) was born in Schuylkill County, Pennsylvania. The Unger family lived in what is today called Pottsville, Pennsylvania, the type area for the Pottsville Formation (Lower and Middle Pennsylvanian). A self-educated scholar, Unger read widely from books that interested him. Professionally, Unger owned and operated a bookstore in Pottsville, and among his specific interests were Pennsylvania Dutch history and ephemera.

Unger was an acknowledged authority on fossils from the Northern Anthracite coalfield. From 1900 to 1930 he collected 7,686 fossil specimens from 186 localities, mostly in Schuylkill County, Pennsylvania. He also served as a guide for a number of distinguished geologists, both American and European, who came to Pennsylvania to study the Anthracite region. In the mid-1930s Unger hosted W. J. Jongmans (see Chapter 5, this volume) and C. B. Read (Read, 1955).

The Reading Public Museum was founded in 1904. In 1929, the Museum purchased Unger's collections from the Anthracite region; the collection then lay dormant for many years. Even so, W. C. Darrah during the late 1960s studied the collection and found the specimens to be invaluable. He found Unger's identifications to be quite adequate (Darrah, 1969).

W. F. Klose studied the Unger Collection in detail in the 1970s and compiled a catalog using the labels, which had partially disintegrated and were lying at the bottoms of the drawers in the specimen cabinets. It was only in 1975 that the original Unger catalog was rediscovered in the Museum safe.

The Reading Museum hired workers to replace the old labels and attach new labels to the specimens. Klose worked on the collection to record the specimens and localities, and the result is his monumental work entitled "Summary of fossils from the Anthracite and Semi-Anthracite basins of Northeastern Pennsylvania (Klose, 1984a) (for further references see Carpenter, 1980; Klose, 1984b).

To assemble a similar collection from the Anthracite region today would be impossible. All the specimens are derived from coal-bearing strata not worked since 1900, and the underground coal mines have long since been abandoned and are currently inaccessible. Klose regards this collection as extremely valuable, because the localities represented in this collection will probably never be available again (W. F. Klose, personal communication, 1992).

## CARR-DANIELS COLLECTION, UNIVERSITY OF ILLINOIS, URBANA, ILLINOIS

### *R. L. Leary*

The Carr-Daniels Collection at the Natural History Museum of the University of Illinois consists almost entirely of specimens from the Mazon Creek area near Morris, Illinois. The collectors, L. E. Daniels and J. C. Carr, both resided in Morris while gathering plant fossils along Mazon Creek in Grundy County, Illinois. The majority of the specimens in these collections are ironstone nodules with only a few compressions.

The Carr-Daniels Collection was acquired by the University of Illinois in 1920. For many years the specimens remained in storage and were inaccessible for study. At one time, part of the Daniels Collection was investigated by A. C. Noé (see Chapter 14, this volume), who published photographs of several specimens (Noé, 1925). In the late 1940s, W. N. Stewart, paleobotanist of the University of Illinois, examined the Carr-Daniels Collection and prepared a catalog of the best-preserved specimens (Stewart, 1950).

As is true of all fossil plants collected from the Mazon Creek area, most specimens are foliage remains of *Pecopteris* and *Neuropteris.* Many of the leaf fragments in the Carr-Daniels Collection are large enough to show something of the high degree of polymorphism within a frond. The Carr-Daniels Collection also contains many fertile specimens of lycopsids, sphenopsids, and pecopterids.

Of the total number of specimens (4,018) in the two collections, 758 belong to the Daniels Collection and 3,260 to the Carr Collection. The combined collections have 103 different species assigned to 51 genera.

## ACADEMY OF NATURAL SCIENCES COLLECTIONS, PHILADELPHIA, PENNSYLVANIA

### *E. D. Morey*

The Academy of Natural Sciences of Philadelphia, Pennsylvania, was founded in 1812. The early professional staff and collaborators included many noted mineralogists and paleontologists of the United States. Among them were Thomas Say, Isaac Lea, H. C. Woods, P. W. Shaefer, J. B. Wetherill, I. M. Mansfield, Angelo Heilprin, and Edward D. Cope.

The early acquisition of specimens in the Academy of Natural Sciences came from collections that were originally part of the holdings of the American Philosophical Society of Philadelphia. The Academy's entire paleobotanical collection remained dormant for many years; to prevent further deterioration, in 1971 Horance G. Richards (1906–1984), then chairman of the Department of Geology and Paleontology of the Academy of Natural Sciences, enlisted the help of William C. Darrah to refurbish the collections and provide a summary catalog of the specimens. After confirming the identifications of specimens and transcribing the original labels, Darrah succeeded in finally compiling a catalog by 1975.

The Academy's Department of Geology and Paleontology was disbanded in the late 1970s. The collection was transferred in 1984 to the Department of Botany (Spamer, 1988). Earle E. Spamer prepared a catalog of types of the fossil plants that made up this collection in 1988 and continues to study its specimens (Spamer, 1989). The collection contains 3,000 cataloged specimens and 59 verified and suspected type specimens that had been published from 1818 to 1983 (Spamer, 1988).

The Lesquereux type material in these collections comes from Nebraska, Colorado, and New Mexico. The Lesquereux specimens were collected by John L. Le Conte and sent to

Lesquereux, who described and identified them. The Lesquereux type specimens were figured in F. V. Hayden's report on the geology of Nebraska (Spamer, 1988).

The H. C. Wood Collection at the Academy is a fine collection containing holotypes that were described in the Proceedings of the Pennsylvania Academy of Science in 1860. Included in the Wood Collection is a collection of European Carboniferous plants assembled before 1860 (Wood, 1860).

The type material of Steinhauer (1818) comes from the Carboniferous of England. The collection was donated to the Pennsylvania Academy by J. P. Wetherill. The Issac Lea specimens were collected in Pennsylvania.

The Academy also has a selection of Mazon Creek fossils of Illinois. There are about 350 pairs (part and counterpart) of specimens in nodules, many of which represent fructifications.

The I. M. Mansfield Collection is mainly a selection of fossil plants from Cannelton, Pennsylvania. It is interesting to note that there are also some fossil plants in this collection that were collected in Greenland during the Peary Expedition of 1892. These specimens were collected by Angelo Heilprin.

In 1958 the Woodbine collection of fossil plants was given to the Pennsylvania Academy of Sciences. The collections, which were made by D. L. MacNeal, contain 20 holotypes from Denton, Denton County, Texas (see MacNeal, 1958).

## PALEOBOTANICAL COLLECTIONS, CARNEGIE MUSEUM, PITTSBURGH, PENNSYLVANIA

***E. D. Morey***

The Carnegie Museum was founded in 1899. From the start, it has had a staff of professional scientists. The most outstanding contributions of the Carnegie Museum are in vertebrate paleontology, but the Museum also has Carboniferous plant and animal fossils from western Pennsylvania. Today the Carnegie Museum collections are in the process of being recurated.

Baron Bayet purchased a beautiful collection of Mazon Creek nodules from the famous locality in Illinois in about 1890. This collection, a gift from Baron Bayet to the Carnegie Museum, contains 1,777 specimens that were studied by W. C. Darrah while he was a student at the University of Pittsburgh (see Chapter 1, this volume).

This collection also contains some fine European material from classic localities that partially are now exhausted. Represented are specimens from the Radstockian of England, Eocene of Belgium (from Heersien Gelinden), Eocene of Italy (from Monte Bolca), and Miocene of Switzerland (from Oeningen).

Between 1906 and 1910, Percy E. Raymond (Raymond, 1909) made local collections of fossil plants. Those collections were small but contained locality and horizon data. Starting from Raymond's collections, Otto E. Jennings worked for the next 20 years to expand the holdings. He tried to obtain specimens from each newly discovered locality in western Pennsylvania. Jennings encouraged W. C. Darrah (Chapter 1, this volume) to collect at the Clarksburg Sandstone of the Conemaugh Formation from 1926 to 1933. Darrah tabulated 16,000 specimens collected from the Mason Shale of the Mahoning Member of the Conemaugh Formation (now considered a group in several states) during 1927 to 1929 (Darrah, 1969).

A recent accession to the Carnegie Museum's collections is a large part (4,850 specimens) of the John Oleksyshyn (1901–1987; Fig. 9) Collection, which is mainly from the Anthracite region but also from the Narragansett basin. This gift to the museum from Oleksyshyn's daughters in 1988, which was facilitated by Paul C. Lyons of the U.S. Geological Survey, came after Oleksyshyn's death in 1987 (see Lyons, 1988). The Oleksyshyn Collection consists of many specimens of pteridosperms such as *Alethopteris* and *Neuropteris,* pteridophytes such as *Pecopteris,* and sphenophytes such as *Sphenophyllum.* There are no types in this part of the Oleksyshyn Collection. Some holotypes and hypotypes of the Oleksyshyn Collection are at Harvard University (see Oleksyshyn, 1976).

The Honorable I. F. Mansfield Collection of Carboniferous plants was obtained from the Darlington (Middle Kittanning) coal bed at Cannelton, Pennsylvania. This bed was mined southwest of Darlington in Beaver County near the boundary of Ohio from 1850 to 1890 (Darrah, 1969). Other

Figure 9. John Oleksyshyn (1901–1987). Many of Oleksyshyn's specimens can be found in the Harvard University Paleobotanical Collection; the State Museum, Harrisburg, Pennsylvania; and the Paleobotanical Collections, Carnegie Museum. Courtesy of Olha Oleksyshyn Holowecky.

significant collections of specimens from the Darlington coal bed can be found in the USNM, the HUPC, and the Pennsylvania State Museum collections (Darrah, 1969).

At the Carnegie Museum, there are 14 types and figured specimens in the collection of Carboniferous specimens, some of which were obtained by means of exchanges with the USNM. The Carnegie Museum also contains representative material from the Cretaceous, Eocene, and Oligocene of Utah and Montana.

## PALEOBOTANICAL COLLECTION, WYOMING HISTORICAL AND GEOLOGICAL SOCIETY, WILKES BARRE, PENNSYLVANIA

*E. D. Morey*

The Wyoming Historical and Geological Society was founded in 1858. R. D. Lacoe (Fig. 2) joined the Society as an active member on March 2, 1882, and became its curator of paleontology in 1884. He held that post for 15 years. Lacoe also served as a trustee of the Society from 1882 to 1889. While acting as curator, he presented to the Society 5,000 specimens, representing 1,200 species of fossil mollusks, as well as a large collection of fossil plants. Many of these specimens were identified by Lesquereux (Hayden, 1901).

Although considering himself an amateur, Lacoe published two catalogs (1883, 1884) for the Society's collections, one on fossil insects and the other on fossil plants. Lacoe had kept excellent records of his mining activities. From J. P. Lesley (1819–1903; Fig. 8), second director of the Pennsylvania Geology Survey (1874–1888), came high praise for Lacoe. Referring to the buried Valley of Wyoming, Lesley said, "I am indebted for most of the records of drilling by the various mining companies to Mr. R. D. Lacoe of Pittston, who has done so much through his magnificent collections to advance our knowledge of the Coal Flora of Pennsylvania and other states" (Hayden, 1901).[9]

When Lacoe resigned his curatorship in 1899, the post was assumed by Joshua L. Welter. Lacoe, together with Drs. Ingham and Wright, constructed a model of the Earth's crust that demonstrated the geology from Precambrian to Cenozoic time. The model was used in the local public school system in the early 1900s.

Unfortunately, the collections of the Wyoming Historical and Geological Society were destroyed by a flood in 1972. Only a few slabs with fossil plants remain on display. At least the data for the collection remain available (Donald Hoff, oral communication, 1991).

## THE WILLIAM CULP DARRAH PALEOBOTANICAL COLLECTION

*E. D. Morey*

The William Culp Darrah Collection was initiated in 1926 when Darrah began gathering Carboniferous plant fossils under the guidance of his scoutmaster, T. Z. Phillips, a Welsh miner who had worked extensively in the Northern Anthracite coalfield of eastern Pennsylvania. During the late 1920s, Darrah collected about 5,000 fossil specimens from the bituminous coal region near Camp Umbstaetter Reservation near Ambridge, Allegheny County, Pennsylvania. Darrah's early collection from about 1925 to 1933 was given to Harvard University when he began working there as a curator of the paleobotanical collection in 1934.

The Darrah Collection is divided into two parts: specimens that Darrah collected or purchased for his teaching at Gettysburg College and specimens that he collected with his daughter, Elsie Darrah Morey, during the 1950s and 1960s. The Darrah Collection contains about 3,000 specimens, mainly of Pennsylvanian plant fossils but also from other geologic periods. The Darrah Collection contains no types. Darrah's type specimens are at Harvard University (HUPC).

In 1953, Joseph R. Lomax (Fig.7), the son of Joseph Lomax, approached Darrah to sell his collection of coal-ball materials. This collection contains primarily specimens from the upper Carboniferous (Shore, Littleborough, Lancashire, Upper Foot Seam, Dulesgate, Lancashire, England) as well as from the lower Carboniferous (Pettycur, Burntisland, Fife, Scotland). This is an important coal-ball collection because it contains much of the material used to prepare thin sections of European coal-ball plants (Scott, 1900). The purchase of the Lomax Collection was completed in 1955.

Darrah's last major collecting of compressional fossils was done in the middle to late 1960s. At that time, he was collecting extensively from the Allegheny Formation; from the Kittanning, the Fairfax, and the Pittsburgh coal beds in West Virginia; from the Middle Kittanning coal bed from Karthaus, Pennsylvania; and from the Mammoth coal bed at St. Clair, Pennsylvania, in the Anthracite region. By the middle 1970s, Darrah's health did not allow him to do much collecting.

The collection has grown from exchanges with paleobotanists over the years. The J. A. Barlow Collection (1968–1969) from the Upper Pottsville of Tennessee became incorporated into the Darrah Collection. The Barlow Collection contains 433 specimens from the following Pennsylvanian coal beds: Top Summitt Coal, High Coal, Wildcat Coal, Lower Splint Coal, Lower Rock Spring Coal, Peewee Splint Coal, Walnut Mountain Coal, Red Ash Coal, Big Mary Coal, and Wind Rock Coal of the Appalachian region of Tennessee.

The Darrah Collection also contains 40 specimens of Late Pennsylvanian (Cantabrian) plant fossils from the Narragansett basin that were collected by C. G. Grant of Easton, Massachusetts, and identified by P. C. Lyons (see Lyons and Darrah, 1978). These specimens were shipped to Darrah by Lyons on December 30, 1975, and acknowledged by Darrah on January 7, 1976, when he completed labeling the specimens.

Currently, this private collection is being curated and recataloged by Elsie D. Morey. Morey's specimens from the Upper Freeport coal bed of West Virginia and her specimens from the Middle Kittanning coal bed at Karthaus, Pennsylvania, are being added to the Darrah Collection.

## ACKNOWLEDGMENTS

The authors were assisted by many people who graciously gave their time and assistance to provide information about the collections discussed in this chapter and the availability of photographs.

We thank Robert W. Baxter, James P. Ferrigno, Dana A. Fisher, Janice F. Goldblum, Donald Hoff, Andrew H. Knoll, William F. Klose, Meredith A. Lane, Alicia Lesnikowska, Paul C. Lyons, Sergius H. Mamay, Patricia V. Pellegrino, F. Charles Petrillo, Gary G. Rosenberg, Susan M. Rossi-Wilcox, and Arthur Watt for information in preparing this paper.

Elsie D. Morey acknowledges the use of the William Culp Darrah Collection of papers and letters. The William Culp Darrah Collection is located at the Morey Paleobotanical Laboratory.

We thank Clifford M. Nelson, Christopher Wnuk, and an anonymous reviewer for their reviews of our manuscript.

## ENDNOTES

1. Ralph Dupuy Lacoe Memorial, 1901, Wyoming Historical and Geological Society, v. 6, p. 45.
2. Ralph Dupuy Lacoe Memorial, 1901, Wyoming Historical and Geological Society, v. 6, p. 46.
3. Letter from C. B. Reed to W. C. Darrah, July 5, 1939. W. C. Darrah Collection.
4. Letter from S. H. Mamay to E. D. Morey, July 22, 1993. W. C. Darrah Collection.
5. Letter from D. White to W. C. Darrah, July 16, 1934. W. C. Darrah Collection.
6. Letter from C. B. Read to W. C. Darrah, August 21, 1934. W. C. Darrah Collection.
7. Letter from R. W. Brown to W. C. Darrah, September 2, 1934. W. C. Darrah Collection.
8. Letter from C. Schuchert to W. C. Darrah, November 6, 1935. W. C. Darrah Collection.
9. Ralph Dupuy Lacoe Memorial, 1901, Wyoming Historical and Geological Society, v. 6, p. 50.

## REFERENCES CITED

Carpenter, F. M., 1980, Studies on North American Carboniferous insects—Upper Carboniferous insects from Pennsylvania: Psyche, v. 87, p. 107–119.

Darrah, W. C., 1934, Leo Lesquereux: Harvard University Botanical Museum Leaflets, v. 2, p. 113–119.

Darrah, W. C., 1936, The Harvard Collections of fossil plants: Harvard University Alumni Bulletin, v. 39, p. 383–385.

Darrah, W. C., 1969, A critical review of the Upper Pennsylvanian floras of the eastern United States with notes on the Mazon Creek flora of Illinois: Gettysburg, Pennsylvania, privately printed, 220 p.

Hayden, H. E., 1901, Mr. Ralph Dupuy Lacoe: Wyoming Historical and Geological Society, v. 6, p. 43–54.

Hitchcock, E., 1833, Report on the geology, mineralogy, botany, and zoology of Massachusetts: Amherst, Massachusetts, Amherst College, 700 p.

Janssen, R. E., 1939, Leaves and stems from fossil forests; a handbook of the paleobotanical collections in the Illinois State Museum: Springfield, Illinois, Illinois State Museum Popular Science Series, v. 1, p. 190.

Janssen, R. E., 1940, Some fossil plant types of Illinois. Part I: A restudy of the Lesquereux types in the Worthen Collection of the Illinois State Museum; Part II: Description of new species from Mazon Creek: Springfield, Illinois, Illinois State Museum Popular Science Series, v. 1, 124 p.

Jeffrey, E. C., 1925, Coal and civilization: New York, MacMillan, 178 p.

Klose, W. E., 1984a, Summary of fossils from the Anthracite and Semianthracite basins of northeastern Pennsylvania: Pennsylvania Geological Survey Open Report, pages unnumbered.

Klose, W. E., 1984b, Summary lists of fossils from the Anthracite fields and nearby areas of northeastern Pennsylvania: Pennsylvania Geology Journal, v. 15, p. 2–16.

Knoll, A., and Barghoorn, E. S., 1984, The paleobotanical collections: Harvard University Botanical Museum Leaflets, v. 30, p. 13.

Lacoe, R. D., 1883, List of Paleozoic fossil insects of the United States and Canada: Wyoming Historical and Geological Society Publication 5, p. 1–21.

Lacoe, R. D., 1884, Catalog of the Paleozoic fossil plants of North America: Pittston, Luzerne County, Pennsylvania.

Langford, G., 1958, The Wilmington coal flora from a Pennsylvanian deposit in Will County, Illinois (second edition): Downers Grove, Illinois, Esconi Associates, 360 p.

Langford, G., 1963, The Wilmington Coal fauna and additions to the Wilmington Coal flora from a Pennsylvanian deposit in Will County, Illinois: Downers Grove, Illinois, Esconi Associates, 280 p.

Leary, R. L., 1976, Inventory of type and figured paleobotanical specimens in the Illinois State Museum: Illinois State Museum, Inventory of Geological Collections, no. 2, pt. 2, p. 1–64.

Leary, R. L., 1985, Fossil noncalcareous algae from Visean (Mid-Mississippian) strata of Illinois (U.S.A.): Congrès international de stratigraphie et de géologie du Carbonifère, 10th, Madrid, Spain, 1983: Compte Rendu, v. 2, p. 307–331, pl. II.

Leary, R. L., 1986, Three new genera of fossil noncalcareous algae from Valmeyeran (Mississippian) Strata of Illinois: American Journal of Botany, v. 37, p. 369–375.

Lesquereux, L., 1854, New species of fossil plants, from the Anthracite and Bituminous coal fields of Pennsylvania, collected and described by Leo Lesquereux, with an introduction by H. D. Rogers: Boston Journal of Natural History, Article 25, v. 6, p. 409–431.

Lesquereux, L., 1858, Fossil plants of the coal strata of Pennsylvania, *in* Rogers, H. D., ed., Geology of Pennsylvania: Pennsylvania State Geological Survey, v. 2, pt. 2, p. 837–884 [2nd ed., 1868].

Lesquereux, L., 1866, An enumeration of the fossil plants found in the Coal Measures of Illinois, with descriptions of the new species, *in* Worthen, A. H., Geology of Illinois: Geological Survey of Illinois, v. 2, Palaeontology, Sect. 111, p. 427–470.

Lesquereux, L., 1870, Report on the fossil plants of Illinois, *in* Worthen, A. H., et al.: Geological Survey of Illinois, v. 4, Geology and Palaeontology, Pt. 2, Sect. 11, p. 375–508.

Lesquereux, L., 1879–1884, Description of the coal flora of the Pennsylvanian and of the Carboniferous formations in Pennsylvania and throughout the United States: Pennsylvania Geological Survey, 997 p. (Atlas, 1879; vol. 1 and 2, 1880; vol. 3, 1884.)

Lesquereux, L., 1887, List of recently identified fossil plants belonging to the United States National Museum, with descriptions of several new species: Proceedings of the United States National Museum, pls. 4, p. 21–46.

Lyons, P. C., 1988, Memorial to John Oleksyshyn: Geological Society of America Memorial Series, p. 65–68.

Lyons, P. C., 1993, Pennsylvanian paleobotany: Geological Society of America, Coal Geology Division, GSA Field Trip 11 Guidebook, p. 81–87.

Lyons, P. C., and Darrah, W. C., 1978, A late Middle Pennsylvanian flora of the Narragansett basin: Geological Society of America Bulletin, v. 89, p. 433–438.

Lyons, P. C., Hatcher, P. G., Brown, F. W., Thompson, C. L., and Millay, M. A., 1985, Coalification of organic matter in a coal ball from the Calhoun coal bed (Upper Pennsylvanian) Illinois Basin, United States of America: Congrès international de stratigraphie et de géologie du Carbonifère, 10th, Madrid, Spain, 1983: Compte Rendu, v. 2, p. 155–159.

MacNeal, D. L., 1958, The flora of the Upper Cretaceous Woodbine Sand in Denton County, Texas: Academy of Natural Sciences of Philadelphia Monograph 10, 152 p., 36 pls.

Merrill, G. P., 1920, Contributions to a history of American state geological and natural history surveys: U.S. National Museum Bulletin 109, p. 459.

Nelson, C. M., and Yochelson, E. L., 1980, Organizing federal paleontology in the United States, 1858–1907: Journal of the Society for the Bibliography of Natural History, v. 9, p. 607–618.

Noé, A. C., 1925, Pennsylvanian flora of northern Illinois: Illinois State Geological Survey Bulletin 52, 113 p.

Oleksyshyn, J., 1976, Fossil plants of Pennsylvanian age, Narragansett basin, *in* Lyons, P. C., and Brownlow, A. H., eds., Studies in New England geology: A memoir in honor of C. Wroe Wolfe: Geological Society of America Memoir 146, p. 143–180.

Oleksyshyn, J., 1982, Fossil plants from the Anthracite coal fields of eastern Pennsylvania: Pennsylvania Geological Survey General Geology Report 72, 157 p.

Raymond, P. E., 1909, Some sections in the Conemaugh series between Pittsburgh and Latrobe, Pennsylvania: Carnegie Museum Annals, v. 1, p. 166–177.

Read, C. B., 1946, A Pennsylvanian florule from the Forkston coal in the Dutch Mountain outlier, north-eastern Pennsylvania: U.S. Geological Survey Professional Paper 210-B, p. 17–26, pls. 1 and 2.

Read, C. B., 1955, Floras of the Pocono Formation and Price Sandstone in parts of Pennsylvania, Maryland, West Virginia: U.S. Geological Survey Professional Paper 263, 32 p., pls. 1–20.

Scott, D. H., 1900, Studies in fossil botany: London, A. and C. Black, 533 p.

Spamer, E. E., 1988, Catalogue of type specimens of fossil plants in the Academy of Natural Sciences: Academy of Natural Sciences of Philadelphia Proceedings, v. 140, p. 1–17.

Spamer, E. E., 1989, A historic piece of petrified wood from the Triassic of Arizona: The Mosasaur, v. 4, p. 149–152.

Steinhauer, H., 1818, On fossil reliquia of unknown vegetables in the coal strata: American Philosophical Society, new series, Transactions, v. 1, p. 265–297.

Stewart, W. N., 1950, Report on the Carr-Daniels collections of fossil plants from Mazon Creek: Illinois Academy of Science Transactions 43, p. 41–45.

Watt, A., 1970, Catalog of the illustrated Paleozoic plant specimens in the National Museum of Natural History: Smithsonian Contributions to Paleobiology 5, 53 p.

Watt, A., 1974, Catalog of specimens illustrated in Lesquereux's "Coal Flora": Bloomington, Indiana University for the Paleobotanical Section of the Botanical Society of America, 42 p.

White, D., 1893, *Sphenopteris lacoei:* U.S. Geological Survey Bulletin 98, p. 56, figs. 5 and 6.

White, D., 1903, Memoir of Ralph Dupuy Lacoe: Geological Society of America Bulletin, v. 13, p. 509–515.

Wood, H. C., 1860, Catalog of Carboniferous plants in the Museum of the Academy of Natural Sciences, with corrections in synonym, descriptions of new species: Academy of Natural Sciences of Philadelphia Proceedings, v. 12, p. 436–443.

Manuscript Accepted by the Society July 11, 1994

Printed in U.S.A.

Geological Society of America
Memoir 185
1995

# *Wilhelmus Josephus Jongmans (1878–1957): Paleobotanist, Carboniferous stratigrapher, and floral biogeographer*

**Robert H. Wagner**
*Jardín Botánico de Córdoba, Avenida de Linneo s/n, 14004 Córdoba, Spain*
**H. W. J. van Amerom**
*Geologisches Landesamt Nordrhein-Westfalen, Postfach 1080, 47710 Krefeld, Germany*

## ABSTRACT

**W. J. Jongmans, a botanist, was drawn into the study of Carboniferous compression floras for stratigraphic purposes. He was the founder and director of the Geological Bureau in Heerlen, Netherlands Limburg, where he organized four highly successful international congresses on Carboniferous stratigraphy and geology. Although starting out with a clear commitment to the monographic description of the Carboniferous flora of The Netherlands, he became increasingly drawn into a worldwide documentation of Carboniferous compression floras. These were also used for floral biogeography. His encyclopedic knowledge of the paleobotanical literature was due in part to his prodigious efforts to compile the Fossilium Catalogus—Plantae, an indispensable work of reference for the paleobotanical taxonomist. He also furthered palynological and coal petrographic studies.**

## GENERAL DATA AND EARLY STUDIES

Born in Leiden (Holland) on August 13, 1878, in a tailor's family, Wilhelmus Josephus Jongmans studied biology at the University of Leiden and afterwards in Munich (under Professor K. Goebel), where he obtained his degree and, in 1906, a doctorate earned with a dissertation on mosses. He also met his wife during his sojourn in Germany, so the time was doubly well spent. They had a long life together and raised a family of eight children.

Jongmans (Fig. 1) returned to Leiden to join the Rijks Herbarium, the Botanical Institute linked to the University of Leiden. Nothing at this point seemed to predispose him for a career in paleobotany. However, as so often happens, chance intervened in the shape of W.A.J.M. van Waterschoot van der Gracht, an eminent geologist with a long experience in the oil industry, primarily in the United States. Waterschoot van der Gracht had been charged by the government of The Netherlands with the exploration, on behalf of the future State Mines, of coal deposits in South Limburg. The area belonged to the same basin as neighboring Aachen in Germany and formed part of the North European paralic coal belt. He invited Jongmans to join the exploration team and to contribute to the investigation by identifying the plant megafossils obtained from fully cored boreholes and using these fossils for correlation. Jongmans met this challenge. He collected all the books on Carboniferous plant fossils he could find in the Rijks Herbarium and established links with paleobotanists abroad, notably with Henry Potonié and his student, Walther Gothan, in Berlin and with Robert Kidston in Scotland.

Potonié involved Jongmans with the Fossilium Catalogus—Plantae, a tremendous task that became a lifelong endeavor. This involvement gave him an encyclopedic knowledge of the paleobotanical literature. The first issue appeared in 1913, the year of Potonié's death, but a forerunner—"Die palaeobotanische Literatur" for the years 1908 to 1911—was published in 1910 to 1913. Jongmans devoted an immense amount of time to the Fossilium Catalogus. At the end of his life he left six shelves of files full of handwritten manuscript notes, being a virtually up-to-date record of paleobotanical taxa published all over the world. Jongmans published quite a few volumes himself (Equisetales, Lycopodiales, Filicales, Pteridospermae, Cycadales, 1913–1957), but

Wagner, R. H., and Amerom, H. W. J. van, 1995, Wilhelmus Josephus Jongmans (1878–1957): Paleobotanist, Carboniferous stratigrapher, and floral biogeographer, *in* Lyons, P. C., Morey, E. D., and Wagner, R. H., eds., Historical Perspective of Early Twentieth Century Carboniferous Paleobotany in North America (W. C. Darrah volume): Boulder, Colorado, Geological Society of America Memoir 185.

Figure 1. Jongmans (seated, with flat cap) in horse and buggy in the Dutch countryside around 1906–1908, as part of a team of botanists mapping the distribution of flora in The Netherlands, an early study of plant ecology. The man in charge was Dr. Goethart, the gentleman with a beard, seated in front. From the photographic archives, Geologisch Bureau, Heerlen, The Netherlands.

certain parts were prepared by colleagues. He did not content himself with just the bare record but also gave a critical assessment that has been generally useful to the monographer of the various groups involved. The enormous quantity of manuscript notes left at the end of his life was finally edited and completed by his successor in this massive undertaking, S. J. Dijkstra, and continued up to the present day by one of the authors of this chapter (H.W.J.v.A.).

Jongmans's love of books and capacity for painstaking work made him particularly apt for the never-ending task of producing the Fossilium Catalogus—Plantae. As a necessary backup for this work, he collected an outstanding paleobotanical library of both books and reprints ranging through all geological periods. A large part of this library was obtained through exchange, but he also bought on behalf of the Geologisch Bureau at Heerlen. The Antiquariat Junk used to send him chests full of reprints to browse through and select from at a standard rate of so many cents per page.

Walther Gothan became a lifelong friend and associate in a large number of papers ranging from stratigraphic reports to the description of floras. The extensive correspondence between Jongmans and Gothan started before December 1914 and ended with Gothan's death in 1954. Both gentlemen were stamp collectors, and the first preserved letter ends with remarks on postage stamps.[1] Although they were virtually the same age, Gothan had made an earlier start in paleobotany, publishing on this subject before his doctoral thesis in 1905. Jongmans and Gothan jointly published the most significant summary paper on the Carboniferous flora of The Netherlands, a sizable contribution published in the "Archiv für Lagerstättenforschung," a publication of the Prussian Geological Survey (1915). In later years they coauthored several reports and papers describing floral remains from such far-flung areas as Sumatra (1925, 1935a, b), the United States (1934), and Brazil (1952). Jongmans also loved to quote his friend (rightly or wrongly) as the man who said "Was man nicht bestimmen kann, sieht man als neue Arten an" (What you cannot identify, you regard as new species), quite a philosophy but only workable with massive experience (and probably said tongue in cheek, although there is a germ of truth in this little rhyme).

Jongmans also made early contact with Robert Kidston in Stirling, Scotland, where he visited for several months in 1908. It is likely that he regarded Kidston as his mentor, given the difference in age and experience at the time. Kidston identified plant megafossil impressions for the Geological Survey of Great Britain (as it was known at the time), and his knowledge would have been vital to Jongmans when the latter was engaged in identifying fossil plants from boreholes in the Carboniferous of The Netherlands. Kidston and Jongmans (1911) jointly published on the ovule of *Neuropteris obliqua*. (Unfortunately, the figured specimen is not quite as convincing as it should be.) They also collaborated in the "Monograph of

Calamites of Western Europe" (115–1917), which remains the most comprehensive treatment ever of *Calamites* pith casts and external impressions.

## ACTIVITIES AT THE GEOLOGISCH BUREAU, HEERLEN

When the borehole campaign in South Limburg finished, Jongmans stayed on in what became the Geological Bureau for the Netherlands Mining District (Geologisch Bureau voor het Nederlandse Mijngebied). Founded in 1908 as a branch of the Rijksopsporing van Delfstoffen (State Exploration Institute for Mineral Resources), it was transferred to a foundation of coal mines in South Limburg in 1924. Jongmans became the director of the Bureau in 1921. At first, the Bureau was in the same building as his home, and it was almost a one-man show, with only one assistant geologist, F. H. van Rummelen, as a companion. Jongmans looked after the Carboniferous, and Van Rummelen dealt with the Cretaceous and Tertiary overburden.

Though not a geologist by training, Jongmans became a stratigraphic paleobotanist, using Carboniferous plant compressions and impressions as stratigraphic indices and establishing correlations with the neighboring areas, particularly the adjoining region around Aachen in Germany. He adapted to the needs of the coal collieries by establishing links with the surveying departments of the individual mines. At each of these departments a surveyor was charged with collecting stratigraphic data, a duty that included fossil collecting so as to determine the correct position within a well-established stratigraphic succession. As late as the 1950s, these geological surveyors would congregate at the Geologisch Bureau in Heerlen on Saturday mornings in order to have the fossils identified and to seek advice on local correlation problems. They would also consult on the aspects of faulting that affected the coal workings. This link with the mine surveyors allowed Jongmans to bring together a vast collection of plant megafossils as well as nonmarine and marine animal remains that he was careful to pass on to specialists abroad (and later also in-house at the Bureau) for identification and description. The marine fossils, identified primarily by G. Delépine at the Catholic University in Lille, France, were particularly important for the proper identification of the widespread, eustatically controlled marine bands used as chronostratigraphic markers throughout the Namurian-Westphalian paralic coal belt of northwestern Europe.

A large selection of fossils, as well as minerals obtained from veins in fracture zones in the coal mines, was put on display at the Bureau. This attracted visitors, mainly local amateurs but also the occasional specialist from abroad. The tradition of combining storage with display has been continued, with more emphasis on the museum aspect, by one of the present writers (H.W.J.v.A) when he became paleobotanist and curator at the Geologisch Bureau in the late 1960s. Jongmans would have been pleased to see his early version converted into a proper museum.

The first curator (S. van der Heide) was appointed after World War II. Prior to his appointment, Van der Heide had already published several monographs on the Carboniferous nonmarine bivalves and the fish remains in the Jongmans Collection from South Limburg (1943a, b, 1946). Later (1951), Van der Heide also published on the arthropods.

The Carboniferous and Cretaceous/Tertiary collections that had accumulated through the efforts of Jongmans and Van Rummelen really paid off during World War II. At that time work had to be found for a number of Dutch geologists who were stranded in The Netherlands when the German armies invaded in May 1940, cutting the links with the outside world. In 1941, Jongmans wrote to Professor Schoute in Groningen that he had more than 30 geologists working at the Geologisch Bureau.[2] This group also included recent graduates for whom work had to be found so they would not be shipped to Germany for forced labor. One of Jongmans's sons was among the latter category (see his letter to W. C. Darrah in October 1945[3]). This large and diverse group of geologists involved themselves not only with the description of the many different groups of fossils from the local Tertiary and Cretaceous deposits as well as from the Carboniferous but also with geological problems related to coal mining. A substantial series of memoirs was published by the Geologisch Bureau as a result of these various investigations. It was clear that these people were not only kept out of the clutches of the German occupation authorities but also paid their way in terms of scientific research with tangible results.

Jongmans took a particular interest in the budding science of palynology and engaged a couple of botanists for the study of Carboniferous megaspores and miospores. The suicide in 1942 of one of these botanists (P. H. van Vierssen Trip) meant that only part of the palynology project came to fruition. S. J. Dijkstra continued and finished the work on megaspores started by Van Vierssen Trip. Dijkstra published one of the standard monographs on this group of Carboniferous fossils (1946) but did not get around to working on the miospores that he was originally intended to investigate. Indeed, Dijkstra became one of the recognized world authorities on megaspores, not only from the Carboniferous but also the Cretaceous. His research benefited from the worldwide contacts established by Jongmans, and he published quite often on samples brought back by Jongmans and his associates on his frequent travels in later years. The Carboniferous miospore studies in The Netherlands did not start until much later, well after Jongmans had left this world.

Another specialist interest developed by Jongmans in The Netherlands concerned coal petrography, now a much-practiced science with both geological and industrial connotations. Jongmans published a few small papers (Jongmans and Koopmans, 1934; Jongmans et al., 1936; Jongmans, 1938a, 1954d) on problems of nomenclature and showing botanical structures in coal. He used an etching technique to demonstrate that vitrinite is structured (as R. Thiessen maintained; see Chapter 11, this volume). However, in due time, the practice of coal petrography for the Dutch coal mines devolved upon Dijkstra, who combined this "chore" with the megaspore work that he enjoyed doing. The techniques of coal petrography were developed at Heerlen during the early 1940s by J. Faber, P. A. Hacquebard, and A.L.F.J. Maurenbrecher. These specialists found employment abroad after World War II finished.

During Jongmans's long sojourn in Netherlands Limburg, he developed a special feeling for this atypical part of the Low Countries. He contributed a number of general articles on topics of local interest, mainly connected with coal mining and the Geological Bureau and its collections. He was particularly active in the preservation of the only important outcrop of Carboniferous strata in The Netherlands, namely, the Heimansgroeve near Epen. A geological and paleontological study of the surroundings of Epen, published in 1925 and written in collaboration with G. Delépine, W. Gothan, P. Pruvost, F. H. van Rummelen, and N. de Voogd, became number 1 in the series of Mededeelingen van het Geologisch Bureau (Contributions from the Geological Bureau). This was a reprint of the journal "Natuurhistorisch Maandblad," issued in 1925 by the Natural History Society of Limburg. Jongmans was particularly interested in the mid-Namurian floral remains obtained from the Epen area, which he described in collaboration with his friend, Walther Gothan. Whenever the outcrop was cleaned up, which happened regularly every five years, he used the opportunity to expand the collection. With the exception of some sandstones, these exposures are no longer accessible.

## WESTPHALIAN FLORA OF THE NETHERLANDS

Jongmans intended to make a full description of the Carboniferous (mainly lower to middle Westphalian) plant megafossils from the South Limburg mining district and set out to do so in a characteristically thorough fashion by organizing the data from the literature in his "Anleitung zur Bestimmung der Karbonpflanzen West-Europas," which he published in 1911. He started with the sphenophytes, which he documented in the style later adopted by the "Traité de Paléobotanique" (E. Boureau, ed.) of the 1960s, reproducing the original illustrations of type material. His Anleitung is still a useful book, and it is regrettable that only the first volume was produced. He followed up this literature survey with an atlas of the different species of *Calamites* and calamitean foliage as found in the Ruhr District of western Germany (Jongmans and Kukuk, 1913) and, above all, with the massive "Monograph of the Calamites of Western Europe" (Kidston and Jongmans, 1915-plates, 1917-text).

Although Jongmans clearly set out to give the coal flora of The Netherlands the full monographic treatment (the work on *Calamites* was intended to be the first in a series devoted to the Carboniferous flora of The Netherlands and adjacent regions), his later papers recorded more incidental finds. The first stratigraphic range chart of Carboniferous plants in South Limburg was published by Jongmans in 1928. This reflected the large amount of fossil plant material that had accumulated at an early date. Another noteworthy contribution that found its way into the textbooks was the record of *Neuralethopteris* in conjunction with *Rhabdocarpus* and *Whittleseya* (Jongmans, 1954b). Immature "seeds" were found in connection, but the microsporangiate organ, which was also reported as being in organic connection, proved, on closer examination by one of the present writers (R.H.W.), to occur in superposition to the foliage. Jongmans (1951a) also recorded the cupules and seeds of *Lyginopteris hoeninghausii* and proved that *Crossotheca* does not constitute the microsporangiate organ of this pteridosperm.

The fossil plant collection in the Geologisch Bureau at Heerlen grew to such proportions that it became an almost impossible task to produce the succession of monographs that would have been required. It is also likely that administrative duties made it increasingly difficult for Jongmans to continue the self-imposed task of monographing this flora. On retirement, at age 68, he obtained the support of the coal mines to continue his work on the full description of the Coal Measure plants of The Netherlands, but by then he could no longer be tied down to a local project in the laboratory. His interests had become worldwide. It was thus left to later workers (Boersma, 1972; Van Amerom, 1975) to monograph different parts of the Jongmans Collection from South Limburg. In fact, the vast collection of plant megafossils assembled from the Carboniferous of The Netherlands still remains to a large extent undescribed. It is to be hoped that this important collection may be used in the future together with the other existing collections from the same geological region (which includes the north of France, Belgium, and western Germany) for a full revision of certain groups of Carboniferous plant megafossils from this classical region of the northwest European paralic coal belt.

## CARBONIFEROUS FLORAL BIOSTRATIGRAPHY WORLDWIDE AND PALEOECOLOGY

Jongmans was increasingly drawn into the study of the stratigraphic aspects of Carboniferous plant fossils and in later years became fascinated by paleogeography in relation to floral distribution. He also took an interest in paleoecology, publishing a paper on the changing composition of a roof shale flora in one of the Dutch coal mines (Jongmans, 1955a). This latter interest may have been sharpened by an early involvement with phytosociology in the Recent flora of The Netherlands (see Fig. 1).

He collected extensively from the Carboniferous in different parts of the world during field trips associated with the International Geological Congress convening, successively in Washington (1933) and in Moscow (1937). Also in the 1930s, Jongmans (Fig. 2) collected from different parts of western Europe. He was convinced that collector's bias tended to distort the record in favor of large, well-preserved specimens, thus giving only a partial view of what made up an assem-

Figure 2. Jongmans in his office at the Geologisch Bureau, Heerlen, The Netherlands (late 1930s). Photograph courtesy of Ms. M. Jongmans.

blage. He therefore tended to collect practically everything susceptible to being identified in due time. When publishing on collections from different parts of the world, particularly from little-known areas, he favored well-illustrated papers even where the identifications might still be regarded as provisional. He thus changed from a monographer into a recorder of floral assemblages.

His increasing interest in Carboniferous floras worldwide and the stratigraphic and paleogeographic conclusions to be drawn from worldwide studies seems to have been at least partly an offshoot of the work he did for the coal mines in South Limburg. This work was aimed at the stratigraphic uses of plant fossils. Jongmans soon realized that the various coal districts in western Europe (mainly the paralic coal belt of northwest Europe but also the various limnic basins in the hinterland area) used a plethora of local stratigraphic names and intervals that had become quite confusing in the absence of detailed correlations and a general stratigraphic framework. In collaboration with two other paleobotanists, W. Gothan and A. Renier, he therefore organized the Congrès pour l'avancement des études de stratigraphie carbonifère, which was held in Heerlen in June 1927. This meeting was a huge success, resulting in a European-wide correlation table and a regional (northwest European) classification of Carboniferous chronostratigraphic units. After further precision was introduced at later Heerlen congresses (1935, 1951, 1958) and at meetings in Paris (1963) and Krefeld (1971), this classification has stood the test of time. Indeed, the West European regional classification of the Carboniferous has become a yardstick for international correlations throughout the paleoequatorial belt of Carboniferous times.

Jongmans remained active in the organization of the International Congress of Carboniferous Stratigraphy and Geology, as it was named in later years. He had the satisfaction of seeing the congress reconvened in 1935 and, after a long recess due to World War II, in 1951. He was also active in the organization of the fourth Carboniferous congress, the last one to be held in Heerlen (in 1958), for which he prepared a questionnaire on the fundamental issue of whether the Carboniferous should stand as a single system or be subdivided into Mississippian and Pennsylvanian, based on the American model. (The Americans were outvoted, but the issue is still alive.) Jongmans did not live to see the fourth Carboniferous congress convened in Heerlen. The congress had grown tremendously since its first beginnings, covering all aspects of Carboniferous geology, while relinquishing its original purpose to the I.U.G.S. Subcommission on Carboniferous Stratigraphy.

Jongmans (1952a) summarized his ideas on correlation and floral distribution in the Carboniferous in a paper presented at the third Carboniferous congress in 1951. In a comparative table (Jongmans, 1952a) he presented the correlations that he regarded as reasonable. It is curious to see how he emphasized the stratigraphic gap between Westphalian and Stephanian strata in the Saar-Lorraine Basin, a gap later to be filled, at least in part, by the Cantabrian Stage (lowermost Stephanian) described from northwest Spain. The Carboniferous of the Iberian Peninsula was poorly known at the time when Jongmans (1952a) presented this paper. He also defended the recognition of a Westphalian E division equivalent in time to the Stephanian (Series) but presenting a gradual transition from the Westphalian D (Stage) upward in terms of floral composition, whereas the classic Stephanian floras of the limnic basins of the French Massif Central would show a different composition. Jongmans had become sensitive to the differences in floral composition in the uppermost part of the Carboniferous by working on sediments of Stephanian age in those parts of the world where these corresponded to marine-influenced basins. Although he was quite right in pointing out the differences in floral composition, his use of a chronostratigraphic term was less fortunate, and the Westphalian E was largely ignored in the literature (see Chapter 26, this volume).

## FLORAL PALEOBIOGEOGRAPHY

Jongmans (1952a) also displayed his interest in floral distribution and was aware of the importance of Carboniferous and Permian floras for checking on aspects of the theory of Continental Drift, now universally admitted as a fundamental aspect of Plate Tectonics. Curiously, he ignored the evidence adduced in an important paper by H. Potonié (1909), who pointed out the strong resemblance between tropical forest mires in Sumatra and Carboniferous peats, and chose to take as his point of departure the generalized belief at the time that the formation of peat on the scale necessary to produce the thick Carboniferous coal seams would be impossible in a tropical environment. This led to the search for a subtropical/tropical flora in an area between the Euramerian and Gondwana floral provinces (Jongmans, 1952a) and gave him a strong incentive for looking at the poorly known North African area. Now that paleomagnetism has led to the repositioning of continental plates at different geological times and the composition of fossil floras and their spatial distribution is generally recognized as being a valuable adjunct to these attempts at continental reconstruction, the early work by Jongmans and others is seen to have been extremely valuable. Although his search for the tropics of Carboniferous times away from the paleoequatorial area that constituted his own locality of origin now seems ironical, there is no doubt that his attitude to the problem was essentially modern and has helped to shape the research attitudes of the present day.

## INDONESIAN FLORAS

Jongmans's worldwide interests were sharpened by access to a flora of considerable paleogeographical importance, namely, that found in eastern Sumatra. A small but significant collection of Paleozoic plants from the Djambi District in Sumatra had been brought together by a geologist, A. Tobler, who had lodged the collection in the Natural History Museum of Basel, Switzerland. It was sent for study to Jongmans through the good offices of Richard Kräusel in Frankfurt-am-Main. Jongmans recognized the importance of this apparently late Carboniferous flora (presently regarded as very early Permian), which looked remarkably European (presently one would say paleoequatorial Amerosinian) despite its general proximity to the *Glossopteris* floras of Australia and India. Jongmans described Tobler's collection in collaboration with Gothan and published the descriptions in 1925. Jongmans's connection with Tobler dated at least from 1919, when he first mentioned the desirability of mounting a paleobotanical expedition to Djambi, then a fairly remote part of the Dutch colonial empire. Jongmans actively promoted the expedition, for which he found official support from the colonial authorities in the Netherlands East Indies (presently Indonesia) as well as the necessary financial backing from a number of institutions. The expedition took place in 1925, the same year when the description of the Tobler Collection appeared in print. Jongmans was not a member of the expedition because his duties did not allow him to be away from The Netherlands during the many months required. The zest for travel was in his blood, so one can well imagine the sense of frustration he must have felt when he could not personally go and collect the fossil plant material in the field. However, a young botanist, O. Posthumus, who had attended Jongmans's paleobotany lectures at the University of Groningen, was found for this work. Posthumus was accompanied by J. Zwierzycki, an experienced geologist who was put in charge of the expedition and who was to write the geological report. The viscissitudes of the paleobotanical expedition to Djambi were told engagingly by Zwierzycki and Posthumus (1926), who went to the area at the worst possible time: The local labor force was profitably engaged in rubber tapping (there was a rubber boom at the time), and the expedition could not possibly compete financially. A small party of laborers was eventually sent out from distant Batavia (now Jakarta), but this proved to consist of big town ne'er-do-wells, so that Zwierzycki and Posthumus ended up doing a good deal of the digging themselves. It sounds familiar!

Jongmans was extremely concerned, as the following excerpt from a letter to Tobler dated August 19, 1925, shows:

> From Dr. Bosscha I got a telegram, that Dr. Zwierzycki did send him a telegram from Djambi, that financial support possibly does not suffice. The costs at Djambi were so high, that he fears, that he could not carry out the plans. I wrote immediately to the Bataafsche (Oil company) if the Company eventually could contribute a bit. But it is not sure, that they shall do it, specially because they have given already a considerable contribution. I do not know, to whom here in the country I could turn for contributions. I do not understand why the expedition cannot get along with nearly 20,000 guilders. Zwierzycki is known to be very reliable. But again it is strange that he did not warn urgently in advance. He might have written to Rutten, that the Djambi nowadays was the expensive part of Indië, but that he should have informed himself more precisely and set up the costs more in detail.[4]

The fossil collection was forwarded to Jongmans in Heerlen, where he described it in collaboration with Gothan. They eventually published their important memoir in 1935 (Jongmans and Gothan, 1935b). (The description of this fairly large, new flora proved more time consuming than had been anticipated, and the intended date of publication, 1930, could not be adhered to.) In the meantime, Posthumus (1927) expressed his disappointment at being excluded from the scientific work that followed from his collecting activity. On the basis of his field notes, Posthumus published the opinion that the Djambi flora was early Permian rather than latest Carboniferous (Stephanian). Jongmans and Gothan (1925, 1935a, b) regarded the Djambi flora as being of Stephanian age, but current opinion tends to support Posthumus's stratigraphic conclusion. It is clear from the correspondence between Jongmans, Zwierzycki, and Posthumus that Jongmans became irritated with the latter and that Zwierzycki kept aloof from the controversy.[5] The expedition was Jongmans's brainchild, and he had obtained the financial backing for it, so the fossil plant collection obviously had to go to him. Posthumus may have been a little too assertive. The relationship between Jongmans and Posthumus ended when the latter obtained a post at the experimental station for sugarcane at Passeroean, Java, and thus left paleobotany. He did not survive the Japanese occupation of the Dutch East Indies during World War II. The paleogeographic significance of the Djambi flora was apparent to all, and its links with the South China subprovince of the Cathaysia Province are generally acknowledged today.

Other fossil floras from Indonesia were sent to Jongmans at later dates. The most important one was a Permian flora from near the south coast of New Guinea (Irian). This showed a mixture of Cathaysian and Gondwana elements. Jongmans (1940a, 1941) compared it with the Djambi flora (in both cases he assigned a late Carboniferous age) and deduced a different paleolatitude by mentioning that the flora from New Guinea came from a more southerly area than that occupied by Djambi, Sumatra. Other Permian floras from New Guinea with a mixed Cathaysian-Gondwana composition were later identified by Hopping and Wagner (in Visser and Hermes, 1962). Jongmans published brief notes on Mesozoic plants from Indonesia in 1951 (1951b) and 1955 (the latter in collaboration with E. Boureau as the senior author). Jongmans rarely ventured outside the Carboniferous and lower Permian, and his sporadic descriptions of Mesozoic plants were prompted by material sent to him for identification.

## CARBONIFEROUS FLORAS OF THE UNITED STATES AND CORRESPONDENCE WITH WILLIAM C. DARRAH

It would appear that the Djambi flora awakened Jongmans's curiosity about floral distribution worldwide and that this spurred him on to collect not only in western Europe but further abroad as well. The International Geological Congress in Washington, D.C. (1933), provided him with the opportunity to collect in the United States, where he visited a number of localities in the Appalachians and in the Midcontinent to which he was guided by American geologists (including Oscar Haught in West Virginia, Gilbert H. Cady of the Illinois State Geological Survey, and W. C. Darrah). This fieldwork allowed Jongmans to apply the recently established regional chronostratigraphy of the Carboniferous in western Europe to the succession of strata with fossil plants in North America. He enjoyed his travels in the United States and also enjoyed the personal contact with the American paleobotanists. His primary contact was David White, a very senior member of the U.S. Geological Survey and clearly the most prominent Paleozoic paleobotanist in the United States. White, who was by then nearing the end of a long and distinguished career, organized Jongmans's travels and recommended him to the people who were to accompany him. As so often happens, Jongmans developed his likes and dislikes. When reminiscing in later years, he would describe one of his American colleagues as "the man who had inherited the mantle of the prophet but who wasn't the prophet" (the prophet being quite clearly David White). One person for whom he developed an obvious regard was William C. Darrah, then an up-and-coming paleobotanist who accompanied both Jongmans and Paul Bertrand (University of Lille, France; see Chapter 7, this volume) in the field. Both European paleobotanists were interested in making comparisons between the successions of Carboniferous floras in Europe and North America.

Although Jongmans, as usual, shared the job of identifying the plant megafossils he had collected in North America with Walther Gothan, one of the papers presented on the American floras during the second Carboniferous congress (Heerlen, 1935) was coauthored by Darrah (Jongmans et al., 1937). The latter was also allowed to present a divergent opinion on the age of some of these floras in a footnote appended to another paper (Jongmans, 1937). Darrah went to Heerlen for the second Carboniferous congress and visited the Jongmans family on that occasion (see Chapter 1, this volume). It appears from the correspondence that Darrah and his wife arrived in Holland in August 1935 and stayed in Heerlen for about a fortnight.[6] On October 9 Darrah wrote from Antwerp, "I wish to express my thanks for the wonderful experiences and treatment we received in Heerlen during and after the Congress. We made excellent collections, received good ideas, and had a most enjoyable time. Above all else, it has been pleasant to work with you again and to meet your family."[7] On November 15, 1935, he wrote to Jongmans again, saying "We arrived safely in America two weeks ago and have settled down to work. The boxes from Holland have arrived but thus far the customs inspectors have not come to open them. My papers are ready to send to you, but I will keep them a little longer until I receive the manuscript of our joint paper and return them all at once."[8]

Despite a measure of frustration due to the slow arrival of some fossil collections that Jongmans expected from America, which finally arrived by ship (he wrote several times to Darrah to remind him), it is obvious that Jongmans had a soft spot for W. C. Darrah and respected his work. This is worth emphasizing because Darrah did not agree with Jongmans's ideas on the Stephanian in North America (Darrah, 1969; see also Chapter 26, this volume) and was more closely in tune with Paul Bertrand (see Chapter 7, this volume). Although wishing to learn from the more-experienced Jongmans, Darrah was independent in his judgment on the age of the American floras and was quick to spot the superior knowledge of Bertrand when it came to Stephanian floras.

The preserved correspondence between Jongmans and W. C. Darrah starts with a letter from America dated January 1, 1934, months after Darrah had accompanied Jongmans on part of his collecting trip in the United States.[9] The most interesting letters, important because they throw light on the lives of both scientists during the traumatic times of World War II, were exchanged on October 1, 1945, and November 14, 1945, respectively. The first letter (by Jongmans) mentions

This morning Jo [one of Jongmans's daughters] received a letter from your wife. We were very glad to read the news, and to see that you all are quite well, especially as we had heard from different sides that you should have died. Fortunately this is not the case and I hope that you are extremely well and that you will have a long life. You know that the best guarantee for a long life is that people had told that you were dead. I hope this will also be the case with you. I learnt from the letter that you did not do much paleobotanical work during the war, but I am sure that your work has been very useful. Please let me know whatever you did. Here, most things are all right. However we lost my son, Rudi, who studied Geology at the University of Amsterdam. He died on February 2nd of this year. His death was not a direct result of the war although he had a difficult time because of the persecution by the Germans. I think you will remember him from the days you spent here as Hans [another one of Jongmans' sons] and he did the administration of the Congress. Here in Heerlen we have been working very much. It was necessary to bring a number of young geologists in security, as the Germans intended to transport them to Germany. So the Direction of Mines opened the possibility to study here the geology of The Netherlands and especially that of the Mining District. I had more than thirty geologists in my Institute, so you can understand that I had a busy time. Fortunately, we could save them all and they did much good work. Besides, I personally have given the opportunity to work here to about ten other geologists who hid here from the Germans [Jongmans used here an untranslatable term dived, which meant to go underground, either without a usable personal documentation or with forged papers, all to avoid being transported to Germany for forced labour]. These young men had no permission so this was illegal work. This work too brought great success. My son Rudi was one of these illegal geologists.[10]

Darrah answered this letter as follows:

I was very glad to hear from you. Especially that you and your family had survived this dreadful war. We were saddened to hear that Rudi lost his life, at least indirectly, through treatment at the hands of the Germans [however, Rudi actually died of cancer]. We remember him well from our visit in Heerlen. Mrs. Darrah has received a long letter from Jo and she too, joins me in sending our greetings to all of the family.

I was somewhat amused and taken aback by the news that I had died two years ago. Though my health was not of the best, I assure you that I am still alive [see Chapter 2, this volume]. During the war I have been engaged in industrial war work—a large part of which was confidential and still cannot be divulged. This much I can tell you. Most of my work was indirectly concerned with radar. It was of a metallurgical and chemical nature. Some of the problems I worked on were particle size of metals, especially powdered zirconium, ultra-violet examination [fluorescence] of metals and more recently the writing of technical manuals and reports. For nearly a year I have administered the activities of a large department of 250 persons, not one of whom had had previous technical experience or training. I have not totally neglected paleobotany, though I have published but a few papers during the war. Most important I have completed a monograph on [the] microstructure of coal which Dr. Frans Verdoorn will publish in the spring of 1946.[11]

The last letter from Darrah dates from September 17, 1968, and was addressed to Jo Jongmans.[12] It deals with the exchange of papers. Among other items he asked whether he could purchase a copy of the forthcoming paper by R. H. Wagner on the genus *Alethopteris.* This memoir did indeed become available in 1968. The work for this doctoral thesis (which was defended in Amsterdam in March 13, 1968) had started in the 1950s under the supervision of Jongmans. An earlier paper (Wagner, 1961) dealt with some rare (partly new) species of alethopterids from South Limburg.

In the course of Jongmans's work on the Carboniferous floras from the United States, he introduced for the first time his concept of "Westphalian E," which Darrah rightly considered to be synonymous with Stephanian. Jongmans was struck by the longer ranges in America of certain Westphalian species (e.g., *Neuropteris ovata*) and wished to convey the Westphalian complexion of the North American Stephanian floras, which he found to be rather different from the classical Stephanian floras of the French Massif Central. He later developed the concept further (Jongmans, 1952a).

Jongmans's last trip to North America took place in June 1952, when he participated in the Second Conference on the Origin and Constitution of Coal, held at Crystal Cliffs, Nova Scotia, Canada. He presented a paper on "Coal Research in Europe" (Jongmans, 1954a) in which he outlined his views on paleobiogeography, expressed in a table showing worldwide correlations of Carboniferous stratigraphy based on floral provinces. He also collected plant fossils from the Sydney Coalfield of Nova Scotia, with particular emphasis on the upper part of the section, which contains *Neuropteris ovata.* The Sydney collecting trips were organized by P. A. Hacquebard and T. B. Haites, who had both worked under Jongmans at the Geological Bureau in Heerlen during World War II.

In the same year Jongmans also managed to get to Texas. He wished to collect from David White's Permian localities and was guided by Arthur Byer and Maxine Abbott to some wild, bush-covered country west of Wichita Falls. On one occasion, when collecting from localities well off the main road, he was surprised by a thunderstorm and was temporarily separated from his companions. They searched for him afterwards but could not find him, and in desperation they decided to report his absence to the sheriff, who organized a search party.

Jongmans did not manage to find the spot where he had left his companions but struck out on his own across country to the highway, which he reached after walking several miles. There he stopped a Greyhound bus and managed to convince the driver to take him to Wichita Falls. This was not easy, since Jongmans had on only a pair of shabby looking shorts and had neither documentation nor money (all that was in his shirt at the collecting site). Arriving in town, he went to Byer's home, where he was staying, but found nobody in (everybody was out looking for him). He probably shrugged his shoulders and went up to bed. It seems there was quite a bit of consternation the next morning, when Jongmans came downstairs for breakfast! His bleary-eyed hosts, rather sheepishly no doubt, called off the sheriff's search party.

It may be that collecting plant fossils in the United States made Jongmans even further aware of paleobiogeography and the differences as well as similarities in floral composition in the different parts of the world. He also became convinced that the differences were often exaggerated because of the geographic isolation of the individual scientists and that different names had been applied to essentially the same species in the different areas. This point was well taken and Darrah followed it up by examining museum collections in Europe when traveling across the ocean to attend the second Carboniferous congress (see Chapter 2, this volume).

## WORK ON THE UKRAINE, BRITAIN, THE ALPS, AND TURKEY

Four years after the Washington, D.C., meeting, the International Geological Congress reconvened in Moscow (1937). Jongmans used the opportunity to collect in the Donets Basin, Ukraine, where he was guided to the localities by M.D. Zalessky. Jongmans became aware of the harsh conditions under which people were forced to work in the Soviet Union and admired Zalessky for his fortitude and his utter dedication to science. Jongmans suspected that Zalessky's assistant, Tschirkova (later his wife), was charged with keeping an eye on this eminent paleobotanist, who had already made a name for himself in czarist times. (During World War II Zalessky was taken as a prisoner of war to Berlin, where he worked at Gothan's laboratory and published a number of paleobotanical papers in German; he returned to Russia after the war, and died in 1946.) Jongmans published an illustrated account of the Carboniferous plants he collected in the Ukraine and provided a critical summary of the data in the literature. In this he did not restrict himself to the Donets Basin but also included the published records from elsewhere in the USSR as well as from China (Jongmans, 1939, 1942). These are useful summaries that show the encyclopedic knowledge that Jongmans gained from his work on the Fossilium Catalogus—Plantae.

More or less at the same time (Jongmans, 1940c), he published an exhaustive literature survey of the Upper Carboniferous plant megafossil records in Great Britain. Considering the fact that the British records were widely scattered among a great many different journals, often better known locally than worldwide, he achieved an impressive coverage. He also did useful work in relating the local stratigraphic units to the more international European classification of the Carboniferous that had been achieved by the two Carboniferous congresses of 1927 and 1935. If Jongmans's review paper has not been used much subsequently, this is not because of any lack in quality but rather because it was written in German, published near the outbreak of World War II, and appeared in the "Jaarverslag" (Annual Report) of the Geologisch Bureau at Heerlen. This may not have been generally available to paleobotanists and Carboniferous stratigraphers in the British Isles. This was the last time for a number of years that Jongmans was to publish in German, the foreign language in which he was most fluent. The German occupation of The Netherlands had put his back up, and although he was less fluent in English, he switched to that language for his more general papers in the postwar period. Whenever possible, he would write in the language of the country from which he described the floras, but English generally replaced German as the vehicle for his scientific publications.

Well before World War II, Jongmans started to collect plant megafossils from the Alps. As early as 1925 he planned to collect in Switzerland (letter to A. Tobler dated May 15, 1925[13]). He wrote in German [as translated by H. W. J. van Amerom] "Do you have the time and inclination to accompany me on Whitsun on a visit to the surroundings of Martigny to look at the Carboniferous of the Alps and to try to collect a bit? I should also like to visit the Botanical Garden at Bourg Saint Pierre and possibly the Grand St. Bernard. I do not like to go alone and your geological guidance would be of great value. Then we might also select the best variety of Wallish wine together." However, the intended visit did not take place, and the correspondence ceased with the death of Dr. Tobler in November 1929. Jongmans then turned to the Carnic Alps, publishing a substantial paper on its Carboniferous floras (1938b). In later years, well after his retirement as director of the Geologisch Bureau at Heerlen, he actually made it to Switzerland, where he collected in 1950 and 1951 and also revised museum collections. This revision became a major undertaking, culminating in the large, well-illustrated memoir that was published posthumously in 1960. Once more, this involved a literature survey, which was particularly important in an area with only incidental paleobotanical records since Heer's major work was published in 1877. The descriptions of taxa in the 1960 memoir are rather sketchy, and the text largely consists of a commentary on synonymies and published information. However, the memoir is important in that it provides a photographic documentation of the Carboniferous floras of Switzerland. (The atlas is exhaustive and of good quality.)

Also before World War II, in 1938, Jongmans became engaged in a major project on the Carboniferous coalfields near the Black Sea in northern Turkey. Little was known about the Carboniferous flora of Zonguldak (the main coalfield area) since Zeiller published his classical 1899/1900 paper on the fossil flora of Heraclea (Ereğli in Turkish) at the turn of the century. Jongmans's involvement started with a small collection of plant fossils sent by geologist W. S. Grancy to Elise Hofmann in Vienna. This collection was briefly described by Jongmans in the Turkish Geological Survey publication "Meteae" in 1939. Jongmans established contact with the geological survey (Maden Tetkik ve Arama Enstitüsü) in Ankara and with the State Coal Mining Company in Zonguldak and collected extensively from stratigraphic sections and more scattered localities in the Zonguldak area as well as in the outlying areas of Amasra, Söğütözü, and other localities in the same general region. This first major collecting trip was in 1938, but he returned after World War II, in 1946.

The collection made in 1938 was forwarded to Heerlen for study in the laboratory. Jongmans prepared an extensive report on this collection, with the intention of having it published by the survey in Ankara. However, this very substantial contribution to the Carboniferous floras of northern Turkey remained in the file of Maden Tetkik ve Arama Enstitüsü and did not see the light of day as a publication. This was a grave disappointment to Jongmans and a loss to science. Some duplicate specimens were kept in Heerlen, together with a large file of photographs. However, the negatives were also sent to Ankara. The plants collected in 1946, on his second field trip, remained in Turkey, where they were lodged in the collections of the Palaeontology Department of Maden Tetkik ve Arama Enstitüsü.

Jongmans (1955b) eventually published a general account of the plant assemblages found in the different localities visited, but this paper was without illustrations and not an adequate substitute for the voluminous report submitted to the survey in Ankara. For a time he worked together with a young Turkish paleobotanist, Recep Egemen, who published a few short papers on elements of the Zonguldak flora. However, Egemen was prevented from coming truly to grips with the large collection because of his untimely death in a car crash while on his way to the Third International Congress of Carboniferous Stratigraphy and Geology in 1951. Jongmans (Fig. 3) hoped that some parts of his voluminous report might be readied for publication when one of the present writers (R.H.W.) joined Maden Tetkik ve Arama Enstitüsü as a paleobotanist in 1957, but the link was cut only a few months later when Jongmans died. Wagner did work on the collections brought together by Jongmans, but he left the survey in Ankara after only one year, which was insufficient time for a proper revision of the large flora involved. The Carboniferous flora of Zonguldak and the various outlying districts in the Black Sea area is still very incompletely known (for a recent summary see Kerey et al., 1986).

In 1946 Jongmans was accompanied in the field by S. J. Dijkstra, who studied the megaspores from Zonguldak and who trained a couple of Turkish palynologists. (For a biography of Dijkstra, see Van Amerom, 1984.)

Figure 3. Jongmans in the garden of the Geologisch Bureau, 1951, hammering a core. Photograph courtesy of A. T. Cross, Michigan State University.

## STRATIGRAPHIC PALEOBOTANY IN SPAIN

Only a year after Jongmans's second trip to Turkey, he was invited in 1947 to study Carboniferous plant fossils from Spain. The meritorious start made by Ruiz Falcó and Madariaga Rojo (1931, 1933) on the description of Carboniferous plants from that country had come to very little by the time the Spanish Civil War broke out, and most of the available records in the literature consisted of lists produced by nonspecialists. These were of little practical value. The few lists that corresponded to specialist work were mainly from the nineteenth century. Jongmans thus moved into a major European area, rich in Carboniferous plant fossils, but poorly known at the time. Only neighboring Portugal had been studied more intensively with regard to Carboniferous palaeobotany, mainly through the efforts of Carlos Teixeira.

The invitation to Spain came from a mining engineer, Ignacio Patac, one of the pioneering geologists in northwestern Spain, who had become involved in a controversy about the dating of a small concealed coalfield near the Asturian coast. This was worked by the La Camocha Mine near Gijón, where coal-bearing strata occurred in a tectonic structure that Adaro (1914) had related to outcrops in beds regarded as equivalent to lower Westphalian. However, Patac (1933) maintained that these were of Stephanian age and backed up his opinion by a list of plant fossils identified as Stephanian taxa. He was naturally interested in having this opinion confirmed by a specialist in Carboniferous paleobotany and managed to obtain financial support for a visit by Jongmans from the Spanish Geological Survey (Instituto Geológico y Minero de España). Jongmans found that the identifications made by Patac were quite spurious and confirmed the opinion of Adaro by finding plant fossils at La Camocha that suggested an early Westphalian age. Jongmans continued his visit by collecting in the Central Asturian Coalfield, at that time the main coal-producing area in Spain; he then went on to southern Spain, where he was guided in the field by Bermudo Meléndez, a paleontologist. This was after the main object of Jongmans's visit to the Asturias had been achieved (with negative results as far as Patac was concerned). Meléndez managed to obtain additional funding from the Scientific Research Council (Consejo Superior de Investigaciones Científicas). The later collecting trips by Jongmans, ranging widely through different parts of Spain, were all organized by Meléndez, with financial support from the Spanish Research Council. On Jongmans's first trip (in 1947) he also visited the Puertollano Coalfield in south-central Spain and the Peñarroya and Valdeinfierno coalfields in Sierra Morena in southwestern Spain. Jongmans also gave a lecture on Carboniferous stratigraphy and the methods to be employed for its study in the University of Granada. Meléndez published this lecture on behalf of Jongmans (1949a), and although it makes curious reading in some places, it provides an insight to the ideas that guided his research on worldwide floral distribution.

Jongmans was most struck by the flora of Valdeinfierno, to which he devoted a preliminary note (Jongmans, 1949b) as well as a more substantial paper (Jongmans and Meléndez, 1950). This was the first time that the presence of early Carboniferous plant fossils, closely similar to those of the classical "Kulm flora" of Germany, had been recognized. Valdeinfierno, a small pull-apart basin of Tournaisian age, is still the most important of the only two published localities with floras of earliest Carboniferous age in the Iberian Peninsula.

A general summary of the results of Jongmans's collecting trip in 1947 was published in 1951 under the title of "Las floras carboníferas de España" (1951c). Meléndez provided the translation into Spanish, a language with which Jongmans was unfamiliar. In this paper Jongmans not only gave the lists of species found in the localities he had visited but also analyzed the available literature and gave his opinion on the published floras. This paper can be regarded as the starting point for modern stratigraphic paleobotany in the Carboniferous of Spain. His most substantial collections were from the Central Asturian Coalfield, for which he published an atlas (Jongmans, 1952c).

He returned to the Central Asturian Coalfield and to La Camocha for more detailed investigations of floral assemblages, which were now collected from measured sections. This was in 1952, when he was assisted by one of the present writers (R.H.W.), who had commenced work on Carboniferous stratigraphy and paleobotany in Spain under Jongmans's supervision. One paper, on the Riosa area of the Central Asturian Coalfield (Jongmans and Wagner, 1957), changed the existing ideas on the tectonic structure. Jongmans wanted to speed up the collecting by getting the miners to provide him with wagonloads of the roof shales of coal seams. This request was complied with, and on the next morning he was presented with a whole row of piles of rock, which he happily tackled with his hammer. Unfortunately, every single pile contained stigmarian rootlets. The sequence was overturned, and he had been presented with all the seat-earths. The immediate practical result was that the local geological map had to be revised. Also in 1952, Jongmans returned to La Camocha for a more systematic study of floral assemblages in stratigraphic succession. His assistant, R. H. Wagner, obtained a detailed stratigraphic section in the La Camocha mine, elucidated its tectonic structure, and sent up wagonloads of fossil material that Jongmans hammered through and selected from under daylight conditions. Jongmans produced a draft manuscript on the floral remains from La Camocha, which he mainly attributed to the Namurian (a list of the taxa identified by Jongmans in that unpublished manuscript was reproduced in Wagner, 1959). Unfortunately, that manuscript, produced near the end of Jongmans's life, turned out to be in need of substantial revision, and the age of this flora is more properly given as Langsettian (ex Westphalian A).

In the same year, 1952, and later in 1953, 1954, and 1955, Jongmans expanded his area of investigation to the Stephanian coalfields of western Asturias, particularly to that of Villablino in the province of León, and visited the Pyrenees, and Cuenca in the Iberian Chain. All these trips were organized by Meléndez, then professor at the Central University of Madrid (now known as Universidad Complutense), and Jongmans was accompanied by R. H. Wagner, J. Talens, and H. M. Helmig as well as B. Meléndez. Jongmans established good relations at Villablino, where geologists from the Geologisch Bureau were subsequently engaged for a stratigraphic study in the mine workings. Together with Meléndez he returned to Sierra Morena, where he collected from several different localities that yielded floras ranging from Viséan to early Permian (Jongmans and Meléndez, 1956).

Jongmans (Figs. 4 and 5) initiated not only the more systematic work on stratigraphic paleobotany in the Carboniferous of Spain but also trained R. H. Wagner (Fig. 6) and J. Talens, who continued his work. It was fortunate that he was actively supported in his investigations by B. Meléndez, who managed to get his travels financed by the Scientific Research Council of Spain, accompanied him on many of his trips, and actively promoted the publication of his results in Spanish journals. Meléndez also sent young paleontologists to Jongmans for training (J. Talens, M. C. Bonet).

Jongmans never neglected collecting coal samples for palynology, and these collections resulted in a megaspore paper published by Dijkstra (1955) and unpublished miospore work by R. H. Wagner. Coal samples collected by Jongmans's assistant in the La Camocha mine workings were investigated by Bonet and Dijkstra (1956) and Neves (1964), and marine and freshwater animal remains were described from the same section by different specialists.

## ADDITIONAL INVESTIGATIONS

At the same time as his commitment to Spain, Jongmans developed an interest in the Carboniferous floras of Morocco and Algeria in North Africa. There he found support from the Geological Survey of Morocco (then still a French Protectorate), which published two of his papers dealing with a late Stephanian or possibly early Autunian flora from the High Atlas region (Jongmans, 1950) and an upper Westphalian one from Djerada (1952b), in the western part of the country, fairly close to Algeria. In the latter country, he teamed up with Paul Deleau, a geologist and paleontologist, to produce a well-illustrated account of the Westphalian plant impressions and nonmarine faunas of the Colomb-Béchar and Djebel Mézarif areas in southern Oran (Jongmans and Deleau, 1951). As happened with most of the publications in this later phase of Jongmans's activities, the emphasis is on a full photographic record, whereas the text is a running commentary on the material collected. The publication reads like a preliminary field report, and it seems that a number of the identifications of species are provisional. However, the documentation provided is invaluable for an area on which little has been published, even to the present day.

Although Jongmans studied plant assemblages and was aware of the paleogeographical and paleoecological implications, the latter aspect would really come into its own much later, with the advent of sedimentology. However, his mind was open for field observations, and he did publish a paper on the density of Carboniferous trees as measured in the roof shales of coal in the Emma deep mine (Jongmans, 1955a). Typical of his method of tackling a major area of investigation was the extensive literature survey of reconstructions of fossil forests through geological time (Jongmans, 1949c). A large, full-color reconstruction of a late Carboniferous forest, done under his auspices, hangs in the Heerlen office of the Geological Survey of The Netherlands.

Figure 4. W. J. Jongmans happy as a sandboy in the field in the middle 1950s. Photograph by R. H. Wagner.

Figure 5. Jongmans at home in the 1950s smoking a small cigar (a characteristic adjunct). Photograph courtesy of Ms. M. Jongmans.

Jongmans derived a great deal of pleasure from his travels, and it is clear that his ideas on Carboniferous paleogeography and floral distribution were based to a large extent on first-hand observation of plant assemblages obtained from the different areas. His unflagging enthusiasm comes through clearly in a letter to W. C. Darrah, written on October 1, 1945:

> Next year I intend to retire from my post as director of the Geologisch Bureau. In August of that year I will have been working here for forty years and at the first of May for twenty-five as director of this institute. I find that date the right one to retire and to leave my place to a younger geologist. But it is not my intention to take a rest. The Netherlands Mines will give me the possibility to prepare my large flora of the Carboniferous and the general stratigraphy. I also intend to write a general flora of the Carboniferous of the world and to try to get at a vertical and horizontal parallelisation of the different regions and to treat especially the question of the variations under climatic and geographical conditions. This is a large job, which I can finish only with the help of colleagues. It will also be necessary to make several journeys to collect the necessary material. As I am working with associations and not with separate species I must see the way of occurrence myself. So I did on my trips through Russia, Turkey and the States. But there is still much that I must see. So a large part of your country and also several parts of Europe (Spain, Portugal) and Russia. I hope to find a possibility to make these journeys after my retirement.[14]

After reading this kind of program, it is clear that he could not possibly finish it, but could document species, a synthesis not being possible in the time available. However, he retained his youthful enthusiasm right to the end and was busily engaged in organizing a trip to South Africa when he went to bed one night and never woke up again (at age 79).

Apart from the floras he collected himself, he was sent two Gondwana floras of great paleogeographic interest. One of these, from Carhuamayo and Paracas in Peru, came his way via an ex-alumnus, N. de Voogd. Jongmans combined it with a collection from Paracas in the British Museum (Natural History) and published a full description (1954). He studied another assemblage from the fringe of the Gondwana area, the Sinai Peninsula (Egypt), in 1940 (1940b) and later, in 1955, in conjunction with a faunal study by Van der Heide (Jongmans and Van der Heide, 1955). Both floras were regarded by Jongmans as being of early Carboniferous age, but in both cases a late Carboniferous age is a distinct possibility.

## GENERAL CONSIDERATIONS

It seems that Jongmans was most happy in the field collecting fossils and getting the feel of the floral assemblages. As time went by he turned from detailed descriptions to broader appreciations of what the floral assemblages would mean in terms of stratigraphic age and paleogeographic distribution. Although a botanist by training, his outlook became increasingly geological. This was to a certain extent detrimental to the quality of his fossil identifications (much of his later work is in need of a systematic revision) but useful for pointing the way toward the geological use of plant impressions of late Paleozoic age. He quickly saw the significance of palynology and coal petrography and created posts in Heerlen for these subjects. However, it is likely that he will be best remembered for the four international congresses of Carboniferous stratigraphy and geology, of which he was the main organizer and which made the name of Heerlen famous among Carboniferous stratigraphers.

Figure 6. W. J. Jongmans sitting in his laboratory in Heerlen, The Netherlands, 1954. The person standing is his student, R. H. Wagner. Photograph courtesy of J. Talens.

Jongmans was quite unconventional and would show this in his garb, which was casual to the extreme. He also liked the good things in life and, wherever possible, would enjoy the service of first-class hotels. A story has circulated that he was more or less forcibly removed from the dining room of the Hotel Principado in Oviedo because the wife of General Franco was due to enter and might take offense at the sight of an elderly gentleman in tatty shorts.

Although primarily a research scientist in what he regarded as important, Jongmans was forced to attend to administrative duties during his years as the director of the Geologisch Bureau. However, one day of the working week he would lock the door of his study and be unavailable for administrative matters. It is easy to sympathize with this attitude, but it may have been just a little frustrating for the people wanting to get a decision from him on the wrong day.

For many years Jongmans was the only paleobotanist in The Netherlands. This was generally known, and eventually the University of Groningen made him a "Professor Extraordinary" in 1932, which meant a long train journey once a week so as to lecture to botany students. Jongmans had few students. He was well aware that his specialty would not present many openings for young professionals. His first Ph.D. student was N. de Voogd, a geologist who worked on the Carboniferous stratigraphy and fossil floras and faunas of the surroundings of Aachen. R. G. Koopmans, a botanist who became Jongmans's assistant, was the second one. He obtained his doctorate on the study of coal balls collected by Jongmans in The Netherlands (this collection had first been offered for a Ph.D. study to O. Posthumus, who declined). World War II made an end to the association between Jongmans and Koopmans because the latter joined the German armed forces and became persona non grata in the Geologisch Bureau after the war ended (see Van Amerom, 1984). When Jongmans was well in his seventies, he became concerned to find a successor to continue the work that he had started and pushed ahead with great energy. It looked as if he had found such a person in one of the present writers (R.H.W.), who was trained in paleobotany by Jongmans and who adopted quite a few of the interests that Jongmans had acquired. However, his putative successor did not stay in Heerlen, and Jongmans did not live to see the publication of the Ph.D. thesis (Wagner, 1968) that he supervised in its early stages. It should be noted, however, that the lines of research developed by Jongmans were continued by this last pupil of his, albeit in different locations. Finally, the large fossil plant collection and important specialist library that Jongmans assembled were curated and used by the second author (H.W.J.v.A.), who thus physically benefited from the Jongmans heritage and continued the Fossilium Catalogus as well as augmenting the collections. Both successors are geologists by training.

For a more or less complete bibliography of Jongmans the reader is referred to Thiadens [A.A.] (1957). For other historical data on the history of the Geological Bureau in Heerlen, the reader should peruse Van Amerom (1991).

## ACKNOWLEDGMENTS

We thank P. A. Hacquebard and A. T. Cross for supplying information on Jongmans's travels to Nova Scotia, Canada, and Texas and A. T. Cross also for supplying a photograph. P. C. Lyons, E. Darrah Morey, A. T. Cross, and J. P. Laveine made helpful suggestions. J. Talens provided photographs and liaised with B. Meléndez in order to clarify organizational details of Jongmans's pioneering studies in Spain. Ms. M. Jongmans kindly supplied the photographs of her father, reproduced as Figures 2 and 5.

## ENDNOTES

1. Letter from W. Gothan to W. J. Jongmans, December 11, 1914. Archives, Rijks Geologische Dienst, Office Heerlen, P. O. Box 126, 6400 AC Heerlen, The Netherlands.
2. Letter from W. J. Jongmans to J. C. Schoute, October 6, 1941. Archives, Rijks Geologische Dienst, Office Heerlen, P. O. Box 126, 6400 AC Heerlen, The Netherlands.
3. Letter from W. J. Jongmans to W. C. Darrah, October 1, 1945. W. C. Darrah Collection.
4. Letter from W. J. Jongmans to A. Tobler, August 19, 1925. Archives, Rijks Geologische Dienst, Office Heerlen, P. O. Box 126, 6400 AC Heerlen, The Netherlands.
5. Letter from W. J. Jongmans to J. Zwierzycki, March 24, 1926. Letter from J. Zwierzycki to W. J. Jongmans, May 14, 1926. Letter from O. Posthumus to W. J. Jongmans, April 1926. Letter from O. Posthumus to W. J. Jongmans, June 16, 1926. Archives, Rijks Geologische Dienst, Office Heerlen, P. O. Box 126, 6400 AC Heerlen, The Netherlands.
6. Letter from W. C. Darrah to W. J. Jongmans, August 26, 1935. Archives, Rijks Geologische Dienst, Office Heerlen, P.O. Box 126, 6400 AC Heerlen, The Netherlands.
7. Letter from W. C. Darrah to W. J. Jongmans, October 9, 1935. W. C. Darrah Collection.
8. Letter from W. C. Darrah to W. J. Jongmans, November 15, 1935. W. C. Darrah Collection.
9. Letter from W. C. Darrah to W. J. Jongmans, January 1, 1934. W. C. Darrah Collection.
10. Letter from W. J. Jongmans to W. C. Darrah, October 1, 1945. W. C. Darrah Collection.
11. Letter from W. C. Darrah to W. J. Jongmans, November 14, 1945. Archives, Rijks Geologische Dienst, Office Heerlen, P.O. Box 126, 6400 AC Heerlen, The Netherlands.
12. Letter from W. C. Darrah to Jo Jongmans, September 17, 1968. W. C. Darrah Collection.
13. Letter from W. J. Jongmans to A. Tobler, May 15, 1925. Archives, Rijks Geologische Dienst, Office Heerlen, P.O. Box 126, 6400 AC Heerlen, The Netherlands.
14. Letter from W. J. Jongmans to W. C. Darrah, October 1, 1945. W. C. Darrah Collection.

## REFERENCES CITED

Adaro, L. de, 1914, Emplazamiento de sondeos para investigar la probable prolongación de los senos hulleros por bajo de los terrenos mesozoicos: Instituto Geológico de España, Boletín, t. 34, p. 9–79, mapa y cortes.

Amerom, H. W. J. van, 1975, Die eusphenopteridischen Pteridophyllen aus der Sammlung des Geologischen Bureaus in Heerlen, unter besonderer Berücksichtigung ihrer Stratigraphie bezüglich des südlimburger Kohlenreviers: Rijks Geologische Dienst, Mededelingen, C-III-1-No. 7, 101 p., Tafn I–XLVIII.

Amerom, H. W. J. van, 1984, Sijbren Jan Dijkstra, 27 February 1906–28 April

1982, Obituary notice, *in* S. J. Dijkstra Memorial Volume: Rijks Geologische Dienst, Mededelingen, v. 37, p. 5–9.

Amerom, H. W. J. van, 1991, Korte geschiedenis van de paleobotanie, koolpetrografie en stratigrafie van het Nederlandse Karboon en het Geologisch Bureau te Heerlen sedert 1906: Natuurhistorisch Maandblad, v. 80, p. 151–156.

Boersma, M., 1972, The heterogeneity of the form genus *Mariopteris* Zeiller. A comparative morphological study with special reference to the frond composition of West-European species [dissertation]: Utrecht, University of Utrecht, privately printed, 172 p., 43 pls.

Bonet, M. C., and Dijkstra, S. J., 1956, Megasporas carboníferas de La Camocha (Gijón): Estudios Geológicos, t. XII, nos. 31–32, p. 245–266, láms XLVIII–LVII.

Boureau, E., and Jongmans, W. J., 1955, *Novoguineoxylon lacunosum* n. gen., n. sp. Bois fossile de cycadophyte de la Nouvelle-Guinée hollandaise: Revue Générale Botanique, t. 62, p. 720–734, pls. L–LII.

Darrah, W. C., 1969, A critical review of the Upper Pennsylvanian floras of eastern United States with notes on the Mazon Creek flora of Illinois: Gettysburg, Pennsylvania, privately printed, 220 p., 80 pls.

Dijkstra, S. J., 1946, Eine monographische Bearbeitung der karbonischen Megasporen. Mit besonderer Berücksichtigung von Südlimburg (Niederlande) (unter Mitarbeitung von Jhr P. H. van Vierssen Trip): Geologische Stichting, Mededeelingen, C-III-1-No. 1, 101 p., Tafn 1–16.

Dijkstra, S. J., 1955, Megasporas carboníferas españolas y su empleo en la correlación estratigráfica: Estudios Geológicos, t. XI, nos. 27–28, p. 277–354, láms XXXV–XLV.

Heer, O., 1877, Flora fossilis helvetiae. Die Vorweltliche Flora der Schweiz: Zürich, Verlag von J. Wurster & Comp., 182 p., Tafn I–LXX.

Heide, S. van der, 1943a, Les lamellibranches limniques du Terrain houiller du Limbourg du Sud (Pays Bas): Geologische Stichting, Mededeelingen. C-IV-3-No. 1, 94 p., pls. 1–6.

Heide, S. van der, 1943b, La faune ichthyologique du Carbonifère supérieur des Pays-Bas: Geologische Stichting, Mededeelingen, C-IV-3-No. 2, 65 p., pls. 1–4.

Heide, S. van der, 1946, Stratigraphie et paléontologie animale du Terrain houiller du Peel: Geologische Stichting, Mededeelingen, C-IV-3-No. 4, 98 p., pl. 1.

Heide, S van der, 1951, Les Arthropodes du Terrain houiller du Limbourg meridional (excepté les Scorpions et les Insectes): Geologische Stichting, Mededeelingen, C-IV-3-No. 5, 84 p., pls. 1–10.

Jongmans, W. J., 1910, Die palaeobotanische Literatur. Bibliographische Übersicht über die Arbeiten aus dem Gebiete der Palaeobotanik. I. Die Erscheinungen des Jahres 1908: Jena, Verlag Gustav Fischer, 217 p.

Jongmans, W. J., 1911a, Die palaeobotanische Literatur. Bibliographische Übersicht über die Arbeiten aus dem Gebiete der Palaeobotanik. II. Die Erscheinungen des Jahres 1909: Jena, Verlag Gustav Fischer, 417 p.

Jongmans, W. J., 1911b, Anleitung zur Bestimmung der Karbonpflanzen West-Europas, mit besonderer Berücksichtigung der in den Niederlanden und den benachbarten Ländern gefundenen oder noch zu erwartenden Arten. I: Thallophytae, Equisetales, Sphenophyllales: Rijksopsporing van Delfstoffen, Mededeelingen, v. 3, 482 p., 390 figs.

Jongmans, W. J., 1913a, Die palaeobotanische Literatur. Bibliographische Übersicht über die Arbeiten aus dem Gebiete der Palaeobotanik. Die Erscheinungen der Jahre 1910 and 1911: Jena, Verlag Gustav Fischer, 569 p.

Jongmans, W. J., 1913b, Fossilium Catalogus. II: Plantae. Pars 1: Lycopodiales I: Berlin, W. Junk, 52 p.

Jongmans, W. J., 1915, Paläobotanisch-stratigraphische Studien im Niederländischen Carbon nebst Vergleichen mit umliegenden Gebieten. Mit Anhang (W. J. Jongmans & W. Gothan): Bemerkungen über einige der in den niederländischen Bohrungen gefundenen Pflanzen: Archiv für Lagerstättenforschung, Heft 18, 186 p., Tafn I–VI.

Jongmans, W. J., 1928, Stratigraphie van het Karboon in het algemeen en van Limburg in het bijzonder; Mijnbouwkundige Vereeniging Delft, Jaarboek, 1926–1927, 50 p. (The same range chart appeared in the Compte Rendu of the first Carboniferous Congress in Heerlen in the same year, but without the documentation of the plates).

Jongmans, W. J., 1937, Contribution to a comparison between the Carboniferous floras of the United States and of Western Europe (with the collaboration of W. Gothan and a postscriptum by W. van Waterschoot van der Gracht): Deuxième congrès pour l'avancement des études de stratigraphie carbonifère, Heerlen, septembre 1935: Compte Rendu, t. I, p. 363–392.

Jongmans, W. J. (with the collaboration of W. Gothan and some additional notes by W. C. Darrah), 1937, Comparison of the floral succession in the Carboniferous of West Virginia with Europe: Deuxième congrès pour l'avancement des études de stratigraphie carbonifère, Heerlen, septembre 1935: Compte Rendu, t. I, p. 393–415, pls. 11–36.

Jongmans, W. J., 1938a, Een en ander over kolenpetrographie: Mijnbouwkundige Vereeniging Delft, Lustrumjaarboek 1936–1937, 17 p.

Jongmans, W. J., 1938b, Die Flora des "Stangalpe" Gebietes in Steiermark: Deuxième congrès pour l'avancement des études de stratigraphie carbonifère, Heerlen septembre 1935: Compte Rendu, t. III, p. 1259–1298, pls. 114–145.

Jongmans, W. J., 1939, Die Kohlenbecken des Karbons und Perms im USSR und Ost-Asien: Geologisch Bureau Heerlen, Jaarverslag over 1934–1937, p. 15–192, Tafn. I–XL.

Jongmans, W. J. (unter Mitwirkung von E. Hofmann, W. Senarclens Grancy und R. Koopmans), 1939, Beiträge zur Kenntnis der Karbonflora in den östlichen Teilen des Anatolischen Kohlenbeckens: "METEAE," Institut für Lagerstaettenforschung Türkei, Veröffentlichung, Ankara, Serie B: Abhandlungen, no. 2, 40 p., Tafn. I–XIV.

Jongmans, W. J., 1940a, Beiträge zur Kenntnis der Karbonflora von Niederländisch Neu-Guinea: Geologisch Bureau Heerlen, Mededeelingen behorende bij het Jaarverslag over 1938–1939, p. 263–273, Tafn. I–III.

Jongmans, W. J., 1940b, Contribution to the flora of the Carboniferous of Egypt: Geologisch Bureau Heerlen, Mededeelingen behorende bij het Jaarverslag over 1938–1939, p. 223–230, pls. I–IV.

Jongmans, W. J., 1940c, Die Kohlenfelder von Gross Britannien: Geologisch Bureau Heerlen, Mededeelingen behorende bij het Jaarverslag over 1938–1939, p. 15–222.

Jongmans, W. J., 1941, Elementen der *Glossopteris*-flora in het Carboon van Nieuw-Guinea: Handelingen 28ste Nederlandse Congres Natuur-en Geneeskunde, p. 267–271.

Jongmans, W. J., 1942, Das Alter der Karbon- und Permflora von Ost-Europa bis Ost-Asien: Palaeontographica, Abt. B, Bd. 87, 58 p.

Jongmans, W. J., 1949a, El problema de la sincronización en el terreno hullero, y los métodos que pueden emplearse: Universidad de Granada, Boletín, p. 179–186.

Jongmans, W. J., 1949b, Note préliminaire sur la flore du Val d'Infierno: Instituto Geológico y Minero de España, Notas y Comunicaciones, no. 19, p. 3–7.

Jongmans, W. J., 1949c, Het wisselend aspect van het bos in de oudere geologische formaties, *in* Boerhave Beekman, W., Hout in alle tijden, Deel I, p. 1–164, 130 figs.

Jongmans, W. J., 1950, Note sur la flore du Carbonifère du versant sud du Haut Atlas: Service Géologique du Maroc, Notes et Mémoires, 7b, B.—Paléontologie, p. 155–172, pls. I-X.

Jongmans, W. J., 1951a, The female fructification of *Sphenopteris hoeninghausi* and the (supposed) relation of this species with *Crossotheca:* The Palaeobotanist, v. 1 (Birbal Sahni Memorial Volume), p. 267–276, pls. 1–21.

Jongmans, W. J., 1951b, Fossil plants of the island of Bintan: Koninklijke Akademie van Wetenschappen, Proceedings, B, v. LIV, p. 183–190.

Jongmans, W. J., 1951c, Las floras carboníferas de España: Estudios Geológicos, t. VII, no. 14, p. 281–330.

Jongmans, W. J., 1952a, Some problems on Carboniferous stratigraphy: Troisième congrès pour l'avancement des études de Stratigraphie et de Géologie du Carbonifère, Heerlen 25–30 juin 1951: Compte Rendu, t. I, p. 295–306.

Jongmans, W. J., 1952b, Note sur la flore du terrain carbonifère de Djerada

(Maroc oriental): Service Géologique du Maroc, Notes et Mémoires, t. 91, 27 p., pls. I–XXI.

Jongmans, W. J., 1952c, Documentación sobre las floras hulleras españolas. Primera contribución: Flora carbonífera de Asturias: Estudios Geológicos, t. VIII, no. 15, p. 7–19, láms. II–XXVIII.

Jongmans, W. J., 1954a, Coal research in Europe: Second Conference on the Origin and Constitution of Coal, Crystal Cliffs, Nova Scotia, June 1952: Halifax, Nova Scotia Department of Mines, p. 3–28.

Jongmans, W. J., 1954b, Contribution to the knowledge of the flora of the seam Girondelle (lower part of the Westphalian A), part I: Geologische Stichting, Mededelingen, C-III-1-No. 4, 16 p., pls. 1–9.

Jongmans, W. J., 1954c, The Carboniferous flora of Peru: British Museum (Natural History), Bulletin, Geology, v. 2, p. 189–224, pls. 17–26.

Jongmans, W. J., 1954d, Vitrinite and the difference between euvitrinite or collinite and the telinite: International Committee for Coal Petrology, Proceedings, no. 1, p. 31, pl. XII.

Jongmans, W. J., 1955a, Quelques remarques sur la présence de forêts fossiles dans le Carbonifère du Limbourg néerlandais, *in* Au Chanoine Félix Demanet en hommage, Bruxelles, Publication Association Etudes Paléontologiques, 21 (hors série), p. 65–69, Figs. 1–3.

Jongmans, W. J., 1955b, Notes paléobotaniques sur les Bassins houillers de l'Anatolie: Geologische Stichting, Mededelingen (Nieuwe Serie), v. 9, p. 55–89, 1 carte.

Jongmans, W. J., 1960 (posthumous), Die Karbonflora der Schweiz: Beiträge Geologische Karte der Schweiz, Bern, (Neue Folge), Bd 108, 97 p., Tafn. 1–58.

Jongmans, W. J., and Deleau, P. C., 1951, Les bassins houillers du Sud-Oranais. Livre II: Contribution à l'étude paléontologique: Service de la Carte géologique de l'Algérie, Bulletin, Série 1, Paléontologie, no. 13, 48 p., pls. I–XVI (flore), pls. I–V (faune).

Jongmans, W. J., and Gothan, W., 1925, Beiträge zur Kenntnis der Flora des Oberkarbons von Sumatra: Gedenkboek Verbeek, Geologisch-Mijnbouwkundig Genootschap Nederland en Koloniën, Verhandelingen, Geologische Serie, VIII, p. 279–303, Tafn. I–V.

Jongmans, W. J., and Gothan, W., 1934, Florenfolge und vergleichende Stratigrafie des Karbons der östlichen Staaten Nord-Amerika's. Vergleich mit West-Europa: Geologisch Bureau Heerlen, Jaarverslag over 1933, p. 17–44.

Jongmans, W. J., and Gothan, W., 1935a, Permo-karbonische Flora auf Sumatra, Niederl. Indien: Pontificiae Academiae Scientiarum Novi Lyncaei (Roma), Acta, Anno LXXXVIII-II Sess., p. 54–63.

Jongmans, W. J., and Gothan, W., 1935b, Die paläobotanischen Ergebnisse der Djambi-Expedition 1925: Mijnwezen Nederlandsch-Indië, Jaarboek 1930, Verhandelingen, Batavia, p. 71–201, Tafn. 1–58.

Jongmans, W. J., and Gothan, W., 1952, Contribução para o conhecimiento de Alethopteris branneri White: Rio de Janeiro, Departamento Nacional da Produção Mineral, Divisão de Geologia e Mineralogia, Notas preliminares e Estudos, no. 55, p. 1–9, Est. 1–3.

Jongmans, W. J., and Heide, S. van der, 1955, Flore et faune du Carbonifère inférieur de l'Egypte: Geologische Stichting, Mededelingen (Nieuwe Serie), v. 8, p. 59–75, pls. 7–17.

Jongmans, W. J., and Koopmans, R. G., 1934, Kohlenpetrographische Nomenklatur: Geologisch Bureau Heerlen, Jaarverslag over 1933, p. 49–63.

Jongmans, W. J., and Kukuk, P., 1913, Die Calamariaceen des rheinisch-westfälischen Kohlenbeckens: Rijks Herbarium, Mededeelingen, Leiden, 20, 89 p. + Atlas (Tafn. 1–22).

Jongmans, W. J., and Meléndez, B., 1950, El hullero inferior de Valdeinfierno (Córdoba): Estudios Geológicos, t. VI, no. 12, p. 191–210, láms XLIII–L.

Jongmans, W. J., and Meléndez, B., 1956, Contribución al conocimiento de la flora carbonífera del SO. de España: Estudios Geológicos, t. XII, no. 29–30, p. 19–58, láms VIII–XXIV.

Jongmans, W. J., and Wagner, R. H., 1957, Apuntes para el estudio geológico de la Zona Hullera de Riosa (Cuenca Central de Asturias): Estudios Geológicos, t. XIII, no. 33, p. 7–26, lám II.

Jongmans, W. J., Delépine, G., Gothan, W., Pruvost, P., Rummelen, F. H. van, and Voogd, N. de, 1925, Geologische en palaeontologische beschrijving van het Karboon der omgeving van Epen (Limb.): Natuurhistorisch Maandblad, v. 14, p. 55–83, pl. 1–14 (reprinted as: Mededeeling 1, Geologisch Bureau voor het Nederlandsche Mijngebied).

Jongmans, W. J., Koopmans, R. G., and Roos, G., 1936, Nomenclature of coal and petrography: Fuel in Science and Practice, v. 15, p. 14–15.

Jongmans, W. J., Gothan, W., and Darrah, W. C., 1937, Beiträge zur Kenntnis der Flora der Pocono-Schichten aus Pennsylvanien und Virginia: Deuxième congrès pour l'avancement des études de stratigraphie carbonifère, Heerlen, septembre 1935: Compte Rendu, t. I, p. 423–444, pls. 43–58.

Kerey, I. E., Kelling, G., and Wagner, R. H., 1986, An outline stratigraphy and palaeobotanical records from the middle Carboniferous rocks of northwestern Turkey: Société Geólogique du Nord, Annales, t. CV, p. 203–216, pls. VII–XI.

Kidston, R., and Jongmans, W. J., 1911, Sur la fructification de *Neuropteris obliqua* Brgt.: Archives Néerlandaises Sciences Exactes et Naturelles, Série III B, v. I, p. 25–26.

Kidston, R., and Jongmans, W. J., 1915–1917, A monograph of the Calamites of western Europe: Rijksopsporing van Delfstoffen, Mededeelingen, 7, 207 p. (1917), pls. 1–158 (1915).

Neves, R., 1964, The stratigraphic significance of the small spore assemblages of the La Camocha Mine, Gijón, Spain. Cinquième Congrès International de Stratigraphie et de Geólogie du Carbonifère, Paris, 9–12 septembre 1963: Compte Rendu, t. III, p. 1229–1238, pls. I–III.

Patac, I., 1933, La Cuenca Carbonífera de Gijón: Riquezas minerales de España: Oviedo, Spain (privately printed), 15 p.

Posthumus, O., 1927, Some remarks concerning the Palaeozoic Flora of Djambi, Sumatra: Koninklijke Akademie van Wetenschappen, Proceedings, vol. XXX, p. 628–634.

Potonié, H., 1909, Die Tropen-Sumpfflachmoor-Natur der Moore des Produktiven Carbons: Königliche Preussische Geologische Landesanstalt, Jahrbuch, Bd. XXX, Teil I, Heft 3, p. 389–443.

Ruiz Falcó, M., and Madariaga Rojo, R., 1931, 1933, Vegetales fósiles del Carbonífero español: Instituto Geológico y Minero de España, Boletín, t. LII, p. 119–143, láms I-V (1931); t. LIII, p. 67–89, láms VI–IX.

Thiadens, A. A., 1957, In memoriam W. J. Jongmans: Geologie en Mijnbouw (Nieuwe Serie), Jaargang 19, p. 417–425.

Visser, W. A., and Hermes, J. J., 1962, Geological results of the exploration for oil in Netherlands New Guinea: Koninklijk Nederlands geologisch mijnbouwkundig genootschap, Verhandelingen, Geologische Serie, XX (speciaal nummer), 265 p., enclosures.

Wagner, R. H., 1959, Flora fósil y estratigrafía del Carbonífero de España NW. y Portugal N.: Estudios Geológicos, t. XV, p. 393–420.

Wagner, R. H., 1961, Some Alethopterideae from the South Limburg Coalfield: Geologische Stichting, Mededelingen (Nieuwe Serie), v. 14, p. 5-13, pls. 1–8.

Wagner, R. H., 1968, Upper Westphalian and Stephanian species of Alethopteris from Europe, Asia Minor and North America: Rijks Geologische Dienst, Mededelingen, C-III-1-No. 6, 188 p., pls. 1–64.

Zeiller, R., 1899/1900, Etude sur la flore fossile du Bassin houiller d'Héraclée (Asie Mineure): Société géologique de France, Mémoires, t. 21, p. 1–91, pls. I–VI.

Zwierzycki, J., and Posthumus, O., 1926, De palaeo-botanische Djambi-Expeditie, Mededeelingen uit de rapporten van Dr. Zwierzycki en Dr. Posthumus: Maatschappij ter bevordering van het Natuurkundig Onderzoek der Nederlandsche Koloniën, Bulletin 81, 16 p.

Manuscript Accepted by the Society July 6, 1994

Printed in U.S.A.

Geological Society of America
Memoir 185
1995

# *Walther Gothan (1879–1954): Botanist, geologist, and teacher*

**Winfried Remy and Renate Remy**
*Forschungsstelle für Paläobotanik, Westfälische Wilhelms-Universität Münster, Hindenburgplatz 57-59, D-48143 Münster, Germany*
**Wolfgang Hartung**
*Weidamm 4, D-26135 Oldenburg, Germany*

## ABSTRACT

**Walther Gothan was one of the leading European paleobotanists between about 1910 and 1954. He was a leader in establishing a floral biostratigraphy of the late Paleozoic. He did not visit North America, but his cooperation with W. J. Jongmans and W. C. Darrah considerably influenced the interpretation of late Paleozoic floras of North America. Gothan published more than 350 papers, including a number of excellent contributions intended for the general public.**

## PERSONAL DATA

Walther Gothan was born on August 16, 1879, in Woldeck (Mecklenburg/Strelitz, Germany). He was the son of a stove manufacturer and potter, Hermann Gothan. Walther attended grammar school in Neu-Strelitz, Doberan, and Goslar/Harz. After he had finished school, he worked one year as a volunteer in the mining industry in Clausthal. Then he started his studies in mining and geology at the mining academies (today technical universities) of Clausthal and Berlin and in botany at the Mining Academy of Berlin and the Universities of Berlin and Jena. His teachers in botany were Henry Potonié, Paul Ascherson, Paul Graebner (senior), Simon Schwendener, and Adolf Engler. Gothan's interest in the anatomy of fossil and modern wood resulted in a doctoral thesis in botany on living and fossil gymnospermous wood (1905), a subject suggested by H. Potonié. Gothan received his doctorate in 1904 from the University of Jena, where he took geology, chemistry, and philosophy as subsidiary subjects.

As a graduate student in 1902, Gothan started to work as a volunteer at the Prussian Geological Survey (Preussische Geologische Landesanstalt), where H. Potonié was director of the paleobotany division. In 1903 he got a temporary position at the Survey; however, the salary was so low that he was forced to seek other jobs in order to finance his graduate studies. From 1903 to 1913 he worked on a series of systematic revisions of fossil plant taxa, mostly from the Carboniferous and Permian (Gothan, 1904–1913; Potonié and Gothan, 1909), prepared in addition to his Ph.D. research project.

In 1908, Gothan completed his "Habilitation" (the second thesis required for a university teaching position) on coal resources and paleobotany at the Mining Academy of Berlin. In that same year he became assistant at the Paleobotany Department of the Prussian Geological Survey in Berlin. Five years later, he was appointed curator. Gothan published a number of papers on wood and wood anatomy that caught the attention of botanists and paleobotanists. A. G. Nathorst asked him to study fossil wood from Spitsbergen (Gothan, 1906, 1907, 1910). In 1906 Gothan visited England and Hungary to conduct paleobotanical studies. His contacts with the North American wood anatomist I. W. Bailey of Harvard University began in 1910. Voigt and Hochgesang in Göttingen sold sets of thin sections of fossil and living woods collected and identified by Gothan with four-page explanatory notes ("Erläuterungen zu der Sammlung von Dünnschliffen lebender und fossiler Hölzer").

Gothan spent World War I as an enlisted man in the Army, first as an infantryman and then as an Army geologist. He was too old to serve in World War II, so he remained in the Geological Survey. In 1943 he arranged to take a Russian pa-

Remy, W., Remy, R., and Hartung, W., 1995, Walther Gothan (1879–1954): Botanist, geologist, and teacher, *in* Lyons, P. C., Morey, E. D., and Wagner, R. H., eds., Historical Perspective of Early Twentieth Century Carboniferous Paleobotany in North America (W. C. Darrah volume): Boulder, Colorado, Geological Society of America Memoir 185.

leobotanist, M. D. Zalessky, out of a prisoner of war camp and allowed Zalessky to do paleobotanical work in Gothan's laboratory in Berlin.

## RESEARCH CAREER

After H. Potonié's death in 1913, Gothan started teaching paleobotany at the Mining Academy of Berlin (1913) and at the Technical University of Berlin (1914). He was the successor of E. Weiss and H. Potonié. He became an adjunct professor at the Prussian Geological Survey in 1919 and slightly later at the Technical University in Berlin. The University of Berlin appointed him as an honorary professor in paleobotany in 1927. He was formally appointed as state geologist at the Prussian Geological Survey in 1929, where he became director of the paleobotany and coal division in 1935. In 1929 Gothan established the Institut für Paläobotanik und Petrographie der Brennsteine, which was a division of the Prussian Geological Survey. Soon after its foundation this institute attracted visiting scientists from various countries, including H. C. Sze and B. Sahni. Several interesting papers of their collaboration were published (Gothan and Sze, 1930, 1931, 1933a, b; Gothan and Sahni, 1937).

After World War II, which interrupted scientific studies in Europe, the research of the Prussian Survey largely concentrated on applied geology. In 1948 dissensions arose with the directors of the strongly political successor organization (Reichsamt für Bodenforschung) of the former geological survey. Gothan, who was then almost 70 years old, retired from the Survey in 1948 after nearly five decades. He remained active as an internationally recognized scientist. He wanted to continue teaching and doing research because paleobotany and coal geology had to be built up again in the postwar years, so in 1948, he transferred to the Department of Geology and Palaeontology of Humboldt University in Berlin, where W. Remy became his teaching assistant.

The former Prussian Geological Survey had very large paleobotanical collections, which had been assembled since Weiss's days and included many type and figured specimens. Although the collections survived the war without much damage, they were seriously threatened. Immediately after the war, space for offices and collections was very scarce because the survey buildings had largely been destroyed. Therefore it was suggested that the collections be dismantled, the various parts to be donated to schools and the rest thrown away. However, when Gothan became a titular member of the German Academy of Sciences in Berlin in 1949, he was in a position to save these paleobotanical collections. In 1951, he founded the Forschungsstelle für Paläobotanik und Kohlenkunde within the Academy of Sciences. He led this research institute until his death on December 30, 1954. He was succeeded in this position by W. Remy in 1955, when his former assistant received his Habilitation.

## GOTHAN'S SCIENTIFIC CONTRIBUTIONS

Gothan's training in geology and mining led him toward applied paleobotany. He followed his predecessors, H. Potonié, who worked in Thuringia and Silesia, and E. Weiss, who was the first geologist who applied Carboniferous and Permian plants from the Saar Basin to biostratigraphy. The studies of R. Zeiller firmly established a floral biostratigraphy for the French coal basins. R. Kidston did the same for the English coalfields. Moreover, as a state geologist Gothan had to cover the stratigraphic aspects from Devonian to Quaternary inclusive. Gothan commenced the systematic work on the coal floras of Silesia (Gothan, 1913a, b).

Like his mentor, H. Potonié, Gothan combined the study of living plants with paleobotany right from the beginning of his scientific work. He used systematics, anatomy, and phytosociology for a better appreciation of fossil floras as living entities, as is shown by the "Vegetationsbilder der Jetzt- und Vorzeit," a series of posters published by Potonié and Gothan (1906, 1912). He applied phytosociological concepts originally established for living plants to fossil floras. Gothan paid special attention to paleophytosociological, paleohydrographical, and ecological aspects of ancient floras. Paleophytogeographical problems were also addressed. Gothan recognized this as an important avenue in paleobotany. He was well aware of the great similarities between Carboniferous floras in Europe and North America, but he also pointed out differences in local geographical and stratigraphical occurrences of individual species. As early as 1913 he noticed the extinction of characteristic, accessory, and endemic species (Gothan, 1913a, p. 248 ff.; 1913b). Very early in his long scientific career he realized that fossil plants could reflect paleofloristical, biostratigraphical, paleohydrographical, paleoecological, and paleoclimatological conditions (Gothan, 1908, 1915a, 1930a, b, 1935b, 1937b, 1951a, b, 1954).

Gothan's studies of the Carboniferous floras of Silesia (1913a, b) showed that they were very similar to the fossil flora of the Zonguldak Basin, near the Black Sea coast, described by Zeiller (1899) as the flora of Héraclée (Ereğli, Northwest Anatolia, Turkey). He realized that paleofloristical and stratigraphical comparisons with other regions in Europe should be made. Therefore, in 1912 Gothan visited many museums and localities in France, Belgium, The Netherlands, England, and Scotland. During his travels abroad he discussed the establishment of a uniformly applicable subdivision and interregional correlation of the Carboniferous, following the leadership of R. Zeiller and R. Kidston. He also became friendly with P. Bertrand (see Chapter 7) and A. Renier. Later these contacts would become important in organizing the Carboniferous Congresses in Heerlen, which he helped to organize with W. J. Jongmans and A. Renier. As early as 1913 Gothan reported on his cooperation with his European colleagues in problems concerning interregional correlations of the Carboniferous (1913a, p. 239).

Around 1913 W. J. Jongmans (see Chapter 5) visited Gothan in Berlin. This would result in a long and warm friendship and in a fruitful cooperation (Jongmans and Gothan, 1915, 1925, 1934, 1935a, b, 1937a, b, 1938, 1951; Jongmans et al., 1925, 1935, 1937a, b; Gothan and Jongmans, 1952). Gothan also extended his trans-Atlantic contacts and started a correspondence with A. C. Noé (Chicago) and W. C. Darrah (Harvard), who both worked on paleofloristics and biostratigraphy.

P. Bertrand was another active member of the European paleobotanical community who was a close friend of Gothan for many years. R. Remy and W. Remy realized that close relationship when they met Bertrand's widow, Cecile Bertrand, at the 5th International Carboniferous Congress in Paris (1963). After she heard from one of our French colleagues that we were both students of Gothan, she immediately gave us all the still available reprints of Bertrand's publications.

Gothan's esteem for and his warm friendship with H. Weyland should be particularly mentioned. He highly valued Weyland's studies on fossil cuticles and his work on the Devonian flora of Elberfeld and appreciated Weyland's modest attitude. Three editions of the "Lehrbuch der Paläobotanik" were published, of which the second and third were edited by Weyland after Gothan's death (Gothan and Weyland, 1954; revised editions 1964, 1973).

Gothan was among the first scientists in Germany to recognize the practical importance of paleobotany for the coal-mining industry. He had very good connections with mining engineers, and therefore he had easy access to many mines where he could collect and study fossil plants and discuss paleobotanical problems with mining geologists. Correlations of coal seams in the Ruhr mining district and comparisons of floras and plant associations from different coal basins in Germany and other regions indicated that plant fossils could be used for regional and interregional correlations, which was something new. In those days, German coal-mining and field geologists correlated coal seams, coal-bearing sequences, and even complete coal basins solely on the basis of coalification data. Gothan could clearly demonstrate that paleobotany offered much better correlations; for example, for the Carboniferous deposits of the Piesberg and Ibbenbüren in Westphalia (Gothan and Haack, 1924a, b; Gothan, 1925). The Piesberg deposits were generally regarded as Namurian, or perhaps earliest Westphalian on the basis of their rank. However, plant fossils definitely indicated a late Westphalian age.

Gothan recognized that plants and plant associations could be used to characterize coal-forming basinal and extrabasinal environments (Gothan, 1913a, b, 1926, 1927, 1928b, 1929, 1931a, b, 1935a, 1937a, c, e, 1938, 1941, 1949, 1952a, 1953; Gothan and Nagalhard, 1922; Gothan and Schlosser, 1924; Gothan and Gimm, 1930, 1952; Picard and Gothan, 1931; Friese and Gothan, 1952; Gothan et al., 1959). He also produced a volume on Carboniferous-Permian index fossils (Gothan, 1923). With his vast knowledge and broad experience of late paleozoic floras from all over Europe, Gothan became one of the leading paleobotanists of his time.

Gothan recognized similar floras from the work of David White (see Chapter 10) and others, leading Gothan to believe that similar plant associations characterized coal basins in North America. Gothan's long-standing wish to establish an interregional and internationally applicable stratigraphy of the Carboniferous and Lower Permian could finally be realized after World War I. Gothan, Jongmans, and Renier organized the First Carboniferous Congress, which was held in 1927 in Heerlen, The Netherlands (Gothan, 1928a). During this congress Gothan proposed an international field trip to the Ruhr and Aachen mining districts, where he wished to show in the field the newly established subdivisions of the Carboniferous. Several famous scientists, including W. N. Edwards, B. Sahni, E. Dix, V. Šusta, K. Oberste-Brink, and P. Kukuk, participated in this excursion, which was held in 1935. Three other Carboniferous Congresses were organized by Jongmans and others in Heerlen (1935, 1951, 1958), and from 1963 onward these International Congresses on Carboniferous Stratigraphy and Geology, which have become an institution, have been held in different places all over the world. It is a pleasure to read the organizers' reports on the aims and what was achieved at the 1927 and 1935 Heerlen Congresses. They show that paleobotany flourished during this period (see also Gothan, 1952b).

Gothan was not only one of the conveners of the first three Carboniferous Congresses but was also actively involved in the 5th and 6th International Botanical Congresses held in Cambridge (1930) and Amsterdam (1935). Paleobotanical excursions were organized in August 1930 in England. Together with T. G. Halle, another personal friend, Gothan served on the committee on nomenclature (Jongmans et al., 1935; Halle, 1951).

Early in Gothan's career he also became interested in North American late Paleozoic floras and biostratigraphy. He wanted to incorporate the North American floras into the European floral biostratigraphy of the Carboniferous, but he did not like to rely exclusively on illustrations from the literature. However, Gothan did not have the opportunity to visit the United States. He corresponded with A. C. Noé (see Chapter 14) and he cooperated with Jongmans, who had collected Carboniferous plant fossils during a visit to the United States in 1933 (Jongmans and Gothan, 1934; Gothan, 1935b). In a letter to W. C. Darrah (1936), which was made available by his daughter, Elsie Darrah Morey, Gothan mentions the 1934 paper he had written with Jongmans (Jongmans and Gothan, 1934) as a basis for further discussion and encouraged Darrah to improve the concepts presented in this publication. He also reminded Darrah of an earlier planned collaboration. Darrah's stay in Europe resulted in two joint publications with Jongmans and Gothan (Jongmans and Gothan, 1937a, b). World War II abruptly interrupted this cooperation. Gothan had a good opinion of Darrah's work, as he assured W. Remy several times (Gothan, personal communication, 1952–1954).

Two problems concerning the comparison of European and North American late Paleozoic floras had Gothan's special attention. Gothan wanted to determine if, in fact, floristic elements characteristic of European late Paleozoic floras were definitely present or were absent in North America. Second, he wanted to resolve the question of the biostratigraphic position of beds attributed to the upper Westphalian by North American workers; their floristic composition more closely resembled ones attributed to the Stephanian or even Early Permian in Europe (Jongmans and Gothan, 1937a, b). Related to this was the question of whether some of the presumed endemic North American Carboniferous plant taxa were in fact synonymous with European taxa. This latter point was of particular importance for his biostratigraphical and phytogeographical interpretations (Gothan, 1937a, b, c). For similar reasons Gothan was greatly interested in T. G. Halle's investigations on the Shansi flora (Halle, 1927) and other floras in China. Together with his student H. C. Sze, Gothan worked and published between 1930 and 1933 on Chinese floras (Gothan and Sze, 1930, 1931, 1033a. b). He also established contacts with A. N. Kryshtofovich and M. D. Zalessky to determine the floral relationships in Russia and worked with Jongmans on the late Paleozoic flora of Sumatra (Jongmans and Gothan, 1935a).

The development of studies on late Carboniferous and Early Permian floras had Gothan's continuous interest. In his later years he interested W. Remy in Stephanian and Early Permian plants and floras, their biostratigraphical applications, and paleoecological implications.

In Germany, Gothan is usually remembered as a paleobotanically oriented stratigrapher. He indeed worked biostratigraphically on Devonian to Quaternary strata. However, he also carried out purely systematic studies. His paper on the genus *Thinnfeldia* (Gothan, 1912) and his systematic description of the Rhaeto-Liassic flora (Gothan, 1914) are important contributions; Gothan's findings were later confirmed by T. M. Harris's studies on floras from East Greenland. However, in Germany Gothan's systematic studies are less well known than his biostratigraphic work. Gothan had studied coal, coal-bearing strata, and the process of coal formation (Potonié 1910, 1920 [sixth revised edition by Gothan]). As for Tertiary floras, he was particularly interested in the recognition of of environmental and peat-forming conditions.

Gothan was able to separate main points from side issues by intuition, and he produced results very quickly. He always told his students that it was better to produce something, even when it was not 100% perfect, than to aim at perfection and then finally end up with nothing. Gothan was very straightforward and down to earth. His approach to paleobotanical work was different from that of Jongmans, with whom he collaborated so often. Jongmans tended to pay more attention to details and often doubted identifications in the literature. He frequently asked Gothan's opinion on identifications. Also, Gothan did not like purely theoretical studies, and he was very critical of speculation. Therefore, Gothan, although familiar with the pre-telome theories (H. Potonié, O. Lignier, or F. O. Bower), did not adhere to the telome theory of W. Zimmermann, which was mentioned only very briefly in his lectures. This was nothing personal against Zimmermann, but Gothan generally disliked theories.

Gothan had a good intuition for relationships that were still unclarified at the time, like the relationships among plants, climate, and environment (for example, paralic versus limnic) (Gothan, 1908, 1951a, b). He continued H. Potonié's work on fossil fuels (Kaustobiolithe), particularly coals. He revised Potonié's textbook on coal and Kaustobiolithe (Potonié, 1920; Gothan, 1937d) and placed much emphasis on microscopy and maceration techniques. Gothan regarded Kaustobiolithe as reflections of once-existing ecotopes, and he viewed coal in relationship to associated strata (Gothan, 1932).

## GOTHAN, THE TEACHER

Throughout his lectures, practicals, and excursions Gothan stimulated his students to be versatile. H. Bode, H. C. Sze, E. Stach, U. Horst, F. Thiergart, R. Daber, G. Roselt, W. Hartung, W. Remy, and R. Remy were all Gothan students; the last three became his research assistants. In the early 1930s, Sze worked together with Gothan in Berlin on Carboniferous, Permian, and Jurassic floras from China. During the same time, W. Hartung set up the paleobotanical exhibit at the Natural History Museum in Berlin. He was the first of Gothan's students who investigated fructifications and their spore contents (Hartung, 1933); W. Remy was the second Gothan student who worked on fructifications and in situ spores (Remy, 1953). R. Remy focused on the petrology of cannel and boghead coals and on maceration techniques. E. Stach's student, M. Teichmüller, who became an internationally renowned expert on coal petrology, was also influenced in paleobotany by Gothan when she was student at the University of Berlin (1935–1937) and later when she worked at the Reichsamt für Bodenforschung (formerly the Prussian Geological Survey).

As a university teacher, Gothan gave a classical type of introductory paleobotany course. Much emphasis was laid on morphology, anatomy, and systematics. By showing examples of fossil plants he interested his students in anatomically preserved material. He revised Potonié's textbook of paleobotany (Potonié, 1899, 1921 [second edition, revised by Gothan]). Gothan macerated and described cuticles of *Cyclopteris, Callipteris,* and neuropterids (1913a, 1915b, c), and he was one of the first who worked on late Paleozoic pteridospermous cuticles. He stimulated E. Hofmann (Vienna) to publish her book "Paläohistologie der Pflanze" (Hofmann, 1934).

Being a botanist, Gothan taught his students to see fossil plants as botanical entities as well as witnesses of the past. Fossil plants document continuously changing environments and paleogeography and can be used biostratigraphically and for paleoecological interpretations as well as for paleoclimatological interpretations. Gothan always regarded terrestrial ecosystems as an interactive complex of biotic and abiotic factors.

Although Gothan himself never worked on fossil microfloras, he recognized the potential danger of a too one-sided evaluation of paleobotanical material. He stimulated studies of in situ spores and pollen from fern and seed-fern fructifications (W. Hartung, R. Remy, and W. Remy).

Gothan's geology students had to identify living plants. Attention was given to the relation of plants to their substrate and to the possible use of plants in geological mapping. He even stimulated his students to work phytosociologically. He led several excursions devoted to modern plants and their relations to fossil ones: for example, "relicts of Pontic plants around Berlin and Frankfurt/Oder," "relicts of Arcto-Tertiary floras," "plants and soil," "arborescent plants during the winter season," "the arboretum of the Sanssouci Palace," "spring vegetation in the beech forest," "pond to mire floral successions (hydroseres)," "fresh-water flora," and "peat vegetation." These botanical excursions, and also his paleobotanical excursions to the brown-coal district on such topics as "brown coal vegetation" and "tree stump horizons in the brown coal," were always very stimulating and most enjoyable since he could recognize birds from their singing and sometimes would even play the flute for his students. Field trips, visits to examine collections, and descents to underground coal mines with him were memorable.

Gothan lectured without a preestablished written concept and sometimes followed spontaneous impulses. Therefore his lectures were mostly rather lively in presentation but could sometimes be difficult to follow. He also tended to present too much data. For young students, it was not always easy to recognize the importance of the many comparisons between living and fossil plants, their growth forms and life habits, and the relevant ecological principles.

Gothan stimulated his students to solve problems, taught them to think as botanists and as paleobotanists, and trained them to make careful observations. He regarded himself as both a botanist and a geologist; he was very proud of his early training as a mining geologist. Gothan managed to find the right balance between pure scientific research and practical applications. To his students and colleagues he made brief but instructive comments. He did not always expect a reply; if there was no response, he would let the subject rest, at least for the time being.

His special concern was also to make his scientific contributions available and above all understandable to the general public. With this intention he once mentioned that he even wrote a general introduction to paleobotany in a trench during World War I.

Gothan never imposed himself, either as a teacher or as a member of the scientific community. That does not mean that he was unaware of his qualities as a scientist. He knew that he and his predecessors E. Weiss and H. Potonié had established paleobotany in Berlin. R. Kräusel spoke about a "Berlin tradition" (see Horst, 1954, p. 499). However, according to Gothan, a tradition did not mean simply following a teacher's views and ideas. He perfectly understood that a teacher strongly influences his students, personally as well as scientifically. A tradition usually determines the further development of a scientific discipline and the specific topics that are emphasized in teaching and research. On the whole, Gothan was more of a researcher than a university teacher. This was probably due to a large extent to his position at the Prussian Geological Survey.

## DIVERSIONS

Gothan found relaxation in playing music, either with colleagues or with his wife and friends. He did not aim at virtuosity but played for pleasure. Gothan was a pipe smoker (Fig. 1); he also liked to drink a glass of white wine and enjoyed billiards. He had a good sense of humor, loved puns, and wrote many witty poems, for example, about his colleagues or field trips. He cultivated his Mecklenburg dialect and wrote letters, which were mostly very direct, as he was in conversation. He often dictated his manuscripts without having an elaborate concept; sometimes papers were printed without his having corrected either the original manuscript or the printer's proofs.

Figure 1. Walther Gothan in a winery at Leubsdorf on the Rhine. Photograph by W. Remy, May 1954.

## RECOGNITION

Gothan received many awards. In 1918, in appreciation of his work on fossil woods from Spitsbergen, a mountain ridge on Spitsbergen was named after him. The Gothankammen is situated south of the Van Keulenfjorden at 77°N, 20°15″E. (The place-names of Svalbard, 80: 159, Norsk Polarinstitutt, Oslo, 1942, n. ed. 1991). In 1933 he was appointed an honorary research fellow of the Academia Sinica and a corresponding member of the Geological Society of China. He received the Leopold von Buch Medal (Deutsche Geologische Gesellschaft Hannover, 1948) and the Orville Derby Medal (Brazil, 1951). He was elected a member of the German Academy of Sciences in Berlin (1949). He was one of the honorary presidents of the 7th and 8th International Botanical Congresses (Stockholm, 1950; Paris, 1954) and also was honored with the Medal of the Société de Botanique, Paris (1954). In the same year he was elected an honorary member of the Paläontologische Gesellschaft in Frankfurt/Main. He was awarded posthumously (April 30, 1955) the Riddare av Kungliga Nordstjärneorden (Knight in the Order of the Swedish Royal North Star).

## ACKNOWLEDGMENTS

Several colleagues, including good friends and former students, published tributes and obituaries of Gothan. We especially want to mention that of his teacher's son, R. Potonié (1955), and those of T. G. Halle (1951), W. J. Jongmans (1955), H. Weyland (1956), H. N. Andrews (1980), and Remy and Remy (1986). We thank Hans Kerp for translating the manuscript into English and for a first review. Patricia G. Gensel, Hermann W. Pfefferkorn, Sergius H. Mamay, Paul C. Lyons, and Robert H. Wagner are gratefully acknowledged for their critic readings of the manuscript and for their helpful comments.

## REFERENCES CITED

Andrews, H. N., 1980, The fossil hunters, in search of ancient plants: Ithaca, New York, Cornell University Press, 421 p. [Gothan: p. 316–318].

Friese, H., and Gothan, W., 1952, Neue Beobachtungen über die Kohlenflora von Dobrilugk-Kirchhain: Geologie, v. 1, p. 6–27, pls. 1–6.

Gothan, W., 1904–1913, *in* Potonié, H., ed., Abbildungen und Beschreibungen fossiler Pflanzen-Reste: Berlin, Königlich Preussische Geologische Landesanstalt und Bergakademie, v. 2–9; 1904, *Rhizodendron oppoliense, Alsophilina* sp., v. 2, no. 31, 12 p.; 1906, *Desmopteris integra*, v. 4, no. 64, 4 p.; *Desmopteris serrata*, v. 4, no. 65, 3 p.; *Neuropteris crenulata*, v. 4, no. 66, 3 p.; *Neuropteris rectinervis*, v. 4, no. 67, 3 p.; *Neurodontopteris obliqua*, v. 4, no. 68, 13 p.; *Piceoxylon Pseudotsugae* als fossiles Holz, *Pseudotsuga* sp. (aff. *Douglasii*) als rezenter Baum, v. 4, no. 80, 5 p; 1907, *Callipteris* resp. *p-Callipteris*, v. 5, no. 84, 6 p.; *Callipteris conferta*, v. 5, no. 85, 18 p.; *Callipteris jutieri*, v. 5, no. 86, 4 p.; *Callipteris naumanni*, v. 5, no. 87, 5 p.; *Callipteris subauriculata*, v. 5, no. 88, 3 p.; *Callipteris oxydata*, v. 5, no. 89, 2 p.; *Callipteris bibractensis*, v. 5, no. 90, 2 p.; *Callipteris curretiensis*, v. 5, no. 91, 3 p.; *Callipteris lyratifolia*, v. 5, no. 92, 4 p.; *Callipteris strigosa*, v. 5, no. 93, 2 p.; *Callipteris flabellifera*, v. 5, no. 94, 4 p.; *Callipteris scheibei*, v. 5, no. 95, 2 p.; *Callipteris martinsi*, v. 5, no. 96, 4 p.; *Neuropteris schlehani*, v. 5, no. 100, 10 p.; 1909, *Lepidopteris*, v. 6, no. 109, 4 p.; *Lepidopteris ottonis*, v. 6, no. 110, 5 p.; *Lepidopteris stuttgardiensis*, v. 6, no. 111, 2 p.; *Callipteris moureti*, v. 6, no. 112, 3 p.; *Callipteris raymondi*, v. 6, no. 113, 3 p.; *Callipteris bergeroni*, v. 6, no. 114, 2 p.; *Callipteris pellati*, v. 6, no. 115, 2 p.; *Lonchopteris*, v. 6, no. 117, 7 p.; *Lonchopteris bricei*, v. 6, no. 118, 6 p.; *Lonchopteris rugosa*, v. 6, no. 119, 4 p.; *Neuropteris praedentata*, v. 6, no. 120, 3 p.; 1910, *Pecopteris aspera, Dactylotheca aspera*, v. 7, no. 121, 9 p.; *Callipteris lodevensis*, v. 7, no. 122, 3 p.; *Callipteris polymorpha*, v. 7, no. 123, 6 p.; *Callipteris nicklesi*, v. 7, no. 124, 4 p.; *Alethopteris valida*, v. 7, no. 125, 5 p.; *Weichselia reticulata*, v. 7, no. 126, 14 p.; *Lonchopteris silesiaca*, v. 7, no. 127, 5 p.; *Lonchopteris haliciensis*, v. 7, no. 128, 2 p.; *Lonchopteris bauri*, v. 7, no. 129, 5 p.; *Lonchopteris westfalica*, v. 7, no. 130, 3 p.; *Lonchopteris conjugata*, v. 7, no. 131, 5 p.; *Lonchopteris eschweileriana*, v. 7, no. 132, 3 p.; *Lonchopteris alethopteroides*, v. 7, no. 133, 2 p.; 1913, *Callipteridium gigas*, v. 9, no. 180, 5 p.

Gothan, W., 1905, Zur Anatomie lebender und fossiler Gymnospermen-Hölzer: Abhandlungen der Königlich Preussischen Geologischen Landesanstalt, Neue Folge, v. 44, p. 1–108, 10 pls.

Gothan, W., 1906, Fossile Hölzer aus dem Bathonien von Russisch-Polen: Verhandlungen der Kaiserlichen Russischen Mineralogischen Gesellschaft zu St. Petersburg, ser. 2, v. 44, p. 435–458.

Gothan, W., 1907, Über die Wandlungen der Hoftüpfelung bei den Gymnospermen im Laufe der geologischen Epochen und ihre physiologische Bedeutung: Sitzungsberichte der Gesellschaft naturforschender Freunde, v. 2, p. 13–26.

Gothan, W., 1908, Die Frage der Klimadifferenzierung im Jura und in der Kreideformation im Lichte paläobotanischer Tatsachen: Jahrbuch der Königlich Preussischen Geologischen Landesanstalt, v. 29, pt. 2, p. 220–242, pls. 16–19.

Gothan, W., 1910, Die fossilen Holzreste von Spitzbergen: Kungliga Svenska Vetenskapsakademiens Handlingar, v. 45, 56 p., 7 pls.

Gothan, W., 1912, Über die Gattung *Thinnfeldia* Ettingshausen: Abhandlungen der naturhistorischen Gesellschaft zu Nürnberg, v. 19, p. 67–80, pls. 13–16.

Gothan, W., 1913a, Die oberschlesische Steinkohlenflora. I: Farne und farnähnliche Gewächse (Cycadofilices bezw. Pteridospermen): Abhandlungen der Königlich Preussischen Geologischen Landesanstalt, N.F., v. 75, 279 p., 23 pls.

Gothan, W., 1913b, Das oberschlesische Steinkohlenbecken im Vergleich mit anderen Becken Mitteleuropas auf Grund der Steinkohlenflora: Glückauf, v. 49, p. 1366–1377.

Gothan, W., 1914, Die unterliassische ("rhätische") Flora der Umgebung von Nürnberg: Abhandlungen der naturhistorischen Gesellschaft zu Nürnberg, v. 19, p. 89–186, pls. 17–39.

Gothan, W., 1915a, Pflanzengeographisches aus der paläozoischen Flora mit Ausblicken auf die mesozoischen Folgefloren: Engler's Botanische Jahrbücher für Systematik und Pflanzengeographie, v. 52, p. 221–271.

Gothan, W., 1915b, Über die Epidermen einiger Neuropteriden des Carbons: Jahrbuch der Königlich Preussischen Geologischen Landesanstalt für 1914, v. 35, pt. 2, p. 373–381.

Gothan, W., 1915c, Über die Methoden und neue Erfolge bei der Untersuchung kohlig erhaltener Pflanzenreste: Sitzungsberichte der Gesellschaft naturforschender Freunde, Berlin, Jahrgang 1915, p. 43–48, pl. 2.

Gothan, W., 1923, Karbon und Perm-Pflanzen, *in* Gürich, G., ed., Leitfossilien, v. 3: Berlin, Gebrüder Borntraeger, 187 p., 45 pls.

Gothan, W., 1925, Ruhrkarbon und Osnabrücker Karbon: Glückauf, Jahrgang 61, p. 777–779.

Gothan, W., 1926, Gemeinsame Züge und Verschiedenheiten in den Profilen des Karbons der paralischen und limnischen (Binnen-)Kohlenbecken: Zeitschrift der Deutschen Geologischen Gesellschaft, v. 77, p. 391–403.

Gothan, W., 1927, Über einige Kulmpflanzen vom Kossberg bei Plauen i. V: Abhandlungen des Sächsischen Geologischen Landesamts, v. 5, p. 3–20,

7 pls.

Gothan, W., 1928a, Der Stand der Vergleichung der mitteleuropäischen Steinkohlenbecken und Vorschläge zur Vereinheitlichung: Congrès pour l'avancement des études de stratigraphie Carbonifère 1st, Heerlen, The Netherlands, 1927: Compte Rendu, p. 259–273.

Gothan, W., 1928b, Die limnischen Becken Deutschlands: Congrès pour l'avancement des études de stratigraphie Carbonifère, 1st, Heerlen, The Netherlands, 1927: Compte Rendu, p. 275–287.

Gothan, W., 1929, Die Steinkohlenflora der westlichen paralischen Carbonreviere Deutschlands: Arbeiten des Instituts für Paläobotanik und Petrographie der Brennsteine, v. 1, 48 p., 16 pls.

Gothan, W., 1930a, Die pflanzengeographischen Verhältnisse am Ende des Paläozoikums: Berichte der deutschen botanischen Gesellschaft, v. 48, p. 63–65.

Gothan, W., 1930b, Die pflanzengeographischen Verhältnisse am Ende des Paläozoikums: Engler's Botanische Jahrbücher für Systematik und Pflanzengeographie, v. 63, p. 350–367.

Gothan, W., 1931a, Die Steinkohlenflora der westlichen paralischen Carbonreviere Deutschlands: Arbeiten des Instituts für Paläobotanik und Petrographie der Brennsteine, v. 1, 47 p., 12 pls.

Gothan, W., 1931b, Der Wert der karbonischen und permischen Flora als Leitfossilien: Palaeontologische Zeitschrift, v. 13, p. 298–309.

Gothan, W., 1932, Coals and associated rock: Proceedings, Third Conference on Bituminous Coal: Pittsburgh, Carnegie Institute of Technology, v. 2, p. 834–837.

Gothan, W., 1935a, Die Steinkohlenflora der westlichen paralischen Steinkohlenreviere Deutschlands: Abhandlungen der Preussischen Geologischen Landesanstalt, Neue Folge, v. 167, 58 p., 20 pls.

Gothan, W., 1935b, Geobotanische Provinzen im Karbon und Perm: Forschungen und Fortschritte, Jahrgang 11, no. 34, p. 437–438.

Gothan, W., 1937a, Vergleich des ost- und westdeutschen Karbons: Congrès pour l'avancement des études de stratigraphie du Carbonifère, 2d, Heerlen, The Netherlands, 1935: Compte Rendu, v. 1, p. 219–224.

Gothan, W., 1937b, Geobotanische Provinzen im Karbon und Perm: Congrès pour l'avancement des études de stratigraphie du Carbonifère, 2d, Heerlen, The Netherlands, 1935: Compte Rendu, v. 1, p. 225–226.

Gothan, W., 1937c, Die Frage des Synchronismus der Perm- und Stephan-Floren und ihre Charakteristika: Congrès pour l'avancement des études de stratigraphie du Carbonifère, 2d, Heerlen, The Netherlands, 1935: Compte Rendu, v. 1, p. 213–217.

Gothan, W., 1937d, Kohle, *in* Beyschlag, F., Krusch, P., and Vogt, J. H. L., eds., Die Lagerstätten der nutzbaren Mineralien und Gesteine, v. 3: Stuttgart, Ferdinand Enke, 432 p.

Gothan, W., 1937e, Zwei interessante Funde von Rotliegendpflanzen in Thüringen: Jahrbuch der Preussischen Geologischen Landesanstalt für 1936, v. 57, p. 507–513, pls. 27–29.

Gothan, W., 1938, Die Bedeutung der Steinkohlenpflanzen für die Stratigraphie des Ruhrkarbons, *in* Kukuk, P., ed., Geologie des niederrheinisch-westfälischen Steinkohlengebietes: Berlin, Julius Springer, p. 141–154.

Gothan, W., 1941, Die Steinkohlenflora der westlichen paralischen Steinkohlenreviere Deutschlands: Abhandlungen der Reichsstelle für Bodenforschung, Neue Folge, v. 196, p. 6–54, 25 pls.

Gothan, W., 1949, Die Unterkarbon-Flora der Dobrilugker Tiefbohrungen: Abhandlungen der Geologischen Landesanstalt Berlin, Neue Folge, v. 217, 32 p., 6 pls.

Gothan, W., 1951a, Die merkwürdigen pflanzengeographischen Besonderheiten in den mitteleuropäischen Karbonfloren: Palaeontographica, Abt. B, v. 91, p. 109–130.

Gothan, W., 1951b, Pflanzengeographisches aus dem mitteleuropäischen Karbon: Geologie, v. 3, p. 219–257, 4 pls.

Gothan, W., 1952a, Die Unterscheidung des (oberen) Unterkarbons vom (unteren) Oberkarbon auf Grund der Pflanzenführung: The Palaeobotanist, v. 1 (Birbal Sahni Memorial Volume), p. 189–206.

Gothan, W., 1952b, Die Heerlener Karbonkongresse: Sitzungsberichte der Deutschen Akademie der Wissenschaften zu Berlin, Klasse für Mathematik und allgemeine Naturwissenschaften, Jahrgang 1952, no. 4, 18 p.

Gothan, W., 1953, Die Steinkohlenflora der westlichen paralischen Steinkohlenreviere Deutschlands. Lief. 5: Beihefte zum Geologischen Jahrbuch, v. 10, 83 p., 44 pls.

Gothan, W., 1954, Geobotanische Provinzen im Karbon und Perm: Forschungen und Fortschritte, v. 28, p. 38–40.

Gothan, W., and Gimm, O., 1930, Neuere Beobachtungen und Betrachtungen über die Flora des Rotliegenden von Thüringen: Arbeiten aus dem Institut für Paläobotanik und Petrographie der Brennsteine, v. 2, p. 39–74, pl. 9.

Gothan, W., and Gimm, O., 1952, Über die verkieselte Kohle des Manebacher Oberflözes: Sitzungsberichte der Deutschen Akademie der Wissenschaften zu Berlin, Klasse für Mathematik und allgemeine Naturwissenschaften, Jahrgang 1952, no. 2, p. 5–13, 1 pl.

Gothan, W., and Haack, W., 1924a, Bericht über die Tiefbohrung Ibbenbüren IV: Jahrbuch der Preussischen Geologischen Landesanstalt für 1923, v. 44, p. 25–27.

Gothan, W., and Haack, W., 1924b, Ruhrkarbon und Osnabrücker Karbon: Glückauf, Jahrgang 60, p. 535–541.

Gothan, W., and Jongmans, W. J., 1952, Contribuição para o conhecimento de *Alethopteris branneri* White: Ministério da Agricultura, Departamento Nacional da Produção mineral, Divisão de Geologia e Mineralogia, Notas preliminares e Estudos, v. 55, p. 1–9, 3 pls.

Gothan, W., and Nagalhard, K., 1922, Kupferschieferpflanzen aus dem niederrheinischen Zechstein: Jahrbuch der Preussischen Geologischen Landesanstalt für 1921, v. 42, p. 440–460, 3 pls.

Gothan, W., and Sahni, B., 1937, Fossil plants from the Po Series of Spiti (N.W. Himalayas): Records of the Geological Survey of India, v. 72, p. 195–206, pls. 16–18.

Gothan, W., and Schlosser, P., 1924, Neue Funde von Pflanzen der älteren Steinkohlenzeit (Kulm) auf dem Kossberge bei Plauen im Vogtland: Leipzig, Kommissionsverlag Max Weg, 13 p., 6 pls.

Gothan, W., and Sze, H. C., 1930, Zu Schenks Publikationen über die Ostasiatische Permokarbon-Flora: Memoirs of the National Research Institute of Geology, Academia Sinica, v. 9, p. 1–55, 1 pl.

Gothan, W., and Sze, H. C., 1931, Pflanzenreste aus dem Jura von Chinesisch Turkestan (Provinz Sinkiang): Contributions from The National Research Institute of Geology, Academica Sinica, v. 1, p. 33–38, pl. 1.

Gothan, W., and Sze, H. C., 1933a, Über die paläozoische Flora der Provinz Kiangsu: Memoir of the National Research Institute of Geology, Academia Sinica, v. 13, p. 1–40, pls. 1–4.

Gothan, W., and Sze, H. C., 1933b, Über "Mixoneura" und ihr Vorkommen in China: Memoirs of the National Research Institute of Geology, Academia Sinica, v. 13, p. 41–57, pl. 5.

Gothan, W., and Weyland, H., 1954, Lehrbuch der Paläobotanik: Berlin, Akademie-Verlag, 535 p. (revised editions 1964, 1973).

Gothan, W. (posthumous), Leggewie, W., and Schonefeld, W. (in cooperation with W. Remy), 1959, Die Steinkohlenflora der westlichen paralischen Steinkohlenreviere Deutschlands: Beihefte zum Geologischen Jahrbuch, v. 36, 90 p., 50 pls.

Halle, T. G., 1927, Palaeozoic plants from Central Shansi: Palaeontologia Sinica, ser. A, v. 2, 316 p., 64 pls.

Halle, T. G., 1951, Zum 70. Geburtstag Walther Gothan's: Palaeontographica, v. 91, Abt. B, p. 93–108.

Hartung, W., 1933, Die Sporenverhältnisse der Calamariaceen: Arbeiten aus dem Institut für Paläobotanik und Petrographie der Brennsteine, v. 3, p. 96–149, pls. 8–11.

Hofmann, E., 1934, Paläohistologie der Pflanze: Wien, Verlag von Julius Springer, 308 p.

Horst, U., 1954, Ein Leben für die Wissenschaft. Zum 75. Geburtstag von Prof. Dr. Walther Gothan: Geologie, v. 3, p. 492–501.

Jongmans, W. J., 1955, Ter herdenking van Prof. Dr. Walther Gothan 1879–1954: Geologie en Mijnbouw (n. ser.), v. 17, p. 88–89, 1 pl.

Jongmans, W. J., and Gothan, W., 1915, Bemerkungen über einige der in den niederländischen Bohrungen gefundenen Pflanzen: Königlich Preussische Geologische Landesanstalt, Archiv für Lagerstättenforschung, v. 18, p. 151–183, 1 pl.

Jongmans, W. J., and Gothan, W., 1925, Beiträge zur Kenntnis der Flora des Oberkarbons von Sumatra: Mededeeling no. 2 van het Geologische Bureau voor het Nederlandsche Mijngebied, Verhandelingen van het Geologisch-Mijnbouwkundig Genootschap voor Nederland en Koloniën, Geologische Serie, v. 8, p. 279–304, 5 pls.

Jongmans, W. J., and Gothan, W., 1934, Florenfolge und vergleichende Stratigrafie des Karbons der östlichen Staaten Nord-Amerika's. Vergleich mit West-Europa: Geologisch Bureau Heerlen, Jaarverslag over 1933, p. 17–44.

Jongmans, W. J., and Gothan, W., 1935a, Die paläobotanischen Ergebnisse der Djambi-Expedition 1925: Jaarboek van het Mijnwezen in Nederlandsch-Indië 1930, Verhandelingen, p. 71–201, 58 pls.

Jongmans, W. J., and Gothan, W., 1935b, Nouvelles corrélations stratigraphiques entre le Carbonifère des Etats-Unis et celui de l'Europe occidentale: Annales de la Société Géologique du Nord, v. 60, p. 1–16.

Jongmans, W. J., and Gothan, W., 1937a, Betrachtungen über die Ergebnisse des zweiten Kongresses für Karbonstratigraphie: Congrès pour l'avancement des études de stratigraphie Carbonifère, 2d, Heerlen, The Netherlands, 1935: Compte Rendu, v. 1, p. 1–40.

Jongmans, W. J., and Gothan, W., 1937b, Contribution to a comparison between the Carboniferous floras of the United States and of Western Europe: Congrès pour l'avancement des études de stratigraphie Carbonifère, 2d, Heerlen, The Netherlands, 1935: Compte Rendu, v. 1, p. 363–392.

Jongmans, W. J., and Gothan, W., 1938, Permo-Karbonische Flora auf Sumatra, Niederl. Indien: Acta Pontificiae Academiae Scientiarum Novi Lyncaei (Roma) Anno 80–20 Sess., p. 54–63.

Jongmans, W. J., and Gothan, W., 1951, Beitrag zur Kenntnis von *Alethopteris branneri* White: Academia Brasileira de Ciências Anals, v. 23, p. 283–290, 4 pls.

Jongmans, W. J., Delépine, G., Gothan, W., Pruvost, P., Van Rummelen, F. H., and de Voogd, N., 1925, Geologische en palaeontologische beschrijving van het Karboon der omgeving van Epen (Limb.): Mededeeling no. 1 van het Geologisch Bureau voor het Nederlandsche Mijngebied, Natuurhistorisch Maandblad, v. 14, p. 55–83.

Jongmans, W., Halle, T. G., and Gothan, W., 1935, Proposed additions to the international rules of botanical nomenclature. Adopted by the Fifth International Botanical Congress (Cambridge, 1930): Heerlen, The Netherlands (privately printed), p. 3–15.

Jongmans, W. J., Gothan, W., and Darrah, W. C., 1937a, Comparison of the floral succession in the Carboniferous of West Virginia with Europe: Congrès pour l'avancement des études de Stratigraphie du Carbonifère, 2d, Heerlen, The Netherlands, 1935: Compte Rendu, v. 1, p. 393–415, 26 pls.

Jongmans, W. J., Gothan, W., and Darrah, W. C., 1937b, Beiträge zur Kenntnis der Flora der Pocono-Schichten aus Pennsylvanien und Virginia: Congrès pour l'avancement des études de stratigraphie du Carbonifère, 2d, Heerlen, The Netherlands, 1935: Compte Rendu, v. 1, p. 423–444, 16 pls.

Norsk Polarinstitutt, 1991, The place-names of Svalbard: Ny-Trykk, Oslo, Skrifter Nr. 80, 539 p., and Nr. 112, 133 p. (supplement 1).

Picard, E., and Gothan, W., 1931, Die wissenschaftlichen Ergebnisse der staatl. Tiefbohrungen bei Dobrilugk N. L. 1927–31: Jahrbuch des Halleschen Verbandes für die Erforschung der mitteldeutschen Bodenschätze und ihrer Verwertung, N.F., v. 10, p. 131–141.

Potonié, H., 1899, Lehrbuch der Pflanzenpalaeontologie: Berlin, Ferd. Dümmlers Verlagsbuchhandlung, 402 p., 3 pls.

Potonié, H., 1910, Die Entstehung der Steinkohle und der Kaustobiolithe überhaupt (fifth edition): Berlin, Gebrüder Borntraeger, 225 p.

Potonié, H., 1920, Die Entstehung der Steinkohle und der Kaustobiolithe überhaupt (sixth edition, revised by W. Gothan): Berlin, Gebrüder Borntraeger, 231 p.

Potonié, H., 1921, Lehrbuch der Paläobotanik (second edition, revised by W. Gothan): Berlin, Gebrüder Borntraeger, 537 p.

Potonié, H., and Gothan, W., 1906, Erklärender Text zu Vegetationsbildern der Jetzt- und Vorzeit, poster 1–3: Esslingen und München, J. F. Schreiber.

Potonié, H., and Gothan, W., 1909, *Palaeoweichselia, in* Potonié, H., ed., Abbildungen und Beschreibungen fossiler Pflanzen-Reste: Berlin, Königlich Preussische Geologische Landesanstalt, v. 6, 4 p.

Potonié, H., and Gothan, W., 1912, Erklärender Text zu Vegetationsbildern der Jetzt- und Vorzeit, poster 4–8: Esslingen und München, J. F. Schreiber.

Potonié, R., 1955, Walther Gothan: Geologisches Jahrbuch, v. 70, p. 27–53, 1 pl.

Remy, W., 1953, Untersuchungen über einige Fruktifikationen von Farnen und Pteridospermen aus dem mitteleuropäischen Karbon und Perm: Abhandlungen der deutschen Akademie der Wissenschaften zu Berlin, Klasse für Mathematik und allgemeine Naturwissenschaften, Jahrgang 1952, no. 2, p. 5–38, 7 pls.

Remy, W., and Remy, R., 1986, Walther Gothan zum Gedächtnis—Die Berliner Schule der Paläobotanik: Argumenta Palaeobotanica, v. 7, p. 1–5, 1 pl.

Weyland, H., 1956, Walther Gothan (1879–1954): Paläontologische Zeitschrift, v. 30, p. 223–225, 1 pl.

Zeiller, R., 1899, Études sur la flore fossile du Bassin Houiller d'Héraclée (Asie mineure): Mémoires de la Société géologique de France, Paléontologie (Paris), Mémoire 21, p. 1–91, pls. 1–6.

MANUSCRIPT ACCEPTED BY THE SOCIETY JULY 6, 1994

Printed in U.S.A.

Geological Society of America
Memoir 185
1995

# Paul Bertrand (1879–1944): French paleobotanist

**Jean-Pierre Laveine**
*Université des Sciences et Technologies de Lille, UFR des Sciences de la Terre, URA CNRS 1365 "Paléontologie et Paléogéographie du Paléozoïque," 59655 Villeneuve d'Ascq Cedex, France, and Curator of the Lille Coal Museum (Musée Charles Barrois) and the Geological Museum (Musée Jules Gosselet)*

## ABSTRACT

**Professor Paul Bertrand was undoubtedly one of the most outstanding French paleobotanists of the twentieth century. He spent most of his professional career at the University of Lille (1903–1938) and ended it in Paris (1938–1944), when he was nominated to the National Museum of Natural History.**

**Paul Bertrand's interests and knowledge were broad, not just restricted to paleobotany. He taught general paleontology, and his research areas included stratigraphy, botany, and coal geology. He was very concerned about the transmission of knowledge to the general public through museum displays and special exhibitions. By the time of his death, in February 1944, more than 125 publications and 25 book reviews attest to the great productivity that gave him international fame. His national and international renown resulted in numerous awards and honorary citations. His death, hastened by wartime conditions, came at a time when he had reached full scientific maturity. It was a great loss to paleobotany, but his impact on this field is indelible.**

## INTRODUCTION

Information concerning Paul Bertrand's work and life is available through several publications, some which he wrote himself (1933b, 1937c, 1943b) as well as a posthumous work (1947) edited under the guidance of the late Professor Louis Emberger. Additional information can be found in memorials by François Morvillez (1918), Emberger (1944), Pierre Pruvost (1945, 1946), and John Walton (1945).

The present portrait presents the main lines of Bertrand's life and scientific activity and thus will heavily rely upon these previous works. However, I will try to present him in a more familiar way and will lay particular stress on his work on Carboniferous plants and his relationship with American paleobotanists, especially William C. Darrah (see Chapters 1 and 2, this volume).

## THE FAMILIAL ENVIRONMENT

Paul Bertrand was born at his father's home on July 10, 1879, at Loos-lez-Lille (Nord, France). As indicated by the name, this locality is a small town close to Lille, the boundary being only administrative.

Phylogenetically Paul Bertrand could hardly escape paleobotany. To fully understand his life and career, it is worthwhile to give a short account about his family background and especially about his father, the botanist Charles-Eugène Bertrand.

Charles-Eugène Bertrand (Fig. 1) was born in 1851 in Paris. His family was poor. Faced with the remarkable abilities of their son, the parents did their best to allow C.-E. Bertrand to attend a secondary school, but he had to work hard to earn money to defray the cost of his education. C.-E. Bertrand was very intelligent and was especially good at mathematics. In 1870 he competed in the very difficult entrance examination for one of the most famous schools of higher education in France, the Ecole Polytechnique, and obtained the second place on the list. It is worth mentioning here that in 1867 the first place was awarded to René Zeiller (1847–1915), another famous paleobotanist. Despite C.-E. Bertrand's successful entrance, his passion for botany was so intense that he decided

Laveine, J.-P., 1995, Paul Bertrand (1879–1944): French paleobotanist, *in* Lyons, P. C., Morey, E. D., and Wagner, R. H., eds., Historical Perspective of Early Twentieth Century Carboniferous Paleobotany in North America (W. C. Darrah volume): Boulder, Colorado, Geological Society of America Memoir 185.

Figure 1. Paul Bertrand's father, Charles-Eugène Bertrand (1851–1917). Paris National Museum of Natural History Collections. Courtesy of C. Blanc.

Figure 2. Paul Bertrand's mother, born Marie Hugonin (1849–1943). Paris National Museum of Natural History Collections. Courtesy of C. Blanc.

not to take up his position at the Ecole Polytechnique. Instead, he went to the National Museum of Natural History in Paris, where he asked Professor Joseph Decaisne to suggest a topic for a thesis. Despite the difficulties resulting from the Franco-Prussian war in 1870, Bertrand's research progressed quickly, owing to his remarkable intelligence and through hard work. He gave an outstanding defense of his thesis (1874) on the comparative anatomy of Gnetaceae and conifers; he was only 23 years old! The same year, he met Bernard Renault (1836–1904). Their long-lasting cooperation and joint papers are well known in the paleobotanical community.

In 1876, C.-E. Bertrand married Marie Hugonin (1849–1943; Fig. 2), who was finishing her scholarship in natural history at Paris University (La Sorbonne). She had also attended the lectures delivered by Adolphe Brongniart (1801–1876) at the National Museum in Paris.

In 1878, the Chair of Botany was created at the Faculty of Sciences of Lille University and C.-E. Bertrand, at age 27, became its first occupant. The young couple moved from Paris to Loos, a suburb of Lille. In his new position, C.-E. Bertrand had to come up with a teaching program, establish a research laboratory, and begin building up botanical and paleobotanical collections.

In 1884, several years after their arrival in Lille, C.-E. Bertrand's wife was promoted to head of the Young Girls Grammar School of Amiens, which required residency. Thus the family had to move to Amiens (the capital of Picardy), approximately 130 km from Lille by railway.

Consequently, C.-E. Bertrand's life became shared between Lille and Amiens. The first half of the week, which was spent in Lille, was entirely devoted to his professorial duties: administrative work, teaching and direction of the work of his students, and organization of the laboratory. The second half of the week was spent in Amiens and was devoted to his private research. He also supervised the education of his children (one son, Paul, and three daughters, one of whom died very young). This shared life lasted almost 30 years, and the numerous difficulties it entailed were mastered by C.-E. Bertrand, who was a remarkable organizer and leader. Scores of renowned botanists and paleobotanists were trained under his guidance. It is worth mentioning that the first assistant of C.-E. Bertrand in Lille was Octave Lignier (1855–1916), later professor at the University of Caen. Lignier—10 years before the discovery of the Rhynie plants (Kidston and Lang, 1917)—developed the Théorie des Cauloïdes (1903, 1908, 1908–1909, 1914), later named the Cladode leaf theory by Francis O. Bower (1923, 1935), which is essentially equivalent to Walter Zimmermann's Telome theory (1930, 1952).

C.-E. Bertrand was an excellent professor. He had a remarkable memory and was able to deliver two-hour lectures without any notes. He was also a severe taskmaster. However, his students appreciated him very much because they could see that he was more severe with himself than with anybody else. They also appreciated the fact that they were allowed to come to the botanical laboratory at any time outside class hours. Each student was personally assigned a microscope for the whole year and had to do some botanical work prepared specifically by C.-E. Bertrand. No one had the same work to do, and they could rapidly learn about botanical variations through a quick look at the preparations and analyses of their fellow students. Such a procedure represented a very heavy load for Professor Bertrand, but his students appreciated his effort.

Concurrently with all his professorial responsibilities, C.-E. Bertrand conducted research on comparative plant anatomy and paleobotany (1883, 1891), achieving considerable fame. Among numerous honors, which included several awards by the Academy of Sciences of Paris, in 1902 he was named Chevalier de la Légion d'Honneur and in 1904 Chevalier de l'Ordre de Léopold by the Belgian authorities. In the same year, he was elected Corresponding Member of the Academy of Sciences of Paris.

By the end of 1913, C.-E. Bertrand's wife retired. At that time he could have reduced his university obligations but did not do so, as he was committed to his work and was a man with a strong sense of duty. Despite the fact that his health was deteriorating, he had declared several times that if war should break out, he wanted to be in Lille to be able to assume his obligations and to protect as much as possible his laboratory and collections. Consequently, the family moved back to Lille at the end of 1913. As it turned out, C.-E. Bertrand became confined to Lille since it was under siege during World War I from October 1914 until his death, and his house was partly destroyed by shelling in August 1916. A few months after this happened he suffered from pulmonary congestion. However, his willpower was so strong that he met his students soon afterward to organize their work. Five days before his death, on August 10, 1917, he was supervising student examinations at the university.

In such a family environment, no doubt one of the main difficulties for his son, Paul, was not to disappoint such a famous father and to bring honor to the Bertrand name. To show clearly how he succeeded in his efforts, and to separate the different events of his life, short accounts of his personal life and the main steps of his administrative career are presented first. An overview of his research is then given, followed by an account of the honors and awards he received. Finally his relationships with the paleobotanists of that time is analyzed, and a special part deals with his contacts with William Culp Darrah.

## PAUL BERTRAND'S PERSONAL LIFE

As already mentioned, the Bertrand family moved from the suburb of Lille to Amiens when Paul Bertrand was a young boy (Fig. 3). No special comments are required about

Figure 3. Paul Bertrand at the age of four, before the family moved to Amiens. Paris National Museum of Natural History Collections. Courtesy of C. Blanc.

his youth, except that he was an excellent pupil. Like his father, he was good at mathematics, physics, and chemistry, and he intended at first to work in one of these fields. For two years, he followed a special preparation in high school for that purpose (Fig. 4). But, like his father 30 years earlier, he was so strongly drawn to natural history and botany that he changed his mind and prepared for a university degree in natural history at Lille, where most of his life was to be spent. At the beginning of his career (Fig. 5), Paul Bertrand was still living with his parents in Amiens, but of course this was not very convenient. After he completed his doctoral thesis, he moved back to Lille in 1909.

In August 1914, with the opening of World War I, Bertrand was mobilized into the auxiliary services. From August 1914 to October 1915, he acted as military corpsman. During the winter he had to face a typhoid epidemic and was in charge of 25 patients.

From November 1915 to April 1917, because of the war needs of the French government, Bertrand served as professor of natural history at the High School of Nantes (western

Figure 4. Paul Bertrand, 18 years old. Paris National Museum of Natural History Collections. Courtesy of C. Blanc.

Figure 5. Paul Bertrand, 24 years old. This is how Bertrand looked at the beginning of his research career. Paris National Museum of Natural Collections. Courtesy of C. Blanc.

France). Then, because of his experience with coal, a mineral of strategic and economic importance, he was put at the disposal of the Houillères de la Loire (that is, mainly the Stephanian coalfield of St. Etienne), from April 1917 to February 1919. Paul Bertrand's father died in August 1917, when Lille was under siege, and the son was not able to pass through the German lines to sustain his family during this difficult time of bereavement.

In 1926, Paul Bertrand married Zénobie Fenaux (1892–1928) at Lille. She was a teacher of literature in Lille's Grammar School. Unfortunately, she died young. In 1929, he married Cécile Touron (1885–1975), and they lived in Lille until 1938.

As a result of his nomination in 1938 as professor at the National Museum of Natural History, Bertrand and his wife moved to Paris. They lived in a flat at 7, rue Pierre Nicole, Paris 5°. The exact personal address is given here because the address is found on the correspondence from Mr. and Mrs. W. C. Darrah, who wrote to Cécile P. Bertrand.

Bertrand died on February 24, 1944. It is noteworthy that he died approximately 15 months before the end of World War II, which again poses a certain parallel with his father, who passed away approximately 15 months before the end of World War I. Paul Bertrand died from complications incurred after contracting a cold during the winter of 1943–1944. His condition was aggravated by the privations due to the war and the fact that, despite his rather elevated position, he refused any special favors and wanted to be treated like everyone else during this difficult period for the French nation. He was buried in Saint Germain en Laye Cemetery near Paris (Fig. 6).

From 1944 until her death in 1975, Cécile P. Bertrand did her best to maintain the memory of her husband, by helping to publish several posthumous works and sending reprints to most of the paleobotanical institutions all over the world. Her strong devotion to her husband is obvious from a letter (Fig. 7) sent to me in 1972. Mrs. Bertrand was then 87 years old, and her handwriting is remarkable. She deserves to be referenced in this portrait in the light of her activities to further the memory of her husband.

## MAIN STEPS IN PAUL BERTRAND'S CAREER

From 1903 to 1905, Paul Bertrand was a young assistant in the chemistry laboratory of Lille University. Bertrand's first publication (disregarding short book reviews and various analyses) was entitled "Sur un persulfate organique" and was coauthored with Professor Richard Fosse, director of the Laboratory of Chemistry (1905).

On February 1, 1906, Bertrand was appointed "prépara-

Figure 6. Paul Bertrand's grave, located at Saint Germain en Laye Cemetery. Cécile P. Bertrand on the left side of the photograph. Photograph taken by Professor B. Sahni on April 19, 1948. The negative was developed by Professor T. G. Halle at Stockholm. Paris National Museum of Natural History Collections. Courtesy of C. Blanc.

teur" at the Coal Museum of Lille under its director, Professor Charles Barrois. On February 27, 1909, he defended his doctoral thesis "Études sur la fronde des Zygoptéridées" (Bertrand, 1909). On November 15, 1910, he was promoted to "maître de conférences de paléontologie" (lecturer in paleontology) and curator of the Lille Coal Museum. On November 1, 1919, he was promoted to full professor (and, within this last grade, promoted from fourth to third class on November 1, 1923, and from third to second class on November 1, 1927). In 1927, the Faculty of Sciences of Lille University, in recognition of Bertrand's accomplishments, decided to create a special Chair of Paleobotany within the Institute of Geology. This Chair subsequently was held from 1944 to 1974 by Professor Paul Corsin (see Chapter 26, this volume), one of Paul Bertrand's students. Since that time I have held the Chair.

On January 1, 1932, Bertrand was promoted to first class professor. On April 1, 1938, he was elected to the Chaire d'Anatomie comparée des Végétaux vivants et fossiles at the National Museum of Natural History in Paris. This prestigious position was held before him by Adolphe Brongniart and by Professor Philippe Van Tieghem (1839–1914), one of the most famous botanical anatomists of the late nineteenth century, who was the promoter of the notion of the stele.

In the French system, the promotion to a Chair at the National Museum is obtained after two ballots, one by the assembly of the Museum professors and another by the Academy of Sciences. Thus, Paul Bertrand could hope to finish his remarkable career with the prospect of being promoted to the Academy of Sciences. Unfortunately, war and fate prevented that final step.

## SCIENTIFIC ACTIVITY

Paul Bertrand's scientific works can be divided into three main categories: (1) comparative anatomy of fossil plants, (2) comparative morphology of living and fossil plants, and (3) studies on Carboniferous compressions, with applications to Carboniferous biostratigraphy.

### *Comparative anatomy of fossil plants*

Bertrand's name is closely associated with the knowledge of the structure of early ferns or fernlike plants. After several short preliminary publications (Bertrand, 1907, 1908a, b), he presented in 1909 an outstanding thesis on the Stauropteridales and Zygopteridales. Of course, important new information has been added about some representatives of these groups in the following decades, but at the beginning of this century, many points were still unclear or unknown about these plants. It may be recalled that the taxa *Metaclepsydropsis, Etapteris,* and *Stauropteris burntislandica,* among others, were defined by Bertrand. He accurately analyzed the kind of ramification of the Stauropteridales, plants of bushy appearance because of the lack of any planated appendages, as mentioned already by Williamson near the end of the nineteenth century (1874). From the knowledge gained by comparing the various modes of ramification existing within the Zygopteridales, Bertrand was able to show that these plants had appendages structurally transitional between branches (radial symmetry) and fronds (bilateral symmetry). Throughout his life, and armed with the observation of other kinds of primitive plants, he studied these appendages. He proposed the term *phyllophore* (Bertrand, 1933a), which is widely used in paleobotanical textbooks, to designate such intermediate appendages present in these primitive taxa.

The Cladoxylales, from the Devonian and the lower Carboniferous, represent another group that was thoroughly studied by Bertrand. A set of 10 publications focused on that group, beginning with a short account (Bertrand, 1908b) and culminating in an outstanding publication in 1935. These plants are characterized by an anatomy with stems containing numerous vascular segments radiating out from the center of a stele and by bearing lateral appendages (*Hierogramma, Syncardia*)

PARIS 115
17 45
8-4
1972
R. EPEE DE BOIS (5e)

LA PROTECTI
UNE NEC
UN DE
0 50
REPUBLIQUE FRANÇAISE

Monsieur
Jean Pierre Laveine
8. Place B. Dorez
Lille
59
Nord

Paris. le 8 Avril 1972
7. Rue Pierre Nicole. V: 75;
Tel: 633.15.76

Cher Monsieur,

Après le plaisir de faire votre connaissance et la remise si délicate de votre magnifique Thèse et autres études, voici que je reçois, par vos soins dévoués,

. . . . . . . . . . . . . . . .

Je reprends à loisir votre important travail sur les Neuroptéridées, le feuilletant ... Si mes connaissances sont plus que faibles en Paléobotanique, j'y retrouve, non sans émotion, bien des plantes au nom familier et aussi celui de mon cher mari...

Soyez donc encore remercié, et que l'avenir vous apporte toujours courage, réussite complète dans ces si importants travaux...

N'oubliez pas, venant à Paris, que le meilleur accueil vous y sera réservé, et veuillez, Cher Monsieur, partager avec Madame J.P. Laveine, l'expression de mes sentiments cordiaux.

C. Paul Bertrand

*Partial translation of Figure 7*

Paris, 8 April 1972
7, Rue Pierre Nicole-V$^{0}$-75
Tel.: 633 15 76

Dear Sir,

. . . I have calmly looked over your important study of the Neuropterids. . . . Although my knowledge of paleobotany is more than weak, I have found, not without emotion, many plants the names of which are familiar to me, and also the name of my dear husband. . . .

Cécile P. Bertrand

Figure 7. Partial reproduction of a letter of Cécile P. Bertrand to Jean-Pierre Laveine. J.-P. Laveine Collection.

showing a bilateral symmetry. Even now the taxonomic position and phylogenetic relationships of this group remain questionable. It is thus not surprising that Bertrand (1941b) changed his mind several times about the supposed relationship of *Clepsydropsis* to *Cladoxylon.* The taxa studied by Bertrand are still regarded as somewhat problematic, which demonstrates that the task was far from easy. Bertrand, with his great scientific honesty, was able to recognize this and to correct his own mistakes in interpreting some of the more dubious structures (for instance 1941c, p. 625). His scientific honesty is further exemplified in the examples in the next section.

### *Comparative morphology of living and fossil plants*

Bertrand focused his attention on two main points. The first one was a defense of some of his father's works concerning the nature of certain kinds of coals, known as bogheads. C.-E. Bertrand and B. Renault interpreted these coals as being due to the accumulation of algae belonging to the Protococcophyceae, an interpretation that had been contested in the early twentieth century. However, following evidence obtained by Paul Bertrand's Russian friend, Michael Zalessky (1914), that the algal genus *Pila* from an Autunian boghead was similar to the present-day *Botryococcus,* Paul Bertrand (1930b) furnished a set of remarkable photographs to demonstrate this relationship.

The second point to which Bertrand paid special attention late in life, mainly after his move to Paris, concerned the general and difficult questions of the origin and relationships of the main groups of plants. With the hope of adding another criterion for the understanding of the general relationships, Bertrand paid special attention to research on the vascularization of the seedlings in various plant groups (for instance, Bertrand, 1937b).

Given the difficulty of attaining these goals, it is not surprising that he changed his mind several times on interpretation but always with great intellectual integrity. To quote but an example, Bertrand criticized (1941a, d) some conclusions of the work of Gustave Chauveaud and Édouard Boureau. Then, just before his death, he recognized (Bertrand 1942, 1943a, p. 245) that he was wrong, thus still giving us a remarkable lesson of intellectual honesty.

### *Carboniferous compressions*

Parallel to his scientific activity on the anatomy and morphology of fossil and living plants, Bertrand became thoroughly familiar with Carboniferous plant compressions, both through the double and inseparable aspect of their morphological characteristics and variability and with regard to their stratigraphic distribution. Approximately 65 of his publications are devoted to this research. Some of them represent a monographic treatment of certain groups (for instance 1930a, on the Neuroptéridées of the Saar-Lorraine coalfield); other papers are devoted to the biostratigraphic uses of compressions (for instance, 1914b). Bertrand was also very interested in the reconstruction of plants (see the background in Fig. 18, which was painted in 1906, or the frontispiece drawing of his 1947 posthumous publication) and in the presentation of botanical landscapes (1950). His knowledge of the anatomy of fossil plants allowed him (1914a) to show the correspondence between the compressional types *Corynepteris-Alloiopteris* and the Zygopteridales.

It is not possible here to comment on all his work on Carboniferous plants and their stratigraphic significance. During more than 30 years (from his appointment in 1906 to the Coal Museum at Lille and in 1938 to the Chair in Paris), he collected in the principal French coalfields. He professed a great admiration for the work of both R. Zeiller and Cyrille Grand'Eury, and after their deaths, in 1915 and 1917, respectively, he became the leading exponent of stratigraphic paleobotany in France. Zeiller and Grand'Eury (see Bertrand, 1920b) had laid the foundation, but Bertrand improved and largely refined the picture, through a fruitful collaboration (for instance, 1937) with his great friend, Pierre Pruvost (Fig. 8) and under the guidance of their venerated master, Professor Barrois.

From 1914 to 1922, Bertrand refined the biostratigraphic successions of the Westphalian of the northern France coalfield (1914b, 1919), the Stephanian of the Loire (1918) and Gard (1920a) coalfields, and then the Westphalian and Stephanian of the Saar-Lorraine coalfield (1922b). The quin-

Figure 8. Paul Bertrand (on right) and his great friend, Professor Pierre Pruvost. The photograph was taken by Helen H. Darrah during the 2nd Carboniferous Congress, Heerlen, 1935. The William Culp Darrah Collection. Courtesy of Elsie Darrah Morey.

tessence of these long-term investigations was presented at several International Congresses: in Brussels (Bertrand, 1922a); in Heerlen in 1927 (Bertrand, 1928a, b, c); and in Heerlen in 1935 (Bertrand, 1937a). A major result of Bertrand's biostratigraphic work is the recognition of the Westphalian D (initially and paleobotanically defined as the Zone à *Mixoneura* (Barrois et al., 1922; Bertrand, 1926), a stratigraphic unit up to then unrecognized, which since has proved fundamental for stratigraphic correlation and tectonic analyses. Some minor corrections concerning the limits of that stratigraphic entity have been added later (see Laveine, 1974, 1977, 1989), and it is possible that the names and number of subdivisions might change in the future, but the fact remains that Bertrand discovered and defined a characteristic interval of the upper part of the Westphalian sequence, which is in general use today. Bertrand worked without any preconceived ideas. He always asked not to be informed of the supposed stratigraphic position of the strata that had yielded the compressions. In that way he could not be influenced, even subconsciously, and could base a determination exclusively on the available characters of the plant fossils at hand.

In 1930 he applied such a procedure to some plant fossils recorded from the Roucourt Conglomerate (Barrois et al., 1930) discovered in the Aniche area, where it lies unconformably on coal-bearing strata. The local geologists assumed this formation to be of Permo-Triassic age, but Bertrand demonstrated that the specimens were of Westphalian age. This proved the early start of orogenic deformation in the northern France coal basin and was of fundamental importance for the unraveling of the structural history of the whole area.

Figure 9. Paul Bertrand wearing his academic gown and his honorary medals. Circa 1939. The Lille Laboratory of Paleobotany Collections.

## HONORS AND AWARDS

Bertrand's activities as teacher and researcher and his outstanding scientific discoveries resulted in many national and international honors.

He was honored by the Society of Sciences, Literature and Arts of Lille (1913, Prix Debray; 1937, Prix Descamps-Crespel); by the Academy of Sciences of Paris (1920, Prix Saintour; 1932, Prix Demolombe; 1935, Prix Henri Wilde; 1947, Prix Guido Triossi); by the Société Géologique de France (1927, Prix Prestwich); and by the Société de l'Industrie Minérale (1933, Prix Henri Fayol).

Bertrand was a member of many French scientific societies and was several times president or vice president of these. In 1920 he was given the title Officier d'Académie and, in 1925, Officier de l'Instruction Publique in recognition of his professorial activities.

In 1933, he was promoted to Chevalier de la Légion d'Honneur (Fig. 9), like his father 30 years earlier. The same year he was nominated by the French government as one of the official delegates to the 16th International Geological Congress in Washington, D.C. (July 1933), where he met for the first time with William C. Darrah. This leads into the next topic, the presentation of a short account of Bertrand's visits to foreign countries and his connection with foreign scientists. A special section is devoted to his relationships with the United States and American scientists, and special attention is given to his friendship with Darrah.

## INTERNATIONAL RELATIONSHIPS

For the preparation of his doctoral thesis, Bertrand was obliged to study the type sections of permineralized specimens described during the nineteenth century. He traveled for the first time to Great Britain in 1905 and to Germany in 1907. He visited several institutions and had the opportunity to meet many scientists whom he acknowledged in his thesis (1909, p. 12–14). In 1910, he participated in the paleobotanical excursion led in Great Britain by Professor Francis W. Oliver. Figure 10 is a copy of two pages of Bertrand's notebook to illustrate his work habits. Figure 11 is a group photograph taken at the end of May 1910 at Manchester. In 1912, Bertrand paid a second visit to Germany to continue his examination of permineralized specimens of the Cladoxylales, but unfortunately circumstances prevented this research for a long time. Apart from a field trip to Morocco in 1932, his later visits to foreign countries were linked with international meetings that he attended: in 1922,

28 Mai 1910.
Visite à M. et Mme Scott
à Oakley Hants.

—

Situé à 2h de Londres, sur la ligne de Southampton à Londres, ou sur une ramification latérale

—

Unger's Sections from the Jermyn Street Museum

—

Il y en a 6 en tout.

15 867. – Kalymma grandis
C'est un mauvais pétiole de Calamopitys

15868. – Calamosyrinx devonica
Pétiole avec tissu mécanique protecteur très épais.
Je ne le comprends pas encore.

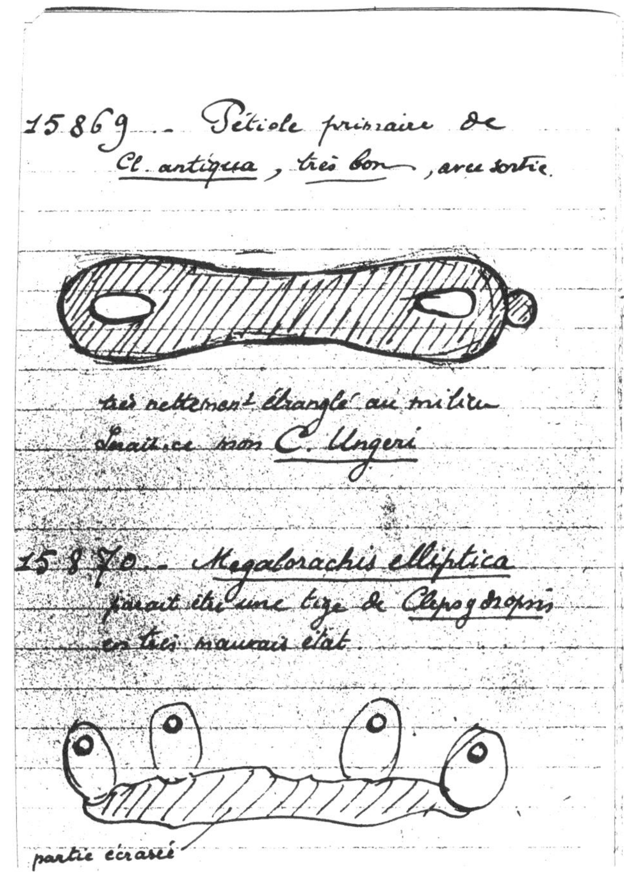

15869. – Pétiole primaire de Cl. antiqua, très bon, avec sortie.

très nettement étranglé au milieu
Serait-ce mon C. Ungeri

15870. – Megalorachis elliptica
paraît être une tige de Clepsydropsis en très mauvais état.

partie écrasée

Figure 10. Copy of some pages of Paul Bertrand's notebook from his visit to Great Britain in 1910. The Lille Laboratory of Paleobotany Collections.

Figure 11. Photograph of a meeting in Manchester, England, May 1910. Seated (left to right): C. Grand'Eury, W. J. Jongmans, F. W. Oliver(?), M. Stopes, P. Bertrand, M. D. Zalessky. Standing (left to right): F. O. Bower, F. E. Weiss, D. H. Scott; all others are unidentified. Paris National Museum of Natural History Collections. Courtesy of C. Blanc.

Brussels, 13th International Geological Congress; in 1927, Heerlen, 1st Congress of Carboniferous Stratigraphy; in 1930, Liège, International Congress of Mines and Applied Geology; Cambridge, 5th Botanical International Congress; in 1933, Washington, D.C., 16th International Geological Congress; in 1935, Amsterdam, 6th International Botanical Congress; Heerlen, 2nd Congress of Carboniferous Stratigraphy. During these years, at all these congresses and through visits from paleobotanists whom Bertrand welcomed in Lille, he met numerous colleagues.

As far as I am aware from letters and rather scarce verbal information, it seems that the judgment given by Professor J. Walton in 1945 nicely summarizes the general opinion of Paul Bertrand held by his colleagues: "Those who were privileged to know him will remember his quiet unassuming character, and his unfailing sympathy and interest in his friends' difficulties and problems. To know the Bertrands was to know what is best and most charming in the French character."

Walton was a young colleague whom Bertrand appreciated very much. As far as it is possible to assess such feelings, it seems to me that Paul Bertrand was sympathetic toward all people. However, there were several scientists with whom he was more friendly, even if at times he would disagree with them scientifically. M. D. Zalessky and Birbal Sahni should be mentioned in this respect.

Zalessky, who professed great admiration for the work of R. Zeiller and C.-E. Bertrand, published his first papers approximately at the same time as Paul Bertrand did. Zalessky met Bertrand in 1910 in Great Britain during the paleobotanical excursion (Fig. 11; see Zalessky, 1910) and kept in touch with him over the years. As a result of this friendship there is, in the library of the Laboratory of Paleobotany in Lille, an almost complete set of Zalessky's publications, most of them with personal dedications to C.-E. Bertrand, and later on to P. Bertrand. Unfortunately the political situation did not allow Zalessky (see Chapter 6, this volume) to visit foreign countries as would have been necessary for keeping close scientific ties.

Birbal Sahni, after graduating from Cambridge University in 1914, began paleobotanical research under Albert Charles Seward's leadership. Because one of Sahni's chief lines of research concerned the study of the Cœnopteridopsida, he of course had to get in touch with Bertrand, an authority on the subject. Despite some scientific controversy about the interpretation of some structures, Bertrand and Sahni became very good friends. There are numerous facts to prove this. For instance, Figure 12 illustrates their research on Cladoxylales, which shows that both friends were in close contact (note that the conclusion about the supposed connection between *Cladoxylon* and *Clepsydropsis* was withdrawn (Bertrand, 1941c). Thore G. Halle, another good friend of both of them, remembered similar examples (Halle, 1952, p. 25–27). A similar fast friendship existed between their wives, who were happy to meet during the congresses that they attended with their husbands. Figure 5 of Lyons and Morey (1995), concerning Birbal Sahni and his North American paleobotanical connection with W. C. Darrah, attests to the close relationship among the Sahnis, Bertrands, and Darrahs.

## PAUL BERTRAND AND THE UNITED STATES

Bertrand had only one opportunity to visit the United States, through the meeting of the 16th International Geological Congress at Washington, D.C., 1933. However, for a long time he had been interested in the possibility of examining the

La position systématique des Cladoxylées était difficile à établir en raison de leur état fragmentaire et de leur structure déconcertante. J'ai dû par deux fois modifier mon opinion à leur sujet.

J'avais annoncé en 1908, que les *Cladoxylon* devaient être les stipes qui avaient porté les pétioles de *Clepsydropsis* ; en 1911 et 1912, j'apportai des arguments très forts à l'appui de cette manière de voir. Mais en 1913 et 1914, faute de preuves décisives, je me vis contraint d'abandonner ma première opinion et je suggérai un rapprochement entre les *Cladoxylon* et les Phanérogames anciennes. En 1930, B. Sahni au cours d'un séjour en Europe reprit mon opinion de 1908 ; malgré tous nos efforts, il ne nous fut pas possible de trouver une seule trace foliaire en *clepsydre,* indubitablement fixée sur un stipe de *Cladoxylon*. Tous les spécimens de *Cladoxylon* ont été décortiqués trop profondément : il ne reste rien des pétioles qu'ils portaient.

Je demeurai donc dans le doute le plus complet. Néanmoins B. Sahni continua à m'affirmer dans plusieurs lettres (1931 et début de 1932) que les *Clepsydropsis* devraient appartenir aux *Cladoxylon*. C'est seulement en Octobre 1932, que je réussis, en me basant sur les caractères histologiques, à démontrer que les pétioles de *Clepsydropsis exigua* et probablement aussi ceux de *Cl. antiqua* appartiennent au *Cladoxylon mirabile*. D'autres pétioles de *Clepsydropsis,* trouvés également à Saalfeld appartiennent les uns au *Cladoxylon tæniatum*, les autres au *Cl. Solmsi*.

Cette belle découverte, due, on le voit, à l'insistance affectueuse de B. Sahni, entraîne des conséquences importantes au point de vue de la classification et des affinités des plantes anciennes :

1° Les *Cladoxylon* sont vraiment des Fougères très primitives ; par les caractères si particuliers de leurs pétioles

Figure 12. Demonstration of the close relationship between Birbal Sahni and Paul Bertrand (from Paul Bertrand's 1933 "Notice Scientifique," *in* Titres et Thavaux Scientifiques: Lille, France, G. Sautai). J.-P. Laveine Collection.

*Translation of Figure 12*

It is difficult to establish the exact position of the Cladoxylées in the classification because of their fragmentation and confusing structure. I have changed my opinion twice on this subject.

I announced in 1908 that the *Cladoxylon* should be the stems that bore the petioles of *Clepsydropsis.* In 1911 and 1912 I produced strong arguments in support of this viewpoint. Yet, in 1913 and 1914, lacking definitive proof, I was forced to abandon my initial opinion and I suggested a closer relationship between the *Cladoxylon* and the ancient Phanerogams. On his visit to Europe in 1930, B. Sahni reconsidered my 1908 opinion; however, in spite of all our efforts, we were unable to find a single foliar trace with a "clepsydre" outline, which was indisputably connected to a stem of *Cladoxylon.* All *Cladoxylon* specimens are too drastically stripped: nothing is left of the petioles which they carried.

I remained, therefore, in complete uncertainty concerning the *Cl.* succession. However, in his numerous letters (1931 and beginning of 1932) B. Sahni still maintained that the *Clepsydropsis* should belong to the *Cladoxylon.* It was not until October of 1932, that I succeeded, by taking into account histological characteristics, to demonstrate that the petioles of *Clepsydropsis exigua* and probably also of *Cl. antiqua* belong to the *Cladoxylon mirabile.* Other petioles of *Clepsydropsis,* also found at Saalfeld, belong to either *Cladoxylon taeniatum,* or to *Cl. solmsi.*

This beautiful discovery, which owes much to the affectionate persistence of B. Sahni, has important consequences in regard to the classification and the affinities of ancient plants . . .

American Carboniferous floras. Figure 13 is a copy of a letter from David White (see Chapter 10, this volume) on July 14, 1926, answering a request from Paul Bertrand on June 25 of the same year and indicating the generosity of Richard A. F. Penrose, Jr. I was unable to find the exact reason why Bertrand could not visit America in 1926, but it was most likely related to the health of his first wife, who died in 1928.

Many Carboniferous workers attended the 16th International Geological Congress in Washington, D.C. (P. Bertrand, W. S. Bisat, Wilhelmus J. Jongmans, P. Pruvost, and Armand Renier, among others; Fig. 14). This marked the beginning of the Bertrand-Darrah relationship. Following the scientific session, the young and talented William C. Darrah was confronted with a rather difficult task. Darrah was asked by D. White to lead separate collecting trips in the Appalachian coalfields for two European paleobotanists: W. J. Jongmans and

UNITED STATES
DEPARTMENT OF THE INTERIOR
GEOLOGICAL SURVEY
WASHINGTON

July 14, 1926.

Dr. Paul Bertrand,
University of Lille,
Lille, F r a n c e.

Dear Doctor Bertrand:

Further in reply to your letter relating to your coming to the Botanical Congress at Ithaca in August and your need for financial assistance to the amount of $500 to enable you to come to America and visit the fossil plant collections in Washington, Chicago, and eastern Museums:

It gives me much pleasure to inform you that Dr. R. A. F. Penrose, jr., well known American geologist, whom you doubtless remember meeting at the Thirteenth International Geological Congress in Brussels, on learning of your desire to attend the Congress and examine collections of Paleozoic plants in this country, has very promptly volunteered to place $500 at your service, for he knows of and highly esteems your paleontological work.

I think Doctor Penrose would prefer not to have you know that it is he who is aiding you, for he seems to wish me to place the money in your hands without revealing the source of the funds. However, since I am leaving Washington for the far West to-morrow, it will not be possible for me to complete the arrangements with Doctor Penrose and transfer the funds to you myself.

I am therefore asking Doctor Penrose to immediately establish contact with you, and he probably will do so in order to forward the funds to you at the earliest possible date.

I am very glad to learn that you will surely be able to attend the Congress in this country, and I share your thanks to Doctor Penrose for his generous aid. Such help to scientists in the promotion of geologic research is rather characteristic of this fine type of American gentleman and geologist, who in this country is highly esteemed for his geological knowledge as well as for his fine personality.

Very truly yours,

David White

David White

Figure 13. Copy of a letter from David White informing Paul Bertrand that Dr. R. A. F. Penrose agreed to fund his visit to the United States. The Lille Laboratory of Paleobotany Collections.

Paul Bertrand. The two gentlemen had rather different personalities. Jongmans's approach was to collect as many specimens as possible. Bertrand was essentially interested in looking at specimens in the field and in the paleobotanical collections in order to analyze the total range of variation. He made extensive entries and many sketches in his notebook. He wanted to collect only those specimens that showed some differences compared to what was known from Europe. No doubt Darrah's brilliant mind and diplomacy made the organization of things run very smoothly in both cases.

Through this first joint field trip, and despite the 30 years difference in age, a sound friendship between Darrah and Bertrand was born. It is clear from the correspondence between Bertrand and Darrah that they held one another in high esteem. Of course, Bertrand was also in touch with many other American paleobotanists, especially those who worked on structurally preserved specimens, as Chester A. Arnold did (Fig. 15; see Chapter 18, this volume).

Through the American field trip, Bertrand strongly supported the views of Darrah as to the existence of Stephanian strata in the Appalachian basin. Several times, Bertrand (see Figs. 16 and 17) insisted on the importance of Darrah's discoveries. They coauthored papers on the American succession (Darrah and Bertrand, 1933, 1934).

In 1935, Bertrand obtained a grant from the University of Lille for Darrah to spend the summer in France and to visit several French coalfields under his guidance. As a result, Darrah became fully acquainted with the characteristics of the Westphalian and Stephanian floras of northwestern Europe. He published a paper (Darrah, 1937) giving the main conclusions of his first-hand comparison between the west European and North American floral successions. During his stay in Lille, Darrah left a profound impression on all people who met him. They were impressed not only by his inborn talents but also by his kindness and dynamism. I can present here an anecdote on this aspect of Darrah's enthusiastic behavior, which I can only now fully appreciate, after preparing this chapter.

I began my research at the Laboratory of Paleobotany at the end of 1961. At that time, the university was still located at 23, rue Gosselet, at Lille, and the Geology Department was directly connected to the Coal Museum. The entrance of the museum was around the corner, on the rue de Bruxelles. In front of the main entrance of the museum was (and still is) located Lille's fire station. Among my duties, I had from time to time to take pupils from the primary and secondary schools to the Coal Museum. At that time, the museum was still as Darrah saw it in 1935 (Fig. 18). One can see in the background the Carboniferous landscape that was created at the beginning of the century under the supervision of Paul Bertrand. The stairs and the lateral intermediate level were added later on. In front of the stairs is the bust of Barrois, in front of which lies a massive boulder from the famous Roucourt Conglomerate that has been mentioned previously. Figure 19 is a close-up of that block, with smaller boulders in the foreground. In 1961, there were still old watchmen (like the one in the photograph) who had worked there for over 30 years; one of them told me the whole story about the big block as follows:

The Conglomerate, discovered in 1875, was reinvestigated in 1929 by Professor Ch. Barrois, in connection with new underground work-

Figure 14. Part of the group photograph of the attendants at the 16th International Geological Congress at Washington, D.C., 1933 (Paul Bertrand shown by the arrow). Original photograph is in the photographic collection of the U.S. Geological Survey Library (Denver, Colorado). The entire group photograph was published in the International Congress Report of the XVI Session, v. 1, Washington, D.C., 1936. The U.S. Geological Survey Collections. Courtesy of P. C. Lyons.

MUSÉUM NATIONAL
D'HISTOIRE NATURELLE

ANATOMIE COMPARÉE
DES VÉGÉTAUX VIVANTS
ET FOSSILES

61, Rue de Buffon, PARIS (Ve)

Paris, le 26 Aout 1940

Cher Docteur Ch. A. Arnold

Je vous remercie de l'envoi de vos publications relatives aux végétaux dévoniens et carbonifères à structure conservé d'Amérique. Toutes vos découvertes offrent un très grand intérêt au point de vue de l'origine et de l'évolution des végétaux vasculaires. Je vous présente mes cordiales félicitations pour vos travaux

. . . . . . . . . . . . . . . . .

*Translation of Figure 15*

Paris, 26 August 1940

Dear Dr. Ch.. A. Arnold:

I thank you for sending your publications on the structurally preserved Devonian and Carboniferous floras from America. All your discoveries are of the greatest interest from the view point of the origin and the evolution of vascular plants. I extend my cordial congratulations for your work. . . .

Figure 15. Partial reproduction of a letter by Paul Bertrand to Chester A. Arnold, written at the beginning of World War II (the letter may not have been sent). Paris National Museum of Natural History Collections. Courtesy of C. Blanc.

**G. Mathieu prépare un important travail sur un bassin houiller de la France.**

Depuis 1933, le Doct^r W. C. Darrah de l'Université de Harvard, a pu, sur nos indications, étendre avec plein succès, aux flores Houillères des États Unis les méthodes appliquées en France, et les résultats acquis.

Figure 16. Handwritten addition to p. 23 of Paul Bertrand's 1933 "Notice Scientifique," referring to William C. Darrah's work concerning the biostratigraphy of the upper Carboniferous. J.-P. Laveine Collection.

*Translation of Figure 16*

G. Mathieu is preparing an important study of a coal basin of France.

Since 1933, Dr. W. C. Darrah, Harvard University, following our direction, extended successfully the methods used and results obtained in France to the coal floras of the United States.

FLORES HOUILLÈRES DES ETATS-UNIS

(Publications : 86, 87, 88, 93)

Lors de mon séjour aux Etats-Unis (juillet-septembre 1933), à l'occasion du XVI[e] Congrès géologique international, j'ai pu grâce à l'obligeance du regretté D. White et grâce à la collaboration du jeune Wm.-C. Darrah me familiariser avec les flores houillères des Etats-Unis. C'était une occasion excellente d'appliquer à une contrée éloignée les méthodes employées en France et les résultats acquis au cours de l'exploration de nos bassins houillers. En compagnie de Wm.-C. Darrah, je visitai quelques gisements houillers de Pensylvanie (régions de Pittsburg et de Wilkes-Barre). C'est de Wilkes-Barre que proviennent plusieurs types houillers décrits par Brongniart.

A la suite de cette exploration, une corrélation précise fut établie entre la série houillère de Pensylvanie et celle de nos gisements français (86, 87). Les travaux de D. White, qui remontaient à la période allant de 1897 à 1908, me permirent d'étendre nos conclusions à tous les bassins houillers explorés par cet auteur qui avait obtenu déjà d'excellents résultats, tant dans ses comparaisons de bassin à bassin en Amérique, que dans les équivalences proposées avec la série stratigraphique de l'Europe occidentale.

Wm.-C. Darrah fut quelques mois plus tard (1934) attaché à l'Université de Harvard avec la double mission de réviser les collections de Lesquereux et d'entreprendre l'étude systématique des flores houillères américaines. Il explora tous les bassins houillers de la région Nord-Est des Etats-Unis. De ses travaux, il y a lieu de retenir surtout : la démonstration de l'existence du Stéphanien aux Etats-Unis, avec plusieurs espèces identiques à celles décrites par Grand'Eury et par Zeiller (flores fossiles de la Loire, du Gard, de Blanzy, de Commentry), et avec zones végétales comparables terme à terme à celles du Stéphanien français (Heerlen, 1935).

*Translation of Figure 17*

CARBONIFEROUS FLORAS OF THE UNITED STATES

During my stay in the United States (July–September 1933) on the occasion of the 16th International Geological Congress, thanks to the efforts of the late D. White and the collaboration of the young Wm. C. Darrah, I was able to familiarize myself with the coal floras of this country. It was an excellent occasion for applying in a remote region the methods used in France, as well as the results acquired during the exploration of our coal basins. Accompanied by Wm. C. Darrah I visited several coal deposits in Pennsylvania (in the Pittsburgh and Wilkes-Barre regions). It was Wilkes-Barre which yielded several type specimens described by Brongniart.

As a result of this exploration, a precise correlation was established between the coal series of Pennsylvania and the French coal deposits (86, 87). The earlier work of D. White, which was dated 1897 to 1908, allowed me to extend our conclusions to all the coal basins explored by this author. White had obtained excellent results, both in his interbasin comparisons in America and in the correlations that he proposed with the stratigraphic series of Western Europe.

Wm. C. Darrah joined Harvard University several months later (1934) pursuing a double mission—to review the Lesquereux collections and to undertake a study of the succession of the American coal floras. He explored all the coal basins of the northeastern United States. Of his works one should remember principally the demonstration of the existence of the Stephanian in the United States, based on numerous type specimens that were identical with those described by Grand'Eury and by Zeiller (fossil flora of the Loire, Gard, Blanzy, and Commentry), and on megafloral zones that were comparable term by term to those of the French Stephanian (Heerlen, 1935).

---

Figure 17. Chapter (p. 43–44) of Paul Bertrand's 1937 "Notice Scientifique" evaluating William C. Darrah's results on upper Carboniferous biostratigraphy. J.-P. Laveine collection.

Figure 18. View of a part of the Lille Coal Museum in its early stages, as seen by William C. Darrah during his 1935 visit to Lille, France. The Lille Laboratory of Paleobotany Collections.

ings by a Coal company. As a direct result of Bertrand's dating of the plant fossils in the thin intercalated silty layers, Barrois decided a few years later to get some blocks of this conglomerate to be exhibited in the Coal Museum of Lille. During one of his visits in 1934 to the coal mine, he marked with a piece of chalk a boulder which was partially visible on the side of the gallery. Barrois was at that time a prominent member of the Administrative Council of the Coal company and any wish he expressed would be carried out immediately. Two miners were assigned to free the boulder. At first glance they thought it would be easy, but as they were digging around the initial rounded area, the boulder grew progressively larger. The two miners were full of admiration for that wonderful Professor who was able to know without direct information which was the largest block! Finally, quite a number of people were required to get the boulder out. At that time, the mining companies had direct access to the railway and the huge block was loaded onto a platform and transferred to Lille by train. It duly arrived at Lille railhead, which was 300 m away from the Institute of Geology. One day, Barrois found on his desk a notifica-

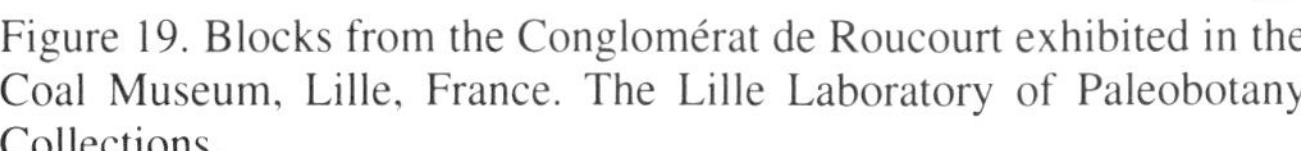

Figure 19. Blocks from the Conglomérat de Roucourt exhibited in the Coal Museum, Lille, France. The Lille Laboratory of Paleobotany Collections.

tion stating: "Rock specimen to collect" without any further information. As Barrois received many rock specimens from various places, he was unaware of what was waiting at the station. He thus sent his office boy to collect the specimen. Imagine how the poor boy was teased by the railway people! At the end of 1934, the boulder was finally transported by horse and wagon and laid on the pavement, on one side of the main entrance of the Museum.

The old watchman then added:

The rock was on the pavement for at least six months with no one specially worrying as to the best and easiest way to get it inside when a young American (and he did not know his name) came for a few months to the laboratory. He inquired why this boulder was lying on the pavement, and was duly informed that people were trying to figure out how to get it inside. The problem was compounded by the fact that the floor of the museum is approximately half a meter above the level of the pavement. The young American took it as a challenge. He went to the fire station and asked the Chief if some firemen might help him at the right moment. As he received a positive answer, he managed to get a ramp prepared by the technicians of the Geological Institute. One day the young people of the institute and the firemen were all inspired by the dynamism of the young American and succeeded in moving the huge block to its present place inside the Museum" [Fig. 19].

The story was so funny that I repeated it many times to the young boys and girls that I was shepherding through the Coal Museum, always using the expression "young American." However, when I was looking for information to prepare this chapter, I suddenly realized that the "young American" must have been William C. Darrah. The "young American" now had a face, and from what I have learned about Darrah's enthusiasm it seems fitting to repeat the story, even though its veracity cannot be checked. It seems to me that it perfectly corresponds to the Darrah spirit. Thus, it is not surprising that he left such a nice memory of himself at Lille and that the Bertrands appreciated Darrah and all his family very much.

Bertrand and Darrah kept in touch over the years and exchanged some specimens (Fig. 20). Even after Bertrand's decease, Helen Hilsman Darrah, W. C. Darrah, and Cecile P. Bertrand maintained their correspondence (Fig. 21).

No doubt such a lasting friendship between P. Bertrand and W. C. Darrah was only possible because they shared many traits: modesty, brilliant mind, scientific integrity, energetic research, and exquisite kindness for everybody. Although they are no longer with us, they still provide us wonderful examples.

Wilkes-Barre
Glen Alden C° — Loomis Colliery
probabl' horizon du Ross Coal
Neuropteris Scheuchzeri

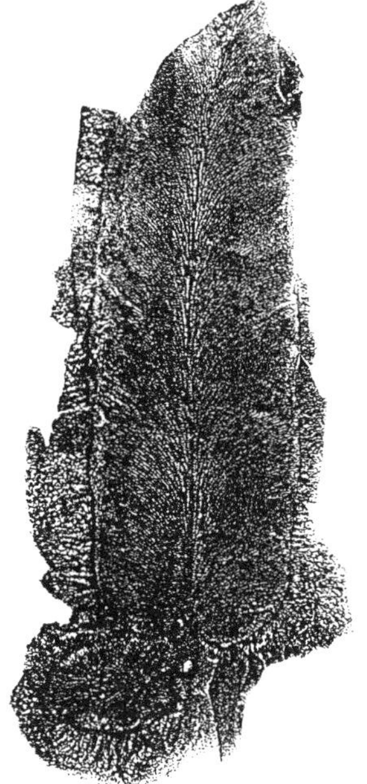

HARVARD UNIVERSITY: BOTANICAL MUSEUM

No. 18946 Orig. No.

Neuropteris decipiens Lesquereux

Carboniferous: Allegheny
Mazon Creek, Illinois

Presented to
the Coal museum of
University of Lille
1936

det. Darrah

Figure 20. Two examples of labels, one written by Paul Bertrand and the other by William C. Darrah. These labels accompany the specimens of two Carboniferous plants in the Lille collections. On the left hand, a copy of a peel made from a Mazon Creek specimen, which was prepared by William C. Darrah. The Lille Laboratory of Paleobotany Collections.

Helen Hilsman Darrah
122 Lincoln Road
Medford, Massachusetts

June 9, 1946.

My dear Madame Bertrand,

I recieved your letter on May 31 - exactly 3 weeks from the time that you had written to us. It was so good to hear from you and to get

. . . . . . . . . . . . . . . .

My dear Mme. Bertrand:

It was indeed a joy, mixed with sadness, that we received your letter, the papers of your distinguished husband, and the statements of his admiring colleagues. We shall always treasure his memory and affectionate courtesies, and his picture which you have so understandingly sent us. I shall consider it an honor to compose a biographical and obituary notice of Professor Bertrand and his work. When copies are available you will receive a number of copies and I shall see personally that his many colleagues will receive copies, also.

It is a double sadness that your husband did not witness the liberation of France. His tender nature must have suffered for his country. We hope some day to visit France and, God willing, to see you and to visit the tomb of Professor Bertrand at St. Germaine.

I shall as you desire, prepare a list of paleobotanists and geologists.

Again may I express affectionate greetings to you and beg that will will tell us what we may send to you to add to your comfort and health.

Sincerely,

Wm. C. Darrah

Figure 21. Copy of the letter sent by William C. Darrah to Cécile P. Bertrand, written on the back of a letter sent by Helen Hilsman Darrah, two years after Paul Bertrand's death. Paris National Museum of Natural History Collections. Courtesy of C. Blanc.

## ACKNOWLEDGMENTS

I am much indebted to C. Blanc, head of the Laboratory of Paleobotany at Paris National Museum of Natural History, for the loan of numerous documents that were deposited by Cécile P. Bertrand. I am also largely indebted to E. D. Morey and to P. C. Lyons for important additional information about the W. C. Darrah–Paul Bertrand relationship.

I express my thanks to C. J. Cleal for his help in correcting a first draft of the manuscript. I am also more than grateful to E. D. Morey, P. C. Lyons, E. L. Zodrow, and R. H. Wagner for their meticulous reviews and helpful comments, which greatly improved the final version of the manuscript.

## REFERENCES CITED

Barrois, C., Bertrand, P., and Pruvost, P., 1922, Observations sur le terrain houiller de la Moselle: Comptes rendus de l'Académie des Sciences (Paris), t. 175, p. 657–660.

Barrois, C., Bertrand, P., and Pruvost, P., 1930, Le Conglomérat houiller de Roucourt: Liège, Congrès International des Mines, Section Géologie, 6th session, p. 147–158, 1 pl.

Bertrand, C.-E., 1874, Anatomie comparée des tiges et des feuilles chez les Gnétacées et les Conifères [Thèse de Doctorat]: Paris, G. Masson éditeur, 149 p., 14 pls.

Bertrand, C.-E., 1883, Recherches sur les Tmésiptéridées: Lille, Imprimerie L. Danel, 350 p.

Bertrand, C.-E., 1891, Remarques sur le *Lepidodendron hartcourtii* de Witham: Lille, Travaux et Mémoires des Facultés, t. 2, mém. 6, 159 p., 10 pls.

Bertrand, P., 1907, Classification des Zygoptéridées d'après les caractères de leurs traces foliaires: Comptes rendus de l'Académie des Sciences (Paris), t. 145, p. 775–777.

Bertrand, P., 1908a, Caractéristiques de la trace foliaire dans les genres *Gyropteris* et *Tubicaulis:* Comptes rendus de l'Académie des Sciences (Paris), t. 146, p. 208–210.

Bertrand, P., 1908b, Sur les stipes de *Clepsydropsis:* Comptes rendus de l'Académie des Sciences (Paris), t. 147, p. 945–947.

Bertrand, P., 1909, Etudes sur la fronde des Zygoptéridées [Thèse de Doctorat]: Lille, Faculté des Sciences, 306 p., 16 pls.

Bertrand, P., 1914a, Relations des empreintes de *Corynepteris* avec les *Zygopteris* à structure conservée: Comptes rendus de l'Académie des Sciences (Paris), t. 158, p. 740–742.

Bertrand, P., 1914b, Les zones végétales du terrain houiller du Nord de la France. Leur extension verticale par rapport aux horizons marins: Annales de la Société géologique du Nord (Lille), t. 43, p. 208–254.

Bertrand, P., 1918, Les grandes divisions paléontologiques du Stéphanien du bassin de la Loire: Comptes rendus de l'Académie des Sciences (Paris), t. 167, p. 689–692.

Bertrand, P., 1919, Les zones végétales du terrain houiller du Nord de la France: Comptes rendus de l'Académie des Sciences (Paris), t. 168, p. 780–782.

Bertrand, P., 1920a, Succession normale des flores houillères dans le bassin houiller du Gard: Comptes rendus de l'Académie des Sciences (Paris), t. 170, p.331–333.

Bertrand, P., 1920b, C. Grand'Eury. Notice nécrologique: Bulletin de la Société géologique de France, 4th ser., t. 19, p. 148–162.

Bertrand, P., 1922a, Succession régulière des zones végétales dans les bassins houillers français: Congrès Géologique International, 13th, Bruxelles, p. 599–608.

Bertrand, P., 1922b, Sur les flores houillères de la Sarre: Comptes rendus de l'Académie des Sciences (Paris), t. 175, p. 770–772.

Bertrand, P., 1926, La zone à *Mixoneura* du Westphalien supérieur: Comptes rendus de l'Académie des Sciences (Paris): t. 183, p. 1349–1350.

Bertrand, P., 1928a, L'échelle stratigraphique du terrain houiller de la Sarre et de la Lorraine: Congrès pour l'avancement des études de stratigraphie Carbonifère, 1st, Heerlen, The Netherlands, 1927: Compte rendu, p. 83–92.

Bertrand, P., 1928b, Stratigraphie du Westphalien et du Stéphanien dans les différents bassins houillers français: Congrès pour l'avancement des études de stratigraphie Carbonifère, 1st, Heerlen, The Netherlands, 1927: Compte rendu, p. 93–101.

Bertrand, P., 1928c, Valeur des flores pour la caractérisation des différentes assises du terrain houiller et pour les synchronisations de bassin à bassin: Congrès pour l'avancement des études de stratigraphie Carbonifère, 1st, Heerlen, The Netherlands, 1927: Compte rendu, p. 103–116.

Bertrand, P., 1930a, Bassin houiller de la Sarre et de la Lorraine. I: Flore fossile, ler fascicule: Neuroptéridées: Etudes des Gîtes minéraux de la France (Lille), p. 1–52, 30 pls.

Bertrand, P., 1930b, Les Charbons d'Algues: Congrès International des Mines, Liège, Section Géologie, 6th session, p. 159–168, 7 pls.

Bertrand, P., 1933a, Observations sur la classification des fougères anciennes (Palaeoptéridales) du Dévonien et du Carbonifère: Bulletin de la Société Botanique de France, t. 80, p. 527–537.

Bertrand, P., 1933b, Titres et travaux scientifiques: Lille, Imprimerie G. Sautai, 35 p.

Bertrand, P., 1935, Contribution à l'étude des Cladoxylées de Saalfeld: Stuttgart, Palaeontographica, Abt. B, v. 80, p. 101–170, 25 pls.

Bertrand, P., 1937a, Tableaux des flores successives du Westphalien supérieur et du Stéphanien: Congrès pour l'avancement des études de stratigraphie Carbonifère, 2d, Heerlen, The Netherlands, 1935: Compte rendu, t. 1, p. 67–69.

Bertrand, P., 1937b, Anatomie et ontogénie comparées des végétaux vasculaires: Paris, Bulletin de la Société Botanique de France, t. 84, p. 515–529.

Bertrand, P., 1937c, Notice sur les travaux scientifiques: Lille, Imprimerie G. Sautai, 64 p.

Bertrand, P., 1941a, Solution du problème posé par l'ontogénie comparée des plantules des Phanérogames: Comptes rendus de l'Académie des Sciences (Paris), t. 212, p. 712–713.

Bertrand, P., 1941b, Observations au sujet d'une note de M. E. Boureau sur les dispositions vasculaires excentriques et pseudo-excentriques: Comptes rendus de l'Académie des Sciences (Paris), t. 212, p. 767–768.

Bertrand, P., 1941c, Nouvelle classification des Filicales primitives: Bulletin de la Société Botanique de France, t. 88, p. 621–635.

Bertrand, P., 1941d, Remarques sur l'organisation générale des Clepsydropsis: Comptes rendus de l'Académie des Sciences (Paris), t. 213, p. 500–503.

Bertrand, P., 1942, Sur l'existence d'une structure primaire exarche dans les plantules des Angiospermes: Comptes rendus de l'Académie des Sciences (Paris), t. 215, p. 284–285.

Bertrand, P., 1943a, Les trois aspects de la loi de récapitulation ontogénique et phylogénique chez les végétaux: Genève, Boissiera VII, p. 232–247.

Bertrand, P., 1943b, Supplément à la notice scientifique: Paris, Centre de Documentation Universitaire, 12 p.

Bertrand, P., 1947 (posthumous), *in* Emberger, L., ed., Les végétaux vasculaires. Introduction à l'étude de l'anatomie comparée, suivie de notes originales: Paris, Masson et Compagnie Éditeurs, 184 p. [Includes a list of Paul Bertrand's publications (p. 174–182).]

Bertrand, P., 1950, Reconstitutions de paysages fossiles. Texte de présentation par P. Corsin: Paris, Annales de Paléontologie, t. 36, p. 126–139, 7 pls.

Bertrand, P. and Pruvost, P., 1937, La question du Westphalien et du Stéphanien en France: Congrès pour l'avancement des études de stratigraphie Carbonifère, 2d, Heerlen, The Netherlands, 1935: Compte rendu, t. 1, p. 81–83.

Bower, F. O., 1923, The ferns (Filicales): Cambridge, England, Cambridge Botanical Handbooks, v. 1, 359 p.

Bower, F. O., 1935, Primitive land plants: London, Macmillan Publishers, 658 p.

Darrah, W. C., 1937, Sur la présence d'équivalents des terrains stéphaniens dans l'Amérique du Nord: Annales de la Société géologique du Nord (Lille), t. 61, p. 187–197.

Darrah, W. C., and Bertrand, P., 1933, Observations sur les flores houillères de Pennsylvanie (régions de Wilkes-Barre et de Pittsburgh): Comptes rendus de l'Académie des Sciences (Paris), t. 197, p. 1451–1453.

Darrah, W. C., and Bertrand, P., 1934, Observations sur les flores houillères de Pennsylvanie: Annales de la Société géologique du Nord (Lille), t. 58, p. 211–224.

Emberger, L., 1944, Paul Bertrand (1879–1944): Bulletin de la Société Botanique de France, t. 91, p. 161–165, 1 pl.

Fosse, R., and Bertrand, P., 1905, Sur un persulfate organique: Comptes rendus de l'Académie des Sciences (Paris), t. 139, p. 600–602.

Halle, T. G., 152, Professor Sahni's palaeobotanical work: The Palaeobotanist (Lucknow), v. 1, Birbal Sahni Memorial Volume, p. 22–41.

Kidston, R., and Lang, W. H., 1917, On Old Red Sandstone plants showing structure, from the Rhynie Chert Bed, Aberdeenshire. Part I: *Rhynia Gwynne-Vaughani,* Kidston and Lang: Transactions of the Royal Society of Edinburgh, v. 51, pt. 3, p. 761–784, 10 pls.

Laveine, J.-P., 1974, Précisions sur la répartition stratigraphique des principales espèces végétales du Carbonifère supérieur de Lorraine: Comptes rendus de l'Académie des Sciences (Paris), t. 278, sér. D, p. 851–854.

Laveine, J.-P., 1977, Report on the Westphalian D, *in* Holub, V. M., and Wagner, R. H., eds., Symposium on Carboniferous Stratigraphy, 1973: Prague, Geological Survey of Prague, p. 71–87, 2 pls.

Laveine, J.-P., 1989, Guide paléobotanique dans le terrain houiller sarrolorrain: Merlebach, Imprimerie des Houillères du Bassin de Lorraine, 154 p., 64 pls.

Lignier, O., 1903, Equisétales et Sphénophyllales. Leur origine filicéenne commune: Bulletin de la Société Linnéenne de Normandie, 5th ser., v. 7, p. 93–137.

Lignier, O., 1908, Essai sur l'évolution morphologique du règne végétal: Congrès de l'Association Française pour l'avancement des Sciences, Comptes rendus, p. 530–542.

Lignier, O., 1908–1909, Essai sur l'evolution morphologique du Règne végétal: Bulletin de la Société Linnéenne de Normandie, 6th ser., 3d v., p. 35–62.

Lignier, O., 1914, Titres et travaux scientifiques: Laval, Imprimerie L. Barnéoud et Compagnie, 120 p.

Lyons, P. C., and Morey, E. D., 1995, Birbal Sahni—His North American connection with William Culp Darrah: Lucknow, Birbal Sahni Centenary Compendium (in press).

Morvillez, F., 1918, Charles-Eugène Bertrand, Correspondant de l'Institut, Professeur de botanique à la Faculté des Sciences de Lille (1851–1917): Caen, Imprimerie H. Delesques, 55 p.

Pruvost, P., 1945, L'œuvre de Paul Bertrand, paléobotaniste (1879–1944). Un portrait: Bulletin de la Société géologique de France, v. 5, p. 229–243.

Pruvost, P., 1946, L'œuvre de Paul Bertrand, paléobotaniste (1879–1944): Annales de la Société géologique du Nord (Lille), t. 65, p. 127–137.

Walton, J., 1945, Prof. Paul Bertrand: Nature, v. 155, p. 419.

Williamson, W. C., 1874, On the organization of the fossil plants of the coalmeasures. Part 6: Ferns: Philosophical Transactions of the Royal Society, London, v. 164, p. 675–703, 8 pls.

Zalessky, M. D., 1910, Excursion paléobotanique en Angleterre: Bulletin du Comité Géologique, Saint Pétersbourg, t. 29, p. 697–713.

Zalessky, M. D., 1914, On the nature of *Pila* of the yellow bodies of boghead and on sapropel of the Ala-Kool Gulf of the Lake Balkhach: Comité Géologique (Saint Pétersbourg), lettre scientifique no. 4, p. 11–14.

Zimmermann, W., 1930, Die Phylogenie der Pflanzen: Jena, G. Fischer Verlag, 454 p.

Zimmermann, W., 1952, Main results of the "Telome theory": The Palaeobotanist , (Lucknow), v. 1, Birbal Sahni Memorial Volume, p. 456–470.

Manuscript Accepted by the Society July 6, 1994

Printed in U.S.A.

Geological Society of America
Memoir 185
1995

# *Carl Rudolf Florin (1894–1965): A pioneer in fossil-conifer studies*

**Elsie Darrah Morey**
*Morey Paleobotanical Laboratory, 1729 Christopher Lane, Norristown, Pennsylvania 19403*

## ABSTRACT

**Carl Rudolf Florin (1894–1965) received his education from the University of Stockholm. During the early years, Florin was a "docent" (lecturer) at the Riksmuseum (Swedish Museum of Natural History) and served under Thore G. Halle, who was director of the Paleobotanical Department. In 1942, Florin received his professorship, and he became the director of the Bergius Foundation in 1944. Florin was a pioneer in fossil-conifer studies. Recognition of his great expertise on primitive conifers and their relationship to the Cordaitales came late in life. His interpretations and concepts are widely utilized today. Florin's magnum opus, entitled "Die Koniferen des Oberkarbons und des unteren Perms," consists of eight parts totaling 729 pages and 186 plates. Florin described about two dozen new genera, of which the best known are *Ernestiodendron* and *Lebachia;* the latter name is now regarded as illegitimate.**

## EARLY YEARS

Carl Rudolf Florin was born on April 5, 1894, in Solna, Sweden. He was the eldest son of Carl Ludwig and Augusta Jansdotter Florin. His father, Carl L. Florin (1869–1933), was the gardener at the Hortus Bergianus Botanical Garden (Stockholm, Sweden) in the 1890s and was appointed head gardener in 1919. This garden atmosphere provided the ideal stimulus for his two sons—Rudolf, who became a paleobotanist, and his younger brother, who became a landscape gardener.

Rudolf began his education at the Beskowska Skolan, a school in Stockholm, and entered the University of Stockholm in 1914. He studied under several prominent professors who influenced him: G. Lagerheim, a pioneer in pollen analysis; O. Rosenberg, a cytologist; and Baron G. de Geer, who founded the Swedish clay-varve geochronology for late Pleistocene glacial deposits.

Rudolf Florin's early work dealt with pomology. One of his papers was coauthored with his father and described a new kind of apple (Florin and Florin, 1918). With his wife, Elsa Borgenstem Florin, Rudolf Florin experimented with fertility and partial sterility of pollen from apples, cherries, and pears to determine the compatibility of several varieties.

Florin passed the Swedish "filosofie kandidat" examination in 1919 and graduated "filosofie licentiat" in 1920. Like her husband, Elsa Florin earned her "filosofie licentiat" in botany, and thus she was an able assistant to her husband, much as Helen Darrah (a botanist) was to William C. Darrah. Later Elsa Florin taught in high school.

Under O. Rosenberg, Florin received instruction in botanical and cytological research techniques. Some of this training was used in relationship to Florin's work on angiosperm embryology, as shown in Florin's work on the Hepatics (Florin, 1918, 1922a) and later in his work with *Ephedra* (Florin, 1932). These studies qualified him for the post of "docent" (lecturer) at the Swedish Museum of Natural History.

## FLORIN'S WORK AT THE SWEDISH MUSEUM OF NATURAL HISTORY

In 1918, Florin (Fig. 1) was appointed assistant keeper to the Paleobotanical Department of the Swedish Museum of Natural History. The director of the Paleobotanical Department was Thore G. Halle (1884–1964). A. G. Nathorst (1850–1921), founder of the Paleobotanical Department, was still working there. Nathorst had wanted Florin to continue Nathorst's work on Tertiary floras, but Florin's principal interest was on the epidermal structure of the conifers. Florin made in-

Morey, E. D., 1995, Carl Rudolf Florin (1894–1965): A pioneer in fossil-conifer studies, *in* Lyons, P. C., Morey, E. D., and Wagner, R. H., eds., Historical Perspective of Early Twentieth Century Carboniferous Paleobotany in North America (W. C. Darrah volume): Boulder, Colorado, Geological Society of America Memoir 185.

Figure 1. Rudolf Florin at the age of 25, sitting behind his microscope in the Riksmuseum (Swedish Museum of Natural History). Courtesy of and permission from the Bergius Foundation, Stockholm, Sweden.

itial comparative studies of the cuticles of some living conifers to find if they had distinct epidermal characteristics. These studies included the shoots of *Sequoia* and *Taxodium,* two genera well represented in the Tertiary floral record. Florin discovered that the two genera could be distinguished by the arrangement of their stomata.

Thus, for his doctorate thesis (1931), Florin chose to investigate the epidermal structures of living conifers and to compare these with the cuticles of certain fossil conifers. Florin's dissertation was entitled "Untersuchungen zur Stammesgeschichte der Coniferales und Cordaitales. I: Morphologie und Epidermisstruktur der Assimilationsorgane bei den rezenten Koniferen." Most important in his thesis are the descriptions of the stomata that can be used to recognize many living and fossil taxa. Florin (1933, 1934a) extended this type of work to the Bennettitales, continuing the work of Thomas and Bancroft (1913) on studies of the cuticles of the Bennettitales and the recent cycads. Florin described the stomatal apparatus of the Cycadales as "haplocheilic" and in the Bennettitales as "syndetocheilic." Syndetocheilic stomata were found only in this extinct group and in the living genus *Welwitschia* (Florin, 1934d).

Florin's (1936a, b, c) work on the fossil Ginkgophytes from Franz-Joseph-Land is monumental. He macerated the fossil cuticles from the rocks, embedded them in paraffin, and sectioned them with the microtome. He recognized and described six new genera in the fossil Ginkgoales and compared the cuticles of modern *Ginkgo biloba* to those of fossil taxa. His study advanced our documentation of the diversity of this group in the Mesozoic.

Florin continued his work on the pollen grains of Paleozoic Cordaitales (Florin, 1936c) and completed an important phytogeographical paper on the Tertiary fossil conifers of South Chile (Florin, 1940b). His study of the silicified cordaitean material from Grand Croix near St. Étienne, France, was completed in 1944. Despite the importance of all of Florin's smaller contributions, they are overshadowed by his monumental monographic work on the Permo-Carboniferous conifers, "Die Koniferen des Oberkarbons und des unteren Perms, which was published in eight parts totaling 729 pages and 186 plates. This work was the result of years of study of fossil material from about 75 European and American localities.

Florin presented an explanation of the nature and evolution of the seed cone of conifers, which was based upon his extensive studies of both modern and fossil taxa. He demonstrated that in modern pines, for example, the seemingly simple and flattened ovuliferous scale was derived phylogenetically from ancestral forms that bore radially arranged small vegetative and seed-bearing leaves on the short fertile shoots in a compound seed cone. These short shoots became progressively reduced, through evolutionary time, and are represented by many related conifer fossils. He demonstrated that this modern conifer's seed-bearing appendage was made simple by reduction and that the modern ovule has its recurved micropyle pointed toward the axis (see Andrews, 1980).

Florin's monographic work (1938, 1939a, 1940a, 1944, 1945) was too large to be published by the Swedish Academy of Science during the years of World War II and was therefore published as a series in *Palaeontographica* by Schweizerbart'sche Verlagsbuchhandlung at Stuttgart, Germany. The monograph sold so well that it saved the firm from economic collapse. Several of the European collections studied by Florin were destroyed during World War II, which makes the careful, thorough description and illustrations within the monograph extremely valuable (Lundblad, 1966).

Florin utilized his work on cuticular characteristics of compressional remains to recognize several new genera of fossil conifers. Sterile remains without cuticles were assigned to the preexisting genus, *Walchia* Sternberg. Florin established two new genera, *Lebachia* and *Ernestiodendron,* for fossils that have preserved cuticle containing certain stomatal features. Clement-Westerhof (1984) reported that the name *Lebachia* is synonymous with *Walchia* and, thus, is nomenclaturally incorrect (see also Mapes and Rothwell, 1991). The ovule-bearing cones of these genera consist of a central axes with spirally inserted bracts that carry a fertile dwarf-shoot in their axils. During evolution, the shoot flattened, and the scales of the dwarf-shoot are simplified by reduction in number and fusion.

The Late Permian genera *Pseudovoltzia* and *Ullmannia* show this feature.

## FLORIN'S TIME IN THE UNITED STATES

Florin and W. C. Darrah (1909–1989) began corresponding after their meeting in The Netherlands, where they both attended the 1935 Sixth International Botanical Congress (Amsterdam) and the Second International Carboniferous Conference (Heerlen).

Florin write Darrah, "It is extremely generous of you to offer me so kindly to borrow your whole material of *Cordaianthus,* sections, transfers and carbonized material."[1] Two months later, Florin wrote Darrah, "It will be a great pleasure to me and of greatest benefit to my work to be able to study also American specimens of cordaitean fructifications."[2]

In 1937, Florin wrote Darrah and indicated that the eight *Walchia* specimens borrowed from the Harvard Botanical Collection were being returned. He said, "I have practically finished the description of all the material of *Walchia* placed at my disposal by different museums. . . . I intend now to visit museums in Central Europe . . . and the United States."[3]

Florin came to the United States for the first time in 1937. When Florin's first visit to the United States and Harvard University was drawing near, Darrah spoke with Professors E. C. Jeffery and R. Wetmore to ask if Florin could speak at Harvard's Biological Colloquium. Florin was invited and notified Darrah that his topic for the lecture would be the structure of pollen grains in the Cordaitales.

Florin sent his itinerary to Darrah, planning arrival on May 13, 1937. First he would go to Washington, D.C., to study the collections of the U.S. National Museum, and from there he would travel to Lawrence, Kansas, to visit with M. K. Elias (see Chapter 13, this volume). The next part of the trip was to be to Urbana and Chicago, Illinois, to visit with A. C. Noé (see Chapter 14, this volume). From Chicago, Florin was to travel to Ithaca, New York, and finally from there to Harvard University in Cambridge, Massachusetts.

The purpose of the visit to the United States was to visit about 10 universities and to lecture at each in hopes that he could find a faculty position at an American university. "Unfortunately, he was a dull, boring lecturer, even with good English. He was unimpressive and [had a] rather condescending, probably 'superior' manner would be more accurate. He was accepted politely and not a single institution was interested even though Yale, Chicago, and the Missouri Botanical Garden were openly seeking a paleobotanist," wrote Darrah to P. C. Lyons.[4]

Andrews (1980, p. 283) wrote of Florin's visit to St. Louis, Missouri: "He was something of a disappointment to my students in that he seemed to have little interest in any fossils other than the conifers, and unfortunately my collections were rather sparse in that direction. But his knowledge of and devotion to the investigation of a great group of plants was so great and so apparent that it was certainly a unique experience."

C. A. Arnold (see Chapter 18, this volume) wrote to Darrah about his meeting with Florin, "I was surprised at Dr. Florin's visit. Didn't even know he was in the country, but one morning while talking to a class I glanced up and he stood in the doorway. Those Europeans, deprecating American paleobotany, have given in to a very prevalent human fallacy of taking the best of one and comparing it to the worst of the other. . . . Dr. Florin, however, I like very much. He may hold the current European views about us wild and obstreperous Americans, but he, at least, is nice about it."[5]

Darrah had two specimens of Appalachian *Walchia* that he had collected in 1930. The first was a small specimen of *Walchia piniformis* from the Clarksburg member of the upper part of the Conemaugh Formation at Rennerdale, Allegheny County, Pennsylvania. The other specimen, *Walchia* (*Lecrosia*), was collected in the arenaceous shale from the lower part of the Greene Formation (Dunkard Group) near Mount Morris, Greene County, Pennsylvania (Darrah, 1975). In 1933, D. White (see Chapter 10, this volume) confirmed the identifications as *Walchia* (see White, 1936). Darrah wrote Lyons, "[White] . . . was at first sight enthusiastic. He said he had been convinced the conifers would be found in the Dunkard because the plant associations in Kansas and Texas included *Walchia.*"[6]

Florin later confirmed Darrah's and White's identifications of *Walchia piniformis* and *Walchia* (*Lecrosia*) from the Appalachian region. In a letter to Lyons, Darrah wrote "Florin many years ago [1937] verified the identifications and said it was probably *Lebachia* a 'rather early occurrence.' He also identified my single specimen from the upper part of the Greene [Dunkard Group] [Formation] as possible *Ernestiodendron.*"[7] In still another letter to Lyons, Darrah explained, "Florin examined the Conemaugh specimen [but he] could not get cuticle, but said it was probably *Lecrosia.*"[8] Darrah's was the first discovery of a walchian conifer in the Dunkard Group (Lyons and Darrah, 1989a, b).

Florin's 1937 stay at Harvard lasted a week. His time was spent studying the extensive fertile cordaitean material, *Cordaianthus,* from the Iowa coal balls that Darrah had been working on (Darrah, 1935, 1936). Prior to his visit, Florin had suggested that when he came to the United States he wished to view all of the Harvard paleobotanical material of *Cordaites* before any of the fossils or preparations would be sent to Stockholm for further study and inclusion in his "Die Koniferen des Oberkarbons und des unteren Perms" (1938, 1939a).

Darrah (1937) also had described Cambrian spores from Sweden. While Florin was in Cambridge, Massachusetts, he examined Darrah's slides from this work. "Florin looked at the preparation under the microscope and said, 'That's an unusual spore—what is it?' When I [Darrah] said 'I don't know; it is upper Cambrian,' he replied, 'Then it can't be a spore.' Later the Russians confirmed this occurrence and Florin was piqued that Swedish paleobotanists had not been asked to investigate the specimen."[9]

At the end of each day's study at Harvard, Florin went to

the Darrahs' apartment for an enjoyable evening of dinner and conversation while they listened to classical music on the phonograph that all enjoyed. On Florin's first acceptance to dinner at the Darrahs', he asked Darrah if they would be having potatoes and said that, if so, he liked them fluffed and that he enjoyed steak. If beer was to be served, Florin liked Budweiser. With amusement, Bill Darrah told his wife, Helen Darrah (1909– ), who knew exactly what fluffed potatoes were. Helen took care of the dinner, and Bill bought the beer.

After returning to Stockholm from Cambridge, Massachusetts, Florin wrote, "I beg to express to you and Mrs. Darrah my heartiest thanks for all the kindness shown to me during my stay at Harvard. I enjoyed the week in Boston-Cambridge very much and I only regret I could not stay longer. Harvard (as well as Cornell) is to me characterized by a most encouraging atmosphere, and I highly appreciate the interest everybody has taken in my present work. The books you were so kind as to send on my behalf have already arrived."[10]

Florin wrote later, "I have now described and figured also all the American material placed at my disposal with the exception of the specimens from Harvard University which have not yet arrived (but will apparently do so in the future). In the meantime, I thank you for your courtesy to send us others in exchange."[11]

Shortly thereafter, Florin wrote Darrah again: "Some few days ago I received in good condition the box with fossil plants you were so kind as to promise to send. It contained two specimens of *Walchia* and one of *Ullmannia* belonging to your Museum—these will soon be returned to you. . . . I myself thank you very much for sending me the conifers, which will be referred to in my monograph on the Upper Carboniferous and Lower Permian Conifers.[12] Florin provided updates as parts of the monograph became completed. "This is to inform you that the three specimens of Paleozoic conifers borrowed from your collection some time ago will now be returned. Two of them have been figured in my monograph on the oldest conifers"[13] (Florin, 1938, 1939a).

Florin had received his position of "docent" (lecturer) at the University of Stockholm in 1931. His salary was a pittance. Darrah's impression of Florin in the 1930s was that he was a somewhat bitter and frustrated man. In Sweden, the only position in paleobotany was that held by Thore G. Halle (Fig. 2), Florin's superior. Florin's reputation at the University of Stockholm was overshadowed by Halle's name and position.

When Florin became director of the Bergius Foundation in 1944 and thus editor of the journal *Acta Horti Bergiani,* he could excel in paleobotany on his own, and his stature was enhanced. Florin was able to purchase equipment for his department that met his needs for paleobotanical research. Along with his administrative work at the Bergius Foundation, Florin continued to study modern and fossil *Taxus.*

## FLORIN'S LATER LIFE

Florin did not reach great prominence as a paleobotanist until late in life. Today we realize the great importance of Florin's work that focused on his research on the evolution of the modern conifer bract-scale complex. It is Florin's early work that today provides the stimulus for paleobotanists to critically examine the morphology and evolution of the conifers (Taylor and Taylor, 1993). For this, Florin is owed a

Figure 2. Thore G. Halle and Rudolf Florin at Stockholm, Sweden, in 1950. Courtesy of and permission from Hunt Institute for Botanical Documentation, Carnegie Mellon University, Pittsburgh, Pennsylvania.

debt of gratitude. His papers of 1950 and 1951 are excellent review articles written in English. Many of our concepts of the primitive conifers were formulated by Florin or influenced by him. Florin recognized the family of Paleozoic voltzialean conifers including the Lebachiaceae, which consists of four genera. *Lebachia* Florin is an illegitimate genus for which *Utrechtia* has been proposed as a substitute genus (Mapes and Rothwell, 1991). These authors retained *Walchia* Sternberg as a form genus for vegetative shoots of late Paleozoic conifers that do not have preserved cuticle detail, in contrast to Clement-Westerhof (1984), who proposed *Walchia* as a natural genus to replace *Lebachia.* The family Utrechtiaceae (Mapes and Rothwell, 1991) replaced the family Lebachiaceae (see also Stewart and Rothwell, 1993). As new conifer fossils are discovered and known fossils are reconsidered, there will be many changes in conifer systematics, but all students of conifers will build on the conceptual framework provided by Florin's investigations.

In 1948 and 1949, Rudolf Florin was a visiting professor at Harvard University. There he presented the John M. Prather Lectures on Biology. His topic at the Prather Lectures was "Evolution in Cordaites and Conifers." Later this lecture series was published in *Acta Horti Bergiani* (Florin, 1951). That paper is an exceptional summary of the relationships between the cordaitean and walchian cones and the evolution of the ovulate conifer cone. H. N. Andrews (1980, p. 282) said of this excellent paper, "Those who are fortunate enough to acquire a copy of this own one of the greatest classics in scientific literature."

Another of Florin's important research works is the paleofloristic investigation of the "Jurassic Taxads and Conifers from North-Western Europe and Eastern Greenland" (Florin, 1958). Florin determined to what extent the class Taxopsida occurred in the northern Jurassic floras and recognized that fossil members of the Taxaceae, Cephalatoxaceae, and Taxodiaceae were present.

In 1960, Florin traveled to the United States for a third time, when he was a visiting professor at the University of California. There he presented lectures for the Charles M. and Martha Hitchcock Foundation of the University of California. His topic was the paleo- and neogeographic distribution of the conifers. In 1963, he published a summary of his life's work—"The Distribution of Conifer and Taxad Genera in Time and Space." It is a synopsis of the lectures that he had presented at the Hitchcock Foundation.

## HONORS AND AFFILIATIONS

Florin has been honored by fellow paleobotanists by having several genera named for him. Some of these are *Florinites* (Schopf et al., 1944), *Ruflorinia* (Archangelsky, 1963), and *Rufloria* (Meyen, 1963).

He was presented with the Linnaeus Gold medal of the Royal Physiographic Society (London), and he also received the corresponding award of the Swedish Academy of Sciences. In 1958, he received the Darwin-Wallace Commemorative Medal of the Linnaean Society in London. Florin (Fig. 3) served as Secretary of the Committee for the Conservation of Nature of the Swedish Academy of Sciences from 1931 to 1947 and later received the Gold Medal from the Conservation of Nature of the Swedish Academy of Sciences.

Florin was a member of many scientific societies. He was president of the Swedish Botanical Society at the time of his death in 1965. Florin served on the Council of the International Association for Plant Taxonomy from 1959 to 1964 and on the Council of the International Organization of Paleobotany, also serving as its first president. He was president of the Paleobotanical Section of the International Union of Biological Sciences from 1954 to 1959 (Lundblad, 1966).

Florin's scientific bibliography includes about 125 papers (see Florin's selected bibliography, this chapter). He described 24 new genera of fossil plants, of which the Paleozoic walchian conifers, *Ernestiodendron* and *Lebachia,* are the best known.

Britta Lundblad reminisced to Morey in a recent letter,

Figure 3. Rudolf Florin, a portrait. Courtesy of and permission from the Hunt Institute for Botanical Documentation, Carnegie Mellon University, Pittsburgh, Pennsylvania.

I remember with pleasure my first year as a paleobotanist, when Florin still worked in the Paleobotanical Department of the Riks museum. It was Florin, who introduced me to a paleobotanical research work. . . . I had a scholarship from Stockholm University and had to give in reports on my work to Florin four times a year. I enjoyed very much having this contact with him. At that time relations between research students and their teachers were still very formal, and I did not "drop titles" with Florin until I defended my thesis. I was never invited to Florin's home as an ordinary guest during his lifetime, but I was there once during the fifties, presenting some professional information. . . . I remember a visit there with H. N. Andrews and his wife. . . . At the University, Florin was known as "Black Rudolf" [B. R.] perhaps because he was severe and exacting. (B. R. is really a charming ballad about a sailor, written by the Swedish Nobel prize winner, E. A. Karlfeldt, but B. R. was really very unlike R. F.).[14]

In another letter to Morey, Lundblad wrote, "Florin was more human than Halle (it was not only my personal opinion but also that of a lady, who for many years was the secretary to the Department of Paleobotany). He [Florin] had for many years a too small salary . . . it seems probable to me . . . [that it may] have acted as a stimulus to Florin."[15]

Again from Lundblad (1966, p. 91): "The impression remains with me that Florin worked in a spirit of calm conscientiousness, yet strangely enough, he never seemed hurried. This is all the more remarkable in view of his great accomplishments, and he must have had extraordinary power of concentration and organization."

Darrah said of Florin (then deceased) that he was "the all time expert on Paleozoic Conifers."[16] It should be noted that 1994 marked the 100th birthday of Carl Rudolf Florin (Fig. 4). From Lundblad, we learn there will be a memorial volume in his honor, which adds additional high praise to his contributions, especially his monumental work on the primitive conifers.

Rudolf Florin, we salute you!

Figure 4. Elsa and Rudolf Florin at Sigtuna, Sweden, 1964. Photograph taken by and courtesy of H. N. Andrews.

## ACKNOWLEDGMENTS

I acknowledge the use of the William Culp Darrah Paleobotanical Collection of papers and letters. Thanks are given to Henry H. Zoller (U.S. Geological Survey Library, Reston, Virginia) for his courtesy in obtaining the references of Florin's publications and to Henry Andrews for letting me reproduce his 1964 photograph of Professor and Mrs. Florin. I thank Paul C. Lyons for permission to quote from letters written to him by W. C. Darrah and for his thought-provoking comments on the manuscript.

The original copies of W. C. Darrah's letters to Rudolf Florin are housed in the Center of the History of Science under the auspices of the Swedish Academy of Sciences, Stockholm, Sweden. Copies of these letters were obtained by courtesy of B. Lundland so that I could refer to them in preparing this chapter.

Hans Kerp, Gar Rothwell, and Gene Mapes are thanked for their reviews of this manuscript.

## ENDNOTES

1. Letter from R. Florin to W. C. Darrah, November 11, 1935. W. C. Darrah Collection.
2. Letter from R. Florin to W. C. Darrah, January 1, 1936. W. C. Darrah Collection.
3. Letter from R. Florin to W. C. Darrah, March 12, 1937. W. C. Darrah Collection.
4. Letter from W. C. Darrah to P. C. Lyons, October 18, 1987.
5. Letter from C. A. Arnold to W. C. Darrah, July 14, 1937. W. C. Darrah Collection.
6. Letter from W. C. Darrah to P. C. Lyons, August 20, 1987.
7. Letter from W. C. Darrah to P. C. Lyons, November 14, 1979.
8. Letter from W. C. Darrah to P. C. Lyons, May 6, 1987.
9. Letter from W. C. Darrah to P. C. Lyons, November 19, 1979.
10. Letter from R. Florin to W. C. Darrah, June 28, 1937. W. C. Darrah Collection.
11. Letter from R. Florin to W. C. Darrah, August 19, 1937. W. C. Darrah Collection.

12. Letter from R. Florin to W. C. Darrah, November 18, 1937. W. C. Darrah Collection.

13. Letter from R. Florin to W. C. Darrah, April 15, 1939. W. C. Darrah Collection.

14. Letter from B. Lundblad to E. D. Morey, May 11, 1992. W. C. Darrah Collection.

15. Letter from B. Lundblad to E. D. Morey, June 16, 1992. W. C. Darrah Collection.

16. Letter from W. C. Darrah to P. C. Lyons, October 18, 1987.

## REFERENCES CITED


Andrews, H. N., 1980, The fossil hunters, in search of ancient plants: Ithaca, New York, and London, England, Cornell University Press, 421 p.

Archangelsky, S., 1963, A new Mesozoic flora from Ticó, Santa Cruz Providence, Argentina: Bulletin of British Museum of Natural History Geology, v. 8, p. 45–92.

Clement-Westerhof, J. A., 1984, Aspects of Permian paleobotany and palynology. IV: The conifer *Ortisea* Florin from the Val Gardena Formation of the Dolomites and the Vicentinian Alps (Italy) with special reference to a revised concept of the Walchiaceae (Goeppert) Schimper: Review of Paleobotany and Palynology, v. 41, p. 51–166.

Darrah, W. C., 1935, Permian elements in the fossil flora of the Appalachian province. 1: *Taeniopteris:* Harvard University Botanical Museum Leaflets, v. 3, p. 137–148.

Darrah, W. C., 1936, Permian elements in the fossil flora of the Appalachian Province. 2: *Walchia:* Harvard University Botanical Museum Leaflets, v. 4, p. 9–19.

Darrah, W. C., 1937, Spores of the Cambrian plants: Science, v. 86, p. 154–155.

Darrah, W. C., 1975, Historical aspects of the Permian flora of Fontaine and White, *in* Barlow, J. A., ed., Proceedings, First I. C. White Memorial Symposium, "The Age of the Dunkard," September 25–29, 1972, Morgantown, West Virginia: Morgantown, Geological and Economic Survey, p. 81–101.

Florin, R. See selected bibliography of Carl Rudolf Florin that follows.

Lundblad, B., 1966, Rudolf Florin: Taxon, v. 15, p. 85–93.

Lyons, P. C., and Darrah, W. C., 1989a, Earliest conifers in North America: Upland and/or paleoclimatic indicators?: Palaios, v. 4, p. 480–486.

Lyons, P. C., and Darrah, W. C., 1989b, Paleoenvironmental and paleoecological significance of walchian conifers in Westphalian (late Carboniferous) horizons of North America: Congrès international de stratigraphie et de Géologie du carbonifère, 11th, Beijing, China, August 31–September 4, 1987: Compte Rendu, v. 3, p. 251–261.

Mapes, G., and Rothwell, G. W., 1991, Structure and relationships of primitive conifers: Neues Jahrbuch für Geologie und Paläontologie Abhandlungen, v. 183, p. 269–287.

Meyen, S. V., 1963, Leaf anatomy and nomenclature of Angaran cordaites: Paleontologichkyeskii Zhurnal, v. 3, p. 96–107.

Schopf, J. M., Wilson, L. R., and Bentall, R., 1944, An annotated synopsis of Paleozoic fossil spores and the definition of generic groups: Illinois Geological Survey Report of Investigations 91, p. 7–72.

Stewart, W. N., and Rothwell, G. W., 1993, Paleobotany and the evolution of plants: New York, Cambridge University Press, 521 p.

Taylor, T. N., and Taylor, E. L., 1993, The biology and evolution of fossil plants: Englewood Cliffs, New Jersey, Prentice-Hall, 982 p.

Thomas, H. H., and Bancroft, N., 1913, On the cuticles of some recent and fossil cycadean fronds: Linnean Society of London Transactions, v. 8, p. 155–204.

White, D., 1936 (posthumous), Some features of the early Permian flora of America: 16th International Geological Congress, Washington, D.C., 1933: Report the 16th Session, United States of America, 1933, v. 1, p. 679–690.


## SELECTED BIBLIOGRAPHY OF CARL RUDOLF FLORIN


Florin, R., 1918, Cytologische Bryophytenstudien. I: Über Sporenbildung bei *Chiloscyphus polyanthus* (L.) Corda: Arkiv för Botanik, v. 15, p. 1–10.

Florin, R., 1920a, Zur Kenntnis der Fertilität und partiellen Sterilität des Pollens bei Apfel und Birnensorten: Acta Horti Bergiani, v. 7, p. 1–39.

Florin, R., 1920b, Über Cuticularstrukturen der Blätter bei einigen rezenten und fossilen Coniferen: Arkiv för Botanik, v. 16, p. 1–32.

Florin, R., 1920c, Zur Kenntnis der jungtertiären Pflanzenwelt Japans: Kungliga Svenska Vetenskapsakademiens Handlingar, v. 61, p. 1–71.

Florin, R., 1922a, Zytologische Bryophytenstudien II and III: Arkiv för Botanik, v. 18, p. 1–58.

Florin, R., 1922b, Zur alttertiären Flora der südlichen Mandschurei: Palaeontologia Sinica A, v. 1, p. 1–45, pls. 1–3.

Florin, R., 1926a, Über eine vermutete Pteridospermen-Fruktifikation aus dem sächsischen Rotliegenden: Arkiv för Botanik, v. 20, p. 1–11, pls. 1–2.

Florin, R., 1926b, Waren Eupodocarpeen (Konif.) in der alttertiären Flora europas vertreten oder nicht?: Senckenbergiana, v. 8, p. 49–62.

Florin, R., 1927a, Pollen production and incompatibilities in apples and pears: Horticultural Society of New York Memoir, v. 3, p. 87–118.

Florin, R., 1927b, Preliminary descriptions of some Palaeozoic genera of Coniferae: Arkiv för Botanik, v. 21, p. 1–7.

Florin, R., 1929, Über einige Algen und Koniferen aus dem mittleren und oberen Zechstein: Senckenbergiana, v. 11, p. 241–266, pls. 1–5.

Florin, R., 1930a, Die Koniferengattung *Libocedrus* Endl in Ostasien: Svensk Botanisk Tidskrift, v. 24, p. 117–131.

Florin, R., 1930b, *Pilgerodendron,* eine neue Koniferen-Gattung aus Süd-Chile: Svensk Botanisk Tidskrift, v. 24, p. 132–135.

Florin, R., 1931, Untersuchungen zur Stammesgeschichte der Coniferales und Cordaitales. I: Morphologie und Epidermisstruktur der Assimilationsorgane bei den rezenten Koniferen: Kungliga Svenska Vetenskapsakademien Handlingar, v. 10, ser. 3, p. 1–588.

Florin, R., 1932, Die Chromosomenzahlen bei *Welwitschia* und einigen *Ephedra*-Arten: Svenska Botanisk Tidskrift, v. 26, p. 205–214.

Florin, R., 1933, Studien über die Cycadales des Mesozoikums nebst Erörterungen über die Spaltöffnungsapparate der Bennettitales: Kungliga Svenska Vetenskapsakademiens Handlingar, v. 12, 134 p., 16 pls.

Florin, R., 1934a, Die Spaltöffnungsapparate der *Williamsonia, Williamsoniella* und *Wielandiella*-Blüten (Bennettitales): Kungliga Svenska Vetenskapsakademiens Handlingar, v. 25, 20 p., 1 pl.

Florin, R., 1934b, Zur Kenntnis der Paläozoischen Pflanzengattungen *Lesleya* Lesquereux und *Megalopteris* Dawson: Kungliga Svenska Vetenskapsakademiens Handlingar, v. 25, 23 p., 3 pls.

Florin, R., 1934c, Über *Nilssoniopteris glandulosa* n. sp., eine Bennettitacee aus der Juraformation Bornholms: Kungliga Svenska Vetenskapsakademiens Handlingar, v. 25, p. 1–19, pls. 1–2.

Florin, R., 1934d, Die Spaltöffnungapparate von *Welwitschia mirabilis* Hook.: Svensk Botanisk Tidskrift, v. 28, p. 264–289.

Florin, R., 1936a, Die fossilen Ginkgophyten von Franz-Joseph-Land nebst Erörterungen über vermeintliche Cordaitales mesozoischen Alters. I: Spezieller Teil: Palaeontographica, Abt. B, v. 81, p. 71–173.

Florin, R., 1936b, Die fossilen Ginkgophyten von Franz-Joseph-Land nebst Erörterungen über vermeintliche Cordaitales mesozoischen Alters. II: Allgemeiner Teil: Palaeontographica, Abt. B, v. 82, p. 1–72.

Florin, R., 1936c, On the structure of the pollen-grains in the Cordaitales: Svensk Botanisk Tidskrift, v. 30, p. 624–651.

Florin, R., 1937, On the morphology of the pollen-grains in some Paleozoic pteridosperms: Svensk Botanisk Tidskrift, v. 31, p. 305–338, pls. 1–3.

Florin, R., 1938, Die Koniferen des Oberkarbons und des unteren Perms., Pt. I: Palaeontographica, Abt. B, v. 85, p. 1–62.

Florin, R., 1939a, Die Koniferen des Oberkarbons und des unteren Perms. Pt. II–IV: Palaeontographica, Abt. B, v. 85, p. 63–242.

Florin, R., 1939b, The morphology of the female fructifications in cordaites and conifers of Paleozoic age: Botaniska Notiser, v. 36, p. 547–565.

Florin, R., 1939c, Über die Calamitaceen-Gattung *Dicalamophyllum* Sterzel aus dem sächsischen Rotliegenden: Kungliga Svenska Vetenskapsakademiens Handlingar, v. 18, p. 3–18, pls. 1–3.

Florin, R., 1940a, Die Koniferen des Oberkarbons und des unteren Perms. Pt. V: Palaeontographica, Abt. B, v. 85, p. 243–364.

Florin, R., 1940b, The Tertiary fossil conifers of South Chile and their phytogeographical significance with a review of the fossil conifers of southern lands: Kungliga Svenska Vetenskapsakademiens Handlingar, v. 19, 107 p., 6 pls.

Florin, R., 1940c, Zur Kenntnis einiger fossiler Salvinia-Arten und der früheren geographischen Verbreitung der Gattung: Svensk Botanisk Tidskrift, v. 34, p. 265–292.

Florin, R., 1940d, On *Walkomia* n. gen., A genus of upper Palaeozoic conifers from Gondwanaland: Kungliga Svenska Vetenskapsakademiens Handlingar, v. 18, p. 3–23, pls. 1–4.

Florin, R., 1940e, Die heutige und frühere Verbreitung der Koniferengattung *Acmopyle pilger:* Svensk Botanisk Tidskrift, v. 34, p. 117–140.

Florin, R., 1940f, Notes on the past geographical distribution of the genus *Amentotaxus pilger* (Coniferales): Svensk Botanisk Tidskrift, v. 34, p. 162–165.

Florin, R., 1940g, Zur Kenntnis einiger fossiler *Salvinia Arten* und der früheren geographischen Verbreitung der Gattung: Svensk Botanisk Tidskrift, v. 34, p. 265–292, pls. 2–3.

Florin, R., 1940h, On Palaeozoic conifers from Portugal: Publicações do Museo e Laboratório de Mineralogia e Geologia, Faculdade de Ciências do Pôrto, v. 18, 20 p., pls. 1–3.

Florin, R., 1944, Die Koniferen des Oberkarbons und des unteren Perms. Pt. VI–VII: Palaeontographica, Abt. B, v. 85, p. 365–654.

Florin, R., 1945, Die Koniferen des Oberkarbons und des unteren Perms. Pt. VIII: Palaeontographica, Abt. B, v. 85, p. 655–729.

Florin, R., 1948a, On the morphology and relationships of the Taxaceae: Botanical Gazette, v. 110, p. 31–39.

Florin, R., 1948b, On *Nothotaxus,* a new genus of the Taxaceae, from eastern China: Acta Horti Bergiani, v. 14, p. 385–395.

Florin, R., 1948c, *Nothotaxus* or *Pseudotaxus?:* Botaniska Notiser, v. 101, p. 270–272.

Florin, R., 1949, The morphology of *Trichopitys heteromorpha* Saporta, a seed-plant of Palaeozoic age, and the evolution of the female flowers in the Ginkgoinae: Acta Horti Bergiani, v. 15, p. 79–109, pls. 1–4.

Florin, R., 1950, Upper Carboniferous and Lower Permian conifers: Botanical Review, v. 16, p. 258–282.

Florin, R., 1951, Evolution in cordaites and conifers: Acta Horti Bergiani, v. 15, p. 285–388.

Florin, R., 1952a, On *Metasequoia,* living and fossil: Botaniska Notiser, v. 105, p. 1–29.

Florin, R., 1952b, Evolution et classification des gymnospermes: Annales de Biologie, v. 28, p. 99–106.

Florin, R., 1953a, Le Jardin botanique de Bergius à Stockholm et autres jardins botaniques suédois: Annales de Biologie, v. 29, p. 263–265.

Florin, R., 1953b, On the morphology and taxonomic position of the genus *Cycadocarpidium* Nathorst (Coniferae): Acta Horti Bergiani, v. 16, p. 257–275.

Florin, R., 1954, The female reproductive organs of conifers and taxads: Biological Reviews, v. 29, p. 367–389.

Florin, R., 1955, The systematics of the Gymnosperms, *in* A century of progress in the natural sciences 1853–1953 (published in celebration of the Centennial of the California Academy of Sciences): San Francisco, California Academy of Science, p. 323–403.

Florin, R., 1956, Nomenclatural notes on genera of living gymnosperms: Taxon, v. 5, p. 188–192.

Florin, R., 1958, On Jurassic taxads and conifers from north-western Europe and eastern Greenland: Acta Horti Bergiani, v. 17, p. 257–402.

Florin, R., 1963, The distribution of conifer and taxad genera in time and space: Acta Horti Bergiani, v. 20, p. 121–312.

Florin, R., 1964, Über *Ortiseia leonardii* n. gen. et sp., eine Konifere aus den Grödener Schichten im Alto Adige (Südtirol): Memorie Geopaleontoligiche dell'Universita di Ferrara, v. 1, p. 1–9.

Florin, R., and Boureau, E., 1956, L'organisation internationale de la paléobotanique, *in* Rapport sur la Paléobotanique dans le monde [I]: Regnum Vegetabile, v. 7, p. 5–7.

Florin, R., and Florin, C., 1918, P. J. Bergius ny applesort: Acta Horti Bergiani, v. 6, p. 1–7.

Florin, R., Boureau, E., and Leclercq, S., 1958, Sur l'organisation internationale de la paléobotanique, *in* Rapport sur la Paléobotanique dans le monde, II: Regnum Vegetabile, v. 7, p. 7–8.

MANUSCRIPT ACCEPTED BY THE SOCIETY JULY 6, 1994

Printed in U.S.A.

Geological Society of America
Memoir 185
1995

# *Marie Stopes (1880–1958): The American connection*

**W. G. Chaloner**
*Geology Department, Royal Holloway, University of London, Egham Hill, Egham, Surrey TW20 OEX, England*

## ABSTRACT

**Marie Charlotte Carmichael Stopes was one of the most remarkable women of the twentieth century. She will be remembered more for her work in pioneering the defense of women's rights and birth control than for her contributions to paleobotany. Three aspects of her research on fossil plants are relevant to the development of that subject in North America; her revision of the Carboniferous "Fern Ledges" flora of New Brunswick, Canada; her work on the origin of coal balls; and her studies on coal petrology.**

## INTRODUCTION

Marie Stopes was described in a biography written nearly 20 years after her death (Hall, 1977) as "one of the twentieth century's most remarkable women." She distinguished herself in at least three quite distinct fields: first as a paleobotanist and coal petrologist of considerable renown; then (chronologically) as a pioneer in proclaiming and defending women's expectations within marriage and in making birth control available and acceptable; and finally as a poet and playwright. But the fame—and some would say notoriety—that she attained through her expositions on sexual technique and in her birth-control work totally overshadowed her activities in the other fields.

In the present context, it is of course her contributions to paleobotany, and particularly to American paleobotany, that primarily concern us, but other aspects of her life inevitably intrude into the story. Four significant biographies of Marie Stopes have been published. The first two were by men (Maude, 1924; Briant, 1962); the later two were by women (Hall, 1977; Rose, 1992). The women, in their very different ways, give much more candid and detailed accounts of her life. A full bibliography was published by Eaton and Warnick (1977). These works give me added reason to concentrate here only on those aspects of Stopes's life that have links with paleobotany and particularly the development of the subject in North America.

Marie Stopes (Fig. 1) was a significant figure in the paleobotanical world between about 1910, after she had returned from a visit to Japan, and 1925, after which her stream of papers on fossil plants and coal rapidly diminished. In those fifteen or so years she added to her earlier work on the structure of living cycad seeds, in contributing significant new information on the Cretaceous Bennettitales; on early angiosperm permineralized woods from the mid-Cretaceous; and, most significantly, on the ways in which coal and coal balls were formed. She developed

Figure 1. Marie Stopes, photographed 1904/1905, at about the time that she obtained her doctorate at Munich. The gas mantle light source is being focused by a water-filled globe acting as a condenser. The slide she was studying is larger than a standard 7.6 × 2.5 cm (3 × 1 in.) and was probably a ground section of a coal ball.

Chaloner, W. G., 1995, Marie Stopes (1880–1958): The American connection, *in* Lyons, P. C., Morey, E. D., and Wagner, R. H., eds., Historical Perspective of Early Twentieth Century Carboniferous Paleobotany in North America (W. C. Darrah volume): Boulder, Colorado, Geological Society of America Memoir 185.

a new and clearly defined terminology for macroscopic coal constituents (Stopes, 1919) and the maceral concept (Stopes, 1935). She also described—or redescribed—the upper Carboniferous "Fern Ledges" flora from New Brunswick, Canada (Stopes, 1914) that was noteworthy for having been thought to be Devonian in age by some earlier workers. This flora was perhaps her only truly North American piece of work. But her research on the terminology of coal constituents and on the process by which coal balls had formed (Stopes and Watson, 1908) probably had a far greater long-term impact on North American paleobotany than her work on the New Brunswick fossils.

I first met Marie Stopes in 1952 at the Geological Society in London, when she attended a meeting at which a lecture was to be presented to the Geologists Association (which, lacking its own rooms, met on the premises of that society). The lecture was to be by anthropologist Kenneth Oakley and was scheduled as the Henry Stopes Lecture—named for Marie Stopes's father. A medal of the Geologists Association, the Henry Stopes Medal is linked to the lecture and is awarded for work of distinction on "the prehistory of man and his geological environment." I was at that time in my second year of paleobotanical research on Carboniferous plants and was introduced to her by Bill Croft, then paleobotanist at the Natural History Museum. I remember (with uncharacteristic precision) the scene as she came into the crowded Geological Society library, where we were having tea prior to the meeting; she was dressed strikingly, with very showy jewelry and a large hat, rather as for a reception at Buckingham Palace than for a scientific meeting. Heads turned, and there were murmurs of recognition. Bill Croft urged that I should not miss this opportunity of meeting her, and introduced me. On learning that I was embarking on paleobotanical research, she announced loudly in a voice well audible above the tea-time chat—"Ah, dear boy, that is wonderful! Of course paleobotany was my first love." That really turned all the heads, and I tried to hide behind Bill Croft. But of course, in later years, I was very proud to have met her then and was able to meet her again on several later occasions.

She sent me offprints of a number of her early papers (slightly gnawed by rats, from prolonged storage in her garage at Leatherhead). I eagerly sent my first reprint to her in return. I kept up a sporadic correspondence with her, and, at her suggestion, she paid a visit in 1957 to the Botany Department at University College London, where I was then working as an assistant lecturer. The purpose was to see the collection of fossil plants there, which had been largely built up by her some 40 years earlier. I think she was disappointed; the memories of striking fossil specimens are often more glowing than the reality. She made that visit the year before she died. Later, after her death, her scientific library was left to University College London. Any books that duplicated their library holdings were then sold off, and I was pleased to be able to acquire several of them. She was a great annotator of printed matter—especially when she disagreed with the author's views—and her books contain some spirited marginal notes (see Fig. 2).

## THE MAKING OF A PALEOBOTANIST

Marie Charlotte Carmichael Stopes was born on October 15, 1880, in Edinburgh. Her father, Henry Stopes, combined a professional career as a civil engineer with an enthusiastic but "amateur" commitment to archeology. Her mother, Charlotte Carmichael, was a teacher and fiction writer, contributing articles to a literary magazine. In her 20s, Charlotte attended classes leading to the M.A. degree at Edinburgh University, but a woman was not allowed at that time to sit for the degree examination. Her writings show that she had strong commitment to what nearly a hundred years later would be characterized as the feminist cause (Rose, 1992). Marie Stopes seems to have combined in her inheritance, in a simple Mendelian way, her father's scientific interest in the record of past life, as preserved in the rocks, and her mother's commitment to women's rights and social reform.

As a young girl, Marie Stopes helped her father wash the products of his archeological digs and label and catalog his archeological collection. She was educated at home by her mother, then started conventional schooling at the age of 12 and progressed to attend the North London Collegiate School. From there, she matriculated and so gained access to University College London. She apparently was not accepted when she sought admission to major in chemistry, but Professor F. W. Oliver, head of the Botany Department, accepted her to major in botany with chemistry and zoology as subsidiary subjects. At that time, 1892–1906, D. H. Scott (honorary keeper of the Jodrell Laboratory, Royal Botanic Gardens, Kew) was giving a series of lectures in paleobotany at Kew, and Oliver was on the verge of his discovery of the pteridosperms, on which he was to publish with Scott (Oliver and Scott, 1904). The rapid advances being made in coal-ball research at that time, revealing some of the major steps in plant evolution, must have given paleobotany something of the excitement of molecular biology in a later age.

With characteristic commitment and confidence in her own abilities, Stopes took night classes in Birkbeck College, another part of the University of London. This enabled her to complete the degree of B.Sc. with honors in botany, with geology as a subsidiary subject, in only two years. While at University College, she became president of the debating society and is reported to have been the innovator of the participation of women in what henceforth became known as "mixed" debates!

Her attainment of a first-class degree evidently formed the basis of Stopes's getting a scholarship for a further year's study, and she took the opportunity of using this to go to Germany to do research in Munich with Professor K. Goebel in the Plant Physiology Institute. She had already worked as assistant to both Professors F. W. Oliver (at University College) and D. H. Scott (at Kew) in the last few months before leaving for Munich. The stay at Munich was to result in her two earliest paleobotanical papers: on a cordaite leaf and on the nature of the epidermis of calamite roots (Stopes 1903a,

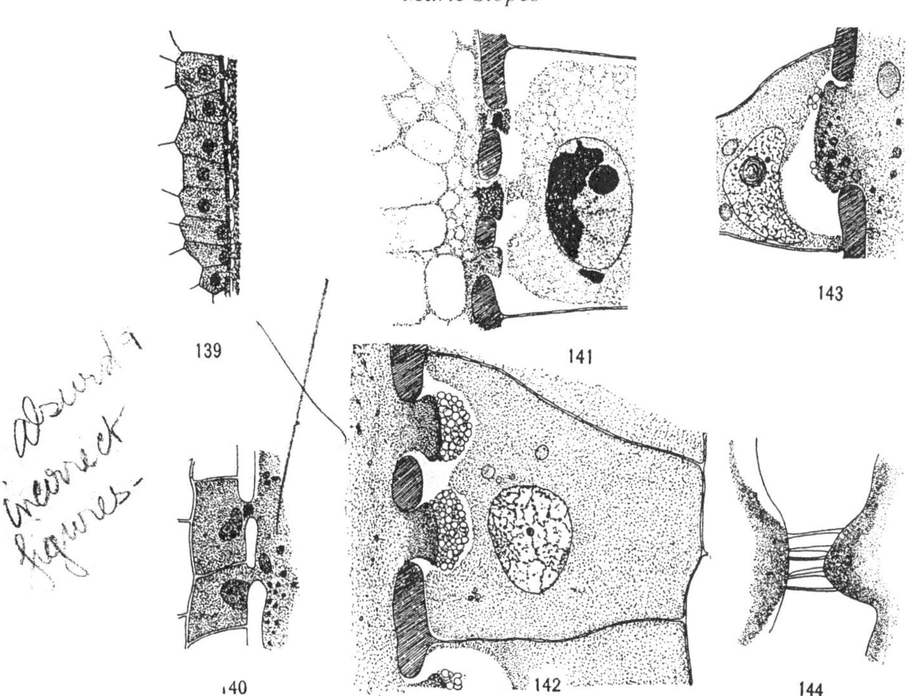


FIGS. 139–144.—Haustoria of cycads: fig. 139, *Cycas revoluta;* ×150; fig. 140, the same ×375; figs. 141–143, *Dioon edule;* ×800; fig. 144, *Encephalartos Lehmanii,* showing the *Plasmodesmen;* from a free-hand section which had been treated with sulphuric acid; ×1,100.—Figs. 139, 140, after IKENO (27); figs. 141–143, after CHAMBERLAIN (46); fig. 144, after STOPES and FUJII (50).

substances pass from the jacket cells into the haustoria as readily as from one part of the cell into another. Doubtless a pit-closing membrane exists here, as elsewhere, during the earlier development of the central cell, but as the haustoria grow larger and project into the cells of the jacket, the closing membrane is ruptured. The situation may be better understood from a series of figures by the various investigators (figs. 139–144).

Oh Lord!

Figure 2. Reproduced from page 135 of Marie Stopes's copy of *Morphology of gymnosperms* (Coulter and Chamberlain, 1910). The annotations indicate her views on Ikeno and Chamberlain's ideas of "haustoria" linking the central cell of the cycad archegonium to the surrounding gametophyte tissue. Her annotations read "absurd and incorrect figures" and "Oh Lord!" opposite the word *haustoria,* underlined by her. From her work with Fujii, she believed that the connections between the central cell (eventually to give rise to the egg) and adjoining female gametophyte were no more than the plasmodesmata that connect the cells within many plant tissues.

b). In Munich University, she was the only woman student among some 5,000 men. She embarked on work on the internal anatomy of cycad seeds, a subject of particular topicality at that time on account of their similarity to various Carboniferous coal-ball seeds then being investigated by Oliver and Scott. Despite the fact that no woman had previously been awarded a doctorate of science at Munich, Stopes persuaded Goebel to have the university regulations altered to allow her to submit her thesis. She not only did this but received her doctorate, having "defended" it in German, and was awarded the degree magna cum laude (Hall, 1977).

The research that had formed the basis of her doctoral degree was published in German (Stopes, 1904, "Contributions to our knowledge of the reproductive organs of the cycads") and was a significant piece of work, reviewing the nature and structure of a wide range of cycadalean seeds. It still stands as an important reference source in that field. She followed it up with a later paper (Stopes and Fujii, 1906) on the nature of the tissue immediately surrounding the egg in a wider range of gymnosperm ovules, published in English and datelined "Manchester and Tokyo" because her coauthor had by that time returned to Japan, and she had moved to Manchester University. That 1906 paper dealt with the degree of autonomy of the egg (and diploid zygote derived from it) from the enclosing haploid gametophyte tissue. In her copy of Coulter and Chamberlain's (1910) *Morphology of gymnosperms*, she made characteristically forthright marginal notes where the interpretation was at variance with hers and Fujii's (reproduced here as Fig. 2).

Some of Stopes's biographers inevitably saw this preoccupation with the (plant) egg and its eventual fertilization as in some way presaging her later interest in the control of fertilization in human beings. Whatever the truth of this, it is a fact that she used her legitimate title of Dr. Stopes (derived from her work on cycad reproductive biology) as a helpful label in opening her birth-control clinic. The libel action that she initiated over the misrepresentation of her work in that clinic and her use of the title of "doctor" are fully explored in Hall's (1977) biography.

In the ensuing sections of this review of American aspects of Marie Stopes's paleobotanical work, I depart somewhat from a straight chronological sequence and consider three subject areas. I believe this arrangement makes the paleobotany more coherent; the chronology of her life is amply dealt with in published biographies.

## MANCHESTER, JAPAN, AND CRETACEOUS PLANTS

Marie Stopes returned in 1905 from her doctoral research in Munich to take up an appointment at Manchester University as a "demonstrator" in botany—approximately equivalent to a teaching assistant in American usage. Professor F. E. Weiss (who had already published a number of papers on coal-ball fossil plants) was then head of the Botany Department and was evidently impressed by this hard-working and dedicated young woman. During her short time at Manchester she published several paleobotanical papers, including one describing the occurrence of Middle Jurassic plants preserved as compression fossils from the east coast of Scotland at Brora (Stopes, 1907). This was her first contact with Mesozoic paleobotany and may well have led her into the contemplation of one of the great mysteries of the Mesozoic—the origin of the angiosperms. It may also have influenced the Natural History Museum in London to invite her to write a catalog of the Cretaceous plants in their collection, although that work was not completed until 1913. While she was in Manchester, she wrote her book *Ancient plants* (Stopes, 1910a)—the first (remarkably successful) book to make the study of fossil plants accessible to a general audience. It still stands as a model of simple exposition of a relatively abstruse subject.

In 1906 she wrote to Fujii, now back in Tokyo, urging him to collect and send to her any promising Cretaceous rock matrix from Hokkaido, which she believed might yield angiosperm fossils. Material that he collected reached her in October 1906, and almost miraculously, as it now seems, the very first section that she cut revealed an angiosperm fruit. This discovery led her to seek money from the Royal Society to get her to Japan, to pursue her search for early permineralized angiosperm fossil flowers. One of her biographers (Hall, 1977) saw this as a barely concealed effort to follow Professor Fujii, to whom she obviously had a strong attachment, to Tokyo. Whatever the truth of this, the fossil material that she had already received from Fujii gave her a very strong scientific case for her projected visit.

This quest for early angiosperms was eventually funded by the Royal Society and Stopes sailed to Japan in 1907. The journey took five weeks. The impact that Japanese culture made upon her—and the impact that this remarkable woman had on the Japanese—are delightfully documented in her *Journal from Japan* (Stopes, 1910b). Her renewed contact with Fujii, and the problems that it brought into both their lives, is given full coverage in rather different terms by Hall (1977) and Rose (1992). Although she did indeed find more permineralized plant material in northern Japan, it was not in a technical sense coal-ball preservation but, rather, structurally preserved plant fragments enclosed within ironstone nodules in a sedimentary sequence and not from within a coal seam. She and Fujii did eventually publish on her Late Cretaceous "flower," *Cretovarium japonicum* (Stopes and Fujii, 1909), but as the name implies it was really just an ovary, rather than a flower, probably of liliaceous affinity. Although the discovery was important at that time, because it extended knowledge of the fossil record of the angiosperms, it was overtaken by subsequent discoveries of more complete Cretaceous angiosperm flowers (see Friis and Crepet, 1987, for a contemporary view of its significance).

I believe that far greater importance should be attached to Stopes's discovery of dicotyledonous permineralized woods in the British marine Cretaceous, which she encountered some years after her return to Britain, while cataloging the Cretaceous plants in the Natural History Museum collection. She described (Stopes, 1912), five new genera of fossil wood in that paper, from various localities in the British Lower Greensand (Aptian, Lower Cretaceous). She emphasized the important point that none of the five showed "primitive" anatomical features; indeed, all were quite advanced and specialized in terms of their anatomical differentiation. It is important to note that she did not herself collect any of the woods; all were in existing (and, in some cases, quite old) museum collections, and all lacked precise locality information. Furthermore, no other records of any angiosperm woods have been obtained from the British Aptian beyond those reported by her. Harris (1956) expressed skepticism about the real provenance of these early Cretaceous angiosperms, emphasizing how easily labels can get misplaced in old museum collections. Hughes (1976) endorsed this skepticism, even though his and others' palynological evidence certainly supports the occurrence of Aptian angiosperms. As with other aspects of Maries Stopes's life and work, we may never know the real truth.

In terms of number of pages, Stopes's largest paleobotanical work was her *Cretaceous flora*, published in two volumes by the Natural History Museum, London (Stopes, 1913, 1915). It was a rather strange assignment, in that Seward (1894–1895) had already written up the Wealden—by far the most important part of the British Cretaceous fossil flora. This left Stopes with a kind of rump collection of permineralized Bennettitales, some conifer seed-bearing cones and woods, and the angiosperm woods that she had made the subject of her earlier paper. The first volume consists of a review of Cretaceous plant occurrences, a worldwide record of all Cretaceous species known at that time, and a literature review. These take up three-quarters of the volume, before progressing to the algae and fungi; the second volume then deals with the remainder, of which noteworthy items are the fern *Tempskya* (of which much better preserved material was later to be described from the United States; see portrait of C. B. Read, Chapter 19, this volume) and a number of British permineralized bennettites and conifers.

Already, the confusion over usage of genera in the bennettites had surfaced, and Stopes remarked (1915): "It is clear that *Cycadeoidea* as used by the American, and *Bennettites* as used by the British palaeobotanists are largely synonymous." This difference of usage still gives an untidiness, particularly in the use of Bennettitales and Cycadeoidales, to transatlantic nomenclature.

## CANADA AND THE "FERN LEDGES" FLORA

Marie Stopes made her first visit to North America en route back to Britain from Japan, arriving by sea in Vancouver in January 1909. It appears that she lectured to women's groups there on her Japanese travels, before continuing overland to the east coast and thence home. On her return to London, she was appointed lecturer in paleobotany in the Botany Department of University College London. Life in Canada must have had some appeal for her, since the following year she apparently applied, unsuccessfully, for a university position in Toronto (Rose, 1992). Then, in the same year, the Canadian government invited her to publish a revision of the so-called Fern Ledges flora of New Brunswick. The plant fossils already reported from that locality had made the age of the flora a matter of some controversy. It had been variously dated as Devonian, Silurian, and Carboniferous, although even before her revision the consensus opinion of R. Kidston, D. White, and R. Zeiller was that the flora was Carboniferous in age.

The "Fern Ledges" plants were not visually exciting—as Stopes herself said, "the plant impressions have been completely graphitised and most of them consist merely of a bright film or streak on the rock. . . .They have no substance, and but little colour contrast with the matrix" (Stopes 1914). She found the "original section was practically worked out," but the nearby Duck Cove exposure to which she was introduced by Mr. McIntosh, curator of the St.John's Natural History Museum, yielded new and useful material. This was the main source of the specimens on which she worked, but she also studied material collected by J. W. Dawson, housed at McGill University and in the St. John museum. Some of this she took back with her to Europe, and she even took specimens across to France to get confirmation of her identifications from Zeiller in Paris. (She writes of having collected in the field in New Brunswick in the summer of 1911, although Rose [1992] reported with good authority that Stopes sailed back to Britain from Canada in April 1911.) In any event, her report was characteristically thorough, with over 140 pages of text and 25 plates of photos of the fossils. The flora proved to include some 28 well-defined species, a number of which were already known from the European upper Carboniferous, and these gave a sound basis for stratigraphic assignment. Her report left no doubt that the age of the "Fern Ledges" flora was Pennsylvanian (Westphalian).

A number of American paleobotanists, from Leo Lesquereux onward, were born in Europe but emigrated to and settled in North America. Equally, a number of other Europeans have taken themselves across the Atlantic to work uninvited on American material—not always with happy results (see Chapter 2, this volume). I believe that Stopes's revision of the "Fern Ledges" flora is remarkable in being perhaps the only case of an Old World scientist's being expressly invited across the Atlantic to write up a North American fossil flora.

The visit to Canada had other more personal significance for Stopes. She had arrived in North America in December 1910 and had gone almost immediately to a meeting of the American Association for the Advancement of Science (AAAS), which was held that year in St. Louis, Missouri. She attended a botanical dinner at the meeting on December 29, where she met Canadian geneticist Ruggles Gates. They became engaged on January 7, 1911, were married in Montreal in March, and in April were sailing back to Britain (Rose, 1992). The marriage was a disaster and ended five years later in a well-publicized nullity suit based on grounds of nonconsummation. There can be no doubt that that experience was the beginning of what became the driving passion in Marie Stopes's life—the preaching of a rational and open attitude toward sex and, with it, the necessity for birth control. In 1918 she published her epoch-making book *Married love,* and within six months had married Humphrey Verdon Roe, brother and partner of A. V. Roe of the Avro aircraft manufacturing firm. They had a son, Harry Stopes-Roe. *Married love* was Marie's most significant publication, which ran to countless editions in several languages. Hall (1977) reported that when a number of American academics were asked in 1935 to list the 25 most influential books of the previous 50 years, *Married love* scored only a little behind Marx's *Das Kapital* and ahead of Einstein's *The Meaning of Relativity* and Hitler's *Mein Kampf.*

## COAL BALLS, COAL, AND ANCIENT FOREST FIRES

One of Marie Stopes's most important contributions to paleobotany, with impact on both sides of the Atlantic, was her pioneering work on coal balls (Stopes and Watson, 1908), carried out when she and D. M. S. Watson were both at Manchester. Watson was at that time working on Carboniferous coal-ball plants, on which he published a number of papers. He later moved away from botany entirely and devoted himself to the animals that had interested themselves in those plants before him—namely, the early tetrapods. He became professor of zoology at University College, London (where Marie had taken her first degree) and achieved international fame for his work in vertebrate paleontology.

Coal balls (calcareous nodules occurring within coal seams, containing plant fossils in three-dimensional preservation) had long been known from Britain and the European continent. Much pioneering work on these Carboniferous fossil plants had been done by W. C. Williamson at Manchester in the latter half of the nineteenth century and later by F. W. Oliver, D. H. Scott, and many other European paleobotanists (see Andrews, 1980).

Stopes and Watson (1908) were concerned with the origin and distribution of coal balls and were able to document convincingly two aspects of their formation: first, that they developed in situ, preserving the plants more or less where they had grown in the coal-forming swamp, and, second, that the carbonate was derived from calcareous shells of marine invertebrates, brought into the swamp environment by marine incursions, depositing a mud that would eventually overlie the coal as a shale. These two basic points dominated the interpretation of coal balls for the ensuing half century. Coal balls were discovered in North America in the late 1920s and, by means of the new "peel" technique, coal-ball research blossomed in the United States, just as the European coal-ball supply was beginning to get "worked out." The idea, implicit in Stopes and Watson's (1908) paper, that coal balls contained the plants directly involved in coal formation lay at the heart of the seminal work of Tom Phillips on the succession of plant communities producing coal seams and on their response to climatic control (see, for example, Phillips et al., 1984, and work therein cited).

The Stopes and Watson (1908) interpretation went essentially unassailed for 50 years, but it came to be significantly challenged by Mamay and Yochelson (1962). Those authors showed that some American coal balls had pellets of marine mud involved in their formation, acting in some cases as a core for the initiation of the coal ball. They both sharpened and broadened the definition of just what constitutes a coal ball and extended the explanation of the circumstances under which such structures formed. But they were essentially building upon the Stopes-Watson model, rather than demolishing it. Scott and Rex (1985) have recently published a review of the mechanism of coal-ball formation, bringing together the strands of European and North American work.

The discovery of coal balls in North America has generally been attributed to A. C. Noé, who demonstrated specimens from Illinois at the International Botanical Congress in Ithaca, New York, in 1926 (see Chapter 14, this volume). However, Henry Andrews (1980) reported in his book *The Fossil Hunters* that Marie Stopes had written to him in 1950 that she herself had found "coal Balls in America, and before Noé, but like so many things I have, I never published about them." She also wrote to S. H. Mamay in the same vein on May 23, 1951, saying "I found American Coal Balls years and years ago, before anyone else I think, but they were not very good structurally, and I have never published on them." Mamay commented on this (personal communication, April 26, 1992)

> If indeed her claim is true, I feel certain that her "discovery" happened at the AAAS episode in St. Louis, which is an easy drive from the Illinois coal mining areas. Perhaps there was a field trip arranged by the AAAS, or, considering Dr. Stopes' interest in coal, perhaps she arranged privately for such an excursion. There were probably plenty of coal balls waiting to be recognized on the spoil piles of many mines in the area, and maybe she did indeed spot some of them.

It is certainly possible, but I remain skeptical. If she found coal balls in the United States in 1910 (the year of the AAAS meeting referred to by Mamay, some 16 years before Noé's announcement), it is astonishing that she never published this. She had just published the Stopes-Watson paper two years previously, and she continued to publish significant papers on paleobotanical topics for some 10 years after this. Clearly, the discovery of American coal balls would have been an item of note at that time, and she was certainly not backward in getting newsworthy items into print. We may never know whether Stopes was the first discoverer of American coal balls, but if that claim is true, she was strangely secretive about the matter until the 1950s.

During World War I, 1914–1918, coal was virtually the sole source of power for industry and the British navy, and coal research was accordingly highly rated in priorities for government-funded research. A laboratory under the aegis of the Department for Scientific and Industrial Research was set up in 1916, headed by R. V. Wheeler. Knowing of Marie Stopes's work in Carboniferous paleobotany, he visited her in University College London in that year, and they agreed to collaborate in research into the nature of coal. This work (Stopes and Wheeler, 1918) was to be Stopes's last contribution to paleobotany, as her concern for the birth-control cause increasingly took over her life. But in terms of the publication's impact on science, it was probably her most important work. The paper and her own publications on coal petrology (Stopes 1919, 1935) laid the foundations for modern coal terminology. At that time, the microscopic investigation of coal involved the laborious and very skilled process of cutting a thin section and carefully polishing it to yield a more or less transparent preparation (see Chapter 11, this volume). This technique came later to be largely superseded by the use of embedded, polished surfaces of coal samples examined by reflected light.

Marie Stopes (1935) coined the term *maceral* as a unit of

coal composition, analagous to *mineral* for a crystalline inorganic rock constituent. The four key rock types (lithotypes of later usage) were her vitrain, clarain, durain, and fusain. In the present context it is interesting to note that at that time there were three distinct "schools" of coal terminology. These were based on Stopes's work in Britain, a German school headed by Robert Potonié (which used a German version of what was in effect the Stopes terminology), and an American school based on the work of Thiessen and Jeffrey (see Chapter 11, this volume). Thiessen argued for only two basic microscopic coal constituents—"anthraxylon" and "attritus." It is fair to say that eventually the Stopes maceral concept and terminology won out and came to be generally adopted internationally with very little modification.

Despite this very general acceptance of Stopes's coal terminology, the *origin* of at least one of her lithotypes, fusain, has had a sustained history of controversy to the present day. Despite her electing to use the French word *fusain,* which means precisely "charcoal" (produced by incomplete combustion of wood), she did not accept that fusain in coal was necessarily the product of paleowildfire. She evidently felt not only that fire was a totally implausible event in the coal swamp vegetation but also that in any case charcoallike preservation could be caused by some kind of aerobic degradation process, and she therefore rejected "the fire origin of fusain." Jim Schopf (see Schopf portrait, Chapter 17, this volume) was a strong supporter of this thesis and continued to deny the general applicability of a fire origin for fusain long after Marie Stopes's death. This controversy, which had been pretty dormant from the 1930s onward, was revived by Tom Harris (1958), who attributed Rhaetic plant debris preserved as fusain to the effect of contemporaneous forest fire. From then on, the subject received growing attention. It formed the subject of a number of papers, and there was an increasing use of physical techniques to discriminate between plant material that had been subjected to pyrolysis by wildfire and other black organic debris (see Cope and Chaloner, 1985; Scott, 1989; and Jones and Chaloner, 1991). Current work on the physical and chemical nature of fusain appears to support a general applicability of the fire origin, and we still have no evidence for a biological process producing fusain, either now or in the past.

Professor Tom Harris of Reading University wrote (personal communication, September 30, 1977), after rereading the Stopes paper of 1919,

> [It] went a long way to making me understand how it was that Stopes led world thought on coal petrography. This had never concerned me much, except only when something hit the evidence on fossil plant structure (mainly fusain). The Ruth Hall book [Hall, 1977, of which he was writing a review at the time] caused me to read quite a lot of Stopes; previously I had only seldom had cause to do this. I must say that her work and writing survive fifty years uncommonly well. I greatly regret that the biography did not include a page or two of appraisal of her scientific work. . . . I am astonished at the difference between Stopes' scientific style—almost too austere, and the excessively juicy and sticky-sweet stuff that she wrote in private letters (but filed to be made public). . . .

He wrote in the same letter, "I am sorry that I only met Marie once, when she turned up at my Linn. Soc. paper on fusain, and opposed my revival of the old fire theory with vigour and in a pleasant voice. If I could have talked it over I doubt if we would have got far because we were thirty years out of phase, but I might have guessed the real basis of her objection to fire origin. And then I might have been convinced."

## CLOSING THOUGHTS

Marie Stopes died at her home in Leatherhead on October 2, 1958. As a paleobotanist she was almost forgotten, but her ideas had brought about a revolution in the quality of life, especially of women, in the western world. As Rayner (1991) has written, "Quite simply, Marie Stopes changed the face of British (and to some extent, Western) society almost beyond recognition . . . she . . . almost single-handedly led women out of the sexual repression of the Victorian Age into an enlightened age of sexual awareness."

## ACKNOWLEDGMENTS

I must record my grateful thanks to June Rose for allowing me to see the manuscript of her perceptive and sensitive biography of Marie Stopes and to Henry Andrews and Sergius Mamay for access to the letters from her that are quoted here. I am also particularly grateful to Harry Stopes-Roe for permission to publish the photo of his mother, reproduced here as Figure 1.

## REFERENCES CITED

Andrews, H. N., 1980, The fossil hunters, in search of ancient plants: Ithaca, New York, Cornell University Press, 421 p.

Briant, K., 1962, Marie Stopes—A biography: London, The Hogarth Press, 286 p.

Cope, M. J., and Chaloner, W. G., 1985, Wildfire, an interaction of biological and physical processes, *in* Tiffney, B. H., ed., Geological factors and the evolution of plants: Hartford, Connecticut, Yale University Press, p. 257–277.

Coulter, J. M., and Chamberlain, C. J., 1910, Morphology of gymnosperms: University of Chicago Press, 458 p.

Eaton, P., and Warnick, M., 1977, Marie Stopes—A preliminary checklist of her writings together with some biographical notes: London, Croom Helm, 59 p.

Friis, E. M., and Crepet, W. L., 1987, Time of appearance of floral features, *in* Friis, E. M., Chaloner, W. G., and Crane, P. R., eds., The origins of angiosperms and their biological consequences: Cambridge, Cambridge University Press, p. 145–179.

Hall, R., 1977, Marie Stopes—A biography: London, Andre Deutsch, 351 p.

Harris, T. M., 1956, The mystery of flowering plants: London, The Listener (26 April), p. 514–516.

Harris, T. M., 1958, Forest fire in the Mesozoic: Journal of Ecology, v. 46, p. 447–453.

Hughes, N. F., 1976, Palaeobiology of angiosperm origins: Cambridge, Cambridge University Press, 242 p.

Jones, T. P., and Chaloner, W. G., 1991, Fossil charcoal, its recognition and palaeoatmospheric significance: Palaeogeography, Palaeoclimatology, Palaeoecology, v. 97, p. 39–50.

Mamay, S. H., and Yochelson, E. L., 1962, Occurrence and significance of marine animal remains in American coal balls: U. S. Geological Survey Professional Paper 354-I, p. 193–224.

Maude, A., 1924, The authorised life of Marie C. Stopes: London, Williams and Norgate, 226 + ii p.

Oliver, F. W., and Scott, D. H., 1904, On the structure of the Palaeozoic seed *Lagenostoma lomaxi*, with a statement of the evidence upon which it is referred to *Lyginodendron*: Royal Society of London Philosophical Transactions, ser. B, v. 197, p. 193–247.

Phillips, T. L., Peppers, R. A., and DiMichele, W. A., 1984, Stratigraphic and interregional changes in Pennsylvanian coal-swamp vegetation: Environmental inferences: International Journal of Coal Geology, v. 5, p. 43–105.

Rayner, R. J., 1991, The private and public life of a palaeobotanist: South African Journal of Science, v. 87, p. 473–478.

Rose, J., 1992, Marie Stopes and the sexual revolution: London, Faber and Faber, 320 p.

Scott, A. C., 1989, Observations on the nature and origin of fusain: International Journal of Coal Geology, v. 12, p. 443–475.

Scott, A. C., and Rex, G., 1985, The formation and significance of Carboniferous coal balls: Royal Society of London Philosophical Transactions, ser. B., v. 311, p. 123–137.

Seward, A. C., 1894–1895, Catalogue of the Mesozoic plants in the Department of Geology, The Wealden flora: British Museum of Natural History, v. 1, p. 1–179, 11 pls; v. 2, p. 1–259, 20 pls.

Stopes, M. C., 1903a, On the leaf structure of Cordaites: The New Phytologist, v. 2, p. 92–98.

Stopes, M. C., 1903b, The "epidermoidal" layer of calamite roots: Annals of Botany, v. 20, p. 792–794.

Stopes, M. C., 1904, Beiträge zur Kenntnis der Fortpflanzungsorgane der Cycadeen: Flora oder Allgemeine Botanische Zeitung, v. 93, p. 435–482.

Stopes, M. C., 1907, The flora of the Inferior Oolite of Brora: Quarterly Journal of the Geological Society of London, v. 63, p. 375–382.

Stopes, M. C., 1910a, Ancient plants: London, Blackie, 198 p.

Stopes, M. C., 1910b, A journal from Japan: London, Blackie, 208 p.

Stopes, M. C., 1912, Petrifactions of the earliest European angiosperms: Royal Society of London Philosophical Transactions, ser. B, v. 203, p. 75–100.

Stopes, M. C., 1913, Catalogue of the Mesozoic plants in the British Museum (Natural History). Part I: Bibliography, Algae and Fungi: London, British Museum of Natural History, 285 p.

Stopes, M. C., 1914, The "Fern Ledges" Carboniferous flora of St. John, New Brunswick: Canada Geological Survey Memoir 41, 142 p.

Stopes, M. C., 1915, Catalogue of the Mesozoic plants in the British Museum (Natural History). Part II: Lower Greensand (Aptian) plants of Britain: London, British Museum of Natural History, 360 p.

Stopes, M. C., 1918, Married love: London, Fifield, 116 p.

Stopes, M. C., 1919, On the four visible ingredients in banded bituminous coal: Royal Society of London Proceedings, ser. B, v. 90, p. 470–487.

Stopes, M. C., 1935, On the petrology of banded bituminous coal: Fuel in Science and Practice, v. 14, p. 4–13.

Stopes, M. C., and Fujii, K., 1906, The nutritive relations of the surrounding tissues to the archegonia in gymnosperms: Beihefte zum Botanischen Centralblatt, v. 20, p. 1–24.

Stopes, M. C., and Fujii, K., 1909, Studies on the structure and affinities of Cretaceous plants: Royal Society of London Philosophical Transactions, ser. B, v. 201, p. 1–90.

Stopes, M. C., and Watson, D. M. S., 1908, On the present distribution and origin of the calcareous concretions in coal seams known as "coal balls": Royal Society of London Philosophical Transactions, ser. B, v. 200, p. 167–218.

Stopes, M. C., and Wheeler, R. V., 1918, The constitution of coal: London, Department of Scientific and Industrial Research, 58 p.

Manuscript Accepted by the Society July 6, 1994

Printed in U.S.A.

Geological Society of America
Memoir 185
1995

# *David White (1862–1935): American paleobotanist and geologist*

**Paul C. Lyons**
*U.S. Geological Survey, MS 956, National Center, Reston, Virginia 22092*
**Elsie Darrah Morey**
*Morey Paleobotanical Laboratory, 1729 Christopher Lane, Norristown, Pennsylvania 19403*

## ABSTRACT

**David White was a model for excellence in the geological profession. From his early years with the U.S. Geological Survey (USGS) under the paleobotanist Lester Ward to his rise to chief geologist of the USGS and chairman of the Division of Geology and Geography of the National Research Council, he displayed a gift for paleobotanical and fossil-fuel research and the ability to combine it with administrative service. His research in paleobotany was focused on the Paleozoic, although his investigations extended into the Cretaceous and the Precambrian. He guided the USGS's geological research program during World War I and directed activity into oil-shale research and geophysical methods in oil exploration. His carbon-ratio theory led to his recognition as an expert on coalification and petroleum geology. He conducted a longtime study of Pottsville floras and was the foremost expert on correlations of coal beds in the basins of the eastern United States. Notably, he discovered seeds attached to the genus *Aneimites.* He also established the early Permian age of the Hermit Shale of the Grand Canyon and was the first to identify and recognize the significance of *Gigantopteris* Schenk in North America.**

**White's enthusiasm for research was contagious, and he inspired a new breed of paleobotanist, coal geologist, and petroleum geologist. His prolific research activity led to his presidency of several scientific societies, membership in the National Academy of Science, honorary membership in foreign societies, and three honorary Doctor of Science degrees. For his research he received several medals from national and international scientific organizations. White, who served under four USGS directors, made an indelible mark as a U.S. government scientist, and his contributions continue to influence the fields of paleobotany, coal geology, and petroleum geology.**

## INTRODUCTION

David White was Leo Lesquereux's (1806–1889) successor to leadership in Carboniferous paleobotany in the United States. White spent his entire 49-year professional career with the U.S. Geological Survey (USGS), from a humble beginning as a draftsman in 1886 under Lester F. Ward to world leadership in the fields of paleobotany, coal geology, and petroleum geology. White's interests extended far beyond paleobotany. His research also included paleoclimatology, stratigraphy, petroleum geology, coal geology, isostasy, geophysical methods in structural geology, and U.S. and world petroleum reserves. White served a decade as chief geologist of the USGS and also held prominent positions in the National Academy of Sciences and the National Research Council. White's creativity, expressed in about 300 publications, won him international fame and awards for his contributions to understanding paleobotany, paleoclimatology, and the origins of coal and petroleum.

David White's contributions to paleobotany and geology and his various leadership roles as scientist, administrator, editor, and public servant have been recorded in many biographies. Most important among them are articles by Miser

Lyons, P. C., and Morey, E. D., 1995, David White (1862–1935): American paleobotanist and geologist, *in* Lyons, P. C., Morey, E. D., and Wagner, R. H., eds., Historical Perspective of Early Twentieth Century Carboniferous Paleobotany in North America (W. C. Darrah volume): Boulder, Colorado, Geological Society of America Memoir 185.

(1935), Mendenhall (1936), and Schuchert (1937), all of whom worked with White. The latter two sources contain White's bibliography. Readers are referred to these two memorials for further details of White's bibliography, which will be only selectively itemized here due to space limitations. Additional biographical data are in Andrews (1980), Berry (1936, 1944), Brown (1935), Campbell (1916), Darrah (1939, 1951, 1975), Mendenhall (1935), Miser (1936), Rabbitt (1980, 1986, 1989), Robertson (1993), Shor (1976), Stadnichenko (1936), and Stanton (1935).

The present biography relies heavily on prior ones but differs in its emphasis on paleobotany and coal geology.

## THE EARLY YEARS

Charles David White was born on July 1, 1862, on his father's farm in Palmyra, Wayne County, New York. He was the youngest of eight children. His parents were of English descent (Berry, 1944). He was encouraged to study the flowering plants of the region by the principal of the Collegiate Institute of Marion, New York, where White went to school. After he had taught in rural schools for two years, at the age of twenty Charles won a state scholarship to Cornell University, where his senior thesis (1886) was on *Ptilophyton vanuxemi* Dawson, an enigmatic Devonian plant fossil that he collected from the shales beneath Cornell's campus.

In 1886, H. S. Williams, White's geology professor at Cornell, recommended him to Lester F. Ward (1841–1913), who was in charge of paleobotanical investigations in the USGS (1881–1906). White's ability as a draftsman and his knowledge of botany won him an appointment as an assistant paleontologist, at a salary of $900 per annum, effective October 1, 1886. His appointment (Fig. 1) was signed by John Wesley Powell, the second director of the USGS.

In February 1888, David White married Mary Elizabeth Houghton of Worcester, Massachusetts, whom he had met at Cornell in 1885. She was a student of literature and history. They lived a long and happy life together in Washington, D.C., where they received many guests, not only geologists but also others interested in societal concerns (Mendenhall, 1936).

## THE SURVEY YEARS (1886–1912)

Although his first duties included only drafting, David White, as he now called himself, undertook on his own, and to Ward's delight, a study of the presumed Tertiary plants at Gay Head, Martha's Vineyard, in offshore Massachusetts. He established their Cretaceous age. This resulted in his first publications (1890a, b). The Gay Head section had been assigned to the Tertiary by Charles Lyell, during his 1845 visit to the United States (Andrews, 1980), and by Professor N. S. Shaler of Harvard University. White established on the basis of "five barrels of very excellent material" (remark by L. F. Ward in discussion of White's 1890b paper) the Middle Cretaceous age of the flora. His conclusion was confirmed by J. S. Newberry (in the same discussion), the foremost expert on Cretaceous floras along the Coastal Plain of the eastern United States.

This excellent biostratigraphic research launched White's paleobotanical career. Ward was pleasantly delighted with White's zeal "during one short season" (Ward, in discussion of White's 1890b paper). It was Ward, whose research interests were in Mesozoic floras, who first accompanied White to collect at the Gay Head section (White, 1890a). However, following Leo Lesquereux's death in 1889, White quickly began to fit into the Carboniferous paleobotanical niche left by Lesquereux.

In the early 1890s, White organized and classified the plant fossils in the R. D. Lacoe Collection, a collection of about 100,000 specimens of Paleozoic plant fossils given to the U.S. National Museum in 1893 (White, 1901a; Campbell, 1916). After White studied the Lacoe specimens, he believed that a careful study of the plant fossils could be used to resolve stratigraphic uncertainties in coal-bed correlations in the Appalachian coalfields. White used the plant fossils to find the "great Pocahontas coal bed on New River, some 30 or 40 mi away from the Pocahontas field . . ." (Campbell, 1916). He also studied the floras of the Lower Coal Measures of Missouri. White discovered a new taeniopterid fern (1893) and described the Pottsville Series along New River, West Virginia (1895). In 1893, White wrote his first systematic paleobotanical paper on *Taeniopteris missouriensis,* the new taeniopterid fern that he had discovered in Henry County, Missouri. Notable in this paper is White's interpretation of the origin of taeniopterid ferns from an Early Devonian *Megalopteris* stock that he suggested also gave rise to the seed ferns—*Neuropteris* and *Alethopteris.* By 1895, he was also using fossil plants to correlate coal beds in the southern Appalachians. White participated in the 1897 U.S. Geological Survey–Johns Hopkins field trip to Potomac River falls, West Virginia (Fig. 2), J. W. Powell's last field trip in the area. Although White continued to be interested in Cretaceous floras, as evidenced by his report with Schuchert on the Cretaceous of Greenland (White and Schuchert, 1898), he directed most of his early paleobotanical studies to the Pottsville in the Appalachian basin and to the coal measures of Missouri (White, 1899).

Charles Schuchert's (1937) biography of White reveals something of the character of this dedicated public servant. White, while working in the Anthracite region of eastern Pennsylvania, discovered a syncline where the coals were only mined on one limb. He recognized that there were millions of tons of coal on the other limb, which were unknown to the coal operators. However, rather than perhaps becoming a coal baron, White merely reported this important discovery to the director of the U.S. Geological Survey.

White's (1900a) paper on the ages of the Kanawha and Allegheny "series" asserted, in contradiction to previous correlations, that only the upper part of the Kanawha is correlative with the Allegheny and that the lower part is "largely of Pottsville floral affinities." This paper represents a major

DEPARTMENT OF THE INTERIOR

UNITED STATES GEOLOGICAL SURVEY

WASHINGTON D. C., Sept. 25, 1886.

The Honorable

The Secretary of the Interior.

Sir:

I have the honor to request that Charles David White, of Palmyra, New York, (27th Congressional District) be appointed an Assistant Paleontologist at $900 per annum, in the temporary force of the Geological Survey, to take effect October 1, 1886, or as soon thereafter as he shall file the oath of office and enter upon duty.

In explanation of the above request I would respectfully state that Mr. White is a graduate of Cornell University, where, owing to his special qualifications, he had performed the duties of instructor in drawing. He has also made the subject of paleontology a branch of study, and is well fitted to resume the work commenced by the late Mr. Pearce, in determining and reproducing fossil plants. His appointment is asked by reason of his qualifications as a scientific expert.

I am, with great respect,

Your obedient servant,

J W Powell

Figure 1. Letter of 1886 appointment of Charles David White as an assistant paleontologist, signed by John Wesley Powell, second director of the U.S. Geological Survey. The letter was reproduced from the original in David White's USGS personnel folder.

A

B
1
2
3
4
5
6
7
8
9
10
11
12
13
14
15
16
17
18
19
20
21
22
23
24
25
26
27
28
29
30
31
32
33
34
35
36
37
38
39
40
41
42
43
44

breakthrough in regional correlation and is great testimony to White's expertise as a stratigraphic paleobotanist.

White's first extensive publications were on the floras of the Lower Coal Measures of Missouri (1899) and on the Pottsville floras in the Southern Anthracite field of Pennsylvania (1900b). In the latter, White made regional correlations from Pennsylvania to Alabama and established the similar age of the variously eroded basins that had been previously thought by the various state surveys to be of different ages. This is the fundamental work on the Pottsville floras, which he proposed were of Early and Middle Pennsylvanian ages.

However, White did not restrict his paleobotanical studies to Cretaceous and Carboniferous floras. He reported a new species of alga in the Upper Silurian of Indiana (1901b) and in the Upper Devonian of New York (1902) and also questioned the validity of algal deposits in the Cambrian of New York (1903). White (1903a, b) published a summary of the Permian plants in Kansas and reported on enigmatic seaweedlike fossils from the Hudson Group (Cambrian). He also developed an interest in the evolution of fossil cycads in North America (1905a) and coauthored an authoritative paper on the geology of the Perry Basin, Maine (Smith and White, 1905).

White took over the leadership of paleobotanical investigations for the USGS when Lester Ward left in 1906 to become professor of sociology at Brown University (Andrews, 1980). Ward, whose paleobotanical publications span the period 1882 to 1905, was a good and influential friend of Powell. Ward was a pioneering sociologist as well as a paleobotanist (Darrah, 1951); his works after 1906 were on his first intellectual interest, sociology—a field that he founded (Andrews, 1980).

One of White's fellow students at Cornell was George H. Ashley, who also joined the USGS around the turn of the century. In 1907, White succeeded Ashley as head of the Section of Eastern Coal Fields, the beginning of White's administrative work, from which he escaped only shortly before his death in 1935. In 1909, after Ashley's departure from the Survey, White also took over Ashley's supervision of cooperative quadrangle mapping in Pennsylvania (Rabbitt, 1986).

By 1908 White had begun investigations on the genesis of coal. He published papers on Paleozoic resins (1909), the origin of coal with R. Thiessen (1913), and a carbon-ratio theory (1915) that related the degree of devolatilization of organic deposits (coal) to the occurrence of petroleum and gas deposits. He maintained that petroleum deposits would not occur where the fixed carbon had exceeded 65 to 70%; White (1935a) revised that to 60% on a pure-coal basis. This important discovery led to White's recognition as an expert on petroleum deposits.

On May 1, 1910, White was put in charge of the Survey's coal and oil work east of the Mississippi River, in addition to his regular work on Paleozoic plant fossils. White's skill in harmonizing the cooperative work of the various state surveys with the federal Survey brought his administrative prowess to the attention of his USGS superiors, M. R. Campbell, geologist-in-charge of Economic Geology of Fuels, and chief geologist Waldemar Lindgren.

## THE SURVEY YEARS (1912–1922)

White became chief geologist of the USGS in 1912 and served in this capacity through 1922. During this time, White (Fig. 3) was in charge of all USGS oil and gas investigations in this country and also supervised the petroleum estimates for the United States (Miser, 1935). In 1913, under White's leadership, a study of the petroleum potential of the shale of the Green River Formation of northwestern Utah was launched (Rabbitt, 1986) Arthur A. Baker (retired, Geologic Division,

Figure 2. Photograph of group on U.S. Geological Survey–John Hopkins University field trip to Potomac River falls in Harpers Ferry, West Virginia, 1897. Original photography by J. S. Diller on occasion of Sir Archibald Geikie's series of lectures at The Johns Hopkins University, 1897. Identifications by Bailey Willis, N. H. Darton, A. C. Spencer, A. C. Lane, and T. W. Stanton, November–December 1947. David White identification by P. C. Lyons. From U.S. Geological Survey photographic files, courtesy of W. R. Reckert, USGS (Reston). *Key:* **1,** C. C. O'Harra, Johns Hopkins University, later president of South Dakota School of Mines; **2,** Sir Archibald Geikie, director, Geological Survey of Great Britain; **3,** W. M. Davis, Harvard University; **4,** H. B. Kummel, Chicago University, later state geologist of New Jersey; **5,** G. B. Shattuck, Johns Hopkins University and Maryland Geological Survey; **6,** R. D. Salisbury, University of Chicago; **7,** A. C. Veatch, Rockport Indiana Schools, later chairman, Land Classification Board, USGS; **8,** L. M. Prindle, USGS; **9,** H. F. Reid, Johns Hopkins University; **10,** C. R. Van Hise, Wisconsin and Chicago Universities, later president, University of Wisconsin; **11,** Cleveland Abbe, Jr., Columbia University; **12,** George W. Stose, USGS; **13,** Thomas Watson, Cornell University, later Virginia state geologist; **14,** E. V. d'Invilliers, consulting geologist, Philadelphia, Pennsylvania; **15,** Dorsey, U.S. Department of Agriculture; **16,** J. W. W. Spencer, Geological Survey of West Indies and Central America; **17,** L. A. Bauer, Maryland Geological Survey, later Carnegie Institution; **18,** A. C. Spencer, USGS; **19,** W. J. McGee, ethnologist-in-charge, Bureau of American Ethnology; **20,** W. B. Clark, Johns Hopkins University, Maryland state geologist; **21,** R. M. Bagg, New York State Museum, later Colorado College; **22,** F. H. Knowlton, USGS; **23,** R. T. Hill, USGS; **24,** H. Ries, Columbia University, later Cornell University; **25,** F. D. Adams, McGill University; **26,** A. P. Coleman, University of Toronto; **27,** T. W. Stanton, USGS, later chief geologist; **28,** O. W. Fassig, Johns Hopkins University, later U.S. Weather Bureau; **29,** S. F. Emmons, USGS; **30,** G. F. Becker, USGS; **31,** A. H. Hoen, USGS, map printer; **32,** G. O. Smith, USGS, later director; **33,** J. F. Kemp, Columbia University; **34,** Bailey Willis, USGS; **35,** unidentified; **36,** E. B. Matthews, Johns Hopkins University; **37,** David White, USGS, later chief geologist, misidentified in original key; new identification by P. C. Lyons; **38,** John Wesley Powell, second director, USGS; **39,** J. Stanley-Brown, editor, GSA Bulletin; **40,** J. A. Holmes, state geologist of North Carolina, later first director, U.S. Bureau of Mines; **41,** C. W. Hayes, USGS, later chief geologist; **42,** F. J. Merrill, director, New York State Museum; **43,** L. C. Glenn, Johns Hopkins University, later Vanderbilt University; **44,** H. S. Williams, Yale University.

Figure 3. David White. circa 1912. USGS photographic files (portrait no. 934), Reston, Virginia, courtesy of L. V. Thompson (USGS, Reston).

USGS), who knew David White quite well during the years following White's 10-year tenure as chief geologist, offered the following remarks regarding White's oil-shale research:

> I believe it was in the spring of 1926 when my field party was moving trucks from Denver to southeastern Utah that David White asked me to stop by the oil shale mine at Rifle, Colorado, to obtain a substantial sample of oil shale that he wished to examine. We climbed the cliffs at the mine site and filled gunny sacks with oil shale, which we then carried back to the trucks and shipped to David White. I never did hear the results of his studies of that sample. He probably turned the samples over to a physical chemist, a Russian woman named Taisia Stadnichenko, who was associated with him for a long time. (A. A. Baker, written communication, 1991).

Some of White's activities as chief geologist are ably summarized by Rabbitt (1986). White focused on petroleum geology; he believed that a fundamental geological approach to the discovery of new oil and gas deposits was necessary. White (1920a) believed that petroleum geology was nothing more than an application of stratigraphy and structural geology to the discovery of oil deposits. As a consequence, Chief Geologist White initiated the USGS's first studies of micropaleontology and gravity measurements as applied to oil exploration. White was a member of the War Minerals Committee during World War I and directed a considerable part of the Geologic Branch's energy to the discovery and development of strategic minerals in short supply (Rabbitt, 1986) and toward oil shale as a potential source of gasoline (Berry, 1944).

White's ideas on the theory of oil distribution were used by the oil industry. While White served as chief geologist, some 90 geologists and 50 support staff worked in the Division of Geology. Many of these geologists were recruited by oil companies after World War I, and a notable proportion of the world's leading petroleum geologists during the 1920s and 1930s were former members of the USGS (Rabbitt, 1986).

During White's administration as chief geologist, "Shorter Contributions to General Geology," a new professional series, was begun so as to draw attention to geological research in the Survey (Rabbitt, 1986). In 1919, White was offered the chairmanship of the newly created Division of Geology and Geography in the National Research Council (NRC). He declined the offer, which would have increased his salary from $4,500 to $6,000 per annum, because he thought he could be of greater service as chief geologist under the fourth USGS director, G. O. Smith. Rabbitt also indicates that a major effort of the USGS's research during White's years as chief geologist was devoted to "energy minerals" and that White was later a founder of the Society of Economic Paleontologists and Mineralogists in the 1920s. In 1920, White (1920b) collaborated with E. Stebinger on estimates of world petroleum resources. White's petroleum-reserve estimates did much to stimulate oil exploration at home and abroad and directed public attention to the need for oil development (Miser, 1935). During these years of administrative work, however, White had to neglect his foremost interest—paleobotany.

One of White's landmark papers (1912) was on the North American occurrence of *Gigantopteris* Schenk, which was only previously reported from south-central and southwestern China. White discovered unattached seeds and polliniferous organs closely associated with this genus and concluded that it should be referred to the cycads or seed ferns. White proposed a common climatic province ("*Gigantopteris* province") connecting China and North America during the early Permian and suggested a North Pacific (Alaskan) route of migration. His work set the stage for the recognition of Permian floral zones in North America (Read and Mamay, 1964) and a line of research on the Gigantopteridaceae of the western United States (for example, Mamay, 1988, 1989).

White compiled a systematic list of all the fossil plant species known from the upper Paleozoic of West Virginia for his "The fossil flora of West Virginia" (1913). In it he referred the Pocahontas and New River Groups to the time periods of the "Millstone Grit" (Namurian) and the Lower Coal Measures of Great Britain ("Westphalian A" of Worsam and Old, 1988), the "Upper Pottsville" to the Middle Coal Measures (Worsam and Old's "Westphalian B") and the "transition" series of Great Britain, and the Allegheny, Conemaugh, and Monongahela (Worsam and Old's "Formations") to the Upper Coal Measures of Great Britain (Worsam and Old's "Westphalian C and D"). Significantly, he correlated the coal beds below the "Kittanning group" to the Westphalian and higher coal beds to the Stephanian. In an earlier paper, White (1904) referred the beds above the "Lower Washington Limestone" to the Permian and the lowermost Dunkard strata to the Late Pennsylvanian (Stephanian). In White (1936, posthumous), he placed the entire Dunkard Group in the Permian.

David White was a frequent speaker at various geological societies, including the Geological Society of Washington (GSW), of which he was president in 1920. During his long career, White gave 31 talks at GSW, a record surpassed by few (Robertson, 1993). Beginning in 1920, White gradually gave up control of the Survey's oil and gas investigations and returned to his research (Rabbitt, 1986).

Figure 4. David White studying the fossil beds in the Hakatai Shale (Precambrian) in the Grand Canyon. Notebook is blurred because White moved his hand. Photograph by George Grant, June 21, 1929, courtesy of S. T. Stebbins, National Park Service. Photograph IM #2864, National Park Service, Grand Canyon National Park.

## THE SURVEY YEARS (1922–1935)

On July 16, 1924, White took leave without pay from the Survey while he served as chairman (1924–1927) of the Division of Geology and Geography of the NRC, and he continued his leadership in oil development and research by raising funds for extensive investigations on the origin and properties of petroleum (Berry, 1944). White also spent several hours per week carrying out his USGS commitments.

In 1926, he became involved in a cooperative effort of the National Park Service, the National Academy of Sciences, and the U.S. National Museum to provide educational material at Grand Canyon National Park (Rabbit, 1986). In 1927, after his resignation from the chairmanship of the Division of Geology and Geography of the NRC, White returned to the Grand Canyon to extend his paleobotanical research to the Precambrian (Fig. 4). His search was fruitful. White (1928) discovered two or three possible fossils, including an algal type referred to Walcott's genus *Chuaria* (Walcott, 1899). In November 1927, he delivered a series of lectures at Columbia University.

In 1929, White participated in a symposium on coal classification at a meeting of the American Institute of Mining and Metallurgical Engineers (Rabbitt, 1986). White chaired the NRC's Committee on Paleobotany (1928–1934), which was composed of E. W. Berry, R. W. Chaney, A. C. Noé, and E. M. Thom. The Committee summarized paleobotanical studies by age, provided status reports on paleobotanists, noted newcomers and deaths, and gave information on museum collections and new techniques. In 1933, David White participated in the 16th International Geological Congress (Washington, D.C.). Figure 5 shows him there with C. B. Read and W. C. Darrah; both were budding paleobotanists who came under White's influence.

A. A. Baker (written communication, 1991) recalled that "White used to climb the stairs at the office regularly. His office being on the fourth floor, he had a stiff climb to get to this office from street level. I was not with him at any time in the field but I remember hearing accounts of his collecting trips in the Grand Canyon and elsewhere in rugged country in the West. I believe he prided himself on remaining in good physical condition right up to the time of his last illness." Sergius Mamay related that the late Lloyd Henbest told him several times that "Mr. White would always take the stairs three steps at a time in his enthusiasm to get to his fourth-floor office" (S. H. Mamay, written communication, 1991).

White spent two months in 1930 at high altitudes in the Grand Canyon. In 1931, he suffered a cerebral hemorrhage that resulted in partial paralysis (Schuchert, 1937; Berry, 1944). On July 1, 1932, White—now 70 years of age—was faced with mandatory retirement from the federal government, but the fifth USGS director, W. C. Mendenhall, requested that White continue to serve. President Herbert Hoover approved Mendenhall's request on June 30, 1932. A medical report in White's personnel folder, which was dated June 4, 1932, reveals some details about White's health. He was 1.8 m in height and weighed 72.7 kg. He had an irregular heartbeat, had a hemiplegic condition on the right side, and had to walk with a cane.

White's greatly diminished physical strength ended his fieldwork. With his mental powers still intact, however, he continued his research after 1931 with great courage and optimism. This research was summarized by White himself in the 1929–1934 reports of the Committee on Paleobotany (of the NRC, Division of Geology and Geography) when he served as chairman. White (1929a) identified Early Cretaceous plants from Alaska and prepared monographs on the Pottsville floras of Illinois (which were to be published by the

Figure 5. Group photograph (left third), 16th International Geological Congress, Washington, D.C., 1933. U.S. Geological Survey photograph. Courtesy of USGS Library (Reston, Virginia). Original in USGS Library (Denver, Colorado); copy in USGS Library (Reston, Virginia). Originally published in the International Geological Congress Report of the XVI Session v. 1, Washington, D.C., 1936. *Key:* **3,** P. Teilhard de Chardin; **4,** W. J. Jongmans; **5,** F. E. Suess; **6,** J. T. Richards; **7,** A. W. Grabau; **8,** Arnold Heim; **9,** V. K. Ting; **10,** L. C. Snider; **11,** G. F. Loughlin; **12,** Therese Fremont; **13,** G. A. Cooper; **14,** W. G. Foye; **15,** E. T. Hodge; **16,** Titus Ulke; **17,** Teiichi Kobayashi; **18,** Mrs. T. Kobayashi; **19,** Mrs. H. G. Ferguson; **20,** R. M. Field; **21,** C. H. Behre, Jr.; **22,** Rudolf Richter; **23,** Mrs. David White; **24,** Lloyd Henbest; **25,** M. W. Shepherd; **26,** W. D. Johnston, Jr.; **27,** E. M. Kindle; **28,** David White; **29,** H. G. Schenck; **30,** J. B. Reeside, Jr.; **30a,** W. H. Bucher; **31,** C. K. Swartz; **32,** W. F. Foshag; **33,** W. T. Gordon; **34,** Shinji Yamane; **35,** Hermann Reisch; **36,** J. S. Williams; **37,** W. S. Bisat; **38,** F. M. Swartz; **39,** L. H. Adams; **40,** Mrs. C. F. Kolderup; **41,** Manuel Santillan; **42,** Friedrich Schumacher; **43,** J.T.J.M. Schmutzer; **44,** M. P. Landman; **45,** A. B. Walkom; **46,** R. E. Dickerson; **47,** Mrs. A. L. du Toit; **48,** G. Marshall Kay; **49,** M. K. Elias; **50,** Charles Schuchert; **51,** C. F. Kolderup; **52,** A. O. Dufresne; **53,** M. Leriche; **54,** W. B. Lang; **55,** A. L. du Toit; **56,** Ernest Obering; **57,** Edwin Kirk; **58,** M. B. Baker; **59,** Charles Butts; **60,** Yoshinosuke Chitani; **61,** R. R. Rosenkrans; **62,** W. S. Bayley; **63,** unidentified; **64,** E. H. Sellards; **65,** H. D. Miser; **66,** C. B. Read; **67,** Fujio Homma; **68,** W. M. Senstius; **69,** H. P. Woodward; **70,** C. F. Bowen; **71,** C. L. Gazin; **72,** M. A. Steinbrock; **73,** W. H. Brown; **74,** F. R. Clark; **75,** Seitarto Tsuboi; **76,** J. T. Sanford; **77,** G. M. Carhart; **78,** Taisia Stadnichenko; **79,** J. Makiyama; **80,** unidentified; **81,** A. C. Spencer; **82,** G. E. Mitchell; **83,** G. B. Richardson; **84,** R. W. Richards; **85,** Mrs. W. C. Mendenhall; **86,** W. C. Smith; **87,** D. F. Hewett; **88,** H. Sommerlatte; **89,** E. S. Moore; **90,** Eugene Callaghan; **91,** R. C. Moore; **92,** Hans Stille; **93,** Mrs. H. Stille; **94,** C. R. Longwell; **95,** G. Betier; **96,** Mrs. V. M. Stadnichenko; **97,** R. S. Knappen; **97a,** V. M. Stadnichenko; **98,** G. W. Stose; **99,** Henry Hinds; **100,** N. I. Svitalsky; **101,** J. J. Sederholm; **102,** Willard Berry; **103,** W. E. Wrather; **104,** M. M. Knechtel; **105,** R. D. Reed; **106,** W. C. Darrah; **107,** Fernando Mouta; **108,** T. Tokuda; **109,** W. C. Mansfield; **110,** W. H. Hass; **111,** L. C. Graton; **112,** Robert Balk; **113,** T. F. W. Barth; **114,** Axel Born; **115,** P. V. Roundy; **116,** Tucker; **117,** I. M. Gubkin; **118,** I. P. Tolmachoff; **119,** A. L. Parsons; **120,** J. M. Bell; **121,** H. L. Fairchild; **122,** James Gilluly.

Illinois State Geological Survey) and on the Mississippian (Pocono) floras of the "Appalachian trough." His studies of a presumed Precambrian genus *Collenia* (which is considered today a chemical fossil) in the Grand Canyon and in Glacier National Park were reported in an annual report of the Carnegie Institution (White, 1929b, p. 392; see also White, 1929c). White believed *Collenia* could be used in correlations with rocks in China. White (1930a) reported the presence of iron bacteria in the Cambro-Ordovician of Virginia and also wrote a paper entitled "Climatic implications of the Pennsylvanian flora" (1931a) for the Quarter Centennial Celebration of the Illinois Geological Survey. He also mentioned in the 1930 NRC Committee's report the publication of his paper on the flora of the Hermit Shale of the Grand Canyon, which he concluded was of early Permian age (the youngest Permian flora then known in the continent), which indicated a semiarid climate. White also noted (1930b) that he was studying a small collection of Gondwanan plants from Brazil. In the same report (1930b, p. 9), White aptly delineated the most important problems in paleobotany:

1. The earliest land plants and their origin.
2. The origin of the angiosperms.
3. The Pleistocene floras, their composition and migrations.
4. Pre-Cambrian plant life.
5. Morphological study of petrified and carbonized plant remains.
6. Early history of the monocotyledons.
7. Study of silicified woods of the Mesozoic in particular, to improve the classification of the extinct types and to better establish their phylogeny.
8. Modern and scientific revision of the Triassic floras of America.
9. Paleobotanical differentiation of the Cretaceous-Tertiary boundary.
10. Floral definition of the Oligocene and Pliocene.
11. Limestones of plant origin of all ages.

White (1931c) mentioned that he and Charles B. Read (USGS) collected algal limestones from the Morrison Formation (Jurassic) of Colorado. In this same report, he also identified a lime-secreting algal deposit in the Maroon Formation (Permian) of Colorado. White also examined the Permian-Pennsylvanian boundary in the Rocky Mountains on the basis of floras, a boundary that he regarded as placed too high in the section on the basis of invertebrate fossils. White (1932) noted that he had nearly finished a study of a Mississippian (Chesterian) flora of Arkansas, and he asked paleobotanists to look at the environmental (paleoecological) and paleogeographic aspects of floras. In the same report, he noted the emergence of algae under subaerial conditions in the petrified peat at Rhynie, Scotland. He also chided paleobotanists for improper citations, lack of locality and age data for new species, and lack of stratigraphic information.

The origin of coal lithotypes (vitrain, fusain, clarain) is discussed in White's (1933a) paper entitled "Role of water conditions in the formation and differentiation of common (banded) coals." White (1933b) indicated that he had completed his paper on the flora of the Wedington Sandstone (Upper Mississippian) of Arkansas and Oklahoma (published posthumously in 1937). In this report, White (1933b) listed American paleobotanists and their affiliations. White's last NRC report (1934f) noted that "Five eminent European paleobotanists [presumably W. J. Jongmans, W. Gothan, P. Bertrand, among others] have taken part in the elaboration of American material and [that] three leading American paleobotanists have reviewed collections from farflung foreign sources." White also noted with gratification the treatment of environment (especially climate) in recent paleobotanical research. In 1934, under White's leadership, the Committee began a bibliography of paleobotany in the United States that has continued to the present day under the auspices of the Paleobotanical Section of the Botanical Society of America. One of White's last papers, on the characteristics of Permian floras, emphasized orogenic movement and resulting Permian climatic change as marked by red beds and floras indicating hot-dry seasons in the Midcontinent and western regions of the United States. He also noted that the middle Conemaugh in the Appalachian basin coincided with the advent of Permian tectonism to the west, a view that is still advocated by some paleobotanists. He advocated glaciation on the Gondwana continent as being responsible for dryness in European and American red-bed deposition. "Some erroneous age records of Paleozoic plant genera (1934a), "The age of the Pocono (1934b)," and the pteridospermic origin of the Permian genus *Supaia* (1934c) are three of White's last contributions. Two other late papers deal with Pocono age, orogeny, and climate (1934d) and the age of the Jackfork and Stanley Formations in the "Ouachita geosyncline (1934e)."

White's NRC report (1934f) also noted his discovery of seeds attached to part of a frond assignable to the genus *Callipteridium.* His last Committee report also includes a section on the paleobotany of coal, an area that had been admirably investigated by his colleague, R. Thiessen (Chapter 11, this volume), and his coworkers, G. C. Sprunk, in their paper "Spores of certain American coals" (1932). After White's death, the chairmanship passed to Roland W. Brown, who formed a new Committee. However, White's influence on the format of the Committee's reports continued for the next few decades, although the succinct summaries of floras by age gradually disappeared, to be replaced by bibliographic items arranged by geologic age. A posthumous work (White, 1943), which was prepared for publication by C. B. Read, deals with Lower Pennsylvanian genera (*Mariopteris, Eremopteris, Diplothema,* and *Aneimites*) in the Appalachian region.

David White prepared a statement that summarized the origin and history of discovery and development of oil that was published in 1934 as part of the USGS's contribution to the hearings before a subcommittee of the Committee on Interstate and Foreign Commerce of the House of Representatives. "This excellent summary [White, 1935b] might not be noted in the usual references" (A. A. Baker, written communication, 1991).

## SOME OTHER REFLECTIONS ON WHITE'S CONTRIBUTIONS AND INFLUENCE

In *The Fossil Hunters,* Andrews (1980) summarized David White's life and contributions. Andrews's account includes White's and Charles Schuchert's trip to study a 30-ton meteorite in Greenland in the summer of 1897 on one of Peary's polar expeditions. They collected plant fossils in what Andrews considers classic Cretaceous areas on the west coast of Greenland (White and Schuchert,1898). Andrews also drew attention to White's (1905b) discovery of seeds attached to the fern genus *Aneimites,* which were previously only known as detached seed forms assignable to the genus *Wardia.* White's discovery of the pteridospermic affinity of *Aneimites* was a major discovery in the history of paleobotany (Darrah, 1939, p. 90, 229).

White, late in his life, did much to encourage the careers of several young paleobotanists. Notable among these were Charles B. Read (1907–1979; Fig. 5; see Chapter 19, this volume) and William C. Darrah (1909–1989; Fig. 5; see Chapter 1, this volume). Read's earliest studies in the late 1920s were on the Tertiary woods (both coniferous and dicotyledonous) of the western United States. Read later (around 1929 or 1930) began a study of Pennsylvanian floras of Colorado and Oklahoma, which White (1931b) regarded as of "Pottsville age." Read went on to become an expert on the late Paleozoic floral biostratigraphy of the United States (see Read and Mamay, 1964; Chapter 19, this volume). White also noted in the 1931 Report of the NRC's Committee on Paleobotany (White, 1931c) the work of William C. Darrah, a student of Professor O. E. Jennings of the Carnegie Museum, Pittsburgh, Pennsylvania, who was studying the Carboniferous floras in the vicinity of Pittsburgh. The Committee's Report (White, 1932) elaborated that Darrah's graduate thesis "The fossil flora of the Mahoning shales" was based on 16,000 specimens containing 55 species correlated with the Radstockian stage (Westphalian D) of Great Britain. In 1932, David White encouraged Darrah to recollect and reappraise the lost Dunkard flora of Fontaine and White (Darrah, 1975); doing so became a lifelong interest of Darrah, who neotyped most of the lost Dunkard species (Darrah, 1975).

White died from a cerebral hemorrhage on February 7, 1935, just one day after he was at his desk at the Survey (Schuchert, 1937), working on his monograph on the Pottsville floras of Illinois (Stanton, 1935). White was one of the USGS's finest scientists and administrators, a truly dedicated public servant whose legacy of fundamental geological research has continued to influence the fields of Paleozoic paleobotany, coal geology, and petroleum geology. He served under four directors of the USGS (Fig. 6; see Robertson, 1993). Few geologists stood close to David White in dedication, loyalty, and world leadership. He was a model for excellence in the geological sciences.

JOHN W. POWELL
1881-1894

CHARLES D. WALCOTT
1894-1907

GEORGE OTIS SMITH
1907-1930

WALTER C. MENDENHALL
1931-1943

Figure 6. The four directors of the U.S. Geological Survey (John W. Powell, Charles D. Walcott, George Otis Smith, and Walter C. Mendenhall) under whom David White served. Taken from U.S. Geological Survey photograph, file no. PIO 83-117 (6A), courtesy of L. V. Thompson and W. R. Reckert (USGS, Reston).

White is buried in a cemetery near the South Rim in Grand Canyon National Park (Figs. 7, 8), the area where he spent his final field days. David White directed some of his last geological research to the Precambrian of the Grand Canyon.

## HONORS AND AWARDS

W. C. Mendenhall (1937), fifth director of the USGS and the last director under whom White served, summarized the memberships and offices held and honors received by David White. White was associate curator and curator of paleobotany at the U.S. National Museum from 1903 to 1935, he served on the "Coal Classification Committee—Subcommittee on Methods of the American Society for Testing and Materials," and he was vice president of the National Academy of Sciences (1931–1933). White was also the president of several societies: Paleontological Society (1909), Washington Academy of Sciences (1914), Geological Society of Washington (1920), and the Geological Society of America (1923). He was elected to the National Academy of Science in 1914 and was an honorary member of the Institute of Petroleum

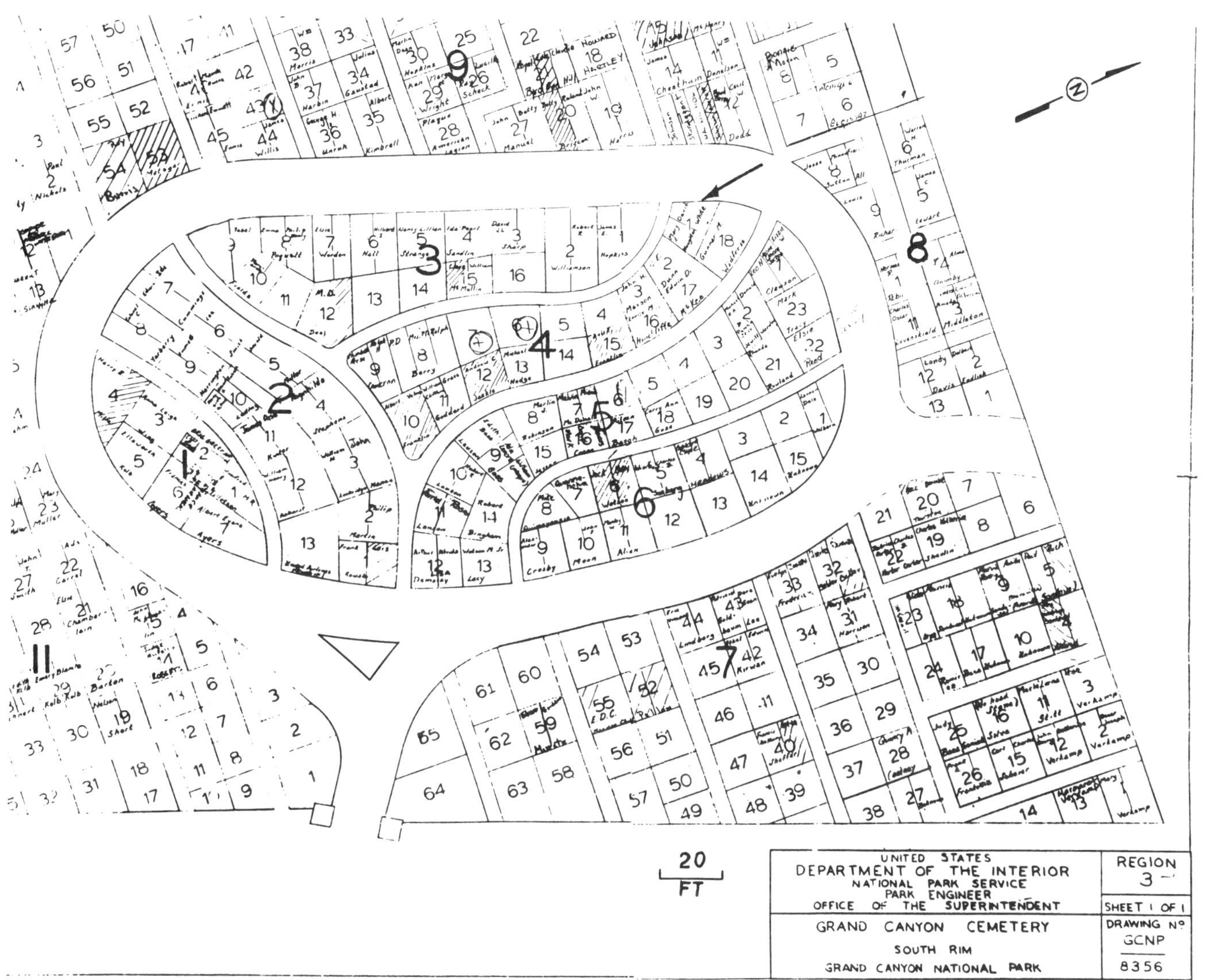


Figure 7. Map showing the grave site of David and Mrs. White (arrow), South Rim, Grand Canyon National Park. Photograph by Jim Tuck, National Park Service, Grand Canyon National Park.

Technologists of London, the Société Géologique de Belgique, and the Geological Society of China.

White received honorary Doctor of Science degrees from the University of Rochester (1924), the University of Cincinnati (1924), and Williams College (1925). He was honored by the National Academy of Sciences with the Thompson and Walcott gold medals, by the Society of Economic Geologists with its Penrose gold medal, and by the Institute of Petroleum Technologists of London with its Boverton Redwood medal (Schuchert, 1937).

## FINAL TRIBUTES

The NRC Committee's 1935 report, now under Roland Brown's chairmanship, noted the passing of David White and included the following short tribute (Brown, 1935, p. 1):

> David White was chairman of this committee until last year when, because of failing health and the pressure of unfinished work, he asked to be relieved entirely from active service. It was then that the present reorganization was arranged. Dr. White's achievements in paleobotany and other phases of geology need not be reviewed in detail here. He himself considered that his structure-carbon ratio theory for the occurrence and distribution of oil and gas was his greatest scientific contribution; but he will also be long remembered for what many of his friends may consider an equally great contribution, namely, his contagious, intelligent enthusiasm by which he encouraged and stimulated his associates, especially the younger group to live an abundant life, filled with zest and productivity. . . .

Mendenhall (1935, p. 381–382) eloquently summarized some of David White's great attributes, which could well serve as his epitaph: "Always an enthusiastic and indefatigable student himself, he was a constant inspirer of others. He took the keenest delight in bringing together a promising student and a

Figure 8. Headstone of the grave of David and Mrs. White, Grand Canyon Cemetery, South Rim, Grand Canyon National Park. Photograph by Jim Tuck, National Park Service, Grand Canyon National Park.

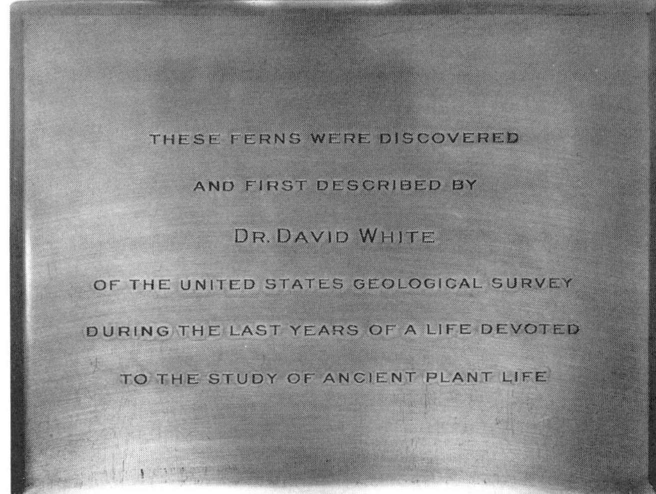


Figure 9. Plaque honoring David White's discovery of early Permian plant fossils in the Grand Canyon. Plaque loaned from Grand Canyon Museum, courtesy of S. T. Stebbins, National Park Service. Photograph by L. V. Thompson, USGS (Reston).

difficult problem, suggesting and counseling out of his own broad experience, helping to overcome difficulties and lending constant encouragement wholly regardless of any drain upon his own energies. His generosity, enthusiasm and keen personal interest, backed by his encyclopedic knowledge and his example of indefatigability, were constant sources of inspiration to his associates."

In 1936, White was honored again by the David White Memorial, a monument on the campus of East Central Oklahoma State Teachers College at Ada (Miser, 1936). The monument is a large tree stump of *Callixylon,* a Devonian tree found in the Woodford Chert (Devonian or Mississippian), which was identified by White in 1930. The fossil tree is 1.37 m in diameter and, according to White, was at the time of identification the largest known tree of so great an age (Miser, 1936). At the dedication ceremony a statement on the paleobotany of *Callixylon* by E. W. Berry, longtime friend and admirer of White, was read (Berry, 1936) as well as various tributes by colleagues, including T. Stadnichenko. She remarked on how "Uncle David" loved and encouraged young people in his late years because he realized the destiny of the country he served so well was in their hands (Stadnichenko, 1936).

In a letter to W. C. Darrah, C. A. Arnold wrote (May 11, 1936): "It seems that when David White died the greater part of our knowledge of the Pocono [flora] went along with him" (from the William Culp Darrah Collection).

At the 544th Meeting of GSW on November 25, 1936, a memorial meeting in honor of David White, the following scientists spoke briefly on Dr. White's life and work: H. D. Miser (introduction), W. C. Mendenhall, J. C. Merriam, Charles Butts, F. E. Wright, K. C. Heald, and E. W. Berry (E. C. Robinson, personal communication, 1991).

The National Park Service honored David White by erecting a plaque (Fig. 9) in the Grand Canyon where he collected his early Permian plant fossils. The plaque is now in safekeeping in the Grand Canyon Museum (S. T. Stebbins, National Park Service, written communication, 1992).

## ACKNOWLEDGMENTS

We have relied heavily on previously published biographies of David White, especially Mendenhall (1936), Schuchert (1937), and Miser (1935, 1936). We tried to contact geologists who knew White as a USGS geologist, but these were very few indeed and their memories were vague on details about him. Only Arthur A. Baker, who started with the USGS's Alaskan Division in 1921, could provide details. David and Mary H. White had no descendants. The William C. Darrah Paleobotanical files were used as a source of some information on David White.

Several geologists, currently or formerly with the USGS, provided valuable leads or information for this biography. These include A. A. Baker, R. M. Kosanke, S. H. Mamay, the late E. McKnight, C. M. Nelson, the late T. B. Nolan, E. C. Robertson, E. L. Yochelson, and A. D. Watt. Also we thank L. V. Thompson, W. R. Reckert, and L. Hadden, USGS, for access to photographic material on David White. Jim Tuck, National Park Service, Grand Canyon National Park, supplied a map of the gravesite of David and Mrs. White and also a photograph of the headstone. Sara Stebbins of the same organization was very helpful in providing photographs of David White, and she loaned the plaque commemorating David White's collections of Permian fossils in the Grand Canyon. Lastly, we are indebted to B. L. Graff, USGS Library, for help in compiling White's bibliography, which is published here only in selected form because of space restrictions. L. C. Matten and W. A. DiMichele reviewed the manuscript and made helpful suggestions for revision.

Any errors of omission or commission are our own. We take full responsibility for them.

## REFERENCES CITED

Andrews, H. N., 1980, The fossil hunters: Ithaca, New York, Cornell University Press, 421 p.

Berry, E. W., 1936, Geology of *Callixylon:* American Association of Petroleum Geologists Bulletin, v. 20, p. 628–630.

Berry, E. W., 1944, White, Charles David: Dictionary of American Biography, v. 21, p. 701–703.

Brown, R. W. (chairman), 1935, Report of the Committee on Paleobotany (April 27): National Research Council, Division of Geology and Geography, 20 p.

Campbell, M. R., 1916, David White stands in the front rank of world's paleobotanists: The Mining Congress Journal (May), v. 2, p. 211–212.

Darrah, W. C., 1939, Principles of paleobotany: Leiden, Holland, Chronica Botanica, 239 p.

Darrah, W. C., 1951, Powell of the Colorado: Princeton, New Jersey, Princeton University Press, 426 p.

Darrah, W. C., 1975, Historical aspects of the Permian flora of Fontaine and White, *in* Barlow, J. A., ed., Proceedings of the First I. C. White Memorial Symposium "The Age of the Dunkard" Morgantown, West Virginia: West Virginia Geological and Economic Survey, p. 81–101.

Mamay, S. H., 1988, Foliar morphology and anatomy of the gigantopterid plant *Delnortea abbotiae,* from the Lower Permian of Texas: American Journal of Botany, v. 75, p. 1409–1433.

Mamay, S. H., 1989, *Evolsonia,* a new genus of Gigantopteridaceae from the Lower Permian Vale Formation, north-central Texas: American Journal of Botany, v. 76, p. 1299–1311.

Mendenhall, W. C., 1935, David White: An appreciation: The Atlantic Monthly (April), p. 380–385.

Mendenhall, W. C., 1937, Memorial of David White: Geological Society of America Proceedings for 1936, p. 271–291, pl. 16.

Miser, H. D., 1935, Memorial, David White: American Association of Petroleum Geologists Bulletin, v. 19, p. 925–932.

Miser, H. D., 1936, David White: American Association of Petroleum Geologists Bulletin, v. 20, p. 625–628.

Rabbitt, M. C., 1980, Minerals, lands, and geology for the common defense and general welfare, Volume 2: 1879–1904: U.S. Geological Survey, 407 p.

Rabbitt, M. C., 1986, Minerals, lands, and geology for the common defense and general welfare, Volume 3: 1904–1939: U.S. Geological Survey, 479 p.

Rabbitt, M. C., 1989, A brief history of the USGS: U.S. Geological Survey, General Information Publication, 48 p.

Read, C. B., and Mamay, S. H., 1964, Upper Paleozoic floral zones and floral provinces of the United States: U.S. Geological Survey Professional Paper 454-K, 35 p., 19 pls.

Robertson, E. C., 1993, Centennial history of the Geological Society of Washington (1893–1993): Washington, D.C., Geological Society of Washington, 165 p.

Schuchert, C., 1937, Biographical memoir of David White, 1862–1935: National Academy of Sciences Biographical Memoirs, v. 17, p. 189–221.

Shor, E. N., 1976, Charles David White: Dictionary of Scientific Biography, v. 14, p. 297–299.

Sprunk, G. C., and Thiessen, R., 1932, Spores of certain American coals: Fuel, v. 11, p. 360–370.

Stadnichenko, T., 1936, Memorial to David White: American Association of Petroleum Geologists Bulletin, v. 20, p. 630–631.

Stanton, T. W., 1935, David White: Journal of Geology, v. 43, p. 778–780.

Walcott, C. D., 1899, Pre-Cambrian fossiliferous formations: Geological Society of America Bulletin, v. 10, p. 199–244, pls. 22–28.

White, D. See Bibliography of David White that follows.

Worsam, B. C., and Old, R. A., 1988, The geology of the country around Coalville: British Geological Survey Memoir for 1:50,000 geological sheet 155 (England and Wales), 161 p.

## SELECTED BIBLIOGRAPHY OF DAVID WHITE

Smith, G. O., and White, D., 1905, The geology of the Perry Basin in southeastern Maine: U.S. Geological Survey Professional Paper 35, 107 p.

White, C. D., 1886, On the nature and systematic classification of *Ptilophyton vanuxemi* [B.S. thesis]: Ithaca, New York, Cornell University (unpublished).

White, D., 1890a, On Cretaceous plants from Martha's Vineyard: American Journal of Science, v. 39, p. 93–101.

White, D., 1890b, On Cretaceous plants from Martha's Vineyard: Geological Society of America Bulletin, v. 1, p. 554–556 (with discussion by J. S. Newberry, L. F. Ward, and F. J. H. Merrill).

White, D., 1893, A new taeniopteroid fern and its allies: Geological Society of America Bulletin, v. 4, p. 119–132.

White, D., 1895, The Pottsville series along New River, West Virginia: Geological Society of America Bulletin, v. 6, p. 305–320.

White, D., 1899, Fossil flora of the lower coal measures of Missouri: U.S. Geological Survey Monographs, v. 37, 467 p., 173 pls.

White, D., 1900a, Relative ages of the Kanawha and Allegheny series as indicated by the fossil plants: Science, new series, v. 11, p. 140–141.

White, D., 1900b, The stratigraphic succession of the fossil floras of the Pottsville formation in the southern anthracite coal field, Pennsylvania: 20th Annual Report of the U.S. Geological Survey, 1898–99, pt. 2, p. 749–930, 14 pls.

White, D., 1901a, Mr. Lacoe's relations to science: Wyoming History of Geology Society Proceedings, v. 6, p. 55–60.

White, D., 1901b, Two new species of algae of the genus *Buthotrephis,* from the Upper Silurian of Indiana: U.S. National Museum Proceedings, v. 24, p. 265–270.

White, D., 1902, Description of a fossil alga from the Chemung of New York, with remarks on the genus *Haliserites* Sternberg: New York State Museum Bulletin, v. 52, p. 593–605, pl. 3.

White, D., 1903a, Summary of fossil plants recorded from the Upper Carboniferous and Permian formations of Kansas: U.S. Geological Survey Bulletin, v. 211, p. 85–117.

White, D., 1903b, Problematic fossils supposed to be seaweeds from the Hudson group: Science, new series, v. 17, p. 264.

White, D., 1904, Permian elements in the Dunkard flora: Geological Society

of America Bulletin, v. 14, p. 538–542.
White, D., 1905a, Fossil plants of the group Cycadofilices: Science, new series, v. 21, p. 664.
White, D., 1905b, The seeds of *Aneimites:* Smithsonian Miscellaneous Collections, v. 47, Quarterly Issue no. 2, p. 322–331.
White, D., 1909, Occurrence of resin in Paleozoic coals: Science, new series, v. 29, p. 945.
White, D., 1912, The characters of the fossil plant *Gigantopteris* Schenk and its occurrence in North America: U.S. National Museum Proceedings, v. 41, p. 493–516.
White, D., 1913, The fossil flora of West Virginia: West Virginia Geological Survey Report, v. 5, p. 390–453.
White, D., 1915, Some relations in origin between coal and petroleum: Washington Academy of Science Journal, v. 5, p. 407.
White, D., 1920a, Genetic problems affecting search for new oil regions: American Institute of Mining and Metallurgical Engineers, Transactions (preprint), no. 994, p. 9–12.
White, D., 1920b, The petroleum resources of the world: American Academy of Political Science Annals, v. 89, p. 111–134, 2 foldout maps.
White, D., 1928, Algal deposits of Unkar Proterozoic age in the Grand Canyon, Arizona: National Academy of Sciences Proceedings (July), v. 14, p. 597–600 (read before the Academy, April 24, 1928).
White, D., 1929a, Description of fossil plants found in some "mother rocks" of the petroleum from northern Alaska: American Association of Petroleum Geologists Bulletin, v. 13, p. 841–848, 8 pls. (July, 1929).
White, D., 1929b, Study of the fossil floras in the Grand Canyon, Arizona: Carnegie Institution of Washington Yearbook, v. 28, p. 392–393.
White, D. (chairman), 1929c, Report of the Committee on Paleobotany (April 27): Washington, D.C., National Research Council, Division of Geology and Geography, 13 p.
White, D., 1930a, Iron bacteria in silicified bog-iron deposits of Cambro-Ordovician age: Science, new series, v. 71, p. 544.
White, D. (chairman), 1930b, Report of the Committee on Paleobotany (May 3): Washington, D.C., National Research Council, Division of Geology and Geography, 10 p.
White, D., 1931a, Climatic implications of Pennsylvanian flora: Illinois Geological Survey Bulletin 60, p. 271–281.
White, D., 1931b, Age of the Maroon and Weber in central Colorado: Pan-American Geologist (February) v. 55, p. 60.
White, D. (chairman), 1931c, Report of the Committee on Paleobotany (April 25): Washington, D.C., National Research Council, Division of Geology and Geography, 13 p.
White, D. (chairman), 1932, Report of the Committee on Paleobotany (April 23): Washington, D.C., National Research Council, Division of Geology and Geography, 20 p.
White, D., 1933a, Role of water conditions in the formation and differentiation of common (banded) coals: Economic Geology, v. 28, p. 556–570.
White, D. (chairman), 1933b, Report of the Committee on Paleobotany (April 22): Washington, D.C., National Research Council, Division of Geology and Geography, 16 p.
White, D., 1934a, Some erroneous age records of Paleozoic plant genera: Science, new series, v. 79, p. 77–78.
White, D., 1934b, The age of the Pocono: American Journal of Science, 5th Series, v. 127, p. 265–272.
White, D., 1934c, The seeds of *Supaia,* a Permian pteridosperm: Science, new series, v. 29, p. 462.
White, D., 1934d, Pocono orogeny, age, climate: Geological Society of America Proceedings for (1933), p. 119.
White, D., 1934e, Age of Jackfork and Stanley formations of Ouachita geosyncline, Arkansas and Oklahoma, as indicated by plants: American Association of Petroleum Geologists Bulletin, v. 18, p. 1010–1017.
White, D. (chairman), 1934f, Report of the Committee on Paleobotany (April 28): Washington, D.C., National Research Council, Division of Geology and Geography, 17 p. (Exhibit A, Annotated Bibliography of Paleobotany of North America, April 1, 1933–April 1, 1934, prepared by E. M. Thom).
White, D., 1935a, Metamorphism of organic sediments and derived oils: American Association of Petroleum Geologists Bulletin, v. 19, p. 589–617.
White, D., 1935b, Outstanding features of petroleum development in America: American Association of Petroleum Geologist Bulletin, v. 19, p. 469–502 (Sections 1 to 6 of history and origin of oil: Pt. 2 of the Hearings before a Subcommittee of the Committee on Interstate and Foreign Commerce, House of Representatives, 73rd Congress [Recess], on H. Res. 441, 1934).
White, D., 1936 (posthumous), Some features of the early Permian flora of America: International Geological Congress, Report of the XVI Session United States of America 1933, v. 1, p. 649–689, 1 pl.
White, D., 1937 (posthumous), Fossil flora of the Wedington sandstone member of the Fayetteville shale: U.S. Geological Survey Professional Paper 186-B, p. 13–41, pls. 4–9.
White, D., 1943 (posthumous), assembled and edited by C. B. Read, Lower Pennsylvanian species of *Mariopteris, Eremopteris, Diplothema,* and *Aneimites* from the Appalachian region: U.S. Geological Survey Professional Paper 197-C, 140 p.
White, D., and Schuchert, C., 1898, Cretaceous series of the west coast of Greenland: Science, new series, v. 7, p. 80.
White, D., and Thiessen, R., 1913, The origin of coal: U.S. Bureau of Mines Bulletin, v. 38, 390 p.

Manuscript Accepted by the Society July 6, 1994

Geological Society of America
Memoir 185
1995

# *Reinhardt Thiessen (1867–1938): Pioneering coal petrologist and stratigraphic palynologist*

**Paul C. Lyons**
*U.S. Geological Survey, MS 956, National Center, Reston, Virginia 22092*

*With personal recollections of Reinhardt Thiessen by:*
**Marlies Teichmüller**
*Geologisches Landesamt Nordrhein-Westfalen, de Greiff Str. 195, D-47799 Krefeld, Germany*

## ABSTRACT

**Reinhardt Thiessen pioneered the thin-section method of coal petrography in the United States as well as the application of spores to the correlation of coal beds. His scientific career, mainly with the U.S. Bureau of Mines, spanned three decades during which he investigated the petrographic and botanical composition of American coals and conducted related research on peat and oil shale. He was the first in North America and, perhaps, worldwide to relate the macroscopic bands in coal to a microscopic classification of coal. Thiessen's published record of about 82 items and the Thiessen coal thin-section slide collection, which consists of about 19,000 slides of both American and foreign coals, are a rich legacy of his monumental work in organic petrology. For his research on coal microscopy, he was awarded the Silver Medal of the Royal Society of Arts (London). The Reinhardt Thiessen Medal of the International Committee for Coal and Organic Petrology honors his name as a pioneering coal petrologist. His research in stratigraphic palynology is recognized by a tree-fern spore, *Thymospora thiessenii* (Kosanke) Wilson and Venkatachala.**

## INTRODUCTION

Reinhardt Thiessen began his career as an educator and spent the last three decades of his life as a U.S. government scientist. He started with the U.S. Geological Survey in 1907. On June 30, 1910, Thiessen was transferred to the newly created U.S. Bureau of Mines in Pittsburgh, Pennsylvania, where he established a microscopy laboratory within the Chemical Division at the Pittsburgh Experiment Station. He spent the rest of his career with that organization. Officially he was a chemist, but unofficially he was a microscopist. His name is intimately associated with the thin-section method of coal petrography, a method that he pioneered in the United States. Thiessen was able to relate the macroscopic properties of coal to a microscopic classification of coal, the first in North America and perhaps worldwide. Thiessen's application of this method, which he adapted for quantitative studies of American coals, laid the foundation for American coal petrology. An alternate method of reflected light microscopy was developed by Erich Stach in Germany, and this method became the standard for western Europe and later the world (Stach et al., 1982). Thiessen was also a pioneering stratigraphic palynologist.

This portrait traces the life and professional career of Thiessen from his birth in Wisconsin in 1867 to his death in Pittsburgh in 1939 and traces his triumph with the development of coal petrological investigations in the United States. Biographical data on Thiessen's life and work are taken from five principal sources (Parks and O'Donnell, 1956; Schopf and Oftedahl, 1976; Thiessen's U.S. government personnel folder; and communications with Linwood Thiessen, his son, and Hugh O'Donnell, his coworker).

Lyons, P. C., and Teichmüller, M., 1995, Reinhardt Thiessen (1867–1938): Pioneering coal petrologist and stratigraphic palynologist, *in* Lyons, P. C., Morey, E. D., and Wagner, R. H., eds., Historical Perspective of Early Twentieth Century Carboniferous Paleobotany in North America (W. C. Darrah volume): Boulder, Colorado, Geological Society of America Memoir 185.

## THE EARLY YEARS

A biographical sketch of Thiessen can be found in B. C. Parks and H. J. O'Donnell (1956). His parents were born in Germany and settled in Wisconsin. Reinhardt Thiessen was born in New Holstein, Wisconsin, on May 1, 1867. In 1895, he received his Ph.B. (chemical physiology) from Lawrence College, Wisconsin, and in 1903, an S.B. (Bachelor of Science) from the University of Chicago. After graduation, he taught science at the Red River Valley University in South Dakota (1898–1901), botany and zoology at Des Moines College in Iowa (1903–05), and science and mathematics at the University High School, Chicago (1906–07). In those days, Thiessen's scientific curiosity led him to study the botanical components of Wisconsin's peat bogs; these early investigations laid the basis for some of his later thinking on the origin of coal. In 1907, Thiessen received his Ph.D. in botany at the University of Chicago under the guidance of the renowned botanist, J. M. Coulter. His Ph.D. thesis (Thiessen, 1908a), on the cycad *Dioon edule,* was published the same year in the Botanical Gazette (Thiessen, 1908b). Thiessen was trained not only in botany but also in biology, mathematics, chemistry, and geology. Early in his career as a scientist, he displayed "great energy and zeal for investigative work" (Parks and O'Donnell, 1956). Thiessen married a Wisconsin native, Clara A. Lindemann. They had three sons: Gilbert (geologist), Linwood (research engineer), and Reinhardt, Jr. (chemist). Gilbert, the oldest son, who died in 1966, followed most closely in his father's research footsteps. At the time of writing the two younger sons are retired in Pittsburgh, Pennsylvania, and Holiday, Florida, respectively.

Thiessen's research career was launched in 1907 with an offer from David White, an established paleobotanist and coal geologist with the U.S. Geological Survey (USGS), who was looking for someone to do laboratory research on the origin of coal. Thiessen joined the Technologic Branch of the USGS in 1907, just after he received his Ph.D. from the University of Chicago.

## THIESSEN'S EARLY WORK ON COAL

In October 1907, Thiessen was appointed as assistant chemist in the Technologic Branch of the USGS (Fig. 1). Working under the direction of David White (see Chapter 10, this volume) and Joseph A. Holmes, chief technologist, his job in Washington, D.C., was to develop a technique for preparing thin sections of coal with a microtome without causing further alteration of the coal (White and Thiessen, 1913). Most of the coals for Thiessen's studies were collected by White in 1907 and 1908 from the western and eastern coalfields of the United States. Thiessen, starting with the lowest rank coal, lignite, and progressing to subbituminous and bituminous coal—each rank requiring special procedures—successfully developed ways to prepare and photograph coals in thin section.

Thiessen and White's collaborative work began with lignite. A thin-section microscopic approach combined with chemical analysis was the foundation of this research.

In 1910, the Technologic Branch of the U.S. Geological Survey in Washington (Fig. 2) became part of the newly created U.S. Bureau of Mines. At this time, Thiessen was transferred from Washington to the Pittsburgh Experiment Station, but White remained with the USGS in Washington. At Pittsburgh, Thiessen was inspired and guided by Dr. A. C. Fieldner (see Parks and O'Donnell, 1956) with whom he coauthored many studies of U.S. coals.

Thiessen and White continued their collaborative work, which led to a monumental contribution to coal science. *The Origin of Coal* (White and Thiessen, 1913) is a classic work on the microscopic nature of coals, a landmark volume including Thiessen's pioneering thin-section work. Thiessen's photomicrographs of coal laid to rest disbeliefs about the woody composition of coal. In 1912, he gave a presentation at the Cosmos Club (Washington, D.C.) entitled "On certain constituents and the genesis of coals," his second publication on coal; his first was published in *Science* the year before (Thiessen, 1911). Thiessen used a modification of Jeffrey's (1910) method for thin-section preparation. Thiessen used weak solutions and mild reagents to minimize alteration of the coal. Minerals were removed from the blocks of coal by acid treatment before the thin sections were made. Thin sections, about 1 to 2 $cm^2$ and 3 μm thick, were produced by cutting the blocks on a rigid microtome, embedding the sections in Canada balsam, and covering them with a cover glass. Some of the treatments of the coal took as long as nine months to get optimum results. Thiessen also used grinding techniques to make larger sections of coal. However, he was successful with grinding techniques for only a few coals because of the problem of adsorption of water and the difficulty of grinding large sections to desired thinness (≈5 μm). To supplement his studies, Thiessen used maceration techniques on coal using a modification of Schulze's (1855) method. Particles of oxidized material were observed in solution under the microscope to determine dissolved and undissolved phases. He also macerated microtome sections in a watch glass and observed the dissolution effects on selected microscopic components.

Thiessen anticipated the development of reflected light microscopy of coal (Stach et al., 1982) in the following statement: "A third method, involving the use of a polished surface, as in metallography, promised good results. This method was not studied fully for want of equipment" (Thiessen, 1931b, p. 206).

Thiessen was able to observe good detail of botanical components in his studies of lignite (White and Thiessen, 1913). He was able to see woody tissue, resin fillings of cells, pollen and spore exines, cuticles, fungal spores, fungal hyphae, bordered pits, and even the middle lamellae of cell walls. Similar components were observed by Thiessen in higher-rank coals.

In reply please refer to CR. and date of this letter.

Address all communications to "Director, U. S. Geological Survey, Washington, D. C."

SUBJECT: Temporary employment of Mr. Thiessen.

Copy for Civ. Ser. Com.

WFM/JAM

DEPT. OF THE INTERIOR, APPOINTMENT DIVISION. RECEIVED OCT 15 1907

DEPARTMENT OF THE INTERIOR

UNITED STATES GEOLOGICAL SURVEY

WASHINGTON, D. C., October 14, 1907.

The Honorable,

The Secretary of the Interior.

Sir:

I have the honor to ask that authority be granted for the employment in Washington, D. C., for ninety (90) days, beginning October 16, 1907, of Mr. Rienhard Thiessen, field assistant at $100.00 per month.

Mr. Thiessen has been employed in the field for some time on investigations into the origin of coal, under the direction of Dr. C. David White, Geologist, and it is desirable that his services be continued in the office in Washington, with a view to his supplementing his field investigations by some laboratory experiments, and also to his preparing a report upon the subject.

I send an extra copy of this, that the Department may use same, if desired, in reporting the authorization to the Civil Service Commission.

Very respectfully,

Geo Otis Smith

Director.

Figure 1. Thiessen's letter of appointment as U.S. Geological Survey field assistant (assistant chemist) in Washington, D.C., signed by George Otis Smith, director of the U.S. Geological Survey. (From the personnel file of R. Thiessen.)

Figure 2. U.S. Geological Survey Coal Testing Plant (circa 1908). U.S. Geological Survey photographic files, Reston, Virginia; courtesy of W. R. Reckert, USGS.

Thiessen demonstrated that cannel coal was composed mainly of spore exines and very minor amounts of cuticular and humic material, rather than being composed of algal remains as proposed by some investigators (Thiessen, 1912; White and Thiessen, 1913). His observations provided very strong evidence of the autochthonous origin of coal and cast considerable doubt on the drift (allochthonous) theory of coal formation.

## THIESSEN'S EARLY WORK WITH BITUMINOUS COAL

After Thiessen completed his early studies of the microstructure of lignite, he spent most of the remaining part of his career studying the botanical and chemical nature of bituminous coal, mainly from coalfields of the eastern United States. In order to work with bituminous coal, Thiessen had to develop new methods of thin-section preparation, which were closely allied to the grinding preparation techniques for noncoal rocks. Thiessen was the first to apply this method to North American coals (Park and O'Donnell, 1956). Thiessen mastered the technique, and a series of papers on the nature of bituminous coal as seen in thin section established him as a pioneer in the field of coal microscopy (Thiessen, 1920a, b, f, 1924; Thiessen and Voorhees, 1922).

Bituminous coals, which are richly illustrated in Thiessen (1920f), were studied in detail mainly by using ground thin sections. Thiessen (1920a, f) observed rodlets that he interpreted to be infillings of mucilaginous canals in medullosan seed ferns. In honor of Thiessen's first discovery of a stem of *Medullosa* in North America (Thiessen, 1920f), Schopf (see Chapter 17, this volume) (1939) named it *Medullosa anglica* var. *thiessenii*. Thiessen was the first to interpret the different laminae in coal. He was able to demonstrate that bright bands of coal consisted of woody tissue, an interpretation that was to lead to considerable debate with his European colleagues, notably M. C. Stopes (1919), who maintained that the bright bands, her vitrain, were structureless. Later Thiessen (1919c, 1920a, b, c, f, 1921c) called the bright bands "anthraxylon" (Greek: *anthrax,* coal; *xylon,* wood), a synonym for vitrite. Anthraxylon is nicely illustrated in Thiessen (1920b, f, 1924). The dull bands in coal were interpreted by Thiessen (1920a) to be composed of laminae of anthraxylon and *attritus,* a microscopic term he used for highly comminuted matter composed of degradation products from cellulose in cell walls, humic matter, spore exines, resins, cuticular remains, rodlets, and some mineral matter. Thus, Thiessen was able to relate the macroscopic properties of coal to a microscopic classification—essentially anthraxylon and attritus—a finding that was probably a first worldwide.

Shortly before 1920, Thiessen (Fig. 3) worked on the forms of sulfur in coal by using microscopic methods (Thies-

Figure 3. Reinhardt Thiessen (circa early 1920s). U.S. Bureau of Mines photograph, courtesy of Linwood Thiessen.

Figure 4. Reinhardt Thiessen at the University of Sheffield, 1926. Courtesy of Linwood Thiessen.

sen, 1919c, d, e). From about 1923 until his death in 1938, he supervised several research fellows at the Carnegie Institute of Technology in their investigations in palynology, coal chemistry, and coal microscopy.

In September 1925, Thiessen went to England with his family as an exchange investigator with the Safety and Mines Research Board of Great Britain to assist in the establishment of a coal research program. It was during these 14 months, at the University of Sheffield (Fig. 4), that he examined the vitrain of Marie Stopes (see Chapter 9, this volume) and determined that it indeed was structured (see discussion in Thiessen and Francis, 1929). The presumed structureless aspect of vitrain was due to a poorer quality of preparation and *not* due to the inherent nature of the material. In Europe, Thiessen met many scientific experts in the field of coal and developed long-standing correspondence with a score of European scientists, including W. Francis, G. H. Hickling, and C. A. Seyler of England; W. J. Jongmans of The Netherlands (see Chapter 5, this volume); E. Stach, R. Potonié, Th. Lange, and K. A. Jurasky of Germany; and A. Duparque of France. He traveled widely in western Europe and visited peat bogs (Fig. 5) and coal mines in England, Ireland, Germany, and other countries

Figure 5. Reinhardt Thiessen in a peat bog in Ireland, 1926, looking for dopplerite and other peat components. Courtesy of Linwood Thiessen.

of western Europe (Linwood Thiessen, personal communication, 1992). He collected coal balls in the English coal mines (Linwood Thiessen, personal communication, 1992) to study the nature of Carboniferous coal-forming plants.

The most important result of Thiessen's stay in England was a paper on a comparison of European and American coal terminology (Thiessen and Francis, 1929a, b). Thiessen concluded that Stopes's (1919) vitrain, as originally applied to British bituminous coals, was essentially anthraxylon, both being characterized by cellular structure. However, he clearly recognized a "structureless vitrain," analogous to dopplerite in peat, that follows the original definition of Stopes (1919). Thiessen and Francis (1929a, b) compared the Glanzkohle of the Germans to Stopes's (1919) vitrain and the Mattkohle of the Germans to the durain of Stopes (1919).

For his work on coal microscopy, in 1926 Thiessen was awarded the Silver Medal of the Royal Society of Arts, Manufacturers and Commerce (London). In 1927, he received an honorary Sc.D. from Lawrence College, his alma mater. Thiessen was a member of Phi Beta Kappa. In 1927 and 1928, English coal investigators came to work in Pittsburgh under Thiessen's direction.

## THIESSEN'S PALYNOLOGICAL WORK

Thiessen was also a pioneer in the application of spores to the correlation of coal beds (Thiessen, 1918, 1920a, b, c). He recognized that the Pittsburgh coal bed was characterized by a single type of spore (Thiessen, 1918) that he later called the "Pittsburgh" spore (Thiessen and Staud, 1923). This spore was later formally named *Thymospora thiessenii* (Kosanke) Wilson and Venkatachala in his honor. Thiessen characterized other coal beds—such as the Brookville, Lower and Middle Kittanning, Elkhorn, Alma, Taggart, and Chilton No. 2 Gas coal beds of the Appalachian basin; the No. 2 and No. 6 coal beds of the Illinois basin; and the Jugger, Mary Lee, and Black Creek coal beds of the Warrior basin—by their spore assemblages (Thiessen, 1918; Thiessen and Wilson, 1924; Thiessen, 1929; Sprunk and Thiessen, 1932). This method of correlation of certain coal beds by spore type established Thiessen as the father of coal stratigraphic palynology. Thiessen and Staud (1923) were the first to classify Carboniferous spores and pollen in the United States (Kosanke, 1969). However, most of the Carboniferous spores in coal thin sections that Thiessen and his coworkers measured and described were simply given numbers for identification. Later, Schopf et al. (1944) produced a coherent classification of genera of Carboniferous and other late Paleozoic palynomorphs.

Darrah (1939, p. 48) credited Thiessen (1925e) with the recognition of the probable algal nature of *Pila* and *Reinschia.* These genera were interpreted as spore exines by Jeffrey (1910) and as bitumen agglomerations by R. Potonié (1924).

## THIESSEN'S LATER WORK ON COAL

In 1921, Thiessen (Fig. 6) was appointed as chief of the Section on Microscopy and Microscopical Research at the Pittsburgh Station. He was allowed to hire two assistants. A new decade of coal microscopy research was launched in the early 1930s with the entire coal-bed column analyzed microscopically, inch-by-inch, for 34 bituminous coals of the eastern and interior coalfields (Parks and O'Donnell, 1956). Chemical analyses were selectively tied to the microscopical categories. Sprunk and Thiessen (1935) reported elemental analyses of anthraxylon (vitrain) from several coal beds.

Thiessen and his coworkers, G. C. Sprunk and H. J. O'Donnell (Fig. 7), recognized many of the paleobotanical constituents of Paleozoic bituminous coals in thin section (see highlights in Thiessen and Sprunk, 1941). Their work was made possible by the superb thin-section preparation techniques of their coworker, H. J. O'Donnell, who brought the art to a high level (Schopf and Oftedahl, 1976) and who later also became a coal microscopist. O'Donnell was solely responsible for the preparation of thousands of thin sections—what his

Figure 6. Reinhardt Thiessen (circa 1937). From the photographic files of Marlies Teichmüller.

Figure 7. Left to right: H. J. O'Donnell, Reinhardt Thiessen, and G. C. Sprunk, June 1937 at the U.S. Bureau of Mines, Pittsburgh, Pennsylvania. Courtesy of H. J. O'Donnell.

later coworkers called "O'Donnell's monument." He vividly remembers Thiessen's great patience and warmly alluded to him as a "nice man" (H. J. O'Donnell, personal communication, 1992). The three coworkers realized that plant constituents could be clearly seen in transmitted light parallel to bedding, the standard method today for coal petrographic studies, as opposed to cross section (Stach et al., 1982). However, this standard practice neglects the three-dimensional aspect of plant remains that is often critical for identification. Thiessen and Sprunk (1941) showed photomicrographs of some of the principal coal-forming plants in Appalachian coals, namely, lycopods, seed ferns, and cordaitean remains. These include fine examples of tracheids with various types of pitting, lycopod periderm, cordaitean leaves with cuticles and tissue, sporangia with preserved spores, and numerous examples of petioles with resinous canals belonging to medullosan seed ferns (cycadophyte of Thiessen).

Thiessen (1928; 1937a, 1947) clearly understood the role of flora in the determination of the type of coal formed. He maintained that coals were bright as a result of the preservation of woody materials or their byproducts, coals were attrital as a result of the preservation of attritus (finely divided plant matter), and banded coals were a mixture of the two. Cannel coals were considered by Thiessen to be composed of "spores, cuticles, and other resistant matter," and boghead coals were mainly derived from oil algae.

Details of Thiessen's classification of coal are in Thiessen, 1930b, 1937a, 1947 (p. 41–47). He distinguished various types of bright coals on the basis of the predominance of constituents such as spores, resins, cuticles, leaves, and barks. Thiessen used the terms *splint coal* and *semisplint coal* for lithotypes having lesser or greater amounts of anthraxylon (vitrite), respectively. Petrographically, he described bituminous coal beds stratigraphically from the bottom to the top of the coal bed, using scores of thin sections. The descriptions included the type and subtype and thickness of the major bands of coals, the amounts of anthraxylon, attritus (both translucent and opaque), and fusain. In addition, each of the major bands of coal types was commonly analyzed chemically, including proximate and ultimate analyses.

In addition to thin sections of coal, thin sections of peatified wood were prepared in Thiessen's laboratory. The wood was compressed under high pressure so that direct comparisons could be made with wood in lignite, brown coal, and higher-rank coals (Thiessen, 1937, 1947).

## THIESSEN'S OIL SHALE RESEARCH

Beginning in the 1920s, under the leadership of David White, Thiessen studied the petroleum potential of oil shales. During World War I, White recognized the need for alternative sources of petroleum and began studies of oil shales from the western and eastern United States. Thiessen's contributions to these efforts resulted in three publications on the subject (Thiessen, 1921c, 1925d).

Thiessen (1921c, d) showed that Devonian oil shales from Illinois were composed mainly of mineral matter and spores. He was able to identify spores of different types (similar to the alga [acritarch] *Tasmanites*; see discussion on the affinities of this genus in Schopf et al., 1944, p. 16, and in Traverse, 1988, p. 119). He also observed comminuted cuticular matter, dark-colored unidentified organic matter, and very little resinous matter. Thiessen made comparisons with similar oil shales in Ohio, Kentucky, and Tennessee. He identified in the Illinois oil shale a spore, *Sporangites huronesis* Dawson (see discussion in Schopf et al., 1944, p. 16). Thiessen also compared oil shale with banded coal, cannel coal and canneloid slate, and boghead coal, all of which differ, according to him, in the amounts of mineral matter, spore-exines, woody organic matter, and resins.

Thiessen published a well-illustrated report on the oil shales of Kentucky (Thiessen, 1925d). He showed that these oil shales consisted of 13 to 31% organic matter. He clearly demonstrated that the spore-rich nature of these oil shales was attributed to autochthonous, resistant plant matter (e.g., spores and cuticles) and admixed drifted or wind-blown organic matter, which accumulated in large lakes marginal to the Appalachian trough. In the same paper he reported and illustrated oil algae in cannel coal from the Freeport coal zone.

## THIESSEN'S PEAT STUDIES

Thiessen's early research was not confined to coal (White and Thiessen, 1913) but included work on the composition and chemistry of peat. Much of this work was conducted on the Holocene peat bogs of Wisconsin (see Thiessen, 1920f, pls. 1–3), which he frequently visited with his coworkers and Carnegie Institute of Technology students. George Sprunk, a chemist by training, in 1930 completed his M.S. thesis on the chemistry of peat under the guidance of Thiessen at the Carnegie Institute of Technology, Pittsburgh. After graduation, he joined the U.S. Bureau of Mines as Thiessen's assistant.

Thiessen (1928) experimented on the activity of bacteria in peat bogs. He found that bacteria are active at least to depths of 7.4 m under acid conditions. Earlier work by Thiessen had generated widespread denial of the existence of bacterial activity under such conditions. Thiessen was able to demonstrate that cellulose was lost from wood in peat bogs and that lignocellulose was more resistant to bacteriological processes than cellulose. In lignite that he investigated, Thiessen observed that all the cellulose was gone. He had earlier recognized that cellulosic components (and lignin) were more reduced in subbituminous coals than in lignites (Thiessen, 1912). These observations are fundamental to our thinking today on the fate of cellulose during early coalification (Hatcher et al., 1981).

Thiessen had a good grasp on the fate of cellulose and lignin during coalification, as revealed in the following statement: "Trees like the cedars, bald cypress, and certain other conifers are full of resins which form preservative substances of the wood. The wood in those trees is not decomposed easily, and the cellulose remains a long time. Those are the trees of which we find the wood intact in the brown coals and lignites. Yet in a number of trees the cellulose has disappeared completely. What has become of that cellulose is a problem" (Thiessen, 1929). He further stated: "All but the lignin decompose almost completely on rotting; the lignin also is changed in some way into a substance of unknown chemical composition known as humin" (Thiessen, 1928). Thiessen (1937a, 1947) also recognized that under aerobic conditions some fungi selectively attack the lignin in wood and *not* the cellulose and cause the so-called white rot.

Hugh O'Donnell relates (personal communication, 1992) an interesting story about a barrel of sectioned peat from Kiel, Wisconsin (see Thiessen 1920f, pls. 1, 3) which was stolen, presumably by Kiel moonshiners, in the summer of 1936. Thiessen put an ad in the paper indicating it was U.S. government property and that the "F.B.I. always gets its man." Thiessen managed to recover the peat, which was then shipped by rail to Washington.

## THIESSEN'S FINAL YEAR

On March 29, 1937, Franklin D. Roosevelt, president of the United States, by executive order gave Thiessen a one-year extension from the mandatory 70-year retirement age because "the public interest so requires." A medical report dated February 12, 1937, shows that Thiessen was 1.80 m tall, weighed 81 kg, and was "in very good physical condition for a man of his age."

Thiessen continued his research on bituminous coals with great enthusiasm until he was stricken with a heart attack on January 29, 1938. He died the next day. A pioneering era of coal microscopic research in the public service ended with his death.

## THIESSEN'S MONUMENTAL LEGACY: THE REINHARDT THIESSEN COAL THIN-SECTION SLIDE COLLECTION

Thiessen's pioneering work on U.S. coal petrology is richly recorded by the Reinhardt Thiessen collection of thin sections of coal. The collection consists of more than 19,000 slides from 682 localities, mainly from U.S. coalfields, including those in Alaska (Schopf and Oftedahl, 1976). Thin sections are material from localities in Canada, Mexico, England, Scotland, France, Belgium, Denmark, Germany, Sweden, Italy, Greece, Yugoslavia, Poland, Morocco, Malaya, Republic of China, Antarctica, Australia, South Africa, Rhodesia, Venezuela, Chile, Peru, Colombia, and Brazil. The collection includes sections of peat and wood from peat; oil shales from the United States and many foreign countries; special collections, such as Lomax's, Jeffrey's, and F. D. Reed's sections and thin-sections of fossil woods (Tertiary?); and J. M. Schopf's western U.S. coal sections studied for uranium during World War II. The Thiessen collection includes some unusual foreign specimens, including a coal sample collected near the South Pole by Admiral Byrd in 1937; lopingite, a cuticle-rich coal from Loping, Kiangsi Province, China; and torbanite from Torbane Hill, Scotland.

After Thiessen's death in 1938, H. J. O'Donnell (1910– ) and G. C. Sprunk (1905– ) continued his great coal petrological research until 1943, when Sprunk, who became an expert coal microscopist (Schopf and Oftedahl, 1976), left the U.S. Bureau of Mines for a job in industry (H. J. O'Donnell, personal communication, 1992). J. M. Schopf (1911–1978) replaced Sprunk at Pittsburgh in 1943 and was in turn replaced by B. C. Parks (1902–?) after Schopf was transferred to the U.S. Geological Survey to take up paleobotanical studies (Schopf and Oftedahl, 1976). In the late 1940s Schopf established a coal petrological laboratory of the U.S. Geological Survey at Ohio State University in Columbus, where he stayed for the rest of his career. O'Donnell carried on Thiessen's work at the U.S. Bureau of Mines until he was transferred from Pittsburgh in 1960.

The Thiessen collection was transferred in the late 1960s or early 1970s to J. M. Schopf of the U.S. Geological Survey in Columbus, Ohio (H. J. O'Donnell, personal communication, 1992); after Schopf's death in 1978, the collection was transferred to the U.S. Geological Survey in Reston, Virginia, where it was housed under the supervision of R. W. Stanton. The collection was transferred to S. S. Sorenson, Department of Mineral Sciences, Museum of Natural History, the Smithsonian Institution, on March 9, 1988 (accession no. 379115; catalog no. 116423). Thiessen's notebooks (15 volumes), which were recovered from the U.S. Bureau of Mines by R. J. Gray, are now at Southern Illinois University at Carbondale and under the care of J. C. Crelling (R. J. Gray, personal communication, 1992). Unfortunately, the slides are now badly cracked as a result of aging, so they have lost much of their research value, but they remain a rich legacy of Thiessen's pioneering work in coal petrology. The thin-section method of transmitted light microscopy is still the best method for paleobotanical studies of coal by microscopical examination, even though the reflected light method is presently the international standard for coal petrological work (M. Teichmüller, in Stach et al., 1982).

Thiessen's pioneering work on American coal petrology is richly recorded both by the Reinhardt Thiessen Coal Thin-Section Slide Collection and by Thiessen's bibliography of about 82 items (see Reinhardt Thiessen complete bibliography, this chapter). The International Committee for Coal and Organic Petrology (ICCP), under the initiative of E. Stach, honored Reinhardt Thiessen by creating an award medal made of briquetted anthracite and showing Thiessen's portrait. The "Reinhardt Thiessen Medal" is presented by the ICCP from time to time to distinguished international coal petrologists and other scientists who are working in the field of organic petrology.

## REINHARDT THIESSEN, THE FATHER OF GENETIC COAL PETROLOGY AND AN UNFORGETTABLE TEACHER: RECOLLECTIONS OF A GERMAN EXCHANGE STUDENT IN THE UNITED STATES, 1938

*Marlies Teichmüller*

Dr. Reinhardt Thiessen was my teacher in coal petrography in January 1938, the year of his untimely death. In September 1937, I was very lucky to study in the United States as a German exchange student. First, I went to Clark University in Worcester, Massachusetts (September–December 1937), to learn more geography and geology. Before I left Germany, Dr. Erich Stach had been my teacher in coal petrography at the University of Berlin. He recommended that I visit Dr. Thiessen who, at that time, worked at the U.S. Bureau of Mines in Pittsburgh, Pennsylvania, and was an internationally known coal petrographer.

In contrast to most coal petrographers in western Europe, who studied coals under the microscope with polished sections under incident light, Thiessen (and other American coal petrographers) worked with thin sections of coal in transmitted light. These different methods had led to different coal microscopic nomenclatures and different results of microscopic analyses. Thus, it seemed reasonable to study the coal thin-section method, which had been highly developed by Thiessen, in order to compare and contrast the various coal microscopic nomenclatures and analytical results. This was the reason why I went to Pittsburgh in January 1938 to visit Thiessen, with the hope that I could work under his supervision for a couple of months.

In a letter of January 19, 1938, I wrote to Rolf Teichmüller, who later became my husband:

> Fourteen days in Pittsburgh have passed. Hail to Pittsburgh, the Bureau of Mines, Thiessen, and work here which makes me so happy! Never before was I able to spend so much time with my books—now it is oftentimes 1 A.M. before I notice that it is time to go to bed. After 5:30 P.M. I remain in the Bureau of Mines pretty much by myself. This is the time when I write down all my new experiences, or go searching for literature in Thiessen's bookcases. I have experienced a lot of inspiration. Thiessen is really great [*ganz prima*]—he always has time for me, he directs my attention to everything that he has noticed or read, leads me through the whole Bureau of Mines, introduces me to the coal people at the Carnegie Institute of Technology; in short, he is a teacher and a leader who could not be surpassed. The other day he suggested that I should remain here and work for him. If he had only known about [my] poor abandoned Rolf!

In other letters written from Pittsburgh in January 1938, I reported that Thiessen was of German descent. His father and grandfather came to the United States around 1850 from Schleswig-Holstein, a German province, lying between the North and the Baltic Seas. Both Reinhardt Thiessen and his wife, Clara, were born in small towns in central Wisconsin where people in 1938 still spoke "Plattdeutsch," a local dialect of northern Germany. Clara Thiessen spent her youth in a village named "Kiel" (in Wisconsin). In Germany, Kiel is the capital of Schleswig-Holstein and is a well-known harbor along the Baltic coast.

During visits to the Thiessen's home, I always enjoyed their great hospitality and the music of Beethoven, which Thiessen especially enjoyed, besides other classical music. According to Stach (1966), Thiessen played various instruments: double bass, violoncello, alto horn, French horn, and trombone. In his youth, he had joined different bands, partly to finance his scientific studies.

Although I stayed until April 1938, my expectations were shattered; three weeks after my arrival, Reinhardt Thiessen died on January 30, 1938, of an unexpected heart attack. He was only 70 years old. It was the greatest disappointment of my life up to that time. In a letter of January 30, 1938, I wrote to Rolf Teichmüller;

> An incomprehensible thing has happened: Thiessen is dead. Yesterday, Saturday, he came to the office, worked as usual, looked good and fresh as always. We had been talking shortly before closing time about travel in Germany. This morning I received a telephone message: he has died from a heart attack. I have been thinking about it all day long—I simply cannot comprehend it. I had great respect for Thiessen; in fact, most recently, something like a friendship had developed between us. He was one of the real great persons whom I met, and I was so tremendously happy that I was allowed to work with him. His desk is covered with outlines and projects. He had just recently started new and important investigations. In spite of his age, no one would have suspected that he was 70 years old; he was as active and creative as a young person.
>
> I have not been able to think about it, how this event will affect my stay here. I believe I can remain here and continue to work with Thiessen's assistant, [G. C.] Sprunk. However, I will miss Thiessen's advice very much, our stimulating conversations, and his assistance in paleobotany.

During the three weeks that I stayed with Thiessen, I learned of his great personality and his unique methodology as a researcher. Thiessen was very broad-minded which, at least partly, was the result of his academic training in various disciplines of science, including botany, palaeobotany, chemistry, geology, and astronomy. He had studied mainly botany at the University of Chicago. His Ph.D. thesis (Thiessen, 1908a) is entitled "The vascular anatomy of the seedling of *Dioon edule,*" and his first nonthesis publication (Thiessen, 1911) is entitled "Plant remains composing coals." Unlike most coal petrographers, Thiessen knew the conditions prevailing in mires and peats. His homeland, Wisconsin, offered many opportunities to investigate the origin of coal in mires of postglacial origin. Later he studied the role of bacterial decomposition and the relationship between cellulose, lignin, and humic substances as well as soluble lipoid substances in peats. Thiessen's main interest was the origin of coal, as is seen from his list of publications, a catalog of his thin-section slide collection, and a detailed report on his activities during the 1907–1910 period at the U.S. Geological Survey in Washington, D.C. (Schopf and Oftendahl, 1976). There he worked with David White (see Chapter 10, this volume) and later at the Bureau of Mines in Pittsburgh, where Dr. A. C. Fieldner was his adviser. One of Thiessen's last publications "What is Coal?" (Thiessen, 1937a, 1947) became a guide for my later work in organic petrology.

In the 1930s, Thiessen became more and more engaged in the field of technologically applied coal petrography. Together with his assistants, George Sprunk, a microscopist who was originally trained as a chemist, and Hugh O'Donnell, who prepared thousands of excellent coal thin sections, as well as with other coal chemists, physicists, and engineers, he published many papers that focused on the detailed petrographic composition of individual U.S. coal seams. The purpose of these studies was to determine their technological behavior, mainly their carbonization properties. These publications contain introductions with detailed geological and paleobotanical information and numerous excellent photomicrographs that were always explained from the viewpoint of a botanist.

No doubt that Thiessen's thin-section method offered notable advantages for botanical-genetical investigations, although it is applicable mainly to low-rank coals up to about 1.2% $R_r$ vitrinite reflectance. Thiessen developed a coal petrographic nomenclature based on his thin-section method. This classification was later introduced into the *International Handbook of Coal Petrography* (International Committee for Coal Petrology, 1963) as the "Thiessen–Bureau of Mines nomenclature."

The major result of my stay with Thiessen in 1938 was my Doctor's thesis. It is a comparison, using an eastern Kentucky coal, of his transmitted-light method and nomenclature with the West European reflected-light method that was based on polished sections and the Stopes-Heerlen nomenclature (Teichmüller, 1941).

I am extremely grateful that I met and learned from Reinhardt Thiessen, whom I regard as the father of genetic coal petrology. He influenced my scientific career more than any other coal petrologist.

## ACKNOWLEDGMENTS

Thanks are given to H. J. O'Donnell and Linwood Thiessen of Pittsburgh for sharing memories of and information on Thiessen and for permission to publish Figures 3–5 and 7. E. D. Morey provided a newspaper memorial of Thiessen and also some correspondence from Thiessen from the William C. Darrah files. E. C. Robertson supplied information on Thiessen's presentation at the Cosmos Club. B. L. Graff of the U.S. Geological Survey Library (Reston, Virginia) assisted with the compilation of Thiessen's bibliography.

Nora Tamberg of the U.S. Geological Survey Library (Reston, Virginia) is thanked for translating the German from two letters to Rolf Teichmüller.

## REFERENCES CITED

Darrah, W. C., 1939, Textbook of paleobotany: New York, D. Appleton-Century, 441 p.

Hatcher, P. G., Breger, I. A., and Earl, W. L., 1981, Nuclear magnetic resonance studies of ancient buried wood. I: Observations on the origin of coal to the brown-coal stage: Organic Geochemistry, v. 3, p. 49–55.

International Committee for Coal Petrology, 1963, International handbook of coal petrology (second edition): Paris, Centre National Recherche Scientifique, 184 p.

Jeffrey, E. C., 1910, Microscopic study of certain coals in relation to the sapropelic hypothesis: Science, new series, v. 32, p. 220–221.

Kosanke, R. M., 1969, Mississippian and Pennsylvanian palynology, *in* Tschudy, R. H., and Scott, R. A., eds., Aspects of palynology: New York, Wiley-Interscience Division of John Wiley and Sons, p. 223–270.

Parks, B. C., and O'Donnell, H. J., 1956, Petrography of American coal: U.S. Bureau of Mines Bulletin 550, 193 p.

Potonié, R., 1924, Einführung in die allgemeine Kohlenpetrographie: Berlin, Borntraeger, 285 p.

Schopf, J. M., 1939, *Medullosa distelica,* a new species of the anglica group of *Medullosa:* American Journal of Botany, v. 26, p. 196–207.

Schopf, J. M., and Oftedahl, O. G., 1976, The Reinhardt Thiessen coal thin-section slide collection of the U.S. Geological Survey, catalog and notes: U.S. Geological Survey Bulletin 1432, 58 p.

Schopf, J. M., Wilson, L. R., and Bentall, R., 1944, An annotated synopsis of Paleozoic fossil spores and the definition of generic groups: Illinois State Geological Survey Report of Investigations 91, 73 p.

Schulze, F., 1855, Ueber das Vorkommen wohlerhaltener Zellulose in Braunkohle und Steinkohle: Sitzungs berichte Königliche Preussische, Akademie der Wissenschaften zu Berlin, p. 676–678.

Stach, E., 1966, Reinhardt Thiessen, 1867–1938, *in* Freund, H., and Berg, A., eds., Geschichte der Mikroskopie, Leben und Werk grosser Forscher, Volume 3: Frankfurt/Main, Umschau Verlag, p. 433–440.

Stach, E., Mackowsky, M.-Th., Teichmüller, M., Taylor, G. H., Chandra, D., and Teichmüller, R., 1982, Stach's textbook of coal petrology (third edition): Berlin, Gebrüder Borntraeger, 535 p.

Stopes, M. C., 1919, On the four visible ingredients in banded bituminous coals: Royal Society of London, Proceedings, ser. B, v. 90, p. 547–564.

Teichmüller, M., 1941, Der Feinbau amerikanischer Kohlen im Anschliff und Dünnschliff—Ein Vergleich kohlenpetrographischer Untersuchungsmethoden: Jahrbuch Reichsstelle für Bodenforschung Berlin, v. 61, p. 20–55, 9 pls., 2 tables.

Thiessen, R. See Bibliography of Reinhardt Thiessen that follows.

Traverse, A., 1988, Paleopalynology: Boston, Massachusetts, Unwin Hyman, 600 p.

White, D., and Thiessen, R., 1913, The origin of coal: U.S. Bureau of Mines Bulletin 38, 390 p.

## BIBLIOGRAPHY OF REINHARDT THIESSEN

Arnold, C. L., Lowy, A., and Thiessen, R., 1934, The isolation and study of the humic acids from peat: U.S. Bureau of Mines Report of Investigations 3258, 9 p.

Arnold, C. L., Lowy, A., and Thiessen, R., 1935, The isolation and study of the humic acids from peat: Fuel, v. 14, p. 107–112.

Fieldner, A. C., Davis, J. D., Thiessen, R., Kester, E. B., and Selvig, W. A., 1931, Methods and apparatus used in determining the gas, coke, and by-product making properties of American coals with results on a Taggart-bed coal from Rhoda, Wise County, Va. U.S. Bureau of Mines Bulletin 344, 107 p.

Fieldner, A. C., Davis, J. D., Thiessen, R., Kester, E. B., Selvig, W. A., Reynolds, D. A., Jung, F. W., and Sprunk, G. C., 1932, Carbonizing properties and constitution of washed and unwashed coal from Mary Lee bed, Flat Top, Jefferson County, Ala. U.S. Bureau of Mines Technical Paper 519, 78 p.

Fieldner, A. C., Davis, J. D., Thiessen, R., Kester, E. B., Selvig, W. A., Jung, F. W., and Sprunk, G. C., 1933, Carbonizing properties and constitution of No. 2 gas bed coal from Point Lick No. 4 Mine, Kanawha County, W. Va.: U.S. Bureau of Mines Technical Paper 548, 52 p.

Fieldner, A. C., Davis, J. D., Thiessen, R., Kester, E. B., Selvig, W. A., Reynolds, D. A., Jung, F. W., and Sprunk, G. C., 1935, Carbonizing properties and constitution of Alma Bed coal from Spruce River No. 4 Mine, Boone County, W. Va.: U.S. Bureau of Mines Technical Paper 562, 41 p.

Fieldner, A. C., Davis, J. D., Thiessen, R., Selvig, W. A., Reynolds, D. A., Sprunk, G. C., and Jung, F. W., 1936, Carbonizing properties and petrographic composition of Pittsburgh bed coal from Pittsburgh Terminal No. 9 Mine, Washington County, Pa.: U.S. Bureau of Mines Technical Paper 571, 33 p.

Fieldner, A. C., Davis, J. D., Thiessen, R., Selvig, W. A., Reynolds, D. A., Sprunk, G. C., and Holmes, C. R., 1937, Carbonizing properties and petrographic composition of Millers Creek bed coal from Consolidation No. 155 Mine, Johnson County, Ky., and the effect of blending Millers Creek coal with Pocahontas bed and Pittsburgh-bed (Warden mine) coals: U.S. Bureau of Mines Technical Paper 572, 50 p.

Fieldner, A. C., Davis, J. D., Thiessen, R., Selvig, W. A., Reynolds, D. A., Brewer, R. E., and Sprunk, G. C., 1938a, Carbonizing properties and petrographic composition of Upper Banner bed coal from Clinchfield No. 9 mine, Dickenson County, Va., and of Indiana No. 4 bed coal from Saxton No. 1 Mine, Vigo County, Ind., and the effect of blending these coals with Beckley bed coal: U.S. Bureau of Mines Technical Paper 584, 81 p.

Fieldner, A. C., Davis, J. D., Thiessen, R., Selvig, W. A., Reynolds, D. A., and Holmes, C. R., 1938b, Carbonizing properties of West Virginia coals and blends of coals from the Alma, Cedar Grove, Dorothy Powellton A, Eagle, Pocahontas, and Beckley beds: U.S. Bureau of Mines Bulletin 411, 162 p.

Sprunk, G. C., and Thiessen, R., 1932, Spores of certain American coals: Fuel, v. 11, p. 360–370.

Sprunk, G. C., and Thiessen, R., 1935, Relation of microscopic composition of coal to chemical, coking, and by-product properties: Industrial and Engineering Chemistry, v. 27, p. 446–451.

Thiessen, R., 1908a, The vascular anatomy of the seedling of *Dioon edule* [Ph.D. thesis]: Chicago, University of Chicago, Department of Botany Contribution No. 119, Hull Botanical Laboratory.

Thiessen, R., 1908b, The vascular anatomy of the seedling of *Dioon edule:* Botanical Gazette, v. 46, p. 357–380, 7 pls. [Thiessen's published Ph.D. thesis].

Thiessen, R., 1911, Plant remains composing coals [abs.]: Science, new series, v. 33, p. 551–552.

Thiessen, R., 1912, On certain constituents and the genesis of coals [abs.]: Washington Academy of Sciences Journal, v. 2, p. 232–233.

Thiessen, R., 1913a, On the constitution and genesis of certain lignites and subbituminous coals [abs.]: 8th International Congress of Applied Chemistry (1912), Original Communications, v. 25, p. 203–204.

Thiessen, R., 1913b, Microscopic study of coal, *in* White, D., and Thiessen, R., The origin of coal: U.S. Bureau of Mines Bulletin 38, p. 187–378.

Thiessen, R., 1918, The determination of the stratigraphic position of coal seams by means of their spore-exines [abs.]: Science, new series, v. 47, p. 469.

Thiessen, R., 1919a, Constitution of coal through a microscope: Coal Mining Institute of America, 33rd Proceedings, p. 34–44; discussion, p. 44–45.

Thiessen, R., 1919b, Constitution of coal through a microscope: Coal Industry, v. 2, p. 558–563.

Thiessen, R., 1919c, Occurrence and origin of finely disseminated sulphur compounds in coal: American Institute of Mining and Metallurgical Engineers Bulletin 153, p. 2431–2444.

Thiessen, R., 1919d, Occurrence and origin of finely disseminated sulphur compounds in coal: American Institute of Mining and Metallurgical Engineers Transactions, v. 63, p. 913–926.

Thiessen, R., 1919e, Occurrence and origin of finely disseminated sulphur compounds in coal: Coal Age, v. 16, p. 668–673.

Thiessen, R., 1920a, Structure in bituminous coals: U.S. Bureau of Mines Report of Inventory 2196, 4 p.

Thiessen, R., 1920b, Structure in Paleozoic bituminous coals: U.S. Bureau of Mines Bulletin 117, 296 p.

Thiessen, R., 1920c, The correlation of coal seams by means of spore-exines [abs.]: Science, new series, v. 51, p. 522.

Thiessen, R., 1920d, Compilation and composition of bituminous coals: Journal of Geology, v. 28, p. 185–209.

Thiessen, R., 1920e, Recent developments in the microscopic study of coal: Coal Mining Institute of America, 34th Proceedings (1920), p. 88–119; discussion, p. 119–120.

Thiessen, R., 1920f, Structure in Paleozoic bituminous coals: U.S. Bureau of Mines Bulletin 117, 296 p.

Thiessen, R., 1920g, Under the microscope coal has already lost much of its former mystery, I: Coal Age, v. 18, p. 1183–1189.

Thiessen, R., 1920h, Under the microscope coal has already lost much of its former mystery, II: Coal Age, v. 18, p. 1223–1228.

Thiessen, R., 1920e, Under the microscope coal has already lost much of its former mystery, III: Coal Age, v. 18, p. 1275–1279.

Thiessen, R., 1921a, Recent developments in the microscopic study of coal: Canadian Mining Journal, v. 42, p. 64–68, 86–91, 109–113, 124–128.

Thiessen, R., 1921b, Under the microscope coal has already lost much of its former mystery, IV: Coal Age, v. 19, p. 12–15.

Thiessen, R., 1921c, Origin and composition of certain oil shales: Economic Geology, v. 16, p. 289–300, pls. 9–10.

Thiessen, R., 1921d, Origin and composition of certain oil shales [abs.]: Geological Society of America Bulletin, v. 32, p. 72–73; discussion by F. R. van Horn, p. 74.

Thiessen, R., 1924, The origin and constitution of coal: Proceedings and collections of the Wyoming Historical and Geological Society for the years 1923–1924, v. 19: Wilkes-Barre, Pennsylvania, The E. B. Yordy Company, p. 3–44, pls. 21A, 21B.

Thiessen, R., 1925a, The constitution of coal: American Institute of Mining and Metallurgical Engineers [A.I.M.E.] Pamphlet 1438-I, 50 p.; discussion, A.I.M.E. Pamphlet 1473-I, p. 12–14.

Thiessen, R., 1925b, The microscopical constitution of coal: American Institute of Mining and Metallurgical Engineers, Transactions, v. 71, p. 35–114; discussion, p. 114–116.

Thiessen, R., 1925c, The microscopical constitution of coal: American Institute of Mining and Metallurgical Engineers Pamphlet 1473; discussion, p. 12–14.

Thiessen, R., 1925d, Microscopic examination of Kentucky oil shales, *in* Thiessen, R., White, D., and Crouse, C. S., Oil shales of Kentucky—A series of four economic and morphological discussions of the Devonian shales of this Commonwealth: Kentucky Geological Survey, ser. 6, v. 21, p. 1–47, 37 pls.

Thiessen, R., 1925e, Origin of the boghead coals, *in* Shorter contributions to general geology, 1923–24: U.S. Geological Survey Professional Paper 132-I, p. 121–137, pls. 27–40.

Thiessen, R., 1925f, Environmental conditions of deposition of coal: Discussion: American Institute of Mining and Metallurgical Engineers, Transactions, v. 71, p. 24–25, 26.

Thiessen, R., 1925g, Microscopical structure of anthracite: Discussion: American Institute of Mining and Metallurgical Engineers, Transactions, v. 71, p. 143–144.

Thiessen, R., 1925h, Resolution of coal by oxidation: Discussion: American Institute of Mining and Metallurgical Engineers, Transactions, v. 71, p. 174.

Thiessen, R., 1926, The micro-structure of coal: Royal Society of Arts Journal (London), v. 74, p. 535–554; discussion, p. 554–557.

Thiessen, R., 1929, Some recent developments on the constitution of coal: Proceedings, International Conference on Bituminous Coal, 2nd, Pittsburgh, November 19–24, 1928, Volume 1: Pittsburgh, Pennsylvania, Carnegie Institute of Technology, p. 695–761; discussion, p. 761–767.

Thiessen, R., 1930a, Classification of coal from the viewpoint of the paleobotanist: American Institute of Mining and Metallurgical Engineers (Coal Division), Transactions, v. 88, p. 419–435; discussion, p. 435–437.

Thiessen, R., 1930b, Splint coal: American Institute of Mining and Metallurgical Engineers Transactions, v. 88, p. 644–671; discussion, p. 671–672.

Thiessen, R., 1930c, Recently developed methods of research in the constitution of coal and their application to Illinois coals: Illinois State Geological Survey, Cooperative Mining Series Bulletin 33, p. 58–89.

Thiessen, R., 1930d, The microscopic structure of coals of the Monongahela series: West Virginia Academy of Sciences Proceedings, v. 3, p. 159–198.

Thiessen, R., 1931a, Recently developed methods of research in the constitution of coal and their application to Illinois coals: Illinois State Geological Survey Bulletin 60, p. 117–147.

Thiessen, R., 1931b, Recently developed methods of research in the constitution of coal and their application to Illinois coals: Fuel, v. 10, p. 72–94.

Thiessen, R., 1932a, Physical structure of coal, cellulose fiber, and wood as shown by Spierer Lens: Industrial and Engineering Chemistry, v. 24, p. 1032–1041.

Thiessen, R., 1932b, Carbonizing properties and constitution of No. 6 Bed Coal from West Frankfort, Franklin County, Ill.: U.S. Bureau of Mines Technical Paper 524, 60 p.

Thiessen, R., 1932c, Carbonizing properties and constitution of Pittsburgh bed coal from Edenborn Mine, Fayette County, Pa.: U.S. Bureau of Mines Technical Paper 525, 60 p.

Thiessen, R., 1932d, Carbonizing properties and constitution of Black Creek Bed Coal from Empire Mine, Walker County, Ala.: U.S. Bureau of Mines Technical Paper 531, 44 p.

Thiessen, R., 1932e, Carbonizing properties and constitution of Chilton bed coal from Boone No. 2 Mine, Logan County, W. Va.: U.S. Bureau of Mines Technical Paper 542, 60 p.

Thiessen, R., 1936a, Carbonizing properties and petrographic composition of Clintwood-bed coal from Buchanan mines Nos. 1 and 2, Buchanan County, Va.: U.S. Bureau of Mines Technical Paper 570, 34 p.

Thiessen, R., 1936b, Fusain content of coal dust from an Illinois dedusting plant: American Institute of Mining and Metallurgical Engineers Technical Publication 664, 12 p.

Thiessen, R., 1937a, What is coal?: 17th Meeting of the Fuel Engineers of Appalachian Coals, Inc. (Cincinnati, Ohio, January 11, 1937), Record of Meetings, v. 4, p. 211–272; discussion, p. 273–279.

Thiessen, R., 1937b, Origin and petrographic composition of the Lower Sunnyside coal of Utah: U.S. Bureau of Mines Technical Paper 573, 34 p.

Thiessen, R., 1938b, Preparation of thin sections of coal: Fuel, v. 17, p. 307–315.

Thiessen, R., 1947, What is coal?: U.S. Bureau of Mines Information Circular 7397, 53 p. Reprinted from 17th Meeting of the Fuel Engineers of Appalachian Coals Inc. (Cincinnati, Ohio, January 11, 1937), Record of Meetings v. 4, p. 211–279; discussion, p. 273–279.

Thiessen, R., and Francis, W., 1928, Classification of coal from the viewpoint of the paleobotanist: American Institute of Mining and Metallurgical Engineers Technical Publication 156, p. 28–44; discussion, p. 44–46.

Thiessen, R., and Francis, W., 1929a, Terminology in coal research: Fuel, v. 8, p. 385–405.

Thiessen, R., and Francis, W., 1929b, Terminology in coal research: U.S. Bureau of Mines Technical Paper 446, 27 p.

Thiessen, R., and Johnson, R. C., 1929, An analysis of a peat profile: Industrial and Engineering Chemistry, Analytical Edition, v. 1, p. 216–220.

Thiessen, R., and Johnson, R. C., 1930, An analysis of a peat profile: Fuel, v. 9, p. 367–372.

Thiessen, R., and Johnson, R. C., 1934, Studies on peat: Alcohol and ether-soluble matter of certain peats: Fuel, v. 13, p. 44–47.

Thiessen, R., and Sprunk, G. C., 1934, The effect of heat on coal as revealed by the microscope: Fuel, v. 13, p. 116–125.

Thiessen, R., and Sprunk, G. C., 1935, Microscopic and petrographic studies of certain American coals: U.S. Bureau of Mines Technical Paper 564, 71 p.

Thiessen, R., and Sprunk, G. C., 1936, Origin of the finely divided or granular opaque matter in splint coals: Fuel, v. 15, p. 304–315.

Thiessen, R., and Sprunk, G. C., 1941, Coal paleobotany: U.S. Bureau of Mines Technical Paper 631, v. 4, 56 p. (presented at Annual Meeting, American Association for Advancement of Science, Indianapolis, Indiana, December 1937).

Thiessen, R., Sprunk, G. C., and O'Donnell, H. J., 1931, Microscopic study of Elkhorn coal bed at Jenkins, Letcher County, Ky.: U.S. Bureau of Mines Technical Paper 506, 30 p.

Thiessen, R., Sprunk, G. C., and O'Donnell, H. J., 1938a, Preparation of thin sections of coal: U.S. Bureau of Mines Information Circular 7021, 8 p., 12 figs.

Thiessen, R., and Staud, J. N., 1923, Correlation of coal beds in the Monongahela Formation of Ohio, Pennsylvania, and West Virginia: Carnegie Institute of Technology, Coal-mining Investigations Bulletin 9, 64 p.

Thiessen, R., and Strickler, H. S., 1934, The distribution of micro-organisms in four peat deposits: U.S. Bureau of Mines, Carnegie Institute of Technology, and Mining and Metallurgical Advisory Boards Cooperative Bulletin 61, 20 p.

Thiessen, R., and Voorhees, A. W., 1922, A microscopic study of the Freeport coal bed, Pennsylvania: Carnegie Institute of Technology, Coal-mining Investigations Bulletin 2, 75 p.

Thiessen, R., and Wilson, F. E., 1924, Correlation of coal beds of the Allegheny Formation of western Pennsylvania and eastern Ohio: Carnegie Institute of Technology, Coal-mining Investigations Bulletin 10, 56 p.

MANUSCRIPT ACCEPTED BY THE SOCIETY JULY 6, 1994

Printed in U.S.A.

Geological Society of America
Memoir 185
1995

# *Elias Howard Sellards (1875–1961): Paleontologist, geologist, and anthropologist*

**Elsie Darrah Morey**
*Morey Paleobotanical Laboratory, 1729 Christopher Lane, Norristown, Pennsylvania 19403*
**Alicia Lesnikowska**
*Department of Biology, Georgia Southwestern College, Americus, Georgia 31709-4693*

## ABSTRACT

**Elias Howard Sellards (1875–1961) was born in Kentucky but moved to Kansas as an adolescent. He received the B.A. (1899) and M.Sc. (1901) degrees from the University of Kansas and a Ph.D. from Yale University in 1903. He taught at Rutgers College, the University of Florida, Gainesville, and the University of Texas at Austin. He was the first state geologist of Florida (1907–1918), and he was staff geologist (1918–1925), acting director (1925–1932) director (1932–1945), and director emeritus (1945–1961) at the Texas Bureau of Economic Geology. From 1938 to 1957, he was simultaneously director of the Texas Memorial Museum. His broad interests included Paleozoic paleobotany and palynology, Paleozoic insects, vertebrate paleontology, early humans in the Americas, and meteorites, to mention several. He made notable contributions to the study of the Permian floras of Kansas and to the study of seed ferns, establishing with David White, Felix Oliver, and others the gymnospermous nature of that group. He named the Permian genus *Glenopteris* and also named the fructification *Codonotheca.* He also advanced the study of fossil cockroaches and of economic geology, vertebrate paleontology, and anthropology of Florida and Texas.**

## EARLY YEARS

Elias Howard Sellards (Fig. 1) was born in Carter, Kentucky, on May 2, 1875, the son of Wiley W. Sellards and Sarah Menix Sellards. He began his schooling in Kentucky, but while he was still in the elementary grades, his family moved to Kansas. They were among the last people to make the trip in a covered wagon (Ellison et al., 1961). Sellards completed high school in Scranton, Kansas, and studied at Washburn Academy in Topeka before enrolling at the University of Kansas, Lawrence. He received both his Bachelor of Arts degree (1899) and his Master of Science degree (1901) from the University of Kansas. His M.Sc. thesis described three species of a new fern genus, *Glenopteris* (Sellards, 1900b).

At the University of Kansas, Sellards studied under the noted vertebrate paleontologist Samuel W. Williston. Sellards was also affiliated with the Kansas Geological Survey (then known as the University Geological Survey of Kansas) and under this aegis collected extensively from the Permian outcrops of east-central Kansas. From deposits near Banner City, in Dickinson County, he collected classic Permian fossils such as *Callipteris* (*Autunia*) *conferta* (Sellards, 1900a), in addition to *Glenopteris* and new species of *Sphenopteris* and *Taeniopteris* (Sellards, 1901a, b). These specimens, collected by Sellards for the University Geological Survey, are currently housed in the Paleobotanical Collections (Figs. 2, 3, and 4) of the McGregor Herbarium at the University of Kansas. Two of Sellards's holotypes are illustrated in Figures 2 and 3.

From the University of Kansas, Sellards moved to Yale University, where he was awarded both a scholarship and a

Morey, E. D., and Lesnikowska, A., 1995, Elias Howard Sellards (1875–1961): Paleontologist, geologist, and anthropologist, *in* Lyons, P. C., Morey, E. D., and Wagner, R. H., eds., Historical Perspective of Early Twentieth Century Carboniferous Paleobotany in North America (W. C. Darrah volume): Boulder, Colorado, Geological Society of America Memoir 185.

Figure 1. Elias Howard Sellards. Photograph courtesy of the Texas Memorial Museum.

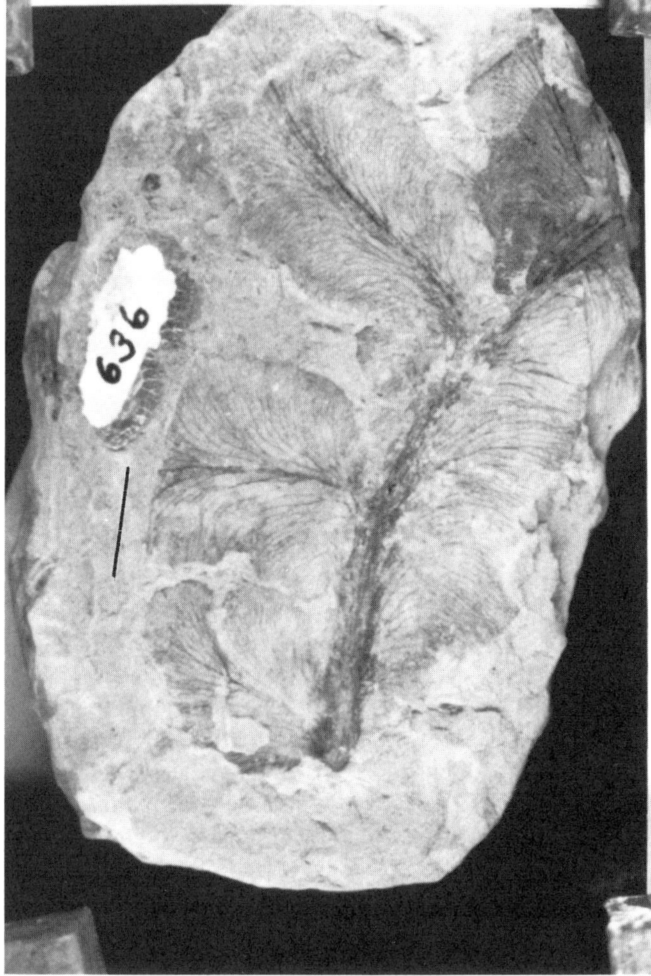


Figure 2. *Odontopteris anomala* Sellards. Kansas specimen number 636 (holotype) currently housed in the Paleobotanical collections of the McGregor Herbarium, University of Kansas. Photograph by A. Lesnikowska. Scale = 1 cm.

fellowship. At Yale, Sellards worked with C. E. Beecher, H. S. Williams, and Charles Schuchert; the latter became a close personal friend of Sellards (Ellison et al., 1961).

During his time at Yale, Sellards worked on the Carboniferous impression flora of the famous Mazon Creek locality in Illinois. He described the fructifications and the spores of *Crossotheca* and *Myriotheca* (Sellards, 1902a) and established the genus *Codonotheca* (Sellards, 1903b). He recognized the affinities of this fructification with what at the time were known as the cycadofilices (cycadlike ferns) and noted its similarity to gymnospermous pollen-producing structures before Oliver and Scott's (1904) famous paper demonstrated that the cycadofilices were, in fact, gymnosperms.

Sellards continued to work on the Permian plants of Kansas for his dissertation, but he also turned to the study of fossil insects. He was the first to describe insects from the Permian deposits of Kansas (Sellards, 1903c) and, together with the Kansas Geological Survey, collected thousands of insect specimens from the Permian and, particularly, Upper Pennsylvanian strata. He described several new species, including *Mylaceris anceps* and *Etoblattina juvenis* (Sellards, 1904).

After receiving the Ph.D. in 1903, Sellards was employed as instructor of geology and mineralogy by Rutgers University for the 1903–1904 academic year, and then he became professor of geology and zoology at the University of Florida, Gainesville. During this time he published on the stratigraphy of Kansas (Sellards and Beede, 1905) and produced several papers on Permian insects (Sellards, 1906a; 1907b; 1909a, b) and Permian plants (Sellards, 1907a).

In 1907, Sellards married Alma Alford; they had two daughters, Helen and Daphne. Alma Sellards died in 1953. In 1907 Sellards became the first state geologist of Florida, a position that he held until 1918. He worked closely with geological surveys in other states (Ferguson, 1981), summarized his work on the paleobotany and biostratigraphy of Kansas and on Kansas insects (Sellards, 1908a, c) in a University Geological

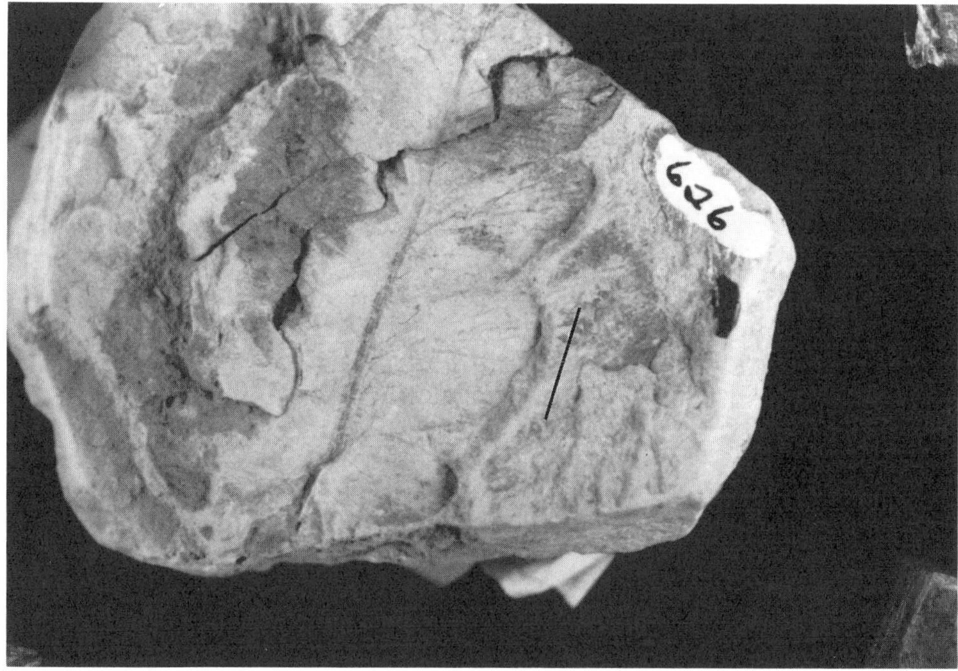


Figure 3. *Odontopteris excellsa* Sellards. Specimen no. 626 (holotype) from the Le Roy Shale, currently housed in the Paleobotanical Collections of the McGregor Herbarium, University of Kansas. Photograph by A. Lesnikowska. Scale = 1 cm.

Survey of Kansas publication, and described two new species of insects from Texas (Sellards, 1911b). However, his attention understandably turned to more practical geological matters, and he wrote numerous articles on the natural resources and general geology of Florida during the 1906–1918 period (see Elias Howard Sellards selected bibliography, this chapter).

During his tenure as state geologist of Florida, an event took place that was to have a major impact on Sellards's scientific career. In 1913, hominid remains were found associated with fossils of extinct mammals in the wall of a canal at Vero (now Vero Beach), Florida. This discovery attracted widespread attention and controversy and resulted in an international symposium in 1916. Sellards's work (Sellards, 1916a, 1917, 1919) was eventually corroborated by J. W. Gridley, a vertebrate paleontologist from the U.S. National Museum. The interest Sellards developed in early humans was to remain with him for the rest of his life. This interest culminated in his book *Early Man in America—A Study of Prehistory* (Sellards, 1952a), which gained him prominence as an authority on the settlement of the Americas. Sellards was in the process of revising the book at the time of his death.

## CONTRIBUTIONS TO SCIENCE IN TEXAS

In 1918, Sellards accepted the post of staff geologist with the Texas Bureau of Economic Resources, where he was to remain for the rest of his career, serving first as acting director, then director, and finally as director emeritus.

He first achieved prominence with the Bureau in 1921 when he was employed by the Texas Attorney-General's office as an expert geologist. Oklahoma claimed land to the southern high-water line of the Red River, whereas Texas asserted that the boundary between the two states was the southern low-water mark. Sellards's data in support of the Texas position were accepted by the U.S. Supreme Court, and the court's ruling in favor of Texas awarded the state oil land worth millions of dollars.

In 1925, Sellards became acting director of the Bureau of Economic Geology, and in 1926, he was named professor of geology and a member of the Graduate Faculty at the University of Texas, Austin.

Sellards served as an able administrator of the Bureau and also for various government projects, such as the Works Project Administration (W.P.A.) mineral resource surveys, the Texas State Planning Board, and the Colorado River Authority. He actively worked to get government money for resource surveys (to aid geologists put out of work by the Great Depression) and was able to obtain sufficient funds to expand the W.P.A. mandate to include vertebrate paleontology and anthropology (Ferguson, 1981).

In 1934, human artifacts were found near Austin, and Sellards dated the sediments as 10,000 to 20,000 years old, thus doubling the age of human occupation of the Americas, which was considered at the time to be 5,000 to 10,000 years. In fact, Sellards became an expert in Pleistocene stratigraphy in order to date human remains (Ferguson, 1981). The surveys directed

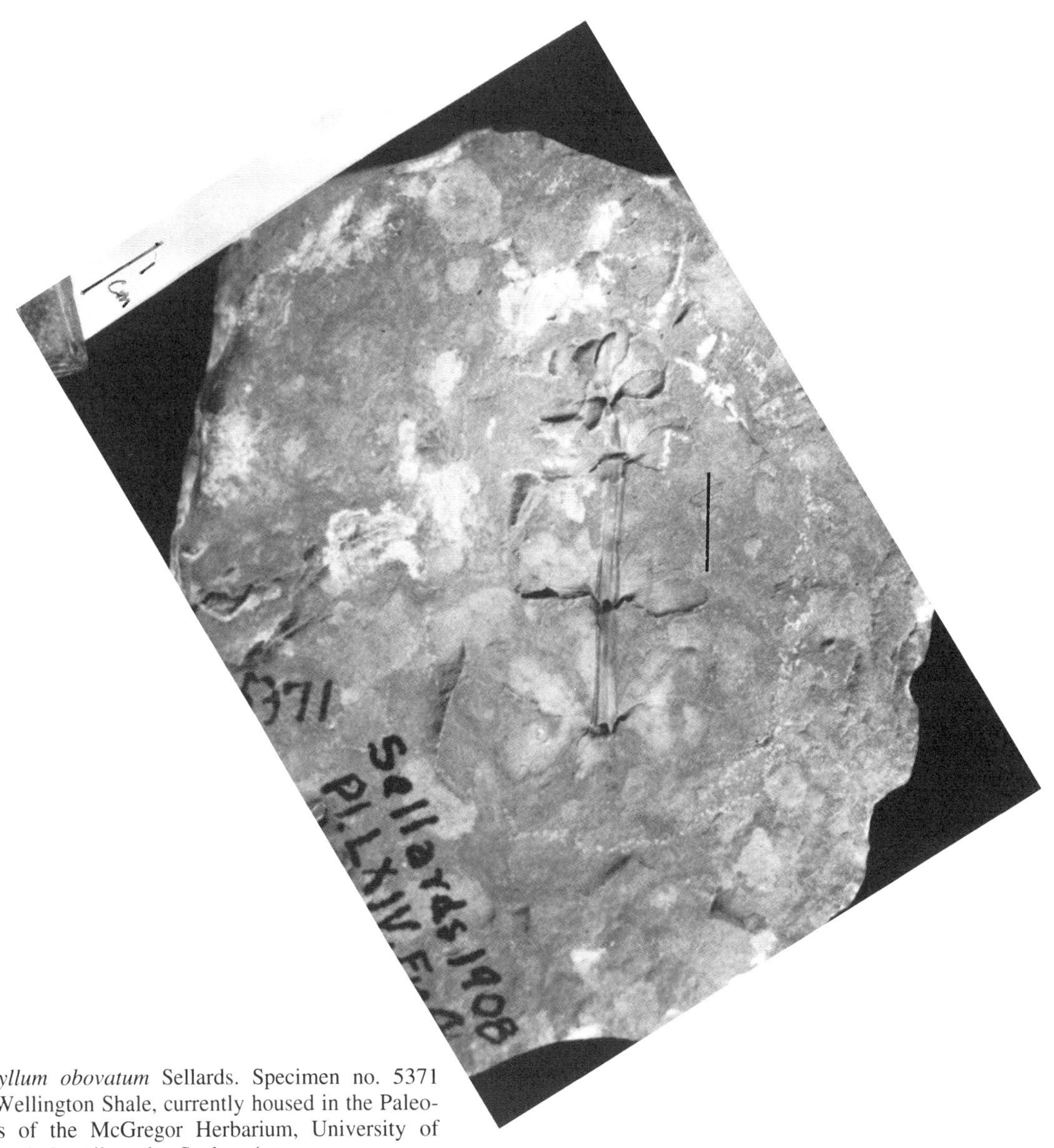


Figure 4. *Sphenophyllum obovatum* Sellards. Specimen no. 5371 (holotype) from the Wellington Shale, currently housed in the Paleobotanical Collections of the McGregor Herbarium, University of Kansas. Photograph by A. Lesnikowska. Scale = 1 cm.

by Sellards uncovered numerous fine specimens from several localities. The fossils eventually became part of the collections of the University of Texas and the Texas Memorial Museum. Sellards was actively involved in creating the latter institution and served as its director from 1938 until his death.

## GENERAL CONTRIBUTIONS

Sellards's career was an unusually varied and productive one. He wrote more than 186 books and articles. Sellards made significant contributions to late Paleozoic stratigraphy and biostratigraphy (Phillips et al., 1973; Darrah, 1969) and the Pleistocene stratigraphy of Texas. He described many fossil plants and monographed the Paleozoic insects. He published extensively on the geology and natural resources of Florida and was instrumental in natural resource exploration and utilization in Texas, although primarily in an administrative role. Finally, Sellards became an authority on the early human settlement of the area of the United States.

Sellards was active in the Geological Society of America and was a charter member of the Paleontological Society, of which he was president in 1931 and vice president in 1943. He served five terms as president of the Southwestern Geological Society. Sellards belonged to the American Association for the Advancement of Science, the Texas Academy of Science (an honorary life member), the Society of Vertebrate Paleontology, the American Association of Petroleum Geologists, the Society of Economic Geologists and Mineralogists, the Society for American Archeology, the Archeological Society, and the Texas Philosophical Society as well as the Texas

Chapter of the National Science Honor Society, Sigma Xi (Evans, 1961).

He was honored by the University of Kansas with the Erasmus Haworth Distinguished Alumnus Award for his contributions to geology and paleontology and by the University of Florida, which named a student geology award for him.

## ACKNOWLEDGMENTS

The authors have relied heavily on Ferguson's 1981 *History of the Bureau of Economic Geology, 1909–1960* in writing this portrait.

We thank Frederic Burchsted of the University Archives, the University of Texas at Austin, and Janice Sorensen, archivist at the Kansas Geological Survey, for invaluable information, and Sally A. Baulch of the Texas Memorial Museum for providing the photograph of E. H. Sellards and some correspondence of Sellards.

We also thank Paul C. Lyons, Keene Ferguson, and J. Thomas Dutro, Jr., for helpful suggestions to improve the manuscript.

## REFERENCES CITED

Darrah, W. C., 1969, A critical review of the Upper Pennsylvanian floras of eastern United States, with notes on the Mazon Creek flora of Illinois: Gettysburg, Pennsylvania, privately printed, 220 p., 80 pls.

Ellison, S. P., Newcomb, W. W., and Eddy, J. R. D., 1961, In memoriam: Elias Howard Sellards, *in* Report of the special Elias Howard Sellards Memorial Resolution Committee: Austin, Texas, Documents and minutes of the General Faculty, The University of Texas, p. 7708–7714.

Evans, G. L., 1961, Elias Howard Sellards (1875–1961): American Association of Petroleum Geologists Bulletin, v. 25, p. 346–347.

Ferguson, W. K., 1981, History of the Bureau of Economic Geology, 1909–1960: Austin, The University of Texas at Austin, Bureau of Economic Geology, 329 p.

Oliver, F. W., and Scott, D. H., 1904, On the structure of the Paleozoic seed *Lagenostoma lomaxi*, with a statement of evidence upon which it is referred to *Lyginodendron:* Royal Philosophical Society of London Transactions, v. 197B, p. 193–247, pls. 13, 14, 23.

Phillips, T. L., Pfefferkorn, H. W., and Peppers, R. A., 1973, Development of paleobotany in the Illinois Basin: Illinois State Geological Survey Circular 480, 86 p.

Sellards, E. H. See selected bibliography of Elias Howard Sellards that follows.

## SELECTED BIBLIOGRAPHY OF ELIAS HOWARD SELLARDS

Sellards, E. H., 1900a, Note on the Permian flora of Kansas: Kansas University Quarterly, v. 9, p. 63–64.

Sellards, E. H., 1900b, A new genus of ferns from the Permian of Kansas: Kansas University Quarterly, v. 9, p. 179–189, pls. 37–42.

Sellards, E. H., 1901a, Fossil plants of the Permian of Kansas: Transactions of the Kansas Academy of Science, v. 17, p. 208–209.

Sellards, E. H., 1901b, Permian plants—*Taeniopteris* of the Permian of Kansas: Kansas University Quarterly, v. 10, p. 1–12, pls. 1–4.

Sellards, E. H., 1902a, On the fertile fronds of *Crossotheca* and *Myriotheca*, and on the spores of other Carboniferous ferns, from Mazon Creek, Illinois: American Journal of Science, 4th ser., v. 14, p. 195–202, 1 pl.

Sellards, E. H., 1902b, On the validity of *Idiophyllum rotundifolium* Lesquereux, a fossil plant from the coal measures at Mazon Creek, Illinois: American Journal of Science, 4th ser., v. 14, p. 203–204.

Sellards, E. H., 1903a, Some new structural characters of Paleozoic cockroaches: American Journal of Science, 4th ser., v. 15, p. 307–315.

Sellards, E. H., 1903b, *Codonotheca*—a new type of spore-bearing organ from the coal measures: American Journal of Science, 4th ser., v. 16, p. 87–95, 1 pl.

Sellards, E. H., 1903c, Discovery of fossil insects in the Permian of Kansas: American Journal of Science, 4th ser., v. 16, p. 323–324.

Sellards, E. H., 1904, A study of the structure of Paleozoic cockroaches with description of new forms from the coal measures: American Journal of Science, 4th ser., v. 18, p. 113–134.

Sellards, E. H., 1906a, Types of Permian insects. Part 1: Odonata: American Journal of Science, ser. 4, v. 22, p. 249–258.

Sellards, E. H., 1906b, Some sink hole lakes of north-central Florida: Science, v. 23, p. 289–290.

Sellards, E. H., 1906c, Geology of Florida in relation to its artesian water supply: Florida State Horticulture Society, p. 117–121.

Sellards, E. H., 1907a, Notes on the spore-bearing organ *Codonotheca* and its relationship with the Cycadofilices: The New Phytologist (London), v. 6, p. 175–178.

Sellards, E. H., 1907b, Types of Permian insects. Part 2: Plecoptera: American Journal of Science, ser. 4, v. 23, p. 345–355.

Sellards, E. H., 1908a, Fossil plants of the Upper Paleozoic of Kansas: Kansas University Geological Survey, v. 9, p. 386–467.

Sellards, E. H., 1908b, Geological investigations in Florida previous to the organization of the State Geological Survey: Florida State Geological Survey, 1st annual report, p. 54–108.

Sellards, E. H., 1908c, Chapter X, Fossil plants of the upper Paleozoic of Kansas (Special Report on Oil and Gas): Bulletin of the University Geological Survey of Kansas, v. 9, p. 386–499.

Sellards, E. H., 1909a, Cockroaches of the Kansas coal measures and of the Kansas Permian: University Geological Survey of Kansas, v. 9, p. 501–535, pls. 70–83.

Sellards, E. H., 1909b, Types of Permian insects. Part 3: Megasecoptera, Oryctoblattinidae and Protorthoptera: American Journal of Science, ser. 4, v. 27, p. 151–173.

Sellards, E. H., 1911a, Two Florida phosphate deposits: American Fertilizer, v. 35, p. 37–47.

Sellards, E. H., 1911b,Two new insects from the Permian of Texas: Carnegie Institution, Washington Publication 146, p. 142–152.

Sellards, E. H., 1913, Classification of soils of Florida: Florida Department of Agriculture, 12th Biannual Report, p. 249–300.

Sellards, E. H., 1915, A few gavial from the Late Tertiary of Florida: American Journal of Science, 4th ser., v. 40, p. 135–138.

Sellards, E. H., 1916a, Human remains from the Pleistocene of Florida: Science, v. 44, p. 615–617.

Sellards, E. H., 1916b, Fossil vertebrates from Florida—A new Miocene fauna; a new Pliocene species; the Pleistocene fauna: Florida State Geological Survey, 8th Annual Report, p. 77-119.

Sellards, E. H., 1917, Further notes on human remains from Vero, Florida: American Anthropologist, v. 19, p. 239–251.

Sellards, E. H., 1919, Literature relating to human remains and artifacts at Vero, Florida: American Journal of Science, 4th ser., v. 47, p. 358–360.

Sellards, E. H., 1920, Geology and natural resources of Bexar County, Texas: University of Texas Bulletin 1932, 202 p.

Sellards, E. H., 1923, The Oklahoma-Texas boundary suit: Science, v. 57, p. 346–349.

Sellards, E. H., 1924, Mineral resources of Texas, the South's development: Manufacturers Record, v. 86, p. 421–423.

Sellards, E. H., 1930, Travis County, *in* Mineral resources of Texas: Austin, University of Texas, Bureau of Economic Geology, p. 41–69.

Sellards, E. H., 1938, Problem of early man in America: Geological Society of America Bulletin, v. 49, p. 1899.

Sellards, E. H., 1940a, Early man in America: Index to localities and selected bibliography: Geological Society of America Bulletin, v. 51, p. 373–431.

Sellards, E. H., 1940b, New Pliocene mastodon (from Texas): Geological Society of America Bulletin, v. 51, p. 1659–1664.

Sellards, E. H., 1940c, Odessa meteor crater: Geological Society of America Bulletin, v. 51, p. 1944.

Sellards, E. H., 1942, Principal war and industrial metals, minerals and mineral substances: University of Texas, Bureau of Economic Geology of Mineral Resources Circular 21, 9 p.

Sellards, E. H., 1944a, Progress in excavating the Odessa, Texas, meteorite crater: Popular Astronomy, v. 51, p. 224–225.

Sellards, E. H., 1944b, Ancient carvings on view in the Texas Memorial Museum: Texas Memorial Museum, Museum Notes 6, p. 23–29.

Sellards, E. H., 1946, The Plainview Texas, fossil bison quarry: Science, v. 103, p. 632.

Sellards, E. H., 1952a, Early man in America—A study in prehistory: Austin, University of Texas Press, 221 p.

Sellards, E. H., 1952b, Age of Folsom man (Texas): Science, v. 115, p. 98.

Sellards, E. H., 1955, Fossil bison and associated artifacts from Milnesand, New Mexico: American Antiquity, v. 20, p. 336–344.

Sellards, E. H., 1960, Some early stone artifact developments in North America: Southwestern Journal of Anthropology, v. 16, p. 160–173.

Sellards, E. H., and Beede, J. W., 1905, Stratigraphy of the eastern outcrop of the Kansas Permian: American Geologist, v. 36, p. 83–111.

Sellards, E. H., Thorp, B. C., and Hill, R. T., 1923, Investigations on the Red River made in connection with the Oklahoma-Texas boundary suit: University of Texas Bulletin 2327, 74 p.

Manuscript Accepted by the Society July 6, 1994

Printed in U.S.A.

Geological Society of America
Memoir 185
1995

# *Maxim Konrad Elias (1889–1982)*

**Alicia Lesnikowska***
*R. L. McGregor Herbarium, University of Kansas, 2045 Constant Avenue, Lawrence, Kansas 66047*

## ABSTRACT

**Maxim Konrad Elias was an important paleobotanist, paleontologist, and biostratigrapher. He was born in Minsk, Russia, as Maxim Konradovich Eliashevich. He received the degree of Engineer of Mines from the Imperial School of Mines, St. Petersburg, in 1917 and worked at the Ural Mining Institute and as a coal company geologist. In 1920 he moved to Vladivostok, where he taught at the Polytechnic Institute and was a member of the Russian Geographical Society. He came to the United States in 1922. Elias was a geologist with the Kansas State Geological Survey from 1927 to 1937 and became a U.S. citizen in 1930. In 1938 he led a geological exploration party in Colombia, South America. He received a Ph.D. from Yale University in 1939. He was paleontologist with the University of Nebraska Conservation and Survey Division from 1939 to 1958 when he retired and became adjunct professor at the Research Institute, University of Oklahoma, Norman. He died in Alliance, Nebraska, May 6, 1982, at the age of 93.**

## INTRODUCTION

Maxim Konrad Elias was not only a paleobotanist of standing but was at the same time a field geologist, an invertebrate paleontologist, and an important biostratigrapher who was concerned throughout his career with the correlation of late Paleozoic strata between Europe and North America. His European training and experience, together with his long career in the United States, made him particularly apt for this work. During his lengthy career he examined pteridophytes, early conifers, prairie angiosperms, and marine fungi and algae; he made extensive studies of bryozoans and cephalopods and at various times considered brachiopods, pelecypods, fusulinids, and conodonts in his biostratigraphic work. He was one of the few persons to study such a tremendous diversity of fossils. His major stratigraphic interest was in the late Paleozoic, but he published articles on fossils from the Cambrian, Ordovician, Silurian, Devonian, Cretaceous, and Tertiary as well as the Carboniferous and Permian.

*Present address: Department of Biology, Georgia Southwestern College, Americus, Georgia 31709-4693.

## EARLY LIFE (1889–1927)

Maxim Konrad Elias was born Maxim Konradovich Eliashevich in Minsk, Russia, on August 12, 1889. He began a career in the Imperial Navy as a cadet at the Naval Academy but transferred to the Imperial School of Mines, St. Petersburg, the center of Russian academic life. In 1915 he married fellow student, Elena Zinovievyev, who was studying poultry science because it was one of the few fields open to women in Tsarist Russia. He graduated from the School of Mines with an advanced degree in geology (Engineering of Mining) in 1917 and moved to Siberia, where he had been involved with geological expeditions as a student. He was docent (instructor) at the Ural Mining Institute in 1917 and associate professor there from 1917 to 1920. Simultaneously he held the position of geologist (1917–1918) and chief geologist (1919–1920) at the Verk-Isetsk Mining Company.

In 1920 the Eliasheviches moved to Vladivostok, where he was professor at the Vladivostok Polytechnic Institute from 1920 until 1923 and a member of the Far Eastern Branch of the Russian Geographical Society from 1920 to 1922. In 1922 he was sent by the Geographical Society and by the Poly-

Lesnikowska, A., 1995, Maxim Konrad Elias (1889–1982), *in* Lyons, P. C., Morey, E. D., and Wagner, R. H., eds., Historical Perspective of Early Twentieth Century Carboniferous Paleobotany in North America (W. C. Darrah volume): Boulder, Colorado, Geological Society of America Memoir 185.

technic Institute to Japan and the United States to conduct geological research. With considerable foresight, he took his fossil collections, his wife, and his infant son, Maxim, with him. The family reached San Francisco, via Japan, on November 22, 1922. Shortly thereafter, Siberia fell under the control of the hard-line communist faction, and Eliashevich, who had gathered intelligence for anticommunist forces during his tenure in Siberia, decided to stay in the United States.

In San Francisco, Eliashevich supported his family as an electrical draftsman while he learned English. He was elected to the California Academy of Science and studied the plant fossils from the coalfields of the midcontinent United States in the possession of that institution, comparing them with his material from the Carboniferous of Siberia. As a result of these investigations, he determined that in the Kuznetz Basin, western Siberia, plant material comes from strata of both Late Carboniferous and Jurassic ages. Much confusion had resulted from the erroneous belief that there was a single plant-bearing horizon, which in a sort of biostratigraphic averaging was assigned to the Permian. He had previously communicated his conclusions to the respected Russian geologist, Dr. Vladimir Obruchev, who had reached the same point of view, and Eliashevitch's data were included in "Geologie von Sibirien" (Obruchev, 1926).

Eliashevitch's comparisons between Siberian and mid-American Carboniferous plant fossils were published as "On a Seed-bearing *Annularia* and on *Annularia* Foliage" in 1931 (1931a). In addition to the biostratigraphic data on the Kuznetz Basin, this paper included a description of a specimen showing seeds in attachment to *Annularia* (generally considered to represent the foliage of spore-producing plants related to the modern horsetails). The conclusion that some horsetail-like plants produced seeds attracted attention even in the popular press (San Francisco Examiner, 1927) and was supported by the prominent paleobotanist David White (see Chapter 10, this volume) but was never widely accepted.

## THE STATE GEOLOGICAL SURVEY OF KANSAS (1927–1937)

In 1927 Eliashevich was able to give up his position with the electrical manufacturing firm to be a geologist for the Kansas Geological Survey, a position he held until 1937. On March 4, 1930, he became a naturalized citizen of the United States as Maxim Konrad Elias, and it was under this name that all his American publications appeared.

At the Kansas Geological Survey Elias (Fig. 1) was involved with detailed mapping in northwestern Kansas and adjacent parts of Colorado. In the foreword to "The geology of Wallace County" (Elias, 1931b), Raymond Moore, head of the Kansas Survey, wrote: "The following report on the geology of Wallace county is more than a county report . . . it gives information that bears on the geology of a large part of the contiguous plains territory. . . . The painstaking field investigations of Mr. Elias and his thorough studies in the preparation of his report have resulted in an important contribution to the knowledge of Kansas geology." Elias's standards never faltered. In this work he set the pattern for the rest of his career by studying the ammonites of the region (Elias, 1933a) as well as the plant fossils. This report was also the beginning of his studies of the grasses and other herbs of the High Plains Tertiary. Elias made major contributions to the systematics and evolutionary biology of grasses and prairie forbs (Elias, 1932a, 1942, 1946c) and developed the first biostratigraphic zonation of the Tertiary based on angiosperm fossils (Elias, 1935; Chaney and Elias,1936). His work on the Tertiary flora extended even to the cryptogams. He described limestones produced by a new kind of alga, which he named *Chlorellopsis* (Elias, 1932b).

Figure 1. Dr. Elias during his years at the Kansas Geological Survey. The photograph was taken about 1935. Courtesy of T. Mylan Stout.

Elias described the now-famous Garnett flora from east-central Kansas (Moore et al., 1936) as well as a scorpion from the same locality (Elias, 1937c). He demonstrated that the "Permian" aspect of the conifer-dominated flora was due to its paleoecology rather than to its age. This is one of the first explicit considerations of the important role of paleoenvironment and paleoecology in plant biostratigraphy in the United States.

For the Kansas Survey, Elias also mapped extensively in north-central Kansas, work that led to the publication of a detailed paleoecological analysis of the upper Carboniferous/Lower Permian marine strata there. Elias concluded, based on the distribution of the marine invertebrates, that the sea had a maximum depth of 55 m (Elias, 1937b). This analysis was one of the first, but by no means the last, of Elias's studies that involved a comparison of fossils and the extant organisms and habitats related to them. Elias continued to investigate ammonoids (Elias, 1938a, b, c) and started to consider in detail Paleozoic bryozoans (Elias, 1937d).

While with the Kansas Survey Elias participated in the Sixteenth International Geological Congress held in Washington, D.C. (Elias, 1933b), and was a leader on the excursion arranged to show European paleobotanists such as W. J. Jongmans (see Chapter 5, this volume) the Carboniferous of Kansas. Elias provided Jongmans with a number of specimens but later took issue with him for assigning European names to the American material (letter to W. C. Darrah dated June 22, 1935). In Elias's presentation to the Second Carboniferous Congress in Heerlen, The Netherlands, to which he was invited by Jongmans, he described the suite of typically Stephanian plants he had collected in Kansas and argued, contrary to Jongmans, for the undeniable Stephanian age of the midcontinent Douglas and Wabaunsee groups (Elias, 1937a).

In addition to his published paleobotanical work, throughout the time he was working for the Kansas Survey he collected many plant fossils and annotated the specimens already in the Survey collection. His handwriting still appears on labels and specimens of this collection, which is currently housed in the University of Kansas McGregor Herbarium (KANU).

In 1937 Elias left the Kansas Survey, and his family moved to Illinois while he led a geological exploration party into Colombia, South America, for the Socony Vacuum Oil Company in 1937 and 1938. On that trip Elias contracted amoebic parasites and was never entirely free of them for the rest of his life.

Elias held an advanced degree from the Imperial School of Mines, but it was not widely recognized in the United States. Thus he spent a year at Yale University and was awarded a Ph.D. in paleontology in 1939 for his monographic treatment of the grasses and other herbs of the Tertiary of the High Plains. This impressive work (Elias, 1942) included a revision of the extant relatives of his Tertiary fossils, based on characters of the palea and lemma, which proved important in his paleobotanical analyses.

## AT THE NEBRASKA SURVEY (1939–1958)

From Yale, Elias moved to the University of Nebraska Conservation and Survey Division in Lincoln, where he was paleontologist until his retirement in 1958. Elias also served as a consulting paleontologist for Standard Oil Company and other firms during his tenure at the Nebraska Survey.

Elias had already worked for the head of the Nebraska Survey, Dr. G. E. Condra, having made collections of Tertiary angiosperms for him in the summers of 1934 and 1935. Upon Elias's arrival at the Nebraska Survey, the two embarked on a long series of investigations of Paleozoic bryozoans (Condra and Elias, 1941, 1944a, b, c, d, e, 1945; Elias and Condra, 1957). They were convinced that many "bryozoans," in particular the well-known *Archimedes* Hall, were symbiotic associations of algae, both red and brown, and fenestellid bryozoans. Although accepted by the well-known paleophycologist J. Harlan Johnson (Johnson and Kenoshi, 1956), this point of view was not widely held.

Dr. Elias also independently examined bryozoans (Elias, 1950b, 1954, 1956) and fossil algae (Elias, 1946a, b, 1947). He further described a unique permineralized specimen of the primitive conifer *Walchia* (Elias, 1948) from Garnett, Kansas; studied fusulinids (Elias, 1950a); and published a series of papers on the Mississippian fauna of the Redoak Hollow Sandstone Member of the Goddard Shale (Elias, 1957a, b, c, 1958) that considered all major taxa including brachiopods, pelecypods, gastropods, and ostracods. Nor did he neglect the ammonoids and their biostratigraphy (Elias, 1952; Colton et al., 1959).

## "RETIREMENT" AND THE UNIVERSITY OF OKLAHOMA

Even in retirement Elias did not give up paleontological research but went to the Research Institute of the University of Oklahoma, Norman, where he was adjunct professor until 1972. In Oklahoma he continued his work on ammonoids (Elias, 1962), fusulinids (Elias, 1959, 1966a), bryozoans (Elias, 1964, 1971), and bryozoan/algal symbioses (Elias, 1965). His work on the last-mentioned topic was well summarized in his last presentation at the Seventh Carboniferous Congress in Krefeld, Germany, 1971 (Elias, 1973). He updated and refined his work on marine paleoecology and on cyclothemic sedimentation (Elias, 1962, 1966b) and on Paleozoic correlations between Europe and North America (Elias, 1960, 1970).

## THE FINAL YEARS (1972–1982)

Elias returned to Lincoln, Nebraska, in the summer of 1972 and worked at the University of Nebraska's Lincoln State Museum until his health failed. In the fall of 1981, he and his wife moved to Alliance, Nebraska, where their daughter, Roxanne Elias Podhaisky, still lives. Dr. Elias died on May 6, 1982, at the age of 93.

## ELIAS THE PERSON

Elias was a man of high principles and firm convictions and very concerned with family. He broke with the head of the Kansas Survey, Raymond C. Moore, when Moore divorced his wife to marry his secretary. The Eliases were both very close to the first Mrs. Moore, and Elias regarded the divorce as an unforgivable betrayal of her (A. T. Cross, personal communication, 1992).

Elias was devoted to his own family—two sons, Maxim and Gregory (a third son, Michael, died in infancy), and a daughter, Roxanne—but Dr. Elias was what we would probably today call a workaholic. Summer vacations with the children were spent collecting fossils, and Mrs. Podhaisky's clearest memories of her father are of seeing him hunched over his microscope in his study at home.

Dr. Elias was a talented artist, generally preparing the illustrations for his publications. He had a strongly practical turn of mind and wanted both his sons to be geologists, which they were, though Maxim was a talented sculptor who initially wanted to pursue art professionally (he did manage several exhibitions with his painter wife). Elias attended the meetings of the professional societies regularly, and his attendance reflects the diversity of his interests. In addition to the Geological Society of America and the International Geological and Carboniferous Congresses, he contributed occasionally to the meetings of the Botanical Society of America and regularly to the Nebraska Academy of Science. In addition to his own presentations, he was always ready with a comment on the other talks. He readily shared his expertise and was active in encouraging young scientists through the science honor societies, Sigma Xi and Sigma Gamma Epsilon.

Elias was a very careful biostratigrapher and did not want to see the least vagueness or carelessness in the use of stratigraphic terms (letter of Elias to W. C. Darrah, dated June 8, 1935). He sometimes criticized younger paleobotanists for taking on too much too soon,as when he took Darrah to task for attempting to make correlations between U.S. and European strata before (in Elias's opinion) Darrah had a sufficient grasp of the American upper Carboniferous biostratigraphy. Nor did Elias hesitate to criticize the works of such well-established paleobotanists as Jongmans and Bertrand when he thought they were hastily done (letter of Elias to Darrah, dated June 22, 1935).

Although Elias was a political refugee, he maintained a positive professional relationship with colleagues from the Soviet Union and was host to Soviet scientists from time to time. On one occasion, Elias was escorting several visiting Soviet scientists to Kansas and then back to Lincoln. It was New Year's Day and the roads were very icy, but Elias was so interested in talking with his guests that he would get carried away and not pay much attention to careful driving. Finally the Russians refused to ride any farther unless Elias's student assistant, Aureal Cross, drove (A. T. Cross, personal communication, 1992).

## HONORS AND AWARDS

Elias received many honors and awards throughout his career. He was the recipient of a Marsh Fund grant from the National Academy of Science in 1933, which supported his work on Tertiary grasses, and of a grant from the Carnegie Institution of Washington in 1938. He served on three committees of the National Research Council: namely, the Committee on Common Problems in Genetics, Paleontology and Systematics; the Committee on a Treatise on Marine Ecology and Paleoecology; and the Subcommittee on Carboniferous correlations of the Committee on Stratigraphy.

He was chosen "a noted man in geology" by the Colorado School of Mines, where a portrait and biography were hung in the School's seminar room. This was in recognition of the diverse contributions he made to the field of geology as a researcher and, in his later years, as a teacher. He was a fellow of the Paleontological Society and the Geological Society of America and a member of the Botanical Society of America, the Nebraska Academy of Science, and the American Association for the Advancement of Science as well as of the American Association of Petroleum Geologists and the Lincoln branch of Rotary International.

## ACKNOWLEDGMENTS

Dr. Elias's biography is still in the process of being compiled. Therefore I have had to rely particularly heavily on unpublished sources and his own publications. I would especially like to thank Roxanne Elias Podhaisky, who patiently answered my numerous questions and provided me with invaluable documents pertaining to her father and her family. Elsie Darrah Morey generously made available to me Elias's letters to her father, William C. Darrah, and Aureal T. Cross provided much information. I would also like to thank T. Mylan Stout (Elias's biographer) for photographs, Tom Phillips and Paul Lyons for helpful suggestions as to sources of information, and Rex Buchanan for providing me with an unpublished manuscript. C. Wnuk and W. H. Pfefferkorn served as reviewers.

## REFERENCES CITED

Chaney, R. W., and Elias, M. K., 1936, Late Tertiary floras from the High Plains: Carnegie Institution of Washington Publication 476, p. 1–46.

Colton, C. C., Elias, M. K., and Amsden, T. W., 1959, Type of *Goniatites choctawensis:* Oklahoma Geology Notes, v. 19, p. 157–164.

Condra, G. E., and Elias, M. K., 1941, *Fenestella* Lonsdale and *Fenestrellina* D'Orbigny: Journal of Palaeontology, v. 5, p. 565–566.

Condra, G. E., and Elias, M. K., 1944a, Study and revision of *Archimedes* (Hall): Geological Society of America Special Paper 53, 243 p.

Condra, G. E., and Elias, M. K., 1944b, Occurrence of the Russian genus *Rhombotrypella* in Utah: Journal of Paleontology, v. 18, p. 148–155.

Condra, G. E., and Elias, M. K., 1944c, Auloporidae and Hederelloidea (and a system to avoid ambiguous use of generic names): Journal of Paleontology, v. 18, p. 529–534.

Condra, G. E., and Elias, M. K., 1944d, *Hederella* and *Corynotrypa* from the Pennsylvanian (U.S.): Journal of Paleontology, v. 18, p. 535–539.

Condra, G. E., and Elias, M. K., 1944e, Carboniferous and Permian ctenostomatous Bryozoa: Geological Society of America Bulletin, v. 55, p. 148–155.

Condra, G. E., and Elias, M. K., 1945, *Bicorbula,* a new Permian Bryozoan, probably a Bryozoan-algal consortium: Journal of Paleontology, v. 19, p. 116–125.

Elias, M. K., 1931a, On a seed-bearing *Annularia* and on *Annularia* foliage: University of Kansas Science Bulletin, v. 20, p. 115–159.

Elias, M. K., 1931b, The geology of Wallace County, Kansas: Kansas Geological Survey Bulletin 18, 254 p.

Elias, M. K., 1932a, Grasses and other plants from the Tertiary rocks of Kansas and Colorado: University of Kansas Science Bulletin, v. 20, p. 333–367.

Elias, M. K., 1932b, Algal limestone of High Plains: The Compass, v. 12, p. 112–115.

Elias, M. K., 1933a, Cephalopods of the Pierre Formation of Wallace County, Kansas, and adjacent area: University of Kansas Science Bulletin, v. 21, p. 289–363.

Elias, M. K., 1933b, Late Paleozoic plants of the Midcontinent region as indicators of time and environment, International Geological Congress, 16th, Washington, D.C.: Reports, Volume 1, p. 691–700.

Elias, M. K., 1935, Tertiary grasses and other prairie vegetation from the High Plains of North America: American Journal of Science, 5th ser., v. 29, p. 24–33.

Elias, M. K., 1937a, Elements of the Stephanian flora in the midcontinent of North America, *in* Congrès pour l'avancement des études de stratigraphie Carbonifère, 2d, Heerlen, The Netherlands, 1935, Compte Rendu, Volume 1: Maastricht, Van Aelst, p. 203–212.

Elias, M. K., 1937b, Depth of deposition of the Big Blue (late Paleozoic) sediments in Kansas: Geological Society of America Bulletin, v. 48, p. 403–432.

Elias, M. K., 1937c, A new scorpion from the Pennsylvanian *Walchia* beds near Garnett, Kansas: Journal of Paleontology, v. 11, p. 335–336.

Elias, M. K., 1937d, Stratigraphic significance of some late Paleozoic fenestrate Bryozoans: Journal of Paleontology, v. 11, p. 303–334.

Elias, M. K., 1938a, Studies of late Paleozoic ammonoids. 1: Methods of drawing sutures, bibliography: Journal of Paleontology, v. 12, p. 86–90.

Elias, M. K., 1938b, Studies of late Paleozoic ammonoids. 2: Revision of *Gonioloboceras* from late Paleozoic rocks of the Midcontinent region: Journal of Paleontology, v. 12, p. 91–100.

Elias, M. K., 1938c, Studies of late Paleozoic ammonoids. 3: *Properrinites plummeri* Elias, n. gen. and sp., from late Paleozoic rocks of Kansas: Journal of Paleontology, v. 12, p. 101–105.

Elias, M. K., 1942, Tertiary prairie grasses and other herbs from the High Plains: Geological Society of America Special Paper 41, 176 p.

Elias, M. K., 1946a, Algae reefs in cap rock of Ogallala formation on Llano Estacado Plateau, New Mexico and Texas: American Association of Petroleum Geologists Bulletin, v. 30, p. 1742–1746.

Elias, M. K., 1946b, Fossil symbiotic algae in comparison with other fossil and living algae: American Midland Naturalist, v. 36, p. 282–290.

Elias, M. K., 1946c, Taxonomy of Tertiary flowers and herbaceous seeds: American Midland Naturalist, v. 36, p. 373–380.

Elias, M. K., 1947, *Permopora keenae,* a new late Permian alga from Texas: Journal of Paleontology, v. 21, p. 46–58.

Elias, M. K., 1948, *Walchia* anatomy of branch and leaf [abs.]: Geological Society of America Bulletin, v. 59, part 2, p. 1319–1320.

Elias, M. K., 1950a, Paleozoic *Ptychocladia* and related Foraminifera: Journal of Paleontology, v. 24, p. 287–306.

Elias, M. K., 1950b, *Fenestella deissi* (new name) from the Middle Devonian of Michigan, and related forms: Journal of Paleontology, v. 24, p. 390–392.

Elias, M. K., 1952, New data on Dinantian-Namurian equivalents in America, *in* Congrès pour l'avancement des études de stratigraphie et de géologie Carbonifère, 3d, Heerlen, The Netherlands, 1951, Compte Rendu, Volume 1: Maastricht, Van Aelst, p. 189–201.

Elias, M. K., 1954, *Cambroporella* and *Coeloclema,* lower Cambrian and Ordovician bryozoans: Journal of Paleontology, v. 28, p. 52–58.

Elias, M. K., 1956, A revision of *Fenestella subantiqua* and related Silurian fenestellids: Journal of Paleontology, v. 30, p. 314–332.

Elias, M. K., 1957a, Late Mississippian fauna from the Redoak Hollow formation of southern Oklahoma, Part 1: Journal of Paleontology, v. 31, p. 370–427.

Elias, M. K., 1957b, Late Mississippian fauna from the Redoak Hollow formation of southern Oklahoma. Part 2: Brachiopoda: Journal of Paleontology, v. 31, p. 487–527.

Elias, M. K., 1957c, Late Mississippian fauna from the Redoak Hollow formation of southern Oklahoma. Part 3: Pelecypoda: Journal of Paleontology, v. 31, p. 737–784.

Elias, M. K., 1958, Late Mississippian fauna from the Redoak Hollow formation of southern Oklahoma. Part 4: Gastropoda, Scaphopoda, Cephalopoda, Ostracoda, Thoracica and Problematica: Journal of Paleontology, v. 32, p. 1–57.

Elias, M. K., 1959, Fusulinid genera *Protricitites, Pseudotricitites,* and *Putrella:* Oklahoma Geology Notes, v. 19, p. 287–289.

Elias, M. K., 1960, Marine Carboniferous of N. America and Europe, *in* Congrès pour l'avancement des études de stratigraphie et de géologie du Carbonifère, 4th, Heerlen, The Netherlands, 1958, Compte Rendu, Volume 1: Maastricht, Van Aelst, p. 151–161.

Elias, M. K., 1962, Studies of Paleozoic ammonoids. Part 4: Differentiation of species in *Gonioloboceras:* Palaeontologische Zeitschrift, Sonderausgabe, Hermann Schmidt Festband: Stuttgart, E. Schweitzerheit, 265 p.

Elias, M. K., 1964, Stratigraphy and paleoecology of some Carboniferous bryozoans, *in* Congrès international de stratigraphie et de géologie du Carbonifère, 5th, Paris, 1963, Compte Rendu, Volume 1: p. 375–382.

Elias, M. K., 1965, Living and fossil algae and fungi, formerly known as structural parts of marine bryozoans: The Palaeobotanist (Lucknow), v. 14, p. 5–15.

Elias, M. K., 1966a, Depth of late Paleozoic sea in Kansas and its megacyclic sedimentation, *in* Symposium on cyclic sedimentation, Champaign-Urbana, Illinois, 1964: Kansas Geological Survey Bulletin 169, v. 1, p. 87–107.

Elias, M. K., 1966b, Late Paleozoic conodonts from the Ouachita and Arbuckle Mountains of Oklahoma: Oklahoma Geological Survey Guidebook 16, 39 p.

Elias, M. K., 1970, Progress in correlation of Carboniferous rocks, *in* Congrès international de stratigraphie et de géologie du Carbonifère, 6th, Sheffield, England, 1967, Compte Rendu, Volume 2: Maastricht, Van Aelst, p. 695–710.

Elias, M. K., 1971, Concept of bud and related phenomena in Bryozoa: University of Kansas Paleontological Contributions 52, 21 p.

Elias, M. K., 1973, Algal-bryozoan symbiosis in the late Paleozoic of America, *in* Congrès international de stratigraphie et de géologie du Carbonifère, 7th, Krefeld, 1971, Compte Rendu, Volume 2: Krefeld, van Acken, p. 449–465.

Elias, M. K., and Condra, G. E., 1957, *Fenestella* from the Permian of West Texas: Geological Society of America Memoir 70, 158 p.

Johnson, J. H., and Kenishi, K., 1956, Studies of Mississippian algae: Quarterly of the Colorado School of Mines, v. 51, 133 p.

Moore, R. C., Elias, M. K., and Newell, N. D., 1936, A "Permian" flora from the Pennsylvanian rocks of Kansas: Journal of Geology, v. 44, p. 1–31. (This is also State Geological Survey of Kansas Contributions to the Paleontology of Kansas 4, 31 p.)

Obruchev, V. R., 1926, Geologie von Sibirien: Soergel's Fortschritte der Geologie und Paläontologie (Berlin), Heft 15, 213 p.

San Francisco Examiner, 1927, Eons ago did plants begin to bear seed?: San Francisco Examiner (May 1), p. K5.


MANUSCRIPT ACCEPTED BY THE SOCIETY JULY 6, 1994

Geological Society of America
Memoir 185
1995

# *Adolf Carl Noé (1873–1939): Pioneer in North American coal-ball studies*

**Elsie Darrah Morey**
*Morey Paleobotanical Laboratory, 1729 Christopher Lane, Norristown, Pennsylvania 19403*
**Paul C. Lyons**
*U.S. Geological Survey, MS 956, Reston, Virginia 22092*

## ABSTRACT

**Adolf Carl Noé von Archenegg was born in Graz, Austria, in 1873. He emigrated to the United States in 1899 and became a naturalized citizen of the United States. In 1899, Noé began his long and distinguished academic career at the University of Chicago where he earned his Doctor of Philosophy degree in 1905. He was an instructor and assistant professor of Germanic languages from 1903 to 1923, when he turned his interest to paleobotany and established the paleobotanical program at the University of Chicago. Noé was vice president of the Paleontological Society in 1931. He was elected as the first chairman of the Paleobotanical Section of the Botanical Society of America in 1936. Noé contributed considerable work on the fossiliferous nodules of the Mazon Creek flora of the Illinois basin, pioneered coal-ball discoveries in North America, and established the Carboniferous stratigraphy of the Eastern Interior basin of the United States. Noé's revision and translation of Stutzer's textbook, *Geology of Coal,* has become a classic.**

## INTRODUCTION

Adolf Carl Noé von Archenegg was born in Graz, Austria, on October 28, 1873 (Phillips et al., 1973). He was the son of Adolf Gustav Noé and Marie Krauss Noé. He attended the University of Graz from 1894 to 1897, studying botany and paleobotany under Baron Constantin von Ettingshausen. Noé was Ettingshausen's scientific assistant or "demonstrator" in paleobotany from 1895 to 1897. When Ettingshausen died in 1897, Noé moved to Göttingen, Germany, where he attended the University of Göttingen from 1897 to 1899 to continue his studies of Germanic languages.

In 1899, at the age of 26, Noé emigrated to the United States (Shull, 1939) and entered the University of Chicago, where he received his Bachelor of Arts degree in 1900 and his Doctor of Philosophy in Germanic Languages in 1905. While attending the University of Chicago, Noé audited science classes and took a course from John M. Coulter in paleobotany. He became a naturalized citizen of the United States.

Noé began teaching German at the Burlington Institute in Burlington, Iowa, in 1900. There he met and married Mary Evelyn Cullatin. They had two daughters, Valerie Noé and Mary Helen Noé (Mrs. Robert S. Mulliken).

Noé left Iowa in 1901 to become an instructor in German at Stanford University in California. From there, he returned to the University of Chicago, where he was an assistant professor of German from 1903 to 1923.

During World War I, Noé did not want to create any suspicion of being German, so he dropped the "von" part of his name. Noé, a tall and distinguished-looking man, had been a cavalry officer in Austria before he came to the United States. For many years, he served as first lieutenant in the First Cavalry Regiment of the Illinois National Guard in Chicago. Before World War I, Noé commanded a volunteer military cadet corps at the University of Chicago that was later absorbed into the infantry of the R.O.T.C. Program.

Noé also commanded the faculty company of the University of Chicago. Walter Loehwing recalled the following

Morey, E. D., and Lyons, P. C., 1995, Adolf Carl Noé (1873–1939): Pioneer in North American coal-ball studies, *in* Lyons, P. C., Morey, E. D., and Wagner, R. H., eds., Historical Perspective of Early Twentieth Century Carboniferous Paleobotany in North America (W. C. Darrah volume): Boulder, Colorado, Geological Society of America Memoir 185.

story. During firing practice at the rifle range at Fort Sheridan, Illinois, Dr. Noé often had trouble "hitting the bull's eye" due to poor eyesight. He would explode into profanity in German or Czech. Loehwing recalled that after such an outburst one of the regular officers at Fort Sheridan was interested in the fact that Noé was a professor at the University of Chicago. He asked him what he taught and received this reply from Noé: "I am a professor of profane language and literature."[1]

Of Noé's drill-sergeant activities for the faculty infantry company, Ralph W. Chaney wrote, "My first acquaintance with Dr. Noé was during the early months of our participation in the [First] World War. A University of Chicago infantry unit drilled on Stagg Field, and the faculty had for its drill sergeant, a tall, angular man who barked his orders with a heavy accent. His efficient and good humored treatment of our rookie squad I shall always remember. It was several years later that I came to know our drill sergeant as Dr. Noé, the paleobotanist" (Croneis, 1939b, p. 223).

## PALEOBOTANICAL STUDIES

By 1923, Noé returned to his original field of interest—paleobotany—and became an assistant professor of paleobotany at the University of Chicago. In 1923, the university created a Chair of Paleobotany, which was supported by the Departments of Botany and Geology (Schopf, 1941). Noé was the first paleobotanist appointed to that chair. In 1924, Noé was promoted to associate professor and "Curator of the Fossil Plants" at the Walker Museum in Chicago. Noé built up the specimens in the museum by his collections and by exchanges with other paleobotanists. The collection consisted of about 500 paleobotanical specimens in 1921 and had grown to over 25,000 specimens in 1939.

Noé had no formal geological training, but, in spite of that, he was on the staff of the Illinois Geological Survey. This provided support for his studies of Pennsylvanian fossil plants.

Noé was a geologist for the 1927 Alan and Garcia Coal Commission to the Soviet Union. There he pioneered geological studies in the Donetz coal basin in Ukraine. His experiences in Russia were the basis for his book, *Golden Days of Soviet Russia* (Noé, 1931d).

Noé was also connected with the Department of Botany of the Field Museum in Chicago. At first his visits were occasional. By 1928, Noé was involved with the reconstructions of Pennsylvanian fossil flora and fauna that were being made at the museum. In recognition of this service, Noé's name was added to the botanical staff of the museum as research associate of paleobotany.

### *Coal-ball studies*

American coal balls were found in Iowa (Bain, 1896; see Chapter 27, this volume) as early as the 1890s but not identified as such. In England, Hooker and Binney (1855) first reported coal balls, which were investigated in detail by Williamson (1871–1882). Zalessky (1910a, b) found coal balls in Ukraine. Noé has been widely credited with the first discovery of a coal ball in North America. The coal ball in question was collected by Gilbert H. Cady, probably in the 1920s or perhaps earlier, who found it in O'Gara Mine Number 9 near Harrisburg, Illinois. Cady gave the "concretion" to the Illinois Geological Survey, and Harold E. Culver showed the concretion to Noé, who identified it as a coal ball (see Noé, 1923b). Noé took the coal ball back to his laboratory to study. J. H. Hoskins (see Chapter 27, this volume), Noé's student, cut the first North American coal ball that was studied as such.

A controversy arose over the question of the discovery of coal balls in North America. Noé once told Croneis that soon after he found the first ones, David White (see Chapter 10, this volume) called upon him. White insisted that there were no coal balls in North America, whereupon Noé went into the next room to return immediately with a "newly discovered specimen" (Croneis, 1939b, p. 220). In the autumn of 1922, Noé wrote a letter to White informing him that coal balls recently had been found in Illinois, and at a conference during the Christmas vacation of the same year, Noé showed a sectioned coal ball to White.

Frederick O. Thompson (see Chapter 22, this volume), amateur collector, knew B. E. Dahlgren, chief curator of botany at the Field Museum in Chicago. In a letter to William C. Darrah (1909–1989), Thompson wrote, "Dahlgren told me a curious story. He recently was told by a man, I believe, by the name of [C. J.] 'Chamberlain' that at one time David White met Dr. Noé at the Field Museum, and David White made the statement that no coal balls had been found in America. Noé said, 'White, I can get you a wheelbarrow full of coal balls.' Noé went to Calhoun, in southern Illinois and came back with the equivalent of a wheelbarrow full of coal balls. This made White so mad that he would never speak to Noé again, and he also interfered with Noé's career, as far as it lay in his powers."[2]

To further the story, Walter Loehwing, professor and head of the Botany Department at the University of Iowa, told F. O. Thompson a version that involved E. W. Berry. After hearing the story, Thompson had Loehwing dictate it to his secretary. Excerpts of this version of the story are quoted below:

> E. W. Berry, geologist [United States Geological Survey] who taught at John Hopkins, gave a lecture to the seminar of the Botany Department of the University of Chicago, about 1920. In the course of the address, he stated that it was regrettable that we did not have coal balls in America, because of the fact that they preserved plant structures so completely that they were the best source of information on detailed tissue structure of plant fossils.
>
> Professor Adolph Carl Noé of the German Department of the University of Chicago was visiting the seminar at the time because of his avocational interest in paleobotany. In his paleobotanical work, Noé had become familiar with the coal balls of the roof shales in the Illinois coal mines. Consequently, after Dr. Berry's lecture, he made the statement during the discussion that he was pretty certain that

coal balls did occur in the United States, particularly in the state of Illinois. His statement was flatly refuted by Dr. Berry before the seminar. Consequently, Professor Noé went down to Morris, Illinois, collected some coal balls, sectioned them and brought the prepared sections to Professor John M. Coulter, Professor and head of Botany at the University of Chicago. Professor Coulter was so much impressed with this discovery, not only because of its importance to geology and botany, but because of the controversial aspect of the question and the irrefutable character of Dr. Noé's evidence that he tendered Noé an appointment as Associate Professor of Paleobotany.

Although Dr. Noé is officially credited as the discoverer of American coal balls, it is interesting to note that their occurrence had been previously observed in the bituminous shales of Polk County, Iowa, and reported by H. F. Bain in the Iowa Geological Survey, Annual Report, Volume VII, 1896, under the heading of "Geology of Polk County," pages 265–412 inclusive.

The reason that the significance of Dr. H. F. Bain's original discovery of coal balls in Polk County, Iowa, was not fully appreciated may have been due to the difficulty of sectioning them to identify and determine the detailed tissue structure of mineralized plant materials which occurred in them. The old process of sectioning a coal ball required the use of diamond or Carborundum saws. Only a fairly thick section could be cut by this process and the slab subsequently had to be ground quite thin by abrasives and then polished to permit microscopic examination. This was a costly and tedious process. It was only [by] means of magnification of transparent fossil tissues that the degree of preservation of plant tissues became fully evident.[3]

In a letter to W. C. Darrah that accompanied the story, Thompson wrote, "Incidentally, Berry was extremely rude to Professor Noé, which does not appear in the mend. He inferred that Noé was a liar, or perhaps even told him he was a liar."[4]

Darrah answered Thompson's letter immediately.

Many thanks for the copy of the interview with Dr. Loehwing. As you will read from my comments, I am at the same time pleased and puzzled. Noé had a very difficult time over his discovery of coal balls in Illinois and instead of pleasing his colleagues, they resented his good fortune. I had heard the story long before from Professor [A. S.] Romer from Harvard but he identified the man who called Noé a liar as David White, who was head [chief geologist] of the U.S. Geological Survey. Certainly I cannot say whether the person was White or Berry. I would like to think it was Berry because the incident does not fit the character of David White, but I hope at some time the truth will [come] out and I can confirm my own records who the guilty one really was. In general, the other details of the story agree very closely with what I had heard from Romer.

There is one aspect of Noé's early discoveries which Loehwing has overlooked. Noé pulled one of those horrible boners which tends to discredit a man's work everafter. He found *Myeloxylon* and in complete ignorance of all that was known of the seed ferns, called it a flowering plant—a corn stem [Noé, 1923a]. Almost without exception his colleagues groveled him in the dirt. Only D. H. Scott of the University of Cambridge in England was gracious enough to imply that Noé might not have been familiar with the recent European literature. It soured Noé a good deal, but he never showed it publicly. Nevertheless, during later years he published exceedingly little giving his students his own ideas and assistance so that their papers would be fully acceptable. You may recall the testy letter of Noé's to Dahlgren when you asked for coal balls before you found them in [Iowa].[5] (See Chapter 22, this volume.)

This paragraph also appears in the Loehwing story:

It was known for example that certain strata were correlated with oil deposits, and Noé's work became a matter of great interest to oil prospectors. For example, if it was known that an oil stratum underlaid a rock stratum containing certain plant fossils, an oil company might continue to drill knowing that they had not yet reached the oil-bearing horizon. If, on the other hand, the reverse relationship prevailed, namely, the oil-bearing stratum overlaid a barren horizon identified by its fossils, drilling could be safely discontinued because a prospector would know that he had already passed through the oil-bearing stratum which the dry well would mark as unproductive in that locality. Oil company officials often came to Noé to show him drill cores and he could frequently tell whether the bottom of the core was above or below oil-bearing stratum or horizons known to bear oil elsewhere.[6]

Darrah related to Thompson,

The other things are a little puzzling. No oil company will trust paleobotanical evidence of impressions or petrifications of plants in stratigraphic work. Paleobotanists have done so much sloppy work that they have been fooled too often. Certainly Noé, despite his attainments, deserves no such credit for assisting petroleum geology. If any man could be given recognition in that direction, it would have been David White. The other point I would like to mention is Loehwing's complete ignorance of modern peel technique. He is talking about the very poor beginnings of the method when [J.] Walton [1928] first worked on it about 20 years ago. The kind of peels he discusses have not been made since 1932 or earlier.[7]

### *Research on compressional plant fossils*

Noé (Fig. 1) spent much research time on the Mazon Creek flora of Illinois. This work was stratigraphically oriented. His main contributions to stratigraphic paleobotany (see Adolf Carl Noé selected bibliography, this chapter) are his papers on the Braidwood flora and plants of the Illinois basin (Noé, 1923c, 1925d, e, 1926b, and 1931a). Noé's students, R. E. Janssen (1937) and F. C. MacKnight (1938), completed their theses on the floras of the Illinois basin.

Noé was an avid correspondent and enjoyed sharing his ideas with colleagues. His correspondence with W. C. Darrah ranges from 1934 to 1939 (the year of Noé's death). Besides the usual letters thanking others for reprint exchanges, Noé wrote, "With regard to *Callipteridium sullivanti,* I feel quite sure that it is not *Callipteridium* but an *Alethopteris.* I have sent specimens of it to various European paleobotanists, including [W.] Gothan (see Chapter 6, this volume), and all agreed with me that it has nothing in common with the European genus *Callipteridium.*"[8]

Noé wrote Darrah of his extensive collecting and added, "I am trying to finance some collecting trips by selling some duplicates from the Wilmington strip mine, the former Braidwood mines near Wilmington and from Mazon Creek. The specimens are fully determined and have been collected at my own personal expense."[9]

Noé answered Darrah's letter,

I have just received your letter of July 26, in which you inquired concerning the seeds of *Neuropteris decipiens.* As far as I know no seeds

Figure 1. Adolf Carl Noé looking at a coal ball in his laboratory at the University of Chicago. Photograph taken in the 1930s. Photograph republished with the courtesy of and permission from the Department of Special Collections, The University of Chicago Archives.

of *Trigonocarpus* have been found in actual connection with *Neuropteris* leaves; but owing to the proximity of the seeds and leaves, I am inclined to consider the seeds of *N. decipiens* to have been of a *Trigonocarpus* type and, therefore, advised the reconstruction in the Field Museum to be along these lines.

I believe Chamberlain [1937] in his book went a step farther and assumed that a connection has been found; I think he is not correct in this. Probably he was influenced by an illustration published by Zeiller in which [a] *Neuropteris* pinna carries a husk with a *Trigonocarpus* seed at its end.[10]

In answering Darrah's September 29, 1936 letter, Noé wrote,

I have no objections to your publishing on *Tingia* [see Darrah, 1938] and *Sphenozamites* which you have collected with the Harvard expedition from the Brazos River, Baylor County, Texas. I have a large collection from this locality which I collected myself in 1933-34, but I do not remember having met Emily Irish whom you mention in your letter. I have prepared an article on Permian Cycadophytes from Texas for the Texas Academy of Science. . . . I appreciate your courtesy in this manner and thank you for giving me this information. . . . I was interested in your paper *The Peel Method in Paleobotany* [Darrah, 1936]. I am especially interested in its application to coal and should be very glad if you could send me for examination a couple of coal peels.[11]

Noé wrote answering Darrah's January 5, 1937 letter: "I gladly accept the honor of being elected President [Chairman] of the Paleobotanical Section of the Botanical Society, and I feel very much flattered."[12]

Later Noé wrote thanking Darrah for his paper on *Codonotheca* and *Crossotheca* (1937a)

"I received [them] before going to the Tropics (West Indies and Central America). . . . Did you give any thoughts to the program of the meeting in Indianapolis. Could we not have a joint meeting with the botanists (taxonomists or morphologists) with a symposium? Let me know your reactions. I collected fossil plants in this part of the world and studied the tropical rain forest in its bearing upon Paleozoic and Tertiary rain forests. . . . I made Barro, Colorado [Panama] my headquarters for a long time, and found it to be an excellent place. Please excuse my scribbling. The moisture in the air is so great that a letter has to be written quickly and sealed soon.[13]

Darrah answered Noé on August 2, 1937, and suggested a symposium on Devonian or Pennsylvanian floras. Darrah was then the secretary of the Paleobotanical Section.

Noé answered Darrah,

I am glad that you have made arrangements for the meetings of the section; I hope you have received offers of papers which will make the meeting attractive. Please put me down on the program for the following subject: "Exploring for Fossil Plants." I have made extensive collecting trips during the last ten years in the middlewest, northwest and southwest of the United States, in Mexico, in Central America, and in Russia. . . . Dr. Reinhardt Thiessen [see Chapter 11, this volume], Research Chemist, Coal Constitution Section, wrote me . . . that he was preparing a paper on "Coal Paleobotany." . . . Please put him on the program.[14]

Later Noé wrote again,

We had a very good meeting of our section with large attendance and with all the prospects of a bright future ahead of us. I am glad that you will continue as secretary and that Dr. Petry accepted the chairmanship. You two ought to make it surely a success. . . . With regard to your question . . . , I do not recollect any specimens of *Neuropteris decipiens* with distinctly odontopteroid pinnules. I am calling *N. decipiens* now *N. scheuchzeri* which I consider the more appropriate name, in accordance with the late David White. . . . I have not found, as far as I recollect any fruiting specimens of *Oligocarpia* [see Darrah, 1937b] nor of *Renaultia* in the Mazon Creek material.[15]

Noé wrote Darrah, "One of my research students, R. E. Janssen, is working up the types of Illinois fossil plants of which there is a collection in the State Museum at Springfield. There are about 50 in that collection. We are anxious to obtain other Illinois types for a revision and redescription which is to be published by the State Museum at Springfield. In case you have any Illinois types of Pennsylvanian age and are willing to have Janssen include them in his revision and redescription, we should be greatly obliged to you for a loan of same."[16]

In another letter, Noé answered Darrah, "Many thanks for your very comprehensive letter of December 15 with enclosed proofs of plates and reprint. . . . I already had advised Mr. Janssen [1939, 1940] to limit himself to a description and illustration of the types from the Illinois State Museum in Springfield. These were sent to him by the chief of the Museum with the request to make this description and it was an arrangement between him and the Museum Chief in which I have no part, except that this description could be accepted as a thesis for a doctors degree."[17]

## NOÉ'S AFFILIATIONS, HONORS, AND RECOGNITION

Noé served as treasurer of the American Commission for Vienna Relief in 1921. Austria bestowed upon him the Cross of the Order of the Austrian Republic for this work.

In 1923, he received the honorary degree of Doctor of Philosophy from Graz and the gold medal from the University of Vienna. Noé received the Doctor of Science degree from Innsbruck in 1929.

He was a member of the Societé Géologique de France and the Sociedad Geológica de México. He was a fellow of the Geological Society of America and a member of the American Association for the Advancement of Science. In 1927, he served as chairman of the Chicago section of the American Institution of Mining and Metallurgical Engineers. In 1931, Noé was vice president of the Paleontological Society and the Illinois Academy of Science. In 1936, he served as the first chairman of the Paleobotanical Section of the Botanical Society of America.

Noé (Fig. 2) was an expert marksman before his eyesight deteriorated and served as an instructor of the University of Chicago Rifle Club. He was a remarkable fencer and an enthusiastic horseman.

Noé felt his contributions to paleobotany were through his students and their productive careers. Among his students were Clayton Bell, Erling Dorf, Roy Graham, J. H. Hoskins, R. E. Janssen, Harriet Krick (Bartoo), F. C. MacKnight, and Fredda Reed (see Chapters 15 and 27, this volume).

Noé will be long remembered for his early studies of the Pennsylvanian floras of Illinois. His colleagues honored him by naming after him a coal-ball plant—*Medullosa noei* (Steidtmann, 1937).

## THE FINAL DAYS

Noé had been interested in translating and revising Otto Stutzer's *Geology of Coal* (see Stutzer and Noé, 1940). On March 11, 1939, Noé went to his office as usual to work on the final chapter of his revision. While at the office he suffered a paralytic stroke from which he never recovered. He died on April 10, 1939. The revision was published posthumously in 1940.

Figure 2. Adolf Carl Noé; photograph taken in the 1930s. Photograph republished with the courtesy of and permission from the Department of Special Collections, The University of Chicago Archives.

## ACKNOWLEDGMENTS

We acknowledge the William Culp Darrah Paleobotanical Collection for the use of the letters from A. C. Noé, the F. O. Thompson letters, and the story by Walter Loehwing.

In writing this portrait we relied heavily on previously written memorials about Noé. They are Croneis (1939a, b), Shull (1939), Schopf (1941), Phillips et al. (1973), and Andrews (1980). These works provided valuable insights into Noé's life and career.

The manuscript was reviewed by T. Delevoryas and G. Mapes.

## ENDNOTES

1. Story about A. C. Noé dictated by Professor Walter Loehwing to the secretary of F. O. Thompson, February 25, 1949. W. C. Darrah Collection.
2. Letter from F. O. Thompson to W. C. Darrah, May 26, 1942. W. C. Darrah Collection.
3. Story about A. C. Noé dictated by Professor Walter Loehwing to the secretary of F. O. Thompson, February 25, 1949. W. C. Darrah Collection.
4. Letter from F. O. Thompson to W. C. Darrah, March 24, 1949. W. C. Darrah Collection.
5. Letter from W. C. Darrah to F. O. Thompson, March 29, 1949. W. C. Darrah Collection.
6. Story about A. C. Noé dictated by Professor Walter Loehwing to the secretary of F. O. Thompson, February 25, 1949. W. C. Darrah Collection.
7. Letter from W. C. Darrah to F. O. Thompson, March 29, 1949. W. C. Darrah Collection.
8. Letter from A. C. Noé to W. C. Darrah, August 25, 1934. W. C. Darrah Collection.
9. Letter from A. C. Noé to W. C. Darrah, December 8, 1934. W. C. Darrah Collection.
10. Letter from A. C. Noé to W. C. Darrah, July 29, 1935. W. C. Darrah Collection.
11. Letter from A. C. Noé to W. C. Darrah, November 25, 1936. W. C. Darrah Collection.
12. Letter from A. C. Noé to W. C. Darrah, January 18, 1937. W. C. Darrah Collection.
13. Letter from A. C. Noé to W. C. Darrah, July 24, 1937, written from Barro Colorado Island Biological Laboratory, Gatun Lake, the Panama Canal. W. C. Darrah Collection.
14. Letter from A. C. Noé to W. C. Darrah, October 29, 1937. W. C. Darrah Collection.
15. Letter from A. C. Noé to W. C. Darrah, February 24, 1938. W. C. Darrah Collection.
16. Letter from A. C. Noé to W. C. Darrah, November 30, 1938. W. C. Darrah Collection.
17. Letter from A. C. Noé to W. C. Darrah, January 2, 1939. W. C. Darrah Collection.

## REFERENCES CITED

Andrews, H. N., 1980, The fossil hunters: Ithaca, New York, Cornell University Press, 421 p.

Bain, H. F., 1896, Geology of Polk County: Iowa Geological Survey, Annual Report, v. 7, p. 265–412.

Chamberlain, C. J., 1937, Gymnosperms structure and evolution: Chicago, Illinois, The University of Chicago Press, 484 p.

Croneis, C. G., 1939a, Adolf Carl Noé: Science, v. 89, p. 379–380.

Croneis, C. G., 1939b, Memorial to Adolf Carl Noé: Geological Society of America Proceedings for 1939, p. 219–227, pl. 11.

Darrah, W. C., 1936, The peel method in paleobotany: Harvard University Botanical Museum Leaflets, v. 4, p. 69–83, 2 pls.

Darrah, W. C., 1937a, *Codonotheca* and *Crossotheca:* Polleniferous structures of pteridosperms: Harvard University Botanical Museum Leaflets, v. 4, p. 153–172, 4 pls.

Darrah,W. C., 1937b, *Oligocarpia* and the antiquity of the Gleicheniaceae: American Journal of Botany, v. 24, p. 743.

Darrah, W. C., 1938, The occurrence of the genus *Tingia* in Texas: Harvard University Botanical Museum Leaflets, v. 5, p. 173–188, 2 pls.

Hooker, J. D., and Binney, E. W., 1855, On the structure of certain limestone nodules enclosed in seams of bituminous coal, etc.: Royal Society of London Philosophical Transactions, v. 145, p. 149–156.

Janssen, R. E., 1937, A key for the identification of plant impressions from the Middle Pennsylvanian of Illinois [M.S. thesis]: Chicago, Illinois, University of Chicago, 54 p.

Janssen, R. E., 1939, Leaves and stems from fossil forests: Illinois State Museum Popular Science Series, v. 1, 190 p.

Janssen, R. E., 1940, Some fossil plant types from Illinois: Illinois State Museum Scientific Papers 1, 124 p.

MacKnight, F. C., 1938, The flora of the Grape Creek Coal at Danville, Illinois [Ph.D. thesis]: Chicago, Illinois, University of Chicago, 118 p.

Noé, A. C. See selected bibliography of Adolf Carl Noé that follows.

Phillips, T. L., Pfefferkorn, H. W., and Peppers, R. A., 1973, Development of paleobotany in the Illinois Basin: Illinois State Geological Survey Circular 480, 86 p.

Schopf, J. M., 1941, Adolf Carl Noé: Chronica Botanica, v. 6, p. 236–237.

Shull, C. A., 1939, Adolf Carl Noé, 1873–1939: Chicago Naturalist, v. 2, p. 51–52.

Steidtmann, W. E., 1937, A preliminary report on the anatomy and affinity of *Medullosa noei* sp. nov. from the Pennsylvanian of Illinois: American Journal of Botany, v. 24, p. 124–125.

Walton, J., 1928, A method of preparing sections of fossil plants contained in coal balls or in other types of petrifications: Nature (London), v. 72, p. 571.

Williamson, W. C., 1871–1882, On the organization of the fossil plants of the coal-measures: Royal Society of London Philosophical Transactions, Parts 1–12, pt. 1 (1871), v. 151, p. 477–510; pt. 2 (1872), v. 152, p. 197–240; pt. 3 (1872), v. 152, p. 238–318; pt. 4 (1873), v. 153, p. 377–408; pt. 5 (1873), Proc. v. 21, p. 394–398; pt. 6 (1874), v. 164, p. 675–703; pt. 7 (1877), v. 166, p. 1–25; pt. 8 (1878), v. 167, p. 213–270; pt. 9 (1879), v. 169, p. 319–364; pt. 10 (1881), v. 171, p. 493–539; pt. 11 (1882), v. 172, p. 283–305; pt. 12 (1882), v. 174, p. 459–475.

Zalessky, M. D., 1910a, On the discovery of well preserved plant remains in one type of rock under Limestone S($I_3$) of a general section of the Donetz Carboniferous deposits: Bulletin de l'Academie Impériale des Sciences, St.-Pétersburg, v. 4, p. 447–449.

Zalessky, M. D., 1910b, On the discovery of the calcareous concretions known as coal balls in one of the coal seams of the Carboniferous strata of the Donetz basin: Bulletin de l'Academie Impériale des Sciences de St.-Pétersburg, v. 4, p. 477–480.

## SELECTED BIBLIOGRAPHY OF ADOLF CARL NOÉ

Fisher, M. C., and Noé, A. C., 1938, A list of coal ball plants from Calhoun, Richland County, Illinois: Illinois State Academy of Science Transactions, v. 31, p. 178–181.

Noé, A. C., 1921a, Cycad-like leaves from the Permian of Texas [abs.]: Geological Society of America Bulletin, v. 32, p. 134.

Noé, A. C., 1921b, New textbooks of paleobotany: Botanical Gazette, v. 71, p. 238–239.

Noé, A. C., 1923a, A Paleozoic angiosperm: Journal of Geology, v. 31, p. 344–347.

Noé, A. C., 1923b, Coal balls: Science, n.s., v. 57, p. 385.

Noé, A. C., 1923c, Fossil flora of Braidwood, Illinois: Illinois State Academy of Science Transactions, v. 15, p. 396–397.

Noé, A. C., 1923d, The flora of the western Kentucky coal fields: Kentucky Geological Survey, ser. 6, v. 10, p. 127–148.

Noé, A. C., 1925a, American coal balls [abs.]: Science, n.s., v. 62, p. 524.

Noé, A. C., 1925b, Coal balls here and abroad: Illinois State Academy of Science Transactions, v. 17, p. 179–180.

Noé, A. C., 1925c, Dakota sandstone plants from Cimarron County, Oklahoma: Oklahoma Geological Survey Bulletin 34, p. 93–107, 7 pls.

Noé, A. C., 1925d, Pennsylvanian flora of northern Illinois: Illinois State Geological Survey Bulletin 52, 113 p., 45 pls.

Noé, A. C., 1925e, The fossil flora of northern Illinois: Illinois State Academy of Science Transactions, v. 18, p. 206–207.

Noé, A. C., 1926a, Applied paleobotany: Botanical Gazette, v. 81, p. 339–340.

Noé, A. C., 1926b, The fossil flora of Harrisburg, Illinois: Illinois State Academy of Science Transactions, v. 19, p. 283–285.

Noé, A. C., 1928, The use of charts in natural sciences: Science, n.s., v. 67, p. 571–574.

Noé, A. C., 1930a, Celluloid films from coal balls (notes for students): Botanical Gazette, v. 89, p. 318–319.

Noé, A. C., 1930b, Correlation of Illinois coal seams with European horizons: Illinois State Academy of Science Transactions, v. 22, p. 470–472.

Noé, A. C., 1931a, Coal-ball floras of Illinois: Illinois State Academy of Science Transactions, v. 23, p. 429–430.

Noé, A. C., 1931b, Evidences of climate in the morphology of Pennsylvanian plants: Illinois State Geological Survey Bulletin 6, p. 283–289, 4 figs.

Noé, A. C., 1931c, Ferns, fossils, and fuel; the story of plant life on Earth: Chicago, Illinois, Thomas S. Rockwell, 128 p., 8 pls.

Noé, A. C., 1931d, Golden Days of Soviet Russia: Chicago, Illinois, Thomas S. Rockwell, 181 p.

Noé, A. C., 1931e, Review of American coal-ball studies: Illinois State Academy of Science Transactions, v. 24, p. 317–320.

Noé, A. C., 1933a, Recent studies of the morphology of plants from the Pennsylvanian of Illinois: Illinois State Academy of Science Transactions, v. 25, p. 129.

Noé, A. C., 1933b, Une forêt de l'époque carbonifère au Musée Field d'histoire naturelle à Chicago: La Nature, no. 2918, p. 510–512, 6 figs.

Noé, A. C., 1935, Some recent attempts to correlate the later Paleozoic of America and Europe: Illinois State Academy of Science Transactions, v. 28, p. 171–172, 1 fig.

Noé, A. C., 1936a, A collecting trip into the Jurassic of southern Mexico: Illinois State Academy of Science Transactions, v. 29, p. 156.

Noé, A. C., 1936b, Fossil palms: Field Museum of Natural History Publication 355, Botanical series, v. 14, p. 439–456.

Noé, A. C., and Janssen, R. E., 1937, Identification key for Illinois plant fossils: Illinois State Academy of Science Transactions, v. 30, p. 236–237.

Stutzer, O., and Noé, A. C., 1940 (posthumous), Geology of coal: Chicago, Illinois, The University of Chicago Press, 461 p.

Manuscript Accepted by the Society July 6, 1994

Printed in U.S.A.

Geological Society of America
Memoir 185
1995

# *Fredda Doris Reed (1894–1988): Educator and paleobotanist*

**Thomas N. Taylor**
*Department of Plant Biology, Ohio State University, 1735 Neil Avenue, Columbus, Ohio 43210*

## ABSTRACT

**Fredda D. Reed was one of the first students of Adolph Noé and was responsible for some of the early work on Carboniferous coal-ball plants in North America. As a faculty member for 35 years at Mount Holyoke College, she touched the lives of generations of students through her strong work ethic, tenacity for detail, and unending sense of humor.**

## INTRODUCTION

The development and continued growth of paleobotany are replete with women who have made significant contributions to the discipline. Through their research Agnes Arber, Eleanor M. Reid, Marie Stopes (see Chapter 9, this volume), Isabel C. Cookson, Marjorie E. J. Chandler, Rina Scott, and Winifred Goldring, to name several, have left an indelible imprint in the study of fossil plants. One who should be added to the list is Fredda Doris Reed from North America.

## PERSONAL DATA AND PROFESSIONAL CAREER

Fredda Doris Reed was born in Parker, Indiana, on August 15, 1894, and was one of four children (she had two brothers and a sister) born to Levi J. and Enola (Cox) Reed. Few details are available regarding her early years, except that she had a keen interest in the plants and animals around her and an infectious enthusiasm in sharing with others the wonders of nature. She enrolled at Earlham College with the singular force of becoming a teacher. She continued her education at the University of Chicago. There she received the M.A. degree in 1923 and her Ph.D. a year later, funded by a fellowship. It was at the University of Chicago that Fredda became associated with Adolph Noé, who had just established a strong paleobotanical research program and was now focusing his attention on Carboniferous plants preserved in calcitic coal balls that had recently been discovered in Illinois. She was one of Noé's first students who mastered the peel technique and initiated a series of papers describing this exceptionally well preserved coal-ball flora.

Fredda returned to Earlham College upon the completion of her degree at the University of Chicago and served as instructor in botany from 1924 to 1926. The following two years were spent back in Chicago as a botanical technician in the General Biological Supply House, then a major supplier of histologic preparations of plant and animal tissues used in laboratory instruction. In 1928, Dr. Reed joined the faculty at Mount Holyoke College, South Hadley, Massachusetts, as an assistant professor of botany. She was promoted to associate professor in 1933 and 10 years later to professor. For a number of years before retiring she chaired the Department of Botany.

## RESEARCH AND TEACHING EXPERIENCE

In 1937 Professor Reed traveled to London, where she spent the summer studying the research collection of fossil plants at the British Museum of Natural History. She was particularly interested in the microscope slides of Carboniferous plants from European coal balls, from the D. H. Scott and W. C. Williamson collections, and spent considerable effort comparing the various genera and species with those that were then being investigated from North American coal balls. There is no doubt that several of her subsequent papers were influenced by her examination of these Carboniferous plants.

Throughout Reed's career she was an active member of several professional societies, including the Botanical Society of America (Paleobotanical Section), International Organization of Paleobotany, Torrey Botanical Club, International Society of Plant Morphologists, Sigma Xi, International Society of Plant Taxonomists, and Sigma Delta Epsilon. She was elected a fellow of the American Association for the Advance-

Taylor, T. N., 1995, Fredda Doris Reed (1894–1988): Educator and paleobotanist, *in* Lyons, P. C., Morey, E. D., and Wagner, R. H., eds., Historical Perspective of Early Twentieth Century Carboniferous Paleobotany in North America (W. C. Darrah volume): Boulder, Colorado, Geological Society of America Memoir 185.

ment of Science. While at Mount Holyoke, Dr. Reed was the recipient of several grants to support her research activities, including ones from the American Philosophical Society and Sigma Xi.

Fredda Reed (Fig. 1) was truly an inspiring teacher who counseled generations of botany majors at Mount Holyoke College. A strong proponent of small, informal classes, she encouraged individual work founded on meticulous accuracy and independent thought. A former student, recounting one of Professors Reed's classes, remarked, "Only afterwards had I realized what a quantity of hard work we put in and how invaluable it is to have learned to make our own decisions on so many matters. I don't think it ever occurred to any of us that one even might ask, how many drawings should we make? or which books should we read?" Another former student, fondly remembering Professor Reed's influence on students at Mount Holyoke, noted at the time of her retirement in 1960, "The single facet for which her students remember her most concerns her meticulous accuracy—both the practice of it and the delight in it, whether in doing her own work or observing that of others." Whether having tea with students, discussing floral hybrids in her garden on Jewett Lane, or telling stories about other botanists, Fredda Reed was a consummate teacher and was forever expanding the horizons of her students at Mount Holyoke College. For her students, "the latchstring was always out."

Figure 1. Fredda Dorris Reed. (Photograph courtesy of Mount Holyoke College Library/Archives).

Reed's research interests ranged from sphenophytes to *Psaronius* and from seed fern seeds to cordaitean leaves. In reading her 1936 paper on *Lepidocarpon,* one is struck by her grasp of the reproductive biology of these plants, a research area in paleobotany that was "rediscovered" nearly 30 years later. Similarly, her paper titled "Coal Flora Studies: Lepidodendrales," which appeared in 1941, was one of the early attempts at what is often referred to today as "whole plant reconstruction." Her detailed line drawings stand as a testament to her love of detail and explicit accuracy and perhaps underscore the influence that paleobotanists like Scott and Williamson had in shaping her research methodology.

## THE LATER YEARS

Fredda Reed retired from Mount Holyoke College after 35 years of service and with increasingly deteriorating eyesight. Throughout her retirement this remarkable woman maintained a positive outlook and disposition despite blindness. Friends and colleagues alike marveled at her at her courage and independence and at her ability to manage sightlessness without complaint. During her retirement she continued to pursued two consuming interests: her life-long love of gardening and a new passion—baseball! In spite of her failed sight Fredda Reed continued to plant bulbs and could tell by feel whether a plant was a weed or flower. Many fulfilling hours were spent sitting on the ground gardening, with baseball on the radio as accompaniment.

She became a true baseball enthusiast and knew each player on every team, despite her never having seen one game. When asked why she was such an ardent fan of the New York Yankees, she responded that she rooted for them because of their abilities. "My father told me to never use the word hate, but when I first started listening everyone I knew around here hated the Yankees. I thought I should dislike them too, but as I listened to their games I thought, they are really good—I should admire them."

One of my earliest papers in paleobotany was concerned with the structure and morphology of the Carboniferous seed *Mitrospermum.* Unfamiliar with the organization of platyspermic seeds, I spent a great deal of time going over Fredda Reed's paper on *Cardiocarpon.* Her careful line drawings and clear explanations of cell types and tissue systems were of great assistance to this paleobotanical neophyte as he attempted to describe the structures in *Mitrospermum.* I never had the opportunity to meet Fredda Reed, but I did talk with her on the telephone and corresponded with her on several occasions. She always had time to talk about fossils and was interested in the progress of my work and what was "going on" in Carboniferous paleobotany.

Such was the character of this educator and paleobotanist who touched the lives of many generations of undergraduate students and who inspired in them a sense of the clarity and symmetry of structure that could be found in both living and fossil plants. Whether in science, gardening, or sport, Fredda Doris Reed always sought excellence.

## ACKNOWLEDGMENTS

Appreciation is extended to Patricia J. Albright and Anne C. Edmonds, Mount Holyoke College Library, for their assistance in preparing this chapter. The chapter was reviewed by L. C. Matten and R. J. Litwin.

## BIBLIOGRAPHY OF FREDDA DORIS REED

Reed, F. D., 1926, Flora of an Illinois coal ball: Botanical Gazette, v. 81, p. 460–469.

Reed, F. D., 1928, Holdfast cells in *Spirogyra:* Proceedings of the Indiana Academy of Sciences, v. 37, p. 339–340.

Reed, F. D., 1936, *Lepidocarpon* sporangia from the Upper Carboniferous of Illinois: Botanical Gazette, v. 98, p. 307–316.

Reed, F. D., 1938, Notes on some plant remains from the Carboniferous of Illinois: Botanical Gazette, v. 100, p. 324–335.

Reed, F. D., 1939, Structure of some Carboniferous seeds from American coal fields: Botanical Gazette, v. 100, p. 769–787.

Reed, F. D., 1941, Coal flora studies—Lepidodendrales: Botanical Gazette, v. 102, p. 663–668.

Reed, F. D., 1946, On *Cardiocarpon* and some associated plant remains from Iowa coal fields: Botanical Gazette, v. 108, p. 51–64.

Reed, F. D., 1949, Notes on the anatomy of two Carboniferous plants, *Sphenophyllum* and *Psaronius:* Botanical Gazette, v. 110, p. 501–510.

Reed, F. D., 1949, A new calamite from American coal fields: American Journal of Botany, v. 36, p. 819–820.

Reed, F. D., 1952, *Arthroxylon,* a redefined genus of calamite: Missouri Botanical Garden Annals, v. 39, p. 173–187.

Reed, F. D., 1956, The vascular anatomy of *Litostrobus iowensis:* Phytomorphology, v. 6, p. 261–272.

Reed, F. D., 1951, and Sandoe, M. T., *Cordaites affinis:* A new species of cordaitean leaf from American coal fields: Torrey Botanical Club Bulletin, v. 78, p. 449–457.

MANUSCRIPT ACCEPTED BY THE SOCIETY JULY 6, 1994

Printed in U.S.A.

Geological Society of America
Memoir 185
1995

# *Walter Andrew Bell (1889–1969): Canadian paleobotanist and earth scientist*

**Erwin Lorenz Zodrow**
*University College of Cape Breton, Subdepartment of Earth Sciences, Sydney, Nova Scotia, B1P 6L2, Canada*

## ABSTRACT

**Walter Andrew Bell (1889–1969) is considered the "Father of Canadian Carboniferous Biostratigraphy." He was one of the most distinguished members of the Geological Survey of Canada.**

**Bell's association with the Canadian Geological Survey spanned almost 60 years, and his research covered a wide geological spectrum but had crystal-clear focus. His greatest goal and accomplishment was the establishment of the major stratigraphic subdivisions of the Carboniferous System of the Canadian Maritime Provinces of Nova Scotia and New Brunswick, and of Newfoundland. He used both plants and invertebrate fossils for correlating the Carboniferous subdivisions of the Maritimes with those of the Carboniferous System of western Europe. This work represented a major advance in the understanding of Carboniferous correlations with western Europe and remains relevant. He also studied the floras of the Mesozoic and Cenozoic Systems of western Canada.**

**From his early years as a paleobotanist he rose to become director of the Geological Survey of Canada (1949–1953), a position he occupied with distinction. Bell was also an educator who shared his knowledge with professionals and the public alike. His life's work was acknowledged by the many rewards he received.**

## INTRODUCTION

Bell's geological talent was recognized early and was nurtured during the 1914–1915 period at Yale University, particularly under Professor Charles Schuchert (McLaren, 1969; see Fig. 1, this chapter; Andrews, 1980) but also in the presence of the Canadian giants F. H. McLearn (see Fig. 1), G. S. Hume, and W. S. McCann and notable American geologists such as W. H. Twenhofel and C. O. Dunbar (McLaren, 1969). In 1910, Bell started work with the Geological Survey of Canada (GSC) as a geological assistant in the Yukon. His lifelong association with the GSC can probably be traced to the influence of Schuchert, because Schuchert in 1920 wrote to the director of the GSC, Dr. R. G. McConnell, in support of Bell's appointment (as well as of Hume's and McCann's). Bell's career as a permanent staff member of the GSC is chronicled as follows, although consulted literature shows some inconsistencies in both dates of appointments and the nature of appointments themselves:

1920–1924: assistant and associate paleobotanist
1925–1936: paleobotanist
1936–1938: geologist
1938–1946: chief, Paleobotany Section
1946–1949: senior geologist
1949–1953: director of the GSC (Zaslow, 1975)
1953–1954: chief-geologist consultant for the GSC
1954–1969: the retirement years, consulting geologist for the GSC and the Nova Scotia Department of Mines.

As a paleobotanist and geologist, Bell applied his talent to several geological systems on the Canadian east and west coasts. First and foremost are his floral, stratigraphic, and structural studies of the lower and upper Carboniferous strata (Mississippian and Pennsylvanian) of the Maritimes and of the

Zodrow, E. L., 1995, Walter Andrew Bell (1889–1969): Canadian paleobotanist and earth scientist, *in* Lyons, P. C., Morey, E. D., and Wagner, R. H., eds., Historical Perspective of Early Twentieth Century Carboniferous Paleobotany in North America (W. C. Darrah volume): Boulder, Colorado, Geological Society of America Memoir 185.

Figure 1. Walter A. Bell at age 71, 1960, Stratigraphic Palaeontology Section, Fuels and Stratigraphy Division. Courtesy of the Geological Survey of Canada, September 23, 1960 (Geological Survey of Canada negative 201790C). *Front row, left to right:* Dr. F. H. McLearn, Miss A. E. Stafford, Dr. A. E. Wilson, and Dr. W. A. Bell. *Second row: left to right:* Mr. B. J. Botte, Mr. G. Prud'homme, Mr. H. Claude, Dr. F. J. E. Wagner, Mrs. L. Shields, Dr. T. T. Uyeno, and Dr. D. C. McGregor. *Third row, left to right:* Dr. D. J. McLaren, Dr. P. Harker, Dr. P. Sartenaer, Dr. T. E. Bolton, Dr. M. J. Copeland, Mr. J. J. Callahan, and Mr. R. Shea. *Fourth or back row, left to right:* Dr. A. W. Norris, Mr. J. E. A. Matte, Dr. J. A. Jeletzky, Dr. E. T. Tozer, Dr. A. W. Sinclair, and Dr. B. S. Norford.

lower Carboniferous of Newfoundland (1948). These studies resulted in the biostratigraphical organization of the Carboniferous strata in eastern Canada and represent his most comprehensive and significant achievement. Bell also worked intensely on floras of Cretaceous and of Tertiary (Paleocene) ages of Alberta (1928b, 1949) and on floras of Early and Late Cretaceous ages of western Canada and British Columbia (1928a, 1962b, 1963, 1965a, b) and investigated the Mesozoic flora of northern Ontario (1928c). Moreover, he concerned himself with exploration of fluorite (Bell, 1924, unpublished record, listed in Bell bibliography this chapter), manganese (Bell, undated, unpublished record), and salt deposits (Bell, 1945, unpublished record), and with oil shale and petroleum potential in Nova Scotia (Bell, 1927b, 1928d, 1932, 1958). After "retirement in 1954" (the effective year was 1956), Bell was fortunate in being able to remain active in a consulting capacity and as a researcher and author of scientific papers, including GSC memoirs and maps, until his death in 1969. Bell's complete publication record is presented here for the first time, together with some of his unpublished reports.

Bell left his mark on the lower and upper Carboniferous biostratigraphy of the Maritimes. This is clear from my personal communications with Permo-Carboniferous specialists (for example, Dr. J.-P. Laveine, written communication, February 22, 1990; Laveine, 1977) and Mesozoic biostratigraphers and paleontologists during the period 1980–1992.

This portrait is written from the perspective of Bell's biostratigraphical synthesis of the coal-bearing strata of the upper Carboniferous strata in the Maritimes, which by and large represents his life's work as writer and geologist from 1912 to 1969. To this end, Bell worked single-handedly and within principles and working hypotheses (Bell, 1921a) that would be considered narrow by today's standards. His most influential works are the accounts of the fossil floras of Nova Scotia, which appeared in three volumes (Bell, 1938, 1943c, 1962a) as well as a minor volume on the Pictou flora of Nova Scotia (Bell, 1940); in these volumes he fully documented more than 200 European fossil-plant species as either hypotype or plesiotype. In today's context, his work certainly confirms and expands the Carboniferous works of Ch. Lyell, W. E. Logan, and J. W. Dawson in Canada and demonstrates European floral affinity with eastern North America, describes correlations with Europe, and confirms plate tectonic theory. Bell's biostratigraphic achievement compares with that of David White and his subdivision of the Carboniferous strata of the United States (see Chapter 10, this volume; see Bell, 1938).

Bell's doctoral dissertation (1920) and two subsequent publications (1921b, 1929) on the lithology and invertebrate fossil assemblages of the Horton-Windsor Group of Nova Scotia (and correlation with age-equivalent rocks in Newfoundland, Bell, 1948) laid the foundation for the current stratigraphy of the lower Carboniferous System in the Maritimes.

As this volume, dedicated to W. C. Darrah, concerns only paleobotany, Bell's work on the invertebrate fauna of the lower Carboniferous is not included. The references to that work are included in Bell's bibliography, this chapter. The documentary material for the present portrait is on deposit with the special section of the library of the University College of Cape Breton.

## PERSONAL CHARACTERISTICS

Walter A. Bell was a rather small, slim, and wiry man with a thin face and a sharp nose (see Fig. 1 and refer to Fig. 3). Records of World War I (WWI), as in his enlistment description on January 25, 1916, show that at the age of 27 he was 1.68 m tall, his girth was 0.81 m, and his eyes and hair were brown. Forty years later, at the age of 67, he would show his nimble foot by vaulting over a fence, after placing one hand on the post (N. Oldale, oral communication, 1991).

An outstanding attribute of Bell was his phenomenal "photographic" memory and encyclopedic knowledge of the Carboniferous history of Maritime Canada. In this regard, Bell's wife, "Budda" (the former Sara Bell Campbell), in a letter to Dr. H. N. Andrews, July 6, 1978, recalls that Walter was able to recite from memory without a pause field notes of surveys done many years ago. Also, N. Oldale (oral communication, 1991) recalled, "Walter would remember the location and details of an outcrop thirty years later," and he had a great "nose" for finding fossils, too. The clarity of Bell's writing and his mastery of English prose are obvious from his publications. His style of writing is particularly noticeable when reading his species' descriptions. The descriptions are vivid and not monotonous, possibly because he did not use the staccato Linnaean style (of botanical Latin).

Dr. S. A. Ferguson (written communication, January 26, 1992) recalls that Bell had a dominating passion for geology and once said that "for any worthwhile geologist, geology is not only his profession, it is his hobby."

Dr. M. J. Copeland (Fig. 1) recalls (written communication, October 30, 1991) that Bell was an incessant pipe smoker and rarely cracked a smile, possibly because of his Presbyterian upbringing. He always had a twinkle in his eye, however, and was never at a loss for conversation. At larger gatherings, however, Bell was not at his best because he was rather deaf from his service years with the artillery (from the same letter of "Budda" Bell to H. N. Andrews in 1978). Bell's verbal parsimony is summed up by the fact that he rarely said "Good morning, it's a nice day" because you could see that the latter part of the greeting was considered superfluous. Any damned fool would know it was a nice day.

In 1992, the Victoria Memorial Museum, Ottawa, Ontario, Canada, was the meeting place for the second Canadian Paleontology Conference to celebrate the GSC's 150th anniversary. At that time an overdue tribute was paid to Bell by von Bitter (1992). In the same museum, Bell, as GSC director, surrounded himself with plant and animal fossils in his large office, which was located in the west gallery.

As GSC director, a position he reluctantly accepted, Bell provided the necessary leadership and "presided over the early stages of a revolution in Survey activities" (Zaslow, 1975, p. 414). Furthermore, he supported the expansion of the GSC into new fields of studies, notably the field of radioactive minerals. As its director, he saw the beginning of all major advances that shaped the Survey for the next two decades.

Although Bell recognized floral and faunal succession as a working hypothesis for biostratigraphy, he stated that "the life-forms first created were the progenitors of all succeeding life" (Bell, 1924c), which implies a theological view expressing faith.

Bell's character can probably be summarized by saying that he was a very thoughtful, sensitive, and polite man who was shy and private about himself and his family (E. S. Belt, written communication, October 19, 1991). The latter aspect may explain the paucity of photographs of Bell. He was of strong and unshakable personal and geological convictions, single-minded of purpose, willing to do his duty for Canada—which he did—and he had a mindset for personal sacrifices. Bell was modest about his achievements, a true gentleman, a man of science, and an excellent field geologist.

## PRE-UNIVERSITY YEARS

Not much seems to be known of the early childhood years of Walter A. Bell. The record for the year 1889 in the Town Hall of St. Thomas, a southwestern Ontario town in Elgin County, Canada, shows that Walter Andrew Bell was born on January 4. His parents were James A. Bell and Kate (née Darroch) Bell. The latter came to the Dominion of Canada from Loch Fyne, Scotland, and Walter A. Bell's father was born in Lobo Township, Middlesex County, Ontario. The parents of James A. Bell originated from Paisley in the Glasgow area, Scotland (J. Bircham, written communication, June 12, 1992).

Walter A. Bell descended from a family of engineers and naturalists. His father was a longtime Elgin County civil engineer who was famous for innovative bridge-building concepts and construction: for example, the construction of the Cantilever Bridge over the Niagara River Gorge (St. Thomas Times-Journal, 1981). His father also instilled the love and respect for nature into Walter and his three brothers. The family background probably influenced Bell's future career and contributed to a lifelong habit of being accurate about geological observations and precise about the descriptions of species. In 1906, Bell graduated from the St. Thomas Collegiate Institute, St. Thomas.

## THE UNIVERSITY YEARS

Bell enrolled at Queen's University in Kingston, Ontario, in 1906 and graduated five years later in April 1911 (Fig. 2) with a Bachelor of Science degree in Geological Engineering—the B.Sc. was an engineering degree only. Prior to 1956, the B.A. was awarded to science graduates (J. M. Dixon, written communication, December 17, 1992). The university records

Figure 2. Walter A. Bell at age 22, in the 1911 science graduating class, Queen's University, Kingston, Ontario, Canada (Bell, see arrow). With permission of Henderson Photographic Studio, Queen's University Archives, Queen's Picture Collection, 1992.

show that while at Queen's, Bell switched programs several times (from Mining to Civil Engineering, Mineralogy, and Geology and to the Arts Faculties), which would explain his five-year study at that university. His B.Sc. thesis, "The Birdseye and Black River formations as occurring on Wolfe Island," was singled out for merit, and the chairman of the University's Committee on 1851 Exhibition Scholarship, Professor Nathan F. Dupuis, recommended Bell as a worthy recipient of the scholarship for the 1911–1913 period. The scholarship (annual value of 150 pounds sterling) was awarded on August 16, 1911, by Her Majesty's Commissioners for the Exhibition of 1851 to Bell, who had given evidence for original research in his thesis, was a British subject, and under thirty years of age. Winners of the scholarship were expected to proceed to an institution other than the one by which they were nominated (J. M. Dixon, written communication, December 17, 1992). Bell chose to go to Yale University. The scholarship no longer exists (E. Farrar, written communication, March 4, 1992).

While still at Queen's in 1910, Bell started to work for the GSC in the Yukon as junior summer assistant to Dr. D. D. Cairnes. This began his lifelong association with the GSC (McLaren, 1969). In 1911, and continuing into 1915 while at Yale University, Bell carried out work as a student assistant for the GSC in Maritime Canada. This commenced his life's work of integrated paleontological and stratigraphical studies of the Maritime basins.

Bell's first publication (1912) was on the famous Joggins Section of Nova Scotia. For that work, Bell mapped the rocks according to the generally accepted classification for the Carboniferous strata, based on analogy with the coalfields of England and Wales, as shown below:

**BRITISH CARBONIFEROUS CLASSIFICATION**

| | | |
|---|---|---|
| G3: | Coal Measures | (youngest) |
| G2: | Millstone Grit | |
| G1: | Carboniferous Limestone Series | |
| GM1: | Carboniferous Conglomerate Series | (oldest) |

The larger problem facing Bell at the very beginning of his geological career was that work on the Carboniferous stratigraphy of the Maritimes had been dormant for more than 50 years and was surrounded by controversy. In many places, Carboniferous strata had been confused with the Devonian strata (see Chapter 25, this volume). Bunbury's (1847) and

Dawson's (1868) attempts at correlating the Carboniferous succession of the Sydney coalfield and of "Acadia" with European strata were limited in scope for lack of an adequately developed biostratigraphic context. However, Bell had the advantage of having at his disposal the detailed, often very accurate, and systematically mapped bedrock geology of the Carboniferous rocks in Nova Scotia by Fletcher (1887, 1892). In particular, Fletcher's map sheets of cliff sections of the Sydney coalfield (Fletcher, 1874), the Windsor and Horton type sections, and the Springhill and the Stellarton coalfields were excellent for locating large stratigraphic sections very much needed for Bell's work (Bell, 1940; Fletcher, 1887, 1892). Without the Fletcher map sheets, Bell might not have been able to accomplish as much as he did (E. S. Belt, oral communication, May 1992). See Gregory (1975) for Fletcher's publications on Nova Scotian Carboniferous geology.

Bell enrolled at Yale University for the Ph.D. degree and became a Dana Fellow, 1914–1915. At that university, his chief academic adviser was Professor Charles Schuchert (Andrews, 1980), whose regional approach to geology greatly influenced Bell and his work. The fieldwork for Bell's doctoral dissertation on the paleontology of the Horton-Windsor District Group was started in 1912 and completed by 1914 (Bell, 1921b). By 1915, all the requirements for the Ph.D. degree were completed, except for the thesis, which was deferred until after WWI. Documents from the National Archives of Canada (war records, attestation paper, discharge certificate, statement of service; J. Russo, written communications, March 12, 1992) show that Bell volunteered as a gunner with the 9th Brigade, 46th Battalion of the Canadian Field Artillery. He served in Canada, in Britain, and in France with the Canadian Expeditionary Force from January 25, 1916, until March 25, 1919, when he was honorably discharged at London, Ontario, Canada. The war records show that Bell was wounded twice in the field (November 11 and 13, 1917) and that he was promoted on June 7, 1918, to the rank of bombardier (corporal in the artillery). He was decorated and received the British War Medal and the Victory Medal. Information about the military operations in which his unit engaged during WWI is given by Nicholson (1962).

After his return to Canada in 1919, Bell received for the second time a temporary staff appointment, because the GSC did not subsidize enlisted assistants (Zaslow, 1975). As a temporary staff member, Bell went back to Europe for study. He was able to spend several months at Cambridge University and some time at the British Museum (Natural History). At Cambridge, he came under the influence of Professors A. C. Seward, a distinguished paleobotanist, and J. E. Marr, paleontologist, and at the University College of London under the influence of Professor E. J. Garwood, paleontologist. Bell also spent some time at the Ecole des Mines, Paris, France, which contained large paleobotanical collections by C. Grand'Eury and R. Zeiller (see Chapter 7, this volume) (Bell, 1929; McLaren, 1969).

In 1919, Birbal Sahni, also a student of Seward, was in London to receive his D.Sc. for research on fossil plants (Sahni, 1952). It is possible that he and Bell met there, which would have established Bell's link with paleobotany in India, although this meeting may not have happened.

In 1920, on his return to Yale, Bell received the Ph.D. degree. A summary of his doctoral thesis, "Stratigraphy of the Horton-Windsor District, Nova Scotia," was published under a different title (Bell, 1921b; see also Bell, 1929). In the publications, Bell warmly thanks Schuchert for his guidance. In 1920, Bell joined the permanent staff of the GSC as assistant paleobotanist (Zaslow, 1975). In the letter of recommendation in support of that appointment Schuchert (May 3, 1920), wrote as follows to Dr. R. G. McConnell, deputy minister and director of the Geological Survey, and unwittingly foretold Bell's career: "Bell is by all odds the best man I ever had, and now with the training in Europe, in the field of honor, and in the temple of learning (Cambridge) he, I prophesy, to be not only a worthy follower of Sir William Dawson in unearthing the geology of the Maritime Provinces, but, let us hope, as great a promoter of the earth sciences in general. He is not only a paleontologist and paleobotanist, but an equally good geologist."

Curiously, Schuchert also promoted W. C. Darrah's career (see Chapter 1, this volume). At the GSC, Bell became associated with two other Canadian paleontologists, Drs. Frank H. McLearn and Alice E. Wilson (see Fig. 1), the first woman professionally employed as a geologist by the GSC (Sarjeant, 1992). The trio made enormous contributions to Canadian biostratigraphy in particular and soft-rock geology in general (McLaren, 1969).

## THE REVISION OF UPPER CARBONIFEROUS STRATA IN THE MARITIMES

### *The upper Carboniferous coal strata: Preliminaries*

Bell (1921a) argued that Carboniferous rocks in one area may represent a similar rock facies in another area but are not necessarily coeval with the same lithological unit found in a different basin (for example, Millstone Grit of the Sydney coalfield is not necessarily time equivalent to the Millstone Grit in England or in Wales). Out of this and based in part on previous stratigraphical work by Robb (1876) and Fletcher (1887, 1892), Bell (1921a) argued that Sydney district would be an ideal basin for studying Carboniferous chrono- and biostratigraphy. Moreover, as a result of accumulated field experience in the Maritime coal basins, Bell (1921a) was the first to hypothesize that the productive coal measures of the Maritimes are (1) of different ages, and not, as it had been tacitly assumed before, "of the same epoch of sedimentation" or coeval according to Fletcher (Hacquebard, 1992), and (2) deposited in downfolded geosynclinal basins. As regards deposition, we now know (Belt, 1965, 1968a) that some are grabens, not geosynclines. Starting in 1924, Bell adopted new nomenclature and named rock units (groups, formations) after geographic type localities (see particularly Bell, 1943c). He further suggested that it is the duty of the paleontologist, given sufficient evidence, to correlate the properly defined formations. As conclusion in the Summary Report for 1920, Bell (1921a) clearly spelled out his regional approach and the working assumptions for subdividing the Carboniferous strata of Nova Scotia.

With the publication of GSC Memoir 133, "The Southern Part of the Sydney Coal Field, Nova Scotia" (Hayes and Bell, 1923, listed in Bell bibliography, this chapter), a proper understanding was achieved of the Carboniferous stratigraphy in the Maritimes, and a subdivision of strata based on well-formulated working assumptions was begun. As Nisbet (1991) put it, definitions of rock units must be in the rocks. In GSC Memoir 133, it is shown how the Morien Basin is subdivided both on faunal and floral grounds—but supporting relevant paleobotanical arguments and data appear later (Bell, 1938). In the subdivision, the oldest floral zone is typified by the presence of the pteridosperm *Neuropteris gigantea*, the middle by the pteridosperm *Linopteris obliqua** with type locality in the Sydney coalfield (Bunbury, 1847), and the youngest zone by the first conspicuous appearance of *Anthracomya* sp.

Later, Bell (Bell, *in* Moore et al., 1944, listed in Bell bibliography this chapter) changed the Time-rock "Series" category—introduced in Hayes and Bell, 1923—to Rock Unit "Group" category in line with earlier thinking to better reflect geographic connotation in naming rock units in the Maritimes. Also, in the 1923 report by Hayes and Bell, a meaningful intercontinental correlation was attempted for the first time since Dawson (1873). They correlated the lower Carboniferous and the Millstone Grit with strata in the eastern United States, the British Isles, and western Europe. In particular, the C5 and C4 Zones of the Morien Series (Table 1) were floristically compared with the Upper and Transition Coal Measures of England, the productive part (C5 Zone) being therefore of latest Westphalian age. The C3 Zone probably corresponds to the upper Middle Coal Measures.

An important consideration for the latter assignment is the presence of the pteridosperm genus *Lonchopteris* in the upper part of the C3 Zone (Hayes and Bell, 1923). Since its discovery then, the species level of identification continues to play a pivotal role as an index for the Westphalian B-C of the Maritimes (Bell in Moore et al., 1944; Bell, 1938). Darrah (1937) noted the significance of Bell's discovery and reference to *Lonchopteris rugosa*, as *Lonchopteris* had not at that time been recorded from the United States. (The genus is questionably present in the Norfolk basin of Massachusetts [Lyons et al., 1976]). Also, in the 1923 report of Hayes and Bell, reference is made to the mode of deposition (flood plain environment in a progressively subsiding river valley) and to the absence of marine horizons (Bell's geosynclinal basin was too far removed from the sea for marine transgression to take place [Bell, 1943c]). Records of agglutinated foraminifera (Thibaudeau, 1987; Wightman et al., 1992), ostracode assemblages (Copeland, 1990), and piscine fossils in limestone (H.-P. Schulze, oral communication, 1985) point to brackish water and marine estuarine conditions for the environment of deposition in parts of the Sydney Basin.

*Bell's revised (1962a) nomenclature is used for his species designation.

**TABLE 1. BIOZONES OF THE CARBONIFEROUS MORIEN GROUP (SERIES), SYDNEY, NOVA SCOTIA, CANADA***

C5: *Anthracomya* zone: Above Emery Seam and equivalent, to the top of the Morien Series

C4: *Linopteris obliqua* zone: Emery to the top of the Tracy Seam

C3: *Neuropteris gigantea* zone: Tracy Seam to bottom of the series

*Hayes and Bell, 1923.

Bell (1924a, 1926a) compared the Minto coal bed with the top zone of the Valenciennes coalfield of northern France and noted (as is accepted today) that the floral affinity between the Minto flora and that of the Appalachians was not as satisfactory as that with Europe. Years later, Arnold (1973) supported Bell by arguing more encompassingly that the floristic elements of the Maritimes are the closest of all American floras to those of Europe. This agrees with modern thinking that the Canadian Maritime coal basins were an integral part of the westerly extension of the area of coal deposition that extended over much of northwestern Europe. Bell's argument of similar floras between the Maritimes and Europe is thus solidly founded from today's perspective (see also Hacquebard, 1992).

Bell continued to unearth species from the Carboniferous strata of the Maritimes heretofore not recorded by Dawson (1868); an example is the loose-meshed veined neuropterid frond of *Reticulopteris (Linopteris) muensteri.* Correlation tables continued to appear (Bell, 1924b), and the regional components of the upper Carboniferous stratigraphy were gradually assembled into a coherent stratigraphy.

In 1924, Bell further refined some components of his working assumptions that were only stated implicitly in his earlier work (Bell, 1921a) by emphasizing the use of fossils in classifying strata and reconstructing geological events (Bell, 1924c). In the practice of that time this meant that for classifying strata using fossils, the range of individual species was used, where appropriate, as an index or a guide fossil. An example was the apparently restricted vertical range of *Linopteris obliqua* in the Sydney coalfield.

Bell (1926a) introduced the Heerlen nomenclature for naming stratigraphic subdivisions of the Carboniferous in the Maritimes (see Jongmans, 1928).

Bell (1927a) envisaged an array of linear geosynclines as recipients of sediments in which the various coal swamps of the Maritime coal basins eventually developed. In this sedimentary scheme, the subsiding geosynclinal floor during a certain time interval would represent the sequence, and repeated subsidence was considered responsible for rhythmic sedimentation. An area of uplift, adjacent to the depositional basin, could be related to crustal movement of an isostatic nature (Bell, 1927a, p. 76); isostasy refers to the theory of geosynclines and mountain building (Wegener, 1966, p. 15). Bell also believed that disconformities, based on records of floral breaks, could be mapped (Bell, 1943c; Bell, *in* Moore et al., 1944; Bell 1945a, p. 203). For mapping purposes and inter- and intrabasin correlations, facies were regarded by Bell as "commonly so diverse laterally, and so distributed vertically in many areas, that he [the geologist] would find it troublesome to make any choice at all, and would have to compromise on some grouping of sediments of purely local significance" (Bell, 1945a, p. 201). Bell's ideas about lithological units and their correlation effectively barred him from recognizing that different facies could be defined and shown to be coeval in the Carboniferous strata of Nova Scotia (Bell, 1962, unpublished).

Apart from published research from 1927 to 1938 (see Bell's bibliography this chapter), Bell's mind in the later part of this time span was also preoccupied with the age equivalence and stratigraphy of the Sydney coalfield, plant-fossil taxonomy, and the practical aspect of fossil-plant identifications. These studies were done to a large extent in preparation for the publication of GSC Memoir 215 (Bell, 1938) and are recorded in letters by Bell to W. C. Darrah from 1935 to 1936 (letters in the William Culp Darrah Collection = WCDC). Bell's letters started (WCDC July 3, 1935), after seeing a copy of Darrah's "American Carboniferous floras" (1937), by pointing out some corrections in Maritime correlations and stating that the Sydney flora is Alleghenian (Middle Pennsylvanian). Bell went on to say in the letter that it was reasonable to assume that Darrah had difficulties correlating floras of the Maritimes because Dawson's determinations were outdated and little of the modern work had been published. Bell and Darrah exchanged study material, and Bell was very much interested in comparative European material that Darrah was able to supply, since he was acquainted with the French Carboniferous coal basins where he collected under guidance from Paul Bertrand (see Chapters 2 and 7, this volume). Bell was also concerned with enhancing his understanding of how to:

(1) separate sphenophyllean species (WCDC, November 21, 1935), particularly specimens of *Sphenophyllum emarginatum* from *S. oblongifolium* (WCDC, November 8, 1935). As is seen (Bell, 1938; Bell, *in* Moore et al., 1944), *S. oblongifolium* was one of the pivotal floral components for distinguishing the Stephanian, and

(2) distinguish particularly among pecopterids with *Asterotheca*-type fructification: *A. arborescens, Eupecopteris* (*Asterotheca*) *cyathea*, and *Pecopteris* (*Asterotheca*) *hemitelioides* (which was a problem for Bell as his working material is fragmentary and larger frond specimens are necessary for accurate determination).

Further, in regard to the second problem, Bell also made it clear that what was needed "on this side" of the Atlantic Ocean was reexamination of *P. hemitelioides* and *A. arborescens* and a diagnosis that did not rely so much on pinnule size.

In WCDC (March 1936), Bell pointed out to Darrah the continuous stratigraphy of the Sydney coalfield from Westphalian B to D. In effect, this is confirmed today and even enlarged to the effect that the Morien Group offers the best opportunity to initiate a detailed reassessment of the Westphalian D for modern northeastern American correlations with the European succession (Zodrow and Cleal, 1985).

Two letters (WCDC, December 10, 1937, and January 21, 1938) discussed floral aspects of the Cumberland and Riversdale Series (Table 2). In conclusion, it is clear from reading the WCDC letters that Bell asked for Darrah's comments and advice and engaged in biostratigraphic and paleobotanic arguments with him. It is assumed, therefore, that Darrah helped shape Bell's biostratigraphical thought in the preparation of GSC Memoir 215 (Bell, 1938), although this is not formally acknowledged in the memoir. For confirmation, Darrah's answers to Bell's letters ought to be examined, which was not possible as I was unable to locate them. However, in Bell's first letter to Darrah (WCDC, July 3, 1935), Bell had written he had completed a study of the Sydney flora.

In accordance with its mandate, the GSC was also engaged in the support of the mineral industry (Winder, 1992). An example of this type of endeavor is Bell's (1925a) stratigraphic unraveling of the unit named the "New Glasgow Conglomerate" by Dawson (1868). On floral evidence, Bell correctly placed the conglomerate unit below the productive coal measures at the base of the Cumberland Group (Bell, 1943c), whereas Dawson (1873) had it placed above the productive coal measures. Bell thus saved the industry money by advising not to drill for coal below the conglomerate unit. Further, Bell's GSC memoirs (1938, 1940) are largely devoted to the economic geology of coal in Pictou County, Nova Scotia, and in the Sydney coalfield, respectively.

### *Biostratigraphical model for coal strata of Westphalian B-D ages: Sydney coal basin*

Bell's GSC Memoir 215 (1938) is one of the finest works on Carboniferous plant-fossil stratigraphy in North America and ranks with Darrah's 1969 monographic work and David White's 1900 work. In particular, GSC Memoir 215 is the stepping-stone into the realm of the Canadian upper Carboniferous paleobotany and the hub for homotaxial correlation of the Maritime with the European coal basins. The memoir features an integrated approach of paleobotany and stratigraphy in the service of the coal industry. Darrah (1969, p. 38) noted the importance of Bell's (1938) GSC Memoir. Nearly 40 years later, C. A. Arnold (see Chapter 18, this volume) also expressed his greatest admiration for that work and the fact that Bell was able to subdivide the Morien Series on fossil plants alone (Arnold,

TABLE 2. BELL'S FLORAL ZONATION SCHEME OF THE UPPER CARBONIFEROUS OF MARITIME CANADA*

| European Chronology | Upper Carboniferous Subdivision | Diagnostic Floral Element |
|---|---|---|
| Westphalian D to late Westphalian B | Pictou Group† | *Linopteris obliqua, Neuropteris rarinervis, N. scheuchzeri, Sphenopteris striata, Sphenophyllum emarginatum, Alethopteris serlii* |
| | Paleontological break (inferred disconformity) | |
| Westphalian B | Cumberland Group | *N. tenuifolia, Sph. valida, Samaropsis baileyi* group |
| Middle Namurian C to Westphalian A | Riversdale Group | *N. smithsii, Mariopteris acuta, Whittleseya desiderata* |
| | Paleontological break (inferred disconformity) | |
| Namurian B to middle Namurian C | Canso Group§ | *Sphenopteridium dawsoni, Mesocalamites cistiiformis.* |

*Bell, in Moore et al., 1944.
†Differentiated into Morien, Stellarton, and Pictou Groups where coal-bearing strata with fossil plants occur.
§Contains oldest known brackish-water clams in Maritime Canada.

1973; C. A. Arnold, oral communication, June 1976). Moreover, Arnold expressed his admiration for Bell's ability to correctly determine fossil plants (C. A. Arnold, oral communication, 1977). This ability can probably be attributed in part to Bell's paleobotanical studies in Europe and to Bell and Darrah's paleobotanical interactions, as shown in the WCDC letters.

In Bell (1938), the 2,100-m-thick Morien Series is subdivided into three floral zones (see Chapter 25, this volume) on the principles embodied in Bell's working assumptions (Bell, 1921a), thereby revising his earlier thought (Table 1). He concluded further that although the floristic zones compare most closely with the English floral succession (see Chapter 25, this volume), they also resemble those of northern France and eastern United States. A problem is the paucity of preserved fossil flora in the *Lonchopteris eschweileriana* zone, a situation that seriously interfered with homotaxial correlation then as now (see also WCDC, July 4, 1935). The Morien subdivision provides a basic model for correlation of the basins of the Maritimes.

Moreover, correlation of the Sydney coal seams, as they occur in "subbasins" of the Morien Series, from the base near Port Morien to the top at Point Aconi, was successfully accomplished for the first time. In this scheme, Bell was, however, unable to track some of the older coal beds (Mullins, Gardiner, and the Tracy Seams) across the coalfield, a problem that remains today (Boehner and Giles, 1986). Historically, Charles Robb (1876) was the first to attempt coal-seam correlation in the Sydney coalfield. Bell (1938) honored him for that work by naming the tree-fern species *Asterotheca robbi* Bell after Robb.

Bell placed the base of the Westphalian D in the roof of the Emery Seam (see Chapter 25, this volume). The discovery by Bell (1938) of *Sphenophyllum oblongifolium* above the Emery Seam in the highest beds in the Morien Series, which according to him carries a Westphalian D flora, caused him to extend downward the stratigraphic range of this species that is well known as a Stephanian index in Europe. The reason is that he was reluctant to consider the presence of the Stephanian in the Sydney coalfield based on only one certain Stephanian index fossil, as the record went at that time. In a way, Bell is vindicated, inasmuch as *S. oblongifolium* is today recognized as a problem species and can no longer unequivocally be used as a Stephanian index fossil (Zodrow and Gao, 1991).

Although it is pointed out by Bell (1938) that the purpose of GSC Memoir 215 was economic and not minute paleobotanical examination, the publication is comprehensive and contains many paleobotanical confirmations. For example, as a result of Bell's observations it was confirmed that hair on *Neuropteris-scheuchzeri* pinnules has taxonomic value, that *Odontopteris subcuneata* is atypical *N. scheuchzeri,* and that forms like *Cyclopteris fimbriata* are part of the frond of *Neuropteris ovata.* In WCDC (March 4, 1936), Bell appeared convinced that his lonchopterid collection was *L. eschweileriana* (and not *L. rugosa* as Darrah had reported in 1937) (see also Bell, 1938). That Bell had problems determining members of the pecopterid group is no surprise and is recorded in two letters (WCDC, November 21, 1935; March 4, 1936). This situation probably contributed to Bell's caution not to publish certain pecopterids, notably *A. arborescens*. However, he did recognize the [apparent] affinity of *A. arborescens* with *A. robbi* (Bell, 1938). *Eupecopteris cyathea* was published (Bell, 1938) despite his earlier opinion in 1935 (WCDC, November 21, 1935) about its doubtful presence in the Sydney coalfield.

In summary, Bell (1938) for the first time in the history of Canadian upper Carboniferous studies gave proper documentation and comparison of 126 plant species from the Maritimes—the majority of the species identified with European taxa. Precise biostratigraphical data, relating to vertical ranges and epiboles for specific taxa, are also given and these form the foundation for the subdivision of the strata of the Sydney coalfield into three floristic zones, correlative with the Westphalian B to D in the European chronology. This work attains additional significance because the uppermost strata of the Morien Series are close to the upper Carboniferous–Lower Permian boundary. Floristic data from there are valuable, as such data are generally not available in North America (W. C. Darrah, written communication, 1986). Bell's memoir also stands as a cornerstone for Maritime Canada plate-tectonic setting, as the Sydney Basin in Carboniferous time was the plate-tectonic intermediate between ancestral Europe and eastern North America.

### *The Biostratigraphy of Westphalian A-B and the Namurian of the Maritimes*

Bell (1943c) published his second-most-important paleobotanical volume on Carboniferous floras of the Maritimes. In this work, Bell described and documented 126 plant species from northern Nova Scotia, about 26 of which are common in the Sydney coalfield. Pecopterids, as expected for the lower Westphalian, are as yet poorly diversified, but Bell was able to collect samples that possibly could represent three species.

Bell (1943c) was also able to propose a comprehensive subdivision of the Carboniferous strata of the Maritimes as a whole and coordinate it with European chronology. On the basis of the vertical distribution of fossil plants, the upper Carboniferous strata are subdivided into four major units: Canso, Riversdale, Cumberland, and Pictou Groups (in ascending stratigraphic order), as shown in Table 2. The ages of the Canso and the Riversdale Groups were established on evidence of brackish-water clams (Bell, 1943c; Bell, *in* Moore et al., 1944). The stratigraphic position of the Canso Group, however, remained uncertain. Bell suggested it might be of late Viséan age of the Mississippian but later designated it as basal upper Carboniferous (Bell, 1943c). Later, Bell participated in the publication on the correlation of Pennsylvanian formations of North America (Bell, *in* Moore et al., 1944) and in the process modified some of his 1943 correlations (Bell, 1943c); the revisions are incorporated in Table 2. The zonations in the Maritimes represent a major contribution to our understanding of Canadian correlations with Europe. (Fig. 3 shows Bell in 1949 at age 60.)

## THE RETIREMENT YEARS, 1954 TO 1969, AND BELL'S SYNTHESIS OF UPPER CARBONIFEROUS COAL STRATA

In his retirement years, Bell settled in New Glasgow, Nova Scotia, with his wife, "Budda," who is still living at the time of this writing. Dr. S. A. Ferguson conjectured (January 26, written communication, 1992) that one of the reasons Bell settled in New Glasgow was that he wished to be close to the Carboniferous coalfields to do geology as long as he was able. And he continued to be occupied with requests for consultation from both the federal and the provincial governments and the Maritime coal industry. The result was that during this pe-

Figure 3. Walter A. Bell at age 60, 1949. Courtesy of the Geological Survey of Canada (Geological Survey of Canada negative 201114).

riod he published 14 papers on topics such as petroleum-reservoir possibilities in Nova Scotia (Bell, 1958) and geological aspects of the Pictou Coalfield (Bell, 1960a). Included in the publications are two GSC memoirs (Bell, 1956, 1957) and one bulletin (Bell, 1963) on the Cretaceous floras of western Canada, in which he summarized the age and the relationship to similar floras in the United States.

Bell found it necessary to amplify lithological results published earlier about the Horton-Windsor District (Bell, 1929). The result was GSC Memoir 314 (Bell, 1960b); see Bell at age 71 in Figure 1. Deficiencies were corrected in this memoir by outlining the stratigraphy of the nonmarine lower Carboniferous Horton Group and by dating the sparse floral remains (*Triphyllopteris-Lepidodendropsis-Rhacopteris*) as Tournaisian in age, based in part on the early Viséan age of the marine fauna of the overlying limestone (of Windsor Group). Interestingly enough, Logan (Bell, 1960b) had assigned this a Triassic age, whereas Lyell (and Bell) considered it correctly as lower Carboniferous. David White (see Chapter 10, this volume) wrote two papers (White, 1926, 1934) from which Bell derived support for his floral-age determination of the Horton Group.

Two years later, Bell (1962a) published the third and concluding volume, GSC Bulletin 87, of the fossil floras of the Maritimes. In it he documented 65 floral taxa (mostly as hypotypes) from the Pictou Group in New Brunswick. Of these, approximately 50% are in common with the fossil flora of the Sydney coalfield. He systematized the flora of the Pictou Group and correlated it with the *Linopteris obliqua* zone of the Sydney coalfield.

With this publication (Bell, 1962a), his revision of the Carboniferous biostratigraphy of the Maritimes was complete. This work represented his final view on the Carboniferous biostratigraphy, stressing, as he always had done, the regional context. He was the very first to do so for the Maritimes and for Newfoundland.

It is puzzling that Bell apparently ignored the theory of continental drift yet consistently argued for far-reaching intercontinental homotaxial correlations without giving a proper reason (the North Atlantic was closed during the late Paleozoic time). For example, an age comparison is made by Bell (1960b, p. 17) between the Horton and the Geigen floras in Bavaria/Germany and between the Horton and that of the Minusinsk Basin in the USSR. In general, he compared the Maritime floras with those of northwestern Europe. Perhaps an answer can be found in Bell's perspective that can be traced back to the influence Schuchert exerted on him, which affected the scope of the work Bell tackled during most of his life. When Bell was at Yale, Schuchert was one of the most famous, and earliest, regional paleogeographers and a dogmatic believer in the existence of land bridges to explain fossil similarities on presently separated continents (Schuchert, 1928). Both Schuchert and Bell were engaged in what today is called "basin analysis." After English publication of Wegener's continental drift theory (1924), Schuchert delivered his "biggest blast" at the theory (John Rodgers, written communication, June 16, 1992) at the 1926 symposium on continental drift in New York (Schuchert, 1928). In his paper, Schuchert (p. 140) implied ridicule of Wegener when he quoted P. Termier, director of the French Geological Survey, that the German theory has "undeniable charm and real beauty." It is "a beautiful dream, the dream of a great poet. One tries to embrace it, and finds that he has in his arms but a little vapor or smoke; it is at the same time both alluring and intangible." To his credit, David White (see Chapter 10, this volume), who participated in the 1926 symposium (White, 1928), appeared to have been more tolerant, although viewing the theory with disfavor.

Bell had the foresight, as a service to the global stratigraphical and paleontological communities, to catalog and curate the fossil-plant type collections that were gathered by him and by others (H. M. Ami, J. W. Dawson, C. D. Walcott's Collections, for example). Bell housed the collections with the Geological Survey of Canada (1962b, 1965a, b, 1969). He also illustrated megaplant fossils of common occurrence in the Carboniferous strata of the Canadian Maritimes (1966).

## EPIEUGEOSYNCLINAL VERSUS RIFTING THEORIES: THE BELT-WEBB MODEL

In the mid-1950s, palynological studies of the Carboniferous strata in the Maritimes were initiated at the Coal Research Section of the GSC situated in Sydney, Nova Scotia (Hacquebard, 1957; Hacquebard et al., 1960). According to Hacquebard, they aroused considerable interest in Bell and by and large confirmed most of his stratigraphic interpretations. The spores were obtained from several areas in which no diagnostic plant megafossils had been found but which could now be assigned to one of Bell's stratigraphic units. This resulted in a most interesting and valuable correspondence between Bell and Hacquebard (in the files of P. A. Hacquebard, November 26, 1956; April 8, 1957). Quotations from these letters by Bell on the spore study of the Pictou coalfield follow: "I consider it an outstanding and extremely valuable contribution . . . you have amply shown convincing evidence for the economic value of correlation by micromethods, including value of such work in interpretation of difficult structures and of stratigraphic anomalies." The point is that the micromethods were used to prove that the Westville coal beds are older than the Albian coal beds and therefore would occur below them.

During the mid-1960s, a workable stratigraphical model relating mid-Carboniferous facies to a tectonic setting was established (Belt, 1964, 1965, 1968a), based on the succession of miospore zones. These zones, which were further developed by M. S. Barss of the GSC, were the first in the Maritimes that related to facies of which the sedimentological aspects were known (Belt, 1964, 1965, 1968b). The use of miospore zones allowed Belt and Webb to originate the model of a complex rift valley containing positive areas and depocenters and having margins developed in some places as a normal fault and in other places as a shear zone (a large pull-apart structure) (Belt, 1968a; Webb, 1968, 1969; Gibling et al., 1987). The new model essentially replaced the epieugeosyncline theory (steep, downfolded margins; Kay, 1951).

Although some geologists have challenged the Belt-Webb model (Keppie, 1989), it is generally accepted today. Perhaps more importantly, the stratigraphic principles (Belt, 1964) are accepted today and applied to strata such as the Cumberland Group (Ryan et al., 1991) or the Horton Group, strata that Belt did not study. Even though the Belt-Webb model did not visualize how the rift scheme fitted the then-developing concepts of the closing of the Iapetus Ocean (Acadian-Orogeny time is too early) or the opening of the Atlantic Ocean (Late Triassic–Early Jurassic is too late), nevertheless, the model has now been integrated with plate tectonic thinking. Neither Belt nor Webb agreed with Keppie's "collage" tectonic schemes (E. S. Belt, written communication, October 19 and December 5, 1991; October 21, 1992; oral communication, May 1992).

In summary, a new paradigm had emerged for analyzing basin tectonics that has ramification in Carboniferous stratigraphy (Belt, 1964, 1965) of the Canadian Maritimes. In the process, Belt revised the middle Carboniferous units of Nova Scotia. Bell's Canso and Riversdale Groups (Bell, 1943c) were no longer being considered valid group terms by Belt (1964). However, Vasey, in a survey of Carboniferous biostratigraphy of the Maritimes for his doctoral dissertation (1984, p. 17), pointed out, that "some authors have opted to redefine Bell's groups as lithostratigraphic units" and that "most authors appear to have retained Bell's original units with some refinement, recognizing that they are not chronostratigraphic units."

## FINAL THOUGHTS

Bell's achievements are an outstanding example of a scientific mind devoted to a geological career that spanned nearly 60 years with one institution. During the time, he labored effectively with relatively few interruptions in a field he had come to understand, left his imprint upon it, and loved it. Moreover, the thought of making personal sacrifices of all sorts lived with him, too, as an integral part of his character. This is exemplified by his volunteering his services in World War I when he did, his late marriage at age 33 (McLaren, 1969), and his willingness to live within small means to be able to serve Canada. As Schuchert put it in the letter of May 3, 1920, to the then GSC director R. G. McConnell in support of Bell's GSC appointment:

> Will the Geological Survey of Canada, in these soaring days of higher salaries, be able to hold these men [Bell, Hume, and McCann] in its service? I hope so, but tempting offers will come to them, and would already have come if I had not advised their exposing themselves to such. . . . The time has come when men of science, and even research men, must be recognized at their full value, and properly paid for their services.

Bell's efforts were tirelessly poured into geological science with determination and resolve—an exemplary stance that inspired future generations of geologists. However, his devotion to science also caused him to live a lonesome life. Through all this, Bell was able to work in stratigraphy, paleontology, biostratigraphy, and oil and mineral geology and to place the investigation of the Carboniferous coal-bearing strata of the Maritime Provinces of Canada on a scientific basis. Moreover, he realized the economic implications of his studies and used them for the benefit of the coal industry of Nova Scotia and New Brunswick and for the economic well-being of Canada in general.

Bell was also an educator and teacher of geology, and he was always interested in passing on his knowledge and expertise to fellow geologists and the public alike. Despite his great achievements and fame, Bell was fundamentally a modest and unassuming gentleman. He was thoughtful, kindly, and a humorous companion as a husband, father (of a daughter, Laura Jean), and friend.

## HONORS, AWARDS, AND ASSOCIATIONS

1911–1913: Recipient of the 1851 Exhibition Scholarship, Queen's University.

1914–1915: Dana Fellow, Yale University.

1920–1968: Member of the Canadian Institute of Mining and Metallurgy, serving as a member of the Council from 1953 to 1955.

1925–1968: Fellow of the Royal Society of Canada, serving as president of Section IX (Geology, 1943–1944).

1944: Recipient of the Gold Medal of the Professional Institute of the Public Service of Canada, honoring Bell for his pioneering work in Carboniferous biostratigraphy and coal geology during the past 30 years.

1945: Recipient of the International Nickel Company of Canada Medal (Inco Medal) of the Canadian Institute of Mining and Metallurgy for "distinguished service to the coal-mining industry of the Maritimes" (Canadian Mining and Metallurgical Bulletin, 1945).

1953: Recipient of an honorary Doctorate of Law degree from St. Francis Xavier University, Antigonish, Nova Scotia, Canada.

1960: A new frond genus was named after Bell, *Bellopteris* (Radforth and Walton, 1960), ". . . who had made, and continues to make, the significant contribution to the palaeontology and age correlation for the geological complex from which all our specimens came" (Radforth and Walton, 1960).

1965: Recipient of the coveted Logan Medal of the Geological Association of Canada. Sir William Edmond Logan founded the Geological Survey of Canada in 1842 (Winder, 1992).

1972: A new Mississippian eocarid taxon, new genus and species, *Bellocaris newfoundlandensis,* was named after

Bell (Fong, 1972).

1982: A new upper Carboniferous fern species, *Oligocarpia bellii,* was named after him (Zodrow and McCandlish, 1982). This is in recognition of Bell's vision for laying the foundation for future Carboniferous studies in the Maritimes.

In a "Montreal Star" newspaper report of 1969, it was supposedly mentioned that Bell was awarded the Dawson Medal. I was unable to confirm this. Bell was also an honorary member of the Mining Society of Nova Scotia and a Fellow of the Geological Association of Canada (McLaren, 1969; Nowlan, 1970).

## ACKNOWLEDGMENTS

As I did not know Bell personally, I consulted many who knew him both professionally and socially during the preparation of this chapter. Without the cooperation, contribution, and dedication of these individuals, an account of the life's work of W. A. Bell and glimpses of his character could not have been written. I thank them very cordially and additionally express my gratitude to persons who contributed other information about Bell: H. N. Andrews (professor emeritus, University of Connecticut); Paul Banfield (assistant archivist, Queen's University); E. S. Belt (Department of Geology, Amherst College); J. Bircham (Local History Department, Elgin County, Ontario); R. C. Boehner and D. MacNeil (Nova Scotia Department of Natural Resources, Energy Branch); A. E. Bourgeois (Geoscience and Information and Communications Division, GSC); Th. E. Bolton and M. J. Copeland (GSC) (see Fig. 1); J. M. Dixon and E. Farrar (Department of Geological Sciences, Queen's University); S. A. Ferguson (Wolfville, Nova Scotia); P. A. Hacquebard (GSC); P. C. Lyons (U.S. Geological Survey); N. Oldale (University College of Cape Breton); J. Rodgers (Silliman Professor of Geology Emeritus, Yale University); J. Russo (Correspondence Section, National Archives of Canada); D. Tedford (GSC Library); and P. H. von Bitter (Royal Ontario Museum). I am also grateful to P. Campbell of the University College of Cape Breton for reading an earlier version of the manuscript, to the reviewers of the manuscript (P. A. Hacquebard and E. S. Belt), and to the editors of the volume for their help.

Specially singled out for praise are E. S. Belt for advice freely given and especially for helping to develop the part of the manuscript that deals with the history and the ramification of the Belt-Webb model; P. A. Hacquebard for critical and constructive advice during the final stages of manuscript preparation; P. H. von Bitter for editorial guidance and for providing additional documentation about Bell; D. Tedford for research on Bell's unpublished reports and bibliography on my behalf; P. C. Lyons for helping compile Bell's bibliography; Th. E. Bolton and M. J. Copeland for written communications and literature about Bell; E. D. Morey for the letters she provided from the William Culp Darrah Collection; and R. C. Boehner for a copy of Bell's thoughts on the Ph.D. thesis of E. S. Belt, additional information about Bell, and constructive advice on the final draft of this manuscript.

I am also grateful to my late wife, Kris; to my daughter, Tanya M. Zodrow; and to Julia O'Blenes for support that I needed and received during the writing of this portrait.

This chapter is dedicated to the Geological Survey of Canada, 1842–1992.

## REFERENCES CITED

Andrews, H. N., 1980, The fossil hunters: Ithaca, Cornell University Press, 421 p.

Arnold, C. A., 1973, Fossil plants and continental drift: Lucknow, First Birbal Sahni Memorial Lecture, Birbal Sahni Institute of Palaeobotany, p. 1–11.

Bell, W. A. See Bibliography of Selected Unpublished and Complete Published Works of Walter Andrew Bell that follows.

Belt, E. S., 1964, Revision of Nova Scotia middle Carboniferous units: American Journal of Science, v. 262, p. 653–673.

Belt, E. S., 1965, Stratigraphy and paleogeography of Mabou Group and related middle Carboniferous facies, Nova Scotia, Canada: Geological Society of America Bulletin, v. 76, p. 777–802.

Belt, E. S., 1968a, Post-Acadian rifts and related facies, eastern Canada, *in* Zen E-An, White, W. S., Hadley, J. B., and Thompson, J. B., Jr., eds., Studies of Appalachian geology: Northern and Maritime: New York, Interscience Publishers, p. 95–113.

Belt, E. S., 1968b, Carboniferous continental sedimentation, Atlantic Provinces, Canada: Geological Society of America Special Paper 106, p. 127–176.

Boehner, R. C., and Giles, P. S., 1986, Geological map of the Sydney Basin, Cape Breton Island, Nova Scotia: Halifax, Nova Scotia Department of Mines and Energy, Map 86-1, scale 1:50,000.

Bunbury, C. J. F., 1847, On fossil plants from the Coal Formation of Cape Breton: Geological Society of London Quarterly Journal, v. 3, pt. 1, p. 423–438, pls. 21–24.

Copeland, M. J., 1990, Report on Ostracoda and Conchostraca from "just above the Harbour Seam near Point Aconi": Geological Survey of Canada unpublished report MP-1-1990-MJC.

Darrah, W. C., 1937, American Carboniferous floras: Congrès pour l'avancement des études de Stratigraphie Carbonifère, 2d (Heerlen, Septembre 1935): Compte Rendu, t. I, p. 109–129.

Darrah, W. C., 1969, A critical review of the Upper Pennsylvanian floras of eastern United States, with notes on the Mazon Creek flora of Illinois: Gettysburg, Pennsylvania, privately printed, 220 p., 80 pls.

Dawson, J. W., 1868, Acadian geology (second edition): London, Mac Millan, 694 p.

Dawson, J. W., 1873, Report on the fossil plants of the lower Carboniferous and Millstone Grit Formations of Canada: Geological Survey of Canada (Montreal), 47 p., 10 pls.

Fletcher, H., 1874, Twenty stratigraphic sections measured along the sea cliffs of the Sydney coalfield, Nova Scotia: Geological Survey of Canada, Ottawa Files, Notebook no. 1533.

Fletcher, H., 1887, Report on geological surveys and explorations in the Counties of Guysborough, Antigonish, and Pictou, Nova Scotia: Geological Survey of Canada, Annual Report for 1886, v. 2, pt. P, p. 5–128.

Fletcher, H., 1892, Report on geological surveys and explorations in the Counties of Pictou and Colchester, Nova Scotia: Geological Survey of Canada, Annual Report for 1890–91, v. 5, pt. P, p. 5–193.

Fong, C. C. K., 1972, *Bellocaris,* a new Mississippian crustacean from Newfoundland: Journal of Paleontology, v. 46, p. 594–597.

Gibling, M., Boehner, R. C., and Rust, B. R., 1987, The Sydney basin of Atlantic Canada: An upper Paleozoic strike-slip basin in a collisional setting, *in* Beaumont, C., and Tankard, A. J., eds., Sedimentary basins and basin-forming mechanisms: Atlantic Geoscience Society Special Publication 5, p. 269–285.

Gregory, D. J., compiler, 1975, Bibliography of the geology of Nova Scotia: Halifax, Nova Scotia Department of Mines, 237 p.

Hacquebard, P. A., 1957, Plant spores in coal from the Horton Group (Mississippian) of Nova Scotia: Micropaleontology, v. 3, p. 301–324.

Hacquebard, P. A., 1992, Reflections on 150 years of coal geology investigations by the Geological Survey of Canada in the Atlantic Provinces: Canadian Institute of Mining Bulletin, v. 85, p. 79–86.

Hacquebard, P. A., Barss, M. S., and Donaldson, J. R., 1960, Distribution and stratigraphic significance of small spore genera in the Upper Carboniferous of the Maritime Provinces of Canada: Congrès pour l'avancement des études de Stratigraphie et de Géologie du Carbonifère, 4th (Heerlen, Septembre 1958): Compte Rendu, t. I, p. 237–245.

Jongmans, W. J., ed., 1928, Discussion générale, Congrès pour l'avancement des études de Stratigraphie Carbonifère, 1st (Heerlen, June 1927), Compte Rendu, p. xxii–xlviii.

Kay, M., 1951, North American geosynclines: Geological Society of America Memoir 48, 143 p.

Keppie, J. D., 1989, Northern Appalachian terranes and their accretionary history: Geological Society of America Special Paper 230, p. 159–192.

Laveine, J.-P., 1977, Report on the Westphalian D, *in* Holub, V. M., and Wagner, R. H., eds., Symposium on Carboniferous stratigraphy, Prague, 1973; Prague, Geological Survey of Prague, p. 71–81.

Lyons, P. C., Tiffney, B., and Cameron, B., 1976, Early Pennsylvanian age of the Norfolk Basin, southeastern Massachusetts, based on plant megafossils, *in* Lyons, P. C., and Brownlow, A. H., eds., Studies in New England geology: Geological Society of America Memoir 146, p. 181–197.

McLaren, D. J., 1969, Walter Andrew Bell 1889–1969: Proceedings of the Royal Society of Canada, series 4, v. 7, p. 44–48.

The Medallists: 1945, Canadian Mining and Metallurgical Bulletin, v. 393, p. 253, 257.

Nicholson, G. W. L., 1962, Canadian Expeditionary Force 1914–1919: Ottawa, Queen's Printer, 621 p.

Nisbet, E. G., 1991, Of clocks and rocks—The four aeons of Earth: Episodes, v. 14, p. 327–330.

Nowlan, J. P., 1970, Memorial to Walter Andrew Bell: Geological Association of Canada Proceedings, v. 21, p. 45–46.

Radforth, N. W., and Walton, J., 1960, On some fossil plants from the Minto Coalfield, New Brunswick: Senckenbergiana Lethaea, v. 41, p. 101–119, pls. 1–5.

Robb, C., 1876, Report on explorations and surveys in Cape Breton, Nova Scotia: Geological Survey of Canada, Report of Progress for 1874–75, p. 166–266.

Ryan, R. J., Boehner, R. C., and Calder, J. H., 1991, Lithostratigraphic revisions of the upper Carboniferous to lower Permian strata in the Cumberland Basin, Nova Scotia and the regional implications for the Maritimes Basin in Atlantic Canada: Bulletin of Canadian Petroleum Geology, v. 39, p. 289–314.

Sahni, M. R., 1952, Birbal Sahni: A biographical sketch of his personal life: The Palaeobotanist, v. 1, p. 1–8.

Sarjeant, W. A. S., 1992, Alice Wilson (1881–1964), first woman geologist with the Geological Survey of Canada [abs.]: Canadian Paleontology Conference, Ottawa, Ontario, September 25–27, Program and Abstracts, no. 2, p. 23–24.

Schuchert, C., 1928, The hypothesis of continental displacement, *in* Van Waterschoot van der Gracht, W.A.J.M., and 13 others, Theory of continental drift, a symposium on the origin and movement of land masses both inter-continental and intra-continental as proposed by Alfred Wegener: Tulsa, Oklahoma, American Association of Petroleum Geologists, p. 104–144.

St. Thomas Times-Journal, 1981, Building bridges major forte for former city engineer: St. Thomas, Ontario, Canada (September 19), p. 12.

Thibaudeau, S. A., 1987, Paleontological evidence for marine incursion in the Sydney Basin deposits (Carboniferous, Nova Scotia) [abs.]: Canadian Paleontology and Biostratigraphy Seminar, London, Ontario, September 25–27, Program with Abstracts, p. 8–9.

Vasey, G. M., 1984, Westphalian macrofaunas in Nova Scotia: Palaeoecology and correlation [Ph.D. thesis]: Glasgow, Scotland, Strathclyde University, 467 p.

von Bitter, P. H., 1992, Walter Andrew Bell (1889–1969): Geological Survey of Canada stratigrapher, palaeontologist, and palaeobotanist *par excellence* [abs.]: Canadian Paleontology Conference, Ottawa, Ontario, September 25–27, Program and Abstracts, no. 2, p. 27–28.

Webb, G. W., 1968, Palinspastic restoration suggesting late Paleozoic North Atlantic rifting: Science, v. 159, p. 875–878.

Webb, G. W., 1969, Paleozoic wrench faults in Canadian Appalachians: North Atlantic—Geology and continental drift: American Association of Petroleum Geologists Memoir 12, p. 754–786.

Wegener, A., 1924, The origin of continents and oceans (translated from the third German edition by J.G.A. Skerl, with an introduction by J. W. Evans): London, Methuen, 212 p.

Wegener, A., 1966, The origin of continents and oceans (translation of the fourth revised edition of Die Entstehung der Kontinente und Ozeane, 1929, Braunschweig, Vierweg und Sohn): New York, Dover Publications, 246 p.

White, D., 1900, The stratigraphic succession of fossil floras of the Pottsville Formation in the southern Anthracite coal field, Pennsylvania: U.S. Geological Survey Annual Report 20, pt. 21, p. 749–930.

White, D., 1926, General features of the Mississippian floras of the Appalachian trough, West Virginia: U.S. Geological Survey County Reports (Mercer, Monroe and Summers Counties), p. 837–843.

White, D., 1928, Discussion of floating continents: *in* Van Waterschoot van der Gracht, W.A.J.M., and 13 others, Theory of continental drift, a symposium on the origin and movement of land masses both inter-continental and intra-continental as proposed by Alfred Wegener: Tulsa, Oklahoma, American Association of Petroleum Geologists, p. 187–188.

White, D., 1934, Pocono, age and climate: Geological Society of America Proceedings for 1933, p. 34.

Wightman, W. G., Scott, D., and Gibling, M. R., 1992, Upper Pennsylvanian agglutinated foraminifers from the Cape Breton Coalfield, Nova Scotia: Their use in the determination of brackish-marine depositional environments [abs.]: Geological Association of Canada and Mineralogical Association of Canada, Joint Annual Meeting, Wolfville, Nova Scotia, May 25–27, Abstracts, v. 17, p. A117.

Winder, C. G., 1992, William Edmond Logan (1798–1875): Canadian Mining and Metallurgical Bulletin, v. 85, p. 13–21.

Zaslow, M., 1975, Reading the rocks: The story of the Geological Survey of Canada, 1842–1972: Toronto, Macmillan, 599 p.

Zodrow, E. L., and Cleal, C. J., 1985, Phyto- and chronostratigraphical correlations between the late Pennsylvanian Morien Group (Sydney, Nova Scotia) and the Silesian Pennant Measures (south Wales): Canadian Journal of Earth Sciences, v. 22, p. 1465–1473.

Zodrow, E. L., and Gao Zhifeng, 1991, *Leeites oblongifolis* nov. gen et sp. (sphenophyllaean, Carboniferous), Sydney Coalfield, Nova Scotia, Canada: Palaeontographica, Abt. B, Band 223, p. 61–80, pls. 1–8.

Zodrow, E. L., and McCandlish, K., 1982, *Oligocarpia bellii,* sp. nov. from the middle Upper Carboniferous of Cape Breton Island, Nova Scotia, Canada: Palaeontographica, Abt. B, Band 181, p. 109–117, pl. 1.

## SELECTED UNPUBLISHED AND COMPLETE PUBLISHED WORKS OF WALTER A. BELL

*Excerpts from Bell's unpublished record, in chronological order*

Bell, W. A., undated, Manganese ore—Loch Lommond District, Inverness Co., N.S. (GSC File 11 F/15-7).

Bell, W. A., 1924, Fluorite deposits of Lake Ainslie (GSC File 11 K/3-1).

Bell, W. A., 1945, Salt reserves in the Malagash Salt Mine, N.S. (GSC File 11 E/14-2 [S2]).

Bell, W. A., 1962, Comments on Dr. E. S. Belt's thesis: Archives of the Nova Scotia Department of Natural Resources, Mineral Resources Division, p. 1–31.

There are an additional 34 unpublished reports by Bell, some of them written as late as 1968, one year before his death.

*Bell's publications*

Bell, W. A., 1912, Joggins Carboniferous section of Nova Scotia: Geological Survey of Canada, Summary Report for 1911, p. 328–333.

Bell, W. A., 1913a, Horton-Windsor (Nova Scotia): Geological Survey of Canada Guidebook 1, pt. 1, p. 136–151.

Bell, W. A., 1913b, The Joggins Carboniferous section (Nova Scotia): Geological Survey of Canada Guidebook 1, pt. 2, p. 326–346.

Bell, W. A., 1913c, Excursion in eastern Quebec and the Maritime Provinces: Horton-Windsor: International Geological Congress, 12th, Canada, 1913, Excursion Guide Book 1, pt. 1, p. 136–144, geological map.

Bell, W. A., 1914, Joggins Carboniferous section, Nova Scotia: Geological Survey of Canada, Summary Report for 1912, p. 360–371.

Bell, W. A., 1915, The Horton-Windsor Carboniferous area, Nova Scotia: Geological Survey of Canada, Summary Report for 1914, p. 106–107.

Bell, W. A., 1920, Stratigraphy of the Horton-Windsor District, Nova Scotia [Ph.D. thesis]: New Haven, Connecticut, Yale University, 303 p.

Bell, W. A., 1921a, The Carboniferous strata of Sydney district, Cape Breton, Nova Scotia: Geological Survey of Canada, Summary Report for 1920, pt. E, p. 17–18.

Bell, W. A., 1921b, The Mississippian Formations of the Horton-Windsor District, Nova Scotia: American Journal of Science, 5th ser., v. 1, p. 153–173.

Bell, W. A., 1922a, Tertiary plant remains collected by G. S. Hume in the Mackenzie River Basin: Geological Survey of Canada Summary Report for 1921, pt. B, p. 76.

Bell, W. A., 1922b, A new genus of Characeae and new Merostomata from the coal measures of Nova Scotia: Royal Society of Canada Proceedings and Transactions, 3d. ser., v. 16, sect. 4, p. 159–168.

Bell, W. A., 1923, Stratigraphy of Great Bras d'Or coal district, Victoria County, Cape Breton, *in* Hayes, A. O., and Bell, W. A., The southern part of the Sydney coal field, Nova Scotia: Geological Survey of Canada Memoir 133, p. 90–104.

Bell, W. A., 1924a, Correlation of the Minto coal horizon: Geological Survey of Canada, Summary Report for 1923, pt. C II, p. 23–32.

Bell, W. A., 1924b, Investigations of coal-bearing formations in Nova Scotia: Geological Survey of Canada, Summary Report for 1923, pt. C II, p. 33–40.

Bell, W. A., 1924c, The subdivision of the Carboniferous rocks of the Maritime Provinces: Canadian Institute of Mining and Metallurgy Bulletin, v. 152, p. 886–894.

Bell, W. A., 1924d, The subdivision of the Carboniferous rocks of the Maritime Provinces: The Transactions of the Canadian Institute of Mining and Metallurgy and of the Mining Society of Nova Scotia, v. 27, p. 607–615.

Bell, W. A., 1924e, The subdivision of the Carboniferous rocks of the Maritime provinces: Canadian Mining Journal, v. 45, p. 1138–1141.

Bell, W. A., 1925a, The New Glasgow conglomerate member of Pictou County, Nova Scotia: Canadian Institute of Mining and Metallurgy Bulletin, v. 158, p. 605–634.

Bell, W. A., 1925b, The New Glasgow conglomerate member of Pictou County, Nova Scotia: The Transactions of the Canadian Institute of Mining and Metallurgy and of the Mining Society of Nova Scotia, v. 28, p. 447–476.

Bell, W. A., 1925c, The New Glasgow conglomerate member of Pictou County, Nova Scotia: Canadian Mining Journal, v. 47, p. 694.

Bell, W. A., 1926a, Carboniferous formations of Northumberland Strait, Nova Scotia: Geological Survey of Canada, Summary Report for 1924, pt. C, p. 142–180, geological map.

Bell, W. A., 1926b, Palaeontology and correlation, *in* Dyer, W. S., Minto Coal Basin New Brunswick: Geological Survey of Canada Memoir 151, p. 18–19.

Bell, W. A., 1927a, Outline of Carboniferous stratigraphy and geologic history of the Maritime provinces of Canada: Royal Society of Canada Proceedings and Transactions, 3rd ser., v. 21, sect. 4, p. 75–108.

Bell, W. A., 1927b, Prospects for petroleum in Lake Ainslie district, Cape Breton Island, with notes on the occurrence of barite and granite: Geological Survey of Canada, Summary Report for 1926, pt. C, p. 100–109.

Bell, W. A., 1928a, Upper Cretaceous plants from Stikine River, Cassiar District, B.C.: Geological Survey of Canada Museum Bulletin 49, Geological series 48, Contributions to Canadian Palaeontology, p. 23–25, 49–55, 4 pls.

Bell, W. A., 1928b, A new Cretaceous conifer from the Blairmore series of Alberta: Geological Survey of Canada Museum Bulletin 49, Geological series 48, Contributions to Canadian Palaeontology, p. 26, 57, 1 pl.

Bell, W. A., 1928c, Mesozoic plants from the Mattagami Series, Ontario: Geological Survey of Canada Museum Bulletin 49, Geological series 48, Contributions to Canadian Palaeontology, p. 27–30, 59–67, 5 pls.

Bell, W. A., 1928d, Oil and gas in the Maritime provinces: Proceedings, Empire Mining and Metallurgical Congress, 2d (triannual), August 22–September 28, 1927: Montreal, Offices of the Congress, p. 3–16.

Bell, W. A., 1929, Horton-Windsor District, Nova Scotia: Geological Survey of Canada Memoir 155, 268 p.

Bell, W. A., 1930, A Mississippian fauna collected by Miss Eleanor T. Long from Windsor, Nova Scotia: Academy of Natural Sciences (Philadelphia), Proceedings, v. 81, p. 617–625.

Bell, W. A., 1932, Oil and gas prospects of the Maritime Provinces; The stratigraphy of bituminous-bearing formations of southern New Brunswick, Nova Scotia, and Prince Edward Island: Geological Survey of Canada, Economic Geology, ser. 9, p. 161–167.

Bell, W. A., 1938, Fossil flora of Sydney coalfield, Nova Scotia: Geological Survey of Canada Memoir 215, 334 p., 1 fig.

Bell, W. A., 1940, The Pictou Coalfield, Nova Scotia: Geological Survey of Canada Memoir 225, p. 161, two maps.

Bell, W. A., 1943a, The St. Rose—Chimney Corner Coalfield, Inverness County, Nova Scotia: Geological Survey of Canada Paper 43-14, 12 p., geological maps.

Bell, W. A., 1943b, The St. Rose—Chimney Corner Coalfield, Inverness County, Nova Scotia: Halifax, Nova Scotia Department of Mines, Annual Report on Mines, 1943, p. 106–122.

Bell, W. A., 1943c, Carboniferous rocks and fossil floras of northern Nova Scotia: Geological Survey of Canada Memoir 238, 277 p.

Bell, W. A., 1944a, Preliminary map, Shinimikas, Nova Scotia: Geological Survey of Canada Paper 44-35, scale 1:63,360.

Bell, W. A., 1944b, Use of some fossil floras in Canadian stratigraphy: Royal Society of Canada Transactions, 3rd ser., v. 38, sect. 4, p. 1–13.

Bell, W. A., 1945a, Use of fossil plants in the coal geology of Eastern Canada: Canadian Institute of Mining and Metallurgy Bulletin, v. 396, p. 199–210.

Bell, W. A., 1945b, Use of fossil plants in the coal geology of Eastern Canada: The Transactions of the Canadian Institute of Mining and Metallurgy and of the Mining Society of Nova Scotia, v. 48, p. 199–210; discussion, p. 622–623.

Bell, W. A., 1945c, Use of fossil plants in the coal geology of Eastern Canada: The Royal Society of Canada Proceedings, 3rd ser., v. 38, p. 155.

Bell, W. A., 1945d, Shinimakas, Cumberland County, Nova Scotia: Geological Survey of Canada, Map 842A, scale 1:63,360.

Bell, W. A., 1946, Age of the Canadian Kootenay Formation: American Journal of Science, v. 244, p. 513–526.

Bell, W. A., 1948, Early Carboniferous strata of St. Georges Bay area, Newfoundland: Geological Survey of Canada Bulletin 10, 45 p.

Bell, W. A., 1949, Uppermost Cretaceous and Paleocene floras of western

Alberta: Geological Survey of Canada Bulletin 13, 231 p.

Bell, W. A., 1951, Much done, but much yet to do in job of mapping Canada: Northern Miner, Annual no., p. 99, 109.

Bell, W. A., 1954, Atomic energy and the Geological Survey: Professional Public Service, v. 33, p. 11–12.

Bell, W. A., 1956, Lower Cretaceous floras of western Canada: Geological Survey of Canada Memoir 285, p. 331.

Bell, W. A., 1957, Flora of the Upper Cretaceous Nanaimo Group of Vancouver Island, British Columbia: Geological Survey of Canada Memoir 293, 84 p.

Bell, W. A., 1958, Possibilities for occurrence of petroleum reservoirs in Nova Scotia: Halifax, Nova Scotia Department of Mines, 177 p.

Bell, W. A., 1960a, Some geological aspects of the Pictou coalfield with reference to their influence on mining operations: Discussion: Third Conference on Origin and Constitution of Coal, Crystal Cliffs, Nova Scotia, 1956: Halifax, Nova Scotia Department of Mines, p. 74–85.

Bell, W. A., 1960b, Mississippian Horton Group of type Windsor-Horton District, Nova Scotia: Geological Survey of Canada Memoir 314, 112 p.

Bell, W. A., 1962a, Flora of Pennsylvanian Pictou Group of New Brunswick: Geological Survey of Canada Bulletin 87, 71 p., 56 pls.

Bell, W. A., 1962b, Catalogue of types and figured specimens of fossil plants in the Geological Survey of Canada Collections: Geological Survey of Canada, Catalogue, 154 p.

Bell, W. A., 1963, Upper Cretaceous floras of the Dunvegan, Bad Heart, and Milk River Formations of western Canada: Geological Survey of Canada Bulletin 94, 76 p., 42 pls.

Bell, W. A., 1965a, Illustrations of Canada fossils: Lower Cretaceous floras of western Canada: Geological Survey of Canada Paper 65-5, 36 p.

Bell, W. A., 1965b, Illustrations of Canadian fossils: Upper Cretaceous and Paleocene plants of western Canada: Geological Survey of Canada Paper 65-35, 46 p.

Bell, W. A., 1966, Illustrations of Canadian fossils: Carboniferous Plants of eastern Canada: Geological Survey of Canada, Paper 66-11, 76 p.

Bell, W. A., 1969 (posthumous), Catalogue of types and figured specimens of fossil plants in the Geological Survey of Canada Collections: Geological Survey of Canada (Megaplant Supplement 1963–67), 36 p.

Bell, W. A., and Goranson, E. A., 1938a, Bras d'Or sheet, Cape Breton and Victoria counties, Nova Scotia: Geological Survey of Canada, Map 359A, scale 1:63,360.

Bell, W. A., and Goranson, E. A., 1938b, Sydney Sheet, west half, Cape Breton and Victoria Counties, Nova Scotia: Geological Survey of Canada, Map 360A, scale 1:63,360.

Hayes, A. O., and Bell, W. A., 1922, Sydney, Cape Breton County, Nova Scotia: Geological Survey of Canada, Map 1767, scale 1:63,360.

Hayes, A. O., and Bell, W. A., 1923, The southern part of the Sydney coal field, Nova Scotia: Geological Survey of Canada Memoir 133, 108 p., plus geological map.

Hayes, A. O., Bell, W. A., and Goranson, E. A., 1938a, Glace Bay sheet, Cape Breton County, Nova Scotia: Geological Survey of Canada, Map 362A, scale 1:63,360.

Hayes, A. O., Bell, W. A., and Goranson, E. A., 1938b, Sydney Sheet, east half, Cape Breton County, Nova Scotia: Geological Survey of Canada, Map 361A, scale 1:63,360.

Kerr, F. A., Jones, I. W., and Bell, W. A., 1938, Springhill sheet, Cumberland and Colchester Counties, Nova Scotia: Geological Survey of Canada, Map 337A, scale 1:63,360.

Kerr, F. A., Jones, I. W., Bell, W. A., Shaw, W. S., and Copeland, M. J., 1959, Cumberland County, west part, Nova Scotia: Geological Survey of Canada, Map 1070A, scale 1:63,360.

Moore, R. C. (chairman), Wanless, H. R., Weller, J. M., Williams Steele, J., Read, C. B., Bell, and 21 others, 1944, Correlation chart of Pennsylvanian formations of North America: Geological Society of America Bulletin, v. 55, p. 657–706.

Norman, G. W. H., and Bell, W. A., 1938a, Oxford sheet, east half, Cumberland and Colchester Counties, Nova Scotia: Geological Survey of Canada, Map 409A, scale 1:63,360.

Norman, G. W. H., and Bell, W. A., 1938b, Oxford sheet, west half, Cumberland and Colchester Counties, Nova Scotia: Geological Survey of Canada, Map 410a, scale 1:63,360.

Weller, J. M. (chairman), Bell, W. A., Caster, K. E., Cooper, C. L., Stockdale, P. B., and Sutton, A. H., 1948, Correlation of the Mississippian formations of North America (Chart 5): Geological Society of America Bulletin, v. 59, p. 91–106.

Manuscript Accepted by the Society July 6, 1994

Printed in U.S.A.

Geological Society of America
Memoir 185
1995

# *James Morton Schopf (1911–1978): Paleobotanist, palynologist, and coal geologist*

**Aureal T. Cross**
*Department of Geological Sciences, Natural Science Building, Michigan State University, East Lansing, Michigan 48824-1115*
**Robert M. Kosanke**
*U.S. Geological Survey, MS 919, Box 25046, Denver Federal Center, Denver, Colorado 80225-0046*
**Tom L. Phillips**
*Department of Plant Biology, 265 Morrill Hall, University of Illinois, 505 S. Goodwin Avenue, Urbana, Illinois 61801*

## ABSTRACT

**The contributions of James (Jim) M. Schopf (1911–1978) to the development of upper Carboniferous and Permian paleobotany, palynology, and coal geology are among the most influential of the mid-twentieth century. Jim's scientific endeavors reached far beyond this scope and extended from the Precambrian to the Pleistocene. His research papers provided many of the integrated benchmarks upon which his and subsequent generations have built. Jim Schopf was respected for his rigorous standards of scholarship and for his candor. He was also one of our most generous and helpful colleagues, and his wise counsel aided many. His graduate studies at the University of Illinois, Urbana, overlapped with pioneering research in the Coal Division, Illinois State Geological Survey (1934–1943). He served with the U.S. Bureau of Mines at Pittsburgh (1943–1947) and the U.S. Geological Survey (1947–1978). In 1949, Jim Schopf established the Coal Geology Laboratory at Ohio State University, Columbus, which became an important center for coal research and scientific exchange—and a special attraction for visiting paleobotanists, palynologists, and coal geologists. The Schopf home was also noted for its gracious hospitality and lively discussions of current research issues. Schopf was especially known as a gregarious and ardent "field tripper," one with a remarkable ability to discover fossil-plant deposits. One of his special challenges was the paleobotany of the Antarctic (1960–1978), in cooperation with the Institute of Polar Studies at Ohio State University. He received many honors, including the Gilbert H. Cady Award of the Coal Geology Division of the Geological Society of America, Mary Clark Thompson Medal of the National Academy of Sciences, and the Paleontological Society Medal. Mt. Schopf in Antarctica was named in honor of his contributions to Antarctic science.**

## INTRODUCTION

The contributions of James M. Schopf to developments in Carboniferous paleobotany of North America during the middle half of the twentieth century are interwoven closely with those of coal-ball studies, palynology, and coal geology. Jim Schopf's scientific endeavors were broad, extending from the paleobiology of the Precambrian to the Pleistocene. He sought the interrelationships within and among subdisciplines and pursued perplexing questions that required a high order of scholarship and provided a lifelong scientific venture.

Brief biographical sketches of James M. Schopf occur in Kosanke (1974, 1979), Andrews (1979, 1980), Cross (1979, 1983), and Phillips et al. (1973). Quotes from some of

Cross, A. T., Kosanke, R. M., and Phillips, T. L., 1995, James Morton Schopf (1911–1978): Paleobotanist, palynologist, and coal geologist, *in* Lyons, P. C., Morey, E. D., and Wagner, R. H., eds., Historical Perspective of Early Twentieth Century Carboniferous Paleobotany in North America (W. C. Darrah volume): Boulder, Colorado, Geological Society of America Memoir 185.

Schopf's prolific correspondence with colleagues through the years are used in this portrait to provide insight into some of his reflections. Most of his publications are listed in the *Bibliography of American Paleobotany for 1978,* compiled by Arthur D. Watt (1979).

James Morton Schopf was born June 2, 1911, in Cheyenne, Wyoming, the son of Ira Morton Schopf and Nettie Bufton Schopf. He attended school at Pine Bluffs, Wyoming, where he graduated from high school in 1927. His college training began at Kemper Military School, Boonville, Missouri, where he learned the orderliness and organization that he next applied to research projects and filing of books, papers, slides, and specimens. After a year at Kemper, Schopf transferred to the University of Wyoming and completed his B.A. (Botany) in 1930. At the university, Dr. Aven Nelson had perhaps the strongest influence on Jim's scholarship.

At the University of Illinois at Urbana (1930–1937) and the Illinois Geological Survey (1934–1943), Schopf was privileged to work with outstanding botanists and geologists. He received a M.S. in Plant Ecology with a minor in Geology (1932). John T. Buchholz in the Botany Department was his Ph.D. adviser at the University of Illinois. Harold R. Wanless was his geology adviser and later his field colleague. Gilbert H. Cady at the Illinois Survey was, like Schopf, a disciplined, constructive critic of the highest scientific integrity. "Doc" Cady, as he was called, had an inestimable influence on Jim Schopf's penetrating studies of the geology of coal and on his whole scientific career. Schopf had quite varied experiences as a teaching assistant in the Botany Department (1930–1933), summer assistant botanist with the Illinois Natural History Survey (1932), and research assistant and then assistant geologist (1935–1943) in the Coal Division of the Illinois Geological Survey, working under Cady.

Esther Julia Nissen, from Cedar Falls, Iowa, was also a graduate student in the Botany Department at the University of Illinois. In 1934 Jim and Esther were married at Toledo, Ohio.

The Board of Trustees of the University of Illinois were probably unaware of the magnitude of the meaning of "philosophy" that Jim, as a young scientist, would impart to science in general and to geology and botany in particular, when he was awarded his Ph.D. in 1937 with a thesis entitled "The Embryology of *Larix*" (Schopf, 1943). Frederick Oliver Bower, the leading plant morphologist, sent Schopf a letter commending him for the significance of his contribution and commenting at length on the various details. One has cause to wonder if that letter might have been inspirational for the Schopfian "Epistles" that so many of us have involuntarily received, especially those that, though constructive, often did not include much identifiable commendation!

In his pursuit of a doctoral degree Schopf had been challenged by the dual responsibilities of his productive Survey research and his thesis. He expressed these thoughts in a letter to Robert Kosanke (October 27, 1952): "There is a considerable question in my mind as to whether the degree can be worth as much as the effort cost, but I believe really that the effort in itself was entirely worthwhile. The diploma then is simply an incidental bit of 'frosting' that signifies that one of our great institutions of learning also believes you have to your credit a job well done."

Schopf was keenly aware of the opportunities within and the influences of the Coal Division of the Illinois Survey during his decade of work there. He considered it a "good training ground" and remarked on the broad backgrounds and breadths of interest necessary for training personnel. Years later he wrote Jack Simon (January 30, 1967) from the McMurdo Naval Base, Antarctica: "The Illinois Survey is one place where a real scientific contribution in coal studies is being made in the U.S.A. I won't say the only place, but the consistency of the Illinois effort has been a great bright spot for those of us that believed it had meaning and value. I want to see that work go on. . . ."

In 1943, Schopf accepted an appointment with the U.S. Bureau of Mines as paleobotanist and was stationed at the Central Experiment Station in Pittsburgh, Pennsylvania. He had become interested in Reinhardt Thiessen's thin-section method (see Chapter 11, this volume) of analysis of coal, and this appointment at the Bureau of Mines provided the opportunity to examine Thiessen's method first hand.

Toward the end of Schopf's work there, he observed in a letter to Robert Kosanke (February 26, 1946): "As things are at present I should be very happy if the U.S. Survey would give *some* attention to coal, or microfossils or anything in that related field. There is such a lot of work that needs doing and it mostly takes just time and application. The sooner they start helping some, the better. It takes men and it takes facilities."

For part of 1947, Schopf was on leave from the Bureau of Mines to serve as a senior research associate with the Council of Scientific and Industrial Research and the South African Geological Survey at Pretoria. His mission there, which was to demonstrate the use of coal petrographic techniques with South African coal, had been arranged by Sydney Houghton, director of the Survey, who had spent a short time with Schopf and associates at the Pittsburgh laboratory.

Upon Schopf's return to the United States in October 1947, he was appointed as a geologist in the U.S. Geological Survey, first with offices in the U.S. National Museum and then in "special facilities" of the Coal Geology Laboratory at Ohio State University in Columbus. His change of address card (November 1949) stated, "The work of this laboratory will be concerned with the many phases of the geology of coal that directly or indirectly rely on study and knowledge of fossil plants." He developed this facility into a world-class laboratory for coal and paleobotanical research. Schopf was recognized throughout his career as an academician of broad interests and scientific pursuits; he held part-time professorships in the Department of Geology and Mineralogy (from 1950) and in the Department of Botany (from 1964) at Ohio State University. Schopf continued his research after his official retirement in 1976 with the most able assistance of his wife, Esther.

## THE SCHOLAR—BREADTH OF INTERESTS AND VISION

The essentials of Schopf's influence and his personal traits are not easily captured by a straightforward presentation of his publications, his curriculum vitae, or even his lifelong activities as we know them (see Andrews, 1979). As a paleobotanical pioneer among pioneers and as a scholar among scholars, Schopf (Fig. 1) was somewhat "larger than life." His sense of scholarship marked his work from the beginning. He was a graduate student by age 19, and six years later he had explored concurrent studies on coal-plant structure, coal spores, and coal balls in Illinois Survey research with Cady, the outstanding American coal geologist.

In retrospect, these beginnings nurtured Schopf's breadth of vision and scholarly contributions that had such enormous impact on Paleozoic paleobotany and coal geology. He was the right man, at the right time, and at the right place. He recognized the necessity for orderly research procedures and reporting of observations and the need for precisely defined technical terms, and he especially appreciated rules of nomenclature that would permit utilization of data and ideas across divergently expanding subdisciplines. He had developed an agenda, and it was both visionary and generous—explore and help influence recognition of what needs to be done—and pass along all the encouragement and direct aid possible.

Figure 1. James M. Schopf in the field during the "The Age of the Dunkard" Symposium in West Virginia (1972). Photograph taken by W. C. Darrah and used courtesy of Elsie Darrah Morey, from the William Culp Darrah Collection.

Lyle Bamber, longtime biology librarian at the University of Illinois, a good friend of the Schopfs from Urbana days, recalled the uncommonly large number of books and journals from "his" library that collected on makeshift bookshelves in their Hill Street home. Indeed, whether at the Bureau of Mines, the Coal Geology Laboratory, or Schopf's den at home in Columbus, reference works, such as geographical dictionaries and atlases, were always a prominent part of the "orderly landscape." Schopf's treatment of the literature on research topics was a model of comprehensiveness, and yet he was always searching for further references.

His efforts with the Russian literature in paleobotany and coal geology were particularly noteworthy at a time when their availability was extremely limited and translations difficult to arrange. This project, particularly at the Coal Geology Laboratory, led to repeated exchanges of translations of major works and an expanded knowledge of the Russian contributions (Schopf, 1964b). In referring to a translation of Zalessky's *Sketch on the Question of Formation of Coal* (1914) in a letter to Tom Phillips (April 28, 1978), Schopf praised Zalessky's scope of the literature and added: "It's not nearly as rough as many that I had prepared by Russian speaking students in years past. Of course, when one gets down to the nitty-gritty of technical detail, I think one must laboriously identify the passage in the original and verify the statement."

Schopf had the exceptional ability to elevate the quality and significance of coal-related research by expectations of rigor that marked his own work. He spared no one, including himself, anytime for any careless or inaccurate action in execution of scientific pursuits. With his knowledge and reputation, he was a formidable questioner at many scientific meetings. His mind was quick and perceptive, and it raced ahead of his words. When Jim Schopf rose to his feet, and he often did, to comment upon a paper or a remark, there was an immediate aura of concentration, not only by him but by the entire audience. We became prepared for a penetrating question, a long comment, or even a mini-lecture. Such remarks were usually prefaced by a genuinely gracious compliment on the better aspects of the paper, if there were any, but none of us were spared the admonition we deserved. In fact, if Schopf made no comment at all, you really had something to be concerned about.

### *Paleobotanical interests*

Schopf's interests in paleobotany had strong roots in his botanical training and in his whole-plant biology perspectives from development to ecology. His gateway to becoming a consummate student of Paleozoic fossil plants in all modes of preservation began with concurrent studies of coal, spores, and coal balls from the upper Carboniferous. His penchant for techniques of preparation and study opened many highly altered fossil plant deposits to observations that some might have overlooked.

Schopf's acquired knowledge of coal-ball plants and their study was like that of other American pioneers of his day—A. C. Noé's students, W. C. Darrah, H. N. Andrews, A. T. Cross, and others (see Chapter 27, this volume). He was mostly self-taught. As a consequence, there was a diverse "American theme" as to the potentials of coal-ball studies. Most of these diverse approaches are evident in Schopf's own early studies, ranging from coal balls as an index to the constitution of coal (Schopf, 1939) to *Mazocarpon* (Schopf, 1941), and to *Dolerotheca* (Schopf, 1948c).

The *Mazocarpon* paper represents a benchmark of assembly, synthesis, and impact. This approach, in combining two major research projects into one with multiple insights, was a sporadic tendency of Jim Schopf to utilize an economy of publications. The stratigraphic chart of North American coal-ball occurrences (Schopf, 1941, p. 9) has been augmented by many of us, and the most comprehensive attempts still reflect the detailed input of Schopf.

About the *Dolerotheca* paper, Henry Andrews, a longtime colleague, wrote to A. T. Cross (February 27, 1978): "Perhaps the most important of all—[of his studies on Carboniferous coal-ball plants]—in his study of *Dolerotheca*—several paleobotanists have been concerned with *Dolerotheca,* but Schopf's account is still by far the most informative and his fine restoration drawings are a model for other workers."

Schopf studied early land plants (Fig. 2) and vascular plants from the Silurian to the Triassic. One of his sustained interests from the 1960s onward was the Precambrian fossil record, and he had an added incentive to keep abreast of this research field. He was delighted to be able to cite the work of his younger son, Bill, in the chapter on "Precambrian microfossils" (Schopf, 1969c).

The Schopf influence on paleobotany was perhaps as much by example as any other means, and this is interwoven with helping to get "things done"! Andrews expressed to A. T. Cross (February 27, 1978): "Quite aside from his own research papers, Jim Schopf has rendered many services to paleobotany. It was largely through his urging and aid that I became engaged in the preparation of the 'Index of Generic Names of Fossil Plants,' the second edition of which appeared in 1970. He deserves a considerable share of the credit for the Index which I believe has served many paleobotanists throughout the world" (see Andrews, 1970).

Wilson N. Stewart, who developed the paleobotanical research programs at the University of Illinois as a result of James M. Schopf's earlier influences, expressed Schopf's status to A. T. Cross (March 13, 1978): "I have known Dr. Schopf since his early days . . . 1941. . . . Since that time Dr. Schopf became the leader—the Dean, if you wish—of North American Paleobotany . . . one has to be impressed not only with the amount and quality of the work, but the scope of Dr. Schopf's expertise in paleobotany. He is 'at home' with studies in palynology or with megafossil plant remains. His work ranges through the geologic column with emphasis on the Paleozoic. Above all, his work is significant and a monument to his enthusiasm and industry. . . ."

Figure 2. James M. Schopf during a field trip into the Devonian of Maine (1966). Photograph provided by H. N. Andrews.

### *Interests in palynology*

Schopf's palynological contributions began at a time when the botanical and geological significance of pollen, spores, and other microscopic-size plant parts, or detritus, were scarcely recognized except by pioneers like Reinhardt Thiessen (1867–1938); see Chapter 11, this volume. At that time, such fossils in peat and coal were generally looked upon only as particulates in coal or mineral-rich sedimentary rocks (e.g., black shales), with little or no reference to animal or plant origin of such entities or degradation products. Schopf's early papers on spores from the Herrin (No. 6) coal of Illinois (Schopf, 1938) and "An annotated synopsis of Paleozoic fossil spores and definition of generic groups" (Schopf et al., 1944) systematized the major groupings of fossil spores.

Some of his inimitable broad reflections incited deeper conceptual thoughts on various subjects. These include phylogeny, botanical origin, significance in geological interpretation of environments of deposition, and general application to ecological considerations of plant-derived detritus in sediments. Some of these appear as papers in journals; a few actually were buried in some otherwise mundane descriptive reports. Tom Schopf remarked in a letter to A. T. Cross (August 8, 1978) that many of these became chapters in books: "Notable among these were Schopf's contributions [1957a, b] to the *Treatise on Marine Ecology and Paleoecology* and in his influential chapter [1964a] in the SEPM Special publication No. 11 on *Palynology in Oil Exploration,* 'Practical problems and principles in study of plant microfossils.' Subsequently, he composed an elegant chapter on 'Systematics and nomenclature in palynology' [1969b], a quarter century after the 'Annotated Synopsis . . .' had appeared."

Schopf's role in nomenclatural matters evolved in large part out of his palynological investigations, namely his role of service and leadership over many years on the editorial board of the International Commission of Botanical Nomenclature (ICBN). Prof. Dr. F. A. Stafleu, internationally recognized leader in plant taxonomy and longtime editor of Taxon wrote of Schopf (to A. T. Cross, March 8, 1978):

> From 1950 onward Dr. Schopf has played a very influential role in the legislation of nomenclature procedures in botany and he has contributed greatly towards clarifying palaeobotanical nomenclature in such a way as to make it compatible with that of actuobotany. Schopf's attitude has always been universal; averse from parochial attempts to fence off a small discipline. On the contrary, he has always been able to approach his own field as a biologist, integrating its findings with those of workers in other subdisciplines. His success in doing so undoubtedly stems from a firm background of excellency in his own subject.

In Schopf's hands, the seemingly dull and sometimes arbitrary-appearing nature of nomenclatural codes and rules became alive with questions or principles of how best to consider, organize, and express our knowledge of living and fossil plants. The ideas he imparted to ICBN have had a lasting effect.

Schopf expressed his philosophy on misconceptions on nomenclature (Schopf, 1978b):

> It would be best for each scientist to concentrate on his own particular problems of correctly expressing plant relationships, and not attempt to set a model that others are bound to follow. Let us all recall that every taxonomic study is a progress report. As new information comes to hand the evidence necessarily deserves a re-evaluation. By careful and correct adherence to the Rules of Nomenclature, we make our understanding of plant relationships more intelligible and easier for the next monographer to revise. In this way, I hope sincerely, that one may contribute to scientific progress.

### *Concerns in coal geology*

The sustained and interwoven scientific interest of Schopf was coal geology. In this field, he consistently argued for proper geological appreciation of the origin, forms, occurrences, conditions of preservation, and alteration products of fossil plants in coals of all ranks. His contributions to the fundamental concepts of the origin, constitution, alteration, and classification of coals were extensive. The classic "Variable coalification: . . ." (Schopf, 1948b) has been copied, reviewed, revised, and rediscovered many times in the ensuing years.

He exhibited great breadth of knowledge and considerable insight in studies of a wide variety of coals and coallike rocks from all over the world. These included cannel coals, bogheads, and torbanites from the United States, Scotland, France, and Australia; natural briquettes from Pennsylvania; bacteria in pyrite concretions from Ohio; structure and plant matter in meta-anthracites of the Narragansett basin, Rhode Island; relation of plant parts as petrographic macerals to coking characteristics of coals from Kentucky, Illinois, West Virginia, Nevada, Colorado, Oregon, Utah, and Washington; coal metamorphism associated with igneous deposits in Antarctica; and the petrography of Willett Range coal deposits, Victoria Land, Antarctica. Schopf analyzed plant sources, rank characteristics, and petrographic quality of various subbituminous coals, lignites, and even some peats (Schopf and Cross, 1947; Schopf, 1967a). He participated in a broad spectrum of studies on uraniferous lignites, oil shales, and black shales during the 1950s and contributed to the paleobiology, petrography, origins, classification, and geochemistry of such deposits.

Schopf's papers, "Was decay important in origin of coal?" (Schopf, 1952) and "A definition of coal" (Schopf, 1956b), were widely influential. His U.S. Geological Survey bulletins "Field description and sampling of coal beds" (Schopf, 1960) and "The Reinhardt Thiessen coal thin-section slide collection of the U.S. Geological Survey—Catalog and notes" (Schopf and Oftedahl, 1976) indicate his deep concern for precise definitions and high quality, diverse reference collections for comparative coal studies.

Dr. Jack Simon, chief of the Illinois State Geological Survey, wrote Robert Carroll (March 1, 1978):

> Dr. Schopf has made important contributions to coal classification and coal petrology. Important among these is the expertise he has brought to working committees on coal for the American Society for Testing and Materials. His work in relating paleobotany to coal formation has had an important impact on coal petrography. By his efforts and publications, the large volume of work of Reinhardt Thiessen of the U.S. Bureau of Mines on coal petrology was analyzed and given modern pertinence in this field that has had large growth throughout the world in the past 25 years. Due solely to his personal interest and recognition of the value of the Thiessen thin section collections, Dr. Schopf organized, catalogued, and was largely responsible for preservation of this valuable collection by his work.

## SCHOPF'S BASES OF OPERATIONS

### *Schopf at Illinois*

Jim Schopf's time at Illinois was enjoyed to the fullest. At the Illinois Geological Survey, his relationship with Gilbert H. Cady was special. Schopf's acceptance speech when he received the first Geological Society of America Coal Division Cady Award in 1973 (Schopf, 1974) expresses this relationship best: "During this period I came to have the greatest respect for Dr. Cady. This is not to say we did not have differences of opinion . . . at the same time I must add that it was a type of pleasure (I did not realize at the time) and a privilege, to have a chance for such frank and open discussion with Dr. Cady. . . . Dr. Cady's encouragement, and above all, his example of courage and integrity, have been a continuing source of strength and inspiration."

Schopf's years at Illinois were very productive, with 14 publications between 1936 and 1943. His description of the spores from the Herrin (No. 6) coal bed (1938) primarily treated megaspores and is a classic—serving as a model of detailed observation and description of spore morphology. His career at the Illinois Survey was greatly influenced by the "training ground" atmosphere and the synergistic interactions with many colleagues who collectively brought the Survey into national and international prominence.

Dr. Cady exerted strong leadership in the Coal Geology Section of the Illinois Survey and over the years selected some remarkably capable geologists, including Schopf. Jack Simon later went on to head this section after Cady's retirement and eventually became chief of the survey. Robert M. Kosanke joined this creative group in January 1943, and Schopf then moved on to the U.S. Bureau of Mines. This period cemented productive, close, professional and personal relationships among Bob Kosanke, Jim Schopf, and Jack Simon that served the science of coal geology actively for nearly four decades. One particularly noteworthy contribution over this period was their combined leadership and physical efforts to develop and sustain the programs, field trips, and symposia of the Coal Geology Division of the Geological Society of America (GSA). The Coal Geology Group (later Division) had its 1945–1955 origins from members of the Society of Economic Geologists (SEG) as a SEG Coal Geology Committee, finally recognized by GSA in 1955. Kosanke was elected the first chairman for the Group, and Schopf was elected vice chairman.

Kosanke, who replaced Schopf at the Illinois Survey, had the pleasure to guide through the Survey review system several notable publications (Schopf et al., 1944; Schopf, 1948c). "An annotated synopsis of Paleozoic fossil spores and definition of generic groups" came at a time when the field of palynology was still in its infancy and chaos ruled. It was in the same year that the term *palynology* was coined by Hyde and Williams (1944). There was considerable diversity in methods of classifying palynomorphs. This work set a standard of excellence and systematized the major groupings of Paleozoic spores for over two decades. Seven guiding principles for treating palynomorphs were discussed in detail. Adherence to the systematic principles embodied in the International Code (at that time it was "Rules") of Botanical Nomenclature was the basis of taxonomic treatment. This publication provided a much-needed degree of stability for taxonomic treatment of fossil palynomorphs.

### *Schopf at the Bureau of Mines*

Schopf was classified as a "paleobotanist" in the Miscellaneous Analysis Section at the U.S. Bureau of Mines Central Experiment Station in Pittsburgh, from 1943 to 1947. His special interests, like his predecessor Thiessen, were in the botanical aspects of coal as viewed by a coal microscopist. World War II placed demands on the Bureau for major efforts on coal and mineral production, processing, and conversion techniques. Therefore, Schopf's energy and scientific knowledge and that of his staff were applied mostly to practical aspects of coal petrology for utilization of coal as coking coal for the steel industry and conversion of coal by the Fischer-Tropsch technology for utilization of hydrocarbons as gas or oil.

In referring to his early Illinois work on studying sections of coal, Schopf wrote to R. M. Kosanke (December 10, 1973): "It was during this period I became frustrated trying to discover just *how* Thiessen performed his method of analysis. . . . This was a real stumbling block and it was one of my most specific personal objectives when I went to the Bureau in 1943. How *did* they regularly perform an anthraxylon determination! Actually, the critical details are finally given for the first time by Teichmüller [1941]. Of course, I was unable to get hold of this until after the war when it only confirmed what I had been a long time finding out." Reinhardt Thiessen had died in 1938 (see Chapter 11, this volume) prior to Schopf's arrival at the Bureau, but Hugh J. O'Donnell, an accomplished preparator who was Thiessen's assistant, served as an aide to Schopf.

Aureal T. Cross came to work with Jim Schopf in the fall of 1943, and this began a lifelong friendship between the two scientists. Their main thrusts were on exploration for coals that were suitable for coking in various less-developed or undeveloped coalfields in the United States. Petrology, palynology, and mineralogy of these coals was also a focus as well as finding sources for extraction of rare minerals (germanium, uranium, etc.). This research direction included studies of bituminous coals in southern West Virginia; eastern Kentucky; Sunnyside, Utah; Kemmerer, Wyoming; Esmeralda County, Nevada; and Paonia, Colorado. Subbituminous coal and lignite were studied in Coos Bay, Oregon; Toledo, Washington; and the Powder River basin.

While at Illinois, Schopf did extensive work in both the laboratory and the field, and there were numerous discussions with Cady, Kosanke, and many other coal geologists on various facets of coal science. Schopf's stay at the Bureau of Mines served as a means of testing and maturing such topics as variable coalification (Schopf, 1948b), a definition of coal (Schopf, 1956b), differential alteration of plant constituents, and others (Schopf, 1947, 1948a, 1949, 1956a).

### *Schopf with the U.S. Geological Survey*

Upon Schopf's return to the United States in 1947 from Pretoria, he was appointed as a geologist in the Paleontology and Stratigraphy Branch of the U.S. Geological Survey and was assigned to offices in the U.S. National Museum, Washington, D.C. He had inadequate laboratory facilities in Washington and wanted to be nearer the coalfields. He had maintained strong interests in coal-ball studies in relation to palynology and coal geology since his apprenticeship days with Cady. This new position offered opportunities to explore these and many other research avenues, after Schopf was placed essentially in an "academic environment."

John Melvin, director of the Ohio State Geological Survey, offered a strong appeal to the U.S. Geological Survey to set up facilities for Schopf at the Ohio Survey headquarters, then located on the "Oval" in Orton Hall on the Ohio State University campus in Columbus. As a result of Melvin's offer, Schopf was transferred in 1949 to the Fuels Branch of the Geological Survey with plans for the Coal Geology Laboratory underway. This turned out to be one of the most fortunate decisions by the U.S. Geological Survey and the Ohio Geological Survey.

The Coal Geology Laboratory became one of the most influential centers for Paleozoic paleobotanical and coal-related research and scientific exchange in the country. Schopf began this enterprise, as always, by going into the field and revisiting or exploring every reported coal-ball locality known to him. This was a historic journey in 1948, many weeks in duration (see Chapter 27, this volume), and it proved to be a crucial link in information-sharing and joint collection opportunities. In effect it was a "jump start" for upper Carboniferous coal-ball studies following World War II and immediately prior to the establishment of the National Science Foundation.

The move to Columbus proved to be very enjoyable and productive. Schopf published a number of papers on various aspects of coal. Schopf (1956a), in petrographic methods for application to coal, discussed various procedures used in analysis of coal and believed coal petrology should be applied more fully to problems of geology and technology. Schopf's (1956b), "A definition of coal," was a product of several years of evaluation: "Coal is a readily combustible rock containing more than 50 percent by weight and more than 70 percent by volume of carbonaceous material formed from compaction or induration of variously altered plant remains similar to those of peaty deposits. Differences in kinds of plant materials (type), in degree of metamorphism (rank), and range of impurity (grade) are characteristic of varieties of coal." Schopf (1966b) reevaluated this definition and indicated a change in the ASTM Standards. This was published after his death by deleting "varieties of" in the last line of the definition and inserting after coal" and are used in classification" (ASTM, 1971, p. 415).

Schopf et al. (1965) reported on the presence of at least two species of well-preserved iron bacteria in pyrite from a concretionary mass associated with a coal of Middle Pennsylvanian age of Ohio. It was a remarkable demonstration that bacteria, even though lacking hard parts, could be fossilized.

Schopf's (1975) "Modes of fossil preservation" is a thoughtful review of 94 publications treating various types of fossil preservation. He considered from this study that there were four basic modes of preservation: (1) cellular permineralization, (2) coalified compression, (3) authigenic preservation or molds and casts, and (4) duripartic (hard part) preservation.

### *Schopf and the Antarctic*

In 1960 Schopf became involved with the investigation of Antarctic paleobotany with the Institute of Polar Studies on the campus at Columbus. During the last two decades of his life, Jim Schopf was challenged by the Antarctic on four very rugged field seasons (1961–1962, 1965–1966, 1966–1967, 1969–1970). Tom Schopf wrote of his father to Cross (August 8, 1978): "Long a drifter, Jim spent several field seasons on the ice continent, in *his* 50's—the oldest person (up to that time) to have done so in that harsh and difficult environment. Returning one afternoon on a several mile hike, he went through the ice into a crevasse, but his pack caught on the edge, and only by swinging his legs back up over the edge was he able to pull himself out!" (Fig. 3).

Publications of his research emanating from this period exhibited greatly broadened vision and penetrating insight into Gondwanan coal accumulations (Schopf, 1966a, 1967b, 1970a, b, 1971). These included synthesis of aspects of plate

Figure 3. James M. Schopf in Antarctica at the Coalsack Bluff Camp, clutching a slab with *Glossopteris*. It was out of this camp that Jim Schopf collected the first Permian and Triassic samples of permineralized peat. Photograph taken in 1969 by James W. Collinson; this use is courtesy of the photographer. The picture was published originally in Andrews (1980, p. 242). Published with permission from H. N. Andrews and Cornell University Press.

tectonics, global paleoclimates, and the plant adaptations, in order to provide substance for speculation on the factors controlling accumulation of coals in the more-exotic paleogeographic patterns.

In a preliminary report on the coal and plant remains of the Horlick Mountains of Antarctica, Schopf (1962) reported the presence of palynomorphs, fusain fragments, preserved woody stems, leaves of *Glossopteris,* seeds of *Samaropsis,* and coal consisting of vitrinite, attrital matter, and resinous bodies. Schopf (1965) depicted the detailed anatomy of the genus *Vertebraria,* a fossil root occurring abundantly with the *Glossopteris* flora in the Ohio Range of Antarctica. The root structures of *Vertebraria* indicate it to be of gymnospermous affinity and attached to large shrubs or small trees. In several short papers he described *Glossopteris* floras. Glossopterid fertile structures were described in considerable detail (Schopf, 1976a).

Writing to Jack Simon (January 30, 1967) from the McMurdo Naval Base, Antarctica, Schopf shared his evaluation of some splendid "fossil finds" and volunteered the following: "But this, let me tell you, was not won without exertion. We had nearly a week living in tents in blizzard conditions about two miles from the outcrop before we got a day in which we could get on the rocks. I was very pessimistic before the 14 hours of good weather finally appeared. And then the weather closed in again while we were breaking camp and we had a slow two hours of toboggan haul back to the pickup strip under more blizzard. Sometimes I could see the flags and sometimes I just ran blind. . . . I am beginning to think this is enough of this sort of stuff for an old man. . . ."

Schopf's paper written with Dr. Rosemary Askin (1980), "Permian and Triassic floral biostratigraphic zones of Southern land masses," was published after Schopf's death in 1978. This paper utilized vertebrate zones, plant megafossils, and palynomorphs in zoning this part of the Gondwanan geologic column.

One other important posthumous paper (Schopf, 1982), reported on the forms and facies of *Vertebraria* in relation to Gondwanan coals. This paper is largely concerned with the forms of preservation and ecologic interpretations as illustrated by *Vertebraria.* These roots are commonly *in situ,* which supports the idea of the autochthonous origin of many Gondwana coal beds.

These fundamental studies of plant morphology (T. J. M. Schopf to A. T. Cross, August 8, 1978): "were balanced by work on the relation of floras as a whole to the question of the how and when of continental drift, including a paper [Schopf, [1969a] on the 'Ellsworth Mountains: position in West Antarctica due to sea-floor spreading'—an idea which occurred to him when flying back from a field season in one of those army transport planes".

Schopf's days at Ohio State University were similar to those at Illinois, very productive, with a warm relationship with his associates. In geology his associates were Walter Sweet, Jim Collinson, Stig Bergstrom, and others. In botany, T. N. Taylor had a close relationship with Schopf. As at Illinois, Schopf's collections at Ohio State were very large. Drs. Bergstrom and Taylor have made arrangements to curate these U.S. Geological Survey collections in the Orton Museum on the campus of Ohio State University. Tom and Edie Taylor have been actively curating the Antarctic collections and adding a considerable amount of new plant-fossil material from their many trips to old and new localities in Antarctica.

## COLLEAGUE, FRIEND, TEACHER, COLLECTOR, AND FAMILY MAN

Jim Schopf was a friend and teacher of students and colleagues alike and a dedicated family man. He promoted his science by sharing his time, his talents, his ideas, even his collections and books. When he was at work he was businesslike and serious minded. At the Bureau of Mines in Pittsburgh Schopf had a hard time arriving at 8:00 A.M., but he had probably worked long into the night on some project he had taken home from the laboratory and had likely forgotten about the time, as he sometimes did, according to his wife. He often came back to work in the evenings. Hugh O'Donnell described Schopf as "wrapped up in his work."

At lunch time (he generally carried his lunch in a heavily taped and retaped, book-mailing box), he was convivial and cajoling, sharing latest game scores, how the pistol team had done in their last match (he even learned to shoot at the Bureau's pistol range and joined their pistol team), how certain words were pronounced, or what was meant by some regional Pittsburghese words or phrases. To demonstrate their lack of correctness Jim Schopf and Aureal Cross bought an old 20-volume set of Encyclopaedia Britannica. It was greeted by fellow workers with derision! Any time they looked up something in it there were defensive cries and ridicule that the encyclopedia was "out-of-date"!

When work was over for the day or at the end of a project, Schopf relaxed. He enjoyed sports, especially football; he was a gracious host; he was eager to go out with the "gang" for a quick beer with generally a mix of secular and professional talk of the events of the day or season. Like his science, everything interested him: a new food, a new book, an impending sports event, a new fossil-plant locality, a burlesque show, a motorcycle, or a newcomer to the group. At times, he was the life of the party.

Schopf once brought back a realistic plaster model of a coiled rattlesnake from a western field trip. Resting on some straw at the bottom of a deep fishing bucket supplied by Hugh O'Donnell, it took on a lifelike quality. Dozens of people in the laboratory, including engineers, secretaries, and custodians, came in to see it. Schopf sat passively nearby, pretending to be deep in work or thought, but he was chuckling inside. Some said they could see it breathe, or move, as they gingerly moved the brick that held down a plate-glass cover, but Jim assured them, "No, it was in hibernation." Their chagrin at learning the truth several days later made them so wary that when a real, live, crated leopard, being shipped to another place, was left in the shipping room by mistake a few weeks later, nobody could be "fooled" into going down to see it—except "Jim" and "Hughie"!

Much of Schopf's teaching was done informally in seminars, one-to-one in laboratory or work setting, or in the field. Esther Schopf wrote to A. T. Cross (January 10, 1990): "The seminars he ran were *not* on a *regular* schedule. It was only if enough students showed an interest. He enjoyed teaching as it *made* him keep up with the literature not *directly* related to his work."

Schopf was noted for his innumerable letters, some of them many pages long, often in longhand. He suggested new ideas, new techniques, or alternative methods. Or he promoted a new program or symposium, or called attention to a problem or to errors or shortcomings in a paper the recipient had written or presented. Sometimes such lessons were "taught" orally over coffee or while collecting or resting at an outcrop. Such a letter or "lecture" was basically "encouragement" and even "teaching" in its thrust, particularly if it was a student or younger scientist. However, if it was addressed to one of his peers or friends, his words could be laudatory (rarely), encouraging, critical but diplomatic, or derisive to scathing in tone (see Schopf, 1978a).

Andrews (1980, p. 241) wrote of Jim Schopf as follows: "He was always generous in sharing what he had and always anxious to place fossil plant collections in the hands of those who he thought could make the most of them. He was known for two other characteristics as well; he was a keen and tireless explorer in the field and could find fossil plants when all others had given up; and he could extract significant information from even the seemingly poorest specimen that most others would ignore."

Andrews (1980) continued the discussion of Schopf: "He discovered an Upper Devonian horizon in West Virginia that contained superbly preserved *Archaeopteris* and *Rhacophyton* material and turned this over to Tom Phillips and myself. Very few men would have 'let go' their claim to such a find, but he had other work to do at the time and was anxious to see the deposit studied."

Jim Schopf's marriage to Esther in 1934 was either a demonstration of great wisdom or an unbelievable stroke of good fortune, perhaps both. Esther Schopf, in her own right, is an exceptional professional scientist, but she carried out her role as mother to the highest level of love, encouragement, and discipline as well as acting as moderating balance wheel in her husband's life and, at times, as his assistant. They had two sons—Thomas J. M. (born 1939) and J. William (born 1941)—who grew up in Pittsburgh and Columbus (Fig. 4). Their exposure to science began as young children in Pittsburgh, where their mother often took them to the Carnegie Museum and other educational facilities. They excelled in public schools throughout their primary and secondary school years. Both boys were very active and talented in different sports and had strong support from home.

Figure 4. James M. Schopf with sons, Thomas J. M. (center) and J. William (right), at the Geological Society of America meeting in Miami (1964). This was the second meeting with all three Schopfs in attendance. Photograph by A. T. Cross.

Schopf, in writing R. M. Kosanke (April 26, 1951) conveyed: "Are you planning to go on the Indiana trip this year? Esther and *the kids* and me are all intending to make it. This may come under the heading of noble experiments not to be repeated, but we will try it at least once. If you are there, you can look out for us!"

Both Tom and Bill Schopf received their college Baccalaureate degrees with honors at Oberlin College. Both became internationally known paleobiologists and geologists, and each was distinguished early in his career, receiving the Paleontological Society's Schuchert Medal for outstanding contributions by a paleontologist under 40 years of age.

Thomas J. M. Schopf (1939–1984) became professor of geology at the University of Chicago. He was leading a group of paleobiology students on a field trip along the Texas coast in March 1984, at the time of his death. He was the author of more than 100 papers; a world authority on multielement taxonomy of conodonts and genetic variation in Bryozoa; the author of a book on paleooceanography and of a book-length manuscript on the history of science; a founder of the journal *Paleobiology;* and the organizer of the memorable symposium in 1971 that led to the classic, *Models in Paleobiology,* which he edited.

The younger son, J. William Schopf, is a professor of paleobiology at the University of California at Los Angeles (UCLA) and an internationally noted Precambrian scholar. He is also the director of the Center for the Study of Evolution and the Origin of Life (CSEOL), a complex institute of nearly 200 member scholars from UCLA and 65 other institutions.

Esther Schopf still resides at the family home in Columbus, Ohio, which the Schopfs bought in 1949. Their address at 140 Oakland Park Avenue is familiar to many who worked and visited with Jim Schopf from time to time at Ohio State University.

## HONORS

Jim Schopf received many awards and honors over his long and exceptional research career. These include a Commendation by the U.S. Department of Interior (1947) for civilian service during World War II, an Antarctic Service Award (1970), The Fellowship Scroll and Medal of the Paleobotanical Society of India (1977), and the Distinguished Paleobotanist Award of the Paleobotanical Section of the Botanical Society of America (1977).

The "highest" honor Schopf received, in the *literal sense,* was having the highest mountain of the Ohio Range in central Antarctica named after him. Mt. Schopf, elevation 2,990 m, was first proposed in his honor by Dr. William E. Long and other Antarctic colleagues. Schopf's four trips there during the Antarctic field seasons and the immense amount of fundamental knowledge that came out of his Antarctic collections and field studies justified this recognition.

The citation to James M. Schopf with the Merit Award of the Botanical Society of America in 1969 read as follows: ". . . in recognition of distinguished achievement in and contributions to the advancement of botanical science. His studies of fossil plants, and especially his work on the petrology of coal, are among the definitive works in the field."

Jim Schopf was also the first recipient of the Gilbert H. Cady Award of the Coal Geology Division, which was presented at the Geological Society of America Meeting, November 13, 1973, ". . . in recognition of his outstanding contributions in the field of coal geology". No more appropriate honor for Jim Schopf could have been found.

It was with "unschopfian pomp and ceremony" that, in 1976, he became the Mary Clark Thompson Medalist of the National Academy of Sciences. That award in stratigraphic geology, given only every third year, is imprinted with the citation: "For leadership in Paleozoic Paleophytology, pioneering in Carboniferous micropaleontology, and origin and constitution of coal, leadership in Antarctican Paleophytology, and as a man of great versatility and synthesizing ability."

The Paleontological Society Medal was privately announced to Schopf in May 1978, when he was ill with leukemia. He was clearly moved and excited about this, and it boosted his spirits. The award was presented posthumously to Esther Schopf and their son Bill at the Paleontological Society Meeting in Toronto six weeks after Jim Schopf's death. Cross (1979) noted at the presentation: "It is the preeminence which he . . . attained as a result of scientific endeavor and the deep and abiding impact on both his peers and his masters which have caused us to seek out . . . Schopf for an award which [was] neither expected nor even hoped for, which could not make him rich or vain, but which is the highest singular honor this great Society can offer . . . as a definitive mark of its esteem and appreciation. . . ."

Jim Schopf was a member of 26 professional societies, which attests to his great range of interests and scientific prowess. He served as president of the International Organization of Paleobotany from 1975 to the time of his death.

A symposium that Jim Schopf would have been delighted to be a part of was "Paleoclimatic Controls on Coal Resources of the Pennsylvanian System of North America" (Phillips and Cecil, 1985), which was organized for the 1982 Annual Meeting of the Geological Society of America Coal Geology Division at Indianapolis. The symposium was dedicated to Schopf (Cross, 1985). It was a fitting remembrance of his impact on all aspects of coal geology as well as indicative of his strong interests in paleoclimate as evidenced by his own writings (Schopf, 1973, 1976b).

President Harold L. Enarson of Ohio State University, in a letter to Esther Schopf (October 6, 1978), expressed some of the thoughts of Schopf's colleagues: "Dr. Schopf's fundamental contributions to our knowledge of fossil plants, his rigorous personal standards of scholarship, his inspired leadership in the field of coal geology, and his influence on students at The Ohio State University and elsewhere in the world have been of the very highest order. He will be remembered as one of the most notable figures in American and world paleobotany of the present century."

Henry Andrews (1980, p. 240) perhaps stated it best: "Jim was a friend and servant to all of us. . . ."

Bill Schopf (J. W. Schopf, 1979) wrote of his father: "Depending on your nomenclatural preference, you might have regarded him as a paleontologist, a paleobotanist, a palynologist, a coal geologist or even as someone once remarked to me at a G.S.A. banquet, 'Oh yes, I know your dad—He's the coal ball man,' in fact . . . Jim . . . defied easy taxonomic description, for he was at the same time both decidedly more geological than his paleobotanical brethren, more whole-plant oriented than practitioners of palynology, and more botanical than his colleagues in coal geology. . . ."

James Morton Schopf died September 15, 1978. His close friend and palynological colleague, Alfred Traverse of Pennsylvania State University, as Episcopal clergyman, officiated at the memorial service in Columbus, Ohio.

## ACKNOWLEDGMENTS

We gratefully acknowledge the assistance of Henry N. Andrews, James W. Collinson, Paul C. Lyons, Elsie Darrah Morey, Hugh J. O'Donnell, and Esther J. Schopf in providing information and photographs. We thank H. N. Andrews, W. G. Chaloner, P. C. Lyons, E. D. Morey, and E. I. Robbins for reviewing the manuscript. We thank all who provided correspondence, copies of which are either in the possession of A. T. Cross or in the University of Illinois Library Archives.

## REFERENCES CITED


American Society for Testing and Materials (ASTM), 1971, Standard definitions related to coal (D2796): Annual Book of ASTM standards, Part 19, p. 415–417.

Andrews, H. N., Jr., 1970, Index of generic names of fossil plants, 1820–1965: U.S. Geological Survey Bulletin 1300, 354 p.

Andrews, H. N., Jr., 1979, Dr. J. M. Schopf, 1911–1978 (IOP President 1975–1978): International Organization of Palaeobotany Newsletter 8, p. 7–10.

Andrews, H. N., Jr., 1980, The fossil hunters, in search of ancient plants: Ithaca, New York, Cornell University Press, 421 p.

Cross, A. T., 1979, Presentation of the Paleontological Society Medal to James Morton Schopf: Journal of Paleontology, v. 53, p. 767–769.

Cross, A. T., 1985, Dedication of symposium to Dr. James Morton Schopf 1911–1978, *in* Phillips, T. L., and Cecil, C. B., eds., Paleoclimatic controls on coal resources of the Pennsylvanian System of North America: International Journal of Coal Geology, v. 5, Special Issue, p. vi–x.

Hyde, H. A., and Williams, D. W., 1944, Right word: Pollen Analysis Circular, v. 8, p. 6.

Kosanke, R. M., 1974, Presentation of the Gilbert H. Cady Award to James M. Schopf—Citation: Geological Society of America Bulletin, v. 85, p. 1349.

Kosanke, R. M., 1979, Memorial to James Morton Schopf 1911–1978: Geological Society of America Memorials, v. x.

Phillips, T. L., and Cecil, C. B., eds., 1985, Paleoclimatic controls on coal resources of the Pennsylvanian System of North America: International Journal of Coal Geology, v. 5, p. i–x, 1–230.

Phillips, T. L., Pfefferkorn, H. W., and Peppers, R. A., 1973, Development of paleobotany in the Illinois Basin: Illinois State Geological Survey Circular 480, 86 p.

Schopf, J. M., 1938, Spores from the Herrin (No. 6) coal bed in Illinois: Illinois Geological Survey Report of Investigations 50, 73 p.

Schopf, J. M., 1939, Coal balls as an index to the constitution of coal: Illinois Academy of Science Transactions, v. 31, p. 187–189.

Schopf, J. M., 1941, Contributions to Pennsylvanian paleobotany: *Mazocarpon oedipternum* sp. nov. and Sigillarian relationships: Illinois State Geological Survey Report of Investigations 73, 40 p.

Schopf, J. M., 1943, The embryology of *Larix:* Urbana, University of Illinois Press, Illinois Biological Monographs, v. 19, 97 p., 5 pl.

Schopf, J. M., 1947, Botanical aspects of coal petrology: Coal from the Coos Bay field in southwestern Oregon: American Journal of Botany, v. 34, p. 335–345.

Schopf, J. M., 1948a, Petrology and type characteristics of coals from the Coos Bay field, p. 10–15, *in* Toenges, A. L., Dowd, J. J., Turnbull, L. A., Schopf, J. M., Cooper, H. M., Abernathy, R. F., Yancy, H. F., and Greer, M. R., Minable resources, petrography, chemical characteristics, and washability tests of coal occurring in the Coos Bay coal field, Coos County, Oregon: U.S. Bureau of Mines Technical Paper 707, 56 p.

Schopf, J. M., 1948b, Variable coalification: The processes involved in coal formation: Economic Geology, v. 43, p. 207–225.

Schopf, J. M., 1948c, Pteridosperm male fructifications: American species of *Dolerotheca,* with notes regarding certain allied forms: Journal of Paleontology, v. 22, p. 681–724.

Schopf, J. M., 1949, Research in coal paleobotany since 1943: Economic Geology, v. 44, p. 492–513.

Schopf, J. M., 1952, Was decay important in origin of coal?: Journal of Sedimentary Petrology, v. 22, p. 61–69.

Schopf, J. M., 1956a, Petrologic methods for application to solid fuels of the future: Mining Engineering, v. 8, p. 629–639.

Schopf, J. M., 1956b, A definition of coal: Economic Geology, v. 51, p. 521–527.

Schopf, J. M., 1957a, Spores and related plant microfossils—Paleozoic, *in* Ladd, H. S., ed., Treatise on marine ecology and paleoecology, Volume 2, Paleoecology: Geological Society of America Memoir 67, v. 2, p. 703–707.

Schopf, J. M., 1957b, "Spores" and problematic plants commonly regarded as marine, *in* Ladd, H. S., ed., Treatise on marine ecology and paleoecology, Volume 2, Paleoecology: Geological Society of America Memoir 67, v. 2, p. 709–717.

Schopf, J. M., 1960, Field description and sampling of coal beds: U.S. Geological Survey Bulletin 1111-B, 70 p., pls. 6–27.

Schopf, J. M., 1962, A preliminary report on the plant remains and coal of the sedimentary section in the Central Range of the Horlick Mountains, Antarctica: Ohio State University Research Foundation, RF 1132, Institute of Polar Studies Report 2, 61 p.

Schopf, J. M., 1964a, Practical problems and principles in study of plant microfossils, *in* Cross, A. T., ed., Palynology in oil exploration—A symposium: Society of Economic Paleontologists and Mineralogists Special Publication 11, p. 29–57.

Schopf, J. M., 1964b, Russian palynology today: Literature and application to exploration, *in* Cross, A. T., ed., Palynology in oil exploration—A symposium: Society of Economic Paleontologists and Mineralogists Special Publication 11, p. 181–200.

Schopf, J. M., 1965, Anatomy of the axis in *Vertebraria, in* Hadley, J. B., ed., Geology and paleontology of the Antarctic: American Geophysical Union, National Academy of Sciences–National Research Council, v. 6, Antarctic Research Series 1299, p. 217–228.

Schopf, J. M., 1966a, Antarctic paleobotany and palynology: Antarctic Journal United States, v. 1, p. 135.

Schopf, J. M., 1966b, Definitions of peat and coal and of graphite that terminates the coal series (graphocite): Journal of Geology, v. 74, p. 584–592.

Schopf, J. M., 1967a, Petrologic examination of East Pakistan peat: U.S. Geological Survey Technical Letter, Pakistan Investigations, PK-29, 14 p. + 7 tables.

Schopf, J. M., 1967b, Antarctic fossil plant collecting during the 1966–1967 season: Antarctic Journal United States, v. 2, p. 114–116.

Schopf, J. M., 1969a, Ellsworth Mountains: Position in West Antarctica due to seafloor spreading: Science, v. 164, p. 63–66.

Schopf, J. M., 1969b, Systematics and nomenclature in palynology, *in* Tschudy, R. H., and Scott, R. A., eds., Aspects of palynology: An introduction to plant microfossils in time: New York, Wiley-Interscience, p. 49–77.

Schopf, J. M., 1969c, Precambrian microfossils, *in* Tschudy, R. H., and Scott, R. A., eds., Aspects of palynology: An introduction to plant microfossils in time: New York, Wiley-Interscience, p. 145–161.

Schopf, J. M., 1970a, Petrified peat from a Permian coal bed in Antarctica: Science, v. 169, p. 274–277.

Schopf, J. M., 1970b, Relation of the Floras of the Southern Hemisphere to continental drift: Taxon, v. 19, p. 657–674.

Schopf, J. M., 1971, Notes on plant tissue preservation and mineralization in a Permian deposit of peat from Antarctica: American Journal of Science, v. 271, p. 522–543.

Schopf, J. M., 1973, Coal, climate, and global tectonics, *in* Tarling, D. H., and Runcorn, S. K., eds., Implications of continental drift, to the Earth Sciences: Corpus Christi, Texas, Academic Press, v. 1, p. 609–622.

Schopf, J. M., 1974, Response to presentation of the Gilbert H. Cady Award to James M. Schopf: Geological Society of America Bulletin, v. 85, p. 1350.

Schopf, J. M., 1975, Modes of fossil preservation: Review of Palaeobotany and Palynology, v. 20, p. 27–53.

Schopf, J. M., 1976a, Morphologic interpretations of fertile structures in glossopterid gymnosperms: Review of Palaeobotany and Palynology, v. 21, p. 25–64.

Schopf, J. M., 1976b, Pennsylvanian climate in the United States, *in* McKee, E. D., and Crosby, E. J., coordinators, Paleotectonic investigations of the Pennsylvanian System. Part II: Interpretive summary: U.S. Geological

Survey Professional Paper 853, p. 23–31.

Schopf, J. M., 1978a, *Foerstia* and recent interpretations of early, vascular land plants: Lethaia, v. 11, p. 139–143.

Schopf, J. M., 1978b, Misconceptions on nomenclature: International Organization of Palaeobotany Newsletter 7, p. 4–5.

Schopf, J. M., 1982 (posthumous), Forms and facies of *Vertebraria* in relation to Gondwana coal: Geology Central Transantarctic Mountains: Antarctic Research Series, v. 36, paper 3, p. 37–62.

Schopf, J. M., and Askin, R. A., 1980, Permian and Triassic floral biostratigraphic zones of southern land masses, *in* Dilcher, D. L., and Taylor, T. N., eds., Biostratigraphy of fossil plants: Successional and paleoecological analyses: Stroudsburg, Pennsylvania, Dowden, Hutchinson & Ross, p. 119–152.

Schopf, J. M., and Cross, A. T., 1947, A glacial-age peat deposit near Pittsburgh: American Journal of Science, v. 245, p. 421–433.

Schopf, J. M., and Oftedahl, P. G., 1976, The Reinhardt Thiessen coal thin-section slide collection of the U.S. Geological Survey—Catalog and notes: U.S. Geological Survey Bulletin 1432, 58 p.

Schopf, J. M., Wilson, L. R., and Bentall, R., 1944, An annotated synopsis of Paleozoic fossil spores and the definition of generic groups: Illinois State Geological Survey Report of Investigations 91, 66 p., 3 pls.

Schopf, J. M., Ehlers, E. G., Stiles, D. V., and Birle, J. D., 1965, Fossil iron bacteria preserved in pyrite: American Philosophical Society Proceedings, v. 109, p. 288–308.

Schopf, J. W., 1979, Response for award of the Paleontological Society Medal to James M. Schopf, October 24, 1978: Journal of Paleontology, v. 53, p. 770.

Teichmüller, M., 1941, Der Feinbau amerikanischer Kohlen im Anschliff und Dünnschliff—Ein Vergleich kohlenpetrographischer Untersuchungs methoden: Jahrbuch Reichsstelle Bodenforschung für 1940, v. 61, p. 20–55.

Watt, A. D., 1979, Bibliography of American paleobotany for 1978: Missoula, University of Montana, 97 p.

Zalessky, M. D., 1914, Sketch on the question of formation of coal: St. Petersburg, Izdanie Geologicheskago Komiteta, 94 p., 12 pls.

Manuscript Accepted by the Society July 6, 1994

Printed in U.S.A.

Geological Society of America
Memoir 185
1995

# *Chester A. Arnold (1901–1977): Portrait of an American paleobotanist*

**Richard A. Scott**
*Morrison Natural History Museum, 501 S. Highway 8, Morrison, Colorado 80465*

## ABSTRACT

**In characterizing Chester A. Arnold, the single word that best describes both the man and the scientist is *substantial.* He was substantial in appearance—a tall, large man whose presence was always felt despite his reticence. His physical size was exceeded by the stature of his contributions to paleobotany; his interests ranged from the Paleozoic to the Tertiary. His impacts upon his colleagues and students were substantive—his terse, measured comments always impaled the moment. Arnold played a critical role in the growth of paleobotany into a major discipline in the United States by way of his introductory textbook on paleobotany (1947) and his numerous fundamental paleobotanical contributions, mainly on Devonian, Mississippian, and Pennsylvanian floras. Arnold was honored by his receipt of the Silver Medal of the Birbal Sahni Institute of Paleobotany (India) and of the Distinguished Service Award of the Paleobotanical Section of the Botanical Society of America and by the naming of several paleobotanical taxa for him.**

## INTRODUCTION

As one of Professor Arnold's students, I recall often being suspended between fear and respect: fear of his penetrating comments on my work and respect for the integrity and insight that produced them. Even now, it is not easy to write about him. Although my student-days apprehension has long since given way to appreciation for a remarkable man and scientist, I think of him as "Chester" for the first time in this portrait.

Information on Arnold's life and work has been presented in Andrews's (1980) book, *The Fossil Hunters.* I shall draw upon that source and supplement it with information from one of Arnold's students (Alan Graham), two of Arnold's postdoctoral fellows (W. G. Chaloner and T. Delevoryas), and two of Arnold's colleagues (C. B. Beck and E. L. Zodrow). From these sources, together with my own remembrances and anecdotes, now faded by the 40-odd years since I was his student, I hope to convey something of the nature and texture of the man. Arnold's bibliography was compiled by Watt (1978) and is included here (in slightly edited form) for convenience (see Chester A. Arnold's bibliography, this chapter).

## EARLY YEARS

Chester Arnold lived from June 25, 1901, until November 19, 1977. His parents were farmers, and his early childhood was on a farm in Leeton, Missouri. In the early 1950s, I recall a drive from Denver to Colorado Springs in which he mentioned traveling that route as a child in a horse-drawn wagon.

That experience suggests that the family may have lived for a time in Colorado. In any case, later the family moved to Ludlowville, New York, not far from Ithaca, where Arnold entered Cornell University to study agriculture. The frugal, taciturn, shy-but-direct nature of the man already was shaped by the austere nature of his farm life during his youth.

At Cornell, Arnold met Professor Loren Petry, who was a student of the Devonian plants of the region. Under this influence, Arnold's interests changed from growing living plants to understanding fossil ones, and in his typically unswerving way, he never looked back. He obtained both his B.S. (1924) and Ph.D. degrees (1929) from Cornell University. Arnold's Ph.D. thesis was on Devonian megafloral paleobotany.

Professor Harley H. Bartlett, then chairman of the Botany

Scott, R. A., 1995, Chester A. Arnold (1901–1977): Portrait of an American paleobotanist, *in* Lyons, P. C., Morey, E. D., and Wagner, R. H., eds., Historical Perspective of Early Twentieth Century Carboniferous Paleobotany in North America (W. C. Darrah volume): Boulder, Colorado, Geological Society of America Memoir 185.

Department at the University of Michigan, brought Arnold to Michigan in 1928 as an instructor of botany. Arnold's entire academic career was spent there. He rose through the academic ranks, becoming professor of botany and geology in 1947 (see Figs. 1 and 2) and serving in both departments and in the University of Michigan Museum of Paleontology until his retirement in 1970.

After retirement, Arnold was appointed professor emeritus of botany and geology and curator emeritus of paleobotany at the University of Michigan Museum of Paleontology, where he continued research until his death in 1977. His wife, the former Jean Elizabeth Davidson, preceded him in death. No one privileged to know the Arnolds and their three children (Eric Bruce, David G., and Patricia Ann) at home, as I was, will forget the warmth of their household.

Very early in Arnold's career he became acquainted with the budding paleobotanist, William C. Darrah, when they both attended the paleobotanical and geological congresses in The Netherlands (Amsterdam and Heerlen, respectively) in 1935. Arnold—beginning with the Fifth International Botanical Congress (1930) in Cambridge, England (see Arnold, 1930a)—regularly attended these international botanical congresses (see Figs. 3 and 4). One of the last that he attended was in Edinburgh, Scotland in 1964. Through the early years of both their careers in 1936 and 1937, they had frequent correspondence about paleobotanical and personal matters. Arnold consulted with Darrah on his discovery of walchian conifers in Colorado (Arnold, 1941c), the Colorado specimens being rather scrappy specimens of *Walchia* and probably *Ernestiodendron* (see Lyons and Darrah, 1989).

Figure 2. Left, R. M. Kosanke; middle, S. H. Mamay; right, C. A. Arnold at Grand Ledge, Michigan, September 1955, AIBS meeting. Photograph courtesy of C. B. Beck, University of Michigan.

Figure 1. Chester A. Arnold, circa 1955. Reproduced from *The Fossil Hunters* (1980) by H. N. Andrews with the permission of Cornell University Press, H. N. Andrews, and C. B. Beck.

In regard to Arnold's family, letters in the William C. Darrah paleobotanical files reveal Arnold's sense of humor about his two sons. Excerpts from these letters show a side of Arnold not known to most paleobotanists. In a letter to Darrah dated May 6, 1937, Arnold concluded, "Our boy is coming up to the Arnold tradition for size. He has recently added 'da-da' to his particular dialect, but I noticed he uses the designation for his celluloid dogs as well as for me." In another letter to Darrah dated September 28, 1937, Arnold noted, "We had a visit from the stork on September 19. Everyone says he looks like me but I don't know whether to be flattered or not as the resemblance seems anthropoid but little else. I can detect a trace of his dad's cantankerous makeup, however."

Alan Graham (December 1993) wrote to Paul Lyons of Arnold's sense of humor:

> Arnold's humor was subtle and often came to the surface, in lieu of anger, in annoying situations. An amateur collector wrote several times to Arnold about a specimen of *Tempskya* he would consider "donating" to the Museum in exchange for a sizable amount of cash, or for a field vehicle. Arnold responded each time that the specimen and its preservation were not exceptional, and the Museum must respectfully decline the offer. The correspondence continued and eventually the owner appeared at the Museum with the threat that if his offer was not accepted he would cut the specimen into pieces "about this thick" (indicating about 5 cm). "What do you think of that?" he demanded. "That would be a little thin for bookends and too thick for toilet paper," Arnold replied in exasperation.

While he was working on the second edition of his textbook, Oxford University Press, which publishes the King James version of

Figure 3. C. A. Arnold in the field in northern France (1954), part of the International Botanical Congress field trip. Front: left to right, Dr. Mahabale (India), C. A. Arnold (USA), Hans Bode (Germany); back, Paul Corsin (France) hammering at an outcrop. Photograph by and courtesy of A. T. Cross, Michigan State University.

Figure 4. Chester Arnold and William Chaloner, 1964, 10th International Botanical Congress, Edinburgh, Scotland. Photograph by and courtesy of Alan Graham, Kent State University.

the Bible, frequently pressured him for a completion date. When the annoyance exceeded his patience, he finally responded, "You waited 2,000 years for the Bible, and you can surely wait a few more months for my book."

Professor Arnold spent the 1958–1959 academic year as visiting scientist at the Birbal Sahni Institute of Palaeobotany, Lucknow, India. He was recipient of the Institute's Silver Medal in 1972.

He served as president of the International Organization of Paleobotany in 1974. Among his honors was the Distinguished Service Award of the Paleobotanical Section of the Botanical Society of America.

Arnold regularly attended the meetings of the Paleobotanical Section, where, in postpresentation discussions of papers, the usual friendly differences of opinion between the taciturn Arnold and the tenacious Jim Schopf (see Chapter 17, this volume) were anticipated with delight by the audience. I recall one such encounter, its point long since forgotten, in which Schopf literally badgered Arnold to admit that there was at least a slight possibility that his (Schopf's) interpretation could be true. Finally, Arnold, tersely and with just a touch of annoyance in his voice, gave in. He silenced Schopf (no mean feat!) with "Yes, Jim, about as much chance as finding a *Dolerotheca* on an oak tree." An in-house joke, but one that brought the house down.

Because the emphasis of this memorial volume is on the Carboniferous, and my background is chiefly in the Tertiary, I can do no better than to quote Henry Andrews's (1980, p. 246) review of some of Arnold's Paleozoic work:

> Chester Arnold's 1930 doctoral thesis dealt with the *Callixylon* woods from the Upper Devonian of central and western New York. It includes descriptions of several new species as well as a historical summary of Devonian plant studies in that general area. As noted elsewhere here, *Callixylon* is now known to be the stem that bore the foliage of *Archaeopteris* and the combination constitutes a plant that presently stands as the best-known member of the progymnosperm group [see Namdoodiri and Beck, 1968] which partially bridges the gap between pteridophytes and seed plants. He added another important link to the progymnosperm story in 1939 [Arnold, 1939c] by demonstrating that a species of *Archaeopteris (A. latifolia),* was almost certainly heterosporous. Since then, several other species have been shown to be heterosporous. In a short paper of 1935 [Arnold, 1935b], he described some "seed-like" compression fossils from the Upper Devonian of northern Pennsylvania. These were later investigated more fully by John Pettitt and Charles Beck [1968] who gave them the name *Archaeosperma arnoldii* [in Arnold's honor] and demonstrated that they consist of a pair of seeds enclosed by a primitive cupule. This is the earliest evidence that we have of seeds in the fossil record.

C. B. Beck (Fig. 5), Arnold's colleague at the University of Michigan, wrote the following concerning Arnold's scientific work:

> The scope and depth of his knowledge were formidable. He had the reputation of being cautious and accurate in his own work and in his references to, and use of the work of, others. Some thought of him as conservative because he was less prone to speculate than they. But he was hardly conservative. His significant paper on classification of gymnosperms [Arnold, 1948a] which had great impact on his field and on systematic botany in general, could not have been written by a conservative. In this paper of 1948, he presented a well-documented rationale for the existence of two independent line of gymnosperms, a viewpoint far ahead of its time and not generally accepted for another decade. (C. B. Beck, written communication, December 1993)

After joining the faculty at the University of Michigan, Chester published papers on fossil plant material from localities ranging from Oklahoma to Greenland; however, his work focused mainly on Upper Devonian and Pennsylvanian plants. In the early 1930s, he began to investigate the local Pennsyl-

Figure 5. Left to right: Professor Charles B. Beck (University of Michigan); Professor John Beaman (Michigan State University); (squatting) Dr. James H. Anderson (Michigan State University) University of Alaska; Mrs. William A. Benninghoff; Professor Aureal T. Cross (Michigan State University); Professor William A. Benninghoff (University of Michigan); Professor Chester A. Arnold (University of Michigan); Professor Ralph W. Chaney (University of California, Berkeley); Dr. Khosron Ebtehadj (Michigan State University), Teheran, Iran; and Dr. Lee R. Parker (Michigan State University), (California Polytechnic State University). Photograph taken May 1969 at Beal Botanical Gardens, Michigan State University, East Lansing by Dr. Ralph Taggart and courtesy of Aureal T. Cross (Michigan State University).

vanian plants from the Michigan Coal Basin. Previously, David White (in Lane, 1902; and in Cooper, 1906) had identified 43 species from the Michigan Basin, the first systematic studies of the Michigan coal flora. Arnold published a preliminary study of the Michigan flora in 1934 (Arnold, 1934b) and a definitive work in 1949 (Arnold, 1949a). In the latter paper he established stratigraphic relationships of the three Michigan Coal Basin floras with Carboniferous floras of the eastern United States, the Maritime Provinces of eastern Canada, and western Europe (see Chapter 8, 25, this volume).

Arnold once told me of a collecting trip during the 1940s to one of the localities of the Michigan Coal Basin in the company of a visitor, Birbal Sahni, the distinguished Indian paleobotanist. Upon reaching the outcrop, Arnold immediately dug in, while Sahni stood back. When Arnold asked why, Sahni explained that in India the scientists always had laborers to do the actual digging. Despite the culture shock, the two men became friends, and their friendship later led to Arnold's year in residence (1958–1959) at the Birbal Sahni Institute of Palaeobotany in Lucknow, India, about a decade after Sahni's death in 1949.

Throughout his career Arnold did not specialize, although he focused on late Paleozoic paleobotany. Nor did he concentrate on a single group of plants, a specific geologic period, or a particular mode of preservation or plant organ. He studied fructifications, wood, bark, megaspores, seeds—whatever paleobotanical material became available. His wide-ranging interests are shown in his summary papers on Paleozoic seeds (Arnold, 1948c), paleobotany and plant classification (Arnold, 1959b), origin and relationships of the cycads (Arnold, 1953d), and the classification of gymnosperms (Arnold, 1948d). By 1950 Arnold ". . . had achieved the well-deserved reputation as one of the foremost paleobotanists in the world" (C. B. Beck, written communication, December 1993).

Arnold's broad interests and his teaching duties led to the writing and publication of his textbook *Introduction to Paleobotany* (1947). This book rapidly became one of the standards in the field and remained in that position for many years. Two generations of paleobotany students met fossil plants for the first time in Arnold's richly illustrated textbook (which sold initially for $5.50). Late in his career Arnold undertook a revision of that book, but it was not completed at the time of his death.

## LATER YEARS

Arnold's 1947 textbook was the beginning of my contact with him. A pre–World War II geology major, after the war ended I was at DePauw University, absorbing enough botany to teach high school science. Professor Truman Yuncker, chairman of the Botany Department at DePauw, who knew Arnold, suggested that I read his then-new textbook. When a question arose, I wrote Arnold and received a gracious reply. Henry Andrews in his book (1980) noted that he, as a young graduate student, also received a cordial reply from Arnold along with the reprints that he requested. Arnold was a warm human being under his brusque exterior.

My inquiry had unexpected consequences. In the spring of that long-ago year, Arnold wrote Professor Yuncker, stating that he needed another field assistant for a collecting trip to Oregon and asking, "Would that Scott student who wrote be suitable and available?" I was, I went, and I became fascinated with some of the materials we collected—Eocene silicified fruits and seeds. I abandoned my high-school teaching goal and, with Professor Arnold's help in skirting formalities after we returned in August, entered graduate school at the University of Michigan in September 1949 to study paleobotany.

Graduate study under Professor Arnold was a curious mixture of the formal and the informal. His laboratory, which also served as his classroom, was a large room in the University Museum. His paleobotany classes met there for his well-organized, definitely formal lectures. At the same time, the laboratory served both for his own and graduate-student research. It usually was cluttered with fossil plants—compressions, coal-ball slices being peeled—all the research materials of Arnold and his two graduate students at the time, J. Stewart Lowther and myself. The disarray was a vivid contrast to the organized mind and manner of the professor. Arnold's graduate students later included Helen Smith, Alan Graham, Hermann Becker (now deceased), and Charles Miller (Arnold's last student). In addition, William Chaloner (Fig. 4) and Ted Delevoryas were "post-docs" under Arnold during the 1950s.

Much of the time a rude sonic background for science was provided in the laboratory by the continuous rasping of a Rube Goldberg device for sectioning large coal balls and even silicified materials. Money for a large diamond saw not being available, Arnold drew upon his farm background to devise a simple substitute. A motor-driven arm pulled a piece of sheet metal back and forth across the specimen, while a hose dripped water into the cut, and a wire pulled to and fro through the bottom of a coffee can supplied carborundum. The infernal machine ran 24 hours a day and took weeks to make a cut. Arnold took pleasure in its elegant simplicity and published a paper about it (Arnold, 1943).

William (Bill) G. Chaloner (Royal Holloway University of London) reminisced to Paul Lyons (written communication, March 1994) about a year in Chester Arnold's Laboratory, 1953–1954:

In the summer of 1953, just after I had got my Ph.D., I obtained a Harkness Fellowship to undertake a year's postdoctoral research in the United States. I sought to take up the fellowship to work in Chester Arnold's laboratory at Ann Arbor, Michigan. Not only was he one of the best-known of American Palaeozoic palaeobotanists, but he had published on Carboniferous megaspores, which were at that time my particular preoccupation.

I was delighted when he accepted me, and I duly took the Cunard ship, the *Coronia,* to New York, arriving on September 24, 1953. From there I went by train directly to Madison, Wisconsin, for the AIBS meeting being held on the University of Wisconsin campus. There was to be a palaeobotanical section meeting, and Chester had suggested that this would be a good start for my year, with a chance of meeting American palaeobotanists. He was right. I met, among others, Henry Andrews [see Chapter 21, this volume] and Jim Schopf [see Chapter 17, this volume], who were most valuable contacts for me, both during that year and far beyond.

On the opening evening of the meeting, there was to be a "smoker" (a term quite foreign to me then) at which it was announced free beer was to be served to those attending the meeting. This was a somewhat surprising development, since I had been led to believe (rightly!) that in the American Midwest alcohol did not flow too freely on university campuses. However, such strictures were waived in a state strongly committed to the brewing industry, and I remarked to Chester what a splendid idea that was. It only came to me later that this was not the right note to have struck, as Chester was a strict teetotaler—and took it seriously!

Chester Arnold was a most gentlemanly, benevolent and scholarly host. I was accommodated in a little office in the ex-Dental Hospital across the road from the Museum of Paleontology. He gave me the run of his own lab, where I could macerate rock samples, cut and section permineralized fossils and take photomicrograph. He was extremely helpful and very generous with his time, although never intrusive into how I was spending mine. He gave me access to a series of Mississippian coal and limestone samples, mainly of his own collecting, from which I was able to obtain an interesting range of new species of lycopod megaspores. This, in turn, led me to seek new material from the same source areas. Not surprisingly, a number of these megaspore species later proved to be present also in the British Lower Carboniferous.

When I wrote up that material, I passed him the manuscript for his criticism; he returned it promptly, having evidently gone over it carefully but making only the most restrained suggestions of a word changed here or there. I was quite nonplussed, being used to radical surgery at the hands of Tom Harris (my Ph.D. supervisor at Reading) on anything I had ever written for publication. However, when at Chester's suggestion I submitted it for publication to the *Contributions from the Michigan University Museum of Paleontology,* it was subjected to a much more vigorous assault by Robert V. Kesling, who then reassembled it with much improvement!

Chester gave a palaeobotanical course for advanced undergraduates—mainly botany majors—and I attended this, although I chickened out on taking the exam. I think I was secretly afraid of the awful disgrace if a post-doc were to fail an undergraduate exam in his chosen field. The course was somewhat catalogic, following the sequence and style of his memorable McGraw Hill 1947 textbook, but I found it captivating and learned a great deal from it. He had a remarkable range of fossil plant material which he laid out lovingly for each of the lab classes, and illustrated his lectures with numerous color slides—which were still a novelty in the fifties.

I went into the field with Chester on several occasions, particularly to the Grand Ledge, Michigan locality which was the nearest spot to Ann Arbor where Pennsylvanian compression fossils could readily be collected. He dressed for field work impeccably, with well-polished shoes, a tie and trilby hat rather as one might expect him to do for a call on the University President. The convention at that time in England was to dress for field work either as a good imitation of a tramp, or, if you could afford it, as though for a final assault on Mt. Everest. Tom Harris was a past master at the former style, and I suppose I tried in a gentle sort of way to imitate him in this. So, Chester's approach to the matter was a novelty for me; indeed, he was not at all what I had expected of an American professor in many ways. For example, he drove his car in a most exemplary manner. My contemporaries at Ann Arbor all drove, vying with one another to break the speed limit to the greatest extent possible without getting caught—a procedure involving the closest attention to the rearview mirror. I had assumed that this was the standard American mode of driving, but Chester disabused me. As we drove through each town, he would slow down to well below the local speed limit, causing (as I later learned) great indignation in the rest of the convoy of student cars which he was leading.

Later in the spring of 1954, he helped me to plan a palaeobotanical journey to the West Coast, taking in the maximum number of fossil-plant collecting spots. He had numerous contacts, and that enabled me to stop off with Ted Delevoryas and Wilson Stewart at Urbana, Illinois; with Henry Andrews in St. Louis, Missouri, Bob Baxter in Lawrence, Kansas; and with Ralph Chaney and Daniel Axelrod in California. I was also to meet a young student of Chaney's by the name of Jane Gray, who had just come across from the East Coast to study Tertiary pollen associated with his leaf floras. Chester's advice and good offices with so many people enabled me to make a wonderful palaeobotanical "Grand Tour" across the West (in a five year old, pea-green Buick) through the early summer of 1954. It was a magical year for me, and I owe Chester Arnold an enormous debt for making it possible.

Ted Delevoryas (University of Texas at Austin) also reminisced (written communication, January 1994) about his postdoctoral fellow days with Arnold in 1954 and 1955:

I sat in on his palaeobotany course and got a good dose of Arnoldian paleobotany. It was a thrilling experience for a fresh, green Ph.D. to be exposed to one of America's most senior paleobotanists. For the lab, however, we were on our own. A lab consisted of Chester's pulling out drawers of fossils that we had just covered in lecture, setting them on a table, and letting us loose to do the best we could with no lab manual or explanations. Perhaps this was good training for real

> life. When you are out in the field and come across fossils, there is no lab manual to direct you. You simply examine the material and extract all the information you can out of it.

Alan Graham also reflected (written communication, December 1993) on his graduate-school days with Arnold:

> As a graduate student, I spent four pleasant and profitable years at a small desk in a corner of Arnold's office/laboratory in the Museum of Paleontology at the University of Michigan (1958–1962). During our daily conversations, I formed several impressions that have remained constant over the years. He certainly was not frivolous, or inane in his humor. He had little patience with those lacking motivation and ability, and virtually no interest in self-aggrandizement. Arnold showed pleasure, surprise and appreciation for the various honors that came his way, enjoyed the unexpected experiences of his travels (as when called upon to explain fossils to the Duke of Edinburgh), and was amused at some of the quaint aspects of his profession (as the claim that ingesting powdered fossils could impart health benefits). He frequently voiced respect and admiration for his colleagues and students, he was uncomfortable with gossip, and I never heard him say anything judgmental or belittling of another person.

Arnold wrote his research papers in that setting and in our presence. Writing was not easy for him. He struggled and fidgeted over each word and phrase to achieve the clarity that characterizes his papers. Yet we students were expected to and welcome to interrupt him at any point for assistance, regardless of the pressure and tension of his own efforts. However, his assistance was often in the form of a counter question or a direction to search; his instructional mode was to encourage inquiry and further investigation.

The fossil material we collected on that Oregon trip in 1949 led to publications by Arnold on both Paleozoic (1953a) and Tertiary (1952a) ferns. Two nonscientific aspects of the long, hot return from the West Coast stand out in memory. We did not pass a single Dairy Queen, a soft-ice-cream shop, without stopping. Arnold was careful with the university expense account, but his meticulous bookkeeping did not extend to denying his two assistants, or himself, frequent portions of ice cream to break the tedious hot journey.

The classic illustration of his care with university funds came on our last night out. We had camped out every night for six weeks, and to prepare us for civilization, Arnold promised that the last night we would stop at a motel, where we could have showers. We two assistants (Robert Lindsley—who later became an invertebrate paleontologist—and I) all day anticipated sleeping in a bed again rather than on the ground. True to Arnold's promise, we had a motel and our showers. But there was only one room and one bed—the field assistants unfolded their bedrolls on the floor—as usual. Befittingly, the professor occupied the bed.

I relate this incident, now if not then humorous to me, to illustrate the basic integrity of the man, not stinginess. In those days, research funds were not easily obtained, and he regarded their expenditure as a "sacred trust." Every penny was to be accounted for, and beds for mere field assistants, themselves a questionable luxury, simply could not be justified. At the same time, the bed—having been paid for—must be slept in. This sense of responsibility from another era may seem naive and quaint today, but it is refreshing now when both universities and researchers have been accused of misapplying research funds. Arnold's spirit of total integrity also pervades and distinguishes his paleobotanical contributions and interactions with his colleagues.

I was privileged to spend a second summer, in 1951, as one of Arnold's field assistants, this time under vastly different circumstances on the Alaskan North Slope. The aftermath of World War II renewed interest in exploration of Naval Petroleum Reserve No. 4, which lies north of the Brooks Range in Alaska. The Office of Naval Research approved a plan for Arnold and two student assistants, Stewart Lowther and me, to perform reconnaissance for fossil plants along the Colville River. At that time, information about fossil plants on the North Slope was limited to rumors among the Native Americans of "plant pictures" in the rocks and knowledge of the existence of a coal mine along the Meade River.

The three of us set out in an open canvas boat down the Colville River, having been deposited by bushplane at Gubic, a well-drilling camp. We had no radio and only the food we could carry; we were alone in the wilderness for a month. Arnold kept a journal from which he later published a narrative account (Arnold, 1952b) from which I quote here: "As our craft began to drift downstream with the current the next morning, I experienced the peculiar feeling that we were going right into the unknown. As far as we knew, there were no human beings except the three of us in an open canvas boat between Gubic and the Arctic Ocean a hundred and fifty miles away. . . . We knew that we could not propel our craft upstream, and there was nothing to do but let the current take us where it would. But after some reflection I realized that is the way life is. One faces the unknown every day of his existence; it is something that cannot be avoided."

Although Arnold published only two papers (Arnold, 1953b; Arnold and Lowther, 1955) on the material we collected, the trip demonstrated the presence of abundant Early and Late Cretaceous plants on the North Slope and led to extensive work by others, among them Charles Smiley of Idaho State University.

Arnold's account of the Colville River trip is testimony to his astute powers of observation. However, there were two incidents that he failed to record. The first of these involved choosing a course among the virtually indistinguishable braids of the meandering Colville River. Stewart Lowther was a Canadian with considerable field experience in canoes due to his work with the Canadian Geological Survey; Arnold was the captain of our fragile craft. Lowther was in the stern; Arnold was in the bow. Each chose a different channel and dug in mightily with his paddle. The canvas boat literally spun in circles while they argued their choices. Lowther persisted and won, but things were tense in camp that evening.

Tension was still high the next morning when Arnold attempted to push our boat from shore into the stream. With one foot on a rock, the other in the boat, he gave a mighty lunge. The boat sprang free and left him spread-eagled between the rock and the boat, struggling to maintain himself upright against the tug of the current. He called for help; we turned, and instead of running to his aid, broke out in laughter. Our dignified, somewhat portly professor was teetering between a rock and a very soft place, in immediate danger of falling into the river; Arnold by sheer strength and determination pulled his legs together and got the boat back to the bank. Once on land, he too saw humor in his predicament and joined us in laughter. The tension was broken. As students we now knew that the professor was human.

C. B. Beck (written communication, December 1993) wrote, "Those of us who knew him well can attest to the fact that he was kind and unselfish, a man of good humor, and a gracious person always ready to accept the successes of others."

Ted Delevoryas recalled (written communication, January 1994) another field trip with Arnold during the middle 1950s when he was a "post-doc" with Arnold:

> He invited me to accompany him on a field trip to Iowa, Wyoming, and Colorado. Chester was normally taciturn and I am not prone to excessive verbalizing; we would drive for hundreds of miles with not a word exchanged between us. But it was a memorable trip nevertheless. We examined petrified cordaitean stumps and logs in an Iowa coal mine, visited the famous Kemmerer locality in Wyoming, saw petrified *"Lepidodendron"* (now known to be *Lepidophloios johnsonii* in Colorado, and collected at the famous Creede, Colorado, site. Other localities were visited as well. Chester was pretty conservative with money and we stayed in some pretty moth-eaten motels. But, of course, that was before we had plush government grants to cover our expenses.

Figure 6. C. A. Arnold, invited scientist (paleobotany), July, 1977, in E. L. Zodrow's laboratory, University College of Cape Breton, Sydney, Nova Scotia, July, 1977. From the archives of the Paleobotanical Collection, University College of Cape Breton. Photograph by Keith McCandlish and courtesy of E. L. Zodrow, University College of Cape Breton.

After retirement, in the summers of 1976 and 1977, Arnold (Fig. 6) was an invited scientist (paleobotany) at Xavier College (component of the present University College of Cape Breton), Sydney, Nova Scotia, and at the Nova Scotia Museum in Halifax, Nova Scotia. The invitation was a result of efforts jointly by R. Grantham as curator (Geology), Nova Scotia Museum, and Professor Erwin L. Zodrow, Department of Geology, who wrote (November 1993):

> In his capacity as visiting scientist, Emeritus Professor Arnold identified a large collection of Carboniferous plant megafossils from Atlantic Maritime Canada, collected jointly by me and my research student, Keith McCandlish, from the Sydney Coalfield on Cape Breton Island since 1974. Arnold not only established the scientific basis of the present Palaeobotanical Collection at the University College of Cape Breton, but also laid the foundation for future palaeobotanical studies by me and co-workers in Maritime Canada; these studies bridge the gap with those of W. A. Bell who died in 1969 [see Bell's portrait, Chapter 16, this volume].

At the Nova Scotia Museum, Arnold in 1977 also identified Carboniferous fossil plants from the Stellarton and Cumberland Basins of Nova Scotia. The data provided by Arnold from studying the Nova Scotian fossil plants, his last scientific contribution, were the basis for an annotated catalogue of Carboniferous fossils of Maritime Canada (Zodrow and McCandlish, 1980) creating a link with Arnold's correlations (1934b, 1949a) between the Michigan Coal Basin and Maritime Canada (see Chapter 25, this volume). Zodrow further reflected, "Professor Arnold was a great teacher, always more than willing to spend time arguing palaeobotanical points with me and Keith McCandlish, was always immaculately dressed for work [Fig. 6], and was a gentleman of strong personal convictions."

## CLOSING REFLECTIONS

Chester A. Arnold was an important figure in twentieth-century paleobotany. He entered the field in the late 1920s, at a time when paleobotany was almost a curiosity in academia.

Arnold played a prominent part as a teacher, textbook writer, and researcher in the growth of paleobotany into a major biological and geological discipline. Although his researches spanned both the continent and the upper Paleozoic and younger geologic column, his scientific roots were in the upper Paleozoic, and his contributions there are most significant.

Professor Arnold's scientific achievements, including about 121 publications, are a matter of record (see Chester A. Arnold's bibliography, this chapter). In this portrait I have attempted, in some small way, to illuminate the personal side of a man who remains indelible not only in twentieth-century paleobotany but in the minds of all those privileged to have known him.

## ACKNOWLEDGMENTS

I sincerely appreciate the courtesy of C. B. Beck, W. G. Chaloner, R. M. Kosanke, C. N. Miller, A. Graham, T. Delevoryas, and E. L. Zodrow in furnishing information and/or photos of Chester Arnold. Cornell University Press, H. N. Andrews, and C. B. Beck are thanked for permission to reproduce the photograph of Arnold (Fig. 1). Information from the William C. Darrah paleobotanical files was used with the permission of Elsie Darrah Morey. C. B. Beck, University of Michigan, graciously provided much help and information that was used in this portrait. Aureal T. Cross, Michigan State University, supplied the photographs shown in Figures 3 and 5. Figure 4 was generously provided by Alan Graham, Kent State University. The bibliography of Arnold was taken with minor changes from Watt (1978). The manuscript was reviewed by T. L. Taylor and Harlan Banks.

## REFERENCES CITED

Andrews, H. N., 1980, The fossil hunters: Ithaca, New York, Cornell University, 421 p.

Arnold, C. A. See Bibliography of Chester A. Arnold that follows.

Cooper, W. F., 1906, Geological report on Bay County [Michigan]: Report of State Board, Michigan Geological Survey, Annual Report for 1905, p. 135–426 (see p. 188).

Lane, A. C., 1902, Coal in Michigan—Its mode of occurrence and quality: Michigan Geological Survey, r. 8, pt. 2, 233 p. (see p. 43–44).

Lyons, P. C., and Darrah, W. C., 1989, Early conifers of North America: Upland and/or paleoclimatic indicators: Palaios, v. 4, p. 480–486.

Namdoodiri, K. K., and Beck, C. B., 1968,A comparative study of the primary vascular system of conifers. I: Genera with helical phyllotaxis: American Journal of Botany, v. 55, p. 447–457.

Pettitt, J. M., and Beck, C. B., 1968, *Archaeosperma arnoldii*—A cupulate seed from the Upper Devonian of North America: Contributions from the Museum of Paleontology, University of Michigan, v. 22, p. 139–154.

Watt, A. D., 1978, Bibliography of American paleobotany for 1977: Botanical Society of America, Paleobotanical Section, 97 p.

Zodrow, E. L., and McCandlish, K., 1980, Upper Carboniferous fossil flora of Nova Scotia, *in* The collections of the Nova Scotia Museum; with special reference to the Sydney Coalfield: Halifax, Nova Scotia, Nova Scotia Museum, 275 p., 150 pls.

## BIBLIOGRAPHY OF CHESTER A. ARNOLD

Arnold, C. A., 1928a, Some Devonian plant localities in Central and Western New York: Science, v. 67, p. 276–277.

Arnold, C. A., 1928b, The development of the perithecium and spermagonium of *Sporormialeqorind* Niessl: American Journal of Botany, v. 15, p. 241–245.

Arnold, C. A., 1929a, On the radial pitting in *Callixylon:* American Journal of Botany, v. 16, p. 391–393.

Arnold, C. A., 1929b, The genus *Callixylon* from the Upper Devonian of Central Western New York: Papers of the Michigan Academy of Science, Arts, and Letters, v. 11, p. 1–50.

Arnold, C. A., 1929c, Petrified wood in the New Albany Shale: Science, v. 70, p. 581–582.

Arnold, C. A., 1930a, The Paleobotanical excursion of the Fifth International Botanical Congress: Science, v. 72, p. 429–430.

Arnold, C. A., 1930b, A petrified lepidophyte cone from the Pennsylvanian of Michigan: American Journal of Botany, v. 17, p. 1028–1032.

Arnold, C. A., 1930c, Bark structure of *Callixylon:* Botanical Gazette, v. 90, p. 427–431.

Arnold, C. A., 1931a, Cordaitean wood from the Pennsylvanian of Michigan and Ohio: Botanical Gazette, v. 91, p. 77–87.

Arnold, C. A., 1931b, On *Callixylon newberryi* (Dawson) Elkins et Wieland: Contributions from the Museum of Paleontology, University of Michigan, v. 3, p. 207–232.

Arnold, C. A., 1932, Macrofossils from Greenland coal: Papers of the Michigan Academy of Science, Arts, and Letters, v. 15, p. 51–60.

Arnold, C. A., 1933a, A lycopodiaceous strobilus from the Pocono Sandstone of Pennsylvania: American Journal of Botany, v. 20, p. 114–117.

Arnold, C. A., 1933b, A sphenopterid fructification from the Pennsylvanian of Michigan: Botanical Gazette, v. 94, p. 821–825.

Arnold, C. A., 1933c, Fossil plants from the Pocono (Oswayo) Sandstone of Pennsylvania: Papers of the Michigan Academy of Science, Arts, and Letters, v. 16, p. 51–56.

Arnold, C. A., 1934a, *Callixylon whiteanum* sp. nov., from the Woodford Chert of Oklahoma: Botanical Gazette, v. 96, p. 180–185.

Arnold, C. A., 1934b, A preliminary study of the fossil flora of the Michigan Coal Basin: Contributions from the Museum of Paleontology, University of Michigan, v. 4, p. 177–204.

Arnold, C. A., 1934c, The so-called branch impressions of *Callixylon newberryi,* (Dawson) Elkins and Wieland and the conditions of their preservation: Journal of Geology, v 42, p. 71–76, 4 figs.

Arnold, C. A., 1935a, Observations on *Alethopteris grandifolia* Newberry, and its seeds: Contributions from the Museum of Paleontology, University of Michigan, v. 4, p. 279–282.

Arnold, C. A., 1935b, On seed-like structures associated with *Archaeopteris,* from the Upper Devonian of northern Pennsylvania: Contributions from the Museum of Paleontology, University of Michigan, v. 4, p. 283–286.

Arnold, C. A., 1935c, Notes on some American species of *Lepidostrobus:* American Journal of Botany, v. 22, p. 23–25.

Arnold, C. A., 1935d, A Douglas fir cone from the Miocene of southeastern Oregon: Washington Academy of Science, v. 25, p. 378–380.

Arnold, C. A., 1935e, Some new forms and new occurrences of fossil plants from the Middle and Upper Devonian of New York State: Bulletin of the Buffalo Society of Natural Science, v. 17, p. 1–12, pl. 1.

Arnold, C. A., 1936a, Observations on fossil plants from the Devonian of eastern North America. I: Plant remains from Scaumenac Bay, Quebec: Contributions from the Museum of Paleontology, University of Michigan, v. 5, p. 37–48.

Arnold, C. A., 1936b, Observations on fossil plants from the Devonian of eastern North America. II: *Archaeopteris macilenta* and *A. sphenophyllifolia* of Lesquereux: Contributions from the Museum of Paleontology, University of Michigan, v. 5, p. 49–56.

Arnold, C. A., 1936c, Some fossil species of *Mahonia* from the Tertiary of eastern and southeastern Oregon: Contributions from the Museum of Paleontology, University of Michigan, v. 5, p. 57–66.

Arnold, C. A., 1936d, The occurrence of *Cedrela* in the Miocene of western America: American Midland Naturalist, v. 17, p. 1018–1021.

Arnold, C. A., 1936e, Paleozoic plants and their environmental relations: Washington, D.C., Report of the Commission on Paleobotany, Natural Resource Council, p. 56–64.

Arnold, C. A., 1937a, The seeds of *Alethopteris* and other pteridosperms from North America: Congrès pour l'avancement des études de stratigraphie Carbonifère, 2d, Heerlen, The Netherlands, 1935: Compte Rendu, v. 1, p. 42–45.

Arnold, C. A., 1937b, Devonian and Mississippian plant-bearing formations in eastern America: Congrès pour l'avancement des études de stratigraphie Carbonifère, 2d, Heerlen, The Netherlands, 1935: Compte Rendu, v. 1, p. 47–62.

Arnold, C. A., 1937c, Observations on the fossil flora of eastern and southeastern Oregon. Part I: Contributions from the Museum of Paleontology, University of Michigan, v. 5, p. 79–102.

Arnold, C. A., 1937d, Observations of fossil plants from the Devonian of eastern North America. III: *Gilboaphyton goldringiae,* gen. et sp. nov., from the Hamilton of western New York: Contributions from the Museum of Paleontology, University of Michigan, v. 5, p. 75–78.

Arnold, C. A., 1938a, Paleobotanical studies in Michigan: Papers of the Michigan Academy of Science, Arts, and Letters, v. 23, p. 95–99.

Arnold, C. A., 1938b, Note on a lepidophyte strobilus containing large spores, from Braidwood, Illinois: American Midland Naturalist, v. 20, p. 709–712.

Arnold, C. A., 1938c, Paleozoic seeds: Botanical Review, v. 4, p. 205–234.

Arnold, C. A., 1939a, Observations on fossil plants from the Devonian of eastern North America. IV: Plant remains from the Catskill delta deposits of northern Pennsylvania and southern New York: Contributions from the Museum of Paleontology, University of Michigan, v. 5, p. 271–314.

Arnold, C. A., 1939b, *Lagenospermum imparirameneum* sp. nov., a seed-bearing fructification from the Mississippian of Pennsylvania and Virginia: Bulletin of the Torrey Botanical Club, v. 66, p. 297–303.

Arnold, C. A., 1939c, A possible missing link in the evolution of the early seed plants: Chronica Botanica, v. 5, p. 360–362.

Arnold, C. A., 1940a, *Lepidodendron johnsonii,* sp. nov., from the Lower Pennsylvanian of central Colorado: Contributions from the Museum of Paleontology, University of Michigan, v. 6, p. 21–52.

Arnold, C. A., 1940b, Structure and relationships of some Middle Devonian plants from western New York: American Journal of Botany, v. 27, p. 57–63.

Arnold, C. A., 1940c, A note on the origin of the lateral rootlets of *Eichhornia crassipes* (Mart.) Solms: American Journal of Botany, v. 27, p. 728–730.

Arnold, C. A., 1940d, Devonian ferns: Chronica Botanica, v. 6, p. 11–12.

Arnold, C. A., 1941a, *Psilophyton* and *Aneurophyton* in the Devonian of eastern North America: Chronica Botanica, v. 6, p. 375–376.

Arnold, C. A., 1941b, Observations on fossil plants from the Devonian of eastern North America. V: *Hyenia banksii* sp. nov.: Contributions from the Museum of Paleontology, University of Michigan, v. 6, p. 53–57.

Arnold, C. A., 1941c, Some Paleozoic plants from central Colorado and their stratigraphic relationships: Contributions from the Museum of Paleontology, University of Michigan, v. 6, p. 59–70.

Arnold, C. A., 1941d, The petrification of wood: The Mineralogist, v. 9, p. 323–324, 353–355.

Arnold, C. A., 1943, Inexpensive saw for large specimens: The Mineralogist, v. 11, p. 226–231.

Arnold, C. A., 1944a, Silicified plant remains from the Mesozoic and Tertiary of western North America. I: Ferns: Papers of the Michigan Academy of Science, Arts, and Letters, v. 30, p. 3–34.

Arnold, C. A., 1944b, A heterosporous species of *Bowmanites* from the Michigan Coal Basin: American Journal of Botany, v. 31, p. 466–469.

Arnold, C. A., 1945a, Instructions to naturalists in the armed forces for botanical field work; No. 10; Directions for collecting plant fossils: Ann Arbor, University of Michigan, Department of Botany.

Arnold, C. A., 1945b, Hugh Miller, patron saint of the rock hounds. The Mineralogist, v. 13, p. 467–471.

Arnold, C. A., 1947, Introduction to paleobotany: New York, McGraw-Hill Book Co., 433 p.

Arnold, C. A., 1948a, Some cutinized seed membranes from the coal-bearing rocks of Michigan: Bulletin of the Torrey Botanical Club, v. 75, p. 131–146.

Arnold, C. A., 1948b, The Mississippian flora: Journal of Geology, v. 56, p. 367–372.

Arnold, C. A., 1948c, Paleozoic seeds, II: Botanical Review, v. 14, p. 450–472.

Arnold, C. A., 1948d, Classification of gymnosperms from the viewpoint of paleobotany: Botanical Gazette, v. 110, p. 2–12, 1 fig.

Arnold, C. A., 1949a, Fossil flora of the Michigan Coal Basin: Contributions from the Museum of Paleontology, University of Michigan, v. 7, p. 131–269.

Arnold, C. A., 1949b, The fossil fern *Osmundites oregonensis:* The Mineralogist, v. 17, p. 233–235.

Arnold, C. A., 1950a, Laboratory guide for plant anatomy: Ann Arbor, Michigan, George Wahr Publishing, 55 p.

Arnold, C. A., 1950b, Fossil dicotyledons, *in* Gunderson, A., Families of dicotyledons: Waltham, Massachusetts, Chronica Botanica Company, p. 3–6.

Arnold, C. A., 1950c, Paleobotany, Chapter 15, *in* Collier's encyclopedia: New York, P. F. Collier and Sons, p. 375–377.

Arnold, C. A., 1950d, Megaspores from the Michigan Coal Basin: Contributions from the Museum of Paleontology, University of Michigan, v. 8, p. 59–111.

Arnold, C. A., 1952a, Fossil Osmundaceae from the Eocene of Oregon: Paleontographica, Abt. B, Band 92, p. 63–78.

Arnold, C. A., 1952b, Paleobotanical investigations in Naval Petroleum Reserve No. 4, Alaska: Science, v. 116, p. 61–62.

Arnold, C. A., 1952c, Observations on fossil plants from the Devonian of eastern North America. VI: *Xenocladia medullosina* Arnold: Contributions from the Museum of Paleontology, University of Michigan, v. 9, p. 297–309.

Arnold, C. A., 1952d, Tertiary plants from North America: The Palaeobotanist, v. 1, p. 73–78.

Arnold, C. A., 1952e, Some observations on the anatomy of the common geranium: Papers of the Michigan Academy of Science, Arts, and Letters, v. 37, p. 3–11.

Arnold, C. A., 1952f, A specimen of *Prototaxites* from the Kettle Point Black Shale of Ontario: Palaeontographica, Abt. B, Band 93, p. 45–56.

Arnold, C. A., 1952g, A paleobotanical excursion in northern Alaska: Asa Gray Bulletin, v. 1, p. 269–282.

Arnold, C. A., 1953a, Fossil plants of Early Pennsylvanian type from central Oregon: Palaeontographica, Abt. B, Band 93, p. 61–68.

Arnold, C. A., 1953b, Silicified plant remains from the Mesozoic and Tertiary of western North America. II: Some fossil woods from northern Alaska: Papers of the Michigan Academy of Science, Arts, and Letters, v. 38, p. 9–20.

Arnold, C. A., 1953c, Searching for fossil plants in Alaska: Research Review, June 1–9.

Arnold, C. A., 1953d, Origin and relationships of the cycads: Phytomorphology, v. 3, p. 51–65.

Arnold, C. A., 1954a, Fossil plants of the Florissant beds, Colorado, by H. D. MacGinite: Review: Science Monitor, v. 78, p. 1.

Arnold, C. A., 1954b, The spores of *Archaeopteris,* with remarks on the affinities of the genus: Proceedings, International Botanical Congress, 7th, Stockholm: Uppsala, Sweden, p. 559–561.

Arnold, C. A., 1954c, Silicified Mesozoic and Tertiary plants of western North America: Proceedings, International Botanical Congress, 7th, Stockholm: Uppsala, Sweden, p. 556.

Arnold, C. A., 1954d, The Michigan Coal Basin: Michigan Alumnus Quarterly Review, v. 60, p. 287–296.

Arnold, C. A., 1954e, Fossil sporocarps of the genus *Protosalvinia* Dawson, with special reference to *P. furcata* (Dawson) comb. nov.: Svensk Botanisk Tidskrift, v. 48, p. 292–300.

Arnold, C. A., 1955a, A Tertiary *Azolla* from British Columbia: Contributions from the Museum of Paleontology, University of Michigan, v. 12, p. 37–45.

Arnold, C. A., 1955b, Tertiary conifers from the Princeton Coal Field of British Columbia: Contributions from the Museum of Paleontology, University of Michigan, v. 12, p. 245–258.

Arnold, C. A., 1955c, "Lehrbuch der Paläobotanik," by W. Gothan and H. Weyland: Review: Science, v. 212, p. 500.

Arnold, C. A., 1956a, Paleobotany: Encyclopedia Americana, v. 21, p. 141–157.

Arnold, C. A., 1956b, Fossil ferns of the Matoniaceae from North America: Palaeontological Society of India Journal (Lucknow), v. 1, p. 118–121.

Arnold, C. A., 1956c, A new calamite from Colorado: Contributions from the Museum of Paleontology, University of Michigan, v. 13, p. 161–173.

Arnold, C. A., 1956d, A new *Tempskya:* Contributions from the Museum of Paleontology, University of Michigan, v. 14, p. 133–142.

Arnold, C. A., 1956e, Petrified cones of the genus *Calamostachya* from the Carboniferous of Illinois: Contributions from the Museum of Paleontology, University of Michigan, v. 14, 149–165.

Arnold, C. A., 1959a, Some paleobotanical aspects of tundra development: Ecology, v. 40, p. 146–148.

Arnold, C. A., 1959b, Palaeobotany and plant classification: 6th Sir Albert Seward Memorial Lecture, Birbal Sahni Institute of Palaeobotany (Lucknow), 5 p.

Arnold, C. A., 1960a, A lepidodendrid stem from Kansas and its bearing on the problem of cambium and phloem in Paleozoic lycopods: Contributions from the Museum of Paleontology, University of Michigan, v. 15, p. 249–267.

Arnold, C. A., 1960b, "The morphology and anatomy of the American species of the genus *Psaronius,*" by Jeanne Morgan, Illinois Biology Monitor, no. 27, Dec. 21, 1959: Review: Paleontology, v. 34, p. 602–603.

Arnold, C. A., 1960c, Archaeopteridales, Volume 1: New York, McGraw-Hill Encyclopedia of Science and Technology, p. 510.

Arnold, C. A., 1960d, Caytoniales, Volume 2: New York, McGraw-Hill Encyclopedia of Science and Technology, p. 578.

Arnold, C. A., 1960e, Nematophytales, Volume 9: New York, McGraw-Hill Encyclopedia of Science and Technology, p. 34.

Arnold, C. A., 1960f, Pleuromeiales, Volume 10: New York, McGraw-Hill Encyclopedia of Science and Technology, p. 417.

Arnold, C. A., 1961a, Cycadales, *in* The encyclopedia of biological sciences: New York, Reinholdt Publishing, p. 287.

Arnold, C. A., 1961b, The Plant Kingdom: The encyclopedia of biological sciences: New York, Reinholdt Publishing, p. 799–803.

Arnold, C. A., 1961c, Re-examination of *Triletes superbus, T. rotatus,* and *T. mamillarius* of Bartlett: Brittonia, v. 13, p. 245–252.

Arnold, C. A., 1963, *Cordaites*-type foliage associated with palm-like plants from the Upper Triassic of south-western Colorado: Journal of the Indian Botanical Society, v. 42A (Maheshwari Commemoration Volume), p. 4–9.

Arnold, C. A., 1965a, Paleobotany—The last 34 years: Taxon, v. 14, p. 12–13.

Arnold, C. A., 1965b, "The Miocene Trapper Creek flora of southern Idaho," by D. Alexrod, University of California Publications in Geology, 51, p. 1–181: Review: Palaeontology, v. 39, p. 510–511 (1964).

Arnold, C. A., 1965c, Rudolf Florin, 1894–1965: Plant Science Bulletin, v. 11, p. 11–12.

Arnold, C. A., 1965d, Some recollections of Percy Train (1876–1942): Huntia, v. 2, p. 111–116.

Arnold, C. A., 1966a, Fossil plants in Michigan: Michigan Botanist, v. 5, p. 3–13.

Arnold, C. A., 1966b, "The morphology of gymnosperms," by K. L. Sporne, London, Hutchinson and New York, Hillary House, 216 p.: Review, Science, v. 152, p. 340.

Arnold, C. A., 1966c, "Spores and pollen of the Potomac Group of Maryland," by G. J. Brenner: Review: Palaeontology, v. 38, p. 1009.

Arnold, C. A., 1967a, "Indian fossil pteridophytes," by K. R., Surange, Botanical Monographs no. 4, Council of Scientific and Industrial Research, New Delhi, 209 p.: Review: Palaeontology, v. 41, p. 1300–1301.

Arnold, C. A., 1967b, "Anatomie des Blattes. I: Blattanatomie der Gymnospermen," by K. Napp-Zinn, Berlin, Gebrüder Borntraeger, 370 p. (1966): Review: Quarterly Review of Biology, v. 42, p. 546–547.

Arnold, C. A., 1967c, The proper designations of the foliage and stems of the Cordaitales: Phytomorphology, v. 17, p. 346–350.

Arnold, C. A., 1969a, Current trends in paleobotany: Earth Science Review, v. 4, p. 283–309.

Arnold, C. A., 1969b, Paleobotany, *in* Ewan, J., ed., A short history of botany in the United States: International Botanical Congress, 11th, Seattle, Washington, 1969: London, Hafner, p. 103–108.

Arnold, C. A., 1970, The fossil-plant record, *in* Tschudy, R. H., and Scott, R. A., eds., Aspects of palynology, Chapter 8: New York, Wiley-Interscience, p. 127–143.

Arnold, C. A., 1971, Some elements of the Gondwanaland flora in North America: Geophytology, v. 1, p. 1–5.

Arnold, C. A., 1973, Fossil plants and continental drift: First Birbal Sahni Memorial Lecture, Birbal Sahni Institute of Palaeobotany, Lucknow, p. 3–11.

Arnold, C. A., and Andrews, H. N., 1970, Orde de Pityales, *in* Boureau, E., ed., Traité de paléobotanique, Volume 4, Filicophyta: Paris, Masson et Cie, p. 422–456.

Arnold, C. A., and Daugherty, L. H., 1963, The fern genus *Acrostichum* in the Eocene Clarno Formation of Oregon: Contributions from the Museum of Paleontology, University of Michigan, v. 18, p. 205–227.

Arnold, C. A., and Daugherty, L. H., 1964, A fossil dennstaedtioid fern from the Eocene Clarno Formation of Oregon: Contributions from the Museum of Paleontology, University of Michigan, v. 18, p. 65–88.

Arnold, C. A., and Lowther, W. J., 1955, A new Cretaceous conifer from northern Alaska: American Journal of Botany, v. 42, p. 522–528.

Arnold, C. A., and Sadlick, W., 1962a, A Mississippian flora from northeastern Utah and its faunal and stratigraphic relations: Contributions from the Museum of Paleontology, University of Michigan, v. 17, p. 241–263.

Arnold, C. A., and Sadlick, W., 1962b, A *Rhexoxylon*-like stem from the Morrison Formation of Utah: American Journal of Botany, v. 49, p. 883–886.

Arnold, C. A., and Stanley, G. M., 1945, Pennsylvanian plants from the glacial drift near Jackson, Michigan: Papers of the Michigan Academy of Science, Arts, and Letters, v. 31, p. 247–250.

Arnold, C. A., and Steidtmann, W. E., 1937, Pteridospermous plants from the Pennsylvanian of Illinois and Missouri: American Journal of Botany, v. 24, p. 644–650.

Manuscript Accepted by the Society July 6, 1994

Geological Society of America
Memoir 185
1995

# *Charles Brian Read (1907–1979): American paleobotanist and geologist*

**Sergius H. Mamay**
*Department of Paleobiology, Smithsonian Institution, Washington, D.C. 20560*
**Sidney R. Ash**
*Department of Geology, Weber State University, Ogden, Utah 84408-2507*
**Paul C. Lyons**
*U.S. Geological Survey, MS 956, National Center, Reston, Virginia 22092*

## ABSTRACT

**Charles Brian Read (1907–1979), paleobotanist and geologist, spent his entire 42-year career with the U.S. Geological Survey. His mentor was the noted geologist and paleobotanist David White, who introduced Read to late Paleozoic floras. In the 1930s, Read became interested in the floral zonation of the Pennsylvanian; this study was to culminate in an authoritative floral zonation scheme for the upper Paleozoic of the United States. Read was also an expert on Devonian floras and the Cretaceous tree fern *Tempskya.* His leadership in studies of the regional geology of New Mexico, including fossil-fuel and uranium reserve estimates, is legendary. Read was mentor to several geoscientists, including S. R. Ash, S. H. Mamay, and the late G. H. Wood, Jr. His continuing influence on the paleobotany of the United States and the geology of New Mexico is high testimony to his multiple talents and the breadth of his accomplishments.**

## INTRODUCTION

When Charles (Charlie) Brian Read (Fig. 1) died on August 30, 1979, there ended a distinguished and influential professional career with the U.S. Geological Survey (USGS), spanning more than four decades of versatile performance in a number of geologically oriented disciplines. Although his scope of competence and experience embraced geologic mapping, military geology, and fuels and mineral explorations, it was paleobotany that was—from his professional beginnings—his primary interest. His paleobotanical studies involved plant anatomy, morphology, floristics, and biostratigraphy. Some of his publications remain standard references in the fields of paleobotany and biostratigraphy.

## THE EARLY YEARS

Read was born on April 26, 1907, in the even-now small town of Dublin, Texas, some 180 km southwest of Dallas—right in the heart of Texas. His early days are not generally well known, but they are partly recorded in a remarkable hand-written, incomplete autobiography of 35 pages, which was retrieved from some odds and ends that were about to be discarded when his last office was being closed in 1972. Probably nobody had seen this chronicle before Sergius Mamay found it, for although the older pages are yellowed and brittle, they show no evidence of prior handling. The story is entitled "Horsing around" and emphasizes Read's fondness of and close association with horses. He wrote:

> My boyhood was spent on a ranch and in a small town in Texas and I can well remember, at the tender age of four years, that I inherited a saddle, bridle and horse from my sister who had outgrown the saddle, as women do, and also her liking for horses. . . . As for the horse, an old mare named Texas, she was twenty some-odd years old and judged by all, including my father and his brothers, as perfectly safe. Looking back, I sometimes wonder why they came to that conclusion.

This account then sketchily relates a few boyhood episodes—terror in the cyclone cellar; skinning skunks too close

Mamay, S. H., Ash, S. R., and Lyons, P. C., 1995, Charles Brian Read (1907–1979): American paleobotanist and geologist, *in* Lyons, P. C., Morey, E. D., and Wagner, R. H., eds., Historical Perspective of Early Twentieth Century Carboniferous Paleobotany in North America (W. C. Darrah volume): Boulder, Colorado, Geological Society of America Memoir 185.

Figure 1. Charles B. Read (circa 1945).

to the house; riding donkeys and calves—and ends many years later (in the early 1940s) in a summer mapping camp in the Sangre de Cristo Mountains of New Mexico. Of this area Read wrote: "The mountains all around were wild and remote and in our mapping we would have to travel by horses for there were no roads." Thus it is clear that horsemanship and the attendant outdoor life were ever-important to Read.

References to Read's education are also sparse, apart from his own meager accounts. He resisted the idea of a college education because he preferred the ranch life. He wrote: "I had been badly hurt while in high school and walked with a limp. I suppose I was not very strong or healthy and they [his family] must have decided that I needed some higher education if I were to ever make a living." He did, however, attend college and received a B.A. degree in botany and geology from Texas Technological College in 1927. His unfinished autobiography relates: "At first I studied agriculture because I wanted to be a rancher. Then I veered in the direction of the sciences, which is where I still am. I ended at the University of California [1927–1930] doing graduate work in some special fields in geology and then was fortunate enough, just at the beginning of the Depression, to gain an appointment with the United States Geological Survey."

The particulars of his curricula are not available, and one wonders from whom Read learned all the botanical/geological specialties with which he was so well versed. He apparently had some contact with Ralph W. Chaney—the Cenozoic paleobotanist, also then in the beginning of his graduate studies at Berkeley—who encouraged Read to pursue paleobotany. Evidently Read also had favorable exposure to W. A. Setchell, the noted phycologist, for Read later named a New Albany Shale (Upper Devonian) species, *Asteroxylon setchelli,* in Setchell's honor. Otherwise, we are led to believe that Read was largely self-taught, with a very valuable part of his education gained through his association with the renowned geologist-paleobotanist, David White (see Chapter 10, this volume), after his college days were behind him.

## THE EARLY SURVEY YEARS

Read's long career with the U.S. Geological Survey began in June 1930, when David White contacted him by mail concerning a possible appointment as his assistant in the U.S. National Museum (USNM), Washington, D.C. The letter reached Read at Yellowstone National Park, where he was beginning a study of the Tertiary plants of the area, under the auspices of the Carnegie Institution of Washington. Read's letter of acceptance, dated June 18, 1930, made it clear that his training in geology had been secondary to plant anatomy and morphology, but he expressed confidence in soon learning to do satisfactory stratigraphic work. Evidently White's letter (not available to us) had suggested that Read consider working on comprehensive taxonomic monographs. Read replied: "Your suggestion suits me exactly. That is what I had in mind with Tertiary plants." (Read's interests were soon to be diverted to the Paleozoic, and his Tertiary studies were only minimal.) Although Read seemed reluctant to abandon both his commitments as a research associate of the Carnegie Institution for that summer as well as "rather definite plans to study under Dr. Bailey [I. W. Bailey, the eminent Harvard University plant anatomist] . . .", he wired his acceptance on June 20, 1930.

Read's appointment as an assistant scientist (Geology) was made official on June 30, 1930, at an annual salary of $2,600, and he was sworn in on August 22 at Fort Collins, Colorado, where he met David White for the first time. He wrote (letter to Mamay, December 9, 1977): "I met Mr. David White in Fort Collins, Colorado, in mid-August, 1930 when I joined the Geological Survey. We spent a few weeks together in the field looking for Permian and Pennsylvanian fossil plants with some success in the case of the Pennsylvanian [Weber Formation of Colorado] and none in the Permian. Mr. White then returned to Washington, D.C., and I went to Oklahoma." White's Report of the Committee on Paleobotany (National Research Council, 1931) indicates that Read was studying a Pennsylvanian flora from the McAlester coalfield, Oklahoma, as a result of his fieldwork there in 1930 and 1931. Read continues: "In mid-November I went to Washington also. There I began my work with Mr. White (Uncle David). I liked him, enjoyed working for him, and consider my years under him very instructive." These remarks clarify the details of Read's hiring by the USGS, while also reflecting his admiration for his mentor/supervisor. His parenthetical reference to White as "Uncle David" is an indication of close rapport with White, although he almost invariably referred to White, even in casual conversation, as "Mr. White." The appellation "Uncle David" was used affectionately by other young assistants of White, including the chemist, Taisia Stadnichenko.

The remainder of that summer (1930) was important to Read's training as a geologist-stratigrapher. Then he began an association with T. M. Hendricks, working as Hendricks's field assistant in the McAlester coalfields of Oklahoma. Read was officially transferred from McAlester to Washington on October 17, 1930, where he took up residence and began work with White in mid-November. However, Read continued to work in Oklahoma with Hendricks for part of 1931. This association led to a lifelong friendship and, in two years (1934–1935), produced five USGS coal maps as well as an important article on correlations of Pennsylvanian rocks in the Arkansas and Oklahoma coal fields (Hendricks and Read, 1934). Read made extensive collections of Pennsylvanian plants during that period; these are now in the USNM Paleobotanical Collections. One of his specimens is a spectacularly well preserved silicified fern stem from the Johns Valley Shale (Upper Mississippian/Lower Pennsylvanian) that Read (1938b) named *Ankyropteris hendricksii* in honor of Hendricks.

Read's career in Washington was mostly spent in cramped quarters on the third floor of the USNM, where he shared an office with Roland Brown ("Brownie"), a Tertiary paleobotanist who was also new on the Washington scene. The paleontologic unit, then known as the Section of Paleontology and Stratigraphy, was headed by T. W. Stanton. Among Read's other USGS associates in his new surroundings were John Reeside, Jr., Lloyd Henbest, and, of course, David White. Read developed close friendships with both Brown and Henbest, learned much from them, and would later collaborate with each—with Henbest, on Pennsylvanian fusulinid distribution in parts of New Mexico (Henbest and Read, 1944), and with Brown, in a pioneering study of American Cretaceous *Tempskya* (a genus of tree ferns), published in 1934.

Many years later the three would surreptitiously joke about each other, and particularly amusing stories are recalled about Brown-Read disagreements on culinary disciplines in their common office space. One of those issues carried over into the government cafeteria, where a group of paleontologists met frequently for lunch. Some of them objected, usually mildly but in agreement with Read, to Brown's practice of carrying with him a bag of sliced onions with which to garnish his bland cafeteria meal. The story has it that one of these paleontologists—a woman—countered Brown's onion-bag opening at one lunchtime by producing from her purse a bottle of AirWick (then a popular household deodorant), which she placed in the middle of the luncheon table. This expression of distaste for the aroma of freshly sliced onions had the intended result, much to Read's satisfaction.

Personal idiosyncracies notwithstanding, Read and Brown shared their office harmoniously for more than a decade. Brown had arrived on the job in 1928 and was specifically charged with paleobotanical studies pertinent to Cretaceous-Tertiary boundary problems; his career terminated with his retirement in 1959, after he had completed his major paper on "Paleocene flora of the Rocky Mountains and Great Plains," published posthumously in 1962. Read's initial interest in Tertiary paleobotany, of course, constituted a potentially disruptive conflict with Brown's mandated research, and it was only appropriate that Read would devote much of his USGS efforts to studies of Paleozoic plants. His sole work on Tertiary plants was published in 1930. Read recognized four species of coniferous wood and provided excellent descriptions and illustrations. He tentatively concluded that the Lamar River flora is late Eocene or early Oligocene in age. In a second paper on coniferous wood, which was based on the Cretaceous *Pinoxylon dakotense,* Read discussed some of the difficulties inherent to phylogenetic interpretation of the early conifers (Read, 1932).

The 1930s were an extraordinarily productive interval in Read's paleobotanical career. It was substantially uninterrupted except for the two seasons of mapping in Oklahoma, which resulted in Read's coauthorship of six coal maps and related articles during 1934 and 1935. During that decade, Read was solely or partly responsible for about 25 papers, which treated subjects ranging from Devonian to Tertiary in age, with important contributions to the Devonian, Pennsylvanian, and Cretaceous—a remarkable span of competence. Two papers, one a short report on *Trichopitys* from the Carboniferous of Colorado (1933) and the other a more substantial account of a Pottsville flora from the Mosquito Range, Colorado (1934), were direct products of Read's introduction to Paleozoic paleobotany by David White in Colorado in 1930.

Aside from the mutual cultural benefits derived from daily exchanges in their shared office, the Read-Brown association was also scientifically productive. Together they successfully completed their sole collaborative study—their *Tempskya* paper (Read and Brown, 1934). This is a superbly illustrated detailed history and discussion of that extinct Early Cretaceous genus and includes descriptions of two new species. Read (1939b) independently published an interesting theoretical article on the evolution of habit in *Tempskya,* involving derivation of radially symmetrical trunk forms from primitive dorsiventral types.

In 1937 Read digressed from his North American investigations in order to study some late Paleozoic floras from South America. A small collection from the Paracas Peninsula of Peru contained sufficient identifiable plant forms to convince Read that the fossiliferous strata were early Carboniferous rather than late Carboniferous (Westphalian), as believed by others (see Read, 1938a). The exact age of the Paracas flora is still not resolved. A second, broader study was based on an extensive collection of late Paleozoic plants that was sent to David White in 1922 by the Brazilian government, with their request for a publishable analysis. White's declining health in the early 1930s precluded his completion of the project, and it fell to Read to finish it (Read, 1941); this was one of several such tasks that Read would subsequently inherit from White. The publication treated plants ranging from Devonian through Permian in age, and for some years to follow it was an authoritative work on the region. This is a very rare library item because, according to Read, the ship carrying the main supply of copies of the volume was lost at sea.

The 1930s also embraced a series of pioneering investigations of Devonian plants, in which Read's penchant for anatomical studies of permineralized material found expression and exerted considerable impact on later North American paleobotany. One of these studies (Read, 1935) reported the first North American occurrence of the genus *Cladoxylon,* which was based on a single, previously overlooked specimen from western New York. In a second and related study, Read (1939a) described two pyritized psilophytic taxa from the upper Middle Devonian Tully Limestone of western New York: *Arachnoxylon* was presented as a new genus resembling *Asteroxylon,* and *Schizopodium mummii* represented only the second known occurrence of the genus, first discovered in Australia.

Meanwhile, Read was channeling considerable time into studies of the permineralized flora of the New Albany Shale of Kentucky and Indiana. His interest in the plant-bearing phosphatic nodules of those beds had been inspired by the work of Scott and Jeffrey (1914), who described six new species. Read's personal involvement began in 1930, when David White presented him with a stem fragment White had collected several years earlier from the vicinity of Junction City, Kentucky. With White's encouragement, Read spent many weeks collecting. Those excursions included several weeks in 1934 and 1935 in the company of Guy Campbell of Corydon, Indiana, an enthusiastic amateur collector and talented stratigrapher who later published a detailed study of the New Albany Shale (Campbell, 1946).

Four important publications resulted from Read's New Albany Shale experience (Read, 1936a, b, 1937; Read and Campbell, 1939). Read's "A Devonian flora from Kentucky" (1936a) discussed 24 species (15 new) among 18 genera (five new). Read concluded that the flora is Late Devonian, rather than Early Mississippian, a possibility entertained earlier by Scott and Jeffrey. Read (1936b) described *Diichnia kentuckiensis,* a new genus of seed ferns characterized primarily by its double leaf trace. Read interpreted that anatomical feature as an evolutionarily advanced modification over related calamopityean species. This paper was followed in 1937 by Read's monographic review of the family Calamopityeae. It contains Read's only venture into theoretical paleobotany involving the New Albany plants and entails an analysis of relationships among the 16 then-known calamopityean species. This monograph is one of Read's more widely cited efforts. Read's final treatment of New Albany Shale plants described 14 new species (Read and Campbell, 1939). The title clearly implied an intent to produce subsequent contributions, but despite Read's many emphatic declarations to that effect, his New Albany Shale series ended with this "preliminary account." With its 33 genera and 42 species, the New Albany flora was, and remains, an unusually diverse Late Devonian assemblage that is based solely on internal anatomy of stems and petioles, with no information whatsoever on leafy or fertile organs. The authors then agreed that "the flora of the upper part of the New Albany can be regarded only as Upper Devonian." Later, Campbell implicitly repudiated his agreement with Read and assigned the plant-bearing beds to the Lower Mississippian (Campbell, 1946). Campbell's views found staunch support in two papers by Hoskins and Cross (1951, 1952), who first revised four of Read's generic identifications that eminently favored a Devonian age assignment and later published a detailed review of the floras of the Devonian-Mississippian black shales, in which they drew heavily on Campbell's largely conodont-based stratigraphic conclusions. Hoskins and Cross tacitly acceded that of nine genera then indigenous to the New Albany Shale, seven were verifiable and attributable to Read (1936a, b, 1937) and Read and Campbell (1939)—their telegraphic descriptions notwithstanding. Read never did yield his side of the New Albany age controversy.

A little-known aspect of the New Albany Shale studies involved the preparation methods used on the material. Because of the low incidence of good, instructive specimens, it was desirable that a minimally kerf-wasteful sectioning method be applied. It was found that the matrix was soft enough to permit accurate sectioning with a hair-thin copper wire and fine abrasive powder. An adaptation of ancient bow-saw rock-cutting techniques was used, in which wire from a spool of "Rea Company Magnet Wire" (an unused spool with a price label of "50¢" in Read's handwriting now rests as an ornament on Mamay's desk) was seesawed and slowly advanced manually. Most of the sectioning was done by K. J. Murata, a clever young technician whose time was shared by Read and Brown; Murata employed this primitive and tedious technique to prepare more than 600 excellent thin section slides. Murata later distinguished himself as a geochemist with the USGS.

This unparalleled slide collection eventually attracted other investigators. In 1949, a request was made and granted for the loan of the entire collection—to the considerable and outspoken annoyance of Read, who ostensibly still intended to return to his interrupted studies. However, he had been transferred from Washington by that time. Other loan requests were to follow. At the time of this writing, about 20% of the slide collection—including Read's original slides of *Diichnia*—is out on loan to three paleobotanists. Now—nearly 60 years after Read described the genus—it is evident that an unpublished reinvestigation of *Diichnia* apparently confirms Read's appraisal of the plant as an evolutionarily advanced form. Thus, Read's pioneering New Albany Shale tracks were closely followed and have been repeatedly trodden upon but are yet far from obliteration!

Read collaborated with C. W. Merriam (1940) on a rich deposit of Early Pennsylvanian plant compressions in the Spotted Ridge Formation of Oregon. They visited the site in 1939, and because of the extreme toughness of the sediments they resorted to dynamite and made a large, selected collection. Later, Read related to Mamay that "There's plenty left on our scrap pile for the next collector." A subsequent paper by Arnold (1953) described the prominent elements in the flora, and in 1956 Mamay and Read elaborated with descriptions of several new taxa. The Spotted Ridge flora is important because of its geographic isolation in western North America where Pennsylvanian plants are extremely rare and because of the apparently mesic facies represented.

Consistent with his love of the outdoors, Read was dedicated to fieldwork and only tolerant of office routine. He would spend as much time in the field as budgetary constraints permitted, often working alone as an economic measure. During the late 1930s his interests in collecting Mississippian and Pennsylvanian floras kept him occupied in the Appalachian and Midcontinent regions. He was essentially a "loner," not known for regular correspondence with his peers. However, he did maintain an intermittent correspondence with his contemporary, William C. Darrah, between 1935 and 1940; the two had met briefly in 1933 at the 16th International Geological Congress in Washington, D.C., which was attended by a number of American and European paleobotanists (P. Bertrand, W. J. Jongmans, and A. Renier; see Chapters 5 and 7). Their letter exchanges mainly concerned loans of specimens and discussions of prospective field and research plans, the latter aimed at avoidance of conflicts relative to their mutually developing interests in paleobotanical work in the southwestern United States. This cordial correspondence was apparently of mutual benefit, and no conflicts developed. One of Read's letters to Darrah, dated July 3, 1939, is particularly interesting in its exposition of Read's field activities, his overall plans, and his general work philosophy. We quote from this letter:

> At one time, I liked best to stay in the office and work. Now I'd rather get out in the field, and generally am out for four or five months each year. I have just returned from a two months trip which was split into about a month in the southern Appalachians and a month in central Oregon. In the last named place I got a very interesting Pennsylvanian flora. Within the week I plan to leave for the Appalachians for 2–3 months work on the Lower Pennsylvanian. . . . My feeling about field work is that while it may slow up production of papers now, I will benefit by it eventually. . . . At present my big aim is to clear up the floral succession of the Lower Pennsylvanian. The Illinois report is in progress of editing. Now I am going at the Appalachian region and have already put in several seasons of fieldwork. The other things that I'm doing are secondary and serve just as relaxations. [!]

These "relaxations" were soon to dominate Read's research prospectives. For reasons not entirely clear to us, Read redirected his attention to the southwestern United States beginning in 1940, when he collected plants and "geologized" in the Pennsylvanian/Permian of Kansas, Oklahoma, Texas, and New Mexico. He never returned to his once-ambitious Appalachian program, and his subsequent publications on the paleobotany of the eastern United States largely reflected his incredible memory and the voluminous data he had assembled during the 1930s.

## THE ALBUQUERQUE YEARS

Early in 1941, Read was detailed, ostensibly on a temporary basis, to the Fuels Branch of the USGS in order to conduct a mapping program in northern New Mexico. His field assistant, Gordon H. Wood, Jr. (1919–1986), was a young and talented geology student from the University of New Mexico (UNM). They became fast and mutually respectful friends who maintained close contact for the rest of their careers. Wood, the protégé, became a high-level surveyor in the U.S. Coast and Geodetic Survey during the early 1940s. He later joined the USGS in 1944, where he became known for his talents as a Pennsylvanian stratigrapher and structural geologist. He received the Geological Society of America Coal Geology Division's Gilbert H. Cady Award in 1985 and the Department of the Interior's highest award—the Distinguished Service Award—in 1986. For two decades, beginning in 1953 while Ralph L. Miller was his branch chief, Wood supervised the Fuels Branch's anthracite field office in Mount Carmel, Pennsylvania. Wood had total faith in Read's Pennsylvanian floral biostratigraphy, and their interaction played a critical role in Wood's pioneering regional analysis of the Anthracite region.

When the United States entered World War II late in 1941, many USGS geologists—including Read—were reassigned to projects that would contribute directly to the national war effort. Read's assignment kept him in New Mexico with a formal transfer to the Fuels Branch. He took up permanent residence in Albuquerque in 1943 and assumed supervisory charge of the Fuels Branch's then new field office. This unit was responsible for a war-related program of exploration and development of fuel and mineral resources of New Mexico and adjacent areas. Read was delighted with this assignment, not only from the standpoint of his elevated responsibilities but also because of the change in geographic locale. He did not enjoy the crowded living conditions in Washington and had often expressed his envy of the choice western field programs in which many of his Washington peers were engaged. Now he was back in the spacious and beautiful Southwest where he could live near, and work in, the mountains—often in the company of horses!—and could apply his talents toward the nation's wartime priorities.

Read's unit was originally housed in one of the old engineering buildings on the UNM campus. This was the beginning of a long and fruitful association with the university, particularly with the faculty and students of the Department of Geology. Several of the geology faculty worked part-time for the Fuels Branch during the war. Many undergraduate and graduate students at UNM worked part-time for Read's group and after graduation became career employees with the USGS in Albuquerque and elsewhere; some of those students completed graduate theses under Read's direction. He seemed to enjoy teaching and soon became a faculty associate in geology at UNM. Whenever anyone brought him a fossil he would expound at great length on the material and its implications and was particularly pleased when he was invited, in 1963, to teach a course in paleobotany. He was quite gregarious and always welcomed visitors graciously. However, he quickly saw through windbags and suffered no fool gladly! In the office he was always neatly dressed in slacks and open-necked sports shirt; in the field, however, he usually wore cowboy boots and a Stetson hat (Fig. 2)—he was, after all, a native Texan! The only jewelry he wore was a handsome silver and turquoise watchband; a pair of antique silver riding spurs permanently decorated his desk. He was clean shaven except for a small, neatly trimmed moustache; his thinning hair was cut short. He was a chain smoker with a persistent cough, the frequency of which seemed a good barometer of his emotional agitation of the moment.

Figure 2. Charlie Read and Serge Mamay—mentor and protégé (New Mexico, May 1955). Photograph from S. H. Mamay files.

Although the Read contingent initially was badly understaffed, their performance was exceptionally effective. In February 1947, Read received a Superior Accomplishment citation from USGS Director W. E. Wrather, worded thus: "I have been advised that despite the handicap of an incomplete staff, you have accomplished superior results in directing and conducting regional studies in New Mexico toward increasing reserves of oil and gas. In addition you have established unusually effective relationships with the numerous agencies which cooperate with the Survey."

By the early 1950s, it was apparent that the USGS was committed to an extended stay in Albuquerque, and an informal agreement was made that culminated in the construction of a three-story building on the campus of UNM to house not only the Geology Department but the offices of the USGS Fuels Branch and Water Resources Division as well. The building was completed in June 1953, and the Fuels Branch office moved into the second floor later that year (Ash, 1955). By then, Read's office personnel had grown to a total of about 15 full-time geologists and support staff. In addition to working closely with the UNM and the New Mexico Bureau of Mines and Mineral Resources, Read's staff were actively cooperating with many coal and oil companies and U.S. federal agencies, including the War Production Board, the Soil Conservation Service, the Bureau of Reclamation, and the Forest Service. Their numerous publications during the 1940s and 1950s involved such resources as coal (a prime example is the 1950 compilation, by Read et al., notably G. H. Wood, Jr., of the coal resources of New Mexico), copper, salt, trace elements in general, and uranium as associated with copper, coal, and black shales. They also produced many geologic maps, structural studies, and stratigraphic correlations and even material on atomic explosions (Read served as a consultant to the Atomic Energy Commission in reference to site selection for underground atomic explosions in Nevada and New Mexico).

His heavily laden schedule of strategic materials studies and related activities notwithstanding, Read continued to participate significantly in paleobotanical projects through the 1940s and 1950s. In 1942 and 1943, he worked with the National Research Council Committee on Paleobotany, assisting in compilation of summaries of American paleobotany. In 1943, the last of David White's works on Lower Pennsylvanian plants from the Appalachian region appeared posthumously with Read as junior coauthor; Read had expedited that important publication by editing and assembling White's notes.

In 1946, Read described a Lower Pennsylvanian florule, including *Lacoeia,* a new genus of pteridosperm fructifications; this was from the Dutch Mountain outlier of northeastern Pennsylvania. Otherwise, the 1940s were highlighted by documentation of Read's thoughts on upper Paleozoic floral zonation, beginning with his invited contribution to the Pennsylvanian correlation chart assembled by Moore et al. (1944).

This entailed nine Appalachian floral zones (See Chapter 8, this volume), most of which were distinguished by a single species; this system drew heavily from the works of David White and to a lesser extent from writings by W. C. Darrah, E. Dix, W. J. Jongmans, and others. Read included no explanatory text, but the columnar format demonstrated for the first time the taxonomic similarities and stratigraphic correspondence between the Pennsylvanian floral zones of North America and those of South Wales as delineated by Dix (1934). Later Read (1947) extended his zonation to include floras of the Midcontinent region and also provided fairly extensive lists of the taxa associated with some of his designated guide fossils. Here he discussed floristic provinciality as observable in the Arcto- and Antarcto-Carboniferous floras and introduced the term *Cordilleran flora* in reference to botanical modifications noted in the Rocky Mountains.

Because of his flair for administration, Read was transferred back to Washington in 1952 for a two-year term as assistant chief of the Fuels Branch under Branch Chief Ralph L. Miller. Read accepted the transfer reluctantly, as the return to the big-city scenario was a matter of considerable annoyance to him. However, he made the best of it and acquitted himself admirably, performing routine and often unattractive tasks while maintaining an outwardly unruffled and dignified mien. It was during this period that an educational and entertaining friendship developed between Read and one of the present writers (SHM). Mamay writes: "Soon after I arrived in Washington in 1951, Charlie showed up for preliminary discussions of his impending transfer, and Roland Brown introduced us. I was slightly apprehensive about this meeting because Brownie had advised me that Read had his brusque moments, and I felt that he might look askance at the young fellow who had arrived to take over his former job as the Paleozoic paleobotanist with the Paleontology and Stratigraphy Branch ("P and S") of the USGS. However, rapport quickly developed and we soon became good friends.

"I remember him as a tall, straight, and slender man who evidently took pride in his personal appearance, always nattily dressed in suit and necktie and 'trademarked' by his trim moustache. He walked gracefully but attributed his characteristic gait to 'limping equally on both legs. . . .' He was ordinarily affable and soft-spoken, but his direct, laconic conversational style bespoke of his impatience with circumlocution; his often blunt deliveries would strike others as overly abrupt and encourage caution in future dealings. Probably this characteristic, which stemmed in part from a pronounced stubborn streak, was largely responsible for getting Charlie saddled with supervisory jobs! During this administrative interlude Charlie kept his hand in paleobotanical endeavor, frequently visiting the National Museum, where he unobtrusively examined specimens and worked on manuscripts. He gave generously of his time, sharing his knowledge and experience with me. He pointed out many interesting aspects of his undescribed collections from the Southwest, and repeatedly urged me to continue with his initial studies of the Paleozoic of that area, particularly the Permian of Texas. Before he returned to Albuquerque in 1954 he had completed a manuscript on *Prosseria grandis* (Read, 1953), a problematical Devonian plant from New York state; his important manuscript on the Mississippian floras of the Appalachian region (Read, 1955); and a collaborative discussion of Permian floras in the southwestern United States, which I presented at the 8th Botanical Congress in Paris (Mamay and Read, 1954). Additionally, we had tentatively agreed to meet in Texas the following spring, when he would introduce me to southwestern Permian floras.

The trip materialized in 1955 as a joint venture with Ellis Yochelson (P and S), whose interests lay in Permian gastropods and who hoped to find some in the general area of my plant collecting. We met Charlie in Dallas, and spent about two weeks on an exhaustive guided tour that touched on many areas in north-central Texas, the Albuquerque area, and points in between and beyond—with Charlie talking almost all the time! He talked mostly about geology, paleontology, and mapping techniques, of course, but interjected many discourses on such varied topics as modern biota, Mexican cuisine (which he loved—the hotter the better!), Spanish-American religious practices in New Mexico, local history and folk lore, meteorology, and flying saucers (he claimed to have observed UFO's twice). His mischievous sense of humor surfaced in his role as arbiter between his two constantly quibbling young colleagues; he would divide his allegiance exactly equally, siding with one or the other on alternate days—no matter how preposterous the argument of the moment (the photograph in Fig. 2 was taken on one of Mamay's days)! After we left Charlie in Albuquerque, we agreed that those memorable two weeks had been educational and inspirational beyond expectation. Charlie had given us many valuable insights into southwestern U.S. geology, and I had been steered into a long and busy career in southwestern Paleozoic paleobotany."

Read devoted most of 1956 and 1957 to a "pet" project—a regional study of the mineral and fuel resources of the Sangre de Cristo Mountains of northern New Mexico and southern Colorado, with A. A. Wanek as his principal collaborator (Read et al., 1964). Their reconnaissance maps and accompanying data were generously made available to G. D. Robinson et al. (1964) for use as an essential ingredient in their book on the Philmont Ranch region (Fig. 3). The Philmont book became one of the USGS's "best sellers," and Read's role in its production gained the commendation of USGS Director T. B. Nolan, whose letter of June 18, 1965, reads:

> I was very happy to see the paper on the Philmont country singled out for greatly deserved recognition. . . . I am much impressed with the way the geologic story is presented in a happy blend of a textbook on principles and a guidebook of the region. . . . In a very short time it has become one of our most popular publications, and thus it will reflect favorably on the Survey in many quarters. . . . Now is a specially appropriate time to recognize your integral role in the Philmont project, visualizing it and preparing it, and contributing indispensable

Figure 3. Charlie Read on the Philmont project (Sangre de Cristo Mountains, New Mexico, 1956). Photograph from C. B. Read files.

knowledge and advice to its authors . . . It is a pleasure to extend to you both my official and warm personal commendation.

The 1950s also saw Read's considerable influence in activities of the New Mexico Geological Society. According to Beaumont et al. (1979), "Mr. Read was a member of the organizing committee that formed the New Mexico Geological Society in 1947. He was the fifth President of the Society (1951–52), and was elected as Honorary Member in 1956. He co-edited the Rio Chama guidebook. He was Chairman of the 1st, 2nd, 7th, 12th and 14th Road Log Committee and Co-Chairman of the 18th field conference [1967]."

Read's bibliography for the 1956–1959 period shows him as a coauthor of about 20 publications, which include several guidebooks, revisions in New Mexico stratigraphic nomenclature, and studies of uranium, salt, and coal. His paleobotanical production had begun to decelerate, however. Except for the Spotted Ridge flora (Mamay and Read, 1956), it was limited to three abstracts precursive of later papers on Paleozoic floral zones and *Tempskya.*

Read's career underwent a drastic change in professional tempo and outlook in 1960. The wartime exigencies responsible for the creation of the Albuquerque office had diminished to the point where maintenance of the unit was no longer economically feasible. The Albuquerque office was disbanded and its personnel reassigned. Charlie was transferred back to the USGS Paleontology and Stratigraphy Branch for administrative reasons, but he was permitted to remain in Albuquerque. During this transition period, one of the present authors (SRA) became acquainted with Read, and yet another career in paleobotany was beginning to form. Ash relates: "I first met Charlie in 1955 after being discharged from active duty with the Seabees. I began working for the Ground Water Branch of the USGS, which was housed in the same building with Charlie's group. At that time my brother was working in eastern New Mexico as a field assistant for one of the geologists in the Fuels office, and his paychecks periodically got misplaced in Washington. I met with Charlie in an attempt to track down the checks and was immediately impressed by his compassion for the well-being of his employees. He made long distance phone calls and wrote several letters on my brother's behalf and eventually got the matter resolved.

"I had planned to obtain a Master's degree in geology under the GI Bill, but as I observed Charlie and his activities my interest in ground water waned and paleontology became increasingly attractive. Thus, when the Fuels Branch office was being closed in 1960 I somehow managed a lateral transfer to the Paleontology and Stratigraphy Branch—a transfer that proved to be a serendipitous opportunity for me!

"My work with Charlie involved several unrelated projects that emphasized his breadth of interests and abilities. Most of these were eventually published, although the principal one was never completed. This was David White's massive monograph—about 1,000 manuscript pages and 100 plates—of the Pennsylvanian (Pottsville) flora of Illinois, which had been set aside during World War II. Charlie was assigned the job of reducing it to manageable size for publication, and through working on it with him I learned much about Pennsylvanian floras. Although we spent considerable time condensing and improving the manuscript, it never reached the publication stage.

"A chance discovery in the fall of 1961 led to our recognition of a rich Upper Pennsylvanian fossil locality in the Kinney Brick Company clay quarry a few miles east of Albuquerque. We had suspected the quarry to be the source of a Carboniferous fern specimen that someone had left at the Geology Department without proper locality data. Therefore, when Charles Felix and an assistant from the Sun Oil Company visited us while prospecting for Pennsylvanian palynological material, Read asked me to guide them through the Manzano Mountains and include a stop at the Kinney quarry. The site was fossiliferous indeed! That day I collected many plant fossils and a few invertebrates; later I found a large insect wing, the first from the quarry. Soon fossil fish were also discovered by John Bradbury, a UNM graduate student. Charlie regarded the site as very important, if not unique, because of the diversity of its fossil biota; he suggested that the pit would eventually warrant as much scientific recognition as the famous Florissant lake beds of Colorado.

" A study of the Cretaceous fern *Tempskya* occupied much of my time in Charlie's laboratory. The genus had previously been unknown in New Mexico, but Robert (Bob) A. Zeller, who taught at UNM in 1950 and 1951, had found silicified plant material in the Albian of southwestern New Mexico and showed it to Charlie, who recognized it as *Tempskya.* Zeller and Read [1956] published a short announcement of this geographically isolated occurrence. Zeller continued to submit specimens to Charlie, and soon I was put to work peeling, sectioning, and photographing Tempskyas (Charlie had taught me all the techniques for thin-sectioning, as well as photography and darkroom procedures). It became apparent that two new species of *Tempskya* were present and that the genus was more restricted, stratigraphically, than previously thought. After two preliminary reports [Read and Ash, 1961a, b], we began work on a monograph of the American *Tempskya,* but its completion was delayed by my departure from Albuquerque in 1964 and by Read's diverted interests and failing health. Ultimately I was enlisted by the USGS to complete the monograph, which I did in 1971 with little assistance from Charlie. Fortunately, he did live to see its publication (Ash and Read, 1976).

"Another chance occurrence was responsible for my career in Triassic paleobotany. Among the collections that Charlie brought to Albuquerque from Washington were a few fossils he had collected in 1941 from Upper Triassic strata—then thought to be Permian—in New Mexico and Texas. Charlie saw close resemblances of these plants to plants previously described from the Upper Triassic Chinle Formation in eastern Arizona [Daugherty, 1941]. Charlie felt that his fossils represented a poorly known flora that warranted some original research. Accordingly, I began a search for Charlie's old localities, particularly in the area near Fort Wingate in western New Mexico. I found several dozen localities there, and with Charlie's continued encouragement I extended my searches through other parts of the West. I accumulated a large number of Triassic specimens, some of which seemed to represent new taxa. The cuticle was preserved on many of my specimens, so Charlie advised me to correspond with the leading Early Mesozoic paleobotanist, Professor T. M. Harris of Reading University in England. As a result of Charlie's suggestions I did correspond with Harris and went to England in the fall of 1964, when I began my studies of the Chinle flora under him. I received my Ph.D. in 1966 and returned to the States to begin my teaching career at Midland Lutheran College in Fremont, Nebraska, and to continue my investigations of Triassic plants."

Read coauthored about 30 publications between 1960 and 1969. These are almost exclusively road logs or reflections of work accomplished before disbandment of the Albuquerque office. However, Read's long-standing interest in Paleozoic floral zonation came to final fruition as his sole notable paleobotanical publication for that decade. He had long considered publishing an illustrated floral zonation for the entire Upper Paleozoic of the United States (see Chapter 8, this volume), and in 1961 he invited Mamay to join in coauthorship of such an undertaking, with principal responsibility for the Permian. The published result (Read and Mamay, 1964) presents a system of 15 floral zones (3 Mississippian, 9 Pennsylvanian, and 3 Permian) with a discussion of Paleozoic floral provinciality in North America. Read also anticipated the eventual recognition of an additional floral zone between the Mississippian and Pennsylvanian—a prediction that was later substantiated by W. H. Gillespie and H. W. Pfefferkorn (U.S. Geological Survey, 1978). This paper (Read and Mamay, 1964), with its extensive illustrations of the key fossils and its glossary of stratigraphic terms, has been received as an extremely useful article. It has been cited frequently and criticized occasionally, but remains unsuperseded.

During most of the 1960s Read exerted a considerable influence on New Mexico paleontology through his persistent publicizing of the Kinney quarry as a potential source of important paleontologic findings. The fossil biota of the lower Virgilian (Upper Pennsylvanian) sediments exposed there, first noted by Ash in 1961, included a truly remarkable association of plants, invertebrates, and vertebrates. Mamay responded to Read's persuasive invitations, visited the site in 1967 and 1969, and obtained a large collection. During these visits, Mamay and his assistant, A. D. Watt, were treated to many helpful and hospitable gestures by Read and Gladys, his second wife. The Reads lived in a modest but tidy and comfortable home on the outskirts of Albuquerque, where Read enjoyed cooking (he won a prize from *Sunset Magazine* for one of his recipes) and vegetable gardening. (He was proud indeed of the powerful pungency quotient of his home-grown Jalapeno peppers, with which he would "surprise" any unsuspecting colleague who might sample Charlie's contribution to the field luncheon!) Mamay relates a particularly memorable example of Charlie's generosity: "Charlie was fond of Scottish terriers and had a pair of those interesting dogs when I visited him in 1967. I offhandedly commented that my children were anxious to have a dog (I didn't say *I* wanted one, because I didn't). Later that summer Charlie telephoned me at home to tell me that he had two puppies left, a male and a female, and which one did I want? Again under the pressure of Charlie's persuasion, I decided to accept the female, and Charlie sent the beautiful pedigreed puppy by air as a gift for my children. "Missie" was a lively and entertaining member of the Mamay household for more than a decade, and my children often expressed their gratitude to the thoughtful friend they never met."

Charlie had grandiose plans for a "Kinney cooperative project," including a chain-link fence around the quarry, Quonset huts for preparation and storage, and other facilities for the large group of participating visiting scientists he envisioned. His plan was not practical, of course, and Charlie's ambition was not realized during his lifetime. He did witness a good beginning, however, in the hundreds of fossil fish collected by David Dunkle of the Smithsonian in the early 1960s and in the collection gathered by Mamay and Watt, which includes not only plants but also fish, insects, myriapods, cephalopods, shrimps, mollusks, and eurypterids.

## THE FINAL YEARS

Two decades after Read's retirement, a volume was published that approximately equates with his vision (Zidek, 1992). This volume, prepared under the joint auspices of the New Mexico Museum of Natural History and the New Mexico

Bureau of Mines and Mineral Resources, contains 23 articles treating all the biologic groups present in and the various geological aspects of the Kinney quarry; an illustrated summary of the flora is among them (Mamay and Mapes, 1992). Mamay had published two earlier articles on Kinney plants, each of which honored Read with a taxonomic patronym—*Dicranophyllum readii* (Mamay, 1981) and *Charliea manzanitanum* (Mamay, 1990). Earlier, Read was similarly recognized for his fossil insect collecting by the taxon *Sandiella readi* (Carpenter, 1970).

Charlie Read's professional career ended in 1969. He had suffered broken hip and leg bones, and a stubborn healing process required intermittent hospitalization and left him confined to a wheelchair. However, his name appeared later as junior coauthor of his two final paleobotanical publications, each of which echoed early, pioneering interests and linked him in publication with one or the other of his two paleobotanical protégés. A 1971 paper with Andrews and Mamay is an anatomical study of *Phytokneme rhodona,* a new genus of Late Devonian plants from Kentucky that recalls Read's New Albany Shale studies; by coincidence, this article also united Mamay in published collaboration with both his mentors. And the comprehensive *Tempskya* study (Ash and Read, 1976) culminated Read's nearly five decades of thoughts on that genus. It remains the definitive *Tempskya* report.

The combination of Read's age and physical decline and the USGS's budgetarily dictated reduction of small satellite offices inexorably resulted in Charlie's retirement in 1972. Mamay writes: "I was given the awkward job of closing Charlie's Albuquerque office. I found him a forlorn figure, seemingly ill beyond his orthopedic problems and hopelessly confined to his wheelchair. However, he was lucidly loquacious, vividly reminiscing on his youth and professional career. He seemed resigned to his prospects, while retaining fierce pride in his past. Although unhappy with some of the events preceding his retirement, he fully understood my role in ending his Albuquerque saga, and tried graciously not to embarrass me. He wasn't able to manage the trip to his office, but would telephone me there every morning to offer suggestions as to the disposition of his books and equipment, and to reminisce some more. Just before our final—and not entirely unemotional—"Goodbye," Charlie offered some unsolicited but prophetic advice regarding developments in my own USGS career. My last contact with my old friend was a letter from Charlie, written December 9, 1977, from his Albuquerque home."

Read left Albuquerque and returned to Texas. There he resided under the watchful eyes of Beverly (Gladys's daughter) and Carl Montera until his death. Predeceased by Gladys, Read died on August 30, 1979, in a Houston nursing home; he is buried in Albuquerque. A 1991 handwritten note by Ralph L. Miller indicates that Charlie had suffered from Alzheimer's disease for several years.

Read is survived by a son by his first marriage, William Alfred Read. In a letter to Ash (June 2, 1992), Al Read fondly recalled his adventuresome boyhood experiences in his father's camps (Fig. 4). He reminisces: "Beginning at Cowles [New Mexico] I learned from my father to love the mountains. . . . It was my job often to feed the horses, round them up, help saddle them and, when I was with my father and Gordon [Wood], to take care of the horses in the mountains when they would leave for much of the day to explore outcrops. . . . He was very kind but quite tough and I was not allowed to complain. . . ." Al Read is now a successful mountaineering entrepreneur who coowns a tour company in San Francisco that specializes in nontraditional tours of the Far East; he also operates a summer mountain-climbing school in Jackson Hole, Wyoming.

## FINAL THOUGHTS

With the passing of Charles B. Read, geology and paleobotany lost a dedicated, illustrious figure. His knowledge of the regional geology of the Southwest, particularly New Mexico, was legendary, and many of the maps produced by Read and his unit during and immediately after World War II are still the only ones available for parts of New Mexico. Of his outdoorsmanship, it is told that he could smell quicksand and even befriend game wardens! His career as a geologist was succinctly summarized by Beaumont et al. (1979). "Charlie Read played a major role in the growth of New Mexico geology both directly through his personal investigations and indirectly through his

Figure 4. A summer 1950 field group, Newcomb, New Mexico. Bottom row, left to right: P. Komalarajun, Ruben Pesquera. Top row, left to right: Don Ziegler, Charlie Read, Ed Beaumont, Al Read, Louis Gardner.

direction of Survey investigations, and the counseling and guidance of numerous students and geologists."

His paleobotanical eminence evinces a talented, versatile, and flexible makeup. He worked comfortably and fluently, pioneering in diverse lines of investigation, while also seeing to the resuscitation of long-neglected areas of paleobotany through his low-keyed but persistent guidance of younger colleagues. Yet Charles B. Read was woefully unsung by his colleagues during his lifetime, despite the astonishing dimensions of his accomplishments and the predictable durability of his influences. This acclamation is long overdue.

## ACKNOWLEDGMENTS

Much of the material presented in this portrait of Charles B. Read derives from the personal recollections, field notes, and correspondence files of Ash and Mamay. We were given access to Read's personnel file by the USGS, Branch of Personnel, and Elsie Darrah Morey provided copies of several letters from Read to William C. Darrah. E. C. Beaumont gave helpful comments on the manuscript and provided the photographs shown in Figure 1 and 4. J. P. Ferrigno and L. V. Thompson gave darkroom assistance. A telephone discussion with W. A. Read was most enlightening.

A bibliography of Read's publications was prepared by Paul Lyons and B. L. Graff (USGS Library, Reston). It includes about 100 references; about 45 are of paleobotanical nature. We doubt that this is a complete list, or even that one exists. However, it provides an important chronological insight to Read's otherwise poorly documented career history. In addition to Read's principal paleobotanical publications, the bibliography included here also refers to other works that impacted on or otherwise reflect Read's career. J. T. Dutro, Jr., N. F. Sohl, and C. Wnuk reviewed the manuscript.

## REFERENCES CITED

Andrews, H. N., Read, C. B., and Mamay, S. H., 1971, A Devonian lycopod stem with well-preserved cortical tissues: Palaeontology, v. 14, pt. 1, p. 1–9.

Arnold, C. A., 1953, Fossil plants of Early Pennsylvanian type from central Oregon: Palaeontographica, Abt B, Band 93, p. 61–68.

Ash, H. O., 1955, The history of the Geology Department of the University of New Mexico: The Compass of Sigma Gamma Epsilon, v. 33, p. 3–8.

Ash, S. R., and Read, C. B., 1976, North American species of *Tempskya* and their stratigraphic significance: U.S. Geological Survey Professional Paper 874, p. 1–42 (with a section on stratigraphy and age of the *Tempskya*-bearing rocks of southern Hidalgo County, New Mexico, by R. A. Zeller, Jr.)

Beaumont, E. C., Kelley, V. C., and Northrop, S. A., 1979, Dedication, *in* New Mexico Geological Society, 30th Field Conference, Guidebook, Santa Fe Country, October 4–6, 1979: Albuquerque, New Mexico Geological Society, p. vi–vii.

Brown, R. W., 1962 (posthumous), Paleocene flora of the Rocky Mountains and Great Plains: U.S. Geological Survey Professional Paper 375, 119 p.

Campbell, G., 1946, New Albany Shale: Geological Society of America Bulletin 57, p. 829–908.

Carpenter, F. M., 1970, Fossil insects from New Mexico: Psyche, v. 77, p. 400–412.

Daugherty, L. H., 1941, The Upper Triassic flora of Arizona: Carnegie Institution of Washington Publication 526, 108 p.

Dix, E., 1934, The sequence of floras in the Upper Carboniferous, with special reference to South Wales: Royal Society of Edinburgh Transactions, v. 57, p. 789–838.

Henbest, L. G., and Read, C. B., 1944, Stratigraphic distribution of the Pennsylvanian Fusulinidae in parts of the Sierra Nacimiento of Sandoval and Rio Arriba Counties, New Mexico: U.S. Geological Survey Oil and Gas Investigations, Chart 2.

Hendricks, T. A., and Read, C. B., 1934, Correlations of Pennsylvanian strata in Arkansas and Oklahoma coal fields: American Association of Petroleum Geologists Bulletin, v. 18, p. 1050–1058.

Hoskins, J. H., and Cross, A. T., 1951, The structure and classification of four plants from the New Albany Shale: American Midland Naturalist, v. 46, p. 684–716.

Hoskins, J. H., and Cross, A. T., 1952, The petrification flora of the Devonian-Mississippian black shale: The Palaeobotanist, v. 1, p. 215–238.

Mamay, S. H., 1981, An unusual new species of *Dicranophyllum* Grand'Eury from the Virgilian (Upper Pennsylvanian) of New Mexico, U.S.A.: The Palaeobotanist, v. 28–29, p. 86–92.

Mamay, S. H., 1990, *Charliea manzanitana,* n. gen., n. sp., and other enigmatic parallel-veined foliar forms from the Upper Pennsylvanian of New Mexico and Texas: American Journal of Botany, v. 77, p. 858–866.

Mamay, S. H., and Mapes, G., 1992, Virgilian plant megafossils from the Kinney Brick Company quarry, Manzanita Mountains, New Mexico: New Mexico Bureau of Mines and Mineral Resources Bulletin 138, p. 61–85.

Mamay, S. H., and Read, C. B., 1954, Differentiation of Permian floras in the southwestern United States, *in* Program, Eighth International Botanical Congress, Paris: Paris-Nice, French/Committee for the Eighth International Botanical Congress, Section 5, p. 157–158.

Mamay, S. H., and Read, C. B., 1956, Additions to the flora of the Spotted Ridge Formation in central Oregon: U.S. Geological Survey Professional Paper 274-I, p. 211–226.

Moore, R. C., and others (the Pennsylvanian Subcommittee of the National Research Council Committee on Stratigraphy), 1944, Correlation of Pennsylvanian formations of North America: Geological Society of America Bulletin, v. 55, p. 657–706.

National Research Council, 1931, Report of the Committee on Paleobotany (Appendix P of annual report of Division): Washington, D.C., Division of Geology and Geography, 13 p.

Read, C. B., 1930, Fossil flora of Yellowstone National Park. I: Coniferous woods of Lamar River flora: Carnegie Institution of Washington Publication 416, p. 1–19.

Read, C. B., 1932, *Pinoxylon dakotense* Knowlton from the Cretaceous of the Black Hills: Botanical Gazette, v. 93, p. 173–187.

Read, C. B., 1933, A new *Trichopitys* from the Carboniferous of Colorado: Washington Academy of Sciences Journal, v. 23, p. 461–463.

Read, C. B., 1934, A flora of Pottsville age from the Mosquito Range, Colorado: U.S. Geological Survey Professional Paper 185-D, p. 79–96.

Read, C. B., 1935, An occurrence of the genus *Cladoxylon* Unger in North America: Washington Academy of Sciences Journal, v. 25, p. 493–497.

Read, C. B., 1936a, A Devonian flora from Kentucky: Journal of Paleontology, v. 10, p. 215–227.

Read, C. B., 1936b, The flora of the New Albany Shale. Part 1: *Diichnia kentuckiensis,* a new representative of the Calamopityeae: U.S. Geological Survey Professional Paper 185-H, p. 149–161.

Read, C. B., 1937, The flora of the New Albany Shale. Part 2: The Calamopityeae and their relationships: U.S. Geological Survey Professional Paper 186-E, p. 81–104.

Read, C. B., 1938a, The age of the Carboniferous strata of the Paracas Peninsula, Peru: Washington Academy of Sciences Journal, v. 28, p. 396–404.

Read, C. B., 1938b, A new fern from the Johns Valley Shale of Oklahoma: American Journal of Botany, v. 25, p. 335–338.

Read, C. B., 1939a, Some Psilophytales from the Hamilton Group in western New York: Torrey Botanical Club Bulletin, v. 65, p. 599–606.

Read, C. B., 1939b, The evolution of habit in *Tempskya:* Lloydia, v. 2, p. 63–72.

Read, C. B., 1941, Plantas fósseis do Neo-paleozóico do Paraná e Santa Catarina: Rio de Janeiro, Brazil, Divisão de Geologia e Mineralogia, Ministereo da Agriculture, Monografia 12, 102 p.

Read, C. B., 1946, A Pennsylvanian florule from the Forkston coal in the Dutch Mountain outlier, northeastern Pennsylvania: U.S. Geological Survey Professional Paper 210-B, p. 17–27.

Read, C. B., 1947, Pennsylvanian floral zones and floral provinces: Journal of Geology, v. 55, pt. 2, p. 271–279.

Read, C. B., 1953, *Prosseria grandis,* a new genus and new species from the Upper Devonian of New York: Washington Academy of Sciences Journal, v. 43, p. 13–16.

Read, C. B., 1955, Floras of the Pocono Formation and Price Sandstone in parts of Pennsylvania, Maryland, West Virginia, and Virginia: U.S. Geological Survey Professional Paper 263, p. 1–31.

Read, C. B., and Ash, S. R., 1961a, The stratigraphic significance of the fossil fern *Tempskya* in the western United States [abs.]: New Mexico Geological Society, 12th Field Conference, Guidebook, Albuquerque County, p. 198.

Read, C. B., and Ash, S. R., 1961b, Stratigraphic significance of the Cretaceous fern *Tempskya* in the western conterminous United States: U.S. Geological Survey Professional Paper 424-D, Art. 383, p. D250–D254.

Read, C. B., and Brown, R. W., 1934, American Cretaceous ferns of the genus *Tempskya:* U.S. Geological Survey Professional Paper 186-F, p. 105–131.

Read, C. B., and Campbell, G., 1939, Preliminary account of the New Albany Shale flora: American Midland Naturalist, v. 21, p. 435–453.

Read, C. B., and Mamay, S. H., 1964, Upper Paleozoic floral zones and floral provinces of the United States (with a glossary of stratigraphic terms, by Grace Keroher): U.S. Geological Survey Professional Paper 454-K, 35 p., 19 pls.

Read, C. B., and Merriam, C. W., 1940, A Pennsylvanian flora from central Oregon: American Journal of Science, v. 238, p. 107–111.

Read, C. B., Duffner, R. T., Wood, G. H., Jr., and Zapp, A. D., 1950, Coal resources of New Mexico: U.S. Geological Survey Circular 89, 24 p.

Read, C. B., Wanek, A. A., Robinson, G. D., Hayes, W. H., and McCallum, M., 1964, Geologic map and sections of the Philmont Ranch region, New Mexico: U.S. Geological Survey Miscellaneous Geological Investigations Map L-425, 2 sheets.

Robinson, G. D., Wanek, A. A., Hays, W. H., and McCallum, M. E., 1964, Philmont country, the rocks and landscapes of a famous New Mexico ranch: U.S. Geological Survey Professional Paper 505, 152 p.

Scott, D. H., and Jeffrey, E. C., 1914, On fossil plants showing structure, from the base of the Wavery Shale of Kentucky: Royal Society of London Philosophical Transactions, Series B, v. 205, p. 315–373.

U.S. Geological Survey, 1978, Geological Survey Research 1976: U.S. Geological Survey Professional Paper 1100, p. 230.

White, D., 1943, Lower Pennsylvanian species of *Mariopteris, Eremopteris, Diplothmema,* and *Aneimites* from the Appalachian region (assembled and edited by C. B. Read): U.S. Geological Survey Professional Paper 197-C, p. 85–140.

Zeller, R. A., and Read, C. B., 1956, Occurrence of *Tempskya minor* in strata of Albian age in southwestern New Mexico [abs.]: Geological Society of America Bulletin, v. 67, pt. 2, p. 1804.

Zidek, J., ed., 1992, Geology and paleontology of the Kinney Brick quarry, Late Pennsylvanian, central New Mexico: New Mexico Bureau of Mines and Mineral Resources Bulletin 138, 242 p.

Manuscript Accepted by the Society July 6, 1994

Printed in U.S.A.

Geological Society of America
Memoir 185
1995

# *Leonard Richard Wilson (1906– ): Palynologist, paleobotanist, and geologist*

**Robert M. Kosanke**
*U.S. Geological Survey, MS 919, Box 25046, Denver Federal Center, Denver, Colorado 80225-0046*
**Aureal T. Cross**
*Department of Geological Sciences, Natural Science Building, Michigan State University, East Lansing, Michigan 48824-1115*

## ABSTRACT

**Leonard Richard Wilson is a pioneering American palynologist. He has contributed extensively through his teaching, research, and publications in palynology, paleobotany, limnology, glacial geology, archeology, and several other areas of geology and botany. His role as an educator has been outstanding from his early days at Coe College in 1934 to his long tenure at the University of Oklahoma, including assignments at several other academic institutions. Many of his students are leaders in palynology and related fields today. His contributions include one of the earliest applications of palynological technology to industrial needs, particularly in oil exploration and coal science.**

## INTRODUCTION

Leonard Richard Wilson was born of sturdy Viking stock in Superior, Wisconsin, July 23, 1906. He traces his roots from the 1500s Gunn Clan in Norway, through three centuries of William Gunn's sons' descendants (hence "Will's son" to "Wilson"), in Viking communities in the Orkney Islands, to a similar community in Thurso, Scotland. Thurso borders the Pentland Firth, the strait that separates the Orkneys from the northern mainland of Scotland.

Richard was the elder of two sons of Ernest and Sara Jane Cooke Wilson. His younger brother, Harold, born in 1909, died of heart complications following a bout with influenza in 1926, when Wilson was a freshman at the University of Wisconsin at Madison.

Richard Wilson married Marian DeWilde, of German descent, September 1, 1930. They have one son, Richard Graham, who lives near Greenland, Arkansas. Wilson's daughter, Marcia Graham, lives near her parents in Norman, Oklahoma.

Wilson is semiretired and lives with Marian in their comfortable home with library, study, microscopes, and memorabilia in Norman, Oklahoma. He still grows a wide variety of flowers and other plants annually and tends a very large, diverse vegetable and fruit garden, complete with grape arbors and fruit trees. He does some research and writing at his office in the Oklahoma Museum of Natural History in Norman.

## YOUTH

Richard Wilson and his younger brother grew up in Superior, Wisconsin. A near neighbor was Dr. George Conklin, a physician, who was an expert on bryophytes and curator of the Sullivant Moss Society Collection. Conklin had a large garden that included many ferns, and from this early exposure to plants, Wilson learned much about ferns, lycopods, and bryophytes. Conklin was also interested in the Boy Scouts, and Richard became the first Eagle Scout in Superior. Conklin was doing research on freshwater sponges, a subject that interested young Wilson and later became valuable to him in his studies of Wisconsin lakes.

Wilson had a paper route for several years and came to know a number of his customers, some of whom were at the Superior State Teachers College (now University of Wisconsin–Superior). One of these was Dr. J. A. Merrill, professor of geology and geography at the Superior State Teachers College, who had done his doctoral dissertation at Harvard Uni-

Kosanke, R. M., and Cross, A. T., 1995, Leonard Richard Wilson (1906– ): Palynologist, paleobotanist, and geologist, *in* Lyons, P. C., Morey, E. D., and Wagner, R. H., eds., Historical Perspective of Early Twentieth Century Carboniferous Paleobotany in North America (W. C. Darrah volume): Boulder, Colorado, Geological Society of America Memoir 185.

versity on a Texas Cretaceous problem. He had also published the first paper on fossil hystrichosphaerids in the United States. This intrigued Wilson. Later Merrill became president of the college, and after he retired, he suggested to the next president of the College that Dr. Wilson be hired away from Coe College, which he had joined in 1934, to bring him "home" to Superior. However, Wilson did not accept the offer because he was just developing his own geological and palynological program at Coe.

Wilson learned to ski cross-country as a boy from Dr. Conklin and later became interested in down-hill skiing and ski jumping. He broke his back at age 19 while practicing ski jumping for an Olympic tryout. This was always some handicap, and many years later it was exacerbated when a tree he was felling in Arkansas, after his retirement, bounced from a nearby stump and fell on him, breaking his back again. He was flown to a medical center in Oklahoma a few days later and has since recovered from that injury relatively well.

Besides hiking, rugged fieldwork in many areas, and occasional mountain climbing, he learned fencing and was on the fencing team at the University of Wisconsin, Madison, and also during his year at the University of Leeds. He later initiated and coached fencing at Coe College for several years. Wilson enjoyed bike riding as a young man; while at Leeds he made a three-week, 1,000-mi "push bike" trip to Land's End, the southwesternmost tip of England, and back up to Robin Hood's Bay, north of Scarborough at the east edge of the North York moors, in time for a Leeds University Biological Station short course for which he was scheduled.

## EDUCATION

Wilson's education and experiences at the University of Wisconsin–Madison were as varied as his mentors, Norman C. Fassett (systematic botany), Chancey Juday (wildlife, limnology), E. A. Birge (zoology, limnology), William H. Twenhofel (sedimentology), and F. T. Thwaites (glacial geology). When Wilson was a freshman student assistant in 1926, Norman C. Fassett, professor of botany, became impressed with Wilson's interest and knowledge of plants and invited him to go along on a botanical foray to southwestern Wisconsin. This hilly region in the "Driftless Area" along the Upper Mississippi River and adjacent regions thinly covered with older glacial drift were ideal for excellent collecting and rough camping as well. Fassett was a brilliant botanist but a penurious leader by reason of extremely limited funds for such work. They survived on fish (mostly canned), bread, and local vegetables. Since "roughing it" was a necessity, their clothes were generally soiled and in poor repair, which accentuated their unkempt appearance. Dogs considered them to be undesirable ragamuffins and attacked them regularly. Wilson reminisced later to some of his advanced students that it was several years before he realized how pathetic and unprofessional they must have looked to the local populace. But carrying his heavy vasculum and attentive to, and entranced by, Fassett's wealth of knowledge of Wisconsin floras, Wilson was unaware of their abject appearance except when reminded by the next angry barking dogs. Fassett later became one of two principal advisers on Wilson's doctoral dissertation.

When Wilson's brother, Harold, died in the fall of 1926, it was necessary for him to drop out and return to Superior to the "Normal" school (now University of Wisconsin–Superior). Wilson returned to the University of Wisconsin for the academic year 1927–1928.

Professor E. A. Birge, former director of the Wisconsin Geological and Natural History Survey and then president of the University of Wisconsin, involved Wilson as a student assistant in conducting studies in zoology and limnology for the Natural History Survey. Professor Birge became a longtime friend and continued some research with Wilson after he went to Coe College in 1934. Birge was principally interested in the physics of light as it affects plant growth in lakes and in other aspects of limnology and zoology.

Wilson's family thought he should study in England a year or two, so for his junior year he selected Leeds University because an exchange student living in the same student house with Wilson at Wisconsin was from Leeds. The exchange student's grandfather was Joseph B. Priestley, who was the discoverer of oxygen. At Leeds, Wilson came under the influence of W. H. Burrell, director of the University Herbarium, who is credited with having published possibly the first palynology paper in England in The Naturalist (1924). Burrell had been working on the Pennine peat bogs when Gunnar Erdtman came to the University to lecture. This sparked Burrell's research on plant remains in peats and other sediments, and Burrell engendered in Wilson a lifelong interest in the study of fossil spores and pollen in peats and coals, an area of research eventually grouped together with the study of other organic remains, under the name "palynology" in 1944. When Wilson returned to Wisconsin he began to weave this new science into the research on his various projects.

Wilson's senior year and graduate school research effort were extremely fruitful and busy. He completed the equivalent of two bachelor's and one master's theses and two doctoral dissertations in combined botanical-geological research. His Ph.B. thesis, 1930, under Dr. Fassett dealt with the spores of the genus *Lycopodium* in the United States and Canada (Wilson, 1934). He also wrote reports on vegetation and plant succession of the Apostle Islands region and compared the flora obtained by pollen analysis with the modern flora there (Wilson, 1935b). The Ph.M. thesis, 1932, and the follow-up papers (Wilson, 1932, 1936), under the direction of Professor F. T. Thwaites and Dr. Fassett, were a combined study of the geology and vegetation of the Two Creeks Forest bed, a major focal point for study of interstadial Late Wisconsinan glacial deposits.

Wilson's Ph.D. dissertation (1936), prepared under the guidance of Dr. Fassett and Professor Thwaites, with some assistance from W. H. Burrell, curator of the Herbarium at Leeds University, and critical reading by Dr. Paul B. Sears, then of Oberlin College, dealt with an analysis of the plant microfossil succession of 10 bogs in Douglas County, Wisconsin (Wilson, 1938). These were selected for their critical positions with reference to the shorelines of Glacial Lakes Duluth and Algonquin and the Nipissing Great Lakes, in order that their fossil spectra could be compared and the vegetational history of the region might be determined.

He established that, as the waters of these lakes receded, there was a clear order of floral succession beginning with the pioneer plant, *Picea*. He was able to associate the effect of soil type as a controlling factor and the influence of certain edaphic factors and fire.

At the same time Wilson prepared another dissertationlike report for the Wisconsin Geological and Natural History Survey on lake development and plant succession in the Highland Lake District and Muskellunge Moraine and outwash of Vilas County, Wisconsin (Wilson, 1935a). This research was conducted under the direction of Professor Chancey Juday and Dr. E. A. Birge, of the Survey, and with guidance by Dr. Fassett. This monograph demonstrated the dynamic ecology and the relationships between aquatic plant associations of several lakes developed on a variety of sites in a complex glacial terrain of terminal and ground moraines, eskers, drumlins, and outwash.

## PROFESSIONAL CAREER

Richard ("Dick") Wilson's professional career has been extensive in both academic and applied areas. He was a teaching research fellow at the University of Wisconsin (1931–1934) and a research associate at the Wisconsin Geological and Natural History Survey (1932–1936). He rose from an instructor of geology to professor of geology at Coe College in Cedar Rapids, Iowa (1934–1946), professor and head of the Geology and Mineralogy Department at the University of Massachusetts (1946–1956), and professor of geology in the Graduate School of Arts and Science at New York University (1956–1957). In 1957, he moved south of the Mason-Dixon Line to become a geologist with the Oklahoma Geological Survey (1957–1977) and professor of geology at the University of Oklahoma at Norman (1957–1962). In 1962, he became curator of micropaleontology and paleobotany of the Stovall Museum of Science and History, now the Oklahoma Museum of Natural History. He was honored by his appointment as the George Lynn Cross Research Professor of Geology and Geophysics at the University of Oklahoma from 1969 until 1977, when he became professor emeritus of geology.

Wilson's expertise in applied palynology resulted in a number of consulting assignments with various oil companies. These include Carter Oil and Humble Oil Companies, subsidiaries of Standard Oil of New Jersey (1945–1962); Union Oil of California (1954), Petrobras and Petroleo Brasileiro, South America (1955); Creole Petroleum Company of Venezuela (1959); Sinclair Research and Refining Company (1964–1966); Continental Oil Company (1954–1972); and numerous independent oil companies.

Wilson (Fig. 1) has directed or was consultant for the research of several graduate students since 1977 and taught a graduate course in palynology at the University of Oklahoma in 1989. He spent the summers of 1950 through 1955 as a visiting professor at the Nova Scotia Bureau of Mines at Antigonish; there, with the late Robert R. Shrock, he conducted geology field camps for graduate students of the Massachusetts Institute of Technology. Wilson recalls that Shrock was the laboratory instructor for his first class in geology at the University of Wisconsin, which was then entitled "Geology I," taught by the renowned W. H. Twenhofel. From 1952 to 1953, Wilson was the director of the Greenland Ice Cap project called Mint Julep.

Wilson was the Melhaupt Scholar at Ohio State University for the academic year 1939–1940, to work on the palynology of Ohio with the eminent ecologist-systematist, Professor E. N. Transeau. Transeau's keen interest in the Postglacial Xerothermic Interval, and the mix of prairie with Ohio woodland floras, was an ideal target for resolution by palynological study. The Wilsons lived in a very historic house in Worthington, about 11 km north of the university. This had been a "station" on the pre–Civil War underground railway: a rest-stop–hiding place for slaves escaping northward toward Cana-

Figure 1. L. R. Wilson, 1992.

da. The "historic atmosphere" of this stately home with its labyrynthian basement and tunnel (and rats!), and massive fireplace with bread-rising ovens, was small comfort for Marian and their young son and infant daughter in the bitter cold winter. However, Wilson, as a botanist-geologist, found solace in the location of the house on a small tributary of the Olentangy River with excellent exposures of the fossiliferous, Devonian-age Ohio Black Shale with its large concretions and plant fossils, *Tasmanites, Protosalvinia* (*Foerstia*), and occasional float pieces of *Callixylon* in the creek and river bed. An exciting place for a paleobotanist!

## WILSON THE TEACHER

Wilson was a born teacher. He was anxious to impart his own broad knowledge of the out-of-doors at every opportunity. He pointed out the flowers, weeds, trees, rocks, fossils, and eskers with equal delight as he herded his classes into the bush or onto the plains. He was like the lead goat in a flock, with his students scrambling after him in some disarray, but trying to keep up, or to hear what he was saying. Or, perhaps, it was more as if Wilson was a soccer ball on the field and the youngsters swarmed after it wherever it chanced to go.

He demanded that students keep notebooks and write field reports within prescribed time frames. The longer the trip, the more pages of report he expected. A zero grade was generally meted out if you missed the field trip or failed to complete a report on time. He would always furnish his own car and seek out local citizens around the community to augment transportation for students who could not crowd into his "Chevy" (Chevrolet automobile). Fossils, rocks, modern plants, and artifacts alike were collected, and even a number of bull snakes or other creatures ended up in gunny sacks—these he carried around in his car or put on the back porch at home to show to his students, the neighborhood kids, or his own children.

Dr. Rudolph Edmund of the Lutheran University of California wrote (1988), in a presentation to the National Association of Geology Teachers, "L. R. Wilson championed the field as the best place and the best way to teach earth processes. . . . He helped students collect Pennsylvanian coal balls and watched them light up when they found spores under the microscope. Students followed him into the field, into the lab, and into research."

The summer geological "reconnaissance" trips that Wilson organized and conducted for Coe College were a major factor in building excitement and wonder in his students. These trips also caught the interest, support, and participation of extroverts of several callings from among faculty and townspeople. The trips involved from seven to more than 20 students plus two or three faculty for over two or three weeks. They ranged across the High Plains into the Middle, Northern, and Canadian Rocky Mountains and were conducted with three to seven personal cars and pickup trucks. Reimbursement to the owners was about one cent per mile, and the students paid about three dollars a day, which included food and travel. Camping was primitive (no motels or research stations), the menu was generally simple and often uninviting, but spirit and enthusiasm were great. Wilson interspersed the long drives with many lectures and many short field studies in various aspects of geology, botany, and ecology. The Black Hills, South Dakota Badlands, Devil's Tower, Glacier National Park, Columbian Ice Fields, Lewis Overthrust, the Beartooth Range, Yellowstone Park, the Tetons, Jackson Hole, the Wind River Mountains, the Green River and other intermontaine basins, Medicine Bows, Big Horn Mountains, and the High Plains were focal points. A select few climbed Mt. Teewinot in 1938 on their "day off," with Wilson leading, and another select group climbed Grand Teton in 1939, but one may wonder about the extent of faith Professor Wilson must have had in taking his novices on such adventures. He required an extensive final report in addition to annotated collections of rocks, fossils, and minerals. Interesting contacts with other geological parties or individuals whetted the appetites of a significant number of these students, including the authors of this portrait, to become professional geologists. One especially high point was a memorable evening spent in 1938 around a campfire in the Tetons with William H. Jackson (then in his 90s), the renowned photographer on the Ferdinand V. Hayden surveys in the American West in the 1870s, who was camped alone in his gray tent near our camp.

Wilson was one of the five founders of the Association of College Geology Teachers, the forerunner of NAGT (the National Association of Geology Teachers), which first met at Augustana College in 1938. A year earlier, in October 1937, we were among the several students from Coe College who sat around or poked at the rock outcrops in a quarry in Wisconsin for an hour or more, at the last stop on the Fifth Tri-State Field Conference, while Wilson, David M. Delo of Knox College, Fritiof M. Fryxell of Augustana College, and Monta E. Wing of Beloit College shared their joys of the exciting field trip just completed and discussed the need for more opportunities to get together and talk about the common problems they shared as lone teachers of the geological sciences in their respective colleges. They needed to share ideas and experiences of the others for developing curricula, course content, innovative techniques, joint field trips, and mitigating financial strictures—problems that each faced.

Though we had greatly enjoyed the field trip, led by Professors Frederick Thwaites and William H. Twenhofel, we were tired and had many hours of driving before reaching Coe College, and Monday morning assignments were yet to be completed. Our Coe contingent of Wilson and three or four students had also camped out for two nights, and on the previous day we had spent a long, but inspiring, evening in front of the great fireplace in Professor Norman C. Fassett's study at Madison, while Fassett and Wilson talked over many things concerning Wilson's research. Dr. Fassett had been to Wilson, for all his college and graduate school years, his mentor in botanical sciences and the principal influence on Wilson's thought—as Wilson was now to us in our brief lives in science. So we waited restlessly while the preliminary foundations of this new organization, which 20 years later would become NAGT, were hammered out. However, when the "quarry conference" was over, we piled into Wilson's "Chevy" and headed west as Wilson talked excitedly about getting together with his colleagues the next year when the first formal meeting of the Association of College Geology Teachers would be held at Augustana on April 13, 1938.

## STUDENTS

Wilson's classes were generally "tough" but extremely interesting, challenging, and fun, resulting in a core of enthusiastic students majoring in geology. Some of the students at Coe College, besides Kosanke and Cross, were the late Arnold Brokaw, Elizabeth Ann Coe (Costello), Eleanor F. Galloway, John Imbrie, Agnes Johnson (Brokaw), H. Douglas Klemme, Charles G. Kos, Sam Patterson, and the late Mart P. Schemel. Two local teachers, Ruth Webster and Iola Tillapaugh, also participated in field and laboratory studies and published jointly with Wilson.

Wilson's years at the University of Oklahoma and the Oklahoma Geological Survey were extremely productive both in the number of published works and in the number of advanced degrees granted. Some publications were coauthored with graduate students working for advanced degrees, among whom were Thomas A. Bond, Kenneth M. Bourdeau, Robert T. Clarke, Phillip N. Davis, James E. Dempsey, Edward D. Dolly, William A. Edwards, Lee B. Gibson, Reginald W. Harris, Jr., Richard W. Hedland, Maurice J. Higgins, James A. Ruffin, James B. Urban, Logan L. Urban, and Virgil Wiggins. B. S. Venkatachala of India was one of several postdoctoral colleagues.

## DIVERSITY OF INTERESTS IN RESEARCH

Wilson's role as a teacher stimulated his interests in broader aspects of geological and botanical sciences and in using palynological or paleobotanical techniques to aid in resolving some geologic problems. He prepared occasional research contributions independently or with his students. An early example was a brief study of a Pennsylvanian-age deposit of leaf and seed fossils found in a sandstone cave filling in Devonian limestone in Linn County, Iowa (Wilson and Cross, 1939). Wilson occasionally utilized palynology in order to interpret the prehistory of Native American sites, such as an Early Caddonian Village in eastern Oklahoma (Wilson, 1977). He also touched upon the recognition of maturation (or alteration) of organic matter in sediments and rocks by color change accompanying or resulting from thermal metamorphism (Wilson, 1971). In addition to a number of studies on diverse topics relevant to Wisconsin glacial geology, botany, and limnology published early in his career, he also published on "Rooted aquatic plants and their relation to limnology of freshwater lakes" (1939).

Wilson's interest in glacial geology has always been on his mind whenever he is in the field in or near glacial terrain. This interest was engendered during his early training under Thwaites and Birge at Wisconsin. He likes to recall that Birge studied glacial geology and zoology at Harvard University about the time of Louis Agassiz's death, in 1873, when the Agassiz influence dominated and associates were still teaching in the Agassiz tradition. So Wilson felt honored to have been in the Agassiz-Birge lineage or, as he said to us in a moment of reflection in 1992, "I stand on the shoulders of Birge who stood on the shoulders of Agassiz."

A research note on discovery of the permafrost structures indicating Arctic climate and tundra conditions in central Iowa, during the Iowan Substage of the Wisconsinan Glacial Stage (Wilson, 1958) is one of several brief studies of glacial features. He included the study and mapping of an esker and other glacial features in class laboratory field trips for us at Coe College during his sojourn there. In another study with his students (Wilson and Kos, 1942), he compared sources and processes influencing accumulation of some cross-bedded, and varvelike, laminated sediments of the Iowan Substage of Wisconsin outwash terraces near Cedar Rapids, Iowa. They conducted stream-table experiments to clarify or reinforce their field conclusions of the origin of the laminated deposits and terraces.

While teaching at Coe College, Wilson's interests broadened further to include Pennsylvanian palynomorph research, with ongoing Pleistocene to Recent studies. This was an outgrowth of his studies at the University of Leeds. Wilson and Coe (1940) described some palynomorphs in "Descriptions of some unassigned plant microfossils from the Des Moines Series of Iowa," and Wilson and Kosanke (1944) noted seven additional species. Wilson and Cross (1943) compared the relative palynological record of a peat bog with that of bottom sediments of a lake in "A study of the plant microfossil succession in the bottom deposits of Crystal Lake, Vilas, Wisconsin, and the peat of an adjacent bog." Wilson (1944) called attention to the value of palynomorphs in stratigraphy and other matters in "Spores and pollen as microfossils."

## MAJOR CONTRIBUTIONS TO POST-PLEISTOCENE PALYNOLOGY

The landmark publication of "An annotated synopsis of Paleozoic fossil spores and the definition of generic groups" (Schopf, Wilson, and Bentall, 1944) was completed at a time when there was considerable diversity in methods of classifying palynomorphs. This publication made a major contribution by proposing seven guiding principles of classification and providing a definition for existing genera.

From 1946 to 1949, Wilson and Ruth M. Webster completed an exhaustive palynologic study of the Bender No. 1

and Griffin No. 5 wells in Montgomery County, Texas (Wilson and Webster, 1946, 1949). The results were compiled by Carter Oil Company (an early subsidiary of Exxon Production Research Company) into two publications in 5 volumes, including 9,600 photomicrographs. These two publications were proprietary but Carter did release a copy to each of several individuals and to the American Museum of Natural History, New York. This study effectively demonstrated the use of palynology as a stratigraphic tool. These two wells were the first in North America to receive such comprehensive palynological analysis and set the stage for subsequent investigations.

Wilson (Fig. 2) made a significant contribution in glaciology with his work on the Greenland Ice Cap during the Mint Julep Expedition, which he directed for the American Geographical Society in 1952 and 1953, including fieldwork by Wilson in 1953. This project was initiated and funded by the Air University Arctic, Desert, and Tropic Information Center, U.S. Air Force. The original problems to be resolved were the origin of large "smooth ice" patches in the Sundstrom Fjord area, which afforded excellent temporary airfields, and the origin of dark bands in the ice that appeared to be rhythmic or episodic. The "smooth ice" was determined to be frozen lakes of meltwater that had collected on the ice surface during the previous summer. The dark bands of a powdery dust, called cryoconite, which had been thought to be of cosmic origin, were demonstrated to be thicker toward the edge of the ice and to have been transported by wind up from the outwash plains below, episodically, as almost continuous cover in places and as partial cover elsewhere (Wilson, 1955).

Figure 2. L. R. Wilson, leader of the American Geographical Society's Greenland Ice Cap project Mint Julep, June 1953. Photograph from the photographic collection of A. T. Cross.

An important palynological study by Wilson and Hoffmeister (1956) was that of the palynomorphs of the Croweburg coal bed of northeastern Oklahoma. In this study nine collections from Okmulgee, Wagoner, Rogers, and Craig Counties were prepared and examined. Four plates with 75 photomicrographs illustrate the significant taxa occurring in this coal bed. Comparisons were made between the taxa of the Croweburg coal bed and that of the Colchester No. 2 coal bed of Illinois as described by Kosanke (1950).

A major Wilson publication of note was the research on "Permian plant microfossils from the Flowerpot Formation of Greer County, Oklahoma" (Wilson, 1962). This report introduced several new taxa and is illustrated by color plates. This was an important early contribution to the study of Permian palynomorphs of the United States. Wilson (1971) published "Palynological techniques in deep basin stratigraphy," in which he reported that factors in palynological stratigraphy are both geologic and biologic in nature. The geologic factors are the time of deposition and the sedimentary and structural history in a given area. The biologic factors are the nature, ecology, preservation, recycling, and degree of alteration of the fossils. Figure 1 in that paper presents graphically the geologic ranges of the principal groups of palynomorphs and presented an early interpretation and technique of determination of level of thermal alteration.

In 1972, Wilson (Fig. 3) presented a paper on Midcontinent Pennsylvanian palynomorphs at the First I. C. White Memorial Symposium, entitled "The Age of the Dunkard." This paper (Wilson and Rashid, 1975) presented compelling evidence on the Pennsylvanian age of the Gearyan Series, thought by some to be Early Permian in age.

Wilson (1984) presented evidence for a revision of the Desmoinesian-Missourian boundary (Middle Pennsylvanian) in a study centered in Tulsa County, Oklahoma. He demonstrated that the palynomorphic genera *Lycospora* and *Thymospora* became extinct after the accumulation of the Dawson coal bed of northeastern Oklahoma and, accordingly, suggested that the Desmoinian-Missourian boundary should be placed between the Dawson and Tulsa coal beds.

Wilson's bibliography includes 65 publications during the eleven-year period, 1957 to 1967. Forty-four of these publications appeared in the "Oklahoma Geology Notes." These are concise, timely reports addressing topics of palynological nomenclature, chitinozoans, dinoflagellates, spores, pollen grains, fungal spores, silicified wood, and compressional remains. Wilson is the first author of most of these publications, many coauthored with colleagues and graduate students.

## HONORS AND PROFESSIONAL SERVICE

Wilson is a member of or has held memberships in and served as an officer or editor of numerous scientific organizations. He is a fellow of the Geological Society of America and

Figure 3. L. R. Wilson (right) listening to a site lecture at the Gentile Wash, Western Book Cliffs, northwest of Price, Utah, October 18, 1975, on the Geological Society of America Coal Geology Division Annual Field Trip; with him are Dr. L. R. Parker (center) and Dr. W. D. Tidwell (left), both former students of A. T. Cross (professional "grandsons" of Wilson).

Figure 4. L. R. Wilson (right), 1988, at 50th anniversary of the founding of the National Association of Geology Teachers, Augustana College, Rock Island, Illinois. Other founders in the photograph are David Delo (left) and Rudy Edmund (center).

the American Association for the Advancement of Science. He has been a member of the Botanical Society of America for over 50 years and was chairman of the Paleobotanical Section, 1947–1948. Wilson was elected as honorary member of the American Association of Stratigraphic Palynologists in 1975 and honorary life member of the National Association of Geology Teachers (NAGT) in 1979; in 1988 Wilson was honored as a founding member at the 50th anniversary of the NAGT (Fig. 4). He is also a member of the American Association of Petroleum Geologists, the Society of Economic Paleontologists and Mineralogists, and the Palaeontological Society of India and was the 1973 recipient of their prestigious award, the Erdtman International Medal for Palynology. He has been a longtime member of the Iowa Academy of Science and served as editor of the Proceedings volumes from 1936 to 1946. Wilson has served on the Commission Internationale de Microflore du Paléozoïque, 1950–1960; the Editorial Board for *Micropaleontology,* 1956–1966; and as research associate (1956–1966) of the American Museum of Natural History. Wilson was elected to the Phi Kappa Phi Scholastic Honorary Society in 1952 at the University of Massachusetts. He was also elected president of the University of Oklahoma Chapter of the Society of Sigma Xi, 1965–1966, and of the Oklahoma Chapter of Phi Beta Kappa Honorary Scholastic Fraternity, 1978–1979, "for scholarly achievement." He was adviser to the Sigma Gamma Epsilon Geology Honorary Fraternity at the University of Massachusetts, 1950–1953. Wilson was elected fellow of the prestigious Explorers Club in 1978.

Wilson's colleagues have also bestowed their respect and affection on him by naming newly discovered or undescribed fossils in his honor. These include: *Wilsonites vesicatus* (Kosanke) Kosanke (Kosanke, 1950), a genus of Pennsylvanian spores; *Wilsonisporites woodbridgei* Kimyai, 1969, a new genus of spores from the Cretaceous Raritan Formation of the eastern United States; *Wilsonipites nevisensis* Srivastava, 1969, a new genus of Maastrichtrian age from western Canada; *Wilsonastrum colonicum* Jansonius, 1962, an asteroid acritarch from the Lower Triassic of western Canada; and recently a new *Pediastrum* fossil from India. A new species of crocodile skull discovered by Wilson in Eocene lignite in 1936 near Wamsutter in southwestern Wyoming (Mook, 1959), has been named *Leidyosuchus wilsoni* Mook after Wilson.

He has had a long, productive, working relationship with many colleagues, including Paul B. Sears, B. S. Venkatachala (India), and R. M. Kosanke, and with several others, now deceased, including James M. Schopf, William S. Hoffmeister, Carl O. Dunbar, and Alex Nicholson. Wilson has played a major role in the field of palynology, both fossil and modern palynomorphs.

At the 1993 Annual Meeting of the American Association of Stratigraphic Palynologists, the Wilsons were brought in to celebrate Marian's 85th birthday, their 63 years of marriage, and Wilson's 65th year in palynological research! For more than a decade, the Leonard Richard Wilson Award has been presented to the student delivering the research paper judged best at the meeting. Wilson was honored at the business luncheon with a large Association Insignia recognizing his 65 years of service to the science of palynology. Capable, persistent, tough and lovable—he keeps going and going and going!

## MOST IMPORTANT CONTRIBUTIONS TO SCIENCE

Wilson's major contributions are threefold: (1) He was among the earliest to demonstrate the value of palynology in resolving stratigraphic correlations and interpreting paleoecology; (2) in the more than 200 scientific papers that he has published on various aspects of palynology, glacial geology, coal geology, paleobotany, paleoecology, limnology, and ecology, he has demonstrated effectively the relationships of plants to sediments and rocks through time; and (3) he has trained or contributed to the inspiration of many students who have become outstanding professional scientists, particularly geologists, serving academia, industry, and government. Without question, Wilson's greatest personal satisfaction comes from the fact that many of his graduate students have contributed so much to their chosen fields and are internationally recognized in their own right.

## ACKNOWLEDGMENTS

Much of the personal information included in this report was provided by Richard and Marian Wilson, for which the authors are most grateful. The assistance of Diane Baclawski of the Michigan State Geological Sciences Library is acknowledged.

## REFERENCES CITED

Burrell, W. H., 1924, Pennine peat: The Naturalist (May), no. 808, p. 145–150, pl. 10.

Kosanke, R. M., 1950, Pennsylvanian spores of Illinois and their use in correlation: Illinois Geological Survey Bulletin 74, 128 p.

Mook, C. C., 1959, A new species of fossil crocodile of the genus *Leidyosuchus* from the Green River beds: American Museum Novitates, no. 1933, p. 1–6.

Schopf, J. M., Wilson, L. R., and Bentall, R., 1944, An annotated synopsis of Paleozoic fossil spores and the definition of generic groups: Illinois Geological Survey Report of Investigations 92, 66 p.

Wilson, L. R., 1932, The Two Creeks Forest bed, Manitowoc County, Wisconsin: Wisconsin Academy of Science, Arts, and Letters Transactions, v. 27, p. 31–46.

Wilson, L. R., 1934, The spores of the genus *Lycopodium* in the United States and Canada: Rhodora, v. 36, p. 13–19.

Wilson, L. R., 1935a, Lake development and plant succession in Vilas County, Wisconsin: Ecological Monographs, v. 5, p. 207–247.

Wilson, L. R., 1935b, The Nipissing flora of the Apostle Islands region: Torrey Botanical Club Bulletin, v. 62, p. 533–535.

Wilson, L. R., 1936, Further fossil studies of the Two Creeks Forest bed, Manitowoc County, Wisconsin: Torrey Botanical Club Bulletin, v. 63, p. 317–325.

Wilson, L. R., 1938, The postglacial history of vegetation in northwestern Wisconsin: Rhodora, v. 40, p. 137–175.

Wilson, L. R., 1939, Rooted aquatic plants and their relation to the limnology of freshwater lakes: American Association for the Advancement of Science Publication 10, p. 107–122.

Wilson, L. R., 1944, Spores and pollen as microfossils: Botanical Review, v. 10, p. 499–523.

Wilson, L. R., 1955, Minor surface features of the southwest Greenland Ice Cap: Mint Julep Reports, Part II: Arctic, Desert, and Tropic Information Center, Air University, U.S. Air Force, p. 73–93.

Wilson, L. R., 1958, Polygonal structures in the soil of central Iowa: Oklahoma Geology Notes, v. 18, p. 4–6.

Wilson, L. R., 1962, Permian plant microfossils from the Flowerpot Formation of Greer County, Oklahoma: Oklahoma Geological Survey Circular 49, 50 p.

Wilson, L. R., 1971, Palynological techniques in deep basin stratigraphy: Shale Shaker, v. 21, p. 124–139.

Wilson, L. R., 1977, Palynological study of feature 4, Mi-63, from an Early Caddonian village in eastern Oklahoma: Oklahoma Highway Archeological Survey, Papers in Highway Archeology 3, p. 243–257.

Wilson, L. R., 1984, Evidence for a new Desmoinesian-Missourian boundary (Middle Pennsylvanian) in Tulsa County, Oklahoma, U.S.A., *in* Evolutionary botany and biostratigraphy, A. K. Ghosh Commemoration Volume: New Delhi, Today and Tomorrow's Printers and Publishers, Current Trends in Life Sciences, v. 10, p. 251–265.

Wilson, L. R., and Coe, E. A., 1940, Descriptions of some unassigned plant microfossils from the Des Moines Series of Iowa: American Midland Naturalist, v. 23, p. 182–186.

Wilson, L. R., and Cross, A. T., 1939, Fossil plants of a Des Moines sandstone cave deposit near Robins, Linn County, Iowa: Iowa Academy of Science Proceedings, v. 46, p. 225–226.

Wilson, L. R., and Cross, A. T., 1943, A study of the plant microfossil succession in the bottom deposits of Crystal Lake, Vilas County, Wisconsin, and the peat of an adjacent bog: American Journal Science, v. 241, p. 307–315.

Wilson, L. R., and Hoffmeister, W. S., 1956, Pennsylvanian plant microfossils of the Croweburg coal in Oklahoma: Oklahoma Geological Survey Circular 32, p. 1–57.

Wilson, L. R., and Kos, C. G., 1942, The laminated Pleistocene sediments of the Cedar Rapids region: Iowa Academy of Science Proceedings, v. 4, p. 359–365.

Wilson, L. R., and Kosanke, R. M., 1944, Seven new species of unassigned plant microfossils from the Des Moines Series of Iowa: Iowa Academy of Science Proceedings, v. 51, p. 329–333.

Wilson, L. R., and Rashid, M. A., 1975, Palynological evidence for a Pennsylvanian age assignment of the Gearyan Series in Kansas and Oklahoma, *in* Barlow, J. A., ed., Proceedings of the First I. C. White Memorial Symposium "The Age of the Dunkard," September 25–29, 1972, Morgantown, West Virginia: Morgantown, West Virginia Geological and Economic Survey, p. 183–193 [includes abstract of paper and discussion].

Wilson, L. R., and Webster, R. M., 1946, Fossil pollen and spores of the Bender No. 1 well, Montgomery County, Texas: Tulsa, Oklahoma, Carter Oil Company Research Division, 150 p.

Wilson, L. R., and Webster, R. M., 1949, Fossil spores and pollen, Bender No. 1 and Griffin No. 5 wells in Montgomery County, Texas: Tulsa, Oklahoma, Carter Oil Company proprietary publications, v. 1, 103 p.; v. 2, 294 p.; v. 3, 152 p.; v. 4, 192 p.

Manuscript Accepted by the Society July 6, 1994

Printed in U.S.A.

Geological Society of America
Memoir 185
1995

# *Henry Nathaniel Andrews, Jr. (1910– ): Paleobotanist, educator, and explorer*

**Tom L. Phillips**
*Department of Plant Biology, 265 Morrill Hall, University of Illinois, 505 S. Goodwin Avenue, Urbana, Illinois 61801*
**Patricia G. Gensel**
*Department of Biology, Coker Hall, University of North Carolina, Chapel Hill, North Carolina 27599*

## ABSTRACT

**From graduate student to "Dean" in the Botany Department at Washington University and as paleobotanist at the Missouri Botanical Garden (1935–1964), Henry N. Andrews, Jr., was one of the pioneers in American coal-ball studies. His contributions and those of his students extended and expanded upper Carboniferous studies in all the major plant groups. Andrews developed a comparably productive research and training program in Devonian paleobotany at the University of Connecticut, Storrs (1964–1975). His quiet way of sharing interests is manifest, in part, by his many collaborative explorations and research projects as well as by diverse written accounts of such adventures. Much of Andrews's broad appreciation of historical, cultural, and scientific heritage comes alive in his numerous books and popular articles. As a native New Englander, he developed intense interests and enjoyment of naturalist opportunities, from seacoast to the mountains. He and his wife, Elisabeth (Lib), have continued their writings, travel, and teaching since "retiring" to the family farm in New Hampshire. *The fossil hunters,* written after retirement, constitutes a benchmark treatment of paleobotanists and conveys much about the author.**

## INTRODUCTION

The contributions of Henry Nathaniel Andrews, Jr., to the historical development of Devonian and Carboniferous paleobotany in North America reflect a lifelong adventure with living and fossil plants as well as with many of the interesting people involved along the way. As a twentieth-century naturalist, explorer, educator, administrator, and historian, Andrews combines intense interests and broad appreciation of the history, geography, culture, and cuisine associated with each exploration in the field. Some of these aspects of his life are apparent in his numerous scientific papers on fossil plants. Equally important, generations of paleobotanists and others will come to know about him largely through his diverse, readable accounts of the study of fossil plants (see Mamay, 1975; Gensel and Andrews, 1984). Throughout his professional career Andrews has enjoyed the written accounts of naturalists and explorers, drawing from them not only additional interests and background for his own explorations but also motivation to share such adventures or, on occasion, misadventures.

Henry and Elisabeth (Lib) Andrews live in the idyllic Lakes Region south of the White Mountains near Laconia, New Hampshire (Fig. 1). Their 20-some-acre farm, with its mainstay of New England furnishings and collections of their handcrafts and mementos of world travel, has been a wellspring of hospitality for visitors from Andrews's boyhood days. The farm and environs bespeak much about his development of outdoorsmanship and woodworking skills as well as his lasting interests in cultural heritage, art, restoration, and conservation projects. In the early 1990s Andrews and a retired professor friend laid out self-guiding history-nature trails through a former Shaker Village in Canterbury, New Hamp-

Phillips, T. L., and Gensel, P. G., 1995, Henry Nathaniel Andrews, Jr. (1910– ): Paleobotanist, educator, and explorer, *in* Lyons, P. C., Morey, E. D., and Wagner, R. H., eds., Historical Perspective of Early Twentieth Century Carboniferous Paleobotany in North America (W. C. Darrah volume): Boulder, Colorado, Geological Society of America Memoir 185.

Figure 1. Henry Andrews at his Laconia, New Hampshire, farm, summer 1992. Photograph taken by his daughter, Nancy.

shire. There tens of thousands of visitors will be able to study the vegetation through the entire year. Lib Andrews continues her variety of charitable affairs, especially those involving cooking, but she has given up "Cooking with Lib," her weekly column of a dozen years in the Laconia newspaper. Any praise of culinary delights to be shared in the Andrews's farm would be an understatement!

That farm is a sustained link interwoven with the activities of Andrews and his family over the decades, as a summer retreat from St. Louis, Missouri, and even from Storrs, Connecticut! This is also where many an expedition was assembled for trips to the north country as well as overseas. Maurice Wonnacott, whom Henry thought of as his "professor" from guidance received at the British Museum of Natural History over the years, once visited the Andrews's farm. Upon surveying the magnificent setting, Maurice asked, "Why do you come to England when you have all of this!" (Andrews, 1990).

Henry Andrews, Jr., was born June 15, 1910, in Melrose, Massachusetts. In 1928 he graduated from Melrose High School, then spent a year at the New Hampton School near the family farm in Laconia, New Hampshire, getting "somewhat better fitted for college," and graduated again in 1929. At the Massachusetts Institute of Technology he majored in food technology and received a B.S. in 1934. He never really regretted doing that because it provided "a very good biological-chemical, etc. education," as Andrews expressed it. During his second year at MIT he passed the paleontology lab several times each week on the way to a chemistry course. He looked in, and it seemed very interesting, so the next year he took a course in paleontology with Hervey W. Shimer. Andrews attributes the serious beginnings of his interests in fossils to that experience (Andrews, 1980, p. 131). He also added a general introductory course in geology.

In 1935, under the tutelage of Professor Ray E. Torrey at the University of Massachusetts (then Massachusetts State College) at Amherst, Andrews's interests and enthusiasm in plant morphology and anatomy developed. He recalled that after his first visit with Torrey, he was very favorably inclined to spend a year just studying what he wanted to—plants! Torrey had worked with fossil coniferous woods, and his teaching was quite influential, as noted by John Hall (in Phillips et al., 1973, p. 32): "His course in morphology really leaned heavily on the fossil record, and he had some of the old Lomax slides which I thought were quite outstanding."

Prior to that time Andrews had been drawn to the West with interests in what he calls "mountain hiking (not real climbing up vertical cliffs!)." The Grand Tetons of Wyoming lured him to such adventures in 1932 and 1934, and later he walked the foothills of the Himalayas (Andrews, 1963d). He also hiked (climbed) extensively closer to home in the White Mountains. It was during the summer hiking in 1934 in the Tetons that Andrews made general fossil collections, among which were a fair number of coniferous woods that he later studied under Torrey's direction.

Of his year at the University of Massachusetts, Andrews recalled in a letter to me: "It was wonderful studying exactly what I wanted to! I really did not know what I was heading into then, but toward the end of the academic year Professor Edgar Anderson [Washington University in St. Louis] paid a visit to Professor Torrey and I met 'Andy.' I did not think very much about it until a few weeks later I had a call from Torrey who told me that Anderson had written saying that he had a half-time graduate teaching position for me if I wanted it." Andrews knew little about Washington University or the Missouri Botanical Garden, but the offer sounded good and that is how he got started in St. Louis. This ultimately was a gateway for his early fossil-plant collecting in the Midwest and further exploration in the West.

There were relatively few universities with paleobotanists in the United States in 1935. At Washington University, Robert E. Woodson, Jr., an angiosperm systematist, served as Andrews's adviser during the course of his studies for an M.S. in 1937. Woodson was exceptionally well grounded in morphology, but as Andrews's interests turned to the extinct pteridosperms, further guidance was sought. Andrews (1990, p. 6) wrote:

> Perhaps, like some other paleobotanists of my own vintage, I never had a formal academic "course" in the subject. As a graduate student I was shipped off to Cambridge University for the year 1937–38 hopefully to absorb some first-hand knowledge about fossil plants. I am not sure that my mentor at Washington University, Prof. Robert E. Woodson, knew exactly what might come of this venture. I was sent to work with Dr. Hamshaw Thomas of Downing College and the Botany School at Cambridge. Dr. Thomas helped me with a research project and in other ways, but there was no formal classroom instruction in paleobotany.

Upon arriving at Cambridge in August 1937, Andrews found the Thomases set for vacation. After a brief research project discussion on seed-fern wood, Andrews was sent off to the British Museum of Natural History to work for a month. It was there that he met Maurice Wonnacott, who kindly provided guidance on specimens to be examined. It was the first of many such visits and the beginning of a treasured friendship. Andrews (1990, p. 6) wrote, "Although I spent some time at other museums and universities, then and over the years, I have always thought of Maurice Wonnacott as my 'Professor.'" At Cambridge, he attended the lectures in the morphology course by Thomas and perhaps most appreciated what he called "short chats in the laboratory" (Andrews, 1980, p. 131). Andrews's doctoral thesis on seed-fern wood anatomy was completed in 1939 after his return to St. Louis (Andrews, 1940).

In 1939, he married Elisabeth (Lib) Claude Ham, a Missourian, whom he had met and courted as a student at Washington University. Lib Andrews recalls that when he returned with a gift of dishes, she knew he was serious! St. Louis was to be "home" (except in the summers) for the next 30 years, and their three children—Hollings, Henry III, and Nancy—were reared there. Their spacious old home in Webster Groves, Missouri, was likened unto a "little New England" (Mamay, 1975) and long noted for its hospitality. Lib and Henry Andrews on occasions hosted many of the participants from the Systematics Symposium, annually held at the Missouri Botanical Garden. This often required the kindly assistance of neighbors who had freezers that could handle the quantities of food prepared.

Andrews was appointed instructor in the Henry Shaw School of Botany at Washington University in 1940. He established a vigorous and productive paleobotanical research program, influencing the lives and careers of generations of students, some near his own age. He joined the Missouri Botanical Garden staff as paleobotanist (1947–1964) and for about five years served as assistant to the director. In the early 1950s he became "The Dean" at the University, a title accompanying the administrative head of the Botany Department. Andrews was also a WAE ("wages as employed"—a temporary status) staff member of the U.S. Geological Survey, 1950–1958. Throughout most of his academic employment, he also served as an administrator, until 1964 at Washington University and then at the University of Connecticut at Storrs. There he was department head, first of Botany (1964–1967) and then of the Systematics and Environmental Section (Biological Sciences Group) (1967–1970). His major paleobotanical research contributions in upper Carboniferous coal-ball studies and then on Devonian plants overlapped during the 1960s, with a progressive shift to the Devonian following his move to Connecticut.

*Ancient plants and the world they lived in* (Andrews, 1947) conveys an agenda of Andrews's developing interests as well as his efforts to entertainingly share them. This delightful book was written during World War II when Andrews was teaching mathematics to servicemen. It was an evening enterprise that kept him in touch with the broader studies of fossil plants. In the preface (Andrews, 1947, p. vii), his view of the relatively young science of paleobotany was explained thusly: "The mortar in the foundation stones is hardly dry, and much of such framework as exists must be altered as our knowledge grows, but it is better to fit the scattered facts together into a temporary edifice than to have no house at all." Some topical chapters, such as "Past epochs of the Arctic," were followed up by expeditions to Ellesmere Island (Andrews, 1963a, b, 1964); others, such as "The fossil hunters," became the benchmark treatment of three centuries of paleobotanists in their best light and an extensive history of our field (Andrews, 1980). Andrews undertook this four-year project upon his formal retirement in 1975. In recent correspondence he stated, "Putting *The Fossil Hunters* together was about the most interesting project that I have ever done." He captures important aspects of colleagues relevant to this volume and conveys some of his philosophy not so easily gained from such a quiet-spoken man (Andrews, 1980, p. 395–396): ". . . younger people who have serious thoughts about devoting their lives to botanical or geological pursuits will be better scientists if they have some knowledge of the struggles and adventures of their predecessors." Andrews also shared his view about mistakes by paleobotanists, and in the context of the Richard Kräusel biography, stated it thus (Andrews, 1980, p. 321): ". . . to me one source of his greatness is that he left them behind where they belonged and continually forged ahead." Accounts of Henry Andrews's contributions to coal-ball studies are included in Phillips et al. (1973), in Phillips and Cross (Chapter 27, this volume), and in the excellent biographical sketch, "anonymously" prepared by Sergius H. Mamay (1975).

Despite never having had a formal academic course in paleobotany, at Washington University Andrews developed an excellent one, with well-illustrated specimens, and wrote a new textbook, *Studies in paleobotany* (Andrews, 1961), including a chapter on palynology by Charles J. Felix. The dedication page reads: "In appreciation of the counsel of three great botanists—Hugh Hamshaw Thomas, Ray Ethan Torrey, Robert E. Woodson, Jr." In the preface Andrews (1961, p. vii) noted, "Perhaps the greatest contribution that paleobotany has made is not in filling gaps in our knowledge of the evolution of the plant kingdom but in showing us how many gaps exist."

Both before and after the preparation of his textbook, Andrews had taken on the prolonged task (at the urging of Jim Schopf) of assembling an index of generic names of fossil plants, based on the Compendium Index of Paleobotany of the U.S. Geological Survey that was first developed by Lester Ward (1841–1913). This index was initiated because of the growing body of literature and a perceived need to have available sources of publication for all fossil plants. Henry limited the index to type species of fossil-plant genera. These enormous undertakings, each representing up to four years of work, are known as the *Index of generic names of fossil plants* (Andrews 1955, 1970). Supplements were prepared by Blazer (1975) and Watt (1982), but currently this work is in abeyance.

## COAL-BALL STUDIES

From the mid-1930s to early 1940s, American coal-ball studies began to move to the forefront as significant research avenues for understanding the development and life history of upper Carboniferous coal-swamp plants or, as stated in Andrews and Mamay, 1952 (p. 66), "the evolution of the pteridophytes and early seed plants of the Palaeozoic." With Andrews's background, the permineralized anatomy of plants in coal balls was ideal for such studies (see Fig. 2). He wrote us: "I believe that I am especially in debt to Jim Schopf and to Eloise Pannell for getting me started. My earliest recollection of Jim is visiting him in Urbana shortly after we both had received our Ph.D.—he very kindly turned over some *Heterangium* specimens to me . . . and shortly after that time he took me to the Berryville locality. . . ."

There are many interesting stories about the Berryville, Illinois, stream-bank locality (Mamay, 1975, p. 5–6), the collectors, and their techniques for gaining permission to collect as well as for excavating the coal balls (see Phillips and Cross, Chapter 27, this volume). What Andrews had in mind at Berryville in 1959 was a bit more than the usual dig. A bulldozer and operator were located through the local highway commission and trucked out to the farm coal-ball locale. As the bulldozer removed the massive overburden of glacial till, the streambed became a mud flow with disheveled coal balls sinking beneath churning dozer tracks. Having rushed to the area from St. Louis, Andrews had not included his field boots. Without hesitation, he donned a cloth bag over each foot, smartly tied at the knee, and waded in to retrieve coal balls (Fig. 3). His attire would not have been so unseemly under the circumstances, except for his double-billed cap of the Sherlock Holmes type!

As one of the pioneers in American coal-ball studies, Andrews, along with his students, made significant contributions on every major plant group present. Many of the early publications appeared in the Annals of the Missouri Botanical Garden during the 1942–1954 period, often as "Contributions to our knowledge of American Carboniferous floras." The coal balls came from the Illinois Basin; from Iowa, particularly due to the generosity of Frederick O. Thompson (see Chapter 22, this volume) (Andrews, 1980, p. 238–239); and from new discoveries in Kansas (Andrews and Mamay, 1952). Examples of the many important studies are those on cordaites (Andrews, 1942a, 1949; Andrews and Felix, 1952), calamites (Andrews, 1952), seed ferns (Andrews, 1942b, 1945; Baxter, 1949), ferns (Mamay, 1950), and lycopsids (Andrews and Pannell, 1942; Felix, 1952, 1954).

Figure 2. Henry Andrews at the St. Wendel, Indiana, creek-bed locality (Parker coal bed), 1958. First an oak tree had to be chopped down; then the *Psaronius* trunk, below, could be bashed into handy carry-home sizes. Photograph taken by Tom Phillips.

Figure 3. Tom Phillips and Henry Andrews, with his "bag boots," retrieving coal balls from Berryville, Illinois, creek-bank locality (Calhoun coalbed), 1959. Photograph provided by Tom Phillips.

In his review of American coal-ball floras, Andrews (1951, p. 464) concluded with this understatement, worthy of a New Englander:

> The tendency to date has been to describe those fossils that are new, especially conspicuous and well preserved, and we are still a long way from the end of this "cream-skimming" stage. With continuation of such studies, which are now in progress in four or five laboratories, and the integration of botanical and geological interests, we should ultimately be able to work out a very interesting picture of the sequence of Upper Carboniferous floras and contribute notably to an

understanding of the evolution of certain pteridophytic groups and early seed plants.

A few selected series of contributions by Andrews and his students give the flavor and significance of research in those and later times. In a letter to us, Andrews stated that ". . . it was Eloise Pannell who introduced me to the Pyramid mine and I still think that her paper on *Lepidodendron* [Pannell, 1942] is one of the best dealing with *Lepidodendron* anatomy." This publication dealt with large populations of stem specimens whereby anatomical composition and size could be related to developmental reconstructions. Such lepidodendrid studies reached a major synthesis in the ontogeny of *Lepidophloios* by Andrews and Murdy (1958, p. 560): "A pattern of determinate growth is postulated for the arborescent lycopods which involves the development of the unbranched primary trunk from a massive apical meristem and, at the onset of dichotomy, a gradually diminishing primary body to a point at which longitudinal growth ceased."

William H. Murdy, dean of Oxford College–Emory University, recently shared this remembrance: "In regard to the paper we did on determinate growth in lycopods, our conclusion was a logical deduction from the diversity of branch sizes and the pattern of their primary tissues. The notion was not well received. I remember giving a paper on the topic at the 1959 International Botanical Congress. All during my talk a couple of French paleobotanists sat in the front row shaking their heads and occasionally murmuring, no, no. It was very disconcerting for a fresh Ph.D."

One of Andrews's keenest interests centered on seed ferns and their evolutionary significance. From his Ph.D. thesis (Andrews, 1940) to invited reviews (Andrews, 1948, 1963c), he projected a high-level visibility to the origins of seed ferns and their evolutionary role toward flowering plants. He championed such related studies as those of Thomas (1925) on the Caytoniales not as solutions but as improved resolutions of evolutionary lines along the way (Andrews, 1980, p. 134–135). This viewpoint of evolutionary paleobotany is still influencing the field today.

Andrews's naturalist appreciations of the present and the past are well identified with living and ancient ferns. Ferns have been conspicuous among the plantings or natural vegetation wherever Andrews has lived, and especially so at the New Hampshire farm. The research record from his laboratory indicates a mixture of both modern and ancient representatives of the Osmundaceae, Gleicheniaceae, Schizaeaceae, and Marattiales as well as the coenopterid-zygopterid complexes of early ferns and fernlike plants. These interests led to a comprehensive treatment (Andrews et al., 1970) of the Filicophyta in Édouard Boureau's *Traité de paléobotanique*. These studies ultimately led to further investigations of those precursors of so-called fernlike plants such as *Rhacophyton* and *Archaeopteris* at a time when progymnosperms and other formative groups of vascular plants were being elucidated from the Devonian. Andrews (1974) captures the breadth of these and other developments in paleobotany from 1947 to 1972.

## DEVONIAN STUDIES

Methodologies and objectives shift as one moves to the Devonian with due appreciation of the three-dimensional burial of the plant specimens. Andrews (1980, p. 345–346) ". . . was introduced to the 'degaging' technique—working under a low-power binocular microscope with a small hammer, needles, and a syringe to blow away the dust." This occurred in 1958 when he collaborated with Suzanne Leclercq at Liège in a study of *Calamophyton*. He further recalls: "I was not allowed to touch any of the precious Middle Devonian plants until I had served a short apprenticeship with some Carboniferous compressions of lesser importance. I have always enjoyed hand-work along with the head-work, and paleobotany requires both in the laboratory as well as in the field. For my choice the degaging technique is the most fun of all; it requires extreme care and patience and every new plant affords a 'microadventure.' "

The Devonian phase of Andrews's research overlaps that of the Carboniferous, but it was closely tied to his explorations of the Appalachians and northward to the United States Mountain Range of Ellesmere Island (Fig. 4). His first *Archaeopteris* studies (Andrews et al., 1965) were, in part, the quest to see the high Arctic and explore all its magnificence (Fig. 5). Andrews has chronicled those adventures (Andrews, 1963a, b, 1964, 1980, p. 270–272), which are well worth the reading. As Andrews suggests (1980, p. 271), ". . . one loves the Arctic very much or not at all." On a sunny day spirits soar, but in

Figure 4. On a summer's hike above Goose Fiord, Ellesmere Island, 1962, Henry Andrews watches the unconcerned arctic hares that remain white year round. Photograph taken by Tom Phillips.

Figure 5. With no sunset, the "evenings" at Lake Hazen in 1963 often found Henry Andrews writing in his journal or, as in this photo, excavating a trench down to the permafrost in search of ancient plants. Photograph provided by Tom Phillips.

storm it is darkness full of foreboding as sand-laden winds whip the mostly barren landscape. In 1962 after such a rare summer storm, Andrews, Norman Radforth, and Tom Phillips found their modest campsite denuded of most lightweight objects, including kitchen implements. After a morning of absence from the camp next day, Andrews returned in triumph with the green plastic washpan and assorted articles, some retrieved from quite a distance up the valley. As the cook on such expeditions, he appreciated what was vital (Fig. 6).

The field season of 1963 on Ellesmere was riddled with transportation problems (Andrews, 1964, 1980, p. 271): downed helicopters, the closure of lake passages by wind-blown ice, and finally hydraulic system malfunction on the south-bound flight of the RCAF Lockheed Hercules transport from Alert. The flight was diverted for an emergency landing at Thule Air Force base in Greenland. Despite an acrid smell of hydraulic fluid in the frigid bay of the plane, the icebergs in the sea below, and attendant uncertainties, an explorer's wanderlust to see Greenland and the great continental ice sheet was granted, if for only 24 hours. Andrews (1980, p. 271–272) explained why he did not return to the Arctic: ". . . in the summer of 1964 I became involved in diggings in northern Maine and later in southeastern Canada, and these areas proved so highly productive that they occupied my time for the next ten years—and there is still much to be done there."

Field trips into West Virginia were numerous from 1965 to almost near the time of Andrew's retirement in 1975. Jim Schopf had discovered the *Archaeopteris* beds and led Andrews and Tom Phillips to the Valley Head site (Fig. 7), where the great finds of *Rhacophyton* also occurred (Andrews, 1980, p. 130). At first the site was just a weathered roadside cut and drainage ditch on the mountain side of the highway. John

Figure 6. Henry Andrews, center, in his cook's apron at the Canadian Defense Research Board camp at Lake Hazen, Ellesmere Island, Northwest Territories, about 81°30′N, 1963. Dish driers are Norman W. Radforth, left, and Tom L. Phillips, right. Photograph provided by Tom Phillips.

Figure 7. Henry Andrews, 1965, contemplating the excavation of *Rhacophyton* slabs at early stages of the dig at the Valley Head, West Virginia site. His "bucket chair" received much service in hauling debris across the highway behind him. Photograph provided by Tom Phillips.

Dennison and Tom Phillips began to expand it, and over the years it grew into a sizable excavation for large slabs; many people participated in the dig, noted for its shadeless exposure. On one occasion Andrews was about to photograph the outcrop with a biology teacher from Clarksburg, working nearby with hammer in hand. The volunteer asked, "What do you want me to do?" Upon being advised, "Just look medium tired," the teacher quipped, "Give me a few minutes to freshen up!" As the dig moved inward, the overburden became excessive, and this recalls a quote out of context from Andrews (1990, p. 7), ". . . I am convinced that it is the collector rather than the locality which is exhausted." Upon hearing of the overburden problem, William H. Gillespie assisted as only he could, and the Valley Head site continued to yield outstanding fossils.

Such digs and the evenings that follow fieldwork are often highlighted by opportunities for students and the professor to share interests. On many such occasions Andrews was primed by inquiries about the paleobotanists he had known and their discoveries. In a sense, Andrews's interests and great storytelling about such acquaintances or those he had read about naturally translated into *The fossil hunters* (Andrews, 1980)—with much scholarly research! At Storrs, Andrews held weekly meetings with students during the academic year on such topics; they were frequently "spiced" by visits from paleobotanists.

In the context of such fieldwork and evening adventures, the extent of Andrews's study of each area of exploration is noteworthy. On one of the Elkins to Valley Head field trips, Andrews confided that there was a "place to dine somewhere along this route." The remote dining establishment, which had been featured in *Forbes Magazine,* was somewhere to be found along a maze of one-lane gravel roads, after he had made reservations by phone. Upon repeated knocking at an ill-lit old farmhouse, Andrews and Phillips were greeted by a dismayed maître d'hôte who admonished them for lacking the required tie and coat but kindly offered to provide the necessary items. "Field tripping" with Henry usually led to adventures in food, whether he was cooking in the Arctic, dining at a remote mountain farmhouse in West Virginia, or returning to home base in Webster Groves, Storrs, or Laconia.

The principal tear in this piece of "wholecloth" mention of food was a misadventure in Newfoundland with Andrew Kasper and Francis Hueber (Andrews et al., 1968; Andrews, 1980, p. 250–252). One mistake—leaving behind the main food supply box—led to others as the three paleobotanists became stranded with few rations in the rough country of the Long Range Mountains. Although Andrews faults his judgment in abandoning the base camp and attempting to walk out, those who read about such an ordeal may draw other lessons about the explorer's wisdom, resourcefulness, and determination under such circumstances.

Andrews's explorations in the Devonian rocks of northern Maine were initiated when he was invited to join Jim Schopf in an assessment of the plants as a means of determining the age of various sediments being studied by geologist Ely Mencher of MIT. Collections were made several times in the Red River and Eagle Lake areas. On one occasion Andrews and Andrew Kasper built a raft to float specimens downriver from the site, but Andrews concluded it was "more work than it was worth due to dragging the raft over rocks in the stream. There was not enough water in the river." His interest extended to Baxter State Park after the evening when he saw several large slabs from there containing extensively preserved plant remains in the basement of William Forbes's home. This interest led to several field trips to Baxter Park by Andrews, his students, and Bill Forbes. Particularly important plants described by Andrews and his student, Andrew Kasper, include several species of *Psilophyton* and a new genus, *Pertica,* both trimerophytes (Andrews et al., 1968; Kasper and Andrews, 1972; Kasper et al., 1974). Several other plants were noted, including a plant similar to lycopsids, *Kaulangiophyton* (Gensel et al., 1969), and the enigmatic *Prototaxites.* Collecting the latter involved working on a steep cliff face, accessible mainly by rope! Jim Schopf climbed up the steep face and sent down *Prototaxites* specimens; Andrews retrieved them from the brook. The locality was accessed by a long hike through the woods that involved navigating with a compass.

In addition to the abundant and generally well preserved fossils to be found in the Trout Brook area, the streams meandering through the forest provided a pleasant means of cooling off on hot days—a resource Andrews took advantage of. This was mitigated a bit by the often voracious black flies. These projects culminated in a summary account of the Trout Valley flora, based both on megafossils and spores (Andrews et al., 1977). Andrews continues to encourage his former students to search for fossils in Maine, true to his long-held belief that outcrops are seldom "worked out."

Andrews's Devonian plant studies were extended in the late 1960s and early 1970s to include exploration for fossil plants north and east of Maine in the classic Devonian localities of the Gaspé peninsula, Quebec, and northern New Brunswick. Again, in collaboration with his students, a series of papers was produced that added considerably to the kinds of plants known from the late Early Devonian of eastern North America. Particularly noteworthy are the occurrence of *Renalia* (Gensel, 1976) and another species of *Pertica* (Granoff et al., 1976) in the Gaspé sequence. Two plants from New Brunswick were important in being more complex than many Early Devonian plants, namely *Chaleuria* (Andrews et al., 1974) and *Oocampsa* (Andrews et al., 1975). Additional information on trimerophytes from New Brunswick appeared just after Andrews's retirement (Doran et al., 1978).

Collecting in the Canadian localities is at times precarious. Andrews added to this on one occasion by employing a bulldozer in New Brunswick, an event still remembered almost with horror by local landowners, who feared they would lose their shoreline holdings. Finding the gentle *Linnea* and Canada dogwood in the woods or the yellow lady-slipper orchids along the New Brunswick roadsides and adding lobster and salmon to the chocolate-cum-butterscotch pudding–based cuisine added to the pleasure of these digs.

It is worth noting that Andrews always actively involved his students in his research program, whether it was coal-ball research or Devonian plant studies. He treated them as valuable contributors and encouraged the best they could produce. He was interested in students personally and listened sympathetically to their woes and joys, often providing succinct but sage advice.

Upon retirement Henry Andrews devoted his efforts to *The fossil hunters* and maintained an interest in Devonian plants. After completion of *The fossil hunters,* he embarked, in collaboration with Patricia Gensel, on writing a book about Devonian plants. The impetus for this was multifold; both Andrews and Gensel had long recognized the need for a compilation of current information about early land plants that would be accessible to botanists and geologists in general as well as to specialists. Both had admired and been inspired by Newell Arber's (1921) *Devonian floras.* This coincided with an interest by a publishing company to produce a series of books about fossil plants. *Plant life in the Devonian* was the only one of the series to be completed (Gensel and Andrews, 1984). Colleagues gave generously of time to offer commentary or photographs, and Andrews proved as usual to be a delightful, insightful collaborator on this endeavor.

## OF BOOKS AND FOSSILS

It is perhaps the hallmark of a lifelong collector of literature and fossil-plant specimens that he treasures the accessibility and utility of such informational sources rather than a sense of ownership. During Andrews's researches and comprehensive writings, he developed one of the very best teaching and reference collections in paleobotany, located at the University of Connecticut. He promptly shipped voucher specimens from large field collections to museums and freely dispersed teaching-quality material where they would be used.

His exceptionally large private library of prized reference works and extensive reprints was no less subject to the same utilitarian care. Kept current and carefully filed, his library was culled for duplicates, promptly passed into welcoming hands. Many general texts were bundled off to India, Russia, or Scandinavia—wherever Andrews sensed they would be used. Binding tape was evident on most of the classic European works: Renault, Scott, Williamson, and others sought out in Cambridge or London bookstores, some from A. C. Seward's library, and some specially treasured gifts.

Andrews's appreciation of such library resources is exemplified by what he implemented during his later years at Connecticut. On one of his visits to the British Museum of Natural History, he arranged for the University of Connecticut library to purchase Wonnacott's reprint collection. Near retirement, as he completed many of the Devonian projects, Andrews invited Andrew Kasper and Patricia Gensel to divide up the bulk of his own reprint collection, retaining what he needed for the development of *The fossil hunters.* He, in turn, donated a substantial portion of the books to the university library. Andrews's sense of generosity and utility about reference works and his passing them along to the next paleobotanists continued into the 1980s as he completed *The fossil hunters* and *Plant life in the Devonian.*

## OF BOATS, BOGS, AND THE SEACOAST

Andrews grew up in the Lakes Region, not far from the seacoast; Cape Cod, in particular, nurtured a lifelong association with water, boats, and beach walking. Henry wrote us:

> . . . we go to the Cape—usually in mid October and April for a week. The ocean is a magnet—if you know it from childhood you can never be happy without getting back now and then. One of my earliest recollections—I guess when I was about 6—is chasing scallops at one of the Cape beaches. The great outer beach is some 30 miles long—it must be one of the loveliest in the world and now well protected as a National Seashore. I have spent considerable time walking the various beaches during the past ten years and have been trying to learn a bit about the various life cycles of the invertebrates—endless and fascinating.

From high school days, when Andrews built and victoriously raced his mahogany pram, to today, when he shares a fiberglass "Seaway," a lobster-boat design, with daughter Nancy and husband, there seems to have been at least one boat in the Andrews family activities. Andrews has described the present craft as rugged although hardly handsome, equipped with a Volvo-Penta engine. By contrast, he had earlier restored a 1940 Chris-Craft, the exact vintage and style as seen in the movie *On Golden Pond.* As far as we know, Andrews, who has an aversion to noisy machines (like dishwashers), did not succumb to motorized boats until after retirement. He still keeps a rowboat and canoe at his island "retreat."

Some of the best shared times with Andrews in leisurely pursuits, both before and after his retirement, were in boating. From a flat-bottomed boat, giving easy, undisturbed access to New England bogs with myriads of insectivorous plants, *Sphagnum* clumps, and deep quiet, to energetic sailing on Lake Winnepesaukee, Andrews shared his enjoyment of such adventures. In the winter the wooden watercraft were his restoration projects, representing long hours of woodworking and maintenance in the large New England–style barn.

To visit that barn, at any time of year, is to share some of Andrews's projects, apart from science. There is an art studio, where selected paintings and drawings from the art colonies are matted and framed. After these were enhanced in value, they were sold to support the restoration of the Lane Tavern in Sanbornton, New Hampshire (Mamay, 1975). Some also found their way to friends and colleagues as reminders of shared experiences. In correspondence with Andrews through some of the extremely heavy snowfalls of past years, it is no wonder that he often mentioned clearing the snow from the barn roof. In one letter he recalled the terrible booming sounds as barn roofs in the distance collapsed under the heavy accumulations of snow. In writing Serge Mamay about the winter snow in 1990, Andrews shared, "There is an old tradition in New Hampshire that we have two seasons—Winter and the 4th of July!"

## EARLY TO RISE AND EARLY TO BED—INTERESTS AND TIMING

It is important to portray a very personal touch of the Henry Andrews that many know and that is reflected in whatever he does. Perhaps from New England upbringing and certainly from his self-motivation, interests, and stamina, he developed a routine from early times of enjoying the quiet of the early morning hours. These were times of planting, trimming, laying a flagstone walkway or brick patio, and other quiet maintenance of the home. Returning to the kitchen, he might clear away dishes from a previous night's hosting and then set about preparation of breakfast for the early risers. He was early to the office—mail read and answered before most others had appeared!

Noontime usually found Andrews at home for lunch. He occasionally spent time sifting through the mail, with some prompt responses. Often there was a current book nearby as a diversion from academic pursuits. In the evenings, he was the attentive host, but when the hour grew late, he bade his farewell no matter how large the number of guests or the occasion. In turn, he rose to regain the next day's momentum with landscaping to launching the current research activities. Photographic sessions at 8 A.M. were common—and this represented a concession on his part to others' less early-bird habits!

Andrews's promptness and self-disciplined modes in these circumstances were indicative of his research momentum as well. The alacrity in his often terse correspondence and revisions moved research projects speedily on their way. In fact, most of his collaborators sensed and appreciated that it was Andrews's interest and momentum that tended to drive each short-term project to completion. It was important to know when his efforts and interests in a project had run their course, as he then promptly shifted to the next research activity at hand. In collaborating with Andrews, one of the key objectives he sought, an obligation of the researcher in his view, was an accurate reconstruction of the plants and/or structures being studied. In this quest, he emphasized acquisition of the most informative specimens possible so that laboratory studies dealt with the best of large quantities of materials.

## GRADUATE STUDENTS

Because Andrews set an admirable tradition that students publish their own theses, it is desirable to mention the students: Eloise Pannell, Lee W. Lenz, Robert W. Baxter, Sergius H. Mamay, Burton R. Anderson, Charles J. Felix, Karen Alt Grant, William H. Murdy, R. Bradley Ewart, Tom L. Phillips, Shripad N. Agashe, Kuldeep Rao, Andrew E. Kasper, Jr., Judy Skog, Bruce Cornet, Jeffrey Doran, Jeffrey Granoff, and Patricia G. Gensel. Henry and Lib Andrews helped these and many other students through the years and, indeed, were instrumental in salvaging the careers of more than one.

## HONORS, SERVICE, AND MORE TRAVEL

The opportunities and accomplishments of Henry Andrews in the broader context of paleobotanical research and service to his profession are marked by many awards. In 1938 he received a fellowship for a one-month study period in Belgium from the Belgium-American Educational Foundation. He is a member of Phi Beta Kappa and Sigma Xi and a fellow in the Geological Society of America and of the American Association for the Advancement of Science. He was a John Simon Guggenheim Memorial Foundation fellow for the years 1950–1951 and 1958–1959 (Belgium). After his year in Belgium, Andrews went to Oslo, Stockholm, Moscow, and Leningrad (Andrews, 1959, 1980), where he visited with paleobotanical colleagues and continued his research. The Guggenheim Foundation gave him a special grant from 1960 to 1965 for exploratory research, such as that in the Arctic. He was a Fulbright lecturer at Poona University, India (1960–1961), special lecturer for Oklahoma State University in Ethiopia (May 1961), and a National Science Foundation post doctoral fellow in Sweden (1964–1965), working at the Natural History Museum in Stockholm.

Andrews served as a secretary for the International Organization of Paleobotany and as a vice president during the 1960s. In 1966 he received the Merit Award of the Botanical Society of America. Andrews also taught for one semester in 1976 at the University of Aarhus in Denmark, after he "retired." In 1977 he was recipient of an award from the Paleobotanical Section of the Botanical Society of America "for Distinguished Service to the Paleobotanical Section and Outstanding Contributions to American Paleobotany." In 1975 Henry N. Andrews was inducted into the National Academy of Sciences. He continues to be an active member of the Academy, traveling each year to the annual meeting in Washington, D.C.

In summary, Andrews's contributions to American paleobotany are numerous, including his recognition of the value of studying American coal balls and his early descriptions of many types of plants from them as well as his many, later contributions to Devonian plant studies. These research efforts are noteworthy for the extent to which they involved students and incorporated aspects of the historical background of the subject matter, the area, and the botanical and evolutionary issues such studies raised. His books and articles will continue to influence many people along the way—as do Henry and Lib Andrews in their personal lies.

## ACKNOWLEDGMENTS

We gratefully acknowledge the assistance of Robert W. Baxter, William H. Murdy, and Sergius H. Mamay in the preparation of the manuscript and of Paul C. Lyons, William H. Gillespie, and Alfred Traverse for the improvements they suggested as reviewers.

## REFERENCES CITED


Andrews, H. N., Jr., 1940, On the stelar anatomy of the pteridosperms with particular reference to the secondary wood: Missouri Botanical Garden Annals, v. 27, p. 51–118 (published Ph.D. thesis).

Andrews, H. N., Jr., 1942a, Contributions to our knowledge of American Carboniferous floras. I: *Scleropteris,* gen. nov., *Mesoxylon* and *Amyelon:* Missouri Botanical Garden Annals, v. 29, p. 1–18.

Andrews, H. N., Jr., 1942b, Contributions to our knowledge of American Carboniferous floras. V: *Heterangium:* Missouri Botanical Garden Annals, v. 29, p. 275–282.

Andrews, H. N., Jr., 1945, Contributions to our knowledge of American Carboniferous floras. VII: Some pteridosperm stems from Iowa: Missouri Botanical Garden Annals, v. 32, p. 323–360.

Andrews, H. N., Jr., 1947, Ancient plants and the world they lived in: Ithaca, New York, Comstock Publishing Company, 279 p. (reprinted 1963).

Andrews, H. N., Jr., 1948, Some evolutionary trends in the pteridosperms: Botanical Gazette, v. 110, p. 13–31.

Andrews, H. N., Jr., 1949, *Nucellangium,* a new genus of fossil seeds previously assigned to *Lepidocarpon:* Missouri Botanical Garden Annals, v. 36, p. 479–504.

Andrews, H. N., Jr., 1951, American coal-ball floras: The Botanical Review, v. 17, p. 431–469.

Andrews, H. N., Jr., 1952, Some American petrified calamitean stems: Missouri Botanical Garden Annals, v. 39, p. 189–218.

Andrews, H. N., Jr., 1955, Index of generic names of fossil plants, 1820–1950: U.S. Geological Survey Bulletin 1013, 262 p.

Andrews, H. N., Jr., 1959, Notes on a visit to Leningrad and Moscow: The American Institute of Biological Sciences Bulletin, v. 9, p. 14–16.

Andrews, H. N., Jr., 1961, Studies in paleobotany: New York, John Wiley & Sons, 487 p.

Andrews, H. N., 1963a, Arctic Island—A botanical trip to Ellesmere Land: Missouri Botanical Garden Bulletin, v. 51, p. 17–23.

Andrews, H. N., 1963b, Botany in Ellesmere Land: Washington University Magazine. Spring Issue, p. 50–55.

Andrews, H. N., 1963c, Early seed plants: Science, v. 142, p. 925–931.

Andrews, H. N., 1963d, Hill stations of India: Appalachia, December Issue, p. 663–677.

Andrews, H. N., 1964, Return to Ellesmere Land: Missouri Botanical Garden Bulletin, v. 52, p. 1–9.

Andrews, H. N., with Hueber, F. M., and Kasper, A. E., Jr., 1968, The long range mountains of Newfoundland: Appalachia, December Issue, p. 288–299.

Andrews, H. N., 1970, Index of generic names of fossil plants, 1820–1965: U.S. Geological Survey Bulletin 1300, 354 p.

Andrews, H. N., 1974, Paleobotany 1947–1972: Missouri Botanical Garden Annals, v. 61, p. 179–202.

Andrews, H. N., 1980, The fossil hunters: Ithaca, New York, Cornell University Press, 421 p.

Andrews, H. N., 1990, Frederick Maurice Wonnacott 1902–1990: International Organization of Palaeobotany Newsletter 43, p. 6–8.

Andrews, H. N., Jr., and Felix, C. J., 1952, The gametophyte of *Cardiocarpus* Graham: Missouri Botanical Garden Annals, v. 39, p. 127–135.

Andrews, H. N., Jr., and Mamay, S. H., 1952, A brief conspectus of American coal ball studies: Palaeobotanist, v. 2, p. 66–72.

Andrews, H. N., Jr., and Murdy, W. H., 1958, *Lepidophloios*—and ontogeny in arborescent lycopods: American Journal of Botany, v. 45, p. 552–560.

Andrews, H. N., Jr., and Pannell, E., 1942, Contributions to our knowledge of American Carboniferous floras. II: *Lepidocarpon:* Missouri Botanical Garden Annals, v. 29, p. 19–34.

Andrews, H. N., Phillips, T. L., and Radforth, N. W., 1965, Palaeobotanical studies in Arctic Canada. I: *Archaeopteris* from Ellesmere Island: Canadian Journal of Botany, v. 43, p. 545–556.

Andrews, H. N., Kasper, A. E., and Mencher, E., 1968, *Psilophyton forbesii,* a new Devonian plant from northern Maine: Torrey Botanical Club Bulletin, v. 95, p. 1–11.

Andrews, H. N., Arnold, C. A., Boureau, E., Doubinger, J., and Leclercq, S., 1970, Filicophyta, *in* Boureau, E., ed., Traité de paléobotanique, v. 4, fasc. 1: Paris, Masson et Cie, 519 p.

Andrews, H. N., Gensel, P. G., and Forbes, W. H., 1974, An apparently heterosporous plant from the Middle Devonian of New Brunswick: Palaeontology, v. 17, p. 387–408.

Andrews, H. N., Gensel, P. G., and Kasper, A. E., 1975, A new fossil plant of probable intermediate affinities (Trimerophyte-Progymnosperm): Canadian Journal of Botany, v. 53, p. 1719–1728.

Andrews, H. N., Kasper, A., Forbes, W. H., Gensel, P. G., and Chaloner, W. G., 1977, Early Devonian flora of the Trout Valley Formation of northern Maine: Review of Palaeobotany and Palynology, v. 23, p. 255–285.

Arber, E. A. N., 1921, Devonian floras: Cambridge, Cambridge University Press, 100 p.

Baxter, R. W., 1949, Some pteridosperm stems and fructifications with particular reference to the Medullosae: Missouri Botanical Garden Annals, v. 36, p. 287–352.

Blazer, A. M., 1975, Index of generic names of fossil plants, 1966–1973: U.S. Geological Survey Bulletin 1396, 54 p.

Doran, J. B., Gensel, P. G., and Andrews, H. N., 1978, New occurrences of trimerophytes from the Devonian of eastern Canada: Canadian Journal of Botany, v. 56, p. 3052–3068.

Felix, C. J., 1952, A study of the arborescent lycopods of southeastern Kansas: Missouri Botanical Garden Annals, v. 39, p. 263–288.

Felix, C. J., 1954, Some American arborescent lycopod fructifications: Missouri Botanical Garden Annals, v. 41, p. 351–394.

Gensel, P. G., 1976, *Renalia hueberi,* a new plant from the Lower Devonian of Gaspé: Review of Palaeobotany and Palynology, v. 22, p. 19–37.

Gensel, P. G., and Andrews, H. N., 1984, Plant life in the Devonian: New York, Praeger Publishers, 380 p.

Gensel, P. G., Kasper, A. E., and Andrews, H. N., 1969, *Kaulangiophyton,* a new genus of plants from the Devonian of Maine: Torrey Botanical Club Bulletin, v. 96, p. 265–276.

Granoff, J. A., Gensel, P. G., and Andrews, H. N., 1976, A new species of *Pertica* from the Devonian of eastern Canada: Palaeontographica, v. 155B, p. 119–128.

Kasper, A. E., and Andrews, H. N., 1972, *Pertica,* a new genus of Devonian plants from northern Maine: American Journal of Botany, v. 59, p. 897–911.

Kasper, A. E., Andrews, H. N., and Forbes, W., 1974, New fertile species of *Psilophyton* from the Devonian of Maine: American Journal of Botany, v. 61, p. 339–359.

Mamay, S. H., 1950, Some American Carboniferous fern fructifications: Missouri Botanical Garden Annals, v. 27, p. 409–476.

Mamay, S. H., 1975, Henry N. Andrews, Jr.: A biographical sketch: Review of Palaeobotany and Palynology, v. 20, p. 3–11.

Pannell, E., 1942, Contributions to our knowledge of American Carboniferous floras. IV: A new species of *Lepidodendron:* Missouri Botanical Garden Annals, v. 29, p. 245–274.

Phillips, T. L., Pfefferkorn, H. W., and Peppers, R. A., 1973, Development of paleobotany in the Illinois Basin: Illinois State Geological Survey Circular 480, 86 p.

Thomas, H. H., 1925, The Caytoniales, a new group of angiospermous plants from the Jurassic rocks of Yorkshire: Royal Society of London, Philosophical Transactions, Series B, v. 213, p. 299–363.

Watt, A. D., 1982, Index of generic names of fossil plants, 1974–1978: U.S. Geological Survey Bulletin 1517, 63 p.


Manuscript Accepted by the Society July 6, 1994

Printed in U.S.A.

Geological Society of America
Memoir 185
1995

# *Frederick Oliver Thompson (1883–1953): Amateur collector and patron of paleobotany*

**Elsie Darrah Morey**
*Morey Paleobotanical Laboratory, 1729 Christopher Lane, Norristown, Pennsylvania 19403*

## ABSTRACT

**Frederick (Fred) O. Thompson spent the last 20 years of his life collecting mainly plant fossils but also trilobites and other invertebrate fossils. He was one of the great amateur fossil collectors in the history of the United States. Thompson shared his knowledge and collections with many well-known paleobotanists, including H. N. Andrews, C. A. Arnold, H. Banks, E. S. Barghoorn, R. W. Baxter, W. C. Darrah, W. L. Fry, S. Leclerq, and S. H. Mamay. Universities in the United States as well as abroad benefited from Thompson's collection efforts and the generosity of his financial support. In recognition of his great contributions to paleontology and paleobotany as an amateur collector, eight species were named in his honor.**

## INTRODUCTION

Frederick (Fred) Oliver Thompson was born in Des Moines, Iowa, in 1883. Following the death of his mother in 1891, the boy moved to Fiddletown, California, where he was tutored during these early years by his stepgrandmother, Charlene Mann Cooper.[1] From California, Thompson went directly to the St. Paul's Preparatory School in Concord, New Hampshire. Upon graduation, he entered Harvard University and graduated in the class of 1907.

While at Harvard, Thompson was active in several social clubs, including the Hasty Pudding Club, the Phoenix Club, and the Institute of 1770. As a student, he had little interest in natural history or science (Barghoorn, 1953).

After completing one year's study at Harvard Law School, Thompson decided to return to Des Moines, where he began his business career in real estate. With time, he became the president of the Utica Realty Company and a trustee of the Thompson Trust.

In 1912, Thompson married Anna Stroh Cram of Des Moines, Iowa. They had four children: Nancy, Ruth, Louise, and Frederick Oliver Thompson, Jr. Anna Thompson was her husband's constant companion and also his field assistant during the later period when Thompson devoted himself to fossil collecting.

## THE EARLY COLLECTIONS

In 1925, a paper appeared entitled "Pennsylvania flora of northern Illinois," by A. C. Noé (see Chapter 14, this volume). This popular guide to the Mazon Creek flora encouraged extensive collecting by amateur collectors, among whom were George Langford, Jr. (see Chapter 23, this volume), and also F. O. Thompson (Darrah, 1969).

Thompson first became interested in collecting fossils in the early 1930s. His friend, Henry Luthe (Fig. 1), first took Thompson to Coal City, Illinois, in 1932 (Fig. 2). There they collected Pennsylvanian plant fossils from the coal strip mines. About 15 boxes of these beautiful, well-preserved, ironstone nodules containing plant fossils were sent to Harvard University in 1933, and it would be only the first of many such shipments.

The early specimens were accepted as a gift to Harvard University by Professor Thomas Barbour, director of the Museum of Comparative Zoology. They were placed in dead storage by Dr. Robert Tracey Jackson (1861–1948),[2] who was the curator of the Paleobotanical Collections.

These unopened boxes of fossils sent by Thompson to the Harvard Botanical Museum were found by William (Bill) C. Darrah (see Chapter 1, this volume) when he arrived at Harvard University in 1934 (Lyons and Morey, 1991, 1993).

Morey, E. D., 1995, Frederick Oliver Thompson (1883–1953): Amateur collector and patron of paleobotany, *in* Lyons, P. C., Morey, E. D., and Wagner, R. H., eds., Historical Perspective of Early Twentieth Century Carboniferous Paleobotany in North America (W. C. Darrah volume): Boulder, Colorado, Geological Society of America Memoir 185.

Figure 1. Henry Luthe, a friend of F. O. Thompson, who in 1932 took Thompson collecting fossils to Coal City, Illinois. William Culp Darrah Collection.

Darrah, a new employee of the Botanical Museum, had as part of his first assignment to unpack, sort, label, and organize the specimens, which became the Thompson Collection of fossil plants. As he unpacked and identified the ironstone nodules, Darrah found them to be more complete and more beautiful than any he had seen before at the Carnegie Museum in Pittsburgh. The specimens had been picked and selected by Thompson with great care. Even so, Darrah observed that further fieldwork would be necessary to provide more data about this Pennsylvanian flora (Darrah, 1969).

Thompson first met Bill Darrah in 1936 when Thompson was attending his 30th class reunion at Harvard. However, before actually meeting Darrah, Thompson had been exchanging letters with him. A long and lasting friendship between the two men would develop.

In 1935, Thompson organized an elaborate search for more specimens to enlarge the collections he would be shipping mainly to Harvard but to other institutions as well. Desirable specimens included seeds, strobili, and fertile fronds as well as associated invertebrate fossils.

The same year, Thompson hired John Herron of Coal City, Illinois, to help him search for more fossiliferous nodules. Herron and Thompson organized high school students (Fig. 3) to help with this endeavor. Herron directed the students during the summer and fall to make the collections.

The intention of these collecting efforts was to follow the coal-stripping operations and to locate the nodules as they became exposed. The students carried the nodules to a central location; from there, two or more experienced collectors split them open and sorted them. The more desirable specimens were then moved to a workshop that had been set up by Thompson (Fig. 4) in Coal City. There they were sorted, packed and labeled for locality in preparation for shipment to Darrah at Harvard (Darrah, 1969).

Like most universities, Harvard had no funding available to build up the fossil collection. It was expected that the individual researcher would receive funding from outside sources. Thompson, in recognizing the young Darrah's need, sent a donation of $2,000 from the Thompson Trust to Harvard University for exclusive use for paleobotany. Oakes Ames, director of the Harvard Botanical Museum, wrote Thompson: "Your generosity is sublime. You can have no conception of what it means to me to have the Botanical Museum take on the aspect of being befriended. . . . I fear we can never look to the University for much aid in the way of allowances because there are so many departments in crying need of assistance."[3]

In 1937, the collecting-for-quantity was discontinued at both Coal City and Wilmington, Illinois, because new species were not found. Twelve thousand specimens were sent to Harvard; 4,000 of them were accessioned into the permanent Harvard Collection (Darrah, 1969). The remaining nodules were used in the exchange program with other universities and institutions, an arrangement established by Darrah.

Bill Darrah developed an improved peel technique (Darrah, 1936b, 1938), a method using cellodian to coat the polished and etched surface of rock that contained plant fossils. When the cellodian dried, it was easily pulled off, resulting in a film containing the replica of the plant fossil. Among the nodules collected by Thompson, Darrah found a new species of *Macrostachya,* a cone. Using the peel technique on the etched surface, Darrah took a small number of peels of the *Macrostachya.* In Fred's honor, the specimen was named *Macrostachya thompsonii* (Darrah, 1936b).

Coal balls (calcareous concretions) were supposedly first discovered in North America by Noé in 1922 at Harrisburg, Illinois (Darrah, 1969). However, Stopes and Watson (1907) mentioned Iowa coal balls (see Chapter 27, this volume). The figured specimen was collected by W. S. Gresley, who loaned the coal ball to Stopes and Watson for their study. In 1938, after a diligent search, Thompson discovered coal balls in Iowa, and these specimens, too, made their way to Harvard.

In a letter to Thompson, Darrah noted: "We have found in one of the coal balls a seed with the embryo plant. This is the *first* Paleozoic embryo ever found. . . . We have been making a discovery a day.

"*Macrostachya thompsonii* has turned up in the Iowa coal balls of all things. You know I think the Lord put those coal balls in your backyard and put your name on them. . . . P.S. They are simply too marvelous for words."[4]

Slowly the Darrah name was coming to be recognized in the field of paleobotany. Ames wrote to Thompson, "Mr. Darrah is doing creditable work in paleobotany and has brought the Harvard Collection of fossils into prominence. Once you were doubtful about the trend of paleobotany here. I am now able to assure you that we are a living force once more."[5]

Again, as Darrah's papers continued to appear, Ames had only warm words for Thompson: "A magnificent thing it is to

Figure 2. The first locality where F. O. Thompson collected fossils in Coal City, Illinois, 1932. William Culp Darrah Collection.

Figure 3. High school students hired by John Herron and F. O. Thompson. John Herron (wearing cap) is checking nodules that had been collected to be taken to the workshop in Coal City, Illinois, 1937. William Culp Darrah Collection.

Figure 4. The workshop as set up by F. O. Thompson. In the workshop the nodules were sorted and packed before they were sent to Harvard University. From left to right in this 1937 photograph: Anna Thompson, John Herron, F. O. Thompson, Jr., David Herron with Buster (the Herron's dog), Bottari Herron, Helen Herron, and F. O. Thompson. William Culp Darrah Collection.

make a young man experience the thrills that come in connection with the substantial recognition of his work. For Mr. Darrah I thank you, for myself I bless you."[6]

Later, Ames wrote Thompson, "Darrah realizes keenly the magnificent opportunity you have spread before him. I feel sure that he will measure up to your expectations and yield the results you have every reason to look for."[7]

Thompson was intrigued by the peel technique, the rock saw, and the polishing equipment, all so important in his new avocation. One of his field assistants, Eugene Kroeger, requested information from Darrah on how to build a rock saw just so that Thompson could use it.[8]

It was not long before Thompson began cutting and polishing the coal balls himself (Fig. 5). Using the peel technique, he made peels to send to Darrah to show him what structures were turning up. From these peels, Darrah could select desirable slabs of coal balls. Thompson made peels for other paleobotanists, for his friends, and also for universities that he enjoyed visiting on his travels around the country.

On occasion, Thompson had trouble leveling the coal balls on his sand trays. Sometimes the peel solution ran over the edge, and sand got on a peel. Said Thompson, "A little sand on a peel looks like a small mountain under a microscope. I also get some tobacco on wet peels. This just shows you that tobacco grew over 300,000,000 years ago. Well fun is fun."[9]

From time to time, Thompson wrote Darrah for further interpretation of words that he found in his correspondence from other scientists. "Just a note to acknowledge receipt of the fine peel of *Cordaianthus shuleri* (Darrah, 1946). I imagine that it is run off one of the types, which makes it doubly valuable. What does he mean by 'one of the types'? Is this what he calls a type? Since I never got through Grammar school, I have a terrible time with you scientists."[10]

Thompson continued to lay the groundwork to collect more and more fossils. He kept busy meeting the mining officials, superintendents of the mine, and the miners themselves, always with the hope they would save the fossils for him when they were found.

Amos Stuver wrote Thompson about the abundance of fossils that were being turned up near Terre Haute, Indiana: "Have quite a number of pieces from Pyramid and Peabody bearing imprints of fern and other plant life. . . . Maumee and Blackhawk are turning out tons and tons of coal balls every day, so you shall not lack material to work on when you decide to visit this summer."[11]

Darrah wrote to Thompson about specific types of fossilized structures that he hoped Thompson would find. When Thompson had none of those requested fossils, he then wrote to potential contacts for another source of material.

Thompson wrote to W. E. Dahlgren of the Field Museum of Natural History (Illinois) for his assistance:

> William C. Darrah, Research Curator of Paleobotany at Harvard University, is anxious to get coal balls which show attachments. For a year I have been trying to locate some but have been unsuccessful. It occurred to me this morning that in your large collection of fossils from the Mazon Creek area there may be some coal balls. If there are,

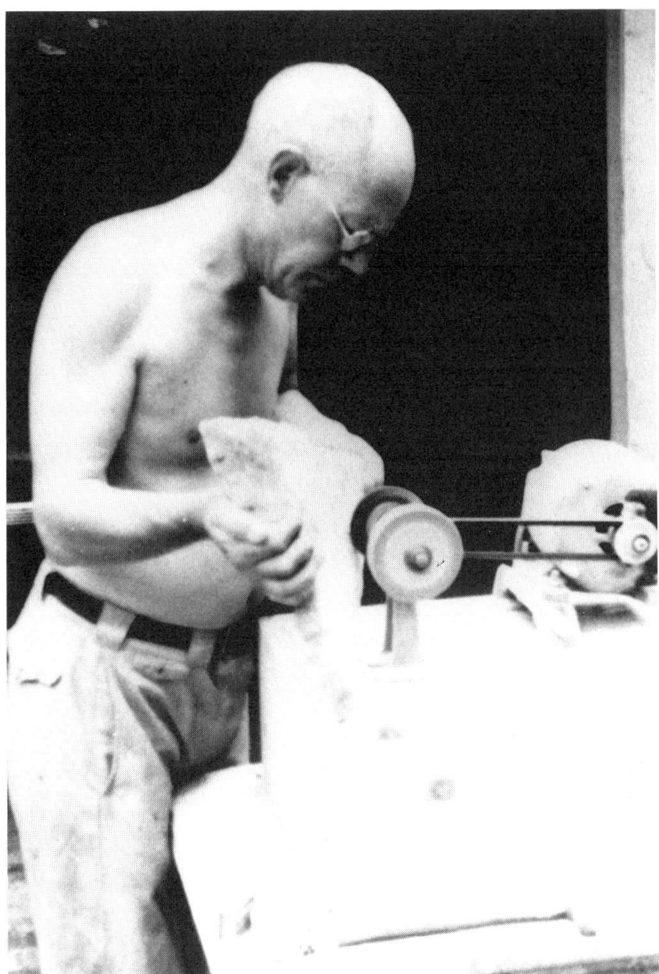

Figure 5. F. O. Thompson using the saw built for him by Eugene Kroeger, 1939. William Culp Darrah Collection.

Mr. Darrah would greatly appreciate the loan of them for study. . . . I am the fellow who wangled a cycad for you from the University of Iowa. That job took me one or two years, but in that case I was successful but in the coal ball business I am a dud.[12]

Dahlgren gave the letter to A. C. Noé, who quickly replied:

Enclosed I am returning Mr. Thompson's letter to you. There are no coal balls in the Mazon Creek area. The coal No. 2 of Northern Illinois does not have any. Furthermore I never saw attachments in any of the hundreds of coal balls which I have sectioned. If I ever found any I would keep them because they would be priceless. Let Dr. Darrah go out in the field and collect coal balls himself. That is the way to do [it]. I have found many hundreds of them by diligent exploring. I have given away some, but I think he who wants them should learn to find them.[13]

That classic remark by Noé was hardly tactful, and Thompson's reaction was typical: "When I heard of the remark 'Let Darrah collect his own coal balls' I was 'hot under the collar' but after finding coal balls by the ton [in the Dallas County, Iowa area], I have had ten thousand laughs over his remark. It is a classic. . . ."[14]

Thompson was meticulous in obtaining the locality data of the mines he visited. Darrah thought that the Shuler (Fig. 6) and Urbandale coal balls might be the same vein. In a letter to Darrah, Kroeger advised: ". . . everyone says they are not. So far we haven't been able to check all the depths of the various veins, but I hope we soon will be."[15]

Ever since you wrote in your letter to Mr. Thompson about your suspicion that the Shuler and Urbandale veins [coal beds] might be the same, Mr. Thompson and I have been gathering information for details. Enclosed I will send a drawing of our efforts. It is all we find out from different places, as the different parties are very peculiar in giving out information of this sort. The best information we always did and will get from the Shuler mine. It comes from Mr. Sanders, the man who drilled the hole, and also the hole for the Des Moines Ice and Fuel Co. This is also the man who told me that the coal from the Des Moines Ice Co. and the Urbandale mines have the same characteristics as the vein from the Shuler mine at 312 ft. And that is why I put them in one line in the drawing [Fig. 7]. Mr. Shuler did give us the information on the yellow paper about a hole drilled in 1919. It was only about one-half mile south of the present Shuler mine in Section 26, Walnut Township, Dallas County.[16]

Mr. Twining, owner of the Urbandale mine, told me that his coal pinched down to a foot in thickness to the west toward the Shuler mine and became too thin to work profitably. I wondered if that was what Charlie Bendixen [official at the Shuler mine] meant when he told me that the Shuler coal pinched out to the east toward the Urbandale mine. Last week Bendixen told me that the Shuler drill holes show that the Shuler coal disappears entirely to the east toward the Urbandale mine. This information should settle the question as to whether or not the two mines worked the same coal vein.[17]

Iowa coal is expensive to mine. It does not lie in level veins. There are "hills"—a raise in the vein; "swamps"—where the vein pitches downward. This bears out your theory of a hill between the Urbandale and Shuler mines.[18]

Darrah was quick to respond to Thompson's letter.

I am not surprised that there are many evidences that the coal beds of Iowa are not flat and that hills or valleys dissect them. I have never meant to imply that my theory of an interruption between the coals of the Urbandale and Shuler mines was absolute but all of the data seem to indicate that this is so. Coal balls occur in such widely scattered places and under such peculiar conditions that it would be very unusual that two or three coal beds lying one above another would each have nodules in it.

In Holland and Belgium [see Chapter 2, this volume] in the horizon known as the Aegir, coal balls are so restricted in their occurrence that the very presence of coal balls is sufficient to identify the horizon. The same thing is true of the Halifax bed in England and the Catherina [Katharina] in Germany. It may well be that in Iowa more than one horizon does have coal balls, but the burden of proof is on the geologist. The plants and other structures found in the many coal balls are so nearly identical that it looks suspiciously like a very widespread stage of coal formation which was essentially the same throughout the region.

Figure 6. Dump truck from the Shuler Mine dumping tons of coal balls into the Thompson yard, 1938. William Culp Darrah Collection.

> Of course, I know that I am on the minority side of this interpretation and, admitting that I may be wrong, I am sitting back waiting for more information.[19]

Always there was more for Thompson to learn about the coal-ball business, about the secrets the nodule could reveal. Some time later, Thompson wrote, "I have become an amateur paleontologist [Fig. 8]. On page 8 of the Annual Report of the Museum of Comparative Zoology (1947–1948), I find the following . . . 'some 2000 Devonian invertebrates from Iowa, presented by Mr. F. O. Thompson of Des Moines'. These were collected by Anna and me at a brick and tile company clay pit near Rockford, Iowa."[20]

But even earlier, Thompson and A. A. Stoyanow had collected invertebrate fossils near Tucson, Arizona (Fig. 9). Thompson was pleased to report to his brother, Oliver Thompson: "Stoyanow got a good laugh out of the way you addressed me: 'Dear Don Quixote Fred'. Stoyanow said the Don Quixote was correct. He said something to the effect that this character never knew where he was going, but was always pushing forward, and that I am doing the same. Of course, Oliver, I just call it 'monkey business', but the result of nine years of scratching around has spread all over the world."[21]

The marvelous collections of invertebrates were sent not only to Harvard University but also to many other universities, including the University of Iowa, University of Arizona, and even Canterbury University College in New Zealand.

## WORLD WAR II

The effects of World War II interrupted Thompson's collecting activities. Once the slow recovery from the war had begun, however, he was quick to resume his fossil collecting, this time at What Cheer, Iowa. As Thompson explained to Darrah, "For years I haven't been able to make collecting trips due to gas rationing, the war, and a car that I couldn't take out of the city limits. I now have a new Dodge car. I intend to go to What Cheer . . . as long as the coal balls hold out."[22]

Darrah was quick to respond to this renewed activity:

> I was delighted to hear that you had found [more] coal balls in Iowa. There has never been any question in my mind that there are a great many more in the ground than above the ground. Despite all of the complications and disappointments, as long as Fred Thompson can drive his car to the workings there is always the possibility that new specimens will be turned up.
>
> The last I saw of the What Cheer balls at the Museum, they were oxidizing rather badly, much more so than from Shuler. I do not think Barghoorn is doing much with them. I hope you are able to give a good series to Henry Andrews. For some curious reason, at no other mine in Iowa does the variety occur as at Shuler. I am waiting for the time when more coal balls are located there, though as you know the Shuler Coal Company could just as soon not contend with them.[23]

## HONORS FOR THOMPSON

Thompson's fossil collections were distributed in many countries. He became well known for his extensive collections and the amount of work that he had done in order to obtain the fossils.

Many scientists recognized Thompson for his splendid accomplishments and named fossil species in his honor. Among them are: *Macrostachya thompsonii* (Darrah, 1936a), *Arctinurus thompsoni* (Miller and Unklesbay, 1944), *Euproops thompsoni* (Raymond, 1944), *Medullosa thomsonii* (Andrews,

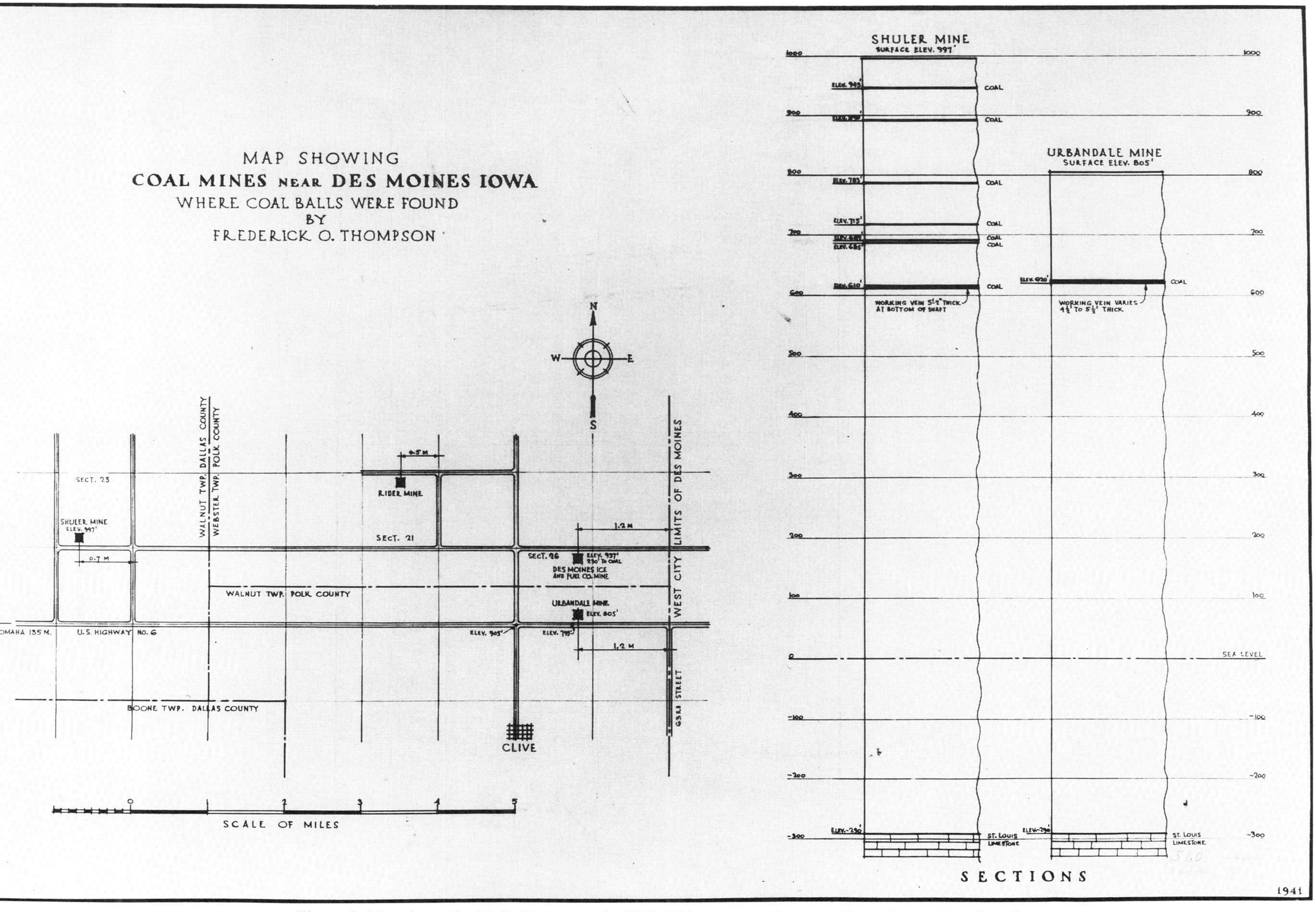


Figure 7. Map drawn by F. O. Thompson in 1941. This map was drawn to show the stratigraphy of the Shuler and Urbandale mines. William Culp Darrah Collection.

Figure 8. F. O. Thompson, 1949, at collecting site of Devonian invertebrates in Iowa. William Culp Darrah Collection.

1945), *Sphenostrobus thompsonii* (Levittan and Barghoorn, 1948), *Mesoxylon thompsonii* (Traverse, 1950), *Neuropteris thompsonii* (Darrah, 1969), and *Mariopteris thompsonii* (Darrah, 1969). It is interesting to note that Thompson had collected all of these specimens over his 20 years of collecting fossils.

Thompson was extremely pleased by the honors bestowed upon him. "This morning I received your two publications on Cephalopods and the one on Trilobites," he wrote to A. K. Miller. "Thank you for tacking my name on to one of them. Now I have my name on a plant fossil and an animal fossil, and feel much puffed up. I fear I shall have to buy myself a new hat."[24] And still reveling in his new "fame," Thompson could not resist a "P.S." in his letter to Darrah, "It is the tail end of the trilobite that gets the new name of *Arctinurus thompsoni*."[25]

Harvard University bestowed upon Frederick Oliver Thompson (Fig. 10) the title of Research Fellow of the Botanical Museum in 1949 (Barghoorn, 1953). The honor was in tribute to the amount of work that Thompson had done amassing the fossil collections at Harvard University. Thompson held the title for the rest of his life.

## THE FINAL YEARS

A time came when Thompson recognized that he was no longer able to work as he had done before, but still he was eager to provide both direction and the means for the new generation of paleobotanists. And as it happened, Thompson and Darrah discovered mutual associations among the aspiring paleobotanists, a matter for some exchange between the two men. Darrah wrote: "I was delighted with your appraisal of Harlan Banks. He visited me frequently at Harvard, . . . and I found him enthusiastic in his quiet way."[26]

Thompson wrote to tell Darrah that he had taken R. W. Baxter, W. L. Fry, and S. H. Mamay to the Atlas Coal Company strip mine. Thompson followed their careers while Mamay was at Cambridge University and Fry at the University of Glasgow.

In later recollections, Mamay wrote, "I was indeed hosted by the Thompsons in their home in 1948, I believe . . . I was with Henry Andrews and Bob Baxter. The Thompsons were very gracious hosts, putting us up in very comfortable sleeping quarters after an elegant dinner."[27]

Henry Andrews also remembered Thompson: "We spent most of the time in the basement going over the many slabs of coal-ball specimens that he [FOT] had gathered. I think he must have sent the bulk of the collections to [Bill Darrah] but he was also quite generous with me. Whatever he had not promised to Harvard he offered to me. I would like to have known him earlier in his 'paleobotanical career' but I am grateful for the time I did have."[28]

W. L. Fry also has memories of Thompson:

> My recollections are all delightful and worthy of remembrance. He was thoroughly fascinated with what we, as paleobotanists, were deciphering from the petrifications found in Iowa coal balls. I do not think he understood the sophisticated and technical data on the fossil plants we were describing. However, that did not detract in any way from his boundless enthusiasm for what we were doing. He would query us relentlessly in his amateurish fashion, yet, there was no question about the honesty of his questions. He was a somewhat brash individual but I welcomed his attitude and confrontation about what we, as students of fossil plants, were trying to accomplish.[29]

R. W. Baxter also offered reminiscences:

> Much of the coal ball material which I worked on for my M.S. and Ph.D. theses were from coal mines of which he had been the original collector or had directed other's attention to. During the early and middle 1940s, U.S. coal ball localities seemed to be relatively rare and certainly many of those which had been discovered were in large part due to Fred Thompson's efforts. The very large deposits of S. E. Kansas did not come to light until late 1949 and early 1950 by which time, largely due to Fred Thompson's laying the groundwork, there was a growing number of trained American paleobotanists available to begin their investigation.[30]

Figure 9. Field trip led by A. A. Stoyanow near Tucson, Arizona, 1938. From left to right: F. O. Thompson, A. A. Stoyanow, and Stoyanow's students. Photograph courtesy of Ruth Thompson Grimes.

Ironically perhaps for a man so meticulous in his efforts at collecting specimens, Thompson kept no record of this 20-year career. On the contrary, he relied on Darrah's penchant for saving every pertinent piece of information that Thompson ever sent to him and then called upon Darrah later on to refresh his memory. Following such a request in 1950, Darrah wrote:

How quickly the years pass! I tried to recall immediately upon receipt of your letter the sequence of events which followed my first letter to you in June of 1934 when I came to Harvard and began reopening the collection which had long been dormant. I remember well the 15 or 20 boxes of Mazon Creek nodules which were piled in the storage room.

How long you were collecting prior to 1934 I do not know, but the largest shipment of nodules which preceded my arrival came to the Museum in the fall of 1933. I have a letter of yours written in June, 1941 in which you say you were collecting "for the past 10 years". Perhaps this is figurative rather than an actual recollection but I have vague notions that you were collecting in 1932 at Coal City.

We met for the first time in 1936, during your 30th reunion and spent our first collecting trip together in 1937. That year, if you remember, we met at Joliet and collected around Coal City.

Your discovery of coal balls [in Iowa], after years of search, occurred in August of 1938 and the following June I came to visit you to see the material first hand. Then in 1941 I returned again—this time we went southward and collected a rich haul at Frontenac, Kansas. You had been to Frontenac with Anna earlier that spring and had written with enthusiasm that you had seen miles of coal balls, though most of them had been rotted by the weather.

Little did you know then, and less did I, the chain of events that would follow the discovery of coal balls in Iowa and Kansas. Scarcely a month passes that some new contribution to paleobotany does not appear in the professional journals. More important than that, the whole approach to the study of coal balls has changed. Noé's idea was to section one ball and to describe its content. It would take a thousand years of puttering to understand the significance of the structures by that method. My notion (of course it was nothing original—I merely happened to be the first in this country to pursue it) was to section a large number of balls and then to describe the species one by one and gradually build up a knowledge of the flora. It is the approach being used at the present time by American paleobotanists.

When I look back over these years, I count among my most cherished recollections, your friendship, your hundreds, indeed thousands of letters, and the boundless enthusiasm with which you threw yourself into the work. I know you do not appreciate the importance of your own work. It is amusing to read in your modest letters how you belittled your own efforts. Yet, you have made it possible, not only with material but with generous contributions of money, for many young men to pursue paleontological research. Just consider the number of institutions and individuals you have benefitted by your interest.[31]

Sadly, Fred Thompson wrote of his continuous ill health in the months that followed. "I had my fun when the going was good," he admitted, "so have no kick about life. If I had [it] to do again I would put in a lot of college botany and geology. . . . At college age a boy cannot foresee what is most useful. If Henry Luthe had not taken me to the Mazon Creek fossil bed I would not regret no knowledge of botany and geology. What a part chance plays in life."[32]

On January 3, 1953, Frederick Oliver Thompson, amateur paleobotanist and paleontologist, died. He will be long remembered for the collections that he made during his final 20 years.

Over the years, Thompson assisted many aspiring paleobotanists. Included among them are W. C. Darrah, E. S. Barghoorn, Suzanne Leclercq, C. A. Arnold, H. N. Andrews, Harlan Banks, R. W. Baxter, S. H. Mamay, and W. L. Fry.

Figure 10. Frederick Oliver Thompson. The photograph was taken in the Thompson home on Forest Drive, Des Moines, Iowa, in the early 1940s. Photograph courtesy of Ruth Thompson Grimes.

Thompson's influence was far reaching. He gave financial assistance for many publications and research projects, in both paleobotany and paleontology.

Barghoorn (1953) wrote, "Frederick Thompson as an amateur paleontologist will be remembered clearly for his selflessness, his generosity and his intense devotion to his chosen field of scientific interest."

Thompson would have been quite pleased in his own modest way. Perhaps he would have gone out again to buy another hat.

## ACKNOWLEDGMENTS

The author wishes to thank Mrs. (Alyce) Frederick O. Thompson, Jr., and Mrs. (Ruth Thompson) Howard Grimes for answering questions about F. O. Thompson, their father-in-law and father, respectively, and for sending photographs and letters that are quoted in this paper. I also thank them both for their kindness and assistance to me during the preparation of the manuscript.

I thank Drs. H. N. Andrews, R. W. Baxter, W. L. Fry, and S. H. Mamay for writing about their recollections of Thompson and their visits to his home. Anne Raymond and Christopher Wnuk are thanked for their reviews of the manuscript.

## ENDNOTES

1. Telephone communication from Ruth Thompson Grimes, F. O. Thompson's daughter. Charlene Mann Cooper was a retired principal of the Des Moines Public School System before her move to Fiddletown, California, with her husband Isaac Cooper.
2. Letter from F. O. Thompson to W. C. Darrah, May 31, 1950. W. C. Darrah Collection.
3. Letter from Oakes Ames to F. O. Thompson, January 30, 1937. W. C. Darrah Collection.
4. Letter from W. C. Darrah to F. O. Thompson, October 18, 1938. Courtesy of Ruth Thompson Grimes.
5. Letter from Oakes Ames to F. O. Thompson, May 11, 1936. W. C. Darrah Collection.
6. Letter from Oakes Ames to F. O. Thompson, November 27, 1936. W. C. Darrah Collection.
7. Letter from Oakes Ames to F. O. Thompson, May 5, 1937. W. C. Darrah Collection.
8. Eugene Kroeger was one of F. O. Thompson's field assistants who helped collect not only the Mazon Creek nodules but also the invertebrate fossils, especially at Winterset, Iowa.
9. Letter from F. O. Thompson to W. C. Darrah, January 30, 1948. W. C. Darrah Collection.
10. Letter from F. O. Thompson to W. C. Darrah, September 12, 1940. W. C. Darrah Collection.
11. Letter from Amos Stuver to F. O. Thompson, April 13, 1942. W. C. Darrah Collection.
12. Letter from F. O. Thompson to W. E. Dahlgren, June 18, 1938. W. C. Darrah Collection.
13. Letter from A. C. Noé to W. E. Dahlgren, July 4, 1938. W. C. Darrah Collection.
14. Letter from F. O. Thompson to W. C. Darrah, January 19, 1940. W. C. Darrah Collection.
15. Letter from Eugene Kroeger to W. C. Darrah, January 8, 1939. W. C. Darrah Collection.
16. Letter from Eugene Kroeger to W. C. Darrah, January 29, 1939. W. C. Darrah Collection.
17. Letter from F. O. Thompson to H. N. Andrews, February 19, 1946. W. C. Darrah Collection.
18. Letter from F. O. Thompson to W. C. Darrah, May 15, 1947. W. C. Darrah Collection.
19. Letter from W. C. Darrah to F. O. Thompson, June 3, 1947. W. C. Darrah Collection.
20. Letter from F. O. Thompson to W. C. Darrah, April 1, 1949. W. C. Darrah Collection.
21. Letter from F. O. Thompson to Oliver Thompson, April 24, 1938. W. C. Darrah Collection.
22. Letter from F. O. Thompson to W. C. Darrah, September 22, 1947. W. C. Darrah Collection.
23. Letter from W. C. Darrah to F. O. Thompson, October 2, 1947. W. C. Darrah Collection.
24. Letter from F. O. Thompson to A. K. Miller, October 19, 1944. W. C. Darrah Collection.
25. Letter from F. O. Thompson to W. C. Darrah, October 19, 1944. W. C. Darrah Collection.
26. Letter from W. C. Darrah to F. O. Thompson, April 21, 1948. W. C.

Darrah Collection.

27. Letter from S. H. Mamay to E. D. Morey, June 20, 1991. W. C. Darrah Collection.

28. Letter from H. N. Andrews to E. D. Morey, July 8, 1991. W. C. Darrah Collection.

29. Letter from W. L. Fry to E. D. Morey, August 18, 1991. W. C. Darrah Collection.

30. Letter from R. W. Baxter to E. D. Morey, September 16, 1991. W. C. Darrah Collection.

31. Letter from W. C. Darrah to F. O. Thompson, March 24, 1950. W. C. Darrah Collection.

32. Letter from F. O. Thompson to W. C. Darrah, January 21, 1951. W. C. Darrah Collection.

## REFERENCES CITED

Andrews, H. N., 1945, Contributions to our knowledge of American Carboniferous floras. Part VII: Some pteridosperm stems from Iowa: Missouri Botanical Garden Annals, v. 32, p. 323–360.

Barghoorn, E. S., 1953, Frederick Oliver Thompson 1883–1953: Harvard University Botanical Museum Leaflets, v. 16, p. 173–178.

Darrah, W. C., 1936a, A new *Macrostachya* from the Carboniferous of Illinois: Harvard University Botanical Museum Leaflets, v. 4, p. 52–63.

Darrah, W. C., 1936b, The peel method in paleobotany: Harvard University Botanical Museum Leaflets, v. 4, p. 69–83.

Darrah, W. C., 1938, A new transfer method for studying fossil plants: Harvard University Botanical Museum Leaflets, v. 7, p. 35–36.

Darrah, W. C., 1940, The fossil flora of Iowa coal balls. III: *Cordaianthus:* Harvard University Botanical Museum Leaflets, v. 8, p. 1–20.

Darrah, W. C., 1969, A critical review of the Upper Pennsylvanian floras of Eastern United States with notes on the Mazon Creek flora of Illinois: Gettysburg, Pennsylvania, privately published, 220 p., 80 pls.

Levittan, E. D., and Barghoorn, E. S., 1948, *Sphenostrobus thompsonii:* A new genus of the Sphenophyllales: American Journal of Botany, v. 35, p. 350–358.

Lyons, P. C., and Morey, E. D., 1991, Memorial to William Culp Darrah (1909–1989): Torrey Botanical Club Bulletin, v. 118, p. 195–200.

Lyons, P. C., and Morey, E. D., 1993, A tribute to an American paleobotanist: William Culp Darrah (1909–1989): Congrès International de Stratigraphie et Géologie du Carbonifère et Permien, 12th, Buenos Aires, September, 1991: Compte Rendu, v. 1, p. 117–126.

Miller, A. K., and Unklesbay, A. G., 1944, Trilobite genera *Goldius* and *Arctinurus* in the Silurian of Iowa and Illinois: Journal of Paleontology, v. 18, p. 363–365.

Noé, A. C., 1925, Pennsylvanian flora of northern Illinois: Illinois State Geological Survey Bulletin 52, 18 p., 45 pls.

Raymond, P. E., 1944, Late Paleozoic Xiphosurans: Museum of Comparative Zoology Bulletin, v. 94, p. 475–508.

Stopes, M. C., and Watson, D. M. S., 1907, On the present distribution and origin of the calcareous concretions in coal seams, known as "coal balls": Royal Society of London Philosophical Transactions, Ser. B, v. 200, p. 167–218.

Traverse, A., 1950, The primary vascular body of *Mesoxylon thompsonii,* a new American cordaitalean: American Journal of Botany, v. 37, p. 318–325.

MANUSCRIPT ACCEPTED BY THE SOCIETY JULY 6, 1994

Printed in U.S.A.

Geological Society of America
Memoir 185
1995

# *George Langford, Sr. (1876–1964): Amateur paleobotanist and inventor*

**Richard L. Leary**
*Illinois State Museum, Research and Collections Center, 1011 East Ash Street, Springfield, Illinois 62703*

## ABSTRACT

**George Langford, Sr., best known for his association with the fossiliferous nodules of the Mazon Creek region of northern Illinois, was also an industrialist, engineer, inventor, archeologist, athlete, artist, and author. He followed in the footsteps of his father and grandfather, graduating with an engineering degree and working in steel mills. He rose up the corporate ladder to the top, becoming president of the McKenna Processing Company. He inherited his father's and grandfather's interest in science, wide-ranging curiosity, and many talents. Following a successful career in industry, George Langford, Sr., retired and began, at age 71, a second career as paleobotanist at the Field Museum of Natural History in Chicago. He continued there until his death in 1964. His works, represented by the two-volume set of popular books on the Wilmington Coal Flora and Fauna, will keep his memory alive for decades to come.**

## INTRODUCTION

George Langford, Sr., is undoubtedly best known for his collections of fossiliferous nodules from the famous Mazon Creek region of northern Illinois and for his two books describing and illustrating Mazon Creek fossils. His long association with fossils from this area assured him a place in the history of paleobotany and paleontology of the early part of the twentieth century.

The fossiliferous nodules of Mazon Creek are known throughout the world. Visit a museum almost anywhere, and you are likely to find a nodule from this area in their exhibits or, if not on display, in their collections. Because the specimens are attractive as well as scientifically significant, they have been prized by museums, universities, and collectors everywhere.

Langford was partly responsible for the fame of the Mazon Creek fossiliferous nodules. Not only was he among the earliest and most active collectors but also he sent specimens to people and institutions all over the world. Sometimes he traded for fossils from elsewhere, and other times he simply sent nodules as gifts. In recognition of his work and generosity, the American Museum of Natural History honored Langford with a life membership, and he was listed as "research associate in anthropology" at the University of Chicago.

## EARLY YEARS

George Langford, Sr., was born in Denver, Colorado, on May 26, 1876. His father was Augustine G. Langford, who grew up in New York, married, and moved west to take advantage of westward expansion.

Augustine Langford had learned the iron trade from his father, who had established an ironworks at Utica, New York—the first west of the Hudson River. Augustine Langford, carrying on family tradition, operated the first blast furnace in the Rockies of the United States.

George Langford's father died when George was nine years old, and his family moved from Denver to St. Paul, Minnesota, to be with the family of his mother (nee Robertson). George received his early education in the public schools of St. Paul.

It was at this time that young Langford began to develop his interest in science and in paleobotany in particular. Both

Leary, R. L., 1995, George Langford, Sr. (1876–1964): Amateur paleobotanist and inventor, *in* Lyons, P. C., Morey, E. D., and Wagner, R. H., eds., Historical Perspective of Early Twentieth Century Carboniferous Paleobotany in North America (W. C. Darrah volume): Boulder, Colorado, Geological Society of America Memoir 185.

his father and grandfather enjoyed natural science as a form of recreation. He began his interest in paleontology at age 10, and this interest remained with him throughout his life. Conversations with his grandfather and readings in his library sparked and nurtured these interests:

> It was in his [maternal] grandfather's library that George first read and studied the botanical monographs of Lesquereux and Brongniart and the paleontological works of Marsh, Cope, and Leidy. Ever after, he shuddered when he recalled that on his grandfather's death in 1895, when George was away at college, the library and his grandfather's files of correspondence with American and European botanists had been sold as scrap paper. (Richardson, 1976, p. 10)

George Langford's uncle, Nathaniel P. Langford, was a member of the Washburn Expedition to Yellowstone in 1870 and later went to Washington, D.C., to urge the creation of a national park. Nathaniel Langford became the first superintendent of Yellowstone National Park. He continued to influence George through his lectures and letters.

George Langford, Sr., was an imposing person, robust and standing 1.9 m tall. According to a letter from his daughter, he was "tall, dark, and handsome."[1] He was an athlete, industrialist, engineer, inventor, archeologist, artist, and author. In many ways, he was a "Renaissance man"—a person of intellect, diverse talents, and a wide-ranging curiosity.

In the fall of 1894, George Langford entered the Sheffield Scientific School as an engineering student. The Sheffield School later became part of Yale University. Langford was unable to take geology and paleontology courses at Yale because of a division between the University and Sheffield. This was a major disappointment to George because he wanted to take classes from the great paleontologist, Professor O. C. Marsh.

While attending the Sheffield Scientific School, George Langford was a member of the Yale rowing crew for three years. In 1895 he went with the crew to the Henley Regatta in England, where Yale raced against Oxford and Cambridge.

## LANGFORD'S RAILROAD YEARS

Following graduation, Langford became a draftsman in the mechanical department of the Chicago Great Western Railroad in St. Paul. Perhaps as a result of having grown up in a family of ironworkers, the following spring he went to work for the McKenna Processing Company in Milwaukee. It was an ironworks that, among other things, refurbished worn-out railroad rails. The plant used a hot-rolling process developed by the plant owner, E. W. McKenna, a former railroad man. Beginning as a laborer, Langford worked his way up through the positions of foreman and assistant superintendent. About 1900, he was sent to the McKenna plant in Joliet, Illinois. He became an engineer and draftsman and spent 45 years in Joliet.

He also established branch plants for McKenna in Kansas City, Kansas, in 1897, and in Elizabeth, New Jersey, in 1898. He spent some time in England in 1904 setting up a plant in Birkenhead, near Liverpool. Langford was promoted to general superintendent, vice president, and then director in 1926. In 1929 he was made president of the McKenna Company (McFadyen, 1937).

George Langford developed many inventions related to the iron industry and railroads. In 1923, he conceived a process of reshaping worn rail-joint bars. Before he retired in 1945, he was awarded nearly 100 patents for his inventions, all in the name of the McKenna Company.

Not all of Langford's experiences with the iron industry were so positive. At the end of his career he was essentially forced out of business in a dispute with the U.S. government over use of his own patents. Early in his career another bad experience befell Langford, but he simply picked up the pieces and continued on to better things, as he was to do years later.

> Early in 1900, in the Joliet plant, George lost his left arm. While he was inspecting a machine, his sleeve was caught between two slow-moving gears and his hand was drawn into the teeth. While he held onto a pillar with his legs and other arm, the entire left arm was pulled off at the shoulder. He was hurried into a cart, which headed for a hospital, across a railroad siding. But a halted freight was blocking the track, and it took twenty minutes for his rescuers to have a coupling opened to let them through. The Joliet hospital patched him up and then sent him to his family in St. Paul, "to die," as he told it.
>
> At this time, George was engaged to a girl he had met a few years earlier in Kansas City, Sydney Holmes. She came to St. Paul to see him while he was recovering and he offered her a chance to end the engagement if she felt she wouldn't be getting full measure. She wouldn't hear of such a thing, and on November 14, 1900, they were married. (Richardson, 1976, p. 11.)

George and Sydney had two children, George Langford, Jr., and Lyda. Both children were as intellectually curious and as active as their father. George Jr. often joined his father in the field collecting fossils and maintained an interest in fossils throughout his life.

They lived in Joliet for many years when George Langford, Sr., worked for the mill there. An article in the Joliet Herald News (May 13, 1979) gives an interesting glimpse into their home life: "A fossilized elephant [mastodon] from Minooka was displayed in the front hall of George and Sydney Langford's Joliet home for many years." Paintings, sculptures, fossils and Indian artifacts decorated their house and the Langford's home was a center for those interested in art and the natural sciences."

Among Langford's talents was a natural writing ability. Not only did he produce the books on fossils (Langford, 1958, 1964) for which he is widely famous, he also wrote children's magazine stories and several children's books.

> While his children were growing up, he amused them with imaginative stories about the extinct animals and their manner of life and changes in their habits and structure. In their summer vacations away from Joliet, his letters contained stories of this kind. They were later published as four books. Informative, based on scientific knowledge

and interestingly told, they were fascinating to young and old. (Deuel, 1964, p. 350)

George Langford was not to be stopped or even slowed by the loss of his arm. In fact, in photographs he tried to obscure the fact that his left arm was missing. If at all possible, he went through life as if it was still present. Whether collecting fossils, doing research, writing, or doing art work, Langford was a match for anyone. He was also an accomplished piano player. His daughter, Lyda Langford Hinrichs, once wrote that his playing sounded "like three hands playing, not one."[2]

George Langford was also something of an artist. He sculpted several bas-reliefs, with natural history subjects, quite naturally. He sculpted models of mammoths based upon studies of the fossils from Europe and the United States (Deuel, 1964).

> Besides various pamphlets, scientific articles, and children's magazine stories, Mr. Langford has written several books pertaining to the natural sciences. Of his books, "Pic, the Weapon Maker," "Kutnar, Son of Pic," and "Stories of First American Animals" were written especially for youthful readers with the view of interesting them in the natural sciences. His poem, "Cave artists of Cro-Magnon," has been published here and abroad.
>
> Mr. Langford is an artist of ability, and has himself illustrated all his books, many of the illustrations being in color. His interest in the art of the cave man led Mr. Langford into the field of sculpture. In the various museums one will note many bas-reliefs of early animals sculptured by Mr. Langford. (McFadyen, 1937, p. 350)

## LANGFORD THE AMATEUR PALEONTOLOGIST AND PALEOBOTANIST

George Langford, Sr., collected fossils from several Ice Age mastodons in northern Illinois and wrote an account of their evolution. His interests in paleontology took him to the Black Hills, where he discovered some noteworthy fossils of the ancestors of the horse (Deuel, 1964).

After recovering from his accident, Langford began to revisit the strip mines in the Coal City, Illinois, area. From 1910 through 1914, he collected fossiliferous nodules along Mazon Creek. These nodules had been known for more than 60 years, and many famous geologists and paleobotanists had visited the area or acquired specimens from there through intermediaries. Leo Lesquereux included many descriptions of fossils from Mazon Creek in his "Pennsylvania Coal Flora" (1879–1884).

Langford paused in his collecting from 1914 until 1925. At that time mining began in the area, bringing fresh nodules to the surface. These "new" specimens renewed interest in the fossil-bearing nodules. Many collectors came to the area, including George Langford and his son. George Sr. wrote in his Wilmington Coal Flora (1958):

> My first fossil-collecting in the Wilmington area was with my son George Jr. We were in the field for about 150 days after which I worked about 400 days alone until 1941. I continued for 100 days or less after that, alone or in company with Dr. Robert H. Whitfield and family, Dr. Eugene Richardson, and Orville Gilpin of the Chicago Natural History Museum, and Mr. and Mrs. John L. McLuckie of Coal City. In 1938–1939 I spent considerable time with Dr. A. E. Noé (see Chapter 14, this volume) and Dr. Raymond Janssen at the University of Chicago, whose library gave me opportunity to study the specimens I was collecting. (Langford, 1958, p. 7)

Bill and June McLuckie, who also amassed an extensive collection of Mazon Creek fossiliferous nodules, had another view of George Langford:

> I was hunting fossils with my dad the first time we met the Langfords. I was 13 or 14 years old (1937 or 1938), but I can still picture him cracking fossils held between his feet because of losing his left arm. He was a tall, slender built man, but very agile for his age and handicap. We were "amateurs" in his field and were tolerated but not much more. He was rather aloof and very difficult to get to know. As time went on and further encounters occurred and different specimens were found there developed a very close friendship between my mom & dad and George.[3]

In his Wilmington Coal Flora (1958), George Langford, Sr., described his collecting technique:

> We gathered up nodules in iron pails, then sat down and split them open, using a stone as an anvil. When placed edge up and struck with a hammer, the nodule split in two halves on the plane of weakness exposing the two plant impressions. One acceptable specimen out of ten nodules split was a good average.
>
> We had no means of knowing what the nodule might contain until we split it open. However if it was well rounded, we could hope for something worth while. A collector becomes familiar with the shapes he likes or dislikes. But he has to do a lot of hammering for the good ones he gets. Splitting 1500 of them was a good day's work. (Langford, 1958, p. 17)

Over the years, in addition to accumulating a large collection of fossils, Langford filled many notebooks in very legible handwriting. He described the fossils in detail and illustrated them with excellent drawings and photographs. These notebooks laid the foundation for the two published volumes on the Wilmington Coal Flora (1958) and Fauna (1964).

He also maintained journals, essentially diaries of visits to strip mines and other sites. These personal accounts describe the weather and other conditions encountered in the field. Langford's scientific precision and completeness are reflected even in these daily logs of activity.

While working in Joliet at the McKenna Processing Company, Langford was able to obtain use of a railroad handcar in order to spend his off-hours collecting fossils along railroad cuts in the area. He also took advantage of his business travels. He collected fossils everywhere he went to establish branch foundries.

## LANGFORD THE AMATEUR ARCHEOLOGIST

George Langford is almost as well known among archeologists as among paleontologists. While exploring in the Joliet area, he became familiar with the Fisher Mounds near Channahon. When the mounds were about to be destroyed by expansion of a gravel pit, Langford contacted Dr. Fay-Cooper Cole, professor of anthropology at the University of Chicago. Convinced of the importance of the site, Cole and students from the University of Chicago worked for two summers, accompanied by Langford. They excavated the mounds and associated village area. Some of the graduate students went on to become leading anthropologists of the twentieth century. Langford published a description of the excavations in the Illinois State Academy of Science Transactions (1927a). He published an excellent report on the artifactual and skeletal material from the site in the *American Anthropologist* (1927b).

As a result of his pioneering work at the Fisher Mounds site, the culture defined by the discoveries there is known as the "Langford Tradition" (Fig. 1).

As if he didn't have enough to do, Langford also became the family historian. He attempted to trace his ancestry—not only his father's line but many of the intermarrying lineages. His son, George Jr., who is still alive at the time this portrait is being written, picked up on this quest and continued for many years.

As part of his search for his roots, George Sr. returned to the family property near Denver.

> In the early 1930s, George returned from Illinois to his childhood haunts. The blast furnace was long gone, but he found the village of Marshall, named for his father's former partner. In that town he also found his old friend Byerley. Before George spoke, the old man asked, "Is your name Langford?"—for the family resemblance was strong.
>
> Byerley showed George the creek on whose banks the foundry had once glowed, but even he had forgotten exactly where it had been. George, employing his fossil-collecting and archeological talents, found the site by following traces of brick and iron in the stream. At a point where no further clues led him upstream, he secured permission to dig, and found three pigs of cast iron, rusted but still sound. (Richardson, 1976, p. 10)

It was well that Langford had many outside interests at the time that his employment in industry came to an early end. He retired at the age of 69 when he still had energy and health to pursue his many activities.

Because of a dispute over the use of patents (for Langford's inventions), the U.S. government forced the plants to cease operations and indicted Langford. He closed the plant in 1945, retired from industry, and left Joliet. The family moved to an apartment in Chicago. Because there was no room for his fossil collections, Langford donated much of his material to the Field Museum of Natural History and the Illinois State Museum.

## LANGFORD'S YEARS AT THE FIELD MUSEUM OF NATURAL HISTORY

However, George Langford could not live without fossils to work on, and he offered to curate the specimens he had donated to the Field Museum of Natural History. In 1947, at age

Figure 1. George Langford, Sr., and unidentified colleagues excavating at Fisher Mounds near Channahon, Illinois. Photograph courtesy of Illinois State Museum Collections.

Figure 2. George Langford, Sr., at the Field Museum of Natural History, Chicago. Photograph courtesy of the Field Museum of Natural History, Chicago, negative GEO80771.

71, he became assistant in fossil plants and began a second career. Three years later he became curator of fossil plants. These positions were not based upon formal education (a handicap far greater than loss of an arm). Langford was given these titles because he had, through personal striving, dedication, hard work, and intelligence, become an expert in the field of paleontology.

Although largely self-taught, George Langford (Fig. 2) was something of an expert on the fossil plants and animals found in the nodules at Mazon Creek. This knowledge was gained through first-hand observation, extensive reading, and many discussions with professional paleontologists who visited the area over the years.

Langford continued to collect, primarily in the Mazon Creek region, until 1959. He also traveled to Tennessee, Mississippi, and Alabama with colleagues from the Field Museum of Natural History to collect fossils. On these travels with Field Museum paleontologists, Langford uncovered a deposit of fossil leaves in clay and developed a technique for preserving the fragile specimens so they could be returned to the laboratory for later study (Deuel, 1964, p. 350). Langford continued as curator until his second retirement in 1962.

George Langford, Sr., died suddenly in Chicago at age 88, on June 16, 1964. His death was a great loss for science, but his life had certainly been its gain. His curiosity and intellect had led him down many paths. Indeed, in many places, he made his own path. His contributions to paleobotany, vertebrate and invertebrate paleontology, and archeology will long remain a monument and beacon. His high standards of scientific knowledge and attention to detail will serve to inspire others to follow. They will also be the standards by which succeeding generations of amateurs (and professionals) will be judged.

The name George Langford may not outlast the fossils still uncollected in his beloved Mazon Creek area. However, the name will remain in the paleontology vocabulary as long as Mazon Creek nodules remain in museums and collections elsewhere.

## ACKNOWLEDGMENTS

Thanks are given to Don Auler, Andrew Hay, Bill and June McLuckie, and Riner Zangerl for their assistance in obtaining information on George Langford, Sr. Any errors or omissions are my responsibility. The manuscript was reviewed by M. D. Henderson and T. L. Phillips; their comments and suggestions are greatly appreciated.

## ENDNOTES

1. Letter from Lyda Langford Hinrichs to Eugene Richardson, February 24, 1976.
2. Letter from Lyda Hinrichs to Eugene Richardson, 1976.
3. Letter from Bill and June McLuckie to Martha Auler, August 6, 1991.

## REFERENCES CITED

Deuel, T., 1964, George Langford: Springfield, Illinois State Museum, The Living Museum, v. 8, p. 349, 350.

Joliet Herald News, 1979, Langford's knowledge rivaled professionals (May 13) [page unknown].

Langford, G., 1927a, Stratified Indian mounds in Will County: Illinois State Academy of Science Transactions, v. 20, p. 247–253.

Langford, G., 1927b, The Fisher Mound Group successive aboriginal occupations near the mouth of the Illinois River: American Anthropologist, v. 29, p. 153–205.

Langford, G., 1958, The Wilmington Coal Flora from a Pennsylvanian deposit in Will County, Illinois: Downers Grove, Illinois, Esconi Associates, 360 p.

Langford, G., 1964, The Wilmington Coal Fauna and additions to the Wilmington Coal Flora from a Pennsylvanian deposit in Will County, Illinois: Downers Grove, Illinois, Esconi Associates, 280 p.

Lesquereux, L., 1879–1884, Description of the Coal Flora of the Carboniferous Formation in Pennsylvania and throughout the United States: Pennsylvania Geological Survey, 2nd Report, Atlas (1879), 87 pls.; parts 1 and 2 (1880), 694 p.; part 3 (1884), p. 695–977, 34 pls.

McFadyen, A. D., 1937, Biographical note (accompanying Paleolithic Patent No. 1 by George Langford): Journal of the Patent Office Society (May) vol. no. unknown, p. 346–350.

Richardson, E., 1976, George Langford 1876–1964: Field Museum of Natural History Bulletin (February), p. 10, 11.

Manuscript Accepted by the Society July 6, 1994

Printed in U.S.A.

Geological Society of America
Memoir 185
1995

# *John Edward Jones (1883–1957): Pioneer in coal-mining safety and generous collector of fossil plants and animals*

**Michael D. Henderson**
*Burpee Museum of Natural History, 737 North Main Street, Rockford, Illinois 61103*

## ABSTRACT

**John Edward Jones (1883–1957) was a major figure in the field of coal-mining safety in the United States. Jones served as an Illinois State mine inspector from 1915 to 1917 and as safety engineer for the Old Ben Coal Corporation from 1917 until his retirement in 1952. He pioneered the practice of rock dusting in U.S. mines early in his career at Old Ben. Jones's contributions to paleobotany were made while he was safety engineer for the Old Ben Coal Corporation. He collected Pennsylvanian fossil plants in southern Illinois, West Virginia, and west-central Arkansas. Jones discovered the first long-leaved specimen of *Lepidodendron* in 1942. Jones passed away on June 30, 1957, at the age of 73.**

## INTRODUCTION

John Edward Jones was one of the great figures in the history of coal-mining safety. He also made a substantial contribution to paleobotany through his fossil-plant collecting achievements. These contributions have remained relatively unknown, overshadowed, perhaps, by the magnitude of his accomplishments in the field of coal-mining safety.

## EARLY YEARS

John E. Jones was born August 20, 1883, at Buckley, Flintshire, Wales. The son of a coal miner, Jones took an interest in geology at an early age and began collecting rocks and fossils that he found in the neighborhood of his father's home.

When Jones was eight years old, his father, Edward Jones, left for the United States in search of work. Upon his arrival in the United States, the elder Jones worked briefly as a coal miner at Sur, West Virginia. He then moved to Lucas, Iowa, where he sent for his family to join him in the United States. After the arrival of his wife and three sons, the Jones family settled in Oglesby, Illinois. Mrs. Jones and the sons became U.S. citizens on April 14, 1894, the date of Edward Jones's naturalization; John was 10 years old.

After finishing grade school, John, at age 14, started working in a coal mine at Oglesby. He was a bright, hard-working young man, and his fellow miners encouraged him to pursue a college education. This he eventually did. Following a mining accident in which both his legs were broken, Jones took lessons from a Scranton, Pennsylvania, correspondence school. He then entered Valparaiso University in Valparaiso, Indiana, where in 1910 he earned his Bachelors degree in civil engineering with a mining engineering option.

Following graduation, Jones worked for five years as a miner, boss, and division engineer for the J. K. Deering Coal Company in Danville, Illinois. It was during this time that Jones married EuTurpe Buntin. The union produced three sons: John Jr., Edward, and Robert.

## JONES THE COAL-MINING SAFETY EXPERT

In 1914, Jones became an Illinois State mining inspector working in southern Illinois. He immediately set to work lobbying mine operators for improved safety measures. In fact, he was nicknamed "Rock Dust Johnny" for his enthusiastic support of the practice of rock dusting.

Coal mining presents special dust hazards, as mining procedures generate high levels of coal dust. Combined with the dust problem is another concern: Coal mines may contain methane, a natural gas, that moves into the mine workings. Due

Henderson, M. D., 1995, John Edward Jones (1883–1957): Pioneer in coal-mining safety and generous collector of fossil plants and animals, *in* Lyons, P. C., Morey, E. D., and Wagner, R. H., eds., Historical Perspective of Early Twentieth Century Carboniferous Paleobotany in North America (W. C. Darrah volume): Boulder, Colorado, Geological Society of America Memoir 185.

to poor ventilation, methane may concentrate and stratify into a methane-air mixture of several percent or more methane, and an explosion may result from ignition by an open lamp, spark, or other source of heat. Methane explosions may be propagated by coal dust, as in the Old Ben No. 11 Mine (Jones, 1955). Thus, Jones suggested that in order to minimize an explosion hazard, an inert rock dust be added to the combustible coal face to prevent the propagation of methane explosions.

Rock dusting refers to the application of finely ground rock dust into mine tunnels, a procedure that effectively coats all solid and dusty walls or surfaces of the coalmine roof, floor, and walls with nonexplosive rock powder. The inert rock dust has the effect of halting the propagation of a small explosion, which could turn into a major disaster, by inhibiting the reactive effects of coal dust. The effectiveness of rock dusting in preventing explosion propagation had been recognized in Britain since the early 1800s. The practice was only sporadically used in the United States because of the high cost and inefficiency of covering miles of mine tunnels with rock dust by hand.

In 1917, Old Ben Coal Corporation's New North mine at Christopher, Illinois, exploded on Thanksgiving night. Seventeen miners on the night shift lost their lives in this explosion. The loss of lives was tragic, but if the accident had occurred during the day shift, as many as 600 miners could possibly have perished.

After learning about the disaster, Jones told D. W. Buchanan, then president of Old Ben, that this type of explosion could be greatly reduced by using rock-dusting procedures. Buchanan immediately offered Jones a position as Old Ben's safety engineer. Jones accepted and resigned his position as a state mine inspector in Illinois.

Jones viewed his position with Old Ben as a chance to introduce the safety measures he had lobbied so hard for. He immediately set to work on the problem of rock dusting. Jones (1955, p. 144) gives some idea of the difficulties of this work:

> The beginning of rock dusting was difficult, too. We began with scraping dust off the highways. Those were horse and buggy days, largely. That dust became wet at once in the mine. Then we tried lime rock dust sweepings from buildings. That was too coarse. Mr. Buchanan then decided to install a mill, after learning our roof shale was a good, though expensive, source for shale dust. Rock dust mills, however, had not yet evolved. He bought a hammer mill to grind the shale lumps into 1½ inch screenings and a ball-mill to grind the screenings into dust, 92% through 255 mesh. This was very good, since the Bureau's requirements were 100% through 150 and 75% through 200 mesh. That mill cost $40,000.00 and that shale dust was used by us and a few of our neighbors for nearly ten years, when lime rock dust became available in paper bags. Very early in that period we supplied the Bureau with five tons of our Mine No. 10 coal and ten tons of shale dust.

The invention and development of a motor and fan rock-dusting machine, for applying rock dust to coal mine ribs and roofs, greatly reduced the labor involved in rock dusting (Fig. 1).

Figure 1. John E. Jones with the first rock-dusting machine, circa 1923. Photograph courtesy of John E. Jones, Jr.

Beginning in the early 1920s, Jones lectured in every major coalfield in the United States on the benefits of rock dusting. Soon rock dusting became standard practice in the United States, thanks to the efforts of Jones and Buchanan.

Jones also played a key role in the introduction of electric battery lamps into coal mines. This was a major safety improvement over the calcium carbide (acetylene gas) lamps, as these closed lights would not ignite methane. Initially, miners resisted the change because of the low candle power and weight of the new lamps. However, Jones finally succeeded in the effort after years of controversy.

## JONES THE AMATEUR FOSSIL-PLANT COLLECTOR

Throughout his life Jones maintained an avid interest in paleontology. Although his most significant collecting efforts were directed toward fossil plants, he collected vertebrate and invertebrate fossils as well.

As safety engineer, Jones routinely visited all of Old Ben's mines. And it was in these mines that he did much of his collecting. During the course of mining operations many fine fossil plant specimens were exposed in the shales overlying the coal seams. Knowing of Jones's interest, miners would bring these specimens to his attention during his visits to the mines.

Assisted by miners or occasionally by one of his three sons, Jones would remove the fossil specimens from the roof of the mine tunnels and rooms. Once removed, the specimens, if large, would be wrapped and placed in a mine car for transportation to the surface.

Jones collected from Old Ben's mines in southern Illinois, West Virginia, and the Paris Basin of west-central Arkansas. While doing some consulting work in Wyoming during the 1920s, he collected dinosaur footprints that he occasionally encountered in sandstone beds capping Late Cretaceous coals. Jones also collected from more-traditional fossil sites when his travels brought him close to a good collecting locality.

Jones was always eager to share his fossils with professional paleontologists. He donated collections of fossils to the Field Museum of Natural History in Chicago, Southern Illinois University at Carbondale, Washington University in St. Louis, and the Illinois State Geological Survey, Urbana (Fig. 2).

Much of the material Jones collected was of exceptional size and quality, and thus it was excellent for display. Because of the high quality of his specimens and his generosity, Jones received frequent requests from paleobotanists for specimens. As a result, whenever Jones encountered a particularly fine or unusual fossil-plant specimen, he would collect and store it, certain in the knowledge that eventually a paleobotanist would call inquiring about his latest finds. In this way Jones sent specimens to individuals and institutions around the country and even as far away as Berlin, Germany.

Figure 2. Dr. M. M. Leighton, late director of the Illinois State Geological Survey (left), with John E. Jones (right). The lycopod stump in the foreground was collected by Jones. The photograph is dated September 4, 1944. Photograph #395, courtesy of the Illinois State Geological Survey.

One of Jones's most interesting discoveries occurred in 1942, when he collected the first long-leaved specimen of *Lepidodendron.* Jones obtained the specimen from the roof shale directly above the Beckley Coal in the Raleigh-Wyoming Mine No. 2 at Glen Rogers, West Virginia. He donated the specimen to the Illinois State Geological Survey, where it was examined and reported by Robert Kosanke (1948, 1979).

It was more than a year after the discovery of the specimen that the roof shale loosened enough to allow its removal. The *in situ* specimen was 3.65 m long and so heavy that it had to be trimmed in order to carry it to the nearest underground transportation area for removal to the hoisting shaft (Kosanke, 1979).

In addition to his collecting efforts, Jones contributed to paleobotany by arranging collecting trips for paleobotanists and other individuals interested in visiting shaft mines in southern Illinois. Professor Henry Andrews and his students visited several deep mines with Jones in the 1940s.

## JONES'S PERSONAL QUALITIES, AFFILIATIONS, AND PUBLICATIONS

Jones possessed an outgoing personality. He was an excellent joke teller and good singer. He also enjoyed the poetry of Kipling and Burns. During his years with Old Ben, he published profusely on issues of coal-mining safety. In 1928 and 1929 he served as president of the Illinois Mining Institute.

Jones retired from Old Ben in 1952 after 35 years of service. During his retirement he occasionally wrote articles on mining issues and continued to collect fossils.

John E. Jones passed away suddenly at his home in Benton, Illinois, on June 30, 1957. He was 73 years old. His substantial contributions to mining safety will be long remembered. His love for fossil collecting produced many fine specimens, which he shared with many institutions, and they will remain a lasting tribute to Jones's memory.

## ACKNOWLEDGMENTS

I thank Matthew Nitecki of the Field Museum, Chicago, who first suggested I write the present manuscript. Jack Simon, chief emeritus of the Illinois State Geological Survey, deserves special thanks for the many leads he provided to information about Jones. Russell Peppers of the Illinois Geological Survey provided me with a list of specimens Jones donated to that institution. Robert M. Kosanke of the U.S. Geological Survey, Denver, provided valuable information about Jones, as did Jones's two surviving sons, John E. Jones, Jr., of St. Albans, West Virginia, and William A. Jones of Benton, Illinois. R. M. Kosanke, A. T. Cross, E. D. Morey, and P. C. Lyons reviewed the manuscript and provided many helpful comments and suggestions.

## REFERENCES CITED

Jones, J. E., 1955, Explosion in Old Ben Coal Corporation's Mine No. 11, *in* A compilation of the reports of the mining industry of Illinois from the earliest records to 1954: Illinois Department of Mines and Minerals, Coal Report, p. 143–145.

Kosanke, R. M., 1948, Unusual specimen of *Lepidodendron* [abs.]: Geological Society of America Bulletin, v. 59, no. 12, pt. 2, p. 1333.

Kosanke, R. M., 1979, A long-leaved specimen of *Lepidodendron:* Geological Society of America Bulletin, v. 90, pt. 1, p. 431–434.

MANUSCRIPT ACCEPTED BY THE SOCIETY JULY 6, 1994

Printed in U.S.A.

Geological Society of America
Memoir 185
1995

# *Early to mid-twentieth century floral zonation schemes of the Pennsylvanian (late Carboniferous) of North America and correlations with the late Carboniferous of Europe*

**Paul C. Lyons**
*U.S. Geological Survey, MS 956, National Center, Reston, Virginia 22092*
**Erwin L. Zodrow**
*University College of Cape Breton, Subdepartment of Earth Sciences, Sydney, Nova Scotia, Canada B1P 6L2*

## ABSTRACT

**Twentieth-century megafloral zonation schemes for the Pennsylvanian System (upper Carboniferous) of North America are partly rooted in the nineteenth-century works of L. Lesquereux and J. W. Dawson in North America and partly in western European floral biostratigraphy. The major contributions to these schemes were by D. White—who independently developed floral correlations with Europe—with later modifications and refinements by W. C. Darrah and C. B. Read of the United States and W. A. Bell of Canada. Minor contributions to European correlations were made by E. H. Sellards, A. C. Noé, and C. A. Arnold of the United States. European paleobotanists—especially P. Bertrand, W. J. Jongmans, and E. Dix—notably influenced the development of Carboniferous floral zonation schemes in North America. Darrah integrated, on the basis of personal collections and examinations of museum collections on both continents, the European and U.S. late Carboniferous floral zonations and correlations. Difficulties with megafloral correlations between Europe and North America occurred because of taxonomic problems and different perspectives on the biostratigraphic value of certain index species and genera that have different ranges in Europe and North America. However, the broad aspects of floral succession and megafloral correlations between the two continents were essentially established by the middle of the twentieth century.**

## INTRODUCTION

Floral zonation schemes of the Pennsylvanian System (upper Carboniferous) of the United States were begun in the nineteenth century by L. Lesquereux (1879–1884) and D. White (1899, 1900). White's work in the early part of the twentieth century (e.g., White, 1903, 1904, 1909, 1912) set the stage for the floral zonation of the Pennsylvanian of North America. W. A. Bell's (1938, 1943, 1944) work in Canada, W. C. Darrah's (1935, 1936, 1937) work in the Appalachian basin, and C. B. Read's (1941, 1944, 1947) work in both the eastern and western United States extended the work of White and set the stage for the late Paleozoic zonation scheme of Read and Mamay (1964).

The first, second, and third Carboniferous Congresses in Heerlen (1927, 1935, 1951) established the subdivisions of the Carboniferous System, as shown in Table 1. The Namurian was established on the basis of marine goniatites, the Westphalian on the basis of plant fossils and to some extent on goniatites, and the Stephanian on the basis of plant fossils. The second Carboniferous Congress Heerlen, 1935, refined the stages of the entire Carboniferous System (Table 1).

The concept of a Westphalian E substage proposed by Jongmans and Gothan (1937) for what was essentially the

Lyons, P. C., and Zodrow, E. L., 1995, Early to mid-twentieth century floral zonation schemes of the Pennsylvanian (late Carboniferous) of North America and correlations with the late Carboniferous of Europe, *in* Lyons, P. C., Morey, E. D., and Wagner, R. H., eds., Historical Perspective of Early Twentieth Century Carboniferous Paleobotany in North America (W. C. Darrah volume): Boulder, Colorado, Geological Society of America Memoir 185.

lower Stephanian was not generally accepted (Darrah, 1969). Fundamental problems occurred because the floral ranges in North America were not exactly the same as in western Europe. Some critical species and genera were missing, and others had different ranges on the two continents. In addition, there were unresolved taxonomic problems relating to such species as *Neuropteris schlehanii* and *N. pocahontas* and to *N. decipiens* and *N. scheuchzeri* (see Darrah, 1969). These differences led to great controversies on the placement of European series and stage boundaries in North America, and even on the placement of systemic boundaries. The proposals for the boundaries of the Pennsylvanian and Permian Systems in North America were radically different (see Barlow, 1975a). These controversies still have not subsided because of the recognition of differing edaphic, climatic, paleogeographic, and plate tectonic factors that influenced the Carboniferous floral succession in Euramerica.

This chapter traces the early evolution of Carboniferous megafloral zonation schemes in North America. These schemes had their roots in the works of J. W. Dawson (1873), L. Lesquereux (1879–1884), and D. White (1899, 1900, 1909).

The Carboniferous floral zonation schemes in this chapter are subdivided into two parts: United States and Canada. Within each country, they will be traced from individual to individual in a historical fashion, as much as possible.

## FLORAL ZONATIONS IN THE UNITED STATES

### *P. C. Lyons*

### *Lesquereux (1879–1884)—The foundation of Carboniferous paleobotany in the United States*

Lesquereux's monumental work (1879–1884) laid the foundation for the taxonomic characteristics of Carboniferous plants and Carboniferous stratigraphic paleobotany in the United States. The three volumes and atlas (1879, 1880, 1884) summarized his three decades of research on the coal floras of the United States. A revision and expansion of part of Lesquereux's work was reported in D. White's many paleobotanical publications (e.g., 1899, 1900, 1903, 1904, 1909, 1912, 1929, 1936) and also by Janssen (1940) and Darrah (1969).

Lesquereux (1879–1884) recognized the value of Carboniferous fossils for identifying individual coal beds and for determining the stratigraphic succession and regional correlations. He compared the American species with those in the Carboniferous strata of Europe and concluded that of 192 species, about one-third were identical to European species and that the most common species are "identical on both continents." He offered this as proof of the Carboniferous age of the coal flora of the United States. However, it is also clear today that Lesquereux erected many new species that were really conspecific with European taxa, which obscured precise European correlations.

Broad correlations of the coal floras and coal beds of the United States and Europe were made by Lesquereux (1879–1884) and White and Ashley (1906). Lesquereux referred the floras to the "lower," "middle," and "upper" Carboniferous Subsystem and the "Permo-Carboniferous."

**TABLE 1. SUBDIVISION OF THE CARBONIFEROUS SYSTEM BY THE FIRST, SECOND, AND THIRD HEERLEN CARBONIFEROUS CONGRESSES**

| Series | Heerlen, 1927 Stage | Heerlen, 1927 Substage | Heerlen, 1935, 1951 Stage | Heerlen, 1935, 1951 Substage | |
|---|---|---|---|---|---|
| Upper Carboniferous | Stephanian | ....... | Stephanian | C<br>B<br>A | |
| | Westphalian | C<br>B<br>A | Westphalian | D<br>C<br>B<br>A | Zone of *Neuropteris ovata* |
| | Namurian | ....... | Namurian | C<br>B<br>A | |
| Lower Carboniferous | Dinantian | ....... | Dinantian | C<br>B<br>A | |

Lesquereux (1879–1884) did not report the presence of the primitive conifer *Walchia* in the Pennsylvanian System of the United States. This is of some stratigraphic importance because of the widespread belief, at the time, that the true conifers (for example, *Walchia*) first appeared in the very uppermost upper Carboniferous or Lower Permian strata. This genus was also *not* reported by Fontaine and White (1880) or White (1904, 1909) in the Dunkard Group flora of late Carboniferous and early Permian age. The absence of *Walchia* was presumed to be an indication of a Carboniferous age, but paleoecologic and paleoclimatic factors were largely ignored in these judgments, except by White (1912), Elias (1936), and, much later, Jongmans (1954).

### *Fontaine and I. C. White's Dunkard Group flora with refinements by D. White and W. C. Darrah*

The late Carboniferous or Permian age of the Dunkard Group flora (Fontaine and White, 1880) has long been controversial (see papers in Barlow, 1975a). The American position is best summarized by Darrah (1975) and the European position by Remy (1975). White (1936, p. 683), in his last paper on the subject, considered the entire Dunkard succession to be of early Permian age on the basis of the occurrence of *Callipteris,* a genus whose representatives are found in great abundance at the level of the Washington coal bed, about 61 m above the base of the Dunkard Group. Lesquereux, unlike

White (1904), considered the red shales and limestones above the Waynesburg Formation (base of Dunkard Group) to represent an "uncertain extent" of the Upper Permian Series (Zechstein) of northwestern Europe. White (1904) initially considered the lower part of the Dunkard Group below the informal Lower Washington Limestone to contain a transitional Carboniferous and Permian flora and the strata above the Lower Washington limestone to be Lower Permian. A similar position on the age of the upper part of the Dunkard was taken by Jongmans and Gothan (1934, 1937), some three decades later; however, because of the recognition of the long ranges of *Neuropteris ovata* and *N. scheuchzeri* in the United States, they assumed that the Stephanian was absent and represented by an unconformity in the Dunkard basin and Midcontinent (see Darrah, 1975). In addition to *Callipteris,* White (1936) considered *Baiera, Saportea,* and *Taeniopteris* as genera additionally indicating an early Permian age of the Dunkard strata. White's position was later supported by Darrah (1975, Figs. 1, 3), who figured *Callipteris conferta* from the basal 61 m of the Dunkard, between the Waynesburg A and Washington coal beds. *Taeniopteris* and *Walchia* were reported by Darrah (1935, 1936) from the upper part of the Conemaugh Group, about 153 m below the base of the Dunkard Group. McComas (1988) and Lyons and Darrah (1989a, b) reported *Walchia* still lower, about 15 m above the base of the Conemaugh Group.

I. C. White (1917) held the view that the red beds in the Conemaugh Group mark the beginning of the Permian System in the Appalachian basin. He used the presence of *Lescuropteris moorii* in the uppermost Conemaugh to suggest that the base of the Conemaugh Group is equivalent to the base of the Lower Permian Series (Rotliegend) in Europe. Apparently, D. White mistakenly believed that *Callipteris* also occurred in the Conemaugh Group (Barlow, 1975b, p. ix, x) which was used by I. C. White to suggest a Permian age for that group.

A similar position on the age of the Dunkard flora was held by Remy (1975) on the basis of investigations in western Europe. He considered the first occurrence of *Walchia* (*Lebachia*) *piniformis* (Darrah, 1936) and *Plagiozamites* sp. in the upper part of the Conemaugh Group (Bassler, 1916) to indicate an early Permian age, even though *Callipteris* has not been reported below the Dunkard Group. Remy (1975) correlated the upper part of the Conemaugh Group (i.e., Connellsville Sandstone) with the Stanton Limestone of Kansas, which contains *Dichophyllum moorei, Taeniopteris* sp., and *Ernestiodendron filiciforme,* all of which first occur in the Lower Permian strata of Europe. Remy's more modern position of using two or more megafloral species to define the Carboniferous-Permian boundary was not in agreement with the accepted practice at that time of using a single species, that is, *Autunia* (*Callipteris*) *conferta* (see Kerp and Haubold, 1988), to define the same boundary. Remy placed the Dunkard beds above the informal Upper Washington limestone, that is, the Greene Formation, in the higher part of the Lower Permian (Saxonian). However, there is broad agreement in the views of D. White (1904, 1909, 1936), Darrah (1937, 1975), and Remy (1975) that all, or almost all, of the Dunkard beds are of Permian age.

The positions of Bode (1975) and Gillespie et al. (1975) represent different points of view. Bode considered the lower part of the Dunkard (i.e., Washington Formation) to be of Westphalian D age and the upper part of the Dunkard (i.e., Greene Formation) to be Stephanian in age, and Gillespie et al. (1975) supported the position that the entire Dunkard is Carboniferous in age and not Permian. Bode's position was based on his observations that *Neuropteris ovata,* the primary index fossil of Westphalian D strata in western Europe, occurs in the Washington Formation up to the base of the Greene Formation. Bode's position was challenged by Darrah (in discussion of Bode's paper, 1975, p. 153), who pointed out that *Neuropteris ovata* and *N. scheuchzeri* in North America were relict species in the same way as some modern floras are relicts of earlier Pliocene floras.

The age of the Dunkard is still open to discussion, especially in light of the fact that there are still several competing definitions for the systemic boundary and that even these "boundaries" have not yet all been correlated with the terrestrial realm. Even a correlation with the European terrestrial sequence of similar age will depend on an evaluation of first occurrences in light of paleoecological and paleoclimatological differences.

### *Sellards's and Elias's Kansas floras*

The Kansas Carboniferous and Permian floras were extensively studied by E. H. Sellards (1901, 1908) and White (1903), and identifications were revised by Cridland et al. (1963). Darrah (1969) summarized the occurrence of Carboniferous floras in Kansas, including a revision of Jongmans's small collections in 1933 from the Lawrence Shale and the now-abandoned LeRoy Shale. Darrah (1969) pointed out that Sellards was the first to recognize the genus *Lescuropteris* in the Lawrence Shale. *Lescuropteris moorii* (see Chapter 26, this volume; a Stephanian plant according to Darrah, 1937; see also Darrah, 1968) was identified by Darrah (1969) from the same formation.

M. K. Elias (1936; see also Moore et al., 1936) reported a *Walchia–Dichophyllum moorei–Taeniopteris–Callipteridium* association in the Garnett Limestone (shale), which he interpreted as representing an upland flora contemporaneous with an upper Carboniferous (Middle Pennsylvanian) lowland flora. The paper was an important contribution to the role of paleoecological factors in the first appearance of Permian-type plants (see Chapter 12, this volume).

Sellards (1908) recognized *Odontopteris genuina, O. minor, Pecopteris candolleana, Alethopteris grandini,* and *Daubreeia pateraeformis* in the Douglas Group and correlated them with the upper Stephanian flora of the Commentry coalfield of south-central France. He also recognized that varieties of *Neuropteris scheuchzeri* and *N. ovata* were associated with these Stephanian forms, a circumstance much like that in the Dunkard Group flora (Darrah, 1969).

The overlying Wellington Shale was referred by Sellards (1908) to the Permian System. He recorded four species of *Callipteris,* including *C. conferta;* five species of *Glenopteris;* three species of *Taeniopteris; Walchia piniformis;* and *Neuropteris permiana,*—an assemblage of undoubted Permian affinity. *C. conferta* was also reported by him in the underlying Wreford and Winfield Limestones, both of probable Permian age.

### D. White's (1899, 1900, 1909) floral correlations with Europe

D. White's (1899, 1900, 1909) intercontinental correlations with Europe represented a milestone in our knowledge of Euramerican upper Carboniferous biostratigraphy. He compared the flora of the Lower Coal Measures of Henry County, Missouri, with the Radstockian [Westphalian D] floras of England. For comparison, White had available a reference collection of Radstockian specimens, which are in the Lacoe Collection at the U.S. National Museum (Washington, D.C.)

White's (1899) flora from the Lower Coal Measures of Missouri consisted of 69 species, half of which were identical, in his opinion, to European taxa. Notable in this flora are *Neuropteris scheuchzeri, N. rarinervis, Alethopteris serlii, Sphenophyllum cuneifolium, S. emarginatum, S. majus, Annularia sphenophylloides, A. stellata, Asterophyllites equisetiformis,* and relatively few *Pecopteris* species; *Neuropteris ovata* is absent. He considered the floral assemblage to be transitional between the Middle and Upper Coal Measures of Great Britain, occupying the position of the "New Rock" and "Vobster Series" of the Bristol and Somerset coalfield and the "Lower Pennant" of the South Wales coalfield of Great Britain.

White (1899) also made a detailed comparison of the Missouri species with floral zones in the Valenciennes area of the Nord/Pas-de-Calais/Charleroi coalfield of northern France and southern Belgium. He found seven identical forms in the lower zone, 19 in the middle zone, and 25 in the upper zone. White concluded that they are assignable to the upper zone, the zone of Bully Grenay, which belongs to the "upper Westphalian." White also recognized the nature of some of the Missouri species of *Pecopteris* (for example, *P. arborescens, P. hemitelioides, P. candolleana*) as being identical to those in the Stephanian of the Saar and Commentry basins. This suggested to him that they may be transitional into the Stephanian as opposed to transitional into the "middle zone" of the Valenciennes basin of northern France.

White (1899) made other notable correlations with Europe. He referred the Pottsville flora to the flora of the Millstone Grit (Namurian) of England and the floras of the Kanawha Formation to the Lower Coal Measures of Great Britain and the "lower zone" of the Valenciennes basin (lower Westphalian). He correlated the Middle Kittanning coal bed of the bituminous coal region and the E coal bed of the Anthracite region of eastern Pennsylvania with the uppermost beds of the Westphalian. Also, he referred the Pittsburgh coal bed of the bituminous coal region and the G coal bed of the Northern Anthracite coalfield of eastern Pennsylvania to the Stephanian (equivalent to the "Ottweiler Schichten" of the Saar-Lorraine Basin in the German-French border area).

In the Pottsville Formation, White (1900) recognized four floral divisions: Lower Lykens, Lower Intermediate, Upper Lykens, and Upper Intermediate. He correlated the Lower Lykens flora with the Upper Culm flora (equivalent to the Ostrau-Waldenburg beds in Silesia), the Upper Lykens with the Millstone Grit of Canada, and the Upper Intermediate with the Lower Coal Measures of Great Britain and Westphalian of Europe. However, in the latter, he reported *Neuropteris ovata,* and *Alethopteris serlii,* which are characteristic of the upper Westphalian. The "Fern ledges" of St. John, New Brunswick (see Chapter 9, this volume), were correlated by White (1900) with the upper part of the Pottsville flora, presumably the upper part of the Westphalian.

White (1909) recognized that many Stephanian species persist into the Dunkard and that *Walchia* was not known in the Appalachian Basin. Nevertheless, he considered the Conemaugh and Monongahela Formations to be Stephanian and placed the Westphalian-Stephanian boundary near the Allegheny-Conemaugh contact on the basis of paleobotanical evidence.

### Noé's (1923, 1930) correlations with Europe

A. C. Noé (1923) reported a floral assemblage from shale beds at Brackwood, Illinois, that contained an unusual mix of different plant genera and species. According to Noé's identifications, the assemblage contained *Annularia sphenophylloides, Lepidodendron* sp., *Sphenophyllum emarginatum, Pecopteris unita, Alethopteris lonchitica, Neuropteris hirsuta* (= *N. scheuchzeri* according to Darrah, 1969), *Callipteris, Odontopteris,* and *Palmatopteris* spp. He also correlated the Braidwood flora with the Mazon Creek flora, the Kittanning flora of Pennsylvania, and the floras of the upper part of the Westphalian and lower part of the Stephanian of Europe. The mix of typical early Westphalian species (e.g., *Alethopteris lonchitica*) with typically Stephanian or Permian genera (*Odontopteris, Callipteris*) seems unlikely. To the contrary, Darrah (1969) considered the Mazon Creek flora to be Westphalian C in age.

Later, following the 1st Carboniferous Congress in Heerlen, 1927, Noé (1930) correlated the Illinois No. 2 (Colchester) coal bed with the Westphalian C, the Pottsville Formation with the Namurian and Westphalian A and B, and the beds above the No. 2 coal bed of the Illinois basin with the Stephanian of Europe. He also correlated the No. 2 coal bed with the Khroustalskaya coal bed (upper Moscovian) of the Donetz basin.

### Jongmans and Gothan's (1934) correlations with Europe

In 1933, W. J. Jongmans collected plant megafossils throughout the Appalachian basin and also in Illinois and Kansas. He reported his findings in collaboration with W. Gothan (Jongmans and Gothan, 1934). They correlated the Pocahontas coal beds with the Namurian B and C of Europe; the New River Group with Westphalian A; the Kanawha Group with Westphalian B; and the Allegheny, Conemaugh, Monongahela, and lowermost part of the Dunkard Group with the Westphalian C. A hiatus, representing the Stephanian, was interpreted by them at the position of the Waynesburg coal bed; almost all of the Dunkard Group they considered Permian. A Stephanian hiatus was also interpreted by them in the Illinois basin, but in Kansas, Stephanian equivalents were considered by them to be questionably present. Later Jongmans (1952) no longer recognized a Stephanian hiatus in the Appalachian basin and considered most of the Conemaugh, Monongahela, and lowermost Dunkard to be of Stephanian age and time equivalent to his Westphalian E facies; the upper Allegheny and lower Conemaugh were considered Westphalian D, which was established in the Heerlen Congress of 1935 (Darrah, 1937).

Jongmans and Gothan (1934) also made correlations between the Appalachian basin and Great Britain. The Pocahontas beds were considered by them to correlate with the Millstone Grit; the New River Group with the Lower Yorkian; the Kanawha Group with the Upper Yorkian; and the Allegheny, Conemaugh, and Monongahela Formations and the beds below the Waynesburg coal bed with the Upper Yorkian and Radstockian.

As detailed elsewhere (Chapter 26, this volume), there is broad agreement between Jongmans and Gothan's (1934) European correlations and those of White (1899) and later biostratigraphers. Notably different, however, is that no hiatus representing the Stephanian was recognized in the Appalachian region by White (1899), who placed the Monongahela beds in the Stephanian.

### *Darrah's (1937) and Moore's (1937) correlations of the Midcontinent and Appalachian-Pennsylvanian sections with the European succession*

The Midcontinent and Appalachian sections of the United States were correlated with the upper Carboniferous of western Europe on the basis of both flora and fauna (Moore and Elias, 1937; see Table 2). A more detailed table that shows essentially the same correlations is in Moore (1937). Of particular interest in Moore (1937) are the plant megafossils in the Des-

**TABLE 2. CORRELATIONS OF THE MIDCONTINENT SECTION OF LATE MISSISSIPPIAN TO EARLY PERMIAN AGE***

| Midcontinent Provincial Series | Appalachian Basin | European Chronology | Basis |
|---|---|---|---|
| Big Blue (Wolfcampian) | Early Permian and Late Pennsylvanian Group (Dunkard Group and uppermost part of Monongahela Formation) | Early Permian | *Callipteris conferta;* also *Taeniopteris, Sphenophyllum fontainianum, Odontopteris, Glenopteris, Walchia* spp. |
| Virgilian | Late Pennsylvanian (Lower to uppermost part of Monongahela Formation) | Stephanian | *Sphenophyllum oblongifolium, Pecopteris* spp., *Sigillaria brardi, Callipteris* (not *C. conferta)* |
| Missourian | Late Pennsylvanian (Upper and middle part of Conemaugh Formation) | Westphalian D | *Mixoneura sarana {= Neuropteris ovata};* disappearance of *Neuropteris heterophylla, Linopteris muensteri* |
| Desmoinesian | Late and Middle (Lower part of Conemaugh Formation, Allegheny Formation, upper part of Pottsville [Kanawha] Formation) | Westphalian B and C | Plant megafossils characteristic of Westphalian B and C; invertebrates (e.g., brachiopods, ammonites) |
| Morrowan | Early Pennsylvanian (Middle and lower parts Pottsville Formation) | Westphalian A and late Namurian | *Neuropteris (Neuralethopteris) schlehani, Sphenopteris hoeninghausii, Zeilleria;* invertebrates (e.g., goniatites) |
| Chesterian | Late Mississippian | Early Namurian and late Visean | *Cardiopteris, Sphenopteris* sp.; goniatites |

*Constructed from data in Moore and Elias, 1937.

moinesian Series: *Sphenophyllum emarginatum, Neuropteris scheuchzeri, N. rarinervis,* and *Linopteris* and *Mariopteris* spp., all of which are considered typical upper Westphalian floral elements, and *Neuropteris loshii, Mariopteris,* and *Walchia* spp. (a rather low occurrence of this last genus in presumably upper Westphalian strata) in the Missourian Series.

Moore (1937) showed the correlations between the Midcontinent and the English succession (Table 3). In a broad sense, Moore's European correlations are similar to those of Darrah (1937).

The first detailed floral zonation scheme of the Pennsylvanian System of the United States was Darrah's (1937), which included correlations with Europe (Tables 4, 5). This work was based mainly on White's studies on the Pottsville Formation and Allegheny Formation floras, Darrah and P. Bertrand's work (1933, 1934) on the Conemaugh Formation and Monongahela Formation floras, and both White's and Darrah's work on the Dunkard flora. Darrah was the first North American paleobotanist to collect Carboniferous plant fossils in the European coalfields and also to study them in European museums. This European experience (see Chapter 2, this volume) gave Darrah a unique biostratigraphic perspective, which led to an integration of the Carboniferous floral zonation schemes and correlations on both continents. The essential parts of Darrah's (1937) zonation scheme and correlations with the Midcontinent and Europe are shown in Table 4. Table 5 is a simplified version of Darrah's scheme.

Darrah's (1937) floral zonation scheme is amplified in Darrah (1969). Of significance in his 1937 scheme is the recognition of *Neuropteris pocahontas,* a form comparable to the alliance of *N. schlehanii,* which occurs in the middle and upper Namurian and Westphalian A of Germany (Josten, 1991, and elsewhere). In Darrah's 1937 zonation scheme, the absence of *Lonchopteris*—a characteristic genus of upper Westphalian A and lower Westphalian B in Europe (compare Josten, 1991)—in the Appalachian basin did not allow precise correlation with Europe. For example, *Paripteris (Neuropteris) gigantea* ranges from upper Westphalian A to Westphalian C in Germany and *Alethopteris lonchitica* and *Sphenophyllum bifurcatum* (= *S. cuneifolium*) range from Namurian B to Westphalian D (Josten, 1991). Thus, one could today conclude that these three species collectively indicate the Westphalian C stage as well as the Westphalian B or even the upper Westphalian A. Darrah recognized the equivalence of the upper Allegheny and lower Conemaugh floras with the Radstockian flora of England (Westphalian D stage) as well as the Rive-de-Gier flora of France with the "middle Conemaugh" (Stephanian A stage of Jongmans and Pruvost, 1950). As to the lower part of the Conemaugh, Darrah (1937) was not completely sure whether it was Westphalian D or lower Stephanian. His large collections from the "Mason shale" (informal name) in the lower part of the Conemaugh (see Darrah, 1969) include abundant specimens of pecopterids, *Sphenophyllum oblongifolium* (considered a reliable indicator of the Stephanian), as well as a number of Allegheny species, such as *Neuropteris ovata, N. scheuchzeri, Annularia sphenophylloides,* and *Alethopteris "serlii-grandini,"* which are now regarded as two separate species (Wagner, 1968).

Darrah's 1937 zonation scheme was a milestone in Pennsylvanian floral zonation schemes in the United States. This scheme was the first detailed and comprehensive comparison of the upper Carboniferous floral successions in Europe and North America. It was based on Darrah's first-hand collecting in North America and Europe and his interaction with W. J. Jongmans and P. Bertrand as well as on an examination of museum collections on both continents.

TABLE 3. MOORE'S CORRELATIONS OF THE AMERICAN MIDCONTINENT AND ENGLAND*

| Midcontinent Provincial Series | European Substage | English "Stages" |
|---|---|---|
| Missourian | ....... | Missing/(unconformity, Asturian orogeny [= late stage of Hercynian orogeny]) |
| Desmoinesian | Westphalian D | Radstockian |
| Atokan | Westphalian C | Staffordian |
| Morrowan | Westphalian B and A, upper Namurian | Yorkian Upper Millstone Grit |
| Chesterian | Middle and lower Namurian | Middle and Lower Millstone Grit |

*Moore, 1937.

### *Read's floral zonation schemes (1944, 1947)*

A somewhat different floral zonation from Darrah's 1937 scheme for the Pennsylvanian strata of the eastern United States was proposed by Read (1944). This scheme (Table 6) follows in a general sense the zonation schemes of Dix (1937) and Darrah (1937) for the Westphalian C and younger Pennsylvanian (late Carboniferous) strata but is significantly different for the older Pennsylvanian strata. Also of significance is Read's recognition of the zone of *Danaeites* spp. in the upper part of the Monongahela Formation.

Dix's (1937) zone of *Pecopteris aspera* (Namurian B and C for south Wales) was correlated by Read (1944) with the basal Pennsylvanian floral zone of *Neuropteris pocahontas* and *Mariopteris eremopteroides,* followed upward by the zone of *Mariopteris pottsvillea* and *Aneimites* (Dix's zone of *Neuropteris schlehani,* Namurian C and Westphalian A) and the zone of *Mariopteris pygmaea* (Dix's zone of *Alethopteris lonchitica,* and *Neuropteris heterophylla,* Westphalian A).

Read's (1947) floral zonation is basically the same as his

**TABLE 4. DARRAH'S FLORAL ZONATION SCHEME FOR THE PENNSYLVANIAN OF THE APPLACHIAN REGION (REORGANIZED)***

| Appalachian Basin | Midcontinent | European Chronology | Floral Assemblage |
|---|---|---|---|
| Greene Formation (upper part of Dunkard Group series) | Marion Formation | Early Permian | *Callipteris conferta, C. lyratifolia, Taeniopteris* spp., *Walchia* cf. *W. piniformis, Baiera* |
| Washington Formation (lower part of Dunkard) | Chase Group | Late Stephanian | *Taeniopteris* spp., *Callipteridium* spp., *Pecopteris schimperiana, P. merianopteroides, Sphenopteris acrocarpa, Sphenophyllum tenuifolium, Sph. thonii* |
| Monongahela Formation (upper part)<br>Monongahela Formation (lower part) | Elmdale Formation | Early Stephanian | *Taeniopteris jejunata, Sphenopteris acrocarpa, S. minutisecta, Sphenophyllum filiculme, Sph. tenuifolium, Odontopteris reichi, O. genuina, Pecopteris truncata, P. arborescens, P. bredovi, P. feminaeformis, P. hemitelioides, P. polymorpha, Sphenopteris minutisecta, Alethopteris magna, A. grandinii, Lescuropteris moorei* |
| Conemaugh Group (upper part)<br>Conemaugh Group (lower part)<br>Allegheny Formation (upper part) | Severy Shale<br>Le Roy Shale<br>Chanute Shale<br>Cherokee Shale (upper part) | Westphalian D | *Pecopteris lamuriana, P. arborescens, Pecopteris* spp., "*Walchia*" cf. *piniformis, Mixoneura neuropteroides* [= *N. ovata*], *Sphenophyllum, Alethopteris grandinii, Sigillaria brardii, Lescuropteris moorei, Sphenophyllum oblongifolium, Neuropteris ovata, N. scheuchzeri, Sphenopteris* spp., *Mariopteris nervosa, Alethopteris serlii, P. vestita* |
| Allegheny Formation (lower part) | No units indicated | Westphalian C | *Neuropteris rarinervis, N. tenuifolia, N. missouriensis, N. scheuchzeri, Alethopteris lonchitica, Mariopteris muricata, A. sullivantii, Pecopteris pseudovestita* |
| Pottsville Formation (upper part)<br>- - - - - - - - -<br>Pottsville Formation (lower part) | No units indicated | Westphalian B<br>Westphalian A | *Zeilleria* cf. *Z. avoldensis, S. hoeninghausi, Eremopteris* spp., *Neuropteris pocahontas, N. gigantea, N. heterophylla, N. missouriensis, Alethopteris grandifolia, A. lonchitica, Sphenophyllum bifurcatum* (= *cuneifolium*), *Mariopteris muricata, M. eremopteroides, Neuropteris pocahontas* (= *N. schlehanii*), *Mariopteris acuta, M. eremopteroides, Sphenopteris hoeninghausii, Sphenopteris asplenioides, Megalopteris* sp. |

*Darrah, 1937.

1944 zonation, but his 1947 work includes amplification of some zones established earlier. Nine zones, first reported by Read (1941), are distinguished in both. In the Midcontinent, Read's (1947) zones 8 and 9, the zones of *Lescuropteris* and *Danaeites*, are not separable and are designated the zone of *Odontopteris* spp. Zone 3 (Read's 1944 zone of *Mariopteris pygmaea*) is amplified to include *Neuropteris tennesseeana, Ovopteris communis, Alloiopteris inaequilateralis*, and *Alethopteris decurrens*. Read's (1947) scheme modified correlations to include the Lykens No. 4 coal bed of the Anthracite region of eastern Pennsylvania in zone 2. This bed was placed in zone 1 in Read's (1944) zonation scheme.

Read (1947) also gave the detailed composition of floral assemblages for zones 3 through 6 and the *Odontopteris* zone in the Midcontinent. In addition, he gave selective correlative floras in the Rocky Mountain region. Notable is his assignment of a *Walchia* florule from Colorado to zone 7, which he correlated with the flora of the upper part of the Allegheny Formation (Westphalian D).

Dix's (1937) zone of *Lonchopteris* (Westphalian B) was

considered by Read (1944) to be partly equivalent to his zone of *"Cannophyllites"* (today *Megalopteris*). It is partly equivalent to his zone of *Neuropteris tenuifolia.*

Read's (1944) basal three zones (Namurian B to Westphalian A) correspond to the Lee Formation in the central Appalachians, the Pottsville Group in the Anthracite region of eastern Pennsylvania, and the Morrow Formation in the Midcontinent, respectively. Westphalian B was considered by Read to be equivalent to the Atokan (= Lampasas) and Kanawha, and the Westphalian C to be equivalent to the Allegheny. The younger floral zones of Read (1944), beginning with the zone of *Neuropteris rarinervis,* approximate those of Darrah (1937), although the stratigraphic boundaries are not exactly the same. The European stages and zones in Read (1944) are not accurately tabulated and correlated, but further mention of this is beyond the scope of this chapter.

TABLE 5. DARRAH'S FLORAL ZONATION SCHEME (SIMPLIFIED)*

| Appalachian Basin | European Chronology | Floral Zone |
|---|---|---|
| Upper part of Dunkard Group (Greene Formation) | Early Permian | Zone of *Callipteris conferta* |
| Lower part of Dunkard Group (Washington Formation) and Monongahela Formation (upper part) | Late Stephanian | Zone of *Callipteridium* spp., *Taeniopteris* spp., *Odontopteris* spp., *Pecopteris arborescens* group |
| Monongahela Formation (lower part) and Conemaugh Group (upper part) | Early Stephanian | Zone of *Sphenophyllum oblongifolium, Lescuropteris moorei, Pecopteris polymorpha, P. lamuriana, Alethopteris grandinii* |
| Conemaugh Group (lower part) and Allegheny Formation (upper part) | Westphalian D | Zone of *Neuropteris ovata* and *Pecopteris* spp. |
| Allegheny Formation (lower part) | Westphalian C | Zone of *Neuropteris rarinervis* and *A. lonchitica;* first appearance of *N. ovata* |
| Pottsville Group (upper part) | Westphalian B | Zone of *Neuropteris gigantea, N. flexuosa, N. tenuifolia, Alethopteris lonchitica, Sphenophyllum bifurcatum (= cuneifolium)* |
| Pottsville Group (lower part) | Westphalian A | Zone of *Neuropteris pocahontas (= schlehanii)* |

*Darrah, 1937.

### *Arnold's (1949) floral zonation for the Michigan coal basin*

A floral succession for the Michigan coal basin (Table 7) was reported by Arnold (1949). This succession was based partly on his earlier investigations (Arnold, 1934). Notable are Arnold's (1949) megafloral correlations with Maritime Canada and western Europe. C. A. Arnold subdivided the Michigan floral succession into the lower, intermediate, and upper floras, which he correlated, respectively, with the floras of the Riversdale (upper part), Cumberland, and Morien (lower part) Groups of Maritime Canada (Tables 8 and 9). He also correlated the Michigan floras with the Yorkian floras of Great Britain and the Westphalian A (upper) and Westphalian B floras of continental Europe. In addition, he correlated his lower, intermediate, and upper floras with Read's (1947) zones 3, 5, and 6(?), respectively. Arnold noted the lack of two biostratigraphically significant genera, *Lonchopteris* and *Linopteris,* which occur in Maritime Canada but not in the Michigan basin (Bell, 1938).

Of particular significance for North American and European correlations is Arnold's (1949) recognition of *Neuropteris (Neuralethopteris) schlehanii* in the lower flora of the Michigan basin. This clearly establishes that his "lower flora" of the Michigan basin (Table 7) is equivalent to the flora of the Lower Pennsylvanian of the central Appalachian basin, Dix's (1934) floral zone C, and the Westphalian A stage of Europe (Arnold, 1949). Arnold's intermediate and upper floras contain some species that are late Westphalian B or younger (see Gillespie and Pfefferkorn, 1979).

### *Read's (1944, 1947) and Read and Mamay's (1964) floral zonation scheme*

Read's (1944, 1947) floral zonation scheme became the central part of the upper Paleozoic floral zonation of Read and Mamay (1964), which has survived with only minor amplification (U.S. Geological Survey, 1979) and minor criticism (see Darrah, 1969; Wagner, 1984) for three decades. This scheme consists of 15 zones, including the nine zones in Read (1944, 1947). Six additional zones were added: three for the Mississippian and three for the Permian.

The basal Mississippian (lower Carboniferous) zone, floral zone 1—the zone of *Adiantites* spp.—is based on the work of Read (1955). It is partly equivalent to the *Lepidodendropsis* flora of Jongmans (1954). Zone 1 also includes species of *Rhodea, Rhacopteris, Alcicornopteris, Lagenospermum, Calathiops,* and *Girtya.* Zone 2, the zone of *Triphyllopteris* spp., is characterized also by *Lepidendropsis* spp. and early forms of *Fryopsis.* Zone 3, the zone of *Fryopsis* spp. and *Sphenopteridium,* is characteristic of the uppermost Mississippian.

Read's (1947) zones 1, 2, and 3 became zones 4, 5, and 6, respectively, in the Read and Mamay (1964) zonation. Zone 4 of Read (1947), which was the zone of *Cannophyllites,* was changed to the zone of *Megalopteris* spp. and became zone 7

**TABLE 6. READ'S FLORAL ZONATION AS COMPARED TO THAT OF DIX'S***

| European Classification | | | | | | American Classification | | |
|---|---|---|---|---|---|---|---|---|
| Series | Stage | Zone | France | England | Floral Zones of South Wales (Dix) | Eastern U.S. | Midcontinent | Floral Zones of Appalachians (Read) |
| Upper Carboniferous | Stephanian | D | | | Zone of *Odontopteris* | Monongahela Series | Virgil Series | Zone of *Odontopteris* and *Danaeites* |
| | | | Assise de St. Etienne | | | | | Zone of *Lescuropteris* |
| | | | Assise de Rive-de-Gier | Radstockian | Zone of *Pecopteris lamurensis* | Conemaugh Series | Missouri Series | |
| | | | | ? | | | | Zone of *Pecopteris* and *Neuropteris flexuosa* |
| | Westphalian | C | Assise de la Houve | Staffordian | Zone of *Neuropteris flexuosa* | Allegheny Series | Des Moines Series | |
| | | | Assise de Bruay | | Zone of *Neuropteris rarinervis* | | | Zone of *Neuropteris rarinervis* |
| | | B | | | Zone of *Neuropteris tenuifolia* | Kanawha Series | Atoka Series (Lampasas) | Zone of *Neuropteris tenuifolia* |
| | | | Assise d'Anzin | Yorkian | Zone of *Lonchopteris* | | | Zone of *Cannophyllites* (*= Megalopteris*) |
| | | A | Assise de Vicoigne | | Zone of *Alethopteris lonchitica* and *Neuropteris heterophylla* | | | Zone of *Mariopteris pygmaea* |
| | | | | ? *Gastrioceras* | Zone of *Neuropteris schlehanii* | Lee Series | Morrow Series | Zone of *Mariopteris pottsvillea* and *Aneimites* |
| | Namurian | C | Assise de Flines | *Reticuloceras* | Zone of *Pecopteris aspera* | | | Zone of *Neuropteris pocahontas* and *Mariopteris eremopteroides* |
| | | B | | *Homoceras* | Zone of *Lyginopteris stangeri* | ? | | |
| | | A | | *Eumorphoceras* | | | Mississippian | |

*Read, 1944; Dix, 1937; modified from Moore et al., 1944.

TABLE 7. ARNOLD'S FLORAL ZONATION FOR THE LOWER AND MIDDLE PENNSYLVANIAN OF THE MICHIGAN COAL BASIN*

| Floral Zone | Characteristic Species | Correlations<br>Eastern U.S. | Maritime Canada | Western Europe |
|---|---|---|---|---|
| Upper Flora | *Neuropteris desorii, N. rarinervis, Sphenophyllum emarginatum* | Kanawha Formation (uppermost part), Tradewater Formation (upper part) | Morien Group (lower part), Cumberland Group (upper part) | Yorkian (upper part), Westphalian B (upper part), Assise d'Anzin, Maurits Group |
| Intermediate Flora | *N. tenuifolia, N. obliqua, N. scheuchzeri, Sph. cuneifolium* | Kanawha Formation (upper part), Tradewater Formation (upper part) | Cumberland Group (lower part) | Yorkian (upper part), Westphalian B |
| Lower Flora | *N. schlehanii, Diplothmema obtusiloba, Alethopteris decurrens, Sph. cuneifolium* | Lee Formation (upper part), New River Formation (upper part) | Riversdale Group (upper part) | Yorkian (lower part), Westphalian A (upper part), Assise de Vicoigne, Baarlo-Wilhelmina Group |

*Arnold, 1949.

TABLE 8. THE SUBDIVISION OF THE MORIEN SERIES, SYDNEY COALFIELD, NOVA SCOTIA, AND CORRELATION WITH THE BRITISH FLORAL "STAGES"*

| European Chronology | Coal Seam | Floral Zone | British "Stage" |
|---|---|---|---|
| Westphalian D | Pt. Aconi<br>Lloyd Cove<br>Hub<br>Harbour<br>Bouthillier<br>Blackpit<br>Phalen<br>(Roof of the Emery Seam) | *Ptychocarpus unitus*<br>366 m | Radstockian |
| Westphalian C | Emery<br>Gardiner<br>Mullins<br>(Roof of the Tracy Seam) | *Linopteris obliqua*<br>701 m | Staffordian |
| Westphalian B | Tracy<br>Shoemaker<br>McAulay | *Lonchopteris eschweileriana*<br>914 m | Yorkian |
| Unconformable base of the Morien Series | | | |

*Bell, 1938.

of Read and Mamay (1964), which was subcharacterized by *Neuropteris lanceolata.* Zones 5, 6, 7, 8, and 9 of Read (1947) became zones 8, 9, 10, 11, and 12, respectively, of Read and Mamay (1964).

Finally, Zones 13, 14, and 15 of Read and Mamay (1964)—which characterize the Lower Permian Series—were based on the works of White (1912, 1929), Read (1941), Mamay and Read (1954), and unpublished data of Mamay (see Mamay, 1976). The Upper Permian Series is absent in North America (S. H. Mamay, personal communication, June 1993). The Lower Permian floral zones are delineated as follows:

| Series | Zone No. | Zone name (Read and Mamay, 1964) | Name Changes (Mamay et al., 1988) |
|---|---|---|---|
| Lower Permian | 15 | Younger *Gigantopteris* flora | *Cathaysiopteris* flora |
| Lower Permian | 14 | Older *Gigantopteris* flora (= *Glenopteris* and *Supaia* floras) | *Gigantopteridium* flora |
| Lower Permian | 13 | *Callipteris* spp. | |

Zone 13, which characterizes the Wolfcampian of the Lower Permian, is not defined by the presence of a single *Callipteris* species, such as *C. conferta,* but by the genus as a whole. Read and Mamay (1964) pointed out that *Callipteris* usually occurs sparingly in early Permian floral assemblages in the United States.

Zone 14, the early Leonardian (early Permian) zone of the older *Gigantopteris* flora, is not composed of a cosmopolitan flora as is Zone 13. Zone 14 can be subdivided into three geographic floras: The *Gigantopteris* flora characterizes Oklahoma and Texas, the *Glenopteris* flora Kansas, and the *Supaia* flora Arizona and New Mexico. The three floras contain *Walchia, Callipteris, Taeniopteris,* and *Sphenophyllum* species, all of which indicate a common floral source (Read and Mamay, 1964).

The uppermost zone in the floral zonation of Read and Mamay (1964) is zone 15, late Leonardian (early Permian), the zone of the younger *Gigantopteris* flora. This zone is based on the absence of *G. americana,* the occurrence of two new *Gigantopteris* species designated A and B, and a form perhaps identical to *Pterophyllum,* a Mesozoic genus (Read and Mamay, 1964). Mamay (1986) formally reported and described these two new gigantopterid species as *Zeilleropteris wattii* and *Cathaysiopteris yochelsonii,* respectively. The "younger *Gigantopteris* flora" and the "older *Gigantopteris* flora" of Read and Mamay (1964) are now known as the

TABLE 9. BELL'S FLORAL ZONATION SCHEME OF THE UPPER CARBONIFEROUS STRATA OF THE MARITIMES OF CANADA*

| European Chronology | Upper Carboniferous Stratigraphic Units | Important Floral Elements |
|---|---|---|
| Westphalian C and D | Pictou Group (Morien and Stellarton Groups) | *Pecopteris* spp.<br>*Linopteris obliqua*<br>*Neuropteris rarinervis*<br>*N. ovata*<br>*Sphenopteris striata*<br>*Ptychocarpus* spp.<br>*Sphenophyllum emarginatum*<br>*Lonchopteris* sp. |
| | Paleontological break (inferred disconformity) | |
| Westphalian B | Cumberland Group (Joggins and Shulie Formations) | *P. plumosa* forma *crenata*<br>*N. tenuifolia*<br>*Sphenopteris valida*<br>*Samaropsis baileyi* group |
| Westphalian A | Riversdale Group (Boss Point, Parrsboro Formation) | *Sphenopteris* spp.<br>*N. smithsii*<br>*Mariopteris acuta*<br>*Whittleseya desiderata* |
| | Paleontological break (inferred disconformity) | |
| Namurian A | Canso Group (Mabou Formation) | *Sphenopteridium dawsonii*<br>*Telangium* sp.<br>*Mesocalamites cistiiformis* |

*Bell, 1943.

"*Cathaysiopteris* flora" and "*Gigantopteridium* flora," respectively, of Mamay et al. (1988). The *Delnortea* beds of north-central Texas represent the youngest Paleozoic floral beds (latest Leonardian, early Permian) in North America (Mamay et al., 1988); Mamay, personal communication, June, 1993).

## FLORAL ZONATIONS IN CANADA

### *E. L. Zodrow*

### *Floral zonations by J. W. Dawson and W. A. Bell, Maritime Provinces of Canada*

The early studies of Carboniferous megafloras in Canada were confined to the coal-producing Maritime Provinces (Nova Scotia and New Brunswick). The development and naming of Carboniferous floral zonation schemes proceeded in the Maritimes and in the United States independently of each other, although in both countries the roots of these schemes are partly western European. However, some age comparisons and paleobotanical interactions did take place between W. A. Bell and W. C. Darrah, and Bell benefited from certain floral works by White (see Chapter 16, this volume).

### *William E. Logan (1843)*

Sir William E. Logan (1843) measured the now-famous Joggins section, which is part of a sea cliff at the Bay of Fundy, Nova Scotia, and determined the Carboniferous section to be 4,441 m thick. Thus began the study of Carboniferous strata in the Maritimes. The most interesting part of the section, an interval of 840 m, was remeasured jointly by Lyell and Dawson in 1852 and 1853. Dawson (1868) conjectured that if the lowermost member of the series were developed and exposed in the Joggins area, the section would be a total of 4,800 m thick.

### *J. William Dawson (1868, 1873)*

Later, Dawson (1868) recognized and assigned to the Carboniferous Series the sequence of 4,800 m at Pictou, Nova Scotia, where he was born in 1820 (Adams, 1899). Richard Brown, Sr., manager of coal mines for the Mining Association of London on Cape Breton Island, Nova Scotia, supplied data to Dawson and indicated that the coal formation on Cape Breton Island (excluding the lower Carboniferous) is estimated at 3,000 m thick. In total then, the upper Carboniferous strata in Nova Scotia, according to Dawson, amounted to about 12,600 m, a figure that was revised by Bell (1943) to about 6,000 m.

Dawson was also the first to attempt to subdivide the Carboniferous System of Maritime Canada, after Lyell in 1843 had assigned both the Windsor and the Horton Groups to the lower Carboniferous (Mississippian), which Dawson later corroborated. Dawson arranged the Carboniferous strata of Nova Scotia, based on his experience with faunal and floral occurrences in the British coalfields, into a stratigraphic sequence (from top to bottom), as follows: (1) the Upper Coal Formation, (2) the Middle Coal Formation, (3) the Millstone Grit Series, (4) the Carboniferous Limestone, and (5) the Lower Coal Measures.

Organic remains were found in all five units. They were identified by Dawson (1868): (1) The Upper Coal Formation contains *Calamites suckowii, Annularia galioides, Cordaites simplex, Alethopteris* (= *Mariopteris*) *nervosa, Pecopteris arborescens, Dadoxylon materiarium, Lepidophloios parvus,* and *Sigillaria scutellata.* (2) The Middle Coal Formation contains species of sigillarian and stigmarian genera that are rather conspicuous. Also abundant are fern, cordaitean, and calamitean species, and all of the genera of Carboniferous plants are represented. Shells of *Naiadites* spp. and *Spirorbis carbonarius* as well as piscine ganoid and placoid remains and *Entomostraca* spp. are present in beds in the vicinity of coal. (3) The Millstone Grit Series contains a sparse record of fossil flora, and Dawson mentioned only *Dadoxylon* wood, preserved as upright trunks in coarse sandstone. (4) The Carboniferous Limestone is characterized by numerous brachiopods and other marine invertebrate fossils. (5) The Lower

Coal Measures have a characteristic assemblage of *Lepidodendron corrugatum,* cyclopterids, *Dadoxylon antiquus,* and *Alethopteris heterophylla;* locally entomostracans are very abundant.

Later, Dawson (1873) enlarged the scope of his earlier stratigraphic inquiry in two important ways. First, he designated type sections for the Lower Coal Measures (= Horton Bluff Group) and the Carboniferous Limestone (= Windsor Group). Second, he attempted intra- and intercontinental correlations of these series and the Millstone Grit Series with the United States, Scotland, Ireland, and Germany, including Silesia.

In his regional approach, Dawson was not able to present a workable biostratigraphy. This, then, is essentially the biostratigraphic picture of the Carboniferous System of the Maritime Provinces that Bell inherited, which provided him with the opportunity for his life's work (see Chapter 16, this volume).

### *Walter A. Bell (1938)*

The upper Carboniferous strata of Cape Breton Island were the first to be carefully examined by Bell. This was no coincidence, as it was earlier realized that the Sydney coal-mining potential was large and needed development.

By considering the vertical distribution of fossil plants (and invertebrates), Bell (1938) succeeded in subdividing the Upper Coal Formation (= Bell's Morien Group) into the *Lonchopteris eschweileriana, Linopteris obliqua,* and *Ptychocarpus unitus* zones (Table 8). To arrive at age assignments, Bell used complex arguments of floral composition, vertical ranges, and epiboles and their intercontinental comparisons. In short, for the *L. eschweileriana* zone, he considered the age to be younger than the early Westphalian B age of the Joggins section and settled for a late Westphalian B age (Yorkian).

The *L. obliqua* zone, which has the richest flora of the three zones, was considered by Bell (1938) to be of Westphalian C age and coeval with the Staffordian Stage of Britain. His argument is based in part on the frequency, or rarity, of occurrences of species that extend in the "British coalfields from the Yorkian into the Staffordian, or higher still into the Radstockian" (Bell, 1938, p. 13). An ancillary argument for a Westphalian C assignment is that the lower part of the *L. obliqua* zone, according to White (1899), is about the same age as that of the Lower Coal Measures of Missouri. Later, Bell (1962) clarified his position that the boundary between the Westphalian C and D (or boundary between *L. obliqua* and *P. unitus* zones) was chosen at the top of the Emery Seam (Table 8) because of the first appearance above the Emery Seam of a number of species not present in the *L. obliqua* zone itself (Table 8). In effect, Bell (1962) disregarded the European usage that one of the best index fossils for Westphalian D age is the first appearance of *Neuropteris ovata* by not pegging the Westphalian C–Westphalian D boundary to the McRury Seam—approximately 46 m below the Emery Seam—where *N. ovata* makes its first appearance.

The bottom of the Westphalian D defines the top of the Westphalian C and, therefore, the *Ptychocarpus unitus* zone is Westphalian D. Bell's homotaxial arguments for the assignment include the observation that of 18 sphenopterid taxa in the *L. obliqua* zone, only six range into the *Ptychocarpus unitus* zone. Moreover, confined to the *Ptychocarpus unitus* zone are several Radstockian taxa, including *Dicksonites plueckenetii, Neuropteris ovata,* and *N. rarinervis. Sphenophyllum oblongifolium* was considered diagnostic of Stephanian age in Europe. Although it occurs in the younger strata of the *Ptychocarpus unitus* zone, Bell was reluctant to consider this species as evidence of a Stephanian age. This may have been based on his impression that it presented a rather low occurrence in an otherwise typical Westphalian D floral assemblage.

### *Walter A. Bell (1943)*

Bell recognized that a purely lithological classification of the Carboniferous strata of Maritime Canada was inadequate. For one thing, it gave the impression that all of the mineable coal beds were deposited in the time interval of the Middle and Upper (Productive) Coal Measures, which they were not.

Bell (1943) described 126 species, of which 24 were reported as new. Of the 24, 15 are sphenopterids, and of these, 10 species are confined to the Riversdale Group. In all, Bell delineated the entire floral succession, including ranges and epiboles, of the upper Carboniferous floral species of Maritime Canada. On comparison, the Maritime floral succession parallels that of western Europe (and not that of the United States), and Bell's recognition of this fact contributed primarily to the development of a subdivision of the Euramerican Floral Province and to the establishment of the Acadian floral subprovince (see summary by Pfefferkorn and Gillespie, 1980).

The continued importance of Bell's work (1943), however, rests on the following: (1) the establishment of a floral zonation for the Namurian and the lower Westphalian—encompassing the Canso, Riversdale, and Cumberland Groups; (2) development of a comprehensive succession of upper Carboniferous floral zonations and their corresponding European ages ("but he named the units independently after the more common plants in the unit" [Pfefferkorn and Gillespie, 1980, p. 94]; and (3) coordination of the subdivisions of the lower and the upper Carboniferous strata of Maritime Canada with European time-rock units: Namurian to Westphalian D ages. Bell's zonation scheme (Table 9) represents a significant advance in our understanding of Canadian correlations with western Europe. It is based mainly on floral data. The ages of the Canso and Riversdale Groups, however, were established partly on the basis of brackish-water clams.

Unlike the Morien Group, in which Bell recognized three

distinct florules, each of the underlying Cumberland, Riversdale, and Canso Groups lacks such floristic potential for subdivision. To this effect, the Canso Group is characterized by sparse macrofloral elements that altogether are not known to extend beyond a particular stratigraphic horizon, marking the boundary between the Canso Group and the overlying Riversdale Group. This aspect, in conformity with Bell's thinking that floral breaks may also signal a stratigraphic break (see Chapter 16, this volume), caused him to postulate a hiatus between these two groups. The age of the Canso Group was established on faunal evidence that indicated to Bell that the basal part is close to the boundary of the lower and upper Carboniferous and that the flora is very early Namurian, which he considered age equivalent to Dix's (1934) Flora A of the South Wales coalfield. However, Bell hedged by admitting the possibility that the Canso Group could also be of Late Mississippian age, but instead he opted for an Early Pennsylvanian age.

The next youngest group, the Riversdale Group, contains a richer flora than the Canso Group. New species records appeared. Among them is a diversified sphenopterid florule with 21 species; this is followed by the equisetalean group with 10 species; the neuropterids with four species; and last, the lycopods, alethopterids, sphenophylls, seeds and fertile organs, mariopterids, and cordaiteans together total 15 species. As far as the Riversdale Group is concerned, Bell thought that, in terms of European chronology, an age not older than late Namurian or younger than Westphalian A was probable. He opted, however, for a Westphalian A age, following Dix's (1934, 1937) age of her Flora C in South Wales. In addition, Bell's assignment was based on European continental floral zonations, if *Neuropteris smithsii* can be considered conspecific with *Neuropteris* (*Neuralethopteris*) *schlehanii* (Namurian to Westphalian A).

The flora in the overlying Cumberland Group is even more diversified than in the two older groups. Notably, it contains a larger number of sphenopterid taxa; pecopterid taxa are recorded for the first time in the Carboniferous strata of Maritime Canada; equisetalean and lycophyte taxa have also increased in number. Bell (1943) summarized as follows: (1) 12 species are present in the Riversdale Group but not in the Pictou Group; (2) 13 species occur in the Pictou Group but not in the older Riversdale Group; (3) 12 species are long ranging and occur in all three groups; and (4) 54 species are confined to the Cumberland Group. After considering the epiboles of the newly emerged taxa and determining that *Lonchopteris eschweileriana* is not present in the Cumberland Group, Bell assigned a Westphalian B age to the Cumberland Group.

Bell (1943) revised age estimates and regarded the Pictou and its equivalents (Morien and Stellarton Groups) as mainly Westphalian C and D. It is noteworthy that Bell (1943) left the boundary line between the Westphalian D and the Stephanian undefined and open in the Morien Group of the Sydney coalfield—possibly anticipating extension of the uppermost strata into the Stephanian. New floral elements in the Pictou Group include: *Alethopteris serlii, Annularia sphenophylloides, Asterotheca miltonii, Pecopteris plumosa* forma *dentata, Hymenotheca bronnii, H. dathei, Linopteris muensteri, Neuropteris scheuchzeri,* and *Sphenopteris striata.* According to Bell, the occurrence of these species marks a major event in the succession of upper Carboniferous floras in Maritime Canada.

### *Walter A. Bell (1944)*

Bell (1944) revised his age correlation for the upper Carboniferous floras as follows: (1) The Canso Group is Namurian B and partly Namurian C, (2) The Riversdale Group is middle Namurian C and also Westphalian A, (3) the Cumberland Group is Westphalian B, and (4) the Pictou Group is late Westphalian B and includes Westphalian C and D (see Chapter 16, this volume).

### *Walter A. Bell (1962)*

The publication on the Pictou Group of New Brunswick, which represents Bell's third and concluding volume on the floral biostratigraphy of Maritime Canada, completed his revision of the Carboniferous stratigraphy of the Maritimes. This report describes all of the identifiable plant fossils from New Brunswick, including four new species and four new combinations, and includes floral comparisons with the Sydney coalfield. He assigned the coal-bearing sequence in New Brunswick to the *Linopteris obliqua* zone (Westphalian C) of the Sydney coalfield and placed it no higher than the lower *Ptychocarpus unitus* zone (Westphalian D). This work concluded Bell's life's work as a worthy follower of Dawson's seminal work on the Carboniferous strata of Maritime Canada.

## SUMMARY

The pioneering work on Pennsylvanian (upper Carboniferous) floral biostratigraphy in North America in the early part of the twentieth century was by D. White (Fig. 1A), W. C. Darrah (Fig. 1B), C. B. Read (Fig. 1C), and W. A. Bell (Fig. 1D). These pioneers set the stage for further floral biostratigraphic refinements in the later part of the century as reported by Read and Mamay (1964) and Gillespie and Pfefferkorn (1979).

## ACKNOWLEDGMENTS

Elsie Darrah Morey is thanked for copies of papers in the William Culp Darrah Collection as well as for her review of the manuscript. S. H. Mamay supplied updated information on Permian taxa and zonation in North America as related to the Permian floral zonation in Read and Mamay (1964). S. H. Mamay, H. W. Pfefferkorn, R. H. Wagner, and S. P. Schweinfurth are thanked for their reviews and thoughtful suggestions for improvement of the paper.

A

B

C

D

Figure 1. The North American pioneers in North American Pennsylvanian (upper Carboniferous) floral biostratigraphy in the early part of the twentieth century. A, David White, 1926, Hermit Shale locality, Cedar Ridge, South Kaibab Trail, Grand Canyon, Arizona. Photograph by F. F. Matthes: Grand Canyon National Park #GRCA 13557, courtesy of Sara T. Stebbins, National Park Service. B, William C. Darrah, 1972, Ames Limestone Member locality, First I. C. White Memorial Symposium field trip. Photograph courtesy of Aureal T. Cross, Michigan State University. C, Charles B. Read, 1955, New Mexico. Photograph courtesy of Sergius H. Mamay, Smithsonian Institution. D, Walter A. Bell, 1952, Antigonish, Nova Scotia. Photograph courtesy of Aureal T. Cross, Michigan State University.

## REFERENCES CITED

Adams, F. K., 1899, Memoir of Sir J. William Dawson: Geological Society of America Bulletin, v. 2, p. 449–579.

Arnold, C. A., 1934, A preliminary study of the fossil flora of the Michigan Coal Basin: Contributions from the Museum of Paleontology, University of Michigan, v. 4, p. 177–204.

Arnold, C. A., 1949, Fossil flora of the Michigan Coal Basin: Contributions from the Museum of Paleontology, University of Michigan, v. 7, p. 131–264.

Barlow, J. A., ed., 1975a, The age of the Dunkard, Proceedings of the First I. C. White Memorial Symposium: Morgantown, West Virginia Geological and Economic Survey, 352 p.

Barlow, J. A., 1975b, Preface, *in* Barlow, J. A., ed., The age of the Dunkard, Proceedings of the First I. C. White Memorial Symposium: Morgantown, West Virginia Geological and Economic Survey, p. vii–xvii.

Bassler, H., 1916, A cycadophyte from the North American Coal Measures: American Journal of Science, v. 42, p. 21–26.

Bell, W. A., 1938, Fossil flora of Sydney Coalfield, Nova Scotia: Geological Survey of Canada Memoir 215, 334 p.

Bell, W. A., 1943, Carboniferous rocks and fossil floras of northern Nova Scotia: Geological Survey of Canada Memoir 238, 277 p.

Bell, W. A., 1944, [Floral zones of Maritime Canada], *in* Moore, R. C., chairman, and others, Correlation of Pennsylvanian formations of North America: Geological Society of America Bulletin, v. 55, p. 657–706.

Bell, W. A., 1962, Flora of Pennsylvanian Pictou Group of New Brunswick: Geological Survey of Canada Bulletin 87, 71 p., 56 pls.

Bode, H. H., 1975, The stratigraphic position of the Dunkard, *in* Barlow, J. A., ed., The age of the Dunkard, Proceedings of the First I. C. White Memorial Symposium: Morgantown, West Virginia Geological and Economic Survey, p. 143–154.

Cridland, A. A., Morris, J. E., and Baxter, R. W., 1963, The Pennsylvanian plants of Kansas and their stratigraphic significance: Palaeontographica, Abt. B, 112, p. 58–92, pls. 17–24.

Darrah, W. C., 1936, Permian elements in the fossil flora of the Appalachian Province. II: *Walchia:* Harvard University Botanical Museum Leaflets, v. 4, p. 9–19.

Darrah, W. C., 1937, American Carboniferous floras: Congrès pour l'avancement des études de stratigraphie du Carbonifère, 2d, Heerlen, The Netherlands, 1935: Compte Rendu, v. 1, p. 109–129.

Darrah, W. C., 1968, The pteridosperm *Lescuropteris*—Characteristics, distribution, and significance: American Journal of Botany, v. 55, p. 725.

Darrah, W. C., 1969, A critical review of the Upper Pennsylvanian floras of eastern United States with notes on the Mazon Creek flora of Illinois: Gettysburg, Pennsylvania, privately printed, 220 p., 80 pls.

Darrah, W. C., 1975, Historical aspects of the Permian flora of Fontaine and White, *in* Barlow, J. A., ed., The age of the Dunkard, Proceedings of the First I. C. White Memorial Symposium: Morgantown, West Virginia Geological and Economic Survey, p. 81–101.

Darrah, W. C., and Bertrand, P., 1933, Observations sur les flores houillères de Pennsylvanie (regions de Wilkes-Barre el de Pittsburgh): Paris, Comptes Rendus de l'Académie des Sciences, v. 197, p. 1451–1453.

Darrah, W. C., and Bertrand, P., 1934, Observations sur les flores houillères de Pennsylvanie: Annals de la Société géologique du Nord, v. 58, p. 211–224.

Dawson, J. W., 1868, Acadian geology: London, Mcmillan, 694 p.

Dawson, J. W., 1873, Report on the fossil plants of the Lower Carboniferous and Millstone Grit Formations of Canada: Geological Survey of Canada (Montreal), 47 p., 10 pls.

Dix, E., 1934, The sequence of floras of the Upper Carboniferous with special reference to South Wales: Royal Society of Edinburgh, Transactions, v. 57, p. 789–838.

Dix, E., 1937, The succession of fossil plants in the South Wales Coalfield with special reference to the existence of the Stephanian: Congrès pour l'avancement des études de stratigraphie du Carbonifère, 2d, Heerlen, The Netherlands, 1935: Compte Rendu, v. 1, p. 159–184.

Elias, M. K., 1936, Late Paleozoic plants of the Midcontinent region as indicators of time and of environment, *in* Report, International Geological Congress, 16th, Washington, D.C., July 1933, Volume 1: Washington, D.C., p. 691–700.

Fontaine, W. F., and White, I. C., 1880, The Permian or Upper Carboniferous flora of West Virginia: 2nd Geological Survey of Pennsylvania, v. PP, 143 p., 38 pls.

Gillespie, W. H., and Pfefferkorn, H. W., 1979, Distribution of commonly occurring plant megafossils in the proposed Pennsylvanian System Stratotype, *in* Englund, K. J., Arndt, H. H., and Henry, T. W., eds., Proposed Pennsylvanian System Stratotype, Virginia and West Virginia: Falls Church, Virginia, The American Geological Institute, p. 87–96.

Gillespie, W. H., Hennen, G. J., and Balasco, C., 1975, Plant megafossils from Dunkard strata in northwestern West Virginia and southwestern Pennsylvania, *in* Barlow, J. A., ed., The age of the Dunkard, Proceedings of the First I. C. White Memorial Symposium: Morgantown, West Virginia Geological and Economic Survey, p. 223–248.

Janssen, R., 1940, Some fossil plant types from Illinois: Illinois State Museum Scientific Papers, v. 1, 124 p.

Jongmans, W. J., 1952, Some problems on Carboniferous stratigraphy: Congrès pour l'avancement des études de stratigraphie du Carbonifère, 3d, Heerlen, The Netherlands, 1951: Compte Rendu, v. 1, p. 295–306.

Jongmans, W. J., 1954, Uniformité et diversité des flores des périodes géologiques, *in* Flores fossiles et climats du Primaire, Congrès international de Botanique, 8e, Paris, 1954: Comptes rendus des séances et rapports et communications, section 5, p. 135–138.

Jongmans, W. J., and Gothan, W., 1934, Florenfolge und vergleichende Stratigrafie des Karbons der östlichen Staaten Nord-Amerika's. Vergleich mit West-Europa: Geologisch Bureau Heerlen, Jaarverslag over 1933, p. 17–44.

Jongmans, W. J., and Gothan, W. J., 1937, Comparison of the floral succession in the Carboniferous of West Virginia and Europe: Congrès pour l'avancement des études de stratigraphie du Carbonifère, 2d, Heerlen, The Netherlands, 1935: Compte Rendu, v. 1, p. 393–415.

Jongmans, W. J., and Pruvost, P., 1950, Les subdivisions du Carbonifère continental: Bulletin Société Géologique de France, 5th série, t. 20, p. 335–344.

Josten, K-H., 1991, Die Steinkohlen-Floren Nordwestdeutschlands: Krefeld, Germany, Geologisches Landesamt Nordrhein-Westfalen, Fortschritte in der Geologie von Rheinland und Westfalen, Band 36, 451 p.

Kerp, J. H. F., and Haubold, H., 1988, Aspects of Permian paleobotany and palynology. VIII: On the classification of the West- and Central-European species of the form-genus *Callipteris* Brongniart 1849: Review of Palaeobotany and Palynology, v. 54, p. 135–140.

Lesquereux, L., 1879–1884, Description of the coal flora of the Carboniferous formation in Pennsylvania and throughout the United States: 2nd Pennsylvanian Geological Survey, Report P, 3 volumes and atlas, 977 p., 111 pls.

Logan, W. E., 1843, Section of the Nova Scotia Coal Measures, developed at the Joggins, on the Bay of Fundy, in descending order, from the neighbourhood of the West Ragged Reef to Minudie, reduced to vertical thickness: Geological Survey of Canada, First Report of Progress for 1843, p. 92–159, appendix.

Lyons, P. C., and Darrah, W. C., 1989a, Earliest conifers of North America: Upland and/or paleoclimatic indicators: Palaios, v. 4, p. 480–486.

Lyons, P. C., and Darrah, W. C., 1989b, Paleoenvironmental and paleoecological significance of walchian conifers in Westphalian (Late Carboniferous) horizons in North America: Congrès International de Stratigraphie et de Géologie du Carbonifère, 11th, Beijing, China, 1987: Compte Rendu, v. 3, p. 251-261.

Mamay, S. H., 1976, Paleozoic origin of the cycads: U.S. Geological Survey Professional Paper 934, 48 p., 5 pls.

Mamay, S. H., 1986, New species of Gigantopteridaceae from the Lower Permian of Texas: Phytologia, v. 61, p. 311–315.

Mamay, S. H., and Read, C. B., 1954, Differentiation of Permian floras in the southwestern United States, *in* Flores fossiles et climats du Primaire, Congrès international de Botanique, 8e, Paris, 1954: Comptes rendus des séances et rapports et communications, section 5, p. 157–158.

Mamay, S. H., Miller, J. M., Rohr, D. M., and Stein, W. E., Jr., 1988, Foliar morphology and anatomy of the gigantopterid plant *Delnortea abbottiae* from the Lower Permian of West Texas: American Journal of Botany, v. 75, p. 1409–1433.

McComas, M. A., 1988, Upper Pennsylvanian compression floras of the 7-11 Mine, Columbiana County, northeastern Ohio; Ohio Journal of Science, v. 88, p. 48–52.

Moore, R. C., 1937, Comparison of the Carboniferous and Early Permian rocks of North America and Europe: Congrès pour l'avancement des études de stratigraphie du Carbonifère, 2d, Heerlen, The Netherlands, 1935: Compte Rendu, v. 1, p. 641–676.

Moore, R. C., chairman, and others, 1944, Correlation of Pennsylvanian formations of North America: Geological Society of America Bulletin, v. 55, p. 657–706.

Moore, R. C., and Elias, M. K., 1937, Paleontologic evidence bearing on correlation of late Paleozoic rocks of Europe and North America: Congrès pour l'avancement des études de stratigraphie du Carbonifère, 2d, Heerlen, The Netherlands, 1935: Compte Rendu, v. 1, p. 677–681.

Moore, R. C., Elias, M. K., and Newell, N. D., 1936, A "Permian" flora from the Pennsylvanian rocks of Kansas: Journal of Geology, v. 44, p. 1–31.

Noé, A. C., 1923, Fossil flora of Braidwood, Illinois: Illinois State Academy of Science, Transactions, v. 15, p. 396–397.

Noé, A. C., 1930, Correlation of Illinois coal seams with European horizons: Illinois State Academy of Science, Transactions, v. 22, p. 470–472.

Pfefferkorn, H. W., and Gillespie, W. H., 1980, Biostratigraphy and biogeography of plant compression fossils in the Pennsylvanian of North America, *in* Dilcher, D. L., and Taylor, T. N., eds., Biostratigraphy of fossil plants: Stroudsburg, Pennsylvania, Dowden, Hutchinson, and Ross, p. 93–118.

Read, C. B., 1941, Pennsylvanian formations and floral zones in the central and northern Appalachian region: Oil and Gas Journal, v. 39, p. 65–66.

Read, C. B., 1944, [Floral zones of the Appalachian region], in Moore, R. C., chairman, and others, Correlation of Pennsylvanian formations of North America: Geological Society of America Bulletin, v. 55, p. 657–706.

Read, C. B., 1947, Pennsylvanian floral zones and floral provinces: Journal of Geology, v. 55, p. 271–279.

Read, C. B., 1955, Floras of the Pocono Formation and Price Sandstone in parts of Pennsylvania, Maryland, West Virginia, and Virginia: U.S. Geological Survey Professional Paper 263, 32 p.

Read, C. B., and Mamay, S. H., 1964, Upper Paleozoic floral zones and floral provinces of the United States: U.S. Geological Survey Professional Paper 454-K, 35 p., 19 pls.

Remy, W., 1975, The floral changes at the Carboniferous-Permian boundary in Europe and North America, *in* Barlow, J. A., ed., The age of the Dunkard, First I. C. White Memorial Symposium: Morgantown, West Virginia Geological and Economic Survey, p. 305–352.

Sellards, E. H., 1901, Permian plants: *Taeniopteris* of the Permian of Kansas: Kansas University Quarterly, v. 10, p. 1–12.

Sellards, E. H., 1908, Fossil plants of the Upper Paleozoic of Kansas: Kansas Geological Survey, v. 9, p. 386–480.

U.S. Geological Survey, 1979, New floral zone in Mississippian of Virginia and West Virginia, *in* Geological Survey Research 1978: U.S. Geological Survey Professional Paper 1100, p. 230.

Wagner, R. H., 1968, Upper Westphalian and Stephanian species of *Alethopteris* from Europe, Asia Minor, and North America: Mededelingen van de Rijks Geologische Dienst, série C, III-1-no. 6, 188 p., 64 pls.

Wagner, R. H., 1984, Megafloral zones of the Carboniferous: Congrès de Stratigraphie et de Géologie du Carbonifère, 9th, Washington, D.C., and Champaign-Urbana, Illinois, 1979: Compte Rendu, v. 2, p. 109–134.

White, D., 1899, Fossil flora of the lower Coal Measures of Missouri: U.S. Geological Survey Monograph 37, 467 p.

White, D., 1900, The stratigraphic succession of fossil floras of the Pottsville Formation in the southern Anthracite coal field, Pennsylvania: U.S. Geological Survey Annual Report, 20th, pt. 2, p. 749–930.

White, D., 1903, Summary of fossil plants recorded from the Upper Carboniferous and Permian formations of Kansas: U.S. Geological Survey Bulletin 211, p. 85–117.

White, D., 1904, Permian elements in the Dunkard flora: Geological Society of America Bulletin, v. 14, p. 538–542.

White, D., 1909, The Upper Paleozoic floras, their succession and range: Journal of Geology, v. 17, p. 320–341.

White, D., 1912, The characters of the fossil plant *Gigantopteris* Schenk and its occurrence in North America: U.S. National Museum, Proceedings, v. 41, p. 493–516, pls. 43–49.

White, D., 1929, Flora of the Hermit Shale, Grand Canyon, Arizona: Carnegie Institute of Washington Publication 405, 221 p., 55 pls.

White, D., 1936 (posthumous), Some features of the Early Permian flora of America, *in* Report, International Geological Congress, 16th, Washington, D.C., 1933, Volume 1: Washington, D.C., p. 679–689.

White, D., and Ashley, G. H., 1906, Correlation of coals: U.S. Geological Survey Professional Paper 49, p. 206–212.

White, I. C., 1917, *in* Hennen, R. V., Braxton and Clay Counties [West Virginia]: Morgantown, West Virginia Geological Survey [Report], 883 p.

Manuscript Accepted by the Society July 6, 1994

Printed in U.S.A.

Geological Society of America
Memoir 185
1995

# *The Stephanian of North America: Early 1900s controversies and problems*

**Paul C. Lyons**
*U.S. Geological Survey, MS 956, National Center, Reston, Virginia 22092*
**Robert H. Wagner**
*Jardín Botánico de Córdoba, Avenida de Linneo s/n, 14004 Córdoba, Spain*

## ABSTRACT

**The recognition of the Stephanian in North America was first made in the early 1900s by D. White and E. H. Sellards on the basis of plant megafossils. In the middle 1930s, their biostratigraphic conclusions were largely confirmed by W. C. Darrah and P. Bertrand (an expert on Stephanian floral biostratigraphy), who together refined the Appalachian correlations with the Westphalian D and Stephanian floras of France. W. J. Jongmans, in collaboration with W. Gothan, considered the Stephanian as a limnic facies that would be absent in the United States, where its time-equivalent in paralic and limno-paralic facies was identified by Jongmans as "Westphalian E." The latter would be characterized by the presence of *Neuropteris ovata* and Stephanian pecopterids. The range zone of *Neuropteris ovata* was equated with Westphalian D by H. Bode, who proposed the incorporation of the Stephanian into an expanded Westphalian D that he subdivided on the presence or absence of associated Stephanian megafloral elements. *N. ovata* ranges into the Dunkard, a unit assigned to lower Permian by D. White, W. C. Darrah, and other workers on the basis of the occurrence of *Autunia (Callipteris) conferta* and other Rotliegend megafloral taxa.**

**Controversies in the literature (still partly unresolved) related to different species concepts, lack of examination of type specimens on both sides of the Atlantic, the phytogeographic distribution of late Carboniferous megafloral taxa in Europe and North America (Amerosinian Realm), the biostratigraphic significance of endemic species, the stratigraphic ranges of key megafloral elements, the role of biofacies, and the presence or absence of stratigraphic breaks and their relative importance in the Carboniferous successions in Europe as well as in North America. Jongmans's Westphalian E Stage has not been accepted because it is essentially a floral facies concept and not a chronostratigraphic unit.**

**A historical analysis of work on the Carboniferous floras of North America, with particular regard to the Stephanian, is accompanied by a brief review of the problems attached to the type Stephanian of Europe, and a full discussion is presented of the views expressed by the European palaeobotanists who traveled to the United States.**

---

Acknowledgments for this chapter are on p. 388.

Lyons, P. C., and Wagner, R. H., 1995, The Stephanian of North America: Early 1900s controversies and problems, *in* Lyons, P. C., Morey, E. D., and Wagner, R. H., eds., Historical Perspective of Early Twentieth Century Carboniferous Paleobotany in North America (W. C. Darrah volume): Boulder, Colorado, Geological Society of America Memoir 185.

## INTRODUCTION

The recognition of Stephanian strata in North America is inextricably linked to megafloral biostratigraphy and, more recently, microfloral biostratigraphy as well. Although it is generally acknowledged that American Carboniferous floras are similar to those found in Europe, the limited amount of interaction between North American and West European paleobotanists has led to taxonomic problems that are still partially unresolved.

Stephanian floras, though widely represented in North America, have not always been recognized as such by European paleobotanists. Certain taxa, regarded as being of stratigraphic importance, show extended ranges in North America. This made the Stephanian floras in North America look different from those of classical Stephanian localities in France and Germany. There have also been problems with the recognition of the Westphalian-Stephanian boundary in western Europe, where the Westphalian and Stephanian stages were typified in different basins of very different characteristics. The role of biofacies and its effect on the composition of floral assemblages have not been taken into account sufficiently, and the use of chronostratigraphic terminology for biostratigraphic concepts has created confusion.

The problems with the correct assignment of stratigraphic age to the Dunkard Group at the top of the Pennsylvanian (upper Carboniferous) succession of the Appalachian basin are the same as those of the Rotliegend and Autunian strata in western Europe. A proper recognition of biofacies is essential to the evaluation of the Dunkard, Rotliegend, and Autunian floras.

## FONTAINE AND I. C. WHITE'S DUNKARD FLORA

Since the publication of the classic work of Fontaine and I. C. White (1880) on the Dunkard flora, few floras have had such a long history of age controversy (see papers in Barlow, 1975). Fontaine and I. C. White (1880) concluded that the Dunkard flora was in part transitional between the Carboniferous and Lower Permian and in part Permian.

David White (1904) undertook a revision of the Dunkard flora but never completed this work (Darrah, 1975). Continuing from D. White, W. C. Darrah in the early 1930s started his work of neotyping the Dunkard flora, a career-long activity, 90% finished at the time of his death in 1989.

The age of the Dunkard Group continued to be a problem for David White (see Chapter 10, this volume), who changed his mind on the Upper Pennsylvanian–Permian boundary over the course of his five-decade career with the U.S. Geological Survey. Early in his career, White (1904) considered the Pennsylvanian-Permian boundary to be at the Lower Washington Limestone—about 70 m above the base of the Dunkard Group—where *Callipteris* (sensu lato) occurs as several species. However, White (*in* Schuchert, 1932; White, 1936, posthumous) later considered the entire Dunkard to be Permian because of the presence of Permian megafloral genera and species in the Cassville Shale, just above the Waynesburg coal bed at the base of the Dunkard. White's (1936) paper contains these illuminating statements:

> There is no clear proof that the Stephanian flora was ever wholly exiled from this part of the trough.
>
> . . . the Dunkard flora and even the Monongahela plant society is marked not so much by species identical with those in the Stephanian of the Old World as by obvious modifications or variations of those species, which for the most part cannot now be specifically identified with Old World forms.
>
> besides representatives of the genera *Callipteris, Taeniopteris, Saportea,* and *Baiera,* the Dunkard carries a number of distinctive species, hardly separable from forms characteristic of the Permian of Europe. A striking example is seen in a giant-toothed *Equisetites.*

Significant in White's last interpretation of the age of the Dunkard is the view that relict floras are not significant as indicators of evolution and time but that the "newly evolved" taxa are (White *in* Schuchert, 1932). In this view then, White takes an antipodal position with regard to that of Jongmans (1937a) and later that of Bode (1958, 1960), who gave great significance to relict taxa such as *Neuropteris ovata* and *N. scheuchzeri* (now assigned to *Macroneuropteris scheuchzeri* [Cleal et al., 1990]) in their interpretations of North American Pennsylvanian floras.

## D. WHITE'S (1903) AND SELLARDS (1908) VIEWS ON THE EUROPEAN CORRELATIONS OF THE MIDCONTINENT U.S. FLORAS

The floras of the Midcontinent of Kansas were investigated by D. White (1903) and Sellards (1908), who extended the work of White. A summary of Kansas floras is in Cridland et al. (1963) and Darrah (1969).

White and Sellards showed that the Douglas Formation contained forms of *Macroneuropteris (Neuropteris) scheuchzeri* and *N. ovata* as well as a number of species of *Pecopteris* (e.g., *P. arborescens, P. candolliana,* and *P. unita)* and several species of *Odontopteris* and *Callipteridium* as well as *Sphenophyllum oblongifolium.* Sellards (1908) concluded: "As compared with the European Coal Measures, the flora of the Douglas Formation presents a close resemblance to that of the Commentry coal field of France, the Stephanian, or distinctly Upper Coal Measures."

Sellards (1908) recognized a new genus, *Glenopteris,* which is related on the one hand to the ". . . callipterid ferns of Permian types, and on the other to the Triassic genera *Cycadopteris* and *Lomatopteris.*" *Glenopteris* as well as *Callipteris* spp. are found in the Wellington shales (overlies Douglas Formation), which Sellards regarded as Permian, as well as in the Wreford and Winfield limestones, which both contain *Callipteris conferta. N. ovata* and *M. (N.) scheuchzeri* were also recognized by Sellards in the Winfield Limestone, a circumstance similar to that in the Dunkard Group, where forms of these species were associated stratigraphically with *Callipteris conferta* (Darrah, 1969, 1975), a Permian index fossil. This species of *Callipteris* is currently assigned to *Autunia* (Kerp and Haubold, 1988).

Darrah (1969), in agreement with White and Sellards, pointed out that the Wellington flora of Kansas has a *Callipteris-Taeniopteris-Walchia-Baiera* association and is also Permian. He also noted the presence of *Danaeites emersoni* and *Lescuropteris moorii*—two North American Late Pennsylvanian megafloral taxa—in the Stranger and Lawrence Shales of the Douglas Group (Table 1) of Kansas, which implies that they are Stephanian in age based on Darrah's (1937) floral scheme (see Chapter 25, this volume). *Sphenophyllum oblongifolium* also occurs in the Stranger Formation (Cridland et al., 1963), a circumstance that supports a Stephanian age assignment.

Jongmans collected in 1933 (see Jongmans, 1937a) from the Lawrence and LeRoy Shales of Kansas and concluded that they were Westphalian C in age (Jongmans and Gothan, 1934). Bertrand (1935), in a critique of their paper, assigned these shales to the Stephanian because of the presence of *Pecopteris*

**TABLE 1. JONGMANS'S CORRELATIONS OF THE CARBONIFEROUS OF THE EASTERN UNITED STATES***

| Europe | West Virginia | Pennsylvania | Illinois | Kansas |
|---|---|---|---|---|
| Rotliegend (Lower Permian) | Permian, according to Fontaine and White, 1880. | Same as West Virginia | ? | Lower Permian (Rotliegend) |
| Westphalian E Highest Upper Carboniferous Beds† | Unknown | Locally present§ | Unknown | Present |
| Westphalian C, D (above the Petit-Buisson [Aegir] marine horizon) | "Dunkard" (pro parte) (Monongahela) (Conemaugh) Allegheny | "?Monongahela" "Conemaugh"** Allegheny | "Monongahela" "Conemaugh" Allegheny | Douglas? Lansing, Kansas City, Marmaton, Cherokee |
| Westphalian B (from Petit-Buisson [Aegir] to Catharina Coal) | Kanawha | Pottsville‡ | Pottsville (Mill Creek) (Crofton, Kentucky) | |
| Westphalian A (from Catharina Coal to Sarnsbank Coal) | New River | | Battery Rock | ? |
| Namurian C (f.i., Lower "Magerkohle" and "Flözleeres" of Germany) | Pocahontas Upper part seam 4-9 | | | |
| Namurian B Younger Ostrau Beds | Pocahontas Middle part seam 1-3 | | | |
| Namurian A Waldenburg Beds | Pocahontas Lower part under seam 1 | | | |

*Jongmans, 1937a.
The footnotes are those published by Jongmans with the table:
†Comprises beds in which *Neuropteris ovata* and similar Westphalian elements are no longer present. These beds can be separated from the underlying as Westphalian E.
§Westphalian "E" from Pennsylvania is present wherever erosion has not completely removed the beds higher than lower Conemaugh. This, of course, is local, but since marine horizons are present and the beds are horizontal, correlation is certain. (Note by Darrah.)
**It has never been my intention to put the whole of the Pennsylvanian Monongahela and Conemaugh into the Westphalian D, but to point out that there are beds, mapped Monongahela and Conemaugh in Pennsylvania, which contain floras, belonging to the Westphalian D. Therefore, the way in which my opinion is abstracted and critized by A. S. Romer, *Early History of Texas Redbeds Vertebrates* (Geological Society of America Bulletin, volume XLVI, 1935, p. 1647), is due to some misunderstanding.
‡It is possible that some beds of the "Pottsville" already belong to the Westphalian C.

*feminaeformis* (presently *Nemejcopteris feminaeformis* [Schlotheim] Barthel), *Pecopteris daubreei, P. densifolia, Alethopteris grandinii* (= *A. zeilleri* Ragot), *Odontopteris genuina* (= *Lescuropteris genuina*), *Sigillaria brardii,* and *Sphenophyllum oblongifolium.* He considered this assemblage similar to that in the St. Étienne and Commentry coalfields of south-central France. Bertrand's views are reflected in Darrah's (1969) correlations between the Appalachian basin and the Midcontinent of Kansas.

## D. WHITE'S VIEWS ON THE AGE OF THE HERMIT SHALE AND CORRELATIVE STRATA OF THE WESTERN UNITED STATES

White's (1929) paper established the age of the Hermit Shale of the Grand Canyon as early Permian. He pointed out (White, 1936) that the Hermit Shale ". . . is marked by the almost total absence of Stephanian genera, with the exception of *Walchia,* which made its appearance in very late Stephanian time, and *Sphenophyllum,* which is represented by a species of Uralian facies, typically Permian in form and habit. The Hermit flora is characterized especially by several species of a fernlike genus, apparently belonging to the *Callipteris* stock. . . ."

According to White (1936), the Hermit Shale of the Grand Canyon correlates with the Greene Formation of the Dunkard basin and the underlying Supai Formation with the Washington Formation of the Dunkard (Table 2, this chapter). In the Rocky Mountain states, White noted the presence of *Walchia* in basal Permian formations that occurred above a regional unconformity. It is clear today that *Walchia* is a facies fossil and has little biostratigraphic value (see Lyons and Darrah, 1989).

White (1936) correlated among the Rocky Mountain, Midcontinent, Appalachian, and European sections (Table 2). He assigned the Monongahela and Conemaugh Formations to the Stephanian, which is the same as in Darrah's (1937) floral zonation scheme (see Chapter 25, this volume). Darrah's 1933 discovery of *Walchia* in the Conemaugh Formation, as noted in White (1936), supports White's (1924) opinion that "Permian types" would be found in the Monongahela Formation or below in the Appalachian basin.

## EUROPEAN VERSUS AMERICAN CARBONIFEROUS SPECIES

In the early 1900s, there was a widespread belief in Europe that many American Carboniferous species as described by Lesquereux and D. White are synonyms of European species (Jongmans, 1937a). The reason for the apparent multiplicity of names was that until W. C. Darrah's visit in 1935 (see Chapter 2, this volume), American paleobotanists did not collect in the Carboniferous of Europe and did not have the opportunity to examine the type specimens in museums in Europe. Consequently, they were forced to erect a large number of "new species" in the American Carboniferous. Conversely, European paleobotanists did not study the types of American species housed in the various museums and institutions in the United States and knew American megafloral taxa chiefly from illustrations in the published record. Some of these illustrations in the North American literature were highly diagrammatic and correspondingly difficult to judge (e.g., Lesquereux's [1879–1884] "Coal Flora" and Fontaine and I. C. White's [1880] "Permian or Upper Carboniferous flora"). Many European workers simply ignored or were suspicious of the North American Carboniferous megafloral record (see Chapter 2, this volume). Jongmans, in particular, maintained that the American species were not adequately illustrated and that many resembled European species. He felt that this was particularly true of White's (1900) Pottsville flora, which became one of the foci of Jongmans's 1933 collecting trip in the United States. Jongmans (1937a) placed *Neuropteris smithsii* and *Neuropteris pocahontas* into his *Neuropteris schlehanii* group (i.e., the *Neuralethopteris* group), which is typical of the middle to upper Namurian and Westphalian A in Europe (see Table 1). He agreed that *Lescuropteris* was a special American genus of local biostratigraphic significance. W. C. Darrah (Fig. 1), in a footnote to Jongmans (1937b), placed *Neuropteris vermicularis* Lesq. into the *N. ovata* group. Darrah recognized this species as distinct from *N. ovata,* a conclusion agreed to by Jongmans (1937b), who considered it probably characteristic of lower Westphalian D, unlike *N. ovata,* which according to Jongmans would range from Westphalian C into highest Westphalian D.

An early, well-founded attempt at revision of controversial identifications from the upper Paleozoic of North America was made by Stopes (1914), who traveled to New Brunswick in eastern Canada and collected from the "Fern Ledges" beds at St. John (see Chapter 9, this volume). This locality had been ascribed to the Devonian by Dawson (1871), but Stopes firmly placed it in the upper Carboniferous, thus confirming the suspicions of several European and American paleobotanists (H. B. Geinitz, R. Kidston, D. White). Bell (1938) considered the "Fern Ledges" to be of Westphalian B age (Middle Pennsylvanian).

The 16th International Geological Congress in Washington (July 1933) provided the opportunity for two eminent European paleobotanists, P. Bertrand (Fig. 2) and W. J. Jongmans (Fig. 3), to collect Carboniferous plant megafossils in the United States. They were guided by W. C. Darrah, among others, because of the poor health of David White, who was their primary contact. Apparently, they agreed with White not to publish on their collections without consulting him first. Being able to examine the material firsthand, Bertrand and Jongmans (the latter in cooperation with W. Gothan, Fig. 4) separately made certain taxonomic revisions (see Chapter 1, this volume), but their main interest was to apply the recently established western European Carboniferous chronostratigraphic units, as based on floral biostratigraphy, to the North American succession.

Much later, in 1956 (after W. Gothan's death in 1954), Gothan's former student, H. Bode (Fig. 5), traveled to the United States for an extensive collecting trip in the Appalachian basin as well as in the American Midcontinent. He evaluated the publications by Darrah and Bertrand (1934), Jongmans and Gothan (1934), Darrah (1937), and Jongmans (1937a) but added his own somewhat unique conclusions on the Carboniferous floral biostratigraphy of North America (Bode, 1958, 1960). Bode identified megafloral taxa and gave European correlations of the Carboniferous megafloras in the Anthracite region of eastern Pennsylvania (Wood et al., 1969, p. 78, 83) and other regions as well. Unfortunately, Bode did not illustrate any specimens, unlike Jongmans (1937a, b; also Jongmans et al., 1937), who presented the first revisions of Carboniferous plant species from the United States in the light of European floral biostratigraphic experience.

The reverse process, namely the recognition of American species in the European Carboniferous, also happened, but rarely. Well-known examples are *Linopteris obliqua* and *Neuropteris rarinervis,* both species described from eastern Canada and widely reported in Europe. However, in this case it is noted that Bunbury's (1847) description of these North American species was published in a European journal.

It is also noted that Brongniart (1828–1838) figured side by side North American and European plant fossils, which thus became established in the European paleobotanical literature. However, the first conscious attempts at recognizing American megafloral taxa in Europe were made by Wagner (1958, 1968). Similarly, the North American genera *Lesleya* and *Supaia* were reported from western Europe by Remy and Remy (1975a) and Doubinger and Heyler (1976), and the European species *Callipteris (Dichophyllum) flabellifera* was recorded from both North America and western Europe by Remy et al. (1980). It is interesting to note that what appears to be geographic parochialism is also practiced in North America. In a general review article on the famous Mazon Creek flora (Horowitz, 1979), no reference is made to Wagner's (1968) unifying monograph on *Alethopteris* in which several specimens from the Mazon Creek flora are identified and illustrated.

There is general agreement that the Carboniferous floras of North America and western Europe are very comparable (Darrah, 1969), with only minor differences, and the taxonomic revisions of such species as *Neuropteris pocahontas* D. W. and *N. (Neuralethopteris) schlehanii* Stur and of *Macroneuropteris (Neuropteris) scheuchzeri* Hoffmann and *N. decipiens* Lesq. will allow an integrated systematic approach that should yield the basic data for a common European and American biostratigraphic zonation using plant megafossils. Indeed, this conviction lies at the root of correlations proposed by different authors, the most recent of which are Gillespie and Pfefferkorn (1979) and Lyons (1984) in the United States and Zodrow and Cleal (1985) and Zodrow (1985) in Canada. Despite these efforts, it is clear that much more comparative work still needs to be done to integrate taxonomic studies on both sides of the Atlantic.

## JONGMANS'S 1933 COLLECTIONS IN THE UNITED STATES AND HIS BIOSTRATIGRAPHIC INTERPRETATIONS

After the 16th International Geological Congress in July 1933, Jongmans collected plant megafossils from the Appalachian and Illinois basins and from the Midcontinent of Kansas. He was guided by O. L. Haught in the Pocahontas coalfield of West Virginia, by R. J. Holden in the Valley coalfield of Virginia, by W. C. Darrah in the bituminous coalfields of Pennsylvania, by C. W. Unger in the Anthracite fields of eastern Pennsylvania, by A. C. Noé (see Chapter 14, this volume) in the Illinois basin, and by M. K. Elias (see Chapter 13, this volume) in the Midcontinent of Kansas. He did not collect in the Dunkard basin but examined Haught's collections from the Dunkard Group (Haught, 1934). Jongmans's collections from the United States were examined conjointly with W. Gothan in Europe, and the two published a joint report (Jongmans and Gothan, 1934). That publication was issued as part of the "Jaarverslag" (Annual Report) of the Geologisch Bureau at Heerlen, The Netherlands, and is little known and often difficult to obtain. Jongmans came to study the North American Carboniferous several years after the First Congress on Carboniferous Stratigraphy (Heerlen, 1927), which was held under his leadership. This congress laid the foundations for a fully integrated stratigraphic classification of the Carboniferous of western Europe. Jongmans and Gothan were interested in finding out whether this classification could also be applied in North America. They presented in their 1934 paper a general correlation chart between the different West European and North American units and related these to the then-current scheme of Dinantian; Namurian A, B, and C; Westphalian A, B, and C; Stephanian; and Permian (see Table 1). The Westphalian D, introduced by Pruvost (1934), was still unknown as such, and it was not yet realized that the Westphalian as characterized by successions in the paralic coal belt of northern Europe (south of the Scandinavian platform) and the Stephanian corresponding to the intermontane (limnic) basins of south-central France, Bohemia, and parts of Germany were well separated in time. The nonrecognition of the important time gap between the standard Westphalian and the standard Stephanian of western Europe, as known in the 1930s, seems to have played a significant role in the errors of stratigraphic judgment with regard to the Stephanian, which are found in the report represented by Jongmans and Gothan (1934).

Jongmans and Gothan's (1934) stratigraphic conclusions were summarized as follows (free translation from the German by R. H. Wagner) (see also Table 1):

1. Part of the so-called Dunkard Group is not Permian but Westphalian C.

TABLE 2. TENTATIVE CORRELATION FROM WHITE (1936) OF PENNSYLVANIAN AND PERMIAN STRATA IN THE ROCKY MOUNTAIN, MIDCONTINENT, AND APPALACHIAN REGIONS ACCORDING TO THE STATE OF KNOWLEDGE OF THE FOSSIL FLORAS*

| Arizona (Grand Canyon) | Central Colorado | Eastern New Mexico | Western Oklahoma | Kansas and Northwestern Arkansas | Appalachian Trough | | Europe |
|---|---|---|---|---|---|---|---|
| Moenkopi Unconformity | Triassic beds | Triassic beds | Triassic shale, sandstone, and gypsum | Triassic shale, etc. | | Triassic | |
| | | Rustler Formation | Quartermaster Formation [Mostly red shale and sandstone] | Cimarron Group | | | |
| Kaibab Limestone, 800± ft | Gypsiferous in upper part [Mostly red beds] | Castile gypsum | Cloud Chief gypsum | | | | |
| | | | Woodward Group | (Wellington at top)* | | Permian | Permian |
| Coconino Sandstone, 400 ft (eolian) | Maroon Formation* (Permian coniferous flora) 2,000±-8,000± ft | [Manzano group, mostly red] San Andres Limestone 500± ft | [Guadalupe Group (limestone in Texas)] Blaine gypsum; Chickasha Fm. (salt); Duncan Sandstone* (salt) | Sumner Group | (Erosional surface) | | |
| Hermit Shale,* 300 ft Unconformity [Red sandstone and shale] | | | Clear Fork Group | Chase Group* | | | |
| | | Yeso Formation, 1,000-1,200 ft (red) (Varied Permian flora at base) | Wichita Group* (Large Permian flora) | Garrison Shale* | Greene Formation* 1,000 ft [Much interbedded red sandstone and shale. Dunkard Group] | | |
| Supai Formation* 800± ft, *Walchia*, *Taeniopteris*, etc. (Thin interbedded algal limestones) | (Varied Permian flora) | | | Cottonwood Limestone | Washington Formation* 200 ft | | |
| | (*Walchia*) | | Pontotoc Group* (*Walchia*) (conglomerates) | Wabaunsee Group [(includes Eskridge Shale (at top). Neva Limestone*, Elmdale Shale*, and Americus Limestone in upper half] | | | |
| (Conglomerate) | Conglomerates and thin algal limestone interbedded | Abo Sandstone* 300-1,000 ft (sandstone with *Walchia*) | ? | | Monongahela Fm,* 300± ft (much fresh-water limestone) [(Red beds interbedded)] | | Stephanian |
| | | | | (Scranton Shale* at top) | (*Walchia* reported in top.) | | |
| | Unconformity | Unconformity | Great unconformity -?- | | | | |

**TABLE 2. TENTATIVE CORRELATION FROM WHITE (1936) OF PENNSYLVANIAN AND PERMIAN STRATA IN THE ROCKY MOUNTAIN, MIDCONTINENT, AND APPALACHIAN REGIONS ACCORDING TO THE STATE OF KNOWLEDGE OF THE FOSSIL FLORAS*** (continued)

| Arizona (Grand Canyon) | Central Colorado | Eastern New Mexico | Western Oklahoma | Kansas and Northwestern Arkansas | Appalachian Trough | | Europe |
|---|---|---|---|---|---|---|---|
| **Diastrophic Base of Permian** | | | | Shawnee Group* | Conemaugh Fm.* 675± ft. Sea expelled by southern uplift. Red beds appear. | Pennsylvanian | Stephanian |
| (Great hiatus) | | | | Douglas Group | | | |
| | | *(Linopteris obliqua)* | | Lansing Group* | (Local conglom- erate) | | |
| | | Magdalena Group*, 0-1,200± ft) | | Kansas City Group* | | | |
| | | | | Pleasanton Group* | Allegheny Fm.*, 275± ft (max. expansion of basin). (Conglomerate) | | Westphalian D |
| | | | | Henrietta Group* | | | |
| (Erosion, with caverning of Mississippian deposits. Pennsylvanian not wholly re- moved at not distant localities.) | Weber (?) Fm.*, 100-2,000± ft (Kerber-Badito) | | Truncated and tilted beds of Pennsyl- vanian and older Paleozoic age in the Arbuckle and Wichita Mountains. | Winslow-Atoka Fm. / Cherokee Shale* Unconformity, | Kanawha Fm.* (Upper Potts- ville, 200-1,800 ft) (Conglomerates) | | Westphalian C |
| | | (Horizon of fireclay at Socorro)* (Sandia horizon)* | | Morrow Group: Bloyd Shale: Sandstone and Shale Member; Kessler Lime- stone Member; Coal-bearing Shale Member*; Shale Member | Sewell Fm.* (several trans- gressive conglom- erates). (Middle Pottsville, 0- 1,200 ft) | | Westphalian B |
| | | | | Morrow Group: Brentwood Lime- stone Member | Pottsville Group: Lee Fm.* (trans- gressive conglom- erates); (Lower Pottsville, 0- 2,000± ft); | | Westphalian A |
| | (Glen Eyrie Member flora) | | | Hale Sandstone | (Uplift, deforma- tion, erosion, and subsidence.) | | Namurian |
| Unconformity | Unconformity | Unconformity (Hiatus) | | Unconformity | | Mississippian | Dinantian |
| Redwall Limestone | Leadville Limestone | Limestones | | Limestone, shale,* and sandstone of Chester age. | Chester and older formations.* | | |

*Stratigraphic nomenclature is from White, 1936, and does not necessarily conform to current usage. Formations marked by asterisks have yielded floras. Diastrophic boundary suggested as base of Permian.

Figure 1. William C. Darrah, 1935, a year after he arrived at Harvard University. Widner Library, Harvard Yard, Harvard University, Cambridge, Massachusetts. William C. Darrah Collection, courtesy of Elsie Darrah Morey.

2. A number of beds attributed to the Stephanian in Illinois, Kansas, and Pennsylvania should be placed in the Westphalian C.
3. Part of the so-called Pottsville Group should be compared with the Westphalian and the other part with the Namurian. Its floral succession matches that of the Namurian and the Westphalian, starting from the equivalent of the Waldenburger Schichten [a unit recognized in Lower Silesia].
4. Part of the so-called Pocono Group can be attributed to the Lower Carboniferous in the European sense.
5. It appears that the floral successions in the United States and Europe are in many respects very similar.
6. The majority of species found in the United States can be identified with European megafloral taxa; however, there are some different varieties as well as some local species. On the whole, the flora of the United States is very similar to the West European one. The differences and similarities can be summarized as follows. In West and Central Europe there are marked differences between the so-called limnic and paralic basins. The exclusively limnic, intermontane basins are separated from the paralic and semiparalic foredeep by the arcuate Variscan Front. It is noteworthy that the tectonic position of the North American Carboniferous areas studied coincides more or less with that of the paralic areas of Europe. It should, therefore, be no surprise that the floras from the paralic and semiparalic basins in Europe resemble so closely those from the United States (with the exception of local species and the understandable differences that occur in such a large area in east-west direction and that are also visible within Europe).

Apparently, Jongmans and Gothan (1934) did not see the Stephanian as far removed in time from the Westphalian, and they regarded the differences in megafloral composition more in the light of different paleogeographic conditions. Jongmans (1937a) later coined the term "Westphalian E" for strata now regarded as being mainly of middle to late Stephanian ages but containing floras with a Westphalian aspect, that is, those corresponding to paralic and semiparalic basins.

Jongmans (1937a) later accepted the Westphalian D, which was introduced by Pruvost (1934) for the Assise de la Houve in Lorraine and which represented Bertrand's (1926) "*Mixoneura* Zone," that is, the zone of *Neuropteris ovata* Hoffmann and related species. Bertrand and Pruvost (1937) mentioned that this zone occurred in the upper part of the Westphalian succession of the Saar-Lorraine Basin, just below the Holz Conglomerate, which is at the base of the Stephanian succession in this basin and which seals an unconformity associated with an important stratigraphic gap. They also regarded the highest part of this zone as occurring in the lowermost unit of the succession in the Gard Basin (Cévennes, France), where a complete Stephanian succession would be present. Thus, Bertrand and Pruvost (1937) and Bertrand (1937) assumed that the Westphalian D = "Zone à *Mixoneura*" would bridge the gap between Westphalian C and Stephanian, with the upper part of the zone being present in the Cévennes. It is not clear whether or not they believed that the basal part of the succession in the Cévennes would correlate with the highest Assise de la Houve in Lorraine (in which case the stratigraphic gap below the Holz Conglomerate in the Saar-Lorraine Basin would comprise only a small part of the lower Stephanian). The lack of precision in the correlations and the equation Assise de la Houve = Westphalian D = Zone à *Mixoneura* (read: *Neuropteris ovata* complex) sowed the seeds of confusion. As the *Neuropteris ovata* range zone expanded upward, with new records outside France and Germany, it went beyond the time limits of the Assise de la Houve.

The Gard Basin on the southeastern margin of the Massif Central in south-central France has been regarded historically as

Figure 2. The "Hall of Discovery," National Museum of Natural History, Paris, France, dedicated in 1967 to Paul Bertrand, whose photograph is on the wall. William C. Darrah Collection, courtesy of Elsie Darrah Morey.

containing a full Stephanian succession. However, it was not this basin that became the type of the various Stephanian subdivisions. Jongmans and Pruvost (1950) defined the Stephanian substages A, B, and C on the three successive lithostratigraphic units present in the St. Étienne (Loire) Coalfield, namely the "Assises" de Rive-de-Gier, St. Étienne, and Avaize. Jongmans (1937a), in reply to a comment from Darrah, stated that the lower part of the French lower Stephanian (Rive-de-Gier) would belong to the Westphalian D because it contained *Neuropteris ovata.* This is apparently incorrect. Later work (Bouroz et al., 1970) placed the basal part of the succession in the Cévennes below the horizon of the Rive-de-Gier (Stephanian A). It seems that Jongmans (1937a) saw the Stephanian (i.e., middle and upper Stephanian = Stephanian B and C of Jongmans and Pruvost, 1950) as a rather short time interval that would be characterized by "Westphalian E" floras (i.e., floras of essentially Westphalian complexion though later in age than Westphalian D) occurring in marine-influenced (coastal) basins that he called paralic, semiparalic, or limno-paralic. Jongmans's apparent refusal to recognize the Stephanian in North America must be seen in the light of this conviction. Darrah (in Jongmans, 1937a) was fully aware of this as he stated that ". . . Professor Jongmans introduces into the literature a new name Westphalian 'E' to designate those beds intermediate or transitional between Westphalian 'D' and Rotliegend. To be sure this is a synonymous term for what is loosely called 'Stephanian.' In my opinion it is not necessary to introduce a new name, and on this matter I disagree with Professor Jongmans."

Jongmans (1937a) summarized his correlations between northwest and central Europe on the one hand and the United States on the other in a table that is reproduced here as Table 1. In his 1937a paper, which differs in some of its conclusions from his 1934 paper with Gothan, Jongmans presents a number of considerations that may be summarized as follows:

A. *West Virginia:* This succession, which goes from the Pocahontas Formation to the Dunkard Group, was dated as ranging from early Namurian (flora equivalent to that of the Waldenburg beds of Lower Silesia) to late Westphalian D. The question of a possible presence of "Westphalian E" and Rotliegend was left open.
B. *Pennsylvania:* Pottsville to Monongahela Formations, ranging from late Westphalian A to late Westphalian D, with the possibility that Westphalian E strata occurred from the middle part of the Conemaugh Formation upward. It is admitted that Fontaine and I. C. White (1880) described Rotliegend (highest Stephanian to lower Permian) floras from Pennsylvania.
C. *Pocono Group in Pennsylvania and Virginia:* The floras collected are unhesitatingly attributed to the Lower Carboniferous, up to and including lower Viséan (i.e., lower Mississippian).
D. *Illinois:* A range from Westphalian A to Westphalian D is reported, with the possibility that "Westphalian E" flora might be present (but was not seen by Jongmans during his collecting trip in 1933).

Figure 3. Wilhelmus J. Jongmans, 1951, at his laboratory at the "Geologisch Bureau," Heerlen, The Netherlands. Photograph by and courtesy of A. T. Cross, Michigan State University.

Figure 4. Walther Gothan, 1951, at the Third Carboniferous Congress, Heerlen, The Netherlands. Photograph by and courtesy of A. T. Cross, Michigan State University.

E. *Kansas:* Westphalian C and D floras are reported, and the presence of "Westphalian E" is admitted on published records (no first-hand information; compare Cridland et al., 1963).

Jongmans et al. (1937) produced a comparison between the floral succession of West Virginia with that known from Europe but did not get beyond Westphalian D. In another paper (1937b), which was devoted specifically to *Neuropteris ovata,* Jongmans admitted that this species as well as *N. scheuchzeri* ranged more extensively in North America. He argued, however, that the beds with *N. ovata* should be called Westphalian D regardless of the differences in the variable top of the range of this species in the different regions and continents. He showed awareness of the fact that this meant an extended Westphalian D range and mentioned that ". . . there remains a relatively small series of strata containing floras which are intermediate between Westphalian D and Lower Rotliegend" (Jongmans, 1937b, p. 421). However, he also mentioned in the same paper that a higher zone could be recognized where *N. ovata* is accompanied by Stephanian pecopterids and that a further subdivision of these higher beds could encompass strata below the Rotliegend but above the uppermost occurrence of *N. ovata.* These would be his "Westphalian E," the name Stephanian being reserved for beds with flora of the intermontane (limnic) basins. He also expressed awareness that this form of correlation would not assure synchroneity.

It is clear that Jongmans's approach to stratigraphy was wholly biostratigraphic and that he consciously sacrificed correlation (i.e., time-equivalence) in order to emphasize paleogeographical and paleoenvironmental differences. It is in this sense that one has to take his concept of "Westphalian E," which, as Darrah (in Jongmans, 1937a) correctly observed, is virtually another name for Stephanian but reflecting a different biofacies. However, Jongmans saw his "Westphalian E" originally as being time-equivalent only to the higher Stephanian and regarded the basal part of the lower Stephanian (i.e., Stephanian A as later proposed) as forming part of the Westphalian D. It is possible, though not explicitly stated, that this point of view was based on Bertrand's inclusion of the basal part of the Stephanian succession in the Gard Basin in his "Zone à *Mixoneura.*" Jongmans, in 1952, modified this view

Figure 5. Hans Bode, 1954, northern France, field trip connected with the 8th International Botanical Congress Paris, France. Photograph by and courtesy of A. T. Cross, Michigan State University.

and stated that the lower Stephanian (i.e., Stephanian A of Jongmans and Pruvost, 1950) of the French Massif Central would constitute a transition between true Westphalian D and the proper Stephanian (meaning, quite clearly, middle and upper Stephanian = Stephanian B and C). Jongmans may well have been influenced by a now-classic paper by Gothan and Gimm (1930), who distinguished between a pecopterid-calamitean association and a *Callipteris-Walchia* association in the Rotliegend of Thüringen in eastern Germany, which are regarded as representative of peat-forming and non-peat-forming environments, respectively. This is a matter that has attracted the interest of paleobotanists in more recent years, with regard to a possible equivalence in time (and difference in biofacies) of upper Stephanian and lower Autunian plant megafossil assemblages (Broutin et al., 1990). A general paper by Remy (1975) should also be cited in this regard. Elias (1936; see Chapter 13, this volume) was the first to describe a similar case in North America. Lyons and Darrah (1989) also discussed this matter. The issue is still with us today, and the fundamental problem is to distinguish the different ages where more apparent differences in fossil assemblages are produced by different climatic and edaphic conditions.

## BERTRAND'S VIEWS ON THE WESTPHALIAN D AND STEPHANIAN IN THE APPALACHIAN BASIN

In 1933, following the 16th International Geological Congress in Washington, D.C., French paleobotanist P. Bertrand (see Chapter 7, this volume) was guided to localities in the Appalachian basin by W. C. Darrah, who had begun collecting in 1926 from the lower part of the Conemaugh Formation (Darrah, 1969). Together, they collected in 1933 from the Conemaugh, Monongahela, and Washington Formations in the bituminous coalfields as well as from the coal-bearing strata in the Anthracite region of eastern Pennsylvania (Darrah, 1969). In contrast to Jongmans, who was mainly familiar with the Namurian and Westphalian A, B, and C of the paralic coal belt in northwestern Europe, Bertrand also had extensive experience with the Westphalian D of Lorraine, France, and the Stephanian of the French Massif Central and the Alps. In a joint paper (Darrah and Bertrand, 1934), credit is given to the correlations proposed by D. White (*in* Handlirsch, 1906), who assigned the Conemaugh and the Monongahela to the Stephanian, a position that is still valid today (see Chapter 25, this volume). Darrah and Bertrand made the point that certain differences in composition as well as in stratigraphic distribution were only to be expected when dealing with different regions, but that, on the whole, the Carboniferous floras of Europe and North America were sufficiently similar to allow correlation. Darrah and Bertrand (1934) presented the following conclusions:

1. The Allegheny Formation of Pennsylvania represents the Westphalian D because it contains the same flora as that of the Assise de la Houve in Lorraine. It is characterized by abundant *Neuropteris ovata* (attributed to *Mixoneura* by Bertrand), the last occurrences of *Mariopteris,* Westphalian forms of *Sphenophyllum,* the presence of cannelate *Sigillaria,* and the first occurrences of a number of Stephanian *Pecopteris* species.
2. The lower part of the Conemaugh Formation does not differ in floral contents from the Allegheny Formation. It corresponds to the transition between Westphalian and Stephanian.
3. In Pennsylvania, the Stephanian is represented by the upper part of the Conemaugh Formation and the Monongahela Formation. The flora from this interval is characterized by *Lescuropteris moorii, Odontopteris* of the *reichii [reichiana]* group, *Mixoneura (Neuropteris) neuropteroides, Pecopteris,* and *Sphenophyllum* of Stephanian types.

Darrah and Bertrand (1934) also accepted White's (1904) attribution of the Dunkard to the lower Permian. As mentioned previously, White's opinion on the age of the Dunkard Group changed from considering the lower part below the Lower Washington Limestone as Late Pennsylvanian (White, 1904), to considering the entire Dunkard as being Permian (White, 1936).

Bertrand (1934) presented a fairly exhaustive analysis of the Carboniferous floras on the basis of data published by David White. He generally agreed with the age relations proposed by the latter but tended to place the Lower Coal Measures of Missouri in Westphalian C rather than Westphalian D, thus implicitly modifying White's correlation between Missouri and Illinois. After reading the 1934 paper by Jongmans and Gothan, Bertrand (1935) presented a critical opinion on the conclusions reached by those authors. He endorsed the correlations made with regard to the floras of Viséan to Westphalian B ages but disagreed on the extended Westphalian C of Jongmans and Gothan, which was later attributed mainly to Westphalian D and Westphalian E in Jongmans's 1937a paper. Indeed, Bertrand stated that the Westphalian C *(senso stricto)* seemed to be absent in Pennsylvania and Virginia, where a stratigraphic gap separated the upper part of the Pottsville Formation from the Allegheny Formation. He reaffirmed the attribution by Darrah and Bertrand (1934) of the Allegheny Formation and lower part of the Conemaugh Formation to the Westphalian D. It should be noted that Darrah, in 1935, after collecting in the Westphalian D of France (see Chapter 2, this volume), clearly made the correlation of Westphalian D with the Mason Shale (in the lower part of the Conemaugh Formation of the Appalachian basin). With regard to the Stephanian, Bertrand (1935) expressed his surprise that certain species, such as *M. (N.) scheuchzeri, N. ovata,* and *Mariopteris,* regarded at that time as exclusively Westphalian in Europe, extended their range into the Stephanian equivalents of North America. However, he placed more emphasis on the occurrence of a large number of Stephanian species in the upper part of the Conemaugh Formation and in the Monongahela Formation and thus operated on the principle that first occurrences (new evolutionary developments) should be regarded as more important than apparent extinctions. He also agreed with Darrah that *Lescuropteris* (Fig. 6) and *Danaeites* (Fig. 7) were Stephanian genera because of their association with Stephanian plant fossils (see Chapter 25, this volume). Bertrand was right in taking this point of view, but one wonders to what extent Jongmans (1937a) agreed with Bertrand when he introduced his Westphalian E as a special biofacies of the middle and upper Stephanian. The list of Stephanian floral elements mentioned by Bertrand (1935, p. 34) is impressive (see also Table 3). The stratigraphic gap between Westphalian D (Lorraine) and lower Stephanian (Loire Basin, Massif Central), underemphasized at the time, is readily apparent from the marked change in floral composition (as shown in Table 3). Among these there is at least one species, *Sphenophyllum tenuifolium,* that, in view of its close similarity with *Sphenophyllum angustifolium* of Europe, may be indicative of Stephanian C. Bertrand (1935) also noted that in Kansas a number of very characteristic Stephanian floral elements occurred intermingled with some Westphalian species that he assumed had a more extended range in North America.

Bertrand (written communication to W. C. Darrah, June 6, 1935), on the basis of photographs received from Darrah, recognized that two species in the Upper Pennsylvanian of the Appalachian basin were excellent Stephanian guide fossils. He stated:

Figure 6. *Lescuropteris moorei* (Lesq.), republished from the "Historical aspects of the Permian flora of Fontaine and White" by W. C. Darrah, *in* Barlow, J. A., ed. Proceedings of the First I. C. White Memorial Symposium, "The Age of the Dunkard" (1975), p. 95, Fig. 5. Republished with the permission of the West Virginia Geological and Economic Survey, Morgantown, West Virginia. Photograph from the W. C. Darrah Collection, courtesy of Elsie Darrah Morey. Scale = 1 cm.

> *There is not the slightest doubt* [emphasis of Bertrand], that you got there in the Monongahela Formation two guide species: *Taeniopteris jejunata* G.'Eury and *Odontopteris genuina* G.'Eury, surely both identical with specimens of St. Étienne, Commentry, Blanzy et le Creusot.
>
> Your *Odontopteris genuina* is evidently the true *genuina* G.'Eury, as you can see immediately if you look at Fig. 2a, Pl. XXIV in Zeiller, Flore fossile du bassin houiller de Blanzy et le Creusot. That species is strictly localized in middle and upper Stephanian, and does not occur at all in lower Stephanian. It is the same thing with *Taeniopteris jejunata.*"

Darrah also recognized another guide species—*Lescuropteris moorii* (Lesq.)—in the upper part of the Conemaugh Formation and lower part of the Monongahela Formation (see Darrah, 1968b, and Chapter 25, this volume). Although this is not a European species, its association with Stephanian index species makes it a good marker for Stephanian equivalents in both the Appalachian basin and Midcontinent of the United States (see Darrah, 1968b).

## STEPHANIAN OF WESTERN EUROPE AND THE WESTPHALIAN-STEPHANIAN BOUNDARY

Some of the confusion in recognizing Stephanian strata and the Westphalian-Stephanian boundary in North America is probably due to the different definitions and variable concepts applied to these units and their mutual boundary in Europe. A brief summary will suffice to point out the problems that have arisen.

Munier-Chalmas and de Lapparent (1893) introduced the Westphalian and Stephanian stages for the terrestrial, coal-bearing deposits in western Europe. (These are currently regarded as series, with constituent Langsettian, Duckmantian, Bolsovian, Westphalian D, Cantabrian, Barruelian, Stephanian B, and Stephanian C stages; see Foreword.) The Westphalian Series was based on its development in the extensive paralic coal belt in northwestern Europe, whereas the Stephanian took its name from the small intermontane basin of St. Étienne in the Massif Central of south-central France. The first Heerlen Congress (1927) subdivided the Westphalian into A, B, and C. A Westphalian D unit was added at the second Heerlen Congress in 1935. It was based on the Assise de la Houve in Lorraine (Pruvost, 1934), that is, the highest unit below the

Figure 7. *Danaeites emersonii* Lesq., roof shale of Pittsburgh coal bed, Fairfax Stone, Tucker County, West Virginia, Appalachian basin, reproduced from Gillespie et al. (1978, pl. 33, Fig. 1) with the permission of the West Virginia Geological and Economic Survey, Morgantown, West Virginia. Photograph courtesy of W. H. Gillespie. Scale = 1 cm.

Holz Conglomerate, which marks a widespread unconformity in the Saar-Lorraine Basin. This is another intermontane basin, intermediate geographically between the northwest European paralic coal belt and the Massif Central and situated partly in northeastern France and partly in western Germany. Jongmans and Pruvost (1950) subdivided the Stephanian into A, B, and C units, on the basis of the succession in the St. Étienne (Loire) Coalfield (Assises de Rive-de-Gier, St. Étienne, and Avaize, respectively). A later proposal to recognize also a Stephanian D unit (Bouroz and Doubinger, 1977) did not gain the approval of the International Union of Geological Sciences (IUGS) Subcommission on Carboniferous Stratigraphy. The "Stephanian D" did not represent a previously unrecognized time interval but was created out of the higher part of Stephanian C and the lowermost part of the succeeding Autunian.

The Westphalian of the paralic coal belt is not followed in continuous succession by Stephanian strata (with the exception of a remnant of very basal Stephanian deposits in South Wales and in the Bristol-Somerset area of southwestern England), and the Stephanian succession of St. Étienne and elsewhere in the Massif Central lies unconformably on basement rocks. Only in the Saar-Lorraine Basin is there superposition of Westphalian and Stephanian strata, but their mutual boundary is an unconformity that represents a fairly substantial time gap even where the Westphalian D succession is most completely preserved (i.e., in Lorraine, on the French side of the basin).

The amount of time missing at the unconformity marked by the Holz Conglomerate in the Saar-Lorraine Basin has been disputed, and the possibility of upper Westphalian D being present in the basal part of the Stephanian succession of the Gard Basin (Cévennes) in the Massif Central has also been considered (Bertrand and Pruvost, 1937). In the 1930s the following was therefore not at all clear: (1) that the stratigraphic break at the Westphalian-Stephanian boundary in Lorraine would be of major importance and (2) that some degree of overlap might not be present between the lowermost Stephanian of the Gard Basin and the highest Westphalian of the Saar-Lorraine Basin. These unsolved questions played a part in the attempts to recognize the Westphalian-Stephanian boundary in North America. A further complication was introduced later by Alpern and Liabeuf (1967, 1969), who proposed lowering the base of the Stephanian to a position *within* the Assise de la Houve (Lorraine), on biostratigraphic (palynological) grounds. This would have changed the definition of Westphalian D, as accepted at the second Heerlen Congress, and although Alpern and Liabeuf's proposal was favorably received to begin with, it was rejected at the meeting of the IUGS Subcommission on Carboniferous Stratigraphy, held in Liège, 1969. Unfortunately, in a general synthesis of palynozones in western Europe (Clayton et al., 1977), the proposal made by Alpern and Liabeuf (1967) was mentioned without stating its subsequent rejection. This is not the place to discuss the various errors of judgment in the palynological synthesis of Clayton et al. (1977), but the uncriti-

**TABLE 3. CHARACTERISTIC PLANT ASSEMBLAGES OF THE WESTPHALIAN D, LOWER, MIDDLE, AND UPPER STEPHANIAN* IN THE MASSIF CENTRAL AND SAAR-LORRAINE BASIN[†]**

| Species | Westphalian D[§] (Zone à Mixoneura) | Lower Stephanian[§] (Flore de Rive-de-Gier) | Middle Stephanian[§] (Flore de St. Etienne) | Upper Stephanian[§] (Flores de Commentry et Blanzy) |
|---|---|---|---|---|
| Alethopteris costei | - | - | - | C |
| A. grand'euryi | - | - | - | C |
| A. grandini | - | P | - | - |
| Alloiopteris heinrichi** | C | - | - | - |
| Asolanus camptotaenia | - | ? | - | - |
| Callipteridium gigas | - | - | P | - |
| C. pteridium | - | P | C | - |
| Cordaites cf. borassifolius | - | P | - | - |
| C. lingulatus-angulosotriatus | - | - | VC | - |
| Diplotmema busqueti | - | - | C | - |
| D. ribeyroni | - | - | R | - |
| Dolerotheca pseudo-peltata | - | - | P | - |
| Dorycordaites aff. palmaeformis | - | C | - | - |
| Eremopteris courtini | - | - | - | C |
| Hapalopteris typica | - | P | - | - |
| Lepidodendron rimosum | - | ? | - | - |
| Lepidophloios crassilepis | P | - | - | - |
| Lep. cf. macrolepidotus | - | P | - | - |
| Linopteris brongniarti | - | - | C | - |
| Linopteris germari | - | - | C | - |
| L. aff. neuropteroides | - | P | - | - |
| Margaritopteris coemansi | C | - | - | - |
| Mariopteris nervosa | C | - | - | - |
| N. (Mixoneura) cordata | - | - | C | - |
| N. dispar | - | - | - | C |
| N. gallica | - | - | - | C |
| N. neuropteroides | - | - | P | - |
| N. ovata | VC | - | - | - |
| N. planchardi | - | - | P | C |
| N. praedentata | - | - | - | C |
| N. pseudo-blissi | - | - | - | C |
| Neuropteris aff. sarana-deflinei | - | P | - | - |
| N. stipulata | - | - | - | C |
| Odontopteris brardi | - | - | C | - |
| O. dufresnoyi | - | - | - | C |
| O. genuina | - | - | R | C |
| O. jean-pauli (= alpina) | C | - | - | - |
| O. lindleyana | R | - | - | - |
| O. peyerimhoffi | R | - | - | - |
| Odontopteris reichi-minor | R | ? | C | - |
| O. schlotheimi | - | ? | P | C |
| Ovopteris goldenbergi | C | - | - | - |
| Ov. pecopteroides | - | - | C | - |
| Pecopteridium armasi[‡] | P | - | - | - |
| Pec. devillei[‡] | P | - | - | - |
| Pec. rubescens-jongmansi[‡] | C | - | - | - |
| Pecopteris arborescens | P | C | VC | - |
| P. bioti-grüneri | - | - | C | - |
| P. bredovi | R | - | - | - |
| P. candollei | R | - | - | - |
| P. crenulata | C | - | - | - |
| P. aff. cyathea | R | - | - | - |
| P. densifolia | - | - | - | C |
| P. elaverica | - | - | - | C |
| P. feminaeformis | - | - | C | - |
| P. hemitelioides | - | P | - | - |
| P. lamuriana | - | C | - | - |
| P. aff. lamuriana | C | - | - | - |
| Pecopteris launayi-monyi | - | - | - | C |
| P. longifolia | C | - | - | - |
| P. micro-miltoni | C | - | - | - |
| P. oreopteridia | - | P | P | - |
| P. pectinata | C | - | - | - |
| P. plückeneti | C | R | - | - |
| P. plumosa-dentata | - | P | - | - |
| P. polymorpha | R | C | C | - |
| P. saraefolia-röhli | C | - | - | - |
| P. saraepontana | P | - | - | - |
| P. sterzeli-plückeneti | - | - | C | - |
| P. unita | C | C | - | - |
| Poacordaites linearis | - | - | C | - |
| Renaultia rutaefolia | C | - | - | - |
| Sigillaria brardi | - | P | C | - |
| Sig. deutschi | - | C | - | - |
| Sig. scutellata | - | C | - | - |
| Sig. tessellata | - | C | - | - |
| Sphenophyllum emarginatum | C? | P | - | - |
| Sph. longifolium | - | - | P | C |

TABLE 3. CHARACTERISTIC PLANT ASSEMBLAGES OF THE WESTPHALIAN D, LOWER, MIDDLE, AND UPPER STEPHANIAN* IN THE MASSIF CENTRAL AND SAAR-LORRAINE BASIN† (continued)

| Species | Westphalian D§ (Zone à Mixoneura) | Lower Stephanian§ (Flore de Rive-de-Gier) | Middle Stephanian§ (Flore de St. Etienne) | Upper Stephanian§ (Flores de Commentry et Blanzy) |
|---|---|---|---|---|
| Sph. majus | - | P | - | - |
| Sph. oblongifolium | - | P | C | - |
| Sph. tenuifolium | - | - | P | - |
| Sph. thoni | - | - | - | C |
| Sphenopteris casteli | - | - | - | C |
| S. fayoli-biturica | - | - | - | C |
| S. matheti | - | - | - | C |
| S. cf. nummularia | C | - | - | - |
| S. quadridactylites | - | P | - | - |
| Taeniopteris carnoti | - | - | - | C |
| T. jejunata | - | - | C | - |
| T. multinervis | - | - | - | C |
| Walchia piniformis | - | - | P | - |
| Zygopteris erosa | - | P | - | - |
| Z. pinnata | - | - | C | - |

* Stephanian A, B, and C of Jongmans and Pruvost, 1950. [Not the sum total of presently recognized Stephanian stages.]
†According to Bertrand, 1937. Table arranged by Paul C. Lyons; due to historical significance, not updated for taxonomy and biostratigraphy.
§P = present; R = rare; C = common; VC = very common; ? = questionably present; - = absent or not recorded.
**Nomen nudum
‡Species of *Praecallipteridium.*

cal use of this European synthesis has led to problems with the recognition of the Westphalian-Stephanian boundary in the more recent American literature.

The problems with the Westphalian-Stephanian transition, and the various correlations involved, could finally be solved in northwest Spain where upper Westphalian D strata were found in continuous succession with lower Stephanian. This allowed gauging the time gap between Westphalian and Stephanian strata in the Saar-Lorraine Basin, which proved to be very substantial, comprising the lower Stephanian Cantabrian Stage (Wagner, 1969) as well as part of the Stephanian A (Barruelian) (Germer et al., 1968). It was also found that the lowermost part of the Stephanian succession in the Gard Basin (Cévennes) correlated with upper Cantabrian (Bouroz et al., 1970). A recent summary of the highest Westphalian and lower Stephanian succession in northwest Spain has been published by Wagner and Winkler Prins (1985). The Spanish area was very poorly known in the 1930s, and it did not play any part in the considerations of Carboniferous stratigraphers at that time.

## RECOGNITION OF THE WESTPHALIAN-STEPHANIAN BOUNDARY IN NORTH AMERICA

The different criteria employed for the recognition of the Westphalian-Stephanian boundary in Europe had their repercussions in the United States. Bertrand (1934), analyzing the megafloral information contained in the publications by David White, correlated the Allegheny Formation of the Pennsylvanian System (Appalachian basin) with the Westphalian D and placed the Conemaugh and Monongahela Formations in the Stephanian. By implication, the Westphalian-Stephanian boundary would approximately coincide with the Allegheny-Conemaugh boundary. Because Bertrand and White agreed that the North American succession would be more continuous than those known for western Europe, where one had to rely on correlations between the northwest European paralic coal belt and the intermontane basins of south-central France, via the independent Saar-Lorraine Basin (with its incomplete succession of Stephanian strata), this first attempt at correlation and the implied position of the Westphalian-Stephanian boundary was obviously important. The same position was accepted by Darrah (1937, p. 128), who noted: "The Stephanian boundaries are conveniently recognized at the Lower Conemaugh at the bottom and at the Washington-Greene division on the top." It should be noted that Darrah placed the Mason Shale in the lowermost part of the Conemaugh Formation in the Westphalian D (see Chapter 1, this volume). Bode (1958) developed his own peculiar ideas about the Westphalian D and Stephanian, which will be discussed separately later in this chapter.

Cridland et al. (1963) discussed the Westphalian-Stephanian boundary at some length. They placed this boundary within the Conemaugh Formation, even though they mentioned the presence of the Stephanian index, *Sphenophyllum oblongifolium,* at the level of the Upper Freeport coal bed, the uppermost bed of the Allegheny Formation. This appears inconsistent but may only reflect that the roof shales of the Upper Freeport coal bed are technically basal Conemaugh. They also accepted the suggestion contained in Moore et al. (1944) that the Westphalian-Stephanian boundary would coincide with the Desmoinesian-Missourian boundary of the American Midcontinent. Peppers (1984) slightly modified this position by placing the Westphalian-Stephanian boundary in the Missourian Series, a little above the Desmoinesian Series (see correlation chart in Foreword). In view of the imprecision of palynology at this level (the data in Clayton et al., 1977, suggest that a single palynozone spans the entire Westphalian D, Cantabrian, and lower Barruelian interval), it is unclear what the factual basis was for this proposed correlation. It is noted that Peppers (1984) has taken the palynological record in the lower part of the Stephanian Series in the Saar-Lorraine Basin as representative of basal Stephanian. He thus ignored the widely acknowledged stratigraphic break at the level of the Holz Conglomerate, which has eliminated most of the lower Stephanian (i.e., Cantabrian as well as lower Barruelian according to Germer et al., 1968). In a correlation chart published by Phillips and Peppers (1984), the Westphalian-Stephanian boundary is again equated to the Desmoinesian-Missourian boundary and also to a position between the Moscovian and Kasimovian in Russia. However, the fusulinid foraminiferal records in the Moscow Basin, set against those in northwest Spain where plant megafossils allow correlations within western Europe, show the Westphalian-Stephanian boundary to lie within the highest Moscovian (Myachkovsky Horizon) (M. Lys *in* Wagner et al., 1977). A later re-adjustment of the Westphalian D–Cantabrian boundary (Wagner and Winkler Prins, 1985) has not substantially affected this correlation.

The famous Mazon Creek flora of Illinois, at the level of th Colchester No. 2 coal bed (in the Francis Creek Shale Member of the Carbondale Formation), has been assigned to Westphalian D by Darrah (1969) and to the upper Westphalian D by Gillespie and Pfefferkorn (1979). These assignments are consistent with an early Cantabrian age (basal Stephanian) for the overlying Herrin No. 6 Coal Member of the Carbondale Formation where the Stephanian indices *Sphenophyllum oblongifolium* and *Alethopteris zeilleri* have been recorded (Wagner, 1984, p. 126). Thus, the Westphalian-Stephanian boundary seems to lie within the upper Desmoinesian and not at the limit between Desmoinesian and Missourian (for which no real evidence has been adduced, as discussed above).

Gillespie and Pfefferkorn (1979) placed the Westphalian-Stephanian boundary within the Conemaugh of the Appalachian basin, but the evidence has not been discussed and appears unclear from the ranges of plant fossils (e.g., *Sphenophyllum oblongifolium*) given in their paper. Their position appears to follow that of Darrah (1969, p. 12), who pointed out a megafloral break at the Ames Limestone (mid-Conemaugh) Formation, although recent evidence indicates that this floral break, as interpreted here, occurs at a position in the basal part of the Conemaugh Formation and below the Brush Creek Limestone (see data in McComas, 1988).

The data presented in McComas (1988) suggest that the Westphalian-Stephanian boundary in Ohio lies between the Mahoning coal bed and the Brush Creek Limestone Member in the lower part of the Conemaugh Formation. Significant taxa reported in this 9-m interval are *Sphenophyllum oblongifolium, Odontopteris brardii, Pecopteris (Lobatopteris) lamuriana,* various species of *Pecopteris,* and two different forms of *Taeniopteris*—all indicative of the Stephanian—as well as two taxa, *Danaeites emersonii* and *Lescuropteris moorii* that are also Stephanian indicators according to Darrah (1937, 1968b, 1969; see also Chapter 25, this volume). The Ohio assemblage (McComas, 1988) as a whole shows a mixture of Westphalian D and Stephanian species. The low occurrence of *Danaeites emersonii* and *Lescuropteris moorii* suggests that they might characterize most of the basal part of the Upper Pennsylvanian rather than just the upper part of the Conemaugh Formation and lower part of the Monongahela Formation.

This interpretation of the Ohio assemblage agrees essentially with Darrah's 1937 opinion. He placed the Westphalian-Stephanian boundary near the base of the Conemaugh Formation (see Chapter 2 and 25, this volume).

## DARRAH'S (1937) CONCLUSIONS ON THE STEPHANIAN IN NORTH AMERICA

Darrah was in essential agreement with Bertrand's (1934) interpretation of floral data as published by David White and obviously paid close attention to the succession of Stephanian floras as summarized by Bertrand (1937). It is here noted that P. Bertrand seems to have been one of the first paleobotanists to provide some idea of frequency by recording megafloral taxa as Rare, Present, Common, or Very Common (see Table 3, this chapter). Bertrand's data on the Stephanian floras of the Massif Central in south-central France and in the Westphalian D of the Saar-Lorraine Basin are reorganized and reproduced in Table 3 of this chapter. It is noted that these data are of historical interest only. Bertrand had no knowledge of the lowermost Stephanian stage, that is, the Cantabrian, and the species mentioned in Table 3 include one nomen nudum and several species that are interpreted differently today. Although Darrah and Bertrand (1934) jointly published data on the American floras, it appears that the latter left the details about the Stephanian floras in the United States to be published by Darrah (1937) in the Compte Rendu of the 2nd Heerlen Congress in 1935.

The contacts established between Darrah and the European paleobotanists, Jongmans and Bertrand, first of all in activities connected with the 16th International Geological Congress in Washington, D.C., in 1933 and afterward during Darrah's 1935 field trips in Europe, allowed him to take an informed view of the resemblances and differences apparent between the North American and European floras. In 1935, Darrah presented a preliminary paper in relation to the Second Carboniferous Congress, in which he stated the most important species characterizing the Pottsville, Allegheny, Conemaugh (lower and upper), Monongahela, and Washington Formations (the latter being the lower part of the Dunkard Group and considered uppermost Pennsylvanian by Darrah, in agreement with White, 1904). Darrah then proceeded to enumerate the characteristic floral elements of the Westphalian, Stephanian, and "Permian" (Rotliegend of Germany or Autunian in France) in western Europe and observed that the floral assemblages in North America and Europe were similar but not identical. For example, *Loncho-*

*pteris,* a common plant in the lower Westphalian of Europe, was unknown in the United States (this genus has subsequently been recorded with some reservation from the Narragansett basin of New England [Lyons et al., 1976]). Darrah (1935) also mentioned *Lescuropteris* as a plant restricted to the Upper Pennsylvanian (= Stephanian) of North America. This genus was then unrecognized in Europe, but a later paper by Remy and Remy (1975b) indicated that *Lescuropteris* could also be recognized from the Central Massif of France in the well-known European Stephanian plant *"Odontopteris" genuina.*

Darrah (1935) also noted the much longer range of *Macroneuropteris (N.) scheuchzeri* in North America, where this species reaches into the upper part of the Dunkard Group. In Europe, it had been considered a typical Westphalian species. Records from Spain, not available when Darrah wrote his 1935 paper, have since extended the range of *M. (N.) scheuchzeri* into the upper part of the Cantabrian, the basal stage of the Stephanian Series. *N. ovata,* at one time considered to be a Westphalian D index in Europe, reaches well into the Dunkard Group in North America (Darrah, 1969). This species has since proved to have an almost equally long range in the Iberian Peninsula, where it is a common element up to and including lower Stephanian C (Wagner, 1965, p. 79).

A more extensive paper was presented by Darrah (1937) at the Carboniferous Congress in Heerlen in 1935. He placed the base of the Stephanian at a position in the middle of the Conemaugh Formation (possibly at the level of the Ames Limestone Member as interpreted here; see also Darrah [1969]) and its top in the Washington Formation of the Dunkard Group. This is similar to the position as stated in Darrah and Bertrand (1934). It is clear that Darrah considered the Mason Shale (in the lower part of the Conemaugh Formation) to be Westphalian D (see Darrah, 1969; Chapter 2 and 25, this volume). The exact position of the Westphalian-Stephanian boundary in the Appalachian basin is still unresolved, but it appears, as interpreted here, to be in the basal part of the Conemaugh Formation, below the Brush Creek Limestone Member. Darrah, like Bertrand, obviously regarded the first occurrences of new Carboniferous species as most important in a biostratigraphic sense and thought that the times of extinction would be diachronous in different regions.

## BODE'S (1958, 1960) CORRELATIONS

In 1956, Hans Bode, a former student of Gothan, made an extensive collecting trip through the United States, visiting exposures in the Appalachian region (Tennessee, Kentucky, West Virginia, Ohio, and Pennsylvania), in the Illinois basin, and in the Midcontinent of Kansas and Missouri. He had the benefit of a more-detailed European floral succession set against more closely defined chronostratigraphic units that included Westphalian A, B, C, and D of the second Heerlen Congress (1935) and Stephanian A, B, and C as introduced by Jongmans and Pruvost (1950). Bode (1958, p. 225) emphasized that the Westphalian D–Stephanian A position in the Saar-Lorraine Basin was associated with a stratigraphic gap and that the differences between the Westphalian D flora and the Stephanian A flora were correspondingly sharp. Of course, the Stephanian type area at St. Étienne (Loire Basin of the Massif Central in south-central France) shows an unconformity below the Stephanian A (Rive-de-Gier beds) and a total absence of Westphalian deposits. It would be more than a decade before a gradual transition between Westphalian and Stephanian strata was discovered in the Cantabrian Mountains of northwest Spain (Wagner, 1969).

Bode (1958) noted that P. Corsin (Fig. 8), Paul Bertrand's successor at Lille (in Pruvost and Corsin, 1949), had assigned the lower part of the succession in the Cévennes (Gard Basin of the Massif Central) to the Stephanian, despite the presence of *N. ovata,* which was still regarded at that time as the primary floral index of Westphalian D in Europe. This is the same case as in the Ohio assemblage of McComas (1988) in the Appalachian basin. Bode disagreed, placed major emphasis on the occurrence of *N. ovata* as indicating Westphalian D, and proposed that Stephanian index species might already occur in the highest Westphalian D. This position was also taken by Bell (1938), who found *Sphenophyllum oblongifolium* in the Sydney Basin

Figure 8. Paul Corsin, 1951, outside his laboratory, Lille, France. Photograph by and courtesy of A. T. Cross, Michigan State University.

of Maritime Canada but probably believed this to be a rather low occurrence of what was ostensibly a Stephanian guide species (see Chapter 25, this volume). Bode, like Jongmans and Bell, preferred to regard extinction as more important stratigraphically than first occurrences and went completely against the conclusions of Darrah and Bertrand. However, this could only lead to a diachronous boundary between Westphalian and Stephanian worldwide, depending on the different top ranges of *N. ovata* in different parts of the world.

The well-known Mazon Creek flora (from the Colchester No. 2 coal bed of the Carbondale Formation) from Illinois was placed by Bode (1958) in the Westphalian D in view of the presence of *N. ovata;* a similar view was taken by Darrah (1969). The same species continued to occur even higher in the Illinois basin succession, that is, well up in the Carbondale Formation, which was correspondingly assigned by Bode in its entirety to Westphalian D. In fact, Bode (1958) denied the presence of Stephanian strata in the Illinois basin and attributed the upper part of the Tradewater Formation and the entire Carbondale and McLeansboro Formations to the Westphalian D. There is no doubt that Bode's insistence on equating Westphalian D with the biozone of *N. ovata* led to faulty correlations and that Jongmans's differentiation of Westphalian E (in which Westphalian floral elements such as *N. ovata* and *M. (N.) scheuchzeri* could still occur) was preferable to this poor application of a definition that was quite at odds with the introduction of the Westphalian D by Pruvost (1934). In contrast, Darrah (1969, 1975) admitted that *Neuropteris ovata* could range throughout the Stephanian and even into the lower Permian.

It is not surprising that Bode (1958) also identified Westphalian D in the Missourian and the lower part of the Virgilian of Kansas, which are regarded by North Americans as Stephanian (Darrah, 1969). As Jongmans and Gothan (1934) had done previously, Bode denied the presence of Stephanian in the European sense in North America.

For the Appalachian basin, Bode (1958) admitted the presence of Westphalian D from the Allegheny Formation up into the lower part of the Dunkard Group. This is the same problem, because Bode equated Westphalian D with the range zone of *N. ovata,* a species that has a diachronous top occurrence. However, Bode was not unaware of the problem involved and tried to solve it by subdividing the Westphalian D (*N. ovata* range zone) into three parts, namely, a lower part in which only Westphalian taxa occurred, a middle part with a mixture of Westphalian and Stephanian species, and an upper part in which *N. ovata* was accompanied by predominantly Stephanian taxa. As interpreted here, the middle and upper parts of the *N. ovata* biozone of Bode (1958) comprise the entire Stephanian as a chronostratigraphic unit. Bode (1952) expressed surprise at the presence of a flora containing *Autunia (Callipteris)* (which was regarded as a lower Permian index) directly overlying a flora with *N. ovata.* He concluded that Permian strata followed immediately on top of Westphalian D in the Appalachian region as well as in Kansas and that an unconformity existed in both regions, a position held also by Jongmans and Gothan (1934). However, no such unconformity exists, as noted by Darrah (1969). Bode (1958) made comparisons with the East Asian Cathaysia flora and saw at least part of the differences with European floras as resulting from paleobiogeography. He mentioned that the higher part of the Dunkard Group might well correspond to the European Stephanian.

Bode (1958) summarized his impressions and conclusions in a free translation from the German as follows by R. H. Wagner:

1. The general character of the Carboniferous flora in the United States is practically the same as that in Europe. Most of the American species can be identified with European taxa, but some local endemisms occur, even though these do not create larger differences than exist between different areas within Europe.
2. The Pennsylvanian floral succession of the United States is not different from that in Europe. The lower Namurian Sudetic flora can easily be recognized in the United States, and the Westphalian floral succession is likewise similar in the United States. The same index species occur. Most widely represented is the *N. ovata* zone, which is more extensive in the United States, containing as it does an upper Westphalian D that is not represented in Europe.
3. The floral break between "Sudetic" and "Westphalian" floras in the European Namurian is not immediately apparent in the United States and may not, in fact, be present.
4. The Pennsylvanian only comprises part of the European Upper Carboniferous, viz., the Namurian and most of the Westphalian. The top of the Westphalian lies within the Dunkard, at the Washington coal bed in the lower part of the Dunkard Group. The concepts of Pennsylvanian and Upper Carboniferous are not identical.
5. The Stephanian in the European sense is not represented in the United States. The flora of the higher Dunkard, that is, that from the shales overlying the Washington coalbed, shows close affinities to the East Asian Cathaysia flora and is Permo-Carboniferous. For the time being, it is impossible to delimit the strata corresponding to the European Stephanian from this "Permo-Carboniferous."

Bode (1958) also provided a general overview of the stratigraphic relationships within the Pennsylvanian (upper Carboniferous of the United States). The same views on the upper Westphalian and Stephanian were repeated in a subsequent paper (Bode, 1960) in which the succession in West Virginia is correlated with those found in different parts of western Europe. Bode used the *N. ovata* biozone subdivided into three parts (as already proposed in his 1958 paper). Surprisingly, Bode proposed here that the base of the Stephanian be linked to the first occurrence of Stephanian floral elements, that is, more or less near the base of the *N. ovata* zone. It is interesting to note that Bode admitted that the succession in Saarland (= Saar-Lorraine Basin) contained a substantial gap below the Holz Conglomerate, which underlies the Stephanian coal-bearing intervals and which lies unconformably on Westphalian D.

The discussions of Bode's (1960) paper show opposition by Wagner (1960) and Remy (1960) to the emphasis placed on *N. ovata* (see also discussion in Darrah, 1975) and, particularly, the top of the range of this species or species complex. Remy (1960, in his discussion of Bode's 1960 paper) insisted that boundaries be linked to first occurrences rather than to extinctions. Wagner (in his 1960 discussion of Bode's 1960 paper) agreed and emphasized the diachronous top occurrences of *N. ovata*. He recalled the observations made to this effect by Darrah (1937). However, Bode maintained that the floral succession associated with *N. ovata* showed an identical development in the different parts of western Europe as well as in West Virginia. The resulting correlations (see Bode, 1960) are currently perceived to be incorrect. Bode's views were reiterated in 1964 and 1975 and were contested by the participants in the discussion (e.g., Wagner, 1964).

## W. A. BELL'S VIEWS ON THE POSSIBLE PRESENCE OF STEPHANIAN-EQUIVALENT STRATA IN MARITIME CANADA

W. A. Bell (1938, 1940, 1944, 1962) did not recognize the Stephanian in Maritime Canada. However, he reported *Sphenophyllum oblongifolium* in the Sydney Coalfield, which prompted the following statement (Bell, 1938, p. 18):

> This raises the question whether the upper subzone (of *Ptychocarpus unitus* zone) with a Stephanian precursor such as *Sphenophyllum oblongifolium* may not actually be Stephanian. The answer appears to be negative if the flora is compared with those of the Sarre Basin. The abundance of *Neuropteris ovata* and *Alethopteris friedeli* and the presence of *Validopteris integra* (= *Alethopteris valida*) indicate a parallelism with the flora of the Assise de la Houve (= Westphalian D), and the absence of *Odontopteris reichi, Pecopteris lamurensis,* and *Callipteridium* does not indicate a stage as late as that of Ottweiler (= Stephanian).

However, recent investigations in the Sydney Coalfield (Zodrow, 1985, 1990) have led to the recognition of Stephanian taxa, including *Odontopteris cantabrica* Wagner, an indicator of basal Stephanian, that is, the Cantabrian Stage. These works and the works of Zodrow and Cleal (1985) and Zodrow and Gao (1991) suggest to them that the uppermost 177 m of strata in the Sydney Coalfield, from the Lloyd Cove Seam upward, are Cantabrian in age.

## DARRAH'S (1969) VIEWS ON THE RECOGNITION OF STEPHANIAN EQUIVALENTS IN THE UNITED STATES

Darrah's reactions to the correlations proposed by Jongmans and Gothan (1934) have already been stated. It is clear that he regarded Bertrand's point of view on Stephanian and Westphalian D floral correlation as much more reasonable than the views expressed by Jongmans and Gothan (1934), which were taken to extremes by Bode (1958, 1960). Darrah (1969) presented a summary of the background on the Stephanian controversy in North America. This monograph also contains descriptions of a number of taxa from Westphalian D and Stephanian strata. It is worthwhile to reproduce here Darrah's (1969, p. 1) opening paragraph, because it is still largely relevant at the present day: "The study of Carboniferous floras of the United States has been long neglected, indeed, no monographic treatments concerning them have been published in the Twentieth Century, although many short contributions have appeared. This inattention is the more remarkable because European paleobotanists have not only continued to develop concepts and methods applicable to the American Carboniferous floras but also have from time to time called attention to the problems and discrepancies in the American literature."

Early in his investigations on the floral biostratigraphy of the Upper Pennsylvanian, Darrah recognized that two U.S. species, *Lescuropteris moorei* and *Danaeites emersonii,* which occur together with typical Stephanian species, were excellent guide fossils to the upper Conemaugh and Monongahela as well as to the Stephanian. These two species were also ". . . useful to regional correlations of the Upper Pennsylvanian in the eastern U.S." These conclusions were confirmed by Bertrand in his 1933 collecting trip in the Appalachian basin and reported by Darrah (in Darrah, 1937), who also made correlations with the Midcontinent of Kansas. Darrah's (1937) floral zonation scheme can be found in Lyons and Zodrow (Chapter 25, this volume) and will not be detailed here. Subsequently, Darrah (1969) reported *L. moorii* from the Lawrence Shales of Kansas, a discovery anticipated by Sellards (1908), according to Darrah. Cridland et al. (1963) also showed the genus *Lescuropteris* in their tabulation from the Lawrence Shale of Kansas.

In Darrah's (1969) discussion of the significance of the *Lescuropteris-Danaeites* association in the upper part of the Conemaugh and Monongahela Formations, he ascribed them to the emergence of new floral elements following the last extensive marine invasion in the Appalachian basin—the Ames sea. He supported his argument with the first occurrence of certain flora taxa in the upper part of the Conemaugh—*Walchia (Lebachia), Callipteridium pteridium,* and *Taeniopteris*—the latter genus confirmed by both White and Bertrand. Bertrand believed that the correlation between the Conemaugh and Monongahela floras and the French Stephanian at St. Étienne was "beautiful" (Darrah, 1969). *Walchia* was subsequently found near the base of the Conemaugh Formation (Lyons and Darrah, 1989), below the position of the Brush Creek Limestone and, presumably, at a stratigraphic position near the Mason Shale to the north. Darrah (1969) noted that the *Lescuropteris-Danaeites* association also occurs in the

Stranger and Lawrence Shales of the Midcontinent of Kansas, a circumstance similar to that in the upper part of the Conemaugh Formation and in the Monongahela Formation.

Darrah (1968a) also reported *L. moorei,* in association with *Alethopteris grandini* (read: *Alethopteris zeilleri*) and *Pecopteris unita,* in the Tracy and Peach Mountain coal beds of the Anthracite region of eastern Pennsylvania. This indicates Stephanian equivalents in that region (Darrah, 1968b; see also Fig. 9, this chapter), where both *Lescuropteris* and *Danaeites* are rare (Darrah, 1969). Wood et al. (1969) considered the highest beds in the Anthracite region to be Conemaugh equivalents. However, the presence of these two genera indicates that beds higher than the Conemaugh (i.e., Monongahela or higher) are present in the Anthracite region (Darrah, 1968a, 1969). The presence of Monongahela floral species in the Anthracite region is a view long ago anticipated by Lesquereux (1879–1884), who correlated the Salem coal bed with the Pittsburgh coal bed, at the base of the Monongahela Formation in the bituminous region of the Appalachian basin.

## CONCLUSIONS

The Stephanian controversy in the United States was a result of a number of misinterpretations, errors of judgment, and different concepts on the use of plant fossils in Carboniferous floral biostratigraphy. Personal differences also played a part in all of this, but W. C. Darrah, following in the steps of D. White, managed to walk a tightrope between the two principal protagonists—P. Bertrand and W. J. Jongmans. Darrah sided with Bertrand in the Stephanian controversy, which turned out to be wise, since the Bertrand-Darrah position is now perceived as the correct one. It ascribes a Stephanian age to the Conemaugh (upper part) and Monongahela Formations and Dunkard Group (lowermost part) (Gillespie and Pfefferkorn, 1979). The finer details on the position of the Westphalian D–Cantabrian (i.e., the Westphalian-Stephanian) boundary in the lower part of the Conemaugh Formation are still not precisely resolved today nor is the age of the uppermost beds in the Anthracite region of eastern Pennsylvania, which Darrah (1968a) also considered to be Stephanian.

Figure 9. *Sphenophyllum oblongifolium,* Upper Pennsylvanian, upper part of Llewellyn Formation, Southern Anthracite coalfield, Anthracite region, Pennsylvania. Photograph by and courtesy of J. R. Eggleston and C. Wnuk, U.S. Geological Survey. Scale = 1 cm.

In spite of the different opinions on the Stephanian in North America, there was broad agreement among the various investigators (e.g., D. White, Darrah, Jongmans, Bertrand, Bode) that the Allegheny Formation should be assigned to the Westphalian D, at least in part. Also, there was essential agreement on the recognition of Westphalian A, B, and C in the Appalachian basin, as noted by a comparison between Jongmans's (1937a) and Darrah's (1937) floral zonation schemes. White's (1936) European correlations (Table 2) are also similar.

## REFERENCES CITED


Alpern, B., and Liabeuf, J. J., 1967, Considérations palynologiques sur le Westphalien et le Stéphanien:—Propositions pour un parastratotype: Académie des Sciences, Paris, Comptes rendus, t. 265, p. 840–843.

Alpern, B., and Liabeuf, J. J., 1969, Palynological considerations on the Westphalian and the Stephanian: Proposition for a parastratotype: Congrès International de Stratigraphie et de Géologie du Carbonifère, 6th (Sheffield, England, 1967): Compte Rendu, v. I, p. 109–114.

Barlow, J. A., ed., 1975, Proceedings of the First I. C. White Memorial Symposium, "The Age of the Dunkard" (September 25–29, 1972): Morgantown, West Virginia Geological and Economic Survey, 352 p.

Bell, W. A., 1938, Fossil Flora of Sydney Coalfield, Nova Scotia: Geological Survey of Canada Memoir 215, 334 p., including 107 pls.

Bell, W. A., 1940, The Pictou Coalfield, Nova Scotia: Canada Department of Mines and Resources, Geological Survey Memoir 225, 141 p., pls. I–X.

Bell, W. A., 1944, Carboniferous rocks and fossil floras in northern Nova Scotia: Canada Department of Mines and Resources, Geological Survey Memoir 238, 119 p., pls. I–LXXIX.

Bell, W. A., 1962, Flora of Pennsylvanian Pictou Group of New Brunswick: Geological Survey of Canada Bulletin 87, 71 p., pls. I–LVI.

Bertrand, P., 1926, La zone à *Mixoneura* du Westphalian supérieur: Académie des Sciences, Paris, Comptes rendus, t. 183, p. 1349–1350.

Bertrand, P., 1934, Les flores houillères d'Amérique d'après les travaux de M. David White: Société Géologique du Nord, Annales, t. LVIII (1933), p. 231–254.

Bertrand, P., 1935, Nouvelles corrélations stratigraphiques entre le Carbonifère des Etats-Unis et celui de l'Europe occidentale d'après MM. Jongmans et Gothan: Société Géologique du Nord, Annales, t. 60, p. 25–38.

Bertrand, P., 1937, Tableaux des flores successives du Westphalien supérieur et du Stéphanien: Congrès pour l'avancement des études de Stratigraphie carbonifère, 2nd (Heerlen, septembre 1935): Compte Rendu, t. I, p. 67–79.

Bertrand, P., and Pruvost, P., 1937, La question du Westphalien et du Stéphanien en France: Congrès pour l'avancement des études de Stratigraphie carbonifère, 2nd (Heerlen, Septembre 1935): Compte Rendu, t. I, p. 81–83.

Bode, H., 1952, Palaeobotanische Feinstratigraphie [with discussion]: Congrès pour l'avancement des études de Stratigraphie et de Géologie du Carbonifère, 3rd (Heerlen, 25–30 juin 1951): Compte Rendu, t. I, p. 39–44.

Bode, H., 1958, Die floristische Gliederung des Oberkarbons der Vereinigten Staaten von Nordamerika: Zeitschrift deutsche geologischen Gesellschaft, Band 110, p. 217–259.

Bode, H., 1960, Die floristischen Verhältnisse an der Westfal/Stefan-Grenze im europäischen und US-amerikanischen Karbon: Congrès pour l'avancement des études de Stratigraphie et de Géologie du Carbonifére, 4th (Heerlen, Septembre 1958): Compte Rendu, t. I, p. 49–56.

Bode, H., 1964, Die Sporengliederung des Oberkarbons aus der Sicht der Megaflora: Congrès International de Stratigraphie et de Géologie du Carbonifère, 5th (Paris, 9–12 Septembre 1963): Compte Rendu, t. III, p. 1145–1150.

Bode, H., 1975, The stratigraphic position of the Dunkard, *in* Barlow, J. A., ed., Proceedings of the First I. C. White Memorial Symposium, "The Age of the Dunkard" (September 25–29, 1972): Morgantown, West Virginia Geological and Economic Survey, p. 143–154.

Bouroz, A., and Doubinger, J., 1977, Report on the Stephanian-Autunian Boundary and on the Contents of Upper Stephanian and Autunian in their Stratotypes, *in* Holub, V. M., and Wagner, R. H., eds., "Symposium on Carboniferous Stratigraphy": Prague, Geological Survey of Prague, p. 147–169.

Bouroz, A., Gras, H., and Wagner, R. H., 1970, A propos de la limite Westphalien-Stéphanien et du Stéphanien inférieur, *in* Streel, M., and Wagner, R. H., eds., "Colloque sur la stratigraphie du Carbonifère": Les Congrès et Colloques de l'Université de Liège, v. 55, p. 205–225.

Brongniart, A., 1828–1838, Histoire des végétaux fossiles: t. I, p. 1–488, pls. I–CLXVI (1828–1836); t. II, p. 1–72, pls. I–XXX (1837–1838), Paris, G. Dufour and Ed. D'Ocagne.

Broutin, J., and eight others, 1990, Le renouvellement des flores au passage Carbonifère Permien: Approches stratigraphique, biologique, sédimentologique: Académie des Sciences, Paris, Comptes rendus, t. 311, série II, p. 1563–1569.

Bunbury, C. J. F., 1847, On Fossil Plants from the Coal Formation of Cape Breton: Geological Society of London, Quarterly Journal, t. III, p. 423–238, pls. XXI–XXIV.

Clayton, G., Coquel, R., Doubinger, J., Gueinn, K. J., Loboziak, S., Owens, B., and Streel, M., 1977, Carboniferous Miospores of Western Europe: Illustration and zonation: Mededelingen Rijks Geologische Dienst, v. 29, p. 1–19, pls. 1–25, charts.

Cleal, C. J., Shute, C. H., and Zodrow, E. L., 1990, A revised taxonomy for late Palaeozoic neuropterid foliage: Taxon, v. 39, p.486–492.

Cridland, A. A., Morris, J. E., and Baxter, R. W., 1963, The Pennsylvanian plants of Kansas and their stratigraphic significance: Palaeontographica, Abt. B, Band 112, p. 58–92, pls. 17–24.

Darrah, W. C., 1935, Some Late Carboniferous correlations in the Appalachian province [abs.]: Geological Society of America, Proceedings for 1934, p. 443.

Darrah, W. C., 1937, American Carboniferous floras: Congrès pour l'avancement des études de Stratigraphie carbonifère, 2nd (Heerlen, septembre 1935): Compte Rendu, t. I, p. 111–129.

Darrah, W. C., 1968a, The age of the highest coals of the Southern Anthracite field: Pennsylvania Academy of Science, Proceedings, v. 43, p. 14.

Darrah, W. C., 1968b, The pteridosperm genus *Lescuropteris:* Characteristics, distribution, and significance: American Journal of Botany, v. 55, p. 725.

Darrah, W. C., 1969, A critical review of the Upper Pennsylvanian floras of eastern United States with notes on the Mazon Creek flora of Illinois: Gettysburg, Pennsylvania, privately printed, 220 p., 80 pls.

Darrah, W. C., 1975, Historical aspects of the Permian flora of Fontaine and White, *in* Barlow, J. A., ed., Proceedings of the First I. C. White Memorial Symposium, "The Age of the Dunkard" (September 25–29, 1972): Morgantown, West Virginia Geological and Economic Survey, p. 81–101.

Darrah, W. C., and Bertrand, P., 1934, Observations sur les flores houillères de Pennsylvanie: Société Géologique du Nord, Annales, t. 58, p. 211–224.

Dawson, J. W., 1871, The Fossil Plants of the Devonian and Upper Silurian Formations of Canada: Geological Survey of Canada, 92 p., 8 suppl. pages, pls. I–XX.

Doubinger, J., and Heyler, D., 1976, Nouveaux fossiles dans le Permien français: Société Géologique de France, Bulletin, v. 17, p. 1176–1180.

Elias, M. K., 1936, Late Paleozoic plants of the Midcontinent region as indicators of time and of environment: International Geological Congress, 16th (Washington, D.C., July 1933): Report, v. 1, p. 691–700.

Fontaine, W. F., and White, I. C., 1880, The Permian or Upper Carboniferous flora of West Virginia: 2nd Geological Survey of Pennsylvania, v. PP, 143 p., pls. I–XXXVIII.

Germer, R., Kneuper, G. K., and Wagner, R. H., 1968, Zur Westfal/Stefan-Grenze und zur Frage der asturischen Faltungsphase im Saarbrücker Hauptsattel: Geologica et Palaeontologica, v. 2, p. 59–71, Tafn. 1–2.

Gillespie, W. H., and Pfefferkorn, H. W., 1979, Distribution of commonly occurring plant megafossils in the proposed Pennsylvanian Stratotype, *in* Englund, K. J., Arndt, H. H., and Henry, T. W., eds., "Proposed Pennsylvanian System Stratotype, Virginia and West Virginia": American Geological Institute, AGI Selected Guidebook Series No. 1, p. 87–96 (including pls. 1–3).

Gillespie, W. H., Clendening, J. A., and Pfefferkorn, H. W., 1978, Plant fossils of West Virginia and adjacent areas: Morgantown, West Virginia Geological and Economic Survey, 172 p.

Gothan, W., and Gimm, O., 1930, Neuere Beobachtungen und Betrachtungen über die Flora des Rotliegenden von Thüringen. Arbeiten aus dem Institut für Paläobotanik und Petrographie der Brennsteine: Preussische Geologische Landesanstalt, Berlin, Bd. 2, p. 39–74, Taf. 9.

Handlirsch, A., 1906, Revision of American Paleozoic insects: U.S. National Museum Proceedings, v. 29, p. 661–820.

Haught, O. L., 1934, Characteristics of the flora of the Greene formation: West Virginia Academy of Sciences, Proceedings, v. 7, p. 83–87.

Horowitz, A. S., 1979, The Mazon Creek Flora: Review of Research and Bibliography: "Mazon Creek Fossils": New York, Academic Press, p. 143–158.

Jongmans, W. J., 1937a, Contribution to a comparison between the Carboniferous floras of the United States and of Western Europe: Congrès pour l'avancement des études de Stratigraphie carbonifère, 2nd (Heerlen, Septembre 1935): Compte Rendu, t. I, p. 363–387.

Jongmans, W. J., 1937b, Some remarks on *Neuropteris ovata* in the American Carboniferous: Congrès pour l'avancement des études de Stratigraphie carbonifère, 2nd (Heerlen, Septembre 1935): Compte Rendu, t. I, p. 417–422, pls. 37–42.

Jongmans, W. J., 1952, Some problems on Carboniferous stratigraphy: Congrès pour l'avancement des études de Stratigraphie et de Geólogie du Carbonifère, 3rd (Heerlen, 25–30 juin 1951): Compte Rendu, t. I, p. 295–306.

Jongmans, W. J., and Gothan, W., 1934, Florenfolge und vergleichende Stratigrafie des Karbons der östlichen Staaten Nord-Amerika's, Vergleich mit West-Europa: Geologisch Bureau Heerlen, Jaarverslag over 1993, p. 17–44.

Jongmans, W. J., and Pruvost, P., 1950, Les subdivisions du Carbonifère continental: Société Géologique de France, Bulletin, 5th série, t. XX, p. 335–344.

Jongmans, W. J., Gothan, W., and Darrah, W. C., 1937, Comparison of the floral succession in the Carboniferous of West Virginia: Congrès pour l'avancement des études de Stratigraphie carbonifère, 2nd (Heerlen, Septembre 1935): Compte Rendu, t. I, p. 393–415.

Kerp, J. H. F., and Haubold, H., 1988, Towards a Reclassification of the West- and Central-European Species of the Form-genus *Callipteris* BRONGNIART 1849: Zeitschrift der geologischen Wissenschaften, Berlin, Band 16, Heft 9, p. 865–876.

Lesquereux, L., 1879–1884, Description of the Coal Flora of the Carboniferous Formation in Pennsylvania and throughout the United States: 2nd Geological Survey of Pennsylvania, Report of Progress, v. 1 (1880), 354 p.; v. 2 (1880), 694 p.; Atlas (1879) pls. A–B, pls. I–LXXXV; v. 3 (1884), p. 695–977, pls. LXXXVIII–CXI.

Lyons, P. C., 1984, Carboniferous Megafloral Zonation of New England: Congrès International de Stratigraphie et de Géologie du Carbonifère, 9th (Washington and Champaign-Urbana, May 17–26, 1979): Compte Rendu, v. 2, p. 503–514.

Lyons, P. C., and Darrah, W. C., 1989, Earliest conifers of North America: Upland and/or paleoclimatic indicators?: Palaios, v. 4, p. 480–486.

Lyons, P. C., Tiffney, B., and Cameron, B., 1976, Early Pennsylvanian Age of the Norfolk Basin, Southeastern Massachusetts, Based on Plant Megafossils, *in* Lyons, P. C., and Brownlow, A. H., eds., Studies in New England Geology: Geological Society of America Memoir 146, p. 181–197.

McComas, M. A., 1988, Upper Pennsylvanian compression floras of the 7-11 Mine, Columbiana County, Northeastern Ohio: Ohio Journal of Science, v. 88, p. 48–52.

Moore, R. C. (chairman), and 26 others, 1944, Correlation of Pennsylvanian formations of North America: Geological Society of America Bulletin, v. 55, p. 657–706.

Munier-Chalmas, E., and Lapparent, A. de, 1893, Note sur la Nomenclature des Terrains sédimentaires: Société Géologique de France, Bulletin, 3e série, t. XXI, p. 438–493.

Peppers, R. A., 1984, Comparison of Miospore Assemblages in the Pennsylvanian System of the Illinois Basin with those in the Upper Carboniferous of Western Europe: Congrès International de Stratigraphie et de Géologie du Carbonifère, 9th (Washington and Champaign-Urbana, May 17–26, 1979): Compte Rendu, v. 2, p. 483–502.

Phillips, T. L., and Peppers, R. A., 1984, Changing patterns of Pennsylvanian coal-swamp vegetation and implications of climatic control on coal occurrence: International Journal of Coal Geology, v. 3, p. 205–255.

Pruvost, P., 1934, Bassin houiller de la Sarre et de la Lorraine. III: Description géologique: Études Gîtes Minéraux de la France, 175 p., carte et coupes.

Pruvost, P., and Corsin, P., 1949, Westphalien supérieur et Stéphanien inférieur: Académie des Sciences, Paris, Comptes rendus, v. 229, p. 1284–1286.

Remy, W., 1960, Discussion of paper "Die floristischen Verhältnisse an der Westfal/Stefan-Grenze im europäischen und US-amerikanischen Karbon" by H. Bode: Congrès pour l'avancement des études de Stratigraphie et de Géologie du Carbonifère, 4th (Heerlen, septembre 1958): Compte Rendu, t. I, p. 56.

Remy, W., 1975, The floral changes at the Carboniferous-Permian boundary in Europe and North America, *in* Barlow, J. A., ed., Proceedings of the First I. C. White Memorial Symposium, "The Age of the Dunkard" (September 25–29, 1972): Morgantown, West Virginia Geological and Economic Survey, p. 305–352.

Remy, W., and Remy, R., 1975a, *Lesleya weilerbachensis* n. sp. aus dem höheren Westfal C des Saarkarbons: Argumenta Palaeobotanica, v. 4, p. 1–11, Taf. 1.

Remy, W., and Remy, R., 1975b, *Lescuropteris* (al. *Odontopteris*) *genuina* Gr. Eury sp. emend. et nov. comb. (Stefan) und Zwischenfiedern bei *Odontopteris* Brongniart: Argumenta Palaeobotanica, v. 4, p. 93–100, Taf. 13.

Remy, W., Remy, R., Leisman, G. A., and Hass, H., 1980, Der Nachweis von *Callipteris flabellifera* (Weiss, 1879) Zeiller 1898 in Kansas, U.S.A.: Argumenta Palaeobotanica, v. 6, p. 1–36, Tafn., 1–6.

Romer, A. S., 1935, Early history of Texas red-beds, vertebrates: Geological Society of America Bulletin, v. 46, p. 1597–1658.

Schuchert, C., 1932, Permian floral provinces and their distribution: American Journal of Science, 5th ser., v. 24, p. 405–413.

Sellards, E. H., 1908, Fossil plants of the Upper Paleozoic of Kansas: Kansas University, Geological Survey, Report, v. 9, p. 386–480, pls. XLIV–LXIX.

Stopes, M., 1914, The "Fern Ledges" Carboniferous flora of St. John, New Brunswick: Canadian Geological Survey Memoir 41, 142 p., pls. I–XXV.

Wagner, R. H., 1958, On *Sphenopteris (Saaropteris?) dimorpha* (Lesq.) nov. comb.: Palaeontographica, Abt. B, Band 104, p. 105–114, pl. 15.

Wagner, R. H., 1960, Discussion of paper "Die floristischen Verhältnisse an der Westfal/Stefan-Grenze im europäischen und US-amerikanischen Karbon" by H. Bode: Congrès pour l'avancement des études de Stratigraphie et de Géologie du Carbonifère, 4th (Heerlen, septembre 1958): Compte Rendu, t. I, p. 57.

Wagner, R. H., 1964, Discussion of paper "Die Sporengliederung des Oberkarbons aus der Sicht der Megaflora" by H. Bode: Congrès International de Stratigraphie et de Géologie du Carbonifère, 5th (Paris, 9–12 Septembre 1963): Compte Rendu, t. III, p. 1150.

Wagner, R. H., 1965, Palaeobotanical Dating of Upper Carboniferous Folding Phases in N.W. Spain: Instituto Geológico y Minero de España, Memorias, t. LXVI, 169 p., 77 pls.

Wagner, R. H., 1968, Upper Westphalian and Stephanian species of Alethopteris from Europe, Asia Minor and North America: Rijks Geologische Dienst, Mededelingen, Serie C, III-1-No. 6, 188 p., 64 pls.

Wagner, R. H., 1969, Proposal for the recognition of a new "Cantabrian" Stage at the base of the Stephanian Series: Congrès International de Stratigraphie et de Géologie du Carbonifère, 6th (Sheffield, England, 11–16 September 1967): Compte Rendu, v. I, p. 139–150.

Wagner, R. H., 1984, Megafloral Zones of the Carboniferous: Congrès International de Stratigraphie et de Géologie du Carbonifère, 9th (Washington and Champaign-Urbana, May 17–26, 1979): Compte Rendu, v. 2, p. 109–134.

Wagner, R. H., and Winkler Prins, C. F., 1985, The Cantabrian and Barruelian stratotypes: A summary of basin development and biostratigraphic information, *in* Lemos de Sousa, M. J., and Wagner, R. H., eds., "Papers on the Carboniferous of the Iberian Peninsula": Faculdade de Ciências, Universidade do Porto, Anais, supplement to v. 64 (for 1983), p. 359-410.

Wagner, R. H., Park, R. K., Winkler Prins, C. F., and Lys, M., 1977, The Post-Leonian Basin in Palencia: A Report on the Stratotype of the Cantabrian Stage, *in* Holub, V. M., and Wagner, R. H., eds., "Symposium on Carboniferous Stratigraphy": Prague, Geological Survey of Prague, p. 89–146.

White, D., 1900, The stratigraphic succession of the fossil floras of the Pottsville Formation in the southern Anthracite coal field, Pennsylvania: U.S. Geological Survey, Annual Report 20, pt. 2, p. 749–918, pls. CLXXX–CXCIII.

White, D., 1903, Summary of fossil plants recorded from the Upper Carboniferous and Permian Formations in Kansas: U.S. Geological Survey Bulletin 211, p. 85–117.

White, D., 1904, Permian elements in the Dunkard flora: Geological Society of America Bulletin, v. 14, p. 538–542.

White, D., 1924, Permian of western America from the paleobotanical viewpoint: Pan Pacific Scientific Congress (Australia, 1923), Proceedings, v. 2, p. 1050–1077.

White, D., 1929, Flora of the Hermit shale, Grand Canyon, Arizona: Carnegie Institution of Washington Publication 405, 118 p., 51 pls.

White, D., 1936 (posthumous), Some features of the early Permian flora of America: 16th International Geological Congress (Washington, D.C., July 1933), Report, v. 1, p. 679–689.

Wood, G. H., Jr., Trexler, J. P., and Kehn, T. M., 1969, Geology of the west-central part of the southern Anthracite field and adjoining areas, Pennsylvania: U.S. Geological Survey Professional Paper 602, 105 p.

Zodrow, E. L., 1985, *Odontopteris* Brongniart in the Upper Carboniferous of Canada: Palaeontographica, Abt. B, Band 196, p. 79–110, pls. 1–3.

Zodrow, E. L., 1990, Revision and emendation of *Pecopteris arborescens* group, Permo-Carboniferous: Palaeontographica, Abt. B, Band 217, p. 1–49, pls. 1–8.

Zodrow, E. L., and Cleal, C. J., 1985, Phyto- and chronostratigraphical correlations between the Late Pennsylvanian Morien Group (Sydney, Nova Scotia) and the Silesian Pennant Measures (south Wales): Canadian Journal of Earth Sciences, v. 22, p. 1465–1473.

Zodrow, E. L., and Gao, Z., 1991, *Leeites oblongifolis* nov. gen. et sp. (sphenophyllaean, Carboniferous), Sydney Coalfield, Nova Scotia, Canada: Palaeontographica, Abt. B, Band 223, p. 61–80, pls. 1–8.

Manuscript Accepted by the Society July 6, 1994

Printed in U.S.A.

Geological Society of America
Memoir 185
1995

# *Early and mid-twentieth century coal-ball studies in North America*

**Tom L. Phillips**
*Department of Plant Biology, 265 Morrill Hall, University of Illinois, 505 S. Goodwin Avenue, Urbana, Illinois 61801*
**Aureal T. Cross**
*Department of Geological Sciences, 206 Natural Science Building, Michigan State University, East Lansing, Michigan 48824*

## ABSTRACT

**Coal balls are concretions containing structurally preserved, coalified peat in coal beds; the plant tissues in coal balls are permineralized by calcite, dolomite, siderite, pyrite, or silica. The earliest known North American coal-ball studies, based on pyritic coal balls from Iowa, were carried out in the late 1890s by W. S. Gresley, a mining engineer from England. He utilized reflected light on polished surfaces to observe the anatomy of coal-ball plant constituents. Sustained development of studies began in the early 1920s at the University of Chicago as a result of calcareous coal-ball discoveries in the Illinois basin, reported by Adolph C. Noé. He was the leading advocate for the discovery and study of American coal balls, receiving assistance from state geological surveys and the coal-mining industry. Interests at the University of Chicago led to the establishment of a chair in paleobotany for Noé and ensuing coal-ball research by his students, J. Hobart Hoskins, Fredda D. Reed, Harriet V. Krick Bartoo, Roy Graham, and others. At the time there were very few paleobotanists in North America. Most pioneers in American coal-ball studies during the 1930s were self taught in paleobotanical research. Consequently, quite diverse perspectives of the significance and use of coal balls were represented. In the 1930s acid etching and the liquid-peel (Darrah solution) technique generally replaced thin sectioning; larger numbers of coal balls were discovered, partly as a result of increased open-pit coal mining. At the Illinois State Geological Survey such discoveries attracted James M. Schopf, who stressed the importance of coal balls as a paleobotanical index for the constitution of coal, a method of identifying the botanical sources of dispersed spores and pollen, as well as their use in a stratigraphic approach to monographic studies of Carboniferous plants. This approach is mirrored in the early monographic works of J. Hobart Hoskins and Aureal T. Cross at the University of Cincinnati. William C. Darrah at Harvard University was particularly interested in the paleofloristic and biostratigraphic information to be gained from coal-ball studies. He instituted mass peel preparations as well as splitting coal-ball specimens to reveal the morphology of associated foliage taxa. At Washington University in St. Louis, Henry N. Andrews emphasized anatomy and development in the evolution of the pteridophytes and early seed plants. American coal-ball studies in the upper Middle and Upper Pennsylvanian expanded by the early 1940s with collections from Iowa, Missouri, Kansas, Texas, Illinois, Indiana, and western Kentucky. Following World War II, research and training continued at Washington University with H. N. Andrews and developed anew at the University of Illinois with**

Phillips, T. L., and Cross, A. T., 1995, Early and mid-twentieth century coal-ball studies in North America, *in* Lyons, P. C., Morey, E. D., and Wagner, R. H., eds., Historical Perspective of Early Twentieth Century Carboniferous Paleobotany in North America (W. C. Darrah volume): Boulder, Colorado, Geological Society of America Memoir 185.

**Wilson N. Stewart. This research led to new generations of paleobotanists during the following decades. In 1949, James M. Schopf established the U.S. Geological Survey's Coal Geology Laboratory in Columbus, Ohio, and continued his influential role in coal-ball studies, encouraging and directly aiding such studies in the United States. In the 1950s the rapid-peel technique with preformed sheets of cellulose acetate expedited studies. The National Science Foundation became a significant source of support for such research, and paleobotany was more broadly represented in universities. Even into the 1960s, however, no one would have predicted how abundant coal-ball deposits in North America would prove to be, as the first occurrences were reported from the Appalachians and New Brunswick, Canada. In this chapter, these developments are described.**

**An epilogue is also provided, which outlines major initiatives and contributions of recent decades that bridge early- and middle-twentieth-century development of coal-ball studies in North America with those of the century to come.**

## INTRODUCTION

Sustained North American coal-ball studies encompass most of the twentieth century and differ significantly from the historical developments in western Europe upon which they were founded. There were relatively few paleobotanists in the 1920s for the vastness of North America, and coal balls were generally not known to have been discovered until the report by Adolph C. Noé (1923a) at the University of Chicago. In the 1920s and 1930s, when contributions by Noé's students were attracting interests in American coal-ball studies, the zenith of such studies in western Europe had already passed. There was an extensive literature on the coal-ball plants from the lower Westphalian, especially from England, as well as on the silicified deposits in the upper Stephanian A and Autunian of France. By the 1930s it was clear that the American coal-ball record was stratigraphically complementary to that of western Europe, principally from the upper Middle Pennsylvanian (Westphalian D) and lower Upper Pennsylvanian (Stephanian A). Five major groups of vascular plants were represented in the coal-ball deposits: lycopsids, ferns, sphenopsids, seed ferns, and cordaites. As research activity in American coal-ball studies grew, contributions from western Europe declined, but there was much interest in the growing body of paleobotanical information on the Euramerican tropical belt. In North America, coal-ball studies developed together with coal palynology and many aspects of coal geology that attracted a number of pioneers with strong backgrounds in botany and drew upon many botanically trained students.

This historical account largely deals with North American coal-ball studies during the 1920s through the 1960s, a period that was marked by the expansion of participants and contributions. The individuals and circumstances are emphasized as much as the consequences. The actual numbers of students who pursued such studies have been relatively small compared to the many people who helped them over the years—miners, field geologists, ardent collectors, and generous friends and colleagues. Our intent is to highlight the circumstances in which individuals and small groups worked and interacted and to show how field discoveries, laboratory techniques, mining operations, research funding, and other identifiable events and changes led to the research and teaching expansions related to coal-ball studies. In focusing mostly on pioneering efforts, it is noted that this history is an interwoven part of upper Carboniferous coal geology and fossil-plant exploration. Most contributors, and especially the American pioneers in the field, had numerous other research activities and, indeed, made significant contributions in many areas not mentioned herein (see biographical sketches of A. C. Noé, J. M. Schopf, and H. N. Andrews, Jr., this volume). Also, many contributors have not been specifically mentioned, as a result of our emphasis on the principal lines of development.

The broad spectrum of paleobotanically oriented history has been captured from personal perspectives in *The Fossil Hunters* (Andrews, 1980). Review articles of American coal-ball studies by Darrah (1941a) and by Andrews (1951) provide comparative viewpoints of the potentials, problems, and stages of progress in relationship to the research hiatus of World War II and in relationship to establishment of the National Science Foundation in 1950. The funding for coal-ball studies was minimal to almost nonexistent through most of the early pioneering days. The activities of state geological surveys and the generous cooperation of mining companies were and still are crucial. Cooperative efforts of such dedicated and generous individuals as Adolph C. Noé, Frederick O. Thompson, and James M. Schopf demonstrated the varied ways in which available sponsorship was translated into reconnoitering, discovery, and research collections of coal balls. *Development of Paleobotany in the Illinois Basin* (Phillips et al., 1973) dealt with some of these aspects and is a reference source for additional historical accounts. We have quoted extensively from this work because of the personal insights provided by correspondence from many of the paleobotanists.

## COAL BALLS

The current concept of coal balls with diverse mineralogic composition and varied shapes and sizes as well as plant, animal, and/or clastic inclusions (Mamay and Yochelson,

1962) has evolved over the last century. Coal balls are concretions containing structurally preserved, coalified peat. The plant tissues have been coalified (Hatcher et al., 1982; Lyons et al., 1984) after permineralization by calcite, dolomite, siderite, pyrite, or silica. Mamay and Yochelson (1962) expanded the coal-ball concept to include those with clastic inclusions: mixed coal balls contain both plant and invertebrate shells, and faunal coal balls contain animal fossils only. Coal balls mainly occur in upper Carboniferous and Permian coals, constituting the most abundant source of anatomically well preserved vascular plants in the fossil record. Essential aspects of and interests in coal balls were stated in the first reported discovery from the Westphalian A of Lancashire, England (Hooker and Binney, 1855, p. 149–150):

> The origin of these nodules may probably be ascribed to the presence of mineral matter, held in solution in water and precipitated upon, or aggregated around certain centres, in the mass of vegetable matter now for the most part turned into coal. . . . The formation of nodules of one mineral in a matrix of another, is one that involves many considerations, and [we] shall therefore confine ourselves to remarking, that the appearances are of these nodules being sealed masses of fossil vegetable remains, and as such are probably a fair sample of the vegetation that has produced the surrounding coal.
>
> From the above observations, it would appear, that the fossils in question are possessed of a double interest; the geologist recognizes in them an association of vegetables that certainly prevailed throughout the epoch of the Coal formation, and in all probability contributed largely, if not almost exclusively, to the formation of that mineral; whilst the botanist detects in them characters of the greatest value as throwing light upon the affinities of the Flora of the period.

The English term *coal ball* was appropriate for some spheroidal concretions found in the coal beds, and many hand specimens are somewhat rounded and lenticular, with a crustal covering of compacted coal. However, there is no characteristic shape of coal balls. The classic study by Stopes and Watson (1909)—"On the present distribution and origin of the calcareous concretions in coal seams, known as coal balls"—was influential in stabilizing the name for such concretions as well as providing substantive insight into their origins. Stopes and Watson (1909, p. 212–213) concluded: "(*a*) The coal balls were formed in the position in which they are now found (and probably also the coal itself was likewise formed *in situ*); (*b*) the sea water was fundamentally important during the coal ball formation in acting both as a temporary preservative and as the source of the calcium and magnesium carbonates required for petrifaction; (*c*) the plants in the roof nodules and shale impressions above the seam represent a different flora from that found in the coal." In their text they clearly indicated that the carbonates could be derived from the degradation of the peat and that the calcium and magnesium sulfates that were critical to the permineralization process were derived from seawater. The process resulted either in the evolution of $H_2S$ or pyrite formation.

Because the most frequently asked questions about coal balls relate to their origins and distribution, these topics are dealt with here out of the general historical sequence of the plant studies of coal balls. The treatise by Stopes and Watson (1909) was so lucid and well documented as to have almost closed the questions on calcareous and pyritic coal-ball origins. Ensuing studies both in Europe and then in the United States determined that where coal balls did occur, there was evidence in the roof rocks of the coal beds of marine conditions during burial of the peat beds. Such evidence was so widespread in the Interior Coal Province of the United States as not to be particularly helpful in prospecting for coal balls. The first American challenge to the marine origins was by José Maria Feliciano (1924), a student of Noé, who emphasized "land waters" as sources of the calcium and carbonates, citing an occurrence of coal balls in Indiana that lacked evidence of marine fossils in its immediate roof. There have been numerous contributions on coal-ball origins, based on American deposits (Kindle, 1934; Evans and Amos, 1961; Cross, 1969; Perkins, 1976; McCullough, 1977; Anderson et al., 1980; Rao, 1985). A general review of coal balls is in Scott and Rex (1985).

Two additional studies stand out as comparable works to that of Stopes and Watson (1909), namely that of Mamay and Yochelson (1962) at the U.S. Geological Survey and that of DeMaris et al. (1983) at the Illinois State Geological Survey. In a real sense these encompass extremes in the range of environments of coal-ball origins, like well-documented bookends.

Following the report of their preliminary analyses of direct evidence of marine influences on some coal-ball origins, Sergius H. Mamay and Ellis L. Yochelson (1953, 1962, p. 193) followed up with a comprehensive faunal survey and chemical analyses of coal balls, including the first for trace elements. They concluded: "The presence of marine animals within coal balls is evidence of transportation of material from a marine environment to a non-marine environment. Though all coal balls have heretofore been considered concretionary in origin, mixed [plant and animal] and faunal coal balls are, at least in part, of clastic origin. The swirled texture of some homogeneous-mixed coal balls is further evidence of transport."

They also suggested that where there was evidence of repeated storm-driven incursions into the peat deposits (Calhoun Coal at Berryville, Illinois), "these temporary marine inundations may have provided a source of calcium carbonate for the formation of some normal [plant-containing] coal balls." Stable isotopic analyses of carbon from some coal-ball carbonates by Weber and Keith (1962, p. 901) were consistent with clastic marine carbonates being introduced into the paludal environment of peat deposition.

The paleobotanist in this research project was S. H. Mamay (Fig. 1), who began his morphological and systematic studies of coal-ball plants with H. N. Andrews at Washington University, notably on the fructifications and anatomy of the ferns (Mamay, 1950, 1952, 1957; Mamay and Andrews, 1950). At the U.S. Geological Survey (USGS) during the early 1950s, Mamay shifted emphasis to upper Paleozoic biostrati-

Figure 1. Sergius H. Mamay, Paleozoic paleobotanist and coauthor of the 1962 paper "Occurrence and significance of marine animal remains in American coal balls." Photograph provided by H. N. Andrews.

graphic studies (Read and Mamay, 1964) and Permian paleofloristic studies, including the Paleozoic origin of the cycads (Mamay, 1976).

Of the origins of the faunal coal-ball project, Mamay (letter to T. L. Phillips, June 24, 1992) shared this:

That happened in 1952, when I made my first official collecting trip with the USGS. I was all fired up about making a sensational name for myself with coal-ball studies and headed out to Kansas, all by myself. I sacked up a couple tons of West Mineral stuff, including a few coal balls that had sea-shells showing on fresh surfaces. In my almost utter naiveté, I thought their presence in a coal ball was curious, but didn't see the ecological implications at the time. After cutting enough coal balls to see that invertebrates were fairly common, I started showing them to the invertebrate specialists in the Museum, and before long here came Ellis Yochelson to my office, sort of demanding to see the snail shells I had found in coal balls. He had been thinking that maybe they were land snails, which in itself could have a very unusual discovery. However, the shells turned out to be marine forms, so that the discovery was indeed unusual, but in a different way, with marine animals included in terrestrial sediments. It only took a few hours (maybe minutes?) to see that a marine fauna was well-represented in those coal balls, and a collaborative effort was born! Ellis and I planned a field trip together, which materialized in 1955, when we toured the southwest—some of the time with Charles B. Read [see Chapter 19, this volume]—looking for Permian plants and snails, and on the way back we hunted for marine coal balls in Kansas and Illinois and collected what would be the meat for our Professional Paper. That trip was a real education for me, and only then did I start thinking seriously about paleoecology. The acid residues were a real thrill to pick. I did most of the picking and preliminary identifications—I never knew what to expect, but soon learned to recognize conodonts, brachiopods, fish scales, soot balls (!), and what have you.

In addition to the clearly established cases of coal-ball origins related to influences of brackish to marine water in storm incursions or in subsequent burial of peat deposits, there are observations of both extremely massive and extensive occurrences of calcareous coal balls without any obvious depositional relationship to marine deposits in the Permian of Australia and Siberia as well as within the Herrin coal bed of the Illinois basin. The Herrin coal bed contains the largest-known occurrences of coal balls worldwide. In some mines coal-ball material extends 4 m in thickness (from seat earth to coal-bed roof) as aggregations of concretions to essentially a "fossiliferous limestone wall." These deposits may extend laterally hundreds of meters. Such occurrences have caused considerable doubt about how marine influences relate to such origins. The study by Philip J. DeMaris and his colleagues at the Illinois State Geological Survey, Robert A. Bauer, Richard A. Cahill, and Heinz H. Damberger (DeMaris et al., 1983), is the most extensive underground study of coal-ball distribution since Kukuk's (1909) analysis in the Ruhr region. The 1983 work also had the advantage of team efforts in detailed geochemical analyses and reconstructional aspects of the depositional environments. This was the most concerted effort to date in coal-ball origins, stressing the causes and predictability of such deposits, and was closely allied to studies of the coal geology (Johnson, 1979) as well as paleobotany (DiMichele and Phillips, 1988). Despite limited availability of the DeMaris et al. study, published as a contract report by the Illinois State Geological Survey, it stands as the most significant contribution on coal-ball origins since that of Stopes and Watson (1909).

The essential findings of this Herrin coal-based study in a mine area within 11.5 km of the Walshville paleochannel were: (1) the coal-ball carbonates in the thickest sequences had average $\delta C^{13}$ values of –23.9‰ (apart from secondary calcite fillings), consistent with $CO_2$ origins from plant degradation—nonmarine; (2) the coal-ball deposits occurred in pods with a semilinear distribution oriented diagonally from the Walshville paleochannel; (3) roof deposits indicated erosional removal of the original Energy Shale (over-the-bank flood deposits of the Walshville as the coal swamp was drowned by rising sea level); and (4) some coal balls were deposited with marine fossils before final burial by the marine Anna Shale (Fig. 2).

The significant hypothesis offered by this study suggests that $CO_2$ retention in rapidly buried peat deposits fed carbonate precipitation along semilinear tracks where degassing occurred. Degassing permitted rapid changes in the pH from acidic to just above neutral, with supersaturation bringing

Figure 2. Multiple zones of coal balls (light gray lenticular pattern) exposed in side of support pillar in Herrin coal bed in underground mine No. 24 of the Old Ben Coal Corporation near Benton, Illinois, where the DeMaris et al. (1983) studies were carried out. Photograph provided by P. J. DeMaris, courtesy of the Illinois State Geological Survey.

Figure 3. An aggregate of calcareous coal balls from the Herrin coal bed, after exposure to weathering for years. The uppermost layers are sheetlike in shape; the middle zone consists of variously rounded to irregular shapes. Scale on right is one yard. Photograph taken by Tom Phillips.

about precipitation. In the study described, it was apparently the removal of the original roof deposit permitting $CO_2$ degassing rather than the subsequent marine roof and associated percolating burial influences that had triggered the formation of coal balls.

Philip J. DeMaris (letter to T. L. Phillips, September 16, 1992) provided the following observations about the purpose of the "Old Ben Mine 24" studies of coal balls in the Herrin coal bed:

> The Old Ben No. 24 longwall study was initiated by Harold Gluskoter and funded by the U.S. Bureau of Mines beginning in early 1976. As mine development in the study area began serious mining problems with coal balls caused a broadening of the research goals to include prediction of coal-ball areas (initially called "pods") [see Fig. 3]. By the winter of 1976–77 Bob Bauer and I had refined our mine mapping methods and realized that local erosion was key to understanding the ultimate distribution of roof units, and also that the most concentrated areas of coal balls were spatially related to the eroded areas. It was not until chemical studies were completed years later that a coherent link was established between the erosion and the permineralization that occurred within the exhumed peat. Multiple collections of stratigraphically-zoned coal balls were made in cooperation with Tom L. Phillips and his students, and we supplemented these with random collections from other areas. The project survived both an unusually long miners' strike and the unexpected closure of the mine, which caused sampling problems. Eventually, a contract report in two parts, several papers and portions of several theses were derived from work among nearly a dozen scientists involved in various ways in this project.

## EARLY COAL-BALL DISCOVERIES AND STUDIES

It was not until the late 1940s that American paleobotanists became generally aware of the earliest American coal-ball discoveries, dating back to those cited in the 1890s in Iowa Geological Survey publications (see Andrews, 1951, p. 439). What called or might have called attention to the reports was the Stopes and Watson (1909, p. 185, 191, and Fig. 15, Plate 17) treatise, which, among other critical observations, pointed out that coal-ball occurrences were not restricted to what Schopf (1941) later designated as the "Great Coal-Ball Horizon" (lower Westphalian A) of western Europe. In doing so, Stopes and Watson (1909, p. 191) not only referred to higher and lower stratigraphic coal-ball occurrences in the Westphalian of Europe but listed "other structures which, though not typical *balls,* are parallel in their geological character. . . ." Among those in the table (p. 191) was listed "Iowa, U.S.A., Plate 17, Photo 5," with the only added clue given in the photo acknowledgment (Stopes and Watson, 1909, p. 217): "Specimen belonging to Mr. Gresley, of Derby" (=Derbyshire, England).

W. S. Gresley was a fellow of the Geological Society (of London), with an intense interest in coal origins. An early paper by Gresley (1887) on Leicestershire and South Derbyshire coalfields conveyed, "My principal object in this paper is to bring forward evidence in opposition to the view now generally accepted that coal-seams were formed from vegetation which grew on the spot" (p. 671). Gresley also noted, "During an extensive experience in the Midland district in connection not only with coal-mining, but also with the working of the underbeds, the fireclays, both underground and in opencast workings, I have had unusual opportunities of studying the relationship of the coal-seams to the underbeds, their fossil contents, etc" (1887, p. 671).

In Gresley's (1893) paper on the "Mammoth Seam" in Pennsylvania, his address was given as Erie, Pennsylvania, as it was in the four papers in which he made observations based on what were undoubtedly coal balls. It is interesting to note that he added to his authorship title of F.G.S. a F.G.S.A. (fellow of the Geological Society of America) and, curiously, in his last coal-ball paper (Gresley, 1901) omitted both.

The first of Gresley's (1899a) papers, based clearly on coal balls, stated about the concretionary masses (p. 77): "My most instructive specimens have come from coal in Iowa, and out of these, by grinding, wetting and polishing, the best plant tissues in them have been brought into view. Binney [Hooker and Binney, 1855] it will be remembered, nearly half a century ago in England, sliced somewhat similar, but much more calcareous masses from coal; and the inferences then drawn were that coal was largely composed of plant remains the same as those seen in the nodules." His interests (Gresley, 1899a, p. 69) were stated this way: "The subject of the origin and formation of coal will always be one of much interest so long as there is anything *new* to be said or *left to be found out* concerning it. While all investigators seem to be agreed that coal (all varieties of coal) is of vegetable origin, they are by no means of one mind as to at least two important points, namely: (1) *the true character of the different kinds of plants involved,* and (2) *the manner or ways in which these vegetable constituents, including residual products, were accumulated or deposited.*"

Gresley (1899b, 1900, 1901) began a three-part series on "Possible new coal-plants in coal," encouraged by the editor of the *American Geologist.* Gresley (1901, p. 13), in referring to the pyritic concretions in the coal bed at What Cheer in Iowa (now a well-known coal-ball locality), stated, ". . . I am greatly indebted to Dr. David White of the United States Geological Survey for suggesting this fossil be figured [*Pecopteris,* pl. 8, figs. 1–9], because [it is] probably of scientific value paleobotanically." Gresley (1901, p. 9) also cited the Iowa Geological Survey report (Bain, 1894) for the "What Cheer" coal bed's being considered to be near the base of the Desmoinesian. This report contains a description of the distribution of the concretions within the coal bed (also see Andrews, 1951). It is important to note that W. S. Gresley was a keen observer and intensely interested in coal origins. His drawings permit identification of most of the coal-swamp plant types, and his papers really constitute a beginning of American coal-ball studies that was not widely recognized until the review by Andrews (1951).

Although David White did not recognize the What Cheer materials as representing coal balls, Marie Stopes (see Chapter 9, this volume) apparently did. In pursuit of this historical puzzle, Henry Andrews received this reply from Marie Stopes (May 31, 1950): "You ask about Gresley's American Coal Balls from Iowa. I am sorry to say I know nothing more about them and never got on the track of Mr. Gresley and his work. I did myself find Coal Balls in America, and before Noé . . . ." In a letter to Sergius Mamay (May 23, 1951), she further stated: "I found American Coal Balls years and years ago, before anybody else I think, but they were not very good structurally and I have never published on them."

Gresley's studies were based on pyritic coal balls, using polished surfaces and reflected light to observe plant anatomy, and it is likely that coal balls were thought to be mostly calcareous at the time. This limited view of coal-ball composition perhaps explains why David White (see Chapter 10, this volume) was so adamant about the lack of coal balls in American coals, because he was aware of the pyritic concretions and other kinds of anatomy-bearing concretions (silicified) in coal beds (White and Thiessen, 1913). Schopf (1971) observed, ". . . later Tilton [1912] and Coulter and Land [1911, 1921] described a *Lepidostrobus* cone from Iowa that showed typical coal-ball preservation, neither White nor anyone else made the connection." The Coulter and Land studies were done in the Hull Botanical Laboratory at the University of Chicago. There were probably other cases of coal-ball deposits that were encountered but not recognized as such, as evidenced by reports by Gilbert H. Cady (1915, p. 76, 1919, p. 55–56), who collected the first coal ball in Illinois in 1922, and by Phillips et al. (1973, p. 12).

## UNIVERSITY OF CHICAGO—DISCOVERIES AND STUDIES

One of the most interesting stories leading up to the first coal-ball discoveries, recognized as such, stems from recollections of a story by C. J. Chamberlain of the University of Chicago as recorded in an unpublished tribute prepared by Fredda D. Reed (see Chapter 15, this volume) and partially quoted in Phillips et al. (1973, p. 12–13):

> David White, Chief of the United States Geological Survey, had come to the University of Chicago to lecture before a joint meeting of the botany and geology departments. During the discussion that followed Dr. Noé . . . ventured to ask why, in America, they did not study the Carboniferous flora as preserved in coal balls. David White replied that there were no coal balls to be found in American coal. And Dr. Noé again, "But you have coal mines, do you not?" Where upon David White reiterated with some asperity, "We have looked for them and there are no coal balls in American coal." Dr. J. M. Coulter and Dr. T. C. Chamberlin hastily brought to an end this turn of the discussion. Later, however, they went to Dr. Noé and asked him for an explanation of his comments. Did he believe there were coal balls in America, and if so, did he think he could find them?

Adolph C. Noé had a broad interest in paleobotany and coal, derived from his European training. From 1921 to 1925 and 1928 to 1936 Noé worked for the Illinois Geological Survey during the summers, and it was through this connection that Illinois coal balls were first discovered. Schopf (1971) wrote of this: ". . . Noé was at the time at the Survey in Urbana and Gilbert H. Cady, who was then working on the Harrisburg Quadrangle, had obtained the specimen from one of the mines in the Harrisburg [Springfield] Coal, now long since closed down. Cady saw enough plant material in it to cause him to show it to Noé and ask about it. Noé, who had studied paleobotany with Ettingshausen in Austria, recognized it as a 'coal ball', similar to those from the Ostrau-Karwin field and in Britain." Concerning this, Noé (1932, p. 317) recorded: "In Autumn, 1922, I wrote a letter to Dr. David White informing him that coal balls had recently been found in Illinois, and at a conference during the Christmas vacation of the same year, I showed a sectioned coal ball to him. . . . In December, 1929, Dr. David White showed me some concretions taken out of a coal seam in 1910 which undoubtedly can be called coal balls, but which were not recognized as such at the time."

Noé's (1923a) report of the discovery of American coal balls in *Science* served as a "lightning rod" for further information about coal-ball occurrences. Schopf (1971) explained about the Newcastle coal balls from Texas: "Coal mining had stopped when oil was discovered in north-central Texas about 1923 . . . , I think that W. E. Wrather himself, when he was doing consulting work, first called attention to this locality and sent material from it to Noé about 1923 or 1924. It is one of the places mentioned by Feliciano [1924], one of Noé's early students, but it never was adequately studied. When I transferred the Illinois Survey collections from Chicago to Urbana after Noé's death [1939], I found nothing that was marked as coming from Newcastle, Texas."

The main series of early discoveries and studies was in the Springfield, Danville, and Calhoun coal beds. At a Pittsburgh meeting at the U.S. Bureau of Mines in 1927, Noé told of coal-ball collecting in the O'Gara Mine:

> In Harrisburg they were found in a mine which had been abandoned. The mine had been shut down for about two years and it was rather hard to get to the coal balls. I had my assistant, a mining engineer of the O'Gara Company, and a foreman with me and we had to climb about 425 feet down the air shaft. I dreaded the idea of carrying the balls up because we had collected over 500 pounds. I could have gotten 5,000 pounds. We let a rope down in the shaft and had one man stay underground. He filled our knapsacks and tied them to one end of the rope and we tied the other end to a car and lifted the balls out that way. It proved to be wonderful material. (Phillips et al., 1973, p. 13–14)

In addition to such collecting, Noé kept in touch with the mining companies and joined organizations that aided in developing personal contacts. Schopf (1971) noted: "During part of the 1920s, he [Noé] usually worked for the Illinois Survey during the summer and, with a Survey driver for he did not drive a car himself, he would make the rounds of the coal company field offices in the state, dispensing good will and advice, and bringing back any available collections of coal balls and plant fossils to Chicago for his students to work on."

Associated with Noé's advocacy to discover and study coal balls, a chair in paleobotany was jointly established for him by the Botany and Geology Departments with his transfer from the language faculty. Perhaps 1923 was one of the best and worst years in early American coal-ball studies. In the 1920s two particular topics were in the forefront of paleobotanical research: the elucidation of an extinct group of gymnosperms known as seed ferns (especially assemblages represented by *Lyginopteris* and *Medullosa*) and the search for ancient flowering plants and their ancestors. These topics were inexplicably intertwined in the first coal-ball study from Noé's laboratory by John Hobart Hoskins (1923) on "A Paleozoic angiosperm from an American coal ball." The monocotlike scattered vascular bundles in medullosan anatomy and the extremely large diameter tracheids led to a particularly embarrassing mistake because Noé (1923b) endorsed the finding in a separate article, only to have A. C. Seward (1923) promptly respond that *Angiospermophyton* was " . . . a specimen of the well-known medullosan petiole *Myeloxylon*." Fredda Reed, who also was a student in Noé's lab at the time, said there were some embarrassed faces, and Croneis (1940, p. 224) wrote of this concerning Noé: "Actually he had an intense dislike, even a fear, of scientific combat. His unfortunate experience with his *Paleozoic* angiosperm, one of his several departures from orthodoxy, is a case in point. The extremely critical reception of this paper not only caused him to fail to make reply, but he even dropped the whole matter completely without any attempt, through further investigations, to substantiate his claims."

Needless to say, it is to the credit of both Noé and Hoskins that they put this behind them and got on with their work. It is noteworthy that the first major monograph on the seeds of the medullosans was by Hoskins and Cross (1946a, b). Nevertheless, the circumstances of those times seem to be best appreciated by G. R. Wieland (1924, p. 233) of Yale University in his opening paragraph on "Recent achievements in paleobotany," which included the controversial contributions from Noé's laboratory: "Until within the last few years a sort of forbidding lonesomeness has seemed to attend the student of fossil plants. Paleobotany is in fact the last of the paleobiologic trio to reach accuracy and finality in the methods of research employed. Above and beyond the discovery of new materials, the great need is men [and women]. There are far too few workers in definitely recognized paleobotanic positions which make active contribution possible—in all the world a few over twenty."

Wieland (1924) went on to list those paleobotanists in the United States: F. H. Knowlton, E. W. Berry, A. C. Noé, D. White, E. C. Jeffrey, G. R. Wieland, W. Goldring, R. E. Torrey, and R. W. Chaney, further stating: "Only in subjects where elaborators are active and their fields somewhat overlap, does criticism become an organic, functioning thing—a force such as can carry the study of ancient plant life to the goal. There is pith in the remark of Dr. Marie Stopes that 'paleobotany requires a serene civilization.' "

Wieland was deeply concerned with the harsh criticism levied on the first American coal-ball papers (see Berry, 1924) and realized that critical peer review of such manuscripts would have avoided the entire matter. He was extremely supportive of the potential of American coal-ball studies and conveyed this in a letter to D. H. Scott, who was, so to speak, the leading English coal-ball paleobotanist (see Oliver, 1935) after the death of W. C. Williamson (see Williamson, 1891, 1893).

Professor Noé gave his first paleobotany course at the University of Chicago in 1923. J. Hobart Hoskins and Fredda D. Reed were students in the first course, and they cut the first coal ball in the basement of the Hull Botanical Laboratory at the University. They also prepared the coal-ball thin sections there.

J. H. Hoskins completed his doctorate at the University of Chicago in 1924 and received a National Research Council Fellowship for paleobotanical studies in Europe, where he had the opportunity to meet two prominent English paleobotanists, D. H. Scott and A. C. Seward. Upon his return from Europe in 1925, Hoskins joined the faculty at the University of Cincinnati, where he spent his entire academic career and continued coal-ball studies with his students, notably Aureal T. Cross.

J. H. Hoskins, Fredda Reed, Harriet Krick, and Roy Graham provided the earliest studies of coal-ball plants from the Illinois Basin from 1926 to 1935. Later, Hoskins coauthored, with Cross, some of the earliest American coal-ball monographic works (Hoskins and Cross, 1943, 1946a, b) that dealt with *Bowmanites* and *Pachytesta.* These studies drew heavily on coal balls from Iowa and were part of an extensive doctoral thesis by Cross. Hoskins and Cross (1941, 1942) collaborated on other coal-ball plants. The last of the coal-ball studies by Hoskins was with Maxine L. Abbott (Hoskins and Abbott, 1956) on *Selaginellites.*

Among his many activities in promoting excellence in teaching, Hoskins served as head of the Department of Botany and Bacteriology from 1931 until his death in 1957. Theodore Just (1957), with whom Hoskins jointly founded *Lloydia* in 1938, wrote: "Dr. Hoskins was an inspiring teacher, whose enthusiasm attracted numerous students to botany. Many of them followed his example and became paleobotanists receiving their training from him".

Among those Hoskins attracted and influenced were Aureal T. Cross and Robert M. Kosanke, who had been fellow students at Coe College, Cedar Rapids, Iowa. Both Cross and Kosanke had been greatly influenced by L. R. Wilson, who *was* the geology department. The links between Cross and Kosanke, both personal and professional, and their respective contributions to the development of coal-ball studies and especially coal palynology are interwoven. Kosanke, after receiving his Masters degree at Cincinnati in 1942, was hired by the Illinois Geological Survey.

Aureal Cross completed his monumental doctoral thesis in 1943 at Cincinnati and received a National Research Council Fellowship in Geology (1943–1944) to study the palynology of Appalachian coals. He subsequently served on the faculty at the University of Notre Dame (1943–1946), University of Cincinnati (1946–1949), and jointly at West Virginia University and with the West Virginia Geological Survey (1949–1957) (see Lyons, 1994). It was during the West Virginia days that "The geology of Pittsburgh coal" (Cross, 1952) was completed with observations of silicified *Psaronius* trunks *in situ* in the coal. From 1957 to 1961 Cross worked at the Pan Am (Amoco) Research Center in Tulsa; he then joined the faculty at Michigan State University, where he spent the rest of his career and became professor emeritus in 1987. Aureal Cross has maintained a strong interest in the origin and distribution of coal balls (Cross, 1969) as well as in their discovery (Phillips, 1980).

Fredda Reed shared the following in a telephone conversation with T. L. Phillips (April 20, 1971). Her accidental entry into paleobotany came about because she planned to take plant morphology, and the professor was on leave. Then she found out that Professor Noé was giving a paleobotany course for the first time. Reed's (1926) doctoral thesis was the "Flora of an Illinois coal ball." She said that she was taken to the coal mines to where the coal balls came from but did not really do any fieldwork.

Fredda Reed and "J. H." Hoskins were great friends. She described him as "rather short, red hair and freckles, blue eyes" and "a brilliant person, graduated summa cum laude." Reed went on to teach at Earlham College for two years and then worked at the General Biological Supply House (Turtox) until joining the botany faculty in 1928 at Mount Holyoke College (see Chapter 15, this volume). She continued her coal-ball studies into the 1950s, drawing upon materials provided by Noé and later by J. M. Schopf, H. N. Andrews, and E. S. Barghoorn. She taught plant morphology but said there was not much paleobotany in the course, but when she became chair of the department she saw to it that more paleobotany was taught in the freshman botany course. Reed retired in 1960; in 1971, reflecting upon paleobotany and her teaching, she said, "It has been the richest life. I could have done nothing more exciting than to go into paleobotany. It opened up a whole new world for me." Reed was a favorite of paleobotanists of several generations because of her supreme kindness and lively responses. Blindness in her later years did not deter her from keeping in touch.

As in the case of Fredda Reed, Harriet Krick also intended to work in plant morphology at Chicago, having taken several correspondence courses with C. J. Chamberlain from 1925 to 1927, but learned to her dismay that Chamberlain was about to retire; he suggested that she take the course in paleobotany. Harriet Krick Bartoo (Phillips et al., 1973, p. 26) provided some observations on those days at Chicago:

> I did not myself collect any of the coal balls with which I worked, since the Harrisburg site was too far away. . . . But he [Noé] had a supply of coal balls which he had collected in large sacks in his basement laboratory workshop of Hull Botanical Laboratory Building . . . next to an outside window stood the large bandsaw—next to it the rotary saw with which we "sliced" the coal balls and across . . . the room a couple of polishing machines. Hoskins and Reed had both sectioned and worked in the same room as far as I know before me. It was before the days of the "peel" method so each section had to be carefully glued with marine glue to the slide and polished until thin enough to see through (and sometimes was lost in the process). I well remember the first time Dr. Noé showed us the new method [liquid-peel technique] but it was too late to save some of the material which had been lost in our grinding away at the wheel.

Harriet noted that Noé liked to have visitors and, in turn, show off his students working. Among the more memorable visitors were Chester A. Arnold of Michigan University (see Chapter 18, this volume) and Erling Dorf (Chicago alumnus) of Princeton University. Arnold and Noé first met at an International Botanical Congress at Cornell in 1926. Arnold recalled (Phillips et al., 1973, p. 14):

> It was about midway in my career as graduate student at Cornell University at the time and we struck up a friendship that lasted until he died. After I came to the University of Michigan in 1928, I visited him a couple of times in Chicago. . . . He always took me home to his apartment for dinner, and the fact that he had two rather attractive daughters made these occasions still more enjoyable. It was during one of these visits that he gave me some coal balls. He led me to the pile and told me to help myself. When I glanced toward him, he had purposely turned his back to me so I would not think I was being watched. It was in that lot of balls that Steidtmann and I later found *Medullosa noei* and *Rotodontiospermum illinoiense* [Steidtmann, 1937, 1944; Arnold and Steidtmann, 1937].

Arnold added regarding Noé's role in coal-ball paleobotany, ". . . Noé apparently was the first to realize that these things were something of significance to paleobotany and that they were worthy of investigation."

Harriet Krick recalled (letter to T. L. Phillips, May 1, 1971) Noé as ". . . a most kindly helpful person, tall in stature and with a rather gruff teutonic exterior . . . truly Gargantuan. . . . " Krick was short. She became involved in the study of seedlike fructifications (Krick, 1932) partly as a result of the propensity of Dr. E. J. Kraus, the new Botany Department head, to ask, What constitutes a seed?, in the examination of prospective doctoral candidates and partly as a result of finding several good specimens among the Harrisburg coal balls. She was perplexed as to the lack of embryos, as we still are. Krick became a botany teacher and an ardent attendee of the national and international meetings with gatherings of paleobotanists. After being widowed, she taught at Western Michigan University, where she enjoyed giving some paleobotany courses that had "field tripping"!

Roy Graham, who followed Krick, was from Staffordshire, England, and made exceptionally important contributions in a promising but unfortunately brief career. Graham was well grounded in anatomy as the result of E. J. Kraus's teachings and was a fellowship student. His first publications (Graham, 1934, 1935a), based on his dissertation drawn from the Calhoun coal bed, constituted pioneering efforts into coal-ball paleobotany of the Upper Pennsylvanian. His study (Graham, 1935b) on lepidodendrid leaves is still a basic reference work. Graham was the first of several American coal-ball paleobotanists to benefit from studies at Cambridge University, in his case with A. C. Seward on a National Research Fellowship. Graham subsequently became an instructor in geology at the University of British Columbia, the first of several experienced coal-ball researchers at such institutions in Canada. Graham worked summers with the Canadian Geological Survey until 1937, when he became mine geologist at the Britannia Mining and Smelting Company in British Columbia. He was killed in a rock fall there in 1939 (Bastin, 1940).

## ILLINOIS STATE GEOLOGICAL SURVEY

The connections between the Illinois Geological Survey at Urbana and the University of Chicago were intricate, in interests as well as people. Harold Culver, head of the Coal Division at Urbana (1920–1925), completed his doctorate at Chicago in 1923 with E. S. Bastin; Noé was employed by the Survey in the summers; Gilbert Cady of the Survey collected the first Illinois coal balls; and Bastin at Chicago was a member of the state board overseeing the Illinois Geological Survey. It's no wonder that W. C. Darrah (1941a, p. 33–34) once suggested that James (Jim) M. Schopf was a student of A. C. Noé. Indeed, in hindsight it is hard to conceive of Jim Schopf as anybody's student; however, his graduate adviser was John T. Buchholz in the Botany Department at the University of Illinois. Then, as in all his later years, Schopf's interests and activities were too diverse to ever label him the "coal-ball man," but in this historical account, indeed he was. From the perspectives of where and when coal-ball studies began in North America, Schopf's contributions are as significantly interwoven here as they are with coal palynology and coal geology.

Jim Schopf became involved in coal-ball studies at the Illinois Geological Survey in the mid-1930s and never ever became uninvolved (see Chapter 17, this volume). In 1971 when Phillips et al. (1973) were preparing a historical account of paleobotany in the Illinois Basin, Schopf provided a 16-page account (Schopf, 1971), with a March 10, 1971, covering letter to T. L. Phillips conveying, "I started writing out a few notes about this on Sunday and probably have come out with a good deal more than you expected. Nonetheless, I am happy for an excuse to set these reminiscences on paper. . . . " As a consequence of this exceptional opportunity, we draw upon a coal-ball pioneer's recollections and other quotes to interconnect the coal-ball history between the mid-1930s and the 1960s, introducing other important contributors as Jim did:

> When I first went with the Illinois Survey in early 1934, my work was concerned with plant microfossils, but, because I had more botany than others on the staff, I also got introduced to a variety of topics. Also, I was sincerely interested in questions about the origin of coal and coal petrography. It seemed to me that coal geology, in general, was the reasonable field of economic interest for someone who was in paleobotany and it has always seemed strange to me that so few paleobotanists have had more than a very generalized acquaintance with coal. David White, of course, was much the exception, but his interest dates back to the period in which there was a Coal Section in the U.S. [Geological] Survey organization. I must say that Cady and the other people at the Survey encouraged me in this broader interest.

In 1935 Schopf's first view of coal balls *in situ* was in the Herrin coal bed of the underground Clarkson Mine near Nashville, Illinois. " . . . I was most enthusiastic about the coal balls, about which I had already heard much, but never had a chance to inspect. We brought back a substantial collection . . . before the mine actually closed, Sam Day [Mine Superintendent] brought out three or four more mine cars of the coal balls for my study."

Coal balls from the Clarkson Mine were sources of Schopf's earliest studies on preservation (Schopf, 1936; see Schopf, 1975) as well as on *Lepidocarpon* and *Medullosa* (Schopf, 1937, 1939). He initiated the first quantitative estimates of plant constituents in American coal balls (Schopf, 1938) and was a sustained advocate of the use of coal balls as a botanical index to the constitution of coal. Of this coal-ball deposit Jim noted:

> . . . it gave me a chance to observe and compare the condition and composition of the pre-coal peat with the compressed top coal that we had represented in coal thin sections, and with the spores, cuticles, and other materials obtained from the same layer by maceration. As a result, I have felt ever since that *I* "knew" the lustrous silky top-coal, commonly found in the 6 or 8 inches of Illinois Coal No. 6, was derived from a dominance of *Psaronius* and *Stigmaria* roots, but it is difficult to find a way to demonstrate the proof. In other words, I was greatly interested in the paleoecology of the deposit, but it was a problem that I never was able to work through to a general conclusion. It seems to me that this is still a very important objective because the coal-ball assemblages stand a better chance of characterizing the coal measures peat swamp environment than almost any other source of information.

It was also in the late 1930s that additional coal-ball localities were found in the Calhoun Coal. In 1937 J. Marvin Weller, then at the Illinois Survey, brought in several sample sacks from a creek bed in Richland County that became known as the "Berryville locality." That spring Schopf organized a trip with "Doc" Cady and Professor John T. Buchholz to investigate. Schopf noted: "A great problem in those days was in determining the stratigraphic position of the coal balls and marine limestone at this location. However, I recall that I had to drop the studies of the material that we had collected here, after a very cursory look, because in 1937 I had to get a thesis written."

In 1938 Don Carroll, an extension geologist with the Survey, told Schopf of a huge mass of coal balls he had found a few kilometers from the Berryville locality, and another trip with the Survey truck was underway. This was the Calhoun-North locality and the source of outstanding specimens of sigillarian *(Mazocarpon)* cones, *Psaronius* stems, *Medullosa noei* and *Dolerotheca*. Jim wrote: "I did the *Mazocarpon* paper first, incorporating in it some stratigraphic notes from a paper I presented in 1940 in Milwaukee [Schopf, 1941; see Fig. 4]. I tried to get a botany student interested in the *Psaronius,* and Bill Stewart decided to do his master's thesis on a comparison of *Isoetes* and *Stigmaria*" [Stewart, 1947].

Schopf's (1941) *Mazocarpon* paper is a classic in numerous respects. His interests in coal palynology were already deeply developed, and it was clear that the Illinois Survey was going to emphasize palynology in biostratigraphic correlations. In the preface to Schopf's 1941 paper, G. H. Cady, senior geologist and head of the Coal Division, wrote: "Spores are not uncommon components of the coal-balls, but what is more important to the paleobotanist is the fact that spores are found associated with the organs from which they were produced and the cones containing such organs. Once certain cones and certain spores are definitely linked together, it becomes much easier to identify the spores in terms of the plants from which they were derived. Thus a step in the process of classifying spores in a natural way is accomplished."

Schopf's (1941) benchmark paper on *Mazocarpon* "reproductive biology," before such a term came into vogue, emphasized the nature of other such studies in a stratigraphic context. His incorporated stratigraphic summary (Fig. 4) constituted the most detailed American treatment of Euramerican coal-ball occurrences and included the first reports from the coal in the Shumway cyclothem of Illinois and the Parker coal bed of Indiana. This chart was updated by Andrews (1951), Phillips (1980), and Phillips et al. (1985). By 1941 (Fig. 5) coal balls were known to occur in Texas, Kansas, Missouri, Iowa, Illinois, Indiana, and western Kentucky; all were upper Middle Pennsylvanian (Desmoinesian) and younger (Darrah, 1941a; Schopf, 1941). Coal-ball studies were literally on the brink of expanding to a new level because of increasing coal-ball accessibility with open-pit coal mining, a nucleus of paleobotanists who were becoming experienced in such research and could train students, and institutions that had acquired significant reference collections. There was, of course, a stratigraphic limitation of where the economic coals are in the Midcontinent region, and it was not until coal beds of the Appalachian and other basins yielded coal balls, starting in the 1950s, that a broader paleogeographic and stratigraphic scope was obtainable in North America (see Phillips, 1980).

From archival correspondence between Robert M. Kosanke, who replaced Schopf at the Illinois Survey in 1943, and Schopf, it seems clear that some of Schopf's penetrating questions of research data and interpretations may have evolved from interactions with Gilbert H. Cady. It was Bob Kosanke's role to help mediate several classic publications by Schopf (Schopf et al., 1944; Schopf, 1948) through the Illinois Survey review system after Schopf moved to the U.S. Bureau of Mines. In such correspondence, as much personal as professional, Schopf acknowledged the exceptional scrutiny and clarity demanded of his Survey "mentor," Gilbert H. Cady. In his brief acceptance remarks of the first Gilbert H. Cady Award for outstanding contributions in coal geology, Jim stated that: "During this period I came to have the greatest respect and admiration for Dr. Cady; this is not to say we did not have differences of opinion. . . . At the same time I must add that it was a type of pleasure (I didn't realize it at the time) and a privilege, to have a chance for such frank and open discussion with Dr. Cady. . . . Dr. Cady's encouragement, and above all, his exam-

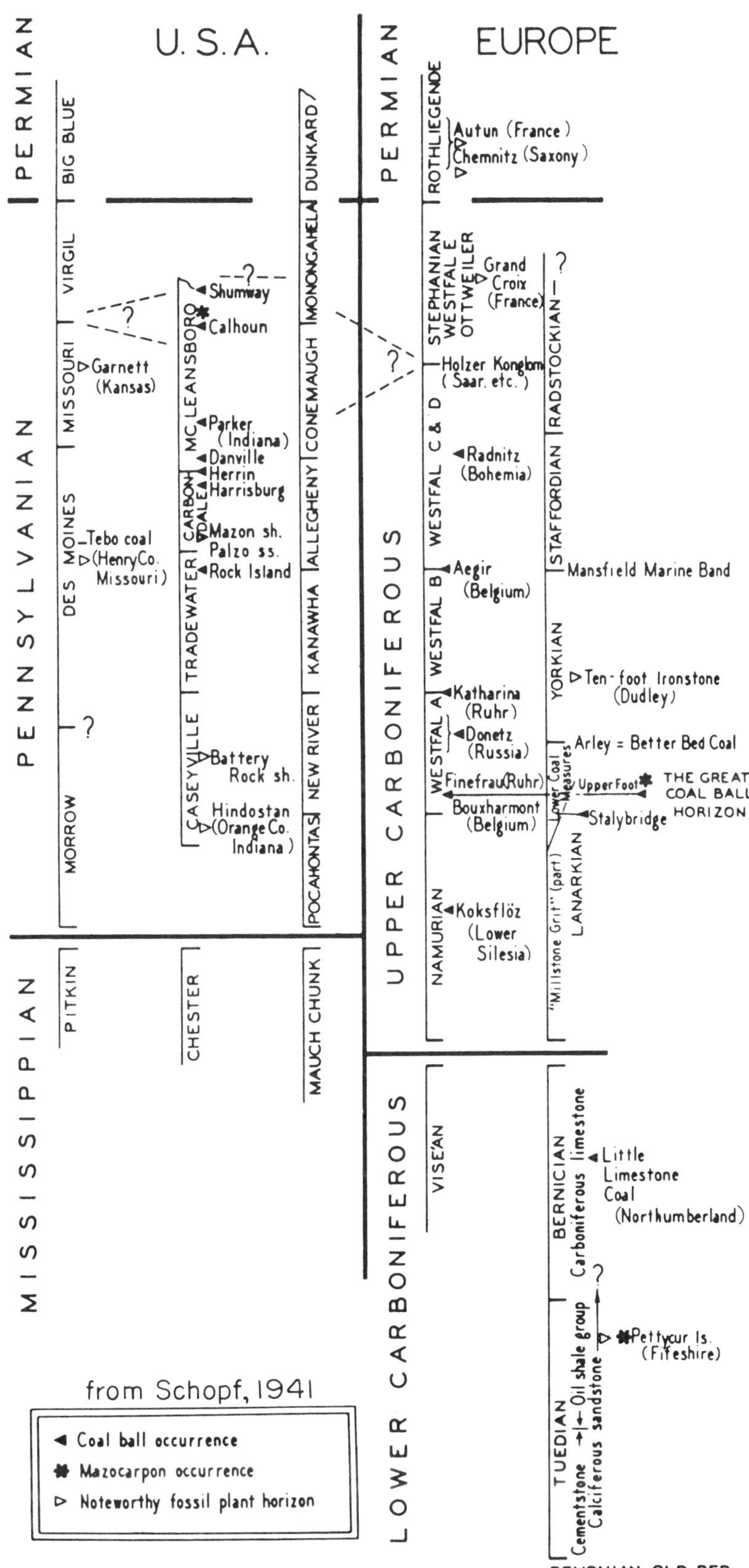


Figure 4. Correlation of Carboniferous beds in America and Europe as published by Schopf (1941).

EUROPEAN STAGES | U.S.A. SERIES | KANSAS GROUPS

AUTUNIAN | LOWER PERMIAN
STEPHANIAN | UPPER PENNSYLVANIAN | WABAUNSEE, SHAWNEE, DOUGLAS, LANSING, KANSAS CITY, PLEASANTON
WESTPHALIAN D | MIDDLE PENNSYLVANIAN | MARMATON
WESTPHALIAN C | MIDDLE PENNSYLVANIAN | CHEROKEE
WESTPHALIAN B | MIDDLE PENNSYLVANIAN | CHEROKEE
WESTPHALIAN A | LOWER PENNSYLVANIAN
NAMURIAN B-C | LOWER PENNSYLVANIAN

## WESTERN INTERIOR COAL REGION

| OKLAHOMA | KANSAS | MISSOURI | IOWA |
|---|---|---|---|
| NEWCASTLE, TEXAS ● | | | |
| IRON POST ○ | | | |
| BEVIER ○ | BEVIER ○ | | |
| | FLEMING ○ | CROWEBURG ○ | |
| MINERAL ○ | MINERAL ○ | | |
| | WEIR-PITTSBURGH ● | TEBO ○ | |
| SECOR ○ | | | |
| | | | LADDSDALE ○ |
| | | | ● |
| | | | CLIFFLAND ● |

MIDCONTINENT SERIES: WOLFCAMPIAN; VIRGILIAN; MISSOURIAN; DESMOINESIAN

ILLINOIS FORMATIONS: MATTOON; BOND; MODESTO; CARBONDALE; SPOON

## EASTERN INTERIOR COAL REGION

| ILLINOIS | INDIANA | KENTUCKY |
|---|---|---|
| SHUMWAY ● | | |
| CALHOUN ● | | |
| OPDYKE ○ | | |
| FRIENDSVILLE ○ | | |
| BRISTOL HILL ○ | | |
| | PARKER ● | |
| DANVILLE ● | DANVILLE ○ | |
| | | BAKER ○ |
| HERRIN ● | | HERRIN ○ |
| BRIAR HILL ○ | COAL V A ○ | |
| SPRINGFIELD ● | SPRINGFIELD ● | SPRINGFIELD ○ |
| SUMMUM (HOUCHIN CREEK) ○ | | |
| COLCHESTER ○ | | |
| | | No. 7 ● |
| | UNNAMED ○ | |
| MURPHYSBORO | ● | |
| | BUFFALOVILLE ○ | |
| ROCK ISLAND | | |

APPALACHIAN GROUPS & FORMATIONS: DUNKARD; MONONGAHELA; CONEMAUGH (CASSELMAN, GLENSHAW); ALLEGHENY (CHARLESTON SANDSTONE); POTTSVILLE (BREATHITT, KANAWHA, NEW RIVER, POCAHONTAS, LEE)

## APPALACHIAN COAL REGION

| PENNSYLVANIA | OHIO |
|---|---|
| | REDSTONE ○ |
| PITTSBURGH ○ | |
| | DUQUESNE ○ |
| | HARLEM ○ |
| | ANDERSON ○ |
| | UPPER FREEPORT ○ |
| MIDDLE KITTANNING ○ | |

| E. KENTUCKY | TENNESSEE |
|---|---|
| | ROCK SPRINGS ○ |
| HAMLIN ○ | |
| HIGNITE ○ | |
| AMBURGY ○ | |
| UPPER PATHFORK ○ | |

| ALABAMA |
|---|
| NEW CASTLE ○ |

## WESTERN EUROPE

| LITHOLOGIC DIVISIONS | GREAT BRITAIN | FORMATIONS | THE NETHERLANDS | ZONES | BELGIUM | SCHICHTEN | RUHR |
|---|---|---|---|---|---|---|---|
| MIDDLE COAL MEASURES | MANSFIELD Marine Band | MAURITS | AEGIR Marine Band | EIKENBERG | MAURAGE Marine Band | HORSTER | AEGIR Marine Band |
| | | | AEGIR ● | | PETIT BUISSON ● | | |
| | TWO-FOOT Marine Band | | | | | | |
| | | HENDRIK | DOMINA Marine Band | ASCH | DOMINA Marine Band | ESSENER | DOMINA Marine Band |
| LOWER COAL MEASURES | CLAY CROSS Marine Band | WILHELMINA | KATHARINA Marine Band | GENK | OUAREGNON Marine Band | BOCHUMER | KATHARINA Marine Band |
| | | | | | | | KATHARINA ● |
| | | | | | | | BLÜCHER ○ |
| | | BARLO | WASSERFALL Marine Band | | WASSERFALL Marine Band | WITTENER | DICKEBANK |
| | UNION SEAM ● | | FINEFRAU-NEBENBANK ● | | BOUXHARMONT ● | | FINEFRAU-NEBENBANK ● |
| | FIRST COAL ● | | SARNSBANK Marine Band | | | | SARNSBANK Marine Band |
| MILLSTONE GRIT / UPPER LS. GP. | *G. subcrenatum* Marine Band | | | | | | |

● Coal-ball occurrences known by 1941
○ Coal-ball occurrences known after 1941

Modified from Phillips and others (1985)

ple of courage and integrity, have been a continuing source of strength and inspiration" (Kosanke, 1979).

All who attended paleobotanical or coal geology meetings with Jim Schopf in the audience felt the importance of his presence, and speakers, especially, were apprehensive of his penetrating questions. Jim, in rising, would usually hitch up his trousers, secured often with red suspenders and a superfluous belt. Usually he kindly prefaced his remarks with a compliment to the efforts of the researcher, hesitating, as if to let his words catch up with his thoughts, and then came the unexpected. His intense interests and questions were such that some speakers often could not tell whether he was just asking questions, adding to their observations, or trying to steer them in a less perilous direction of interpretations. This distinctive quality in Schopf's own researches, as mirrored by his exemplary contributions, was historically important in setting paleobotanical and coal geology standards. From the 1949 establishment of the USGS Coal Geology Laboratory in Columbus, Ohio, Jim was noted for encouraging and directly helping many students and colleagues with their projects; he regarded this as part of his overall mission.

## A 1948 FIELD TRIP WITH JAMES M. SCHOPF

Schopf wrote (1971):

> I think I should add one more note now, that may prove of some historical interest. In 1948, I was able to obtain a little money for USGS field work and I decided to visit all the American coal-ball localities that I knew of. I got 6¢ a mile for use of my own Jeep. I visited Art Blickle in Athens, Ohio, and spent a day or two probing about occurrences of the Athens County silicified *Psaronius* material. This is still a puzzler, but it is not far off stratigraphically from similar, but more poorly preserved, silicified tree-fern material in the Illinois Basin [coal bed in the Shumway Cyclothem; see Scheihing, 1978].
>
> I had made arrangements with Bob Kosanke to meet me near Boonville, Indiana. I had never had a chance to see the coal balls from Indiana V = Harrisburg 5 Coal, although I knew Noé had made collections, and we were lucky. We both got good collections and then I called Henry Andrews, made arrangements to meet at Berryville, Illinois, and told him about the strip mine area which would again be covered in a few days. I also told the people at the Indiana Survey. Henry at least got a good big truck load from there.
>
> Bob and I then went to St. Wendel. I had been there once before, but at that time had been unable to get much material. We did better this time though the exposure was still difficult. This is another locality that Marvin Weller had initially discovered and told me about.
>
> Then, we went to Berryville. This time I had a Jeep and we drove through the gate up close to the outcrop. Henry Andrews joined us there—[Henry and Bob Baxter] went to Boonville a little later in the summer. We all got quite a lot of material and, as you know, Bob Kosanke later told Wilson Stewart about it and the locality was considerably revisited later.

Figure 5. Coal-ball occurrences in the United States and western Europe as known in 1941 and as of 1993. See symbols on chart. Modified from Phillips et al. (1985).

After visits to old mine entries of the Springfield Coal south of Danville, Illinois, pursuing collection notes of Noé, the trip resumed from Urbana. "Then, Bob [Kosanke] and Doc [Cady] and Jack Simon and I went down to a new locality they knew about at the Pyramid Mine in No. 6 Coal in southern Illinois. Nearly the whole bed had been permineralized and good plant material was abundant. I know the Illinois Survey got a couple truck loads and I collected about 600 pounds."

Schopf went on to Missouri, Kansas, Oklahoma, and Texas, commenting about the Young County, Texas, occurrence:

> The other locality in which I was interested was near Newcastle, Texas. These should be the youngest coal balls in U.S.A. I think, in fact, that the base of the Permian has been drawn a little lower now and these may now be stratigraphically located in the Lower Permian. W. E. Wrather, at that time, Director of the Survey [USGS], put me in touch with a friend of his who gave me a few specimens. I spent several days going over all the old mine dumps and outcrops in the locality.
>
> I thought I had done well enough collecting and the next couple of years I spent, among other things, investigating paleobotanical laboratory facilities and possibilities at the National Museum. I finally got essential facilities here in Ohio [USGS Coal Geology Laboratory, Orton Hall, Ohio State University, Columbus], but I have not yet completed the study of my 1948 collections.

## THE IOWA CONNECTIONS

Much as there was an early Iowa "connection" in the first discovery of American coal balls, there were subsequently important Iowa "connections" in the history of coal-ball discovery, collecting, and innovations by William (Bill) C. Darrah (see Chapter 1, this volume) and Frederick O. Thompson (see Chapter 22, this volume). In the late 1930s Bill Darrah at Harvard University developed one of the largest coal-ball processing operations and collections for that time. Darrah's primary interests were compressional floras and biostratigraphy at a time when European paleobotanists with substantial expertise were interested also in "resolving" American biostratigraphic "problems" in the upper Carboniferous (see Chapters 25 and 26, this volume).

Darrah's interests in coal balls developed from his interactions with Paul Bertrand, first on Bertrand's visit to the United States in 1933 and then during Darrah's visit to Europe in 1935. Of Paul Bertrand's influence, Darrah (Phillips et al., 1973, p. 23) wrote, "We got along well and I learned a great deal from him too, especially my first introduction to histologically preserved material." Darrah had the splendid opportunity to study the anatomically preserved Stephanian and Autunian plants from France and considered Paul Bertrand his mentor during the important experiences gained there. He also had the opportunity to collect coal balls from the Bouxharmont seam in Belgium. The Bouxharmont coal balls were black, both inside as well as outside, coming from an anthracite bed (Lyons et al., 1985), and Darrah cautioned Frederick Thompson about the color differences between the Iowa (American) brown versus black, in sending a specimen to him for aid in recognition of coal balls.

F. O. Thompson, a graduate of Harvard University, had become acquainted with Darrah as a result of Darrah's interests in the Mazon Creek nodule flora of northeastern Illinois and the collections Thompson had sent to Harvard even before Darrah's arrival in 1934. The Darrah-Thompson friendship led to discoveries of coal-ball localities and enormous collections at a time when access to and acquisition of such coal-ball material was a critical factor in the development of coal-ball studies. Keeping in mind Darrah's penchant for compression paleofloristics and biostratigraphic correlations with western Europe, as well as his other interests (see Lyons and Morey, 1991), it is clear that he was faced with an incredible opportunity to obtain more coal-ball specimens than could be processed and more on the way.

Thompson, who was a successful businessman dedicated to helping paleontology long the way, kept Darrah informed with telegraphiclike letters and notes about his exploration of the now classic localities known as the Shuler and Urbandale mines of Des Moines, Iowa: "An official of the mine said he would fill a car with coal balls and have it pulled out by mules to save carrying the stuff up" (Shuler Mine, F. O. Thompson letter to W. C. Darrah, August 31, 1938, p. 1). "I know the time it takes to cut, polish, etch and peel coal balls. If I did not feel that I would destroy many specimens, I would be glad to do the work for you" (same letter, p. 2).

Several years later, after tons of coal balls had been shipped to Harvard University, Thompson in a note (April 4, 1941) conveyed, "Brought in more coal balls today from the Urbandale mine. This was my third trip there this spring, and a few months ago I thought no more coal balls would be found in Iowa. It's a great state, Bill!" Bill Darrah was then 32 years of age, having come to Harvard University in 1934 as a research assistant, from the Carnegie Museum, and later becoming research curator of paleobotany at Harvard University (Lyons and Morey, 1991).

Of the Thompson collecting, Darrah wrote (letter to T. L. Phillips, March 16, 1971):

> I went with Thompson to Iowa in 1939 in May and we found what appeared to be coal balls, though pyritized, at the Urbandale Mine, just outside of the city limits of Des Moines. We had no saw but we broke a few nodules and were satisfied that *Mesoxylon* and some seeds were present. I returned to Cambridge and within three weeks Thompson sent me about a dozen very excellent small coal balls from the Shuler mine, also in the immediate vicinity of Des Moines. . . . I have photographs of Thompson's backyard with a 5-ton coal truck dumping a full load of coal balls [see Chapter 22, this volume]. He sent, literally, more than 3,000 to Harvard and sawed at least a thousand more.

Of Thompson's disposition of coal-ball specimens, Elsie D. Morey, Darrah's daughter, wrote (letter to T. L. Phillips, December 4, 1992): "Fred Thompson would send slabs of the same coal ball to many places not realizing the importance of keeping it together in spite of W. C. Darrah's efforts to explain this. In a few cases part of the same coal ball went to the Harvard Paleobotanical Collections, to Henry [Andrews] at Washington University and to my father's personal collection sent to his home and elsewhere depending on Fred's whims."

This is what happened to the coal-ball slabs containing *Medullosa thompsoni* described by Andrews (1945), leading to extensive correspondence between H. N. Andrews and W. C. Darrah. The letters were marked by salutations of "Dear Darrah" and "Dear Andrews" as was the custom of those times.

If necessity be "the mother of invention," it is noteworthy that Darrah's (1936) first contribution related to coal balls was "The peel method in paleobotany." This laboratory technique evolved in Europe (Walton, 1928; Walton and Koopmans, 1928), was modified in the United States (Noé, 1930; Graham, 1933) and, with further experimentation, became known as the "Darrah solution." This technique replaced thin sectioning by etching coal-ball surfaces with hydrochloric acid and re-embedding the cells walls of plants in a parlodion solution (not so unlike embedding techniques of living plant materials). When the parlodion dried, it was lifted (often assisted with a razor blade) from the surface, and the "peel" retained the cell walls of the fossil plants. This technique, used widely into the 1950s and occasionally today, was a major step up from thin sectioning. Nevertheless, the cutting and liquid peeling of vast numbers of coal balls were logistically time-consuming activities, requiring much help and attendant expenses.

F. O. Thompson helped in these regards too. In his first publication of the fossil flora of Iowa coal balls, Darrah (1939, p. 136) acknowledged such support, not only for the essential lapidary equipment provided but also by saying, "a word of recognition is due to the ten students who have cheerfully labored to prepare the thousands of slides now available for study." The late William S. Benninghoff, noted ecologist at Michigan University, was one of those undergraduate assistants in Darrah's laboratory at Harvard University. He published an early report of the coal-ball flora from the Springfield Coal flora of Indiana (Benninghoff, 1942).

Darrah's (1941a) summary of "Studies of American coal balls" constitutes a benchmark overview, based on incredible amounts of sectioned and peeled coal balls for the time, almost 3,500, mostly but not exclusively from Iowa. His approach was that of a census of the paleofloras, using both peels and split layering coal-ball surfaces for identifications. His estimates of higher species diversity in some Iowa coals than in the Illinois basin are probably still close to the evidence at hand. He first noted the dominance of cordaites qualitatively in certain Iowa coal balls (Shuler and Urbandale mines) and remarked on the lycopsid-dominated deposits in some Iowa deposits and especially in Illinois and Indiana coal beds. Darrah (1941a, p. 44) observed, "We cannot choose between temporal and ecological explanations for the *Lepidodendron* association at certain localities and *Cordaites* at others."

One of the driving inquiries by Darrah concerning biostratigraphy caused him to take advantage of the abundance of coal balls, particularly the some 4,000 from the Shuler Mine. Hundreds of these were split for morphological identification of the fern and seed-fern foliage so important in biostratigraphy (Darrah, 1941a, p. 45). Two particular conceptual biases were prevalent at the time with regard to coal-swamp floras. The first was that the peat-forming plants represented by coal balls were the same as those in the roof-shale floras (see Chapter 28, this volume); Darrah embraced this concept because he could match, or nearly so, such compressions of seed-fern foliage with those on split coal-ball layers. The second bias was that cordaites were mainly "upland" or non–coal swamp plants. It was probably not until the late 1970s that both notions were dispelled, but the Iowa coal-ball deposits still constitute the highest level of cordaitean dominance known within the Euramerican tropical belt (Phillips et al., 1985; Raymond, 1988). It is of historical significance, also, that the most "uncertain" stratigraphic position of the most abundant sources of Iowa coal balls, the Shuler and Urbandale mines, has not changed from the earliest correlations with the Rock Island coal bed of Illinois (Schopf, 1941), and that these sources are now designated the Cliffland coal bed of Iowa (Ravn et al., 1984; Ravn, 1986) and placed in the Westphalian C-D transition as just below the boundary.

Darrah (letter to T. L. Phillips, April 2, 1971) wrote of his other acquisitions of coal balls: "I do not think that one would ever confuse American or European coal balls on the basis of appearance alone. In the second place, all European coal balls that I have seen contain a relatively large ratio of dolomite, whereas in American coal balls dolomite is minimal. I have collected coal balls in Belgium (Wérister Mine) and in Holland (Finefrau-Nebenbank). I have never collected coal balls in England, but . . . I purchased the Lomax collection between 1953 and 1956."

Darrah then explained about the Lomax collection (same letter): "After trying to sell unsuccessfully some 550 choice blocks (some of which he cut pieces for Scott and Williamson as a very young man), he [J. Lomax] wrote me and asked if I would buy them. . . . With it, without charge, he gave me his complete file of reprints on English coal balls since the days of Carruthers and Williamson. Lomax served as a preparator for Williamson in the 1890s. To make my story complete, and I do not want you to publish this until after I die, I entered the slide making business with peels from Lomax nodules in order to recoup the [costs]. It is as simple as that."

As in the case of Hoskins and Noé, Darrah and others of the pioneering generations made mistakes in the taxonomic identification of coal-ball plants. This is an understandable part of American researchers finding their way in a growing environment, as Wieland might have expressed it, "where elaborators are active and their fields somewhat overlap." This involved the report of embryos from supposed *Lepidocarpon* (*Lepidophloios* trees) (Darrah, 1941c). One of the pointed searches of early coal-ball studies in the Pennsylvanian of the United States was for embryos in seeds and seedlike structures such as *Lepidocarpon.* Considering the phenomenally good preservation of the peat deposits, including "ovules" and lepidocarps, it was puzzling (and still is) why embryo stages of the seed plants were not encountered. Darrah's (1941c) report of embryos stimulated studies both of *Lepidocarpon* (Hoskins and Cross, 1941) as well as of the actual *Nucellangium* (cordaitean ovule specimens), which Andrews (1949) studied from coal balls given him from the same Iowa locality by F. O. Thompson. It is ironic that the first authentic upper Carboniferous embryo discovered in *Lepidocarpon* was described as a gametophyte by Andrews and Pannell (1942) because it was avascular (Phillips et al., 1975) and the "cellular remains of a young embryo" in *Nucellangium* were subsequently reported in the study of Stidd and Cosentino (1976).

Before turning from the significant coal-ball contributions of Darrah, it is important to emphasize the breadth of his interests and discoveries in coal balls (Darrah, 1941b) prior to his departure from Harvard University in 1946. He was responsible for the reports of other coal-ball discoveries in Missouri and Kansas (Darrah, 1941a) and the acquisition of enormous fossil-plant collections for Harvard, including those from the Illinois basin. His influence on numerous students there and later at Gettysburg College in Pennsylvania was very important in the continued development of paleobotany. Darrah's (1969, p. 71–76) book on Pennsylvanian floras of the eastern United States includes some summaries of coal-ball floras from the Midcontinent as well as from western Europe (see Phillips, 1980).

According to Elsie D. Morey (letter to T. L. Phillips, December 24, 1992), of all his colleagues and mentors, W. C. Darrah was probably influenced the most by Oakes Ames, the then director of the Harvard Botanical Museum. Darrah was, in turn, influential in the training of W. S. Benninghoff, G. W. Dillon, John Burbank, R. E. Schultes, and E. S. Barghoorn. Elso Barghoorn, who completed his doctorate with I. W. Bailey in 1941, returned to Harvard in 1946 to assume the position that Darrah had vacated. In turn, Barghoorn shifted the paleobotanical interests in reopening the Brandon lignite deposits as well as pursuing the Precambrian microbiota with his students. However, there were also studies of coal-ball plants (Roberts and Barghoorn, 1952; Brush and Barghoorn, 1955). Also, as noted by Lyons and Morey (1991, p. 196): "His Iowa coal-ball peels, prepared by his own improved technique [Darrah, 1936], developed with his botanist wife, Helen Hilsman Darrah, were shipped all over the world and did much to popularize and promote paleobotany, especially coal-ball studies in the United States. Darrah's peels of Iowa coal balls were produced in such prodigious numbers that A. Traverse, while a Ph.D. student at Harvard during the late 1940s, was employed for a summer sorting and labeling them."

The late 1940s were a critical resumption and expansion period in paleobotany, as previously noted in the activities and influences of J. M. Schopf. Alfred Traverse, Pennsylvania State University, recalls:

I began immediately to work on the pollen and spores of the Brandon lignite, using at first a sample from the Harvard Museum that had been collected by, I believe Jeffrey, in the 1890s and samples collected by Barghoorn and his first student, William Spackman, the previous summer. I took the paleobotany course that year (1947–48). Spackman was the T.A. [Teaching Assistant], and it was he who taught me to make peels from the vast Thompson collection of coal balls, using the Darrah [1936] technique. In the summers of 1949–50 I worked part time in the Thompson collection, sorting and organizing the large collection of slides made by Darrah, and under his direction, by retinues of helpers. I got quite good at the Darrah liquid peel techniques, and my first publication [Traverse, 1950] was based on such slides of *Mesoxylon.*

Also in Barghoorn's first paleobotany course was James E. Canright, a student of I. W. Bailey, who went to Indiana University in 1949, where he taught plant anatomy and morphology and, later, paleobotany. Jim Canright (in Phillips et al., 1973, p. 24) shared his further involvement with paleobotany and coal balls: "After I gave a talk to the Geology Colloquium on the paleobotany of the Paleozoic, Dr. Charles Deiss [director, Indiana Geological Survey and head, Geology Department, Indiana University] invited me to be a Field Party Chief during the summers of 1953 and 1954 to study the plant megafossils associated with the Indiana coals. Joe Wood was my field assistant during the 1953 season and C. F. Shutts assisted me in the 1954 season. The fossil plants and coal balls collected during these two field seasons were accessioned into the collections of the Coal Section of the Indiana Geological Survey."

These included coal balls from many Indiana localities, especially in the Springfield coal (see Canright, 1958, 1959). Canright noted, "The most impressive coal ball find was the one in the Parker Coal of Posey County (Henry Andrews told me about the site.). A huge calcified mass is embedded in a creek bed near St. Wendel—roots of *Psaronius* are most abundant and beautifully preserved."

## WASHINGTON UNIVERSITY AND DERIVED STUDIES

Henry N. Andrews, Jr., (Fig. 6) established one of the most productive coal-ball research programs in the United States at Washington University, St. Louis, from 1940 to 1964, when he moved to the University of Connecticut (see Chapter 21, this volume). As a graduate student in St. Louis, beginning in 1935, Henry Andrews's paleobotanical interests were essentially self-nurtured until, as he put it, he was "shipped off" to Cambridge University for a year by his adviser, Robert Woodson, an angiosperm systematist.

Henry Andrews wrote from Cambridge (letter to T. L. Phillips, April 1, 1971) of this: "I believe the first time that I actually saw a coal ball and made peel preparations was when I was a student here at Cambridge in 1937–38. Hamshaw Thomas suggested that I look into the anatomy of the secondary wood of the Pteridosperms and I obtained some specimens of *Lyginopteris oldhamia* from Hemingway."

Figure 6. Henry N. Andrews in 1965 at the greenhouse facilities at the University of Connecticut, Storrs. Photograph taken by Tom Phillips.

Andrews spent much time at the British Museum of Natural History that academic year and over the following decades, studying the plant collections at the Museum. From the first visit, he struck up a lifelong friendship with Maurice Wonnacott, who helped administer the paleobotanical collections and advised him on specimens to examine; Henry considered Wonnacott as his "professor" (Andrews, 1990). Upon return to St. Louis, Andrews (1940) completed his doctorate in 1939 along the lines suggested by Hamshaw Thomas. Andrews (1942a, b) began a series of "Contributions to our knowledge of American Carboniferous floras," mostly on coal-ball plants, and turned out his first graduate student, Eloise Pannell (1942; Andrews and Pannell, 1942), with the first of numerous important contributions on the coal-swamp lepidodendrids from his laboratory. Pannell joined the WAVES (U.S. Navy) and, as elsewhere, graduate training in paleobotany was placed mostly on hold for the duration of the war.

Andrews's intense interests in all the major plant groups of the upper Carboniferous coal swamps were primarily from the evolutionary viewpoint, with his research forte in plant anatomy and development. He pursued these as time permitted while teaching servicemen at Washington University, where he began preparation of broader studies (Andrews, 1946, 1947, 1948), including an extensive review of "American coal-ball floras" (Andrews, 1951). By 1951, the graduate training program in coal-ball studies at Washington University had expanded with an influx of veterans representing the U.S. Navy, Army, and Marines, namely Robert W. Baxter (pteridosperms), Sergius H. Mamay (ferns), Charles J. Felix (lycopsids), and Burton Anderson (calamites). Andrews always included his graduate students in the fieldwork, and he was an ardent field tripper and explorer himself (see Chapter 21, this volume). Thus, observations shared in Andrews's (1951, p. 433) review reflected a very broad acquaintance with coal-ball occurrences (Fig. 7): "The largest unit mass of coal balls that we have yet encountered turned up in a strip mine near Boonville in southern Indiana. An

Figure 7. Henry N. Andrews in 1948 examining the coal-ball aggregate in the Springfield coal bed near Boonville, Indiana. Photograph provided by H. N. Andrews.

aggregation of some 30–40 tons occurred in a single huge lump, so large in fact that mining operations had been diverted around it. The quality appeared to be exceptionally good; the pit superintendent indicated that it was the only "rock" that had been encountered in recent years in the coal [Springfield], and that continued operations would cover up the coal balls within a few weeks. Under such conditions it was necessary to act immediately and on a rather large scale or not at all."

Robert W. Baxter (letter to T. L. Phillips, April 29, 1971), who established coal-ball research studies at the University of Kansas, Lawrence, in 1949 shared this view of the trip: "This was a major coal ball collecting trip for Henry and me. Henry had gotten word [from Schopf] that a large quantity of coal balls had been uncovered at the Wasson Coal Co., north of Boonville, Indiana and arranged for the rental of a truck in order to bring back a full load. Since neither of us had ever done much truck driving before, 'getting there was half the fun.'" Andrews (letter to T. L. Phillips, April 1, 1971) admitted: "This was quite a trip because the truck apparently had some sort of carburetor trouble and every few miles it would go into a terrible fit of 'bucking' and we had to bring it to a near stop to stop this."

Indeed, field tripping with Henry's own station wagon was considered more precarious than working in the strip mines, particularly the trips to southeastern Kansas. It was said that the graduate students (perhaps from Charles J. Felix's stories) had to be fairly good mechanics in order to get there and back. The discovery of coal balls by paleobotanists at the Pittsburg-Midway Mine No. 19 near West Mineral, Kansas (see Andrews and Mamay, 1952) was a major find. Of this locality, best known as "West Mineral," Andrews (1951, p. 433) wrote:

> . . . coal balls of good quality occur in an abundance that a paleobotanist hopes for but does not often expect. They have been taken out of the coal here by the truck loads, and because they constitute the hardest rock available in the immediate vicinity they have been put to use as road ballast. Many tons of specimens are exposed along the sides of the roads leading to one of the pits. The specimens at this locality are also unique in that they contain an exceptionally diverse flora, especially large plant fragments, and of particular significance is the fact that they have weathered in such a way that renders possible the identification of many plant organs in the field; thus one can be quite selective in the choice of specimens.

The "West Mineral" locality became one of the best known for coal balls for decades, and Robert W. Baxter, who pioneered in their study, showed several generations how to heft and sort the highly pyritic from the more calcareous, especially at a road ballast supply pile (Fig. 8). This was an enormous strip mine, and at the time the stripping shovel, named "Brutus," placed in operation there in 1963, was one of the two largest shovels in the world. Baxter and Hornbaker (1965) provided an important guidebook on the "Pennsylvanian fossil plants from Kansas coal balls."

Continuing with Andrews's "culling techniques" as shared in his 1951 review, large coal balls had to be broken into smaller masses in order to be cut with slow-speed lapidary saws with industrial diamonds embedded in the blade. Large (3-ft [0.9-m] diameter blades) saws did not come into coal-ball cutting until the 1960s. The smoothed and hydrochloric acid–etched cut surfaces were examined under a low-power binocular microscope. As he states (Andrews, 1951, p. 436): "In this way many worthless specimens or ones containing well-known plant remains are immediately eliminated. After peeling the selected specimens, peels were studied with a binocular dissecting microscope using reflected light or some mounted for compound microscopy with transmitted light. In this way two or three tons of coal balls constituting an impressive if unattractive, dirty

Figure 8. Robert W. Baxter, paleobotanist at the University of Kansas, Lawrence, standing in the enormous coal-ball pile from the Fleming coal bed at the Peabody Pittsburg-Midway Mine No. 19 near West Mineral, Kansas, in 1958. Many specimens were collected for study but most served as "road ballast." Photograph taken by Tom Phillips.

and space-absorbing pile of 'rocks' are ultimately reduced to a few small drawers of envelopes."

His strategy of culling is further conveyed (Andrews, 1951, p. 437) regarding coal balls dominated by *Psaronius* roots arborescent lycopsid periderm, and cordaitean wood where these were the obvious bulk components: "Such specimens may be weeded out to a considerable extent in the field." As an experienced and almost indefatigable collector, Andrews culled at every stage of the field and lab work, knowing it took large collections of the best-preserved material to study and reconstruct the plants. His laboratory and library reflected the same space-saving tendencies to minimize excess clutter—although his garage invariably contained fossil plant collections "to be sorted"!

Andrews's own contributions to coal-ball studies were diverse, involving every one of the five major vascular plant groups represented. In coal-ball studies, one of the most significant breakthroughs with continued important ramifications is the concept of determinate growth in arborescent lepidodendrids (Andrews and Murdy, 1958, p. 560), ". . . which involves the development of the unbranched primary trunk from a massive apical meristem and, at the onset of dichotomy, a gradually diminishing primary body to a point at which longitudinal growth ceased." Many of the important contributions on lepidodendrids were developed in Andrews's laboratory, from those of Pannell (1942), Andrews and Pannell (1942), and Felix (1952, 1954) to such a conceptual synthesis (Andrews and Murdy, 1958). Although not well received at first, the subsequent studies of Donald A. Eggert (1961, 1962), a student of Theodore Delevoryas at Yale University, left little doubt of the documentation of determinate growth in both lepidodendrids and calamites.

Beyond the individual and typical graduate student–colleague studies that mark the career of Andrews, there are two additional contributions that should be mentioned in relationship to coal-ball studies. Andrews is, at heart, an explorer-naturalist, and from this vantage point he became an extremely effective synthesizer, writer, and story teller of adventures in paleobotany, coal-ball studies, and even the history of paleobotany, among other topics. He passed along much more than the literature and methodologies to his students—to all students—and perhaps accomplished more in generating interests in historical accounts of this type than anyone. His students developed quite diverse research interests with his minimal pedantic role as an adviser. The contributions of Mamay and Baxter, close friends through the years, have been mentioned briefly. Charles J. Felix, after two years at the USGS Coal Geology Laboratory in Columbus with James M. Schopf, joined the Sun Oil Company in Richardson, Texas, as a palynologist and later went to Abilene Christian College, from which he recently retired. The last two coal-ball–oriented students of Andrews were Tom L. Phillips, who joined the faculty at the University of Illinois, and Shripad Agashe, who returned to India, where he is department head in botany at Bangalore.

Andrews's interest in vascular plant evolution naturally led him to Devonian paleobotanical studies. The last part of Andrews's career at the University of Connecticut was followed by major synthetic works, after retirement, on paleobotanical history (Andrews, 1980) and Devonian paleobotany (Gensel and Andrews, 1984).

## UNIVERSITY OF ILLINOIS—MULTIPLE LEGACIES IN COAL-BALL STUDIES

The most complex and sustained network of expanded 1950s–1960s coal-ball studies in North America developed from the University of Illinois at Urbana with the original influences of James M. Schopf and the teaching enthusiasm of Wilson N. Stewart and his colleagues, Theodore Delevoryas, Robert M. Kosanke, and Tom L. Phillips. If Schopf ever inspired a student to interests in coal-ball studies, Stewart more than justified the investment in time. This aspect of North American coal-ball studies history stands out because it reflects the power and influence of outstanding teaching and facilitating of graduate research and training with large coal-ball collections.

From the cryptic recording of history of the Botany Department at Urbana by A. G. Vestal (1987) comes the following: "James Schopf, who started in ecology, switched to morphology (embryology of *Larix*), and wound up playing with coal balls. Bill Stewart came back to Illinois in September 1947. He later developed a lab of paleobotany. In earlier years, paleobotany had been taught partly as incidental to advanced morphology courses, partly as research."

Wilson N. Stewart (letter to T. L. Phillips, September 27, 1971 (Fig. 9) recalled how he became interested in paleobotany:

> I had developed an interest in *Isoetes* when I was an undergraduate in Botany at the University of Wisconsin. At that time I completed a senior thesis on the morphology of the plant. Much of the reading I did related to the possible phylogenetic relationships and origin. I remember reading Williamson's [1887] monographic work on *Stigmaria ficoides*.
>
> After I graduated from Wisconsin I received an assistantship at the University of Illinois (1940–41) and signed up with John Buchholz to do a master's thesis. When we sat down to discuss the problem I told him of my interest in the evolutionary study of *Isoetes*. He said he was not well enough versed in the subject, but thought Jim Schopf, who had just gotten his degree with Buchholz the previous year, might help out. So that is how I got interested in paleobotany. At the time, Jim Schopf was working with the Illinois State Geological Survey. It was in Jim's laboratory that I saw my first coal ball. As I recall, they were from the Nashville, Illinois, locality. I never had a course in paleobotany or one in geology. So you might say I came into paleobotany through the back door—a door that Jim Schopf opened for me. (Phillips et al., 1973, p. 31)

And Bill Stewart in turn helped open the door for many students in paleobotany, especially in coal-ball studies, with theses published by Theodore Delevoryas (1955), E. Jeanne Morgan (Willis) (1959), Thomas N. Taylor (1965), Margaret

Figure 9. Wilson N. Stewart, former professor of botany at the University of Illinois, Urbana, where he established the largest coal-ball collections and training program in the 1950–1960 era of paleobotanical expansions. Photograph provided by W. N. Stewart.

Kain Balbach (1962, 1965, 1966a, b, 1967), Julian M. Frankenberg (Frankenberg and Eggert, 1969), and later Gar W. Rothwell (1975).

The "Stewart days" at Urbana (1947–1965) were quite exciting, and there were numerous other students with paleobotanical interests, including Grace Somers Brush, Robert A. Evers, Jean H. Langenheim, and David L. Dilcher, in addition to Robert M. Kosanke and his students, Frederick W. Cropp, Katherine M. Schaeffer, Russel A. Peppers, and Lewis R. Gray. Kosanke, at the Illinois Geological Survey, had earlier (1952) finished his doctorate under Stewart and also had appointments in the Departments of Botany and Geology. His role in training the students in palynology and in the many coal-ball acquisitions was particularly important for the growth of the research programs.

Regarding teaching, Stewart (Phillips et al., 1973, p. 32) wrote: "The paleobotany course just grew like Topsy. If students like the course, it may have something to do with a natural enthusiasm I have for the material and the fact that I have tried to build a teaching collection that will illustrate many of the things we talk about. If I have any objective in teaching paleobotany or any other course, it is to stimulate students' interest in their world. Some want to investigate it in detail and become graduate students in paleobotany." After Stewart's early retirement from the University of Alberta that enthusiastic and well-illustrated style of teaching was captured in his textbook, *Paleobotany and the Evolution of Plants* (Stewart, 1983; Stewart and Rothwell, 1993), which draws heavily on the studies of coal-ball plants.

From his first course in paleobotany in 1949, Stewart's influence was noted by John W. Hall:

> When Bill Stewart began teaching paleobotany, I was also well along in my thesis [with Oswald Tippo] so that it was not feasible to change. There is no question about Stewart's influence, however. In my final year at Illinois, Bill was interviewed for a position at Minnesota, teaching general botany. He turned it own, and in turn recommended me for the position. I accepted it, realizing that there was already one morphologist-anatomist there (Ernst Abbe), and that a paleobotanist (H. P. Banks) had once been there, but left two years before I arrived, it was a logical move to declare myself a paleobotanist. The summer after graduating, I scrambled around collecting coal balls at a number of localities and also got a number of peels from Stewart's material. These formed the nucleus of my paleobotany course. (Phillips et al., 1973, p. 32–33)

Hall was particularly noted for the painstaking observations evident in his coal-ball studies (Hall, 1952, 1954, 1961; Melchior and Hall, 1961; Hall and Stidd, 1971) and for his generosity with coal-ball specimens. In the later 1950s he focused on Cretaceous palynology.

Gilbert A. Leisman of Emporia Kansas State College, like John Hall, "discovered" coal-ball paleobotany late in his graduate-school studies and shared this of Hall's teaching:

> I first became interested in paleobotany while taking Hall's course at Minnesota. In cutting coal balls from Iowa, I had the good fortune to find a new pteridosperm male fructification and I also tried my hand at macerating coal-ball fragments to obtain leaf cuticles. Both of these research experiences really fired my enthusiasm. However, by this time I was so far along in my Ph.D. research in ecology that it was impractical to change. Probably the lack of Ph.D. thesis research in paleobotany was my biggest drawback when I started serious paleobotanical research at Kansas State Teachers College. I almost literally had to train myself in the basics of research and literature search. Conversely, many ecological concepts and principles have proven useful to me in paleobotanical research. (Phillips et al., 1973, p. 33)

The network of connections from Stewart to Hall to Leisman ultimately led to a complete circuit with Benton M. Stidd of Western Illinois University. Stidd was attracted into coal-ball paleobotany by Leisman at Emporia, went on to study with Stewart at Urbana (Stidd, 1971), finishing with Tom Phillips, and, in turn, joined Hall at Minnesota for two years before establishing his own program at Western Illinois University.

Leisman also noted (letter to T. L. Phillips, May 3, 1971): "As far as influencing my paleobotanical thinking, Al Traverse probably was the first. As a neophyte at my first few AIBS [American Institute of Biological Sciences] meetings, Al went out of his way to compliment my papers and to encourage further work on my part. However, the person who probably has influenced me the most is Jim Schopf. Through conversation and correspondence he has consistently guided my thinking on basic concepts in paleobotany."

Leisman considered the paleofloristic and paleoecological comparisons between the Kansas coal-ball sources (Weir-Pittsburg, Mineral and Fleming, and Bevier coal bed) and the Herrin coal bed (as revealed at the Sahara Coal Co. Mine No. 6 in southern Illinois) a major project. Like Darrah's comparisons from Iowa, Leisman observed the great abundance of cordaites in Kansas coal balls compared to their paucity in the Herrin coal bed. He stated (Leisman, letter to T. L. Phillips, May 3, 1971). "There is also a notable increase in arborescent lycopod material as you progress from the oldest to the youngest stratum, the Herrin being the youngest." Leisman made many contributions on the coal-ball plants of Kansas and is noted for the comparative observations of specimens within the Interior coal province, especially lycopsids and sphenopsids (Leisman, 1962a, b, 1970; Leisman and Graves, 1964; Leisman and Stidd, 1967; Schlanker and Leisman, 1969; Leisman and Bucher, 1971; Leisman and Phillips, 1979).

Coal-ball studies at the University of Illinois with Stewart developed from some of the largest collections and from the most diverse American sources of coal balls, reflecting his philosophy and initial objectives (Fig. 10). Stewart, with his appreciation of morphological variation, wanted to have large numbers of plant specimens for studies. He recalled (letter to T. L. Phillips, September 27, 1971): "The first coal ball collecting trip was made with Jim Schopf and Bob Kosanke. We visited the New Delta mine which is only a couple of miles from the Sahara mine [Herrin coal bed, Illinois]. The coal balls had the same flora. We went to the area around Harrisburg and West Frankfort and ended up at Berryville. I am still completely fascinated by the degree of preservation that there is in coal balls. It's almost as though 'someone' had seen to it that the structure of Pennsylvanian plants was preserved just so we would have material to look at today."

Figure 10. Theodore Delevoryas (left) and Chester A. Arnold (right) examining a coal-ball fragment from the Calhoun coal bed of Illinois in the mid-1950s. Arnold (letter to T. L. Phillips, March 14, 1971) wrote, "Wilson Stewart took Ted Delevoryas and me to the Berryville locality. It was then that I found the coal ball containing the three *Calamostachys americana cones* [Arnold, 1958]."

Stewart and his students and colleagues went on to amass the largest coal-ball collections in the United States. This work was motivated, in part, because of his early aims to establish the coal-ball floras of the entire American coal belt then available in the 1950s. This proved too extensive a project with Stewart's administrative and teaching commitments. However, the initial surveys, aided by visits to Elso Barghoorn's laboratory at Harvard University (Thompson and Darrah Collections), Ted Delevoryas's laboratory at Yale University, Fredda Reed's at Mount Holyoke College, and Maxine Abbott's at the University of Cincinnati (Hoskins Collections), gave Stewart an excellent overview of the paleofloristic differences.

The initial interests, engendered by Schopf, on the medullosans was evident in the complementary and comprehensive studies of Stewart and his students (Stewart, 1951, 1954, 1958; Delevoryas, 1955; Stewart and Delevoryas, 1952, 1956; Taylor and Delevoryas, 1964; Taylor, 1965). In the early 1950s, two of these students, Theodore Delevoryas and Jeanne Morgan, were known among some of us as the "coal [gold!] dust twins" because of their pioneering studies on the coenopterid ferns (Delevoryas and Morgan, 1952, 1954a, b; Morgan and Delevoryas, 1954). Their many other studies included their benchmark monographs on *Medullosa* (Delevoryas, 1955) and *Psaronius* (Morgan, 1959).

Jeanne Morgan Willis joined the faculty at Otterbein College in Westerville, Ohio, where she went on to become a department chair in the Division of Science and Mathematics. Delevoryas became the paleobotanist at Yale University and, in between his two tenures there, briefly returned to the faculty at Urbana from 1960 to 1962. At that time Robert M. Kosanke was also teaching in the Botany and Geology Departments. Donald A. Eggert spent a postdoctoral year at Urbana before establishing the first paleobotany program at Southern Illinois University, Carbondale, and later at the University of Iowa. In 1961, a fourth paleobotanist, Tom L. Phillips, joined the Urbana faculty. It was an extremely interesting time with a diversity of graduate studies, including those of Thomas N. Taylor, Margaret Kain Balbach, David L. Dilcher (from John Hall's laboratory), and Julian M. Frankenberg, among others. Taylor, after a postdoctorate at Yale with Delevoryas, went on to establish highly productive coal-ball research programs at the University of Illinois at Chicago Circle (later joined by Eggert, presently there), then Ohio University (replaced by Gar Rothwell), and now Ohio State University where both T. N. Taylor and Edith Smoot Taylor have followed in Schopf's pioneering path of Antarctic studies in paleobotany. Margaret K. Balbach became an outstanding teacher at the University of Illinois, subsequently moving to Eastern Illinois and then Illinois State University as a noted horticulturalist and newspaper writer. David Dilcher, after finishing his doctorate at Yale with Delevoryas, replaced Canright at Indiana University and became a leading force in the studies of fossil angiosperms.

In the mid-1960s Stewart moved to the University of Alberta, Canada, and there established a broad paleobotanical program still with a significant coal-ball aspect. In 1965 Andrews moved to the University of Connecticut, shifting his emphasis to Devonian paleobotany. In 1961, Aureal T. Cross joined the faculty at Michigan State University. Chester A. Arnold was still at Michigan University; John Hall at the University of Minnesota; Robert W. Baxter at the University of Kansas; Gilbert A. Leisman at Kansas State Teachers College, Emporia; Thomas N. Taylor (Fig. 11) and Donald A. Eggert at the University of Illinois, Chicago Circle; Tom L. Phillips at the University of Illinois, Urbana; James M. Schopf at Columbus, Ohio; Sergius H. Mamay at the U.S. Geological Survey in Washington, D.C.; Theodore Delevoryas at Yale University; and Elso Barghoorn at Harvard University. These paleobotanists and their programs were particularly important in research and/or training that incorporated potentials for coal-ball studies. They represented only a part of the overall growth of American paleobotany from the time of A. C. Noé and constitute an important heritage of guidance for the later generations that have followed.

Figure 11. Coal balls in the Herrin coal bed at Sahara Coal Company Mine No. 6 near Carrier Mills, Illinois, were collected over several decades (1960s–1980s). Among the collectors, shown left to right in this 1962 photograph, Henry (Hank) Harris, who always helped with paleobotanical field trips from the University of Illinois, Thomas N. Taylor (front center), Marie Leisman (back center), and Gilbert A. Leisman. Photograph provided by Tom Phillips.

## EPILOGUE

Several important characteristics of North American coal-ball studies are evident in hindsight since the pioneering days of Noé, Hoskins, Reed, Graham, Schopf, Darrah, Andrews, Cross, and others. The paleogeographic and stratigraphic potentials for such research have expanded in the Midcontinent and especially within the Appalachian basin, making "permanent" coal-ball collections and extensive reference sources a necessity for future systematic and evolutionary studies; these collections are required also for basic paleoecologic research related to plants and peat-coal relationships. The precedence of Darrah, Stewart, Baxter, and others has been followed in amassing such coal-ball collections, and the research and teaching importance of such collections will continue to increase with time as has that of collections in Europe.

In Andrews's (1951) review of "American coal-ball floras" he captured the enthusiasm of the time: "The tendency to date has been to describe those fossils that are new, especially conspicuous and well preserved and we are still a long way from the end of this 'cream skimming' stage." To be sure, much of the potential of North American coal-ball studies has been developed by establishing the nature of the plants, or at least of their parts, which is essential for reconstruction of whole plant assemblages and life history biology. The initiatives of the past several decades indicate both the interests and capabilities of doing this well into the next century—another type of "cream skimming." However, one should not overlook the labor and insight required to get to this stage. The progress in coal-ball studies thus far is essential in establishing the plants, their diversity, stratigraphic distribution, and lineages. This botanical emphasis has been paramount to reconstructing a major cast of characters in a different world of tropical biology in the late Carboniferous.

With the increased emphasis on whole plant biology, the challenges of paleobiology are becoming more and more integrative with the development of paleoecology, which links both long-term evolutionary studies and evolutionary theory with ecosystem dynamics of the ancient tropics. The growth of studies in paleoecology, including vegetational analyses, growth and reproductive biology, evidence of plant-animal interactions, litter history, and the roles of microorganisms, collectively represents thresholds of questions and problems now within reach of younger generations of paleobotanists. Concurrently, the distinctiveness of coal-ball studies is likely to diminish by the continued interconnections with coal palynology and compression plant studies—all of which help frame the reconstructed landscapes of the late Carboniferous. In turn, the historical lineages of "coal-ball paleobotanists" will be blurred further by the diversity of their contributions. Coal-ball studies provide an anatomical reference base that is being interconnected botanically and geologically at numerous levels of resolutions from plants and coal to paleoclimate.

Coal-ball studies of the past and the future represent a strong influence on botanical studies, reflecting the interests and training of several generations of paleobotanists. This serves paleobiology well. However, relatively few students have been involved in the aspects of coal balls that relate to coal geology, coalification, coal anatomy, coal palynology, and coal-ball origins and distribution. These avenues of coal-ball research also offer substantial opportunities to contribute to reconstructive studies with both basic and applied consequences.

## ACKNOWLEDGMENTS

We gratefully acknowledge the following contributors who generously provided information and permitted use of quotes from their correspondence and other personal communications: Henry N. Andrews, Chester A. Arnold, Harriet V. Krick Bartoo, Robert W. Baxter, James E. Canright, Philip J. DeMaris, John W. Hall, Gilbert A. Leisman, Paul C. Lyons, Sergius H. Mamay, Elsie D. Morey, Fredda D. Reed, James M. Schopf, Wilson N. Stewart, and Alfred Traverse. We especially thank manuscript reviewers H. N. Andrews, P. C. Lyons, T. N. Taylor, and Christopher Wnuk. We also thank Elsie Darrah Morey for providing correspondence of F. O. Thompson to W. C. Darrah and for permission to quote from additional correspondence of W. C. Darrah from the William Culp Darrah Collection. Carol Kubitz rendered the stratigraphic chart.

## REFERENCES CITED

Anderson, T. F., Brownlee, M. E., and Phillips, T. L., 1980, A stable isotope study on the origin of permineralized peat zones in the Herrin Coal: Journal of Geology, v. 88, p. 713–722.

Andrews, N. H., Jr., 1940, On the stellar anatomy of the pteridosperms, with particular reference to the secondary wood: Missouri Botanical Garden Annals, v. 27, p. 51–118.

Andrews, H. N., Jr., 1942a, Contributions to our knowledge of American Carboniferous floras. Part 1: *Scleropteris,* gen. nov., *Mesoxylon,* and *Amyelon:* Missouri Botanical Garden Annals, v. 29, p. 1–18.

Andrews, H. N., Jr., 1942b, Contributions to our knowledge of American Carboniferous floras. Part 5: *Heterangium:* Missouri Botanical Garden Annals, v. 29, p. 275–316.

Andrews, H. N., Jr., 1945, Contributions to our knowledge of American Carboniferous floras. Part 7: Some pteridosperm stems from Iowa: Missouri Botanical Garden Annals, v. 32, p. 323–360.

Andrews, H. N., Jr., 1946, Coal balls—A key to the past: Scientific Monthly, v. 62, p. 327–334.

Andrews, H. N., Jr., 1947, Ancient plants and the world they lived in: Ithaca, New York, Comstock Publishing, 279 p. (reprinted 1963).

Andrews, H. N., Jr., 1948, Some evolutionary trends in the pteridosperms: Botanical Gazette, v. 110, p. 13–31.

Andrews, H. N., Jr., 1949, *Nucellangium,* a new genus of fossil seeds previously assigned to *Lepidocarpon:* Missouri Botanical Garden Annals, v. 36, p. 479–505.

Andrews, H. N., Jr., 1951, American coal-ball floras: Botanical Review, v. 17, p. 430–469.

Andrews, H. N., Jr., 1980, The fossil hunters: Ithaca, New York, Cornell University Press, 421 p.

Andrews, H. N., Jr., 1990, Frederick Maurice Wonnacott 1902–1990: International Organization of Palaeobotany Newsletter 43, p. 6–8.

Andrews, H. N., Jr., and Mamay, S. H., 1952, A brief conspectus of American coal ball studies: The Palaeobotanist, v. 1, p. 66–72.

Andrews, H. N., Jr., and Murdy, W. H., 1958, *Lepidophloios*—and ontogeny in arborescent lycopods: American Journal of Botany, v. 45, p. 552–560.

Andrews, H. N., Jr., and Pannell, E., 1942, Contributions to our knowledge of American Carboniferous floras. Part 2: *Lepidocarpon:* Missouri Botanical Garden Annals, v. 29, p. 19–34.

Arnold, C. A., 1958, Petrified cones of the genus *Calamostachys* from the Carboniferous of Illinois: University of Michigan Museum Paleontology Contribution, v. 14, p. 149–164.

Arnold, C. A., and Steidtmann, W. E., 1937, Pteridospermous plants from the Pennsylvanian of Illinois and Missouri: American Journal of Botany, v. 24, p. 644–650.

Bain, H. F., 1894, Geology of Keokuk Co.: Iowa Geological Survey Annual Report 1894, v. 4, p. 255–311.

Balbach, M. K., 1962, Observations on the ontogeny of *Lepidocarpon:* American Journal of Botany, v. 49, p. 984–989.

Balbach, M. K., 1965, Paleozoic lycopsid fructifications. 1: *Lepidocarpon* petrifactions: American Journal of Botany, v. 52, p. 317–330.

Balbach, M. K., 1966a, Paleozoic lycopsid fructifications. 2: *Lepidostrobus takhtajanii* in North America and Great Britain: American Journal of Botany, v. 53, p. 275–283.

Balbach, M. K., 1966b, Microspore variation in *Lepidostrobus* and comparison with *Lycospora:* Micropaleontology,v. 12, p. 334–342.

Balbach, M. K., 1967, Paleozoic lycopsid fructifications. 3: Conspecificity of British and North American petrifactions: American Journal of Botany, v. 54, p. 867–875.

Bastin, E. S., 1940, Obituary—Roy Graham (1908–1939): Science, v. 91, p. 87–88.

Baxter, R. W., and Hornbaker, A. L., 1965, Pennsylvanian fossil plants from Kansas coal balls: Field Conference Guidebook for the Annual Meetings, Geological Society of America and Associated Societies, p. 24.

Benninghoff, W. S., 1942, Preliminary report on a coal ball flora from Indiana: Indiana Academy of Science Proceedings, v. 52, p. 62–68.

Berry, E. W., 1924, Paleobotany at the New York State Museum: Science, v. 59, p. 336–337.

Brush, G. S., and Barghoorn, E. S., 1955, *Kallostachys scottii:* A new genus of sphenopsid cones from the Carboniferous: Phytomorphology, v. 5, p. 346–356.

Cady, G. H., 1915, Coal resources of District I (Longwall): Illinois Geological Survey, Cooperative Coal Mining Inventory Bulletin 10, 149 p.

Cady, G. H., 1919, Geology and mineral resources of the Hennepin and La Salle Quadrangles: Illinois Geological Survey Bulletin 37, 136 p.

Canright, J. E., 1958, History of paleobotany in Indiana: Indiana Academy of Sciences Proceedings, v. 67, p. 268–273.

Canright, J. E., 1959, Fossil plants of Indiana: Indiana Geological Survey Report of Progress 14, 45 p.

Coulter, J. M., and Land, W. J. G., 1911, An American *Lepidostrobus:* Botanical Gazette, v. 51, p. 449–453.

Coulter, J. M., and Land, W. J. G., 1921, A homosporous American *Lepidostrobus:* Botanical Gazette, v. 72, p. 106–108.

Croneis, C. G., 1940, Memorial to Adolf Carl Noé [1873–1939]: Geological Society of America Proceedings (1939), p. 219–227.

Cross, A. T., 1952, The geology of Pittsburgh Coal, *in* Second Conference on the Origin and Constitution of Coal: Crystal Cliffs, Nova Scotia Department of Mines, p. 32–99.

Cross, A. T., 1969, Coal balls: Their origin, nature, geological significance and distribution [abs.]: Geological Society of America Abstracts with Programs, Part 7, p. 42.

Darrah, W. C., 1936, The peel method in paleobotany: Harvard University Botanical Museum Leaflets, v. 4, p. 69–83.

Darrah, W. C., 1939, The fossil flora of Iowa coal balls. Part 1: Discovery and occurrence: Harvard University Botanical Museum Leaflets, v. 7, p. 125-136.

Darrah, W. C., 1941a, Studies of American coal balls: American Journal of Science, v. 239, p. 33–53.

Darrah, W. C., 1941b, The coenopterid ferns in American coal balls: American Midland Naturalist, v. 24, p. 233–269.

Darrah, W. C., 1941c, The fossil flora of Iowa coal balls. Part 4: *Lepidocarpon:* Harvard University Botanical Museum Leaflets, v. 9, p. 85–100.

Darrah, W. C., 1969, A critical review of the upper Pennsylvanian floras of eastern United States with notes on the Mazon Creek flora of Illinois: Gettysburg, Pennsylvania, privately printed, 220 p., 80 pls.

Delevoryas, T., 1955, The Medullosae—Structure and relationships: Palaeontographica, v. 97B, p. 114–167.

Delevoryas, T., and Morgan, E. J., 1952, *Tubicaulis multiscalariformis*—A new American coenopterid: American Journal of Botany, v. 39, p. 160–166.

Delevoryas, T., and Morgan, E. J., 1954a, A further investigation of the morphology of *Anachoropteris clavata:* American Journal of Botany, v. 41, p. 192–198.

Delevoryas, T., and Morgan, E. J., 1954b, Observations on petiolar branching and foliage of an American *Botryopteris:* American Midland Naturalist, v. 52, p. 374–387.

DeMaris, P. J., Bauer, R. A., Cahill, R. A., and Damberger, H. H., 1983, Geologic investigation of roof and floor strata: Longwall demonstration, Old Ben Mine No. 24. Prediction of coal balls in the Herrin Coal: Final Technical Report, Part 2, Illinois State Geological Survey Contract/Grant Report 1983-2, 69 p.

DiMichele, W. A., and Phillips, T. L., 1988, Paleoecology of the Middle Pennsylvanian–age Herrin Coal swamp (Illinois) near a contemporaneous river system, the Walshville paleochannel: Review of Palaeobotany and Palynology, v. 56, p. 151–176.

Eggert, D. A., 1961, The ontogeny of Carboniferous arborescent Lycopsida: Palaeontographica, Abt. B, v. 108, p. 43–92.

Eggert, D. A., 1962, The ontogeny of Carboniferous arborescent sphenopsida: Palaeontographica, Abt. B, v. 110, p. 99–127.

Evans, W. D., and Amos, D. H., 1961, An example of the origin of coal balls: Geologists' Association Proceedings, v. 72, p. 447–454.

Feliciano, J. M., 1924, The relations of concretions to coal seams: Journal of Geology, v. 32, p. 230–239.

Felix, C. J., 1952, A study of the arborescent lycopods of southeastern Kansas: Missouri Botanical Garden Annals, v. 39, p. 263–288.

Felix, C. J., 1954, Some American arborescent lycopod fructifications: Missouri Botanical Garden Annals, v. 41, p. 351–394.

Frankenberg, J. M., and Eggert, D. A., 1969, Petrified *Stigmaria* from North America. Part I: *Stigmaria ficoides,* the underground portions of Lepidodendraceae: Palaeontographica, Abt. B, v. 128, p. 1–47.

Gensel, P. G., and Andrews, H. N., 1984, Plant life in the Devonian: New York, Praeger Publishers, 380 p.

Graham, R., 1933, Preparation of palaeobotanical sections by the peel method: Stain Technology, v. 8, p. 65–68.

Graham, R., 1934, Pennsylvanian flora of Illinois as revealed in coal balls: Botanical Gazette, v. 95, p. 453–476.

Graham, R., 1935a, Pennsylvanian flora of Illinois as revealed in coal balls. Part 2: Botanical Gazette, v. 97, p. 156–168.

Graham, R., 1935b, An anatomical study of the leaves of the Carboniferous arborescent lycopods: Annals of Botany, v. 49, p. 587–608.

Gresley, W. S., 1887, Notes on the formation of coal-seams, as suggested by evidence collected chiefly in the Leicestershire and South Derbyshire coalfields: Geological Society of London Quarterly Journal, v. 43, p. 671–674.

Gresley, W. S., 1893, Note on anthracite "coal-apples" from Pennsylvania: American Institute of Mining Engineers Transactions, v. 21, p. 824–832.

Gresley, W. S., 1899a, Side-light upon coal formation: American Geologist, v. 23, p. 69–80.

Gresley, W. S., 1899b, Possible new coal plants in coal: American Geologist, v. 24, p. 199–204.

Gresley, W. S., 1900, Possible new coal-plants in coal. Part II: American Geologist, v. 26, p. 49–59.

Gresley, W. S., 1901, Possible new coal-plants etc., in coal. Part III: American Geologist, v. 27, p. 6–14.

Hall, J. W., 1952, The phloem of *Heterangium americanum:* American Midland Naturalist, v. 47, p. 763–768.

Hall, J. W., 1954, The genus *Stephanospermum* in American coal balls: Botanical Gazette, v. 115, p. 346–360.

Hall, J. W., 1961, *Anachoropteris involuta* and its attachment to a *Tubicaulis* type of stem from the Pennsylvanian of Iowa: American Journal of Botany, v. 48, p. 731–737.

Hall, J. W., and Stidd, B. M., 1971, Ontogeny of *Vesicaspora,* a late Pennsylvanian pollen grain: Palaeontology, v. 14, p. 431–436.

Hatcher, P. G., Lyons, P. C., Thompson, C. L., Brown, F. W., and Maciel, G. E., 1982, Organic matter in a coal ball: Peat or coal?: Science, v. 217, p. 831–833.

Hooker, J. D., and Binney, E. W., 1855, On the structure of certain limestone nodules enclosed in seams of bituminous coal, with a description of some trigonocarpons contained in them: Royal Society of London Philosophical Transactions, v. 145, p. 149–156.

Hoskins, J. H., 1923, A Paleozoic angiosperm from an American coal ball: Botanical Gazette, v. 75, p. 390–399.

Hoskins, J. H., and Abbott, M. L., 1956, *Selaginellites crassicinctus,* a new species from the Desmoinesian Series of Kansas: American Journal of Botany, v. 43, p. 36–46.

Hoskins, J. H., and Cross, A. T., 1941, A consideration of the structure of *Lepidocarpon* Scott based on a new strobilus from Iowa: American Midland Naturalist, v. 25, p. 523–547.

Hoskins, J. H., and Cross, A. T., 1942, New interpretation of *Sphenophyllostachys* based on a petrified specimen from an Iowa coal ball: Illinois, Academy of Sciences Transactions, v. 35, p. 68–69.

Hoskins, J. H., and Cross, A. T., 1943, Monograph of the Paleozoic cone genus *Bowmanites* (Sphenophyllales): American Midland Naturalist, v. 30, p. 113–163.

Hoskins, J. H., and Cross, A. T., 1946a, Studies in the Trigonocarpales. Part 1: *Pachytesta vera,* a new species from the Des Moines Series of Iowa: American Midland Naturalist, v. 36, p. 207–250.

Hoskins, J. H., and Cross, A. T., 1946b, Studies in the Trigonocarpales. Part 2: Taxonomic problems and a revision of the genus *Pachytesta:* American Midland Naturalist, v. 36, p. 331–361.

Johnson, P. R., 1979, Petrology and environments of deposition of the Herrin (No. 6) Coal Member,Carbondale Formation, at the Old Ben Coal Company Mine No. 24, Franklin County, Illinois [M.S. thesis]: Urbana-Champaign, University of Illinois, 169 p.

Just, T. K., 1957, John Hobart Hoskins (1896–1957): Lloydia, v. 20, p.ii–vi.

Kindle, E. M., 1934, Concerning "lake balls," "*Cladophora* balls" and "coal balls": American Midland Naturalist, v. 15, p. 752–760.

Kosanke, R. M., 1979, Memorial to James Morton Schopf 1911–1978: Geological Society of America Memorials, v. 9, p. 1–4.

Krick, H. V., 1932, Structure of seedlike fructifications found in coal balls from Harrisburg, Illinois: Botanical Gazette, v. 93, p. 151–172.

Kukuk, P., 1909, Über Torfdolomite in den Flözen der niederrheinisch—westfälischen Steinkohlenablagerung: Glückauf Berg-und Hüttenmännische Zeitschrift, No. 32 (7 August 1909), p. 1137–1150.

Leisman, G. A., 1962a, A *Spencerites* sporangium and associated spores from Kansas: Micropaleontology, v. 8, p. 396–402.

Leisman, G. A., 1962b, *Spencerites moorei* comb. nov. from southeastern Kansas: American Midland Naturalist, v. 68, p. 347–356.

Leisman, G. A., 1970, A petrified *Sporangiostrobus* and its spores from the middle Pennsylvanian of Kansas: Palaeontographica, Abt. B, v. 129, p. 166–177.

Leisman, G. A., and Bucher, J. L., 1971, Variability in *Calamocarpon insignis* from the American Carboniferous: Journal of Paleontology, v. 45, p. 494–501.

Leisman, G. A., and Graves, C., 1964, The structure of the fossil sphenopsid cone, *Peltastrobus reedae:* American Midland Naturalist, v. 72, p. 426–437.

Leisman, G. A., and Phillips, T. L., 1979, Megasporangiate and microsporangiate cones of *Achlamydocarpon varius* from the Middle Pennsylvanian: Palaeontographica, Abt. B, v. 168, p. 100–128.

Leisman, G. A., and Stidd, B. M., 1967, Further occurrences of *Spencerites* from the Middle Pennsylvanian of Kansas and Illinois: American Journal of Botany, v. 54, p. 316–323.

Lyons, P. C., 1994, Aureal T. Cross, Gordon H. Wood, Jr., Memorial Award: Northeastern Geology, v. 15, p. 238–240.

Lyons, P. C., and Morey, E. D., 1991, Memorial to William Culp Darrah (1909–1989): Torrey Botanical Club Bulletin, v. 118, p. 195–200.

Lyons, P. C., Thompson, C. L., Hatcher, P. G., Brown, F. W., Millay, M. A., Szeverenyi, N., and Maciel, G. E., 1984, Coalification of organic matter in coal balls of the Pennsylvanian (Upper Carboniferous) of the Illinois Basin, United States: Organic Geochemistry, v. 5, p. 227–239.

Lyons, P. C., Hatcher, P. G., Brown, F. W., Krasnow, M. R., Larson, R. R., and Millay, M. A., 1985, Role of static load (overburden) pressure in coalification of bituminous and anthracite coal, *in* Proceedings, International Conference on Coal Science, Sydney, New South Wales, October 1985: Sydney, Australia, Pergamon Press, p. 620–623.

Mamay, S. H., 1950, Some American Carboniferous fern fructifications: Missouri Botanical Garden Annals, v. 37, p. 409–476.

Mamay, S. H., 1952, An epiphytic American species of *Tubicaulis* Cotta: Annals of Botany, new series, v. 16, p. 145–163.

Mamay, S. H., 1957, *Biscalitheca,* a new genus of Pennsylvanian coenopterids, based on its fructification: American Journal of Botany, v. 44, p. 229–239.

Mamay, S. H., 1976, Paleozoic origin of the cycads: U.S. Geological Survey Professional Paper 934, p. 1–48.

Mamay, S. H., and Andrews, H. N., Jr., 1950, A contribution to our knowledge of the anatomy of *Botryopteris:* Torrey Botanical Club Bulletin, v. 77, p. 462–494.

Mamay, S. H., and Yochelson, E. L., 1953, Floral-faunal associations in American coal balls: Science, v. 118, p. 240–241.

Mamay, S. H., and Yochelson, E. L., 1962, Occurrence and significance of marine animal remains in American coal balls: U.S. Geological Survey Professional Paper 354-I, p. 193–224.

McCullough, L. A., 1977, Early diagenetic calcareous coal balls and roof shale concretions from the Pennsylvanian (Allegheny Series): Ohio Journal of Science, v. 77, p. 125–134.

Melchior, R. C., and Hall, J. W., 1961, Calamitean shoot apex from the Pennsylvanian of Iowa: American Journal of Botany, v. 48, p. 811–815.

Morgan, E. J., 1959, The morphology and anatomy of American species of the genus *Psaronius:* Urbana, Illinois, University of Illinois Press, Illinois Biological Monographs 27, 108 p.

Morgan, E. J., and Delevoryas, T., 1954, An anatomical study of a new coenopterid and its bearing on the morphology of certain coenopterid petioles: American Journal of Botany, v. 41, p. 198–203.

Noé, A. C., 1923a, Coal balls: Science, v. 57, p. 385.

Noé, A. C., 1923b, A Paleozoic angiosperm: Journal of Geology, v. 31, p. 344–347.

Noé, A. C., 1930, Celluloid films from coal balls: Botanical Gazette, v. 89, p. 318–319.

Noé, A. C., 1932, Review of American coal-ball studies: Illinois Academy of Sciences Transactions, v. 24, p. 317–320.

Oliver, F. W., 1935, Dukinfield Henry Scott, 1854–1934: Annals of Botany, v. 49, p. 823–840.

Pannell, E., 1942, Contributions to our knowledge of American Carboniferous floras. Part 4: A new species of *Lepidodendron:* Missouri Botanical Garden Annals, v. 29, p. 245–274.

Perkins, T. W., 1976, Textures and conditions of formation of Middle Pennsylvanian coal balls, central United States: The University of Kansas Paleontological Contributions Paper 82, p. 1–13.

Phillips, T. L., 1980, Stratigraphic and geographic occurrences of permineralized coal-swamp plants—Upper Carboniferous of North America and Europe, *in* Dilcher, D. L., and Taylor, T. N., eds., Biostratigraphy of fossil plants: Stroudsburg, Pennsylvania, Dowden, Hutchinson and Ross, p. 25–92.

Phillips, T. L., Pfefferkorn, H. W., and Peppers, R. A., 1973, Development of paleobotany in the Illinois Basin: Illinois State Geological Survey Circular 480, 86 p.

Phillips, T. L., Avcin, M. J., and Schopf, J. M., 1975, Gametophytes and young sporophyte development in *Lepidocarpon* [abs.]: Corvallis, Oregon, Botanical Society of America, p. 23.

Phillips, T. L., Peppers, R. A., and DiMichele, W. A., 1985, Stratigraphic and interregional changes in Pennsylvanian coal-swamp vegetation: Environmental inferences: International Journal of Coal Geology, v. 5, p. 43–109.

Rao, C. P., 1985, Origin of coal balls of the Illinois Basin, *in* Cross, A. T., ed., Economic Geology—Coal, Oil and Gas: Congrès International de Stratigraphie et de Géologie du Carbonifère, 9th, Washington, D.C., and Urbana-Champaign, Illinois, 1979: Compte Rendu, v. 4, p. 393–406.

Ravn, R. L., 1986, Palynostratigraphy of the Lower and Middle Pennsylvanian coals of Iowa: Iowa Geological Survey Technical Paper 7, 245 p.

Ravn, R. L., Swade, J. W., Howes, M. R., Gregory, J. L., Anderson, R. R., and Van Dorpe, P. E., 1984, Stratigraphy of the Cherokee Group and revision of Pennsylvanian stratigraphic nomenclature in Iowa: Iowa Geological Survey Technical Information Series 12, 76 p.

Raymond, A., 1988, The paleoecology of a coal-ball deposit from the Middle Pennsylvanian of Iowa dominated by cordaitalean gymnosperms: Review of Palaeobotany and Palynology, v. 53, p. 233–250.

Read, C. B., and Mamay, S. H., 1964, Upper Paleozoic floral zones and floral provinces of the United States: U.S. Geological Survey Professional Paper 454-K, p. K1–K35.

Reed, F. D., 1926, Flora of an Illinois coal ball: Botanical Gazette, v. 81, p. 460–469.

Roberts, D. C., and Barghoorn, E. S., 1952, *Medullosa olseniae:* A Permian *Medullosa* from north central Texas: Harvard University Botanical Museum Leaflets, v. 15, p. 191–200.

Rothwell, G. W., 1975, The Callistophytaceae (Pteridospermopsida). I: Vegetative structures: Palaeontographica, Abt. B, v. 151, p. 171–196.

Scheihing, M. H., 1978, A paleoenvironmental analysis of the Shumway Cyclothem (Virgilian), Effingham County, Illinois [M.S. thesis]: Urbana-Champaign, University of Illinois, 184 p.

Schlanker, C. M., and Leisman, G. A., 1969, The herbaceous Carboniferous lycopod *Selaginella fraiponti* comb. nov.: Botanical Gazette, v. 130, p. 35–41.

Schopf, J. M., 1936, The paleobotanical significance of plant structure in coal: Illinois Academy of Sciences Transactions, v. 28, p. 106–110.

Schopf, J. M., 1937, Two new lycopod seeds from the Illinois Pennsylvanian: Illinois Academy of Sciences Transactions, v. 30, p. 139–146.

Schopf, J. M., 1938, Coal balls as an index to the constitution of coal: Illinois Academy of Sciences Transactions, v. 31, p. 187–189.

Schopf, J. M., 1939, *Medullosa distelica,* a new species of the Anglica group of *Medullosa:* American Journal of Botany, v. 26, p. 196–207.

Schopf, J. M., 1941, Contribution to Pennsylvanian paleobotany: *Mazocarpon oedipternum* sp. nov., and sigillarian relationships: Illinois Geological Survey Report of Investigations 75, 53 p.

Schopf, J. M., 1948, Pteridosperm male fructifications—American species of *Dolerotheca,* with notes regarding certain allied forms: Journal of Paleontology, v. 22, p. 681–724.

Schopf, J. M., 1971, Notes on paleobotany related to Illinois: unpublished, sent to T. L. Phillips with covering letter, March 10, 1971, designated LR-341, 16 p.

Schopf, J. M., 1975, Modes of fossil preservation: Review of Palaeobotany and Palynology, v. 20, p. 27–53.

Schopf, J. M., Wilson, L. R., and Bentall, R., 1944, An annotated synopsis of Paleozoic fossil spores and the definition of generic groups: Illinois Geological Survey Report of Investigations 91, 74 p.

Scott, A. C., and Rex, G., 1985, The formation and significance of Carboniferous coal balls: Royal Society of London Philosophical Transactions, v. 311B, p. 123–137.

Seward, A. C., 1923, A supposed Paleozoic angiosperm: Botanical Gazette, v. 76, p. 215.

Steidtmann, W. E., 1937, A preliminary report on the anatomy and affinity of *Medullosa noei* sp. nov. from the Pennsylvanian of Illinois: American Journal of Botany, v. 24, p. 124–125.

Steidtmann, W. E., 1944, The anatomy and affinities of *Medullosa noei* Steidtmann, and associated foliage, roots, and seeds: University of Michigan Museum of Paleontology Contributions, v. 6, p. 131–166.

Stewart, W. N., 1947, A comparative study of stigmarian appendages and *Isoetes* roots: American Journal of Botany, v. 34, p. 315–324.

Stewart, W. N., 1951, *Medullosa pandurata* sp. nov. from the McLeansboro Group of Illinois: American Journal of Botany, v. 38, p. 709–717.

Stewart, W. N., 1954, The structure and affinities of *Pachytesta illinoense* comb. nov.: American Journal of Botany, v. 41, p. 500–508.

Stewart, W. N., 1958, The structure and relationships of *Pachytesta composita* sp. nov.: American Journal of Botany, v. 45, p. 580–588.

Stewart, W. N., 1983, Paleobotany and the evolution of plants (first edition): New York, Cambridge University Press, 405 p.

Stewart, W. N., and Delevoryas, T., 1952, Bases for determining relationships among the Medullosaceae: American Journal of Botany, v. 39, p. 505–516.

Stewart, W. N., and Delevoryas, T., 1956, The medullosan pteridosperms: Botanical Review, v. 22, p. 45–80.

Stewart, W. N., and Rothwell, G. W., 1993, Paleobotany and the evolution of plants (second edition): New York, Cambridge University Press, 521 p.

Stidd, B. M., 1971, Morphology and anatomy of the frond of *Psaronius:* Palaeontographica, Abt. B, v. 134, p. 87–123.

Stidd, B. M., and Cosentino, K., 1976, *Nucellangium:* Gametophytic structure and relationship to cordaites: Botanical Gazette, v. 137, p. 242–249.

Stopes, M. C., and Watson, D. M. S., 1909, On the present distribution and origin of the calcareous concretions in coal seams, known as coal balls: Royal Society of London Philosophical Transactions, series B, v. 200, p. 167–218.

Taylor, T. N., 1965, Paleozoic seed studies: A monograph of the American species of *Pachytesta:* Palaeontographica, Abt. B, v. 117, p. 1–46.

Taylor, T. N., and Delevoryas, T., 1964, Paleozoic seed studies: A new Pennsylvanian *Pachytesta* from southern Illinois: American Journal of Botany, v. 51, p. 189–195.

Tilton, J. L., 1912, The first reported petrified American *Lepidostrobus* is from Warren County, Iowa: Iowa Academy of Sciences Proceedings, v. 19, p. 163–165.

Traverse, A., 1950, The primary vascular body of *Mesoxylon thompsonii,* a new American Cordaitalean: American Journal of Botany, v. 37, p. 318–325.

Vestal, A. G., 1987, History of Botany Department, University of Illinois (unpublished notes): Botany Newsletter, Department of Plant Biology, Urbana, p. 7.

Walton, J., 1928, A method of preparing fossil plants: Nature, v. 22, p. 571.

Walton, J., and Koopmans, R. G., 1928, Preparation of cellulose films and their use in making serial sections of coal ball plants: Glasgow, British Association for the Advancement of Science Report, p. 615,688.

Weber, J. N., and Keith, M. L., 1962, Carbon-isotope composition and the origin of calcareous coal balls: Science, v. 138, p. 900–901.

White, D., and Thiessen, R., 1913, The origin of coal: Washington, D.C., U.S. Government Printing Office, U.S. Bureau of Mines Bulletin 38, 390 p.

Wieland, G. R., 1924, Recent achievements in paleobotany: Science, v. 60, p. 233-235.

Williamson, W. C., 1891, General, morphological, and histological index to the author's collective memoirs on the fossil plants of the Coal Measures. Part 1: Memoirs and proceedings of the Manchester Literary and Philosophical Society, ser. 4, v. 4, p. 53–68.

Williamson, W. C., 1893, General, morphological, and histological index to the author's collective memoirs on the fossil plants of the Coal Measures. Part 2: Memoirs and proceedings of the Manchester Literary and Philosophical Society, ser. 4, v. 7, p. 91–127.

Manuscript Accepted by the Society July 6, 1994

Printed in U.S.A.

Geological Society of America
Memoir 185
1995

# *Taphonomic and sedimentologic characterization of roof-shale floras*

**Robert A. Gastaldo**
*Department of Geology, Auburn University, Auburn, Alabama 36849-5305*
**Hermann W. Pfefferkorn**
*Department of Geology, University of Pennsylvania, Philadelphia, Pennsylvania 19104-6316*
**William A. DiMichele**
*Department of Paleobiology, National Museum of Natural History, Washington, D.C. 20506*

## ABSTRACT

**Roof-shale floras have been a major source of data for the understanding of Carboniferous vegetation. Early debate on their origin centered around the question of whether these megafloral assemblages are autochthonous or allochthonous. In these discussions, the sedimentological context in which the preserved fossil assemblage (taphoflora) occurred was largely ignored. W. C. Darrah saw the complexity of these issues, presented helpful starting points for further investigations, and influenced the thinking of the next generation. This chapter characterizes the sedimentological and taphonomic features of a spectrum of roof-shale floras. There are three levels at which the preservation of plant parts can be viewed: (1) early taphonomic processes and earliest diagenesis can destroy or preserve plant parts in a given clastic depositional setting; (2) those plant parts that are preserved can be autochthonous, parautochthonous, or allochthonous in relationship to their original place of growth; (3) with respect to a peat layer (coal bed), the overlying clastic material can be deposited in a continuous transition, after a short temporal break (discontinuity), or after a significant hiatus of time. Characterization of roof-shale floras must take into consideration the sedimentological interpretation of the associated lithologies, the stratigraphic sequence, and the taphonomic processes involved in their formation. Characterization is essential before such floras can be used in higher-level interpretations, such as paleoecological reconstructions.**

## INTRODUCTION

A majority of all late Carboniferous adpression assemblages (taphofloras) are found as roof-shale floras, those preserved in the rock stratum overlying a coal seam. Many authors over the past nearly 100 years have attempted to clarify the relationships of these fossil assemblages to the bounding strata. In some instances these assemblages are genetically related to the underlying coal bed and represent the final phases of peat accumulation (for example, Scott, 1978; DiMichele and DeMaris, 1987) or thanatocoenoeses of peat swamp forest (death assemblages resulting from burial [for example, Demko and Gastaldo, 1992]). In other instances, these assemblages bear little or no genetic relationship to the underlying coal bed (for example, Peppers and Pfefferkorn, 1970; Baird et al., 1985).

The fossil floras under consideration have been variously called compression-impression floras, adpression[1] floras, fossil-plant asssemblages, megafloral assemblages, or taphofloras. We will use several of these terms interchangeably depending on the aspect of the fossil flora that we want to emphasize. The

*[1]Adpression—A plant fossil specimen showing a mixture of compression (plant parts compressed by sediment where some original or chemically altered plant tissue is still preserved) and impression (an imprint of the fossil plant on sediment or rock surface) states. Modified from Shute and Cleal (1987).

Gastaldo, R. A., Pfefferkorn, H. W., and DiMichele, W. A., 1995, Taphonomic and sedimentologic characterization of roof-shale floras, *in* Lyons, P. C., Morey, E. D., and Wagner, R. H., eds., Historical Perspective of Early Twentieth Century Carboniferous Paleobotany in North America (W. C. Darrah volume): Boulder, Colorado, Geological Society of America Memoir 185.

term *roof-shale flora* expresses the stratigraphic position of a plant-bearing bed above a coal seam but contains no taphonomic information in itself, as we will demonstrate in this chapter. Roof rocks themselves can be siltstones or sandstones, even though shales are most common. Adpression floras occur not only in roof rocks but also in floor rocks, often referred to as underclays (Scheihing and Pfefferkorn, 1984; Wnuk and Pfefferkorn, 1987), or in beds that have no direct or indirect relationship to coal beds. However, roof-shale floras are more common and represent a very specific setting in which the relationship to the coal bed begs for an explanation.

Paleobotanical diversity characterized the scientific interests of the late William C. Darrah (1909–1989). During the more than 50 years of his career he investigated topics ranging from pure systematics (for example, Darrah, 1940, 1969), developmental paleobiology (for example, Darrah, 1941), and plant evolution (Darrah, 1938) to plant biogeography (for example, Darrah and Bertrand, 1933; Darrah et al., 1937), plant biostratigraphy (for example, Darrah and Barlow, 1972; Lyons and Darrah, 1979), and what would now be categorized as plant taphonomy (Darrah, 1969, p. 66–71, and written communications to Gastaldo, 1986, 1987). The one question that Darrah pursued throughout his career, though, focused on the interpretation and significance of roof-shale floras, those fossil-plant assemblages from which he collected a multitude of data. Through his teaching and writing he influenced the thinking of several members of the next generation of paleobotanists, including the first and second authors of this chapter.

Earlier interpretations of roof-shale assemblages were generally based upon observations restricted to limited in-mine and outcrop exposures. Too often explanations to account for their genesis were made without regard to complex sedimentological and stratigraphical relationships that may have played a significant role in development of the fossil deposits. The purpose of this contribution is to summarize the spectral characteristics of roof-shale floral assemblages found in the late Carboniferous. It is beyond the scope of this chapter to discuss in detail the large-scale roles that tectonics, eustacy, climate, and sometimes volcanism play in the development of such assemblages (see Gastaldo et al., 1993).

## PREVIOUS INTERPRETATIONS OF ROOF-SHALE FLORAS

As is the case in many scientific discussions, early debate on the origin of roof-shale floras was polarized. The polarity of modern interpretations traces its origin to debate during the late nineteenth and early twentieth centuries concerning the autochthony or allochthony of coal itself. Dawson (1878) described in detail the section in Joggins, Nova Scotia, that contains many upright tree trunks of substantial height. Therefore, he advocated at least for these specific occurrences an autochthonous origin and could prove it beyond reasonable doubt. White and Thiessen (1913) brought a similar set of observations from the American coalfields and advocated the autochthonous origin of most coal beds and at least of those roof rocks containing erect trunks. However, most localities do not show such clear evidence. In an attempt to settle the debate about autochthony versus allochthony, Stevenson (1911a, 1912, 1913) reviewed the arguments of the opposing sides, detailed transport processes of plant detritus and its accumulation, surveyed the Carboniferous stratigraphy of the Appalachian basin and the occurrence of preserved plant material, and elucidated the relationships of clastic strata bounding the coal deposits. He presented a discussion that utilized the preservational state of vegetation within the coal-bearing sequence to argue for autochthony of the coal (Stevenson, 1913). Part of his discussion focused on the roof rock. He noted that a variety of lithologic types may overlie a coal bed, ranging from black shale to sandstone and even conglomerate (Stevenson, 1913), and described the variety of plant fossil assemblages encountered in these rocks, ranging from erect cast trees to fragmentary detritus. In some instances, he attempted to compare the preservational mechanisms to modern analogues. Stevenson did not take a dogmatic position but rather concluded that the plant-part compositional and preservational state of the assemblage in the roof shale could be used as an indicator of autochthony or allochthony. He stated that erect trees are found only in deposits on which there was never more than a shallow cover of water (Stevenson, 1913, p. 114).

Stevenson's arguments present a range of possible developmental processes for roof shales and the associated floras. In contrast, Davies (1929) took a position at one end of the spectrum. He asserted that the vegetational elements found within roof-shale assemblages directly reflected the plants that comprised the underlying coal bed and that spatial variation in species diversity and quantity of plant material represented the heterogeneity of the original mire community. Hence, he presumed that roof-shale assemblages were all autochthonous. North (1935), on the other hand, argued that the material was not *in situ* and that nearly all specimens originate in topogenous lows, and the prevailing wisdom asserted that drifted material overlay coal seams. Therefore, the systematic affinity of plants in the roof shale was not necessarily identical to those found in the underlying coal. Roof-shale floras, then, could only have been comprised of allochthonous megafloral elements, and preserved plants represented vegetation sampled from a variety of habitats. Autochthony was proposed for all well-preserved roof-shale floras by some authors in Germany (Keller, 1931; Draegert, 1964); others discussed various origins but emphasized hypoautochthonous occurrences (Josten, 1961; Hartung, 1966—equivalent term to parautochthonous as used herein).

The natural world rarely operates at either extreme of a polarized spectrum, and Gothan and Gimm (1930) and Havlena (1961) recognized that both autochthonous (flöznah, lycopsid-dominated assemblages) and allochthonous (flözfern, fern- and pteridosperm-dominated assemblages) associations could be identified above a variety of coals within the same sedimentary basin. Similarly, Oshurkova (1974) recognized autochthonous and allochthonous assemblages (phytoryctocoenoses of lepidodendrid stems and pteridospermous rachises, respectively) in central Kazakhstan, and Scott (1979) reported them from Great Britain. Attempts at identifying the gradient between these extremes have been alluded to (for example, Gastaldo, 1987), but never, to our knowledge, characterized.

Darrah (1969) summarized his experience and thinking about geological problems in the interpretation of fossil floras, including questions of paleoecology and plant taphonomy. He made the point that sedimentology had been neglected. He called compression floras coal floras and interpreted most of them as allochthonous or what we would call parautochthonous. However, he also pointed out the complexity of these interpretations and how provisional his interpretation had to be. The strength of his discussion was the recognition of what had to be done and in which direction research had to go. We present in this chapter a short summary of work and ideas that resulted from studies of the nature he suggested.

Jennings (1986) has taken a different approach to the question. Acknowledging that roof-shale floras may differ in quantitative composition from coal-ball floras and palynofloras, he argued (p. 304) that the plant fossils in the roof shales of the Mississippian and Pennsylvanian represent primarily the same taxa that comprise the coal itself. This is certainly true to a certain degree of taxonomic resolution. The overlap between coal-ball floras and adpression floras is very high at the family or genus level, as any survey of the literature will reveal. However, it appears that many of the species known from compressions do not occur in coal balls. This is most notably the case for medullosan pteridospermous foliage (see DiMichele et al., 1985) and for many small ferns and other pteridosperms. Other groups, such as the arborescent lycopsids, clearly had lineages ecologically and evolutionarily centered in mires. The edaphic differences between peat and mineral substrates undoubtedly underlie these distributional patterns. By generalizing the argument to one of broad taxonomic similarity, Jennings overlooked the importance of understanding the depositional environment of roof shales to the interpretation of the provenance of the flora and the degree to which peat-forming floras can be understood by studies of adpressions. The latter is basically a paleoecological, not a taxonomic, issue.

## THE CHARACTERIZATION AND CLASSIFICATION OF ROOF-SHALE FLORAS

Terrestrial clastic deposits overlying coals display a variety of sedimentological features that directly reflect the processes responsible for their accumulation. The presence or absence of plant parts within these rocks is a direct function of processes operating prior to, during, and after deposition. The understanding of a roof-shale flora requires the observation of the small-scale stratigraphy below and above the plant-bearing bed, the sedimentologic characteristics of these rocks, and the taphonomic characteristics of the plant fossils themselves. We can approach the taphonomic-sedimentologic-stratigraphic interpretation at three different levels of processes or relationships: (1) The preservation or nonpreservation of plant matter in the stratum, (2) the relationship between the place of growth of the flora and the site of its preservation, and (3) the relationship of the flora in the overlying sediment to the underlying peat. As discussed below, these three levels are largely, but not in every case, independent of each other.

Preservation or nonpreservation can depend on processes operating in the environment itself or during early diagenesis (Fig. 1). As paleobotanists we tend to concentrate on cases in which preservation took place. Barren roof rocks are frequently observed but rarely reported because of the absence of megascopic plant parts. Sedimentological structures characteristic of these roof shales run the gamut from pin-stripe laminations to megaforms, depending upon the depositional setting of the terrestrial clastics. Often these roof shales are massive mudstone in which primary sedimentary structures are absent, but cases have been reported where tidal rhythmites overlie the coal bed (Scott, 1979; DiMichele and Beall, 1990; Gastaldo et al., 1990). Palynomorphs and dispersed cuticles may be recoverable from macerations, but visually observable detritus is missing from this roof-shale type.

We recognize three basic categories of taphofloras if we want to express the relationship between site of growth and location of burial: autochthonous, parautochthonous, and allochthonous fossil-plant assemblages. The principal processes and features are summarized in Figure 2 and Table 1, and examples are discussed below. This classification is valid for any

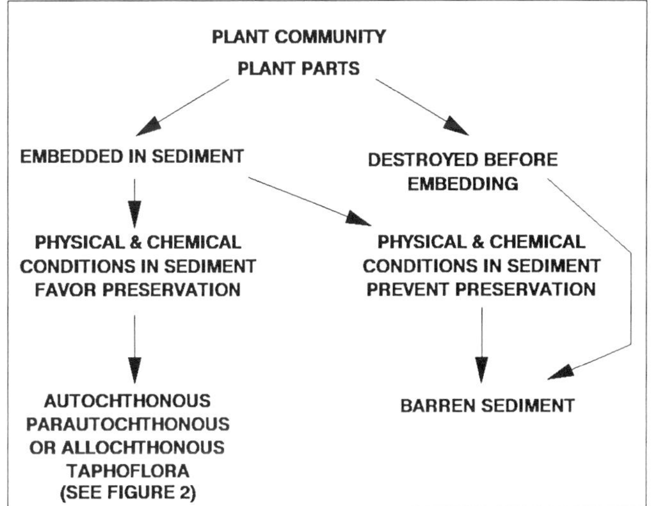


Figure 1. Pathways leading to the preservation or nonpreservation of plant parts in clastic sediments. Barren sediments in general and barren roof shales in particular are common. Phytodebris and/or palynomorphs may or may not be present in barren sediments.

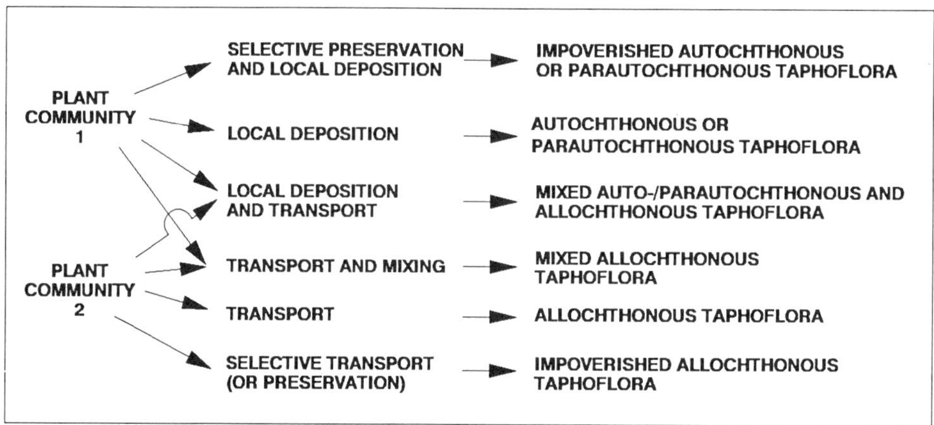


Figure 2. Generalized schematic diagram of the transformation of plant communities into fossil-plant assemblages (taphofloras). Only the major and most common processes are illustrated, demonstrating some of the factors that must be understood before roof-shale floras can be interpreted. These taphonomic processes must be combined with knowledge of the sedimentological system to come to a credible interpretation of roof-shale floras.

TABLE 1. SEDIMENTOLOGIC AND PLANT-TAPHONOMIC CHARACTERISTICS OF ADPRESSION FLORAS ADDRESSING THE RELATIONSHIP BETWEEN SITE OF GROWTH AND DEPOSITION OF THE PLANT PARTS FOUND

| | Sedimentological Characteristics | Paleobotanical Characteristics |
|---|---|---|
| Autochthonous | **Catastrophic:** Massive mudstone up to several meters thick; gray to light gray in color; primary sedimentary structures generally absent; where present are large scale. **Noncatastrophic:** Coaly shale matrix; black, carbonaceous, fissile. Bedding horizontal, particularly tidalite facies. | **Catastrophic:** Erect trees of various systematic affinity; *Calamites* may show regenerative features; less robust vegetation subhorizontal to subvertical orientation; forest floor litter concentrated; Gaussian size distribution restricted to basal ± 10 cm; adpressions and partial casts. Second canopy litter accumulation may be present at top of event deposit. **Noncatastrophic:** Erect trees of various systematic affinity may or may **not** be present; less robust vegetation horizontal; aerial axes represented by vitrain bands, partial to completely filled casts; higher proportion of woody detritus with foliage rarely preserved; distance between bedded litters increases up section. |
| Parautochthonous | Coaly shale matrix, generally a gray to dark gray shale, varying degrees of fissility; bedding horizontal; other primary sedimentary features where present reflect depositional regime (splay deposits). | Horizontally bedded detritus of varying systematic affinity, mostly composed of pteridosperm, pteridophyte, and sphenophyte parts; concentrated assemblages with well-preserved detail; generally Gaussian size distribution of plant parts; may be up to several meters in thickness. |
| Allochthonous | May range from mudstones to sandstones of varying composition and texture depending upon environment of deposition of final burial. | Plant detritus of varying size attributes and preservational features; mixture of floral elements sometimes of isolated pinnules, pinnae, seeds, branchlets, etc.; floras may be impoverished or enriched. Plant "hash" common. |

plant-bearing deposit, but we will consider it from the viewpoint of roof-shale floras.

Stratigraphic categories have been used in evaluating the specific relationship between a plant-bearing roof shale and the underlying coal (Table 2; Fig. 3). Sedimentation can be continuous while changing from organic to clastic, a short temporal break can occur between the cessation of organic sedimentation and the onset of clastic deposition, or a hiatus of significant duration can occur.

### *Autochthonous taphofloras*

Autochthonous fossil-plant assemblages are those in which remains are preserved at the death site of the organism or the site where parts were discarded either by physiological or trauma-induced loss (Bateman, 1991; Behrensmeyer and Hook, 1992; Gastaldo, 1992a). The genesis of these assemblages may be the result of catastrophic burial (event deposition) or slow but regular sedimentation; for instance, on a daily basis with less than a millimeter deposited each day (Table 1). The taphonomic aspect of the preserved plant matter is quite distinct in these two cases (Figs. 3A, 3B). Roof-shale floras may clearly represent the last vestiges of peat-swamp vegetation in cases where no diastem separates the roof shale from the coal bed. The plants are preserved *in situ,* and observed floral assemblages often consist of erect trees rooted within and preserved for some distance above the coal bed. Prostrate and subhorizontal plants and plant parts that represent not only the canopy but also understory vegetation are found to occur between the erect vegetation (DiMichele and DeMaris, 1987; DiMichele and Nelson, 1989; Gastaldo, 1990a). These include complete stems, leaves, aerial branches, and reproductive structures and their propagules.

Several taphonomic features allow for the delimitation of this subcategory of autochthonous peat-swamp assemblages if burial resulted from catastrophic high-discharge, low-frequency flood events (Demko and Gastaldo, 1992). Erect trees are surrounded and cast by gray mudstone in which primary sedimentological structures generally are rare (if present they are a function of the relative proximity of the buried forest to the channel[s] transecting the area). Erect lycopsids, calamites, and tree ferns may be preserved for heights up to 8 m (Gastaldo, 1986, 1990b). Particular taxa, especially *Calamites* sp., show signs of regenerative behavior after burial (Gastaldo, 1992b). Understory vegetation may be preserved in place, within the basal 10 cm, and is found generally in a subvertical to subhorizontal orientation, cross-cutting the encasing rock. The preserved megaflora is heterogeneous with respect to plant-part types and plant-part sizes and is generally taxonomically diverse (ecological parameters will constrain the degree of heterogeneity or homogeneity within the floral assemblage; DiMichele and Nelson, 1989). Most plant parts are preserved as concentrated adpressions, although incompletely filled mud-cast logs may be present (Gastaldo et al., 1989). The incomplete casting of these hollow voids is the result of a very short duration (on the order of up to several weeks) of available suspension-load sediment within the area. Litter may be deposited on top of the flood deposit when erect buried trees undergo canopy part abscission, forced reproduction, or death (Gastaldo, 1987, 1990a). Consequently there may be a barren interval between the buried forest floor and the final litter layer produced.

Slow but continuous burial of a peat swamp may occur due to local/regional subsidence and/or peat compaction generating accommodation space. This may result in the slow death and drowning of mire taxa, on occasion accompanied by a transition to a different assemblage adapted to a mineral substrate (clastic swamp; Wnuk and Pfefferkorn, 1987). Characteristics of this assemblage include the presence of discrete bedded detritus within a coaly shale matrix (dark gray to black, fissile and carbonaceous) or within the basal portion of tidalite sequences. The thickness of the lithology reflects the depth of generated accommodation space (usually >0.5 m). Erect trunks may occur on hummocks, whereas isolated hori-

TABLE 2. SEDIMENTOLOGIC AND PLANT-TAPHONOMIC CHARACTERISTIC OF ROOF-SHALE FLORAS ADDRESSING THE RELATIONSHIP OF THE CLASTIC BED IN WHICH THE PLANTS ARE PRESERVED AND THE UNDERLYING COAL BED

| | | |
|---|---|---|
| Continuous sedimentation | Sedimentation changes from organic to clastic; bony coal and carbonaceous shales may occur; boundary could be sharp in some cases. | Paleobotanical assemblages may be autochthonous, parautochthonous, or allochthonous; however, only in this setting can floras be found that are autochthonous with respect to underlying coal. |
| Short temporal break in sedimentation (discontinuity) | Chemically induced alteration of sediments above coal; flooding of swamps provides impetus for reaction of peat with overlying water column; includes pyritization and authigenic cementation by siderite. | Paleototanical assemblages may be autochthonous, parautochthonous, or allochthonous with respect to site of clastic deposition; assemblages are always unrelated to underlying coal. |
| Significant hiatus often with erosional event | Sedimentary environments may represent any of the settings mentioned in Table 3; however, channel deposits are frequently typical for this setting. | Paleobotanical assemblages may be autochthonous, parautochthonous, or allochthonous with respect to site of clastic deposition; assemblages are always unrelated to coal. |

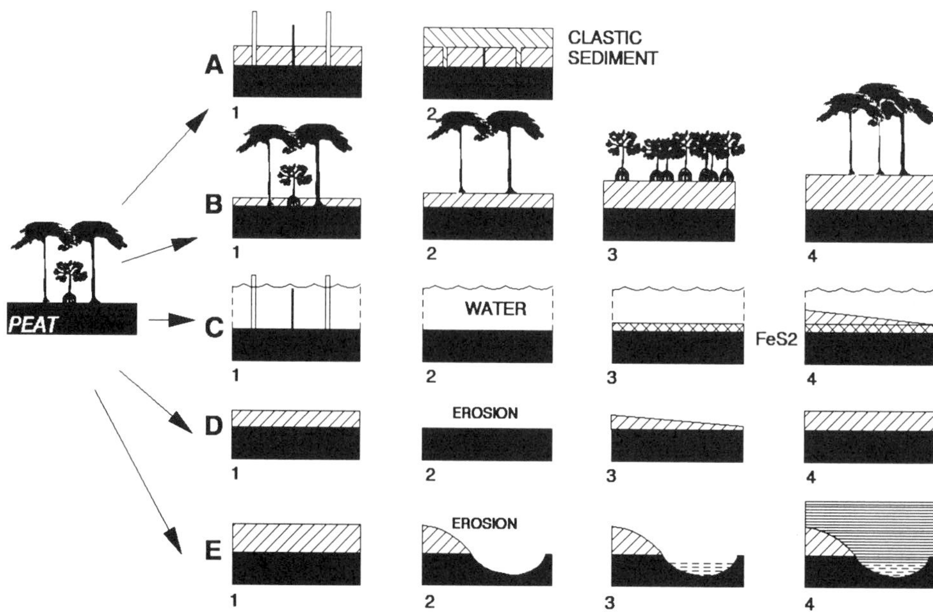


Figure 3. Schematic diagram of some of the processes responsible for roof-shale formation over peat. The illustrated processes must be considered in conjunction with the information in Figures 1 and 2 and Tables 1 and 2 to interpret a particular roof-shale flora. **A,** Catastrophic burial by clastic sediment of forest growing on peat results in plant death and decay above sediment interface (A1) followed by subsequent burial and infilling (A2). **B,** Mire forest is subjected to low rate of clastic sedimentation with part of the litter layer preserved; stumps decay and may be infilled (B1); subsequent colonization and continued sediment accretion may result in successive communities rooted in mineral substrates (B2–B4). **C,** Mire is permanently flooded without accompanying sediment input (C1); tree death and decay result, leaving only fallen flattened stems in the upper peat layer (C2); water circulates through the peat; uppermost part of peat is oxidized, pyritized, phosphatized, or otherwise mineralized (C3); clastic sedimentation occurs at a later point in time (C4). In this scenario, water chemistry may range from fresh, to brackish, to marine. **D,** Erosion removed sediment cover down to the peat (D1–D2); unknown interval of time passes before new sediments are deposited (D3–D4); earlier sediment may remain on peat in irregular lenses. **E,** Erosion removes sediment and cuts into the peat body (E1–E2); new sediment is deposited (E3–E4); earlier sediment may remain present in irregular lenses or may be absent.

zontal trunks may be preserved either as discrete thick vitrain bands, shale-cast logs with periderm structure preserved as enveloping vitrain bands, or elliptical to circular mud-cast logs. The proportion of woody axial material may be higher than that of foliage (Scott, 1978; Wnuk and Pfefferkorn, 1984; DiMichele and DeMaris, 1987) as a result of long-term exposure at the sediment surface and/or secondary rooting. Litter is concentrated with resistant plant parts, particularly cuticles, which are often abundant. Plant-part distribution along bedding planes may display features reflecting physical processes that acted upon the site during accretion (for example, Wnuk and Pfefferkorn, 1987).

### *Parautochthonous taphofloras*

Parautochthonous floral assemblages are those composed of remains that are transported from the death or discard site but reside within the original habitat (Bateman, 1991; Behrensmeyer and Hook, 1992). These are the classic roof-shale floras derived from vegetation that colonized slightly higher topographies (for example, channel margins and levees) adjacent to the clastic depositional setting in which the plant parts were preserved (see Scott, 1978, 1979; Scheihing and Pfefferkorn, 1984; Pfefferkorn et al., 1988). Plant parts may have been introduced by a variety of traumatic (overbank flooding and splay development; Gastaldo, 1987, demonstrates the relationship) or physiological mechanisms. Preservation of the predominantly pteridophyte-pteridosperm and herbaceous sphenophyte assemblages typically is excellent (for example, Scott, 1978; DiMichele et al., 1991). The assemblages are often concentrated (sensu Krasilov, 1975) in horizontally bedded gray shales or silty shales. Concentration does not imply an overlapping of individual plant parts as one would find in autochthonous floras (representing true forest-floor litter). Parts are often discretely isolated spatially from each other. Plant parts generally reflect no specific abiotically generated orientation pattern. The total thickness of a parautochthonous accumulation may be up to a few meters (depending upon available accommodation space).

### *Allochthonous taphofloras*

Allochthonous floral assemblages represent the other end of the transport spectrum from autochthonous assemblages. Such remains have been moved from the site of death or discard and moved out of their original site of habitat (Bateman, 1991; Behrensmeyer and Hook, 1992). Deposition of this plant detritus may occur many kilometers away from the actual site of growth (for example, Gastaldo et al., 1987; Gastaldo and Huc, 1992), and the detritus is emplaced over peat bodies that have been subjected to relative change in elevation. This usually follows a hiatus in sedimentation following the termination of peat accumulation. Plant parts are generally a mixture of floral elements originating from a variety of habitats naturally sampled from along the transect of a feeder channel (for example, Scott and Chaloner, 1983; Peppers and Pfefferkorn, 1970; Pfefferkorn, 1979; Scheihing and Pfefferkorn, 1984). Plant parts are commonly found dispersed in the sediment and represent isolated pinnules, pinnae, seeds, branchlets, and woody detritus (for example, Scott, 1977). The depositional site may be any one of many found within depositional settings in fluvial plains, coastal plains, deltas, or estuaries (Table 3). The sedimentological features characterizing these settings, then, would be reflected in the lithology (for example, Walker and James, 1992). The thickness of sediments in which allochthonous megafloras may accumulate may be greatly dependent upon available accommodation space generated by compaction, tectonics, or a rise in base level. Allochthonous floras may be impoverished by selective transport (for example, Mosbrugger, 1989; Cunningham et al., 1993) or by sorting of taxa through hydrologic processes. Conversely, floras may be enriched by the amalgamation of taxa sampled from a variety of ecological settings (for example, Darrah, 1969, 1972; Pfefferkorn, 1979). In general, plant parts exhibit a variety of size extremes and preservational features that are dependent upon the freshness of the plant part at the time of emplacement on the sediment-water interface and subsequent physical processes that may have mechanically fragmented this detritus (see Spicer, 1989; Gastaldo, 1992c). Where physical destruction has been taken to the extreme, plant hash assemblages predominate (Gastaldo, 1994).

TABLE 3. PRINCIPAL DEPOSITIONAL ENVIRONMENTS AND THEIR SUBENVIRONMENTS COMMONLY ENCOUNTERED ABOVE COAL BEDS

| Major Environments | Subenvironments or Deposits |
|---|---|
| Fluvial channel | Bedload sands and lags; point-bar sediments; longitudinal and lateral bar sands*; bank deposits and slumps*. |
| Natural levee | Levee sediments*; crevasse splay channels*; crevasse splay levee*; crevasse splay shield*; clastic backswamp*. |
| Flood plain | Intermittently "dry" flood plain; flood plain lake; oxbow lake. |
| Coastal plain | Flood deposits*; tidal deposits under various salinities*; tidal channels*; wetlands*; estuarine and lagoonal deposits*; tidal flats*; beach sands, washover fans. |
| Delta | Interdistributary bay; delta-front sheet sand; those listed with an asterisk above. |
| Nearshore marine | Tidal flats; barrier-bar sands; tidal channels. |

The categories discussed so far apply to the fossil-plant assemblage and are applicable to roof-shale and other adpression floras. The categories that follow address the stratigraphic relationship between the clastic plant-bearing bed and the underlying coal bed. This relationship has to be recognized in order to interpret taphonomy and origin of the flora properly. These categories can be combined with the other categories discussed above in many different combinations, but it must be kept in mind that some combinations are logically impossible and others rare.

### *Continuous sedimentation across peat–to–clastic-sediment boundary*

Continuous sedimentation is a precondition for the preservation of an autochthonous or parautochthonous peat flora in the overlying clastic sediment. Therefore, this case has been discussed under the heading autochthonous taphofloras. In the two cases discussed below where short or long interruptions in the sedimentation occur, the taphoflora in the clastic material cannot be autochthonous with respect to the underlying coal.

### *Short temporal break in sedimentation between peat and clastic sediment*

We recognize this category based on a very sharp but otherwise conformable boundary at the coal-clastic interface. This happens where a relative rise in base level floods the swamp and kills the peat-forest vegetation (Fig. 2C). The resulting flood plain lake or lagoon may not receive any sediment for several years. This break may be accompanied by chemical reactions of the peat with the overlying water column, which may result in precipitation of a thin layer of pyrite, phosphate, or carbonate within the very top of the peat body. Some coal-ball occurrences may have a similar origin (Scott and Rex, 1985). Other pore water reactions may result in the development of phosphatic-sideritic nodules (Woodland and Stenstrom, 1979; Curtis, 1977). After some time sediment will start to reach the site, often in the form of the distal ends of crevasse splays. The type of paleobotanical assemblage found under these circumstances in the roof shale varies considerably and has no relationship to the underlying peat.

However, the flora may represent one or several subenvironments of the same larger depositional system (i.e., coastal plain, fluvial plain, or delta).

### *Significant hiatus between peat and clastic sediment*

Although it can be argued that the presence of any terrestrial clastic lithology above a coal bed is, in itself, representative of a disconformity, we refer to this category those cases in which a clastic unit has been secondarily emplaced on top of a peat body after erosion of the original overlying sediments (Figs. 3D, 3E). These floral assemblages may appear to be similar to other roof-shale floras (Fig. 3D) or may have been deposited within channels of various dimensions that have eroded down to the top or even into the peat (Fig. 3E). Hence, they represent material that has been introduced above the peat some time after relative base-level rise. The contact between the plant-bearing lithology and the coal bed is unconformable. The plants in these deposits can represent a different depositional system and even a different chronostratigraphic stage from that of the peat.

Often these sediments are channel deposits, and this particular setting requires some further discussion. Because the coal is a saturated flaccid accumulation, it may act as a localized trap for bedload-transported detritus, and basal-lag accumulations therefore develop. Other in-channel litter deposits may develop in a variety of bar structures. Channelization may be a function of fluvial (see Ferm and Horne, 1979; DeMaris et al., 1979) or tidal activity (Gastaldo et al., 1993), and sedimentary structures will reflect the prevailing physical depositional conditions (see Walker and James, 1992). Plant assemblages are generally composed of the most resistant aerial parts, including trunks and branches, which may be compressed into vitrain bands or cast in the prevailing bedload sediment. Depending upon the physical processes operating in the channel, macroscopic plant detritus may occur as concentrated bedded litter or drapes overlying cross beds. Where degraded and physically fragmented detritus has been redistributed into channels via fluctuating water conditions, plant parts may occur as plant hash.

## DISCUSSION

It is evident that a wide variety of complex interactive taphonomic and sedimentologic processes affect the genesis of roof-shale floras. The diversity of interacting factors results in an assortment of assemblage characters. Local and/or regional physical processes ultimately result in the rise of the relative position of base level over short (for example, rapid subsidence in response to tectonics) or longer time intervals (for example, prolonged subsidence in response to tectonics, sea-level rise, or sediment compaction). This relative rise in water level inversely relates to a relative descent in the elevation of the peat-forming forest. Where the mire forest has developed a planar geomorphic form, the entire region would then be placed in jeopardy of increased flooding by low-frequency and high-magnitude floods (Demko and Gastaldo, 1992). Where ombrogenous peat swamps exist (McCabe, 1984), the once-elevated interior portions of these swamps may or may not be placed in jeopardy of being flooded, which may explain the distribution of some coal partings. Once a peat-accumulating forest has been placed at risk, it is only a matter of time before catastrophic burial may occur. Pfefferkorn et al. (1988) noted in the Orinoco delta, Venezuela, that during the recorded 150-year flood (regarded by them as a moderate-frequency and moderate-magnitude flood), longitudinal bars (up to 5 m in height) had been deposited on top of levees that are up to several meters above stream-channel waters. This means that sediment normally restricted to bedload transport is carried in suspension load, and normal suspension-load sediment is transported into forested areas adjoining these channels, a consequence of water rise on the order of greater than 10 m. Flood waters regress from the forest over a period of about a few weeks or months. Under circumstances that may generate the 1,000- or 10,000-year-magnitude flood, an increased sediment load would remain resident in the waters overlying the forests for longer periods of time. This would result in the catastrophic burial of the forest by alluvium, leaving only those trees erect that exhibit the greatest structural integrity. Evidence for such floods has been well documented in the Carboniferous (for example, Liu and Gastaldo, 1992a).

Base-level changes occurring over the longer term can be related to tectonics in intramontane basins and a combination of tectonics and eustatic changes in paralic basins. This type of change may slowly subject the peat swamp to incursion not only be fresh to brackish waters (the presence of orbiculoid and/or lingulid brachiopods in roof shales is indicative of such settings) but also by marine waters. Slow rise in base level accompanied by intermittent overbank sedimentation or freshwater tidal deposition (Scheihing and Pfefferkorn, 1984) provides the setting in which wetland vegetation not only can survive but also can contribute canopy parts to the accumulating sediment in which it is growing. Concentrated plant litter accumulating within mud may ultimately be preserved as an organic-rich shale.

The above scenarios are not intended to imply that mud-cast erect vegetation can only be formed under a subsidence and catastrophic burial scenario, and we do not intend to imply that roof shales with well-bedded autochthonous litter can only be formed under a slow base-level rise or by localized peat compaction. These are merely the most conspicuous of simple mechanisms to explain autochthonous roof-shale floras. It is possible that mud-cast erect vegetation could have been formed within a setting controlled by longer-term processes (for example, DiMichele and DeMaris, 1987). But such interpretations cannot be made subjectively based solely upon the characteristics of the fossil-plant assemblage. A thorough understanding of the sedimentological characters associated with the assemblage is essential. As fluvial, coastal, or deltaic plain settings are heterogeneous with respect to geomorphological features at any one point in time, so are the laterally correlative potential depositional sites that may preserve plant litter. It is imprudent to place an unequivocal interpretation on the genesis of an autochthonous megafloral assemblage, or in fact any assemblage, without the sedimentological context within which it is preserved.

Most Carboniferous biostratigraphy is based upon those megafloral assemblages interpreted by us to have been preserved as parautochthonous accumulations. Such assemblages include a high proportion of pteridophyte and pteridospermous elements that typically inhabited slightly better drained peat or clastic soils in coastal wetland settings (for example, Gastaldo, 1987; Wnuk and Pfefferkorn, 1984). Their presence may indicate some transport into a site with higher preservation potential. Their mere presence alone, though, does not justify the interpretation of parautochthony. Wnuk and Pfefferkorn (1984) and Gastaldo (1990b) have demonstrated that there may be significant numbers of autochthonous pteridophyte and pteridospermous elements in a particular shale. It is therefore important to understand not only lateral lithofacies distribution of the roof shale but also the sedimentological features in it. The taxonomic diversity, size of preserved plant parts, and plant-part disposition in the rock cannot provide a definitive answer as to whether the assemblage is parautochthonous.

The character of an allochthonous assemblage relates to the mode of transport and emplacement of the plant detritus in the depositional environment that has become established over the former peat-forming swamp. The degree to which plant parts are recognizable is dependent upon several factors that include the freshness of the part when introduced to the water column (traumatic loss or dehiscence following dysfunction), the residency time of the part in the water column and the resistance to degradation of any particular plant part (Gastaldo, 1992c), the physical processes to which the parts have been subjected (Gastaldo, 1994), and the biogeochemical conditions within the substrate. Plant parts are normally evenly distributed throughout the deposit; as new material is transported into the depositional environment, it settles at the sediment-water interface and is incorporated through continued burial. In some instances, plant detritus may be reentrained and transported to beaches of lakes, lagoons, or the ocean, where it may accumulate as detrital peat. Reentrainment and mechanical fragmentation of allochthonous plant detritus result in the development of plant hash often found distributed along bedding surfaces.

Megafloras deposited in sediments that follow a short time of nondeposition represent an inherent change in geochemical conditions that reflect a perturbation within the depositional setting. For example, an increase in activity of sulphate-reducing bacteria and preservation of plant parts by pyrite signal a widespread change in concentrations of pore-water oxygenation. This change may be induced by one or more chemical processes operating in the shallow subsurface, at the sediment-water interface, or within the water column. The same can be applied to nodule floras found throughout the Midcontinent. The development of such assemblages reflects a postdepositional alteration of the chemical environment that may, or may not, have been induced by the introduction of the phytoclasts. It does imply, though, not only that the organic substrates were in place prior to the chemical alteration but also that they had not yet been subjected to lithification.

Fossil-plant assemblages preserved in sediments deposited on the peat bed only after a long hiatus are often found in small, localized deposits that represent their introduction above the peat through erosion of the overlying sediment. These floras may be allochthonous (detritus transported into the accumulation site), parautochthonous (contributed from vegetation growing along the margins of the channel), or autochthonous (representing vegetation growing in a shallow or abandoned channel). An understanding of the megaflora in sedimentological context within these channel deposits is the only way that interpretations can be resolved between the three possible accumulations. In instances where the flora is regionally distributed, its presence may be in response to local subsidence or regional transgression accompanied by shoreface erosion. In the latter instance, ravinement processes may contact the underlying peat, and this detritus may become part of the ravinement bed (Liu and Gastaldo, 1992b).

Many roof shales lack megafloral remains. Although rarely discussed, the lithologies that comprise this category are varied, representing a spectrum of depositional settings that provide information about basinal history. The absence of megafloral elements may be due to the hydrological regime in which the sediments were deposited, geochemical conditions (Eh and pH relationships) that prevailed during accumulation that may have prevented plant-part preservation, or extended distance of the depositional setting from a source area. Often, those roof shales considered to be barren are designated as such only because concentrated megafloras are not preserved. It is common, though, to find scattered bits and pieces that may provide information not readily available in other roof-shale types (Tiffney, 1990).

## CONCLUSIONS

Roof-shale floras have been, and continue to be, a primary source of information concerning Carboniferous vegetation. Interpretations as to what these floras actually represent have often sought simplistic absolutes ranging from autochthony to allochthony. An evaluation of the types of roof-shale floras encountered and reported in the Carboniferous results in the recognition of several categories that fall into three different groups. There is preservation resulting in a taphoflora versus nonpreservation resulting in a barren roof shale. Depending on the amount of transport plant parts experience before embedding, one can distinguish autochthonous, parautochthonous, and allochthonous fossil-plant assemblages. Finally one can distinguish the relationship of the peat bed and the overlying clastic sediments. Deposition can be continuous but changing from organic to clastic or discontinuous with a short interval of nondeposition, or a long hiatus can intervene that could include an erosional event.

In addition, we understand today that we are dealing with two broad groups: those that represent the final vegetation of the mire forest that formed the underlying peat and those from lowland mineral substrate habitats that were deposited in the same site after the cessation of peat formation. Roof-shale floras that exemplify the former are autochthonous or parautochthonous assemblages with regard to the underlying peat representing the final vegetational stand of the peat-forming forest. In essence, this is a geologically instantaneous picture of the forest that has undergone either catastrophic burial and death or death from edaphic stress followed by slow burial. The final peat-swamp forest may be gradually replaced by clastic swamps if colonization occurs within the mineral substrate, resulting in a succession of autochthonous clastic-swamp accumulations in the shale above the coal bed. These clastic accumulations have no genetic relationship to those established in the peat substrate. Their establishment may occur over a relatively short stratigraphic distance (<1 m) or over several meters of section. Roof-shale floras that do not represent the final mire vegetation may rest conformably or unconformably on the peat. Megafloras that are conformable were deposited after a short break in sedimentation. Plant parts are locally introduced via flooding and splays into standing bodies of water that are resident above the peat body. The transported plant material may come from a local source (parautochthonous) or from outside the immediate surroundings of the site of deposition (allochthonous) or both. Plant parts that have been subjected to bedload transport and/or mechanical degradation will result in the development of "plant hash" assemblages (sensu Gastaldo, 1994) distributed along bedding planes. Where these overlying terrestrial clastic sediments are colonized, wetland plant communities may develop, but these bear no relationship to the underlying peat-swamp community. Roof-shale floras preserved within channels that have eroded into an underlying peat body are unconformable and may represent the range of allochthonous to autochthonous assemblages. Again, though, these floras bear no direct relationship to the underlying peat-swamp vegetation.

Differentiation of any category of roof-shale floras must be made in conjunction with the sedimentological features of the lithologies in which they are preserved. Roof-shale floras cannot be interpreted in a sedimentological void, and an understanding of the depositional context of a preserved flora is essential before attempts can be made at interpreting its significance.

## ACKNOWLEDGMENTS

The first author acknowledges the guidance and support from William (Bill) C. Darrah during his undergraduate career at Gettysburg College and the years thereafter. Bill's direction, encouragement, and true friendship influenced my career choice as a paleobotanist. Chapter Four in Darrah's (1969) book influenced the thinking of the second author at a time in the early 1970s when he was first trying to reconstruct Carboniferous floras. The second author would also like to acknowledge the intellectual influence of Vaclav Havlena through his lectures as a visiting professor in Muenster during the 1960s. The development of this chapter is the result of a variety of financial support, including NSF EAR 8407833 and 8618815 to RAG, NSF EAR 7622562, 8019690, and 8916826 and DFG 98/3 to HWP, and NSF EAR 8313094 (to T. L. Phillips) and DEB 8210475 to WAD. We thank W. H. Gillespie, Paul Lyons, and Chris Wnuk for comments on the manuscript. This is contribution no. 11 from the Evolution of Terrestrial Ecosystems Consortium, which has also provided partial support.

## REFERENCES CITED

Baird, G. C., Shabica, C. W., Anderson, J. L., and Richardson, E. S., Jr., 1985, Biota of a Pennsylvanian muddy coast: Habitats within the Mazonian Delta complex, northeast Illinois: Journal of Paleontology, v. 59, p. 253–281.

Bateman, R. M., 1991, Palaeoecology, *in* Cleal, C. J., ed., Plant fossils in geological investigation, The Palaeozoic: New York, Ellis Harwood, p. 34–116.

Behrensmeyer, A. K., and Hook, R. W., rapporteurs, 1992, Paleoenvironmental contexts and taphonomic modes in the terrestrial fossil record, *in* Behrensmeyer, A., Damuth, J., DiMichele, W. A., Potts, R., Sues, H.-D., and Wing, S., Terrestrial ecosystems through time: Chicago, Illinois, University of Chicago Press, p. 15–138.

Cunningham, C. R., Feldman, H. R., Franseen, E. K., Gastaldo, R. A., Mapes, G., Maples, C. G., and Schultze, H-P., 1993, The Upper Carboniferous (Stephanian) Hamilton Fossil-Lagerstätte (Kansas, U.S.A.): A valley-fill, tidally influenced depositional model: Lethaia, v. 26, p. 225–236.

Curtis, C. D., 1977, Sedimentary geochemistry: Environments and processes dominated by involvement of aqueous phases: Royal Society of London Philosophical Transactions, series A, v. 286, p. 353–372.

Darrah, W. C., 1938, Fossil plants and evolution: Evolution, v. 4, p. 5–6.

Darrah, W. C., 1940, The fossil flora of Iowa coal balls. III: *Cordaianthus:* Harvard University Botanical Museum Leaflets, v. 8, p. 1–20.

Darrah, W. C., 1941, Fossil embryos in Iowa coal balls: Chronica Botanica, v. 6, p. 388–389.

Darrah, W. C., 1969, A critical review of the Upper Pennsylvanian floras of eastern United States with notes on the Mazon Creek flora of Illinois: Gettysburg, Pennsylvania, privately printed, 220 p., 80 pls.

Darrah, W. C., 1972, Historical aspects of the Permian flora of Fontaine and White, *in* Arkle, T., ed., The Age of the Dunkard: Symposium: Abstracts and Reference Papers, Morgantown, West Virginia Geological Survey, p. 1–4.

Darrah, W. C., and Barlow, J. A., 1972, Fossil plants of the Waynesburg Coal and Cassville Shale at Mount Morris, Pennsylvania, *in* Arkle, T., ed., The Age of the Dunkard: Symposium: Abstracts and Reference Papers, Morgantown, West Virginia Geological Survey, p. 14–16.

Darrah, W. C., and Bertrand, P., 1933, Observations sur les flores houilleres de Pennsylvanie: Annales Societé Géologique du Nord, v. 58, p. 211–224.

Darrah, W. C., Jongmans, W. J., and Gothan, W., 1937, Comparison of the floral succession in the Carboniferous of West Virginia and Europe: Congrès pour l'avancement des études de Stratigraphie Carbonifère, 2nd (Heerlen, The Netherlands, 1935): Compte Rendu, v. 1, p. 393–415.

Davies, D., 1929, Correlation and palaeontology of the Coal Measures in East Glamorganishire: Royal Society of London Philosophical Transactions, series B, v. 217, p. 91–154.

Dawson, J. W., 1878, Acadian geology (third edition): London, Macmillan, 694 p.

DeMaris, P. J., DiMichele, W. A., and Nelson, W. J., 1979, A compression flora associated with channel-fill sediments above the Herrin (No. 6) coal, *in* Abstracts, International Congress of Carboniferous Stratigraphy and Geology, 9th, Urbana-Champaign: Urbana, Illinois, American Geological Institute, p. 50.

Demko, T. M., and Gastaldo, R. A., 1992, Paludal environments of the Lower Mary Lee coal zone, Pottsville Formation, Alabama—Stacked clastic swamps and peat mires: International Journal of Coal Geology, v. 20, p. 23–47.

DiMichele, W. A., and Beall, B. S., 1990, Flora, fauna, and paleoecology of the Brazil Formation of Indiana: Rocks and Minerals, v. 65, p. 244–250.

DiMichele, W. A., and DeMaris, P. J., 1987, Structure and dynamics of a Pennsylvanian-age *Lepidodendron* forest—Colonizers of a disturbed swamp habitat in the Herrin (No. 6) Coal of Illinois: Palaios, v. 2, p. 146–157.

DiMichele, W. A., and Nelson, W. J., 1989, Small-scale spatial heterogeneity in Pennsylvanian-age vegetation from the roof-shale of the Springfield Coal (Illinois Basin): Palaios, v. 4, p. 276–280.

DiMichele, W. A., Phillips, T. L., and Peppers, R. A., 1985, The influence of climate and depositional environment on the distribution and evolution of Pennsylvanian coal-swamp plants, *in* Tiffney, B., ed., Geological factors and the evolution of plants: New Haven, Connecticut, Yale University Press, p. 223–256.

DiMichele, W. A., Phillips, T. L., and McBrinn, G. E., 1991, Quantitative analysis and paleoecology of the Secor Coal and roof-shale floras (Middle Pennsylvanian, Oklahoma): Palaios, v. 6, p. 390–409.

Drägert, K., 1964, Pflanzensoziologische Untersuchungen in dem Mittleren Essener Schichten des nördlichen Ruhrgebietes: Forschungsberichte Land Nordrhein-Westfalen, no. 1363, 111 p.

Ferm, J. C., and Horne, J. C., eds., 1979, Carboniferous depositional environments in the Appalachian region: Columbia, South Carolina, Carolina Coal Group, University of South Carolina, p. 509–516.

Gastaldo, R. A., 1986, Implications on the paleoecology of autochthonous Carboniferous lycopods in clastic sedimentary environments: Palaeogeography, Palaeoclimatology and Palaeoecology, v. 53, p. 191–212.

Gastaldo, R. A., 1987, Confirmation of Carboniferous clastic swamp communities: Nature, v. 326, p. 869–871.

Gastaldo, R. A., 1990a, Early Pennsylvanian swamp forests in the Mary Lee coal zone, Warrior Basin, Alabama, *in* Gastaldo, R. A., Demko, T. M., and Liu, Yuejin, eds., Carboniferous coastal environments and paleocommunities of the Mary Lee coal zone, Marion and Walker Counties, Alabama: Southeastern Section of the Geological Society of America, 39th Annual Meeting, Guidebook for Field Trip 6, Tuscaloosa, Alabama: Tuscaloosa, Geological Survey of Alabama, p. 41–54.

Gastaldo, R. A., 1990b, Earliest evidence for helical crown configuration in a Carboniferous tree of uncertain affinity: Journal of Paleontology, v. 64, p. 146–151.

Gastaldo, R. A., 1992a, Taphonomic considerations for plant evolutionary investigations: The Palaeobotanist, v. 41, p. 211–223.

Gastaldo, R. A., 1992b, Regenerative growth in fossil horsetails *(Calamites)* following burial by alluvium: Historical Biology, v. 6, p. 203–220.

Gastaldo, R. A., 1992c, Plant taphonomic character of the late Carboniferous Hamilton quarry, Kansas, USA—Preservational modes of walchian conifers and implied relationships for residency time in aquatic environments, *in* Kovar-Eder, J., ed., Palaeovegetational development in Europe and regions relevant to its palaeofloristic evolution: Vienna, Museum of Natural History, p. 393–399.

Gastaldo, R. A., 1994, The genesis and sedimentation of phytoclasts with examples from coastal environments, *in* Traverse, A., ed., Sedimentation of Organic Particles: Cambridge, Cambridge University Press, p. 103–127.

Gastaldo, R. A., and Huc, A. Y., 1992, Sediment facies, depositional environments, and distribution of phytoclasts in the Recent Mahakam River delta, Kalimantan, Indonesia: Palaios, v. 7, p. 574–591.

Gastaldo, R. A., Douglass, D. P., and McCarroll, S. M., 1987, Origin, characteristics and provenance of plant macrodetritus in a Holocene crevasse splay, Mobile delta, Alabama: Palaios, v. 2, p. 229–240.

Gastaldo, R. A., Demko, T. M., Liu, Yuejin, Keefer, W. D., and Abston, S. L., 1989, Biostratinomic processes for the development of mud-cast logs in Carboniferous and Holocene swamps: Palaios, v. 4, p. 356–365.

Gastaldo, R. A., Demko, T. M., and Liu, Yuejin, 1990, Carboniferous coastal environments and paleocommunities of the Mary Lee coal zone, Marion and Walker Counties, Alabama: Southeastern Section of the Geological Society of America, 39th Annual Meeting, Guidebook for Field Trip 6, Tuscaloosa, Alabama: Tuscaloosa, Geological Survey of Alabama, 139 p.

Gastaldo, R. A., Demko, T. M., and Liu, Yuejin, 1993, The application of sequence and genetic stratigraphic concepts to Carboniferous coal-bearing strata: An example from the Black Warrior basin, USA: Geologische Rundschau, v. 82, p. 212-234.

Gothan, W. and Gimm, O., 1930, Neuere Beobachtungen und Betrachtungen ueber die Flora des Rotliegenden von Thuringen: Institut für Paläobotanik und Petrographie ver Brennsteine, v. 2, p. 39–74.

Hartung, W., 1966, Fossilführung und Stratigraphie im Aachener Steinkohlengebirge: Fortschritte Geologie Rheinland und Westfalen, v. 13, p. 339–564.

Havlena, V., 1961, Die flöznahe und flözfremde Flora des Oberschlesischen Namurs A und B: Palaeontographica, Abt. B, v. 108, p. 22–38.

Jennings, J. R., 1986, A review of some fossil plant compressions associated with Mississippian and Pennsylvanian coal deposits in the central Appalachians, Illinois Basin, and elsewhere in the United States: International Journal of Coal Geology, v. 6, p. 303–325.

Josten, K.-H., 1961, Pflanzensoziologische Beobachtungen an Steinkohlenbohrungen im Ruhrgebiet: Palaeontographica, Abt. B, v. 108, p. 39–42.

Keller, G., 1931, Ueber die Pflanzenhorizonte Sarnsbank I und Finefrau im Essener Gebiet: Jahrbuch der Preussischen Geologischen Landesanstalt, v. 52, p. 425–440.

Krasilov, V. A., 1975, The palaeoecology of terrestrial plants: Basic principles and methods: New York, J. Wiley and Sons, 283 p.

Liu, Yuejin, and Gastaldo, R. A., 1992a, Characteristics and provenance of log-transported gravels in a Carboniferous channel deposit: Journal of Sedimentary Petrology, v. 62, p. 1072–1083.

Liu, Yuejin, and Gastaldo, R. A., 1992b, Characteristics of a Pennsylvanian ravinement surface: Sedimentary Geology, v. 77, p. 197–214.

Lyons, P. C., and Darrah, W. C., 1979, Fossil floras of Westphalian D and early Stephanian age from the Narragansett Basin, Massachusetts and Rhode Island, Guidebook for Field Trip 5, *in* Cameron, B., ed., Carboniferous basins of southeastern New England: International Congress of Carboniferous Stratigraphy and Geology, 9th, p. 81–89.

McCabe, P. J., 1984, Depositional environments of coal and coal-bearing strata, *in* Rahmani, R. A., and Flores, R. M., eds., Sedimentology of coal and coal-bearing sequences: International Association of Sedimentologists Special Publication 7, p. 147–184.

Mosbrugger, V., 1989, Zur Gliederung und Benennung von Taphozönosen: Courier Forschunginstitut Senckenberg, v. 109, p. 17–28.

North, F. J., 1935, The fossil and geological history of the South Wales Coal Measures: Cardiff Nature Society Transactions, v. 62, p. 16–44.

Oshurkova, M. V., 1974, A facies-paleoecological approach to the study of fossilized plant remains: Palaeontological Journal, v. 3, p. 363–370.

Peppers, R. A., and Pfefferkorn, H. W., 1970, A comparison of the floras of the Colchester (No. 2) coal and Francis Creek Shale, *in* Smith, W. H., and others, eds., Depositional environments in parts of the Carbondale Formation—Western and Northern Illinois: Illinois State Geological Survey Guide Book Series, v. 8, p. 61–74.

Pfefferkorn, H. W., 1979, High diversity and stratigraphic age of the Mazon Creek flora, *in* Nitecki, M. H., ed., Mazon Creek fossils: New York, Academic Press, p. 129–142.

Pfefferkorn, H. W., Fuchs, H., Hecht, C., Hofmann, C., Rabold, J. M., and Wagner, T., 1988, Recent geology and plant taphonomy of the Orinoco Delta—Overview and field observations: Heidelberger Geowissenschaftliche Abhandlungen, v. 20, p. 21–56.

Scheihing, M. H., and Pfefferkorn, H. W., 1984, The taphonomy of land plants in the Orinoco Delta—A model for the incorporation of plant parts in clastic sediments of late Carboniferous age of Euramerica: Review of Palaeobotany and Palynology, v. 41, p. 205–240.

Scott, A. C., 1977, A review of the ecology of Upper Carboniferous plant assemblages, with new data from Strathclyde: Palaeontology, v. 20, p. 447–473.

Scott, A. C., 1978, Sedimentological and ecological control of Westphalian B plant assemblages from West Yorkshire: Yorkshire Geological Society Proceedings, v. 41, p. 461–508.

Scott, A. C., 1979, The ecology of Coal Measure Floras from northern Britain: Geologists' Association Proceedings, v. 90, p. 97–116.

Scott, A. C., and Chaloner, W. G., 1983, The earliest fossil conifer from the Westphalian B of Yorkshire: Royal Society of London Proceedings, v. 220B, p. 163–182.

Scott, A. C., and Rex, G., 1985, The formation and significance of Carboniferous coal balls: Royal Society of London Philosophical Transactions, series B, v. 311, p. 123–137.

Shute, C. H., and Cleal, C. J., 1987, Palaeobotany in museums: Geological Curator, v. 4, p. 553–559.

Spicer, R. A., 1989, The formation and interpretation of plant fossil assemblages: Advances in Botanical Research, v. 16, p. 95–191.

Stevenson, J. J., 1911a, The formation of coal beds. I: An historical summary of opinion from 1700 to the present time: American Philosophical Society Proceedings, v. 50, p. 1–116.

Stevenson, J. J., 1911b, The formation of coal beds. II: Some elementary problems: American Philosophical Society Proceedings, v. 50, p. 519–643.

Stevenson, J. J., 1912, The formation of coal beds. III: The rocks of the Coal Measures: American Philosophical Society Proceedings, v. 51, p. 423–553.

Stevenson, J. J., 1913, The formation of coal beds. IV: American Philosophical Society Proceedings, v. 52, p. 31–162.

Tiffney, B. H., 1990, The importance of little bits and pieces: Palaios, v. 5, p. 497–498.

Walker, R. G., and James, N. P., eds., 1992, Facies models—Response to sea level change: St. Johns, Newfoundland, Geological Association of Canada, 409 p.

White, D., and Thiessen, R., 1913, The origin of coal: U.S. Bureau of Mines Bulletin, v. 38, 390 p.

Wnuk, C., and Pfefferkorn, H. W., 1984, The life habits and paleoecology of Middle Pennsylvanian medullosan pteridosperms based on an *in situ* assemblage from the Bernice Basin (Sullivan County, Pennsylvania, U.S.A.): Review of Palaeobotany and Palynology, v. 41, p. 329–351.

Wnuk, C., and Pfefferkorn, H. W., 1987, A Pennsylvanian-age terrestrial storm deposit—Using plant fossils to characterize the history and process of sediment accumulation: Journal of Sedimentary Petrology, v. 57, p. 212–221.

Woodland, B. G., and Stenstrom, R. C., 1979, The occurrence and origin of siderite concretions in the Francis Creek Shale (Pennsylvanian) of northeastern Illinois, *in* Nitecki, M. H., ed., Mazon Creek fossils: New York, Academic Press, p. 69–103.

Manuscript Accepted by the Society July 6, 1994

Printed in U.S.A.

Geological Society of America
Memoir 185
1995

# *History and development of Carboniferous palynology in North America during the early and middle twentieth century*

**Aureal T. Cross**
*Department of Geological Sciences, Natural Sciences Building, Michigan State University, East Lansing, Michigan 48824*
**Robert M. Kosanke**
*U.S. Geological Survey, MS 919, Box 25046, Denver Federal Center, Denver, Colorado 80226*

## ABSTRACT

**Three main roots of upper Paleozoic palynology in North America date from the opening of the twentieth century. These are Gresley's recognition of spores in Iowa coal balls in 1901, analyses of spores by Sellards from Mazon Creek compressions in 1902, and Thiessen's analyses of dispersed spores from coal macerations and thin sections in 1913. The *Pollen Analysis Circular* brought workers with dual interests in Holocene and ancient spore/pollen analyses into contact in the 1940s and generated interest in older fossils. Two umbrella organizations—the Paleobotanical Section of the Botanical Society of America (1936) and the Coal Geology Division of the Geological Society of America (1955)—encouraged palynologists to participate in meetings and field trips.**

**Fundamental papers by Schopf et al. in 1944 and Kosanke in 1950 established Carboniferous palynology in North America. Active teaching and research centers at the University of Chicago in the 1920s and the University of Illinois and Coe College in the 1930s spawned new palynological centers, particularly throughout the Midwest. Palynological contributions on dispersed spores, mainly from coals and associated rocks, appeared from educational centers from 1929 through the 1950s (in approximate succession) from the University of Michigan, University of Cincinnati, Harvard University, Washington University (St. Louis), West Virginia University, Pennsylvania State University, Nova Scotia Research Foundation, Indiana University, University of Missouri, and the University of Oklahoma. Limited reviews of early researches at these early palynologic centers are here included by regions: Maritime Canada, the Appalachian, Illinois and Michigan basins, the Midcontinent and Texas, western North America, and the Arctic islands. Palynology applied to petroleum exploration appeared in the 1940s, and major petroleum companies had palynology laboratories in place by 1960. The first international palynology journals appeared in the 1950s and catalogs first appeared in the mid-1960s, except the Catalog of Fossil Spores and Pollen, which began in 1957. The first specific palynology organization, the American Association of Stratigraphic Palynologists, was founded in 1968.**

Cross, A. T., and Kosanke, R. M., 1995, History and development of Carboniferous palynology in North America during the early and middle twentieth century, *in* Lyons, P. C., Morey, E. D., and Wagner, R. H., eds., Historical Perspective of Early Twentieth Century Carboniferous Paleobotany in North America (W. C. Darrah volume): Boulder, Colorado, Geological Society of America Memoir 185.

## INTRODUCTION

### *Palynology, paleopalynology, and palynomorphs*

Palynology is the study of organic particles of microscopic size, up to 1 mm and occasionally up to several millimeters in greatest dimension, found dispersed in sediments and rocks. These dispersed organic particulates are called "palynomorphs" following the suggestion of Tschudy (1961, p. 53).

Paleopalynology is the study of fossil palynomorphs and its application to the resolution of problems of paleobiology, paleoecology, paleoenvironments of life, and environments of sedimentary deposition. Schopf (1964, p. 30, 31) summarized paleopalynology succinctly: "In its broadest sense, it includes the study of practically all plant particles which are subject to normal processes of dispersal and distribution . . . , and . . . plant microfossils include any type of fossil plant fragment of microscopic dimensions that conveys meaningful information about the plant represented."

Palynofloral assemblages may be constituted by a true microflora such as whole, microscopic-size plants or protists, organic-walled acritarchs, dinoflagellates, or other algae or their cysts; they may be dispersed reproductive organs (pollen, spores, cysts, or eggs); or they may be fragments of larger plant bodies as diverse as fungal hyphae, charophyte oogonia, insect scales, or cuticular, vascular and other tissue fragments, and organic or mineral inclusions of more complex plant bodies. This review will concentrate primarily on those palynomorphs found in the Carboniferous coals or coal-bearing rocks or in associated stratigraphic sequences. We will only briefly refer to algae or microplankton of lakes or seas of the late Paleozoic. Schopf (1964, p. 44, 50) wrote, "Most effective use of plant microfossils will link their occurrences with [their] contemporaneous megafossils which, taken all together, constitute the available floral assemblage. . . . Since, in applied palynology the primary intent is to provide a practical tool for stratigraphic interpretation. . . . it would simplify some problems . . . for students to regard nomenclatural rules nonapplicable." He also "suggested use of a symbolic coding system which seems most convenient and advantageous for many practical applications."

### *Six main stages of development*

There have been six main "thrusts" in the development of palynological studies in North America. All include some aspects of, or relationship to, the development of palynology as a branch of paleontology or the paleopalynology of Carboniferous and associated Upper Devonian and Lower Permian rocks. The first thrust was the early contributions in the twentieth century in North America, which began with paleobotanical studies of spores and pollen in fructifications. A second thrust was the recognition of spores in coals and other carbonaceous rocks and their study in both maceration residues and thin sections. The third was the widespread development of "pollen analysis," mainly on Pleistocene and Holocene peats and lake-bottom sediments, which occasionally do contain late Paleozoic recycled plant microfossils. Pollen analysis has impacted paleopalynological studies through common principles and techniques of study and the mutual interest of a number of early workers, including E. H. Sellards, L. R. Wilson, J. M. Schopf, R. M. Kosanke, A. T. Cross, W. Spackman, and others. The fourth thrust was the very extensive development of coal-ball studies, beginning with the "recognition" of coal balls in North America by A. C. Noé (1923), and the earliest studies of fructifications and their spores and pollen in coal balls by Noé's students, Hoskins (1923, 1926) and Reed (1926).

The fifth thrust and probably the major impetus to paleopalynology research in North America was generated by fundamental research on coal and associated rocks coming from the Illinois State Geological Survey Coal Section under the leadership of G. H. Cady. This research began with the publications of McCabe (1932), Henbest (1933, 1936), and Schopf (1936a, b, 1938), and the pivotal works of Schopf, Wilson, and Bentall (1944) and of Kosanke (1950). The sixth thrust has been the paleopalynology of rocks of various ages and types, including Carboniferous, for general application to geological exploration and interpretation of the origin of sedimentary rocks and their environments of deposition, biostratigraphy, and other aspects of oil exploration. This application was first developed in North America by L. R. Wilson and W. S. Hoffmeister in the early 1940s.

These six main thrusts now overlap or have been integrated with others at various levels as each expanded. The only upper Paleozoic paleopalynological thrusts not to be dealt with in this chapter are those that include studies and utilization of microplankton of marine and freshwater environments, including acritarchs and other organic cysts, charophytes, chitinozoa, microforaminifera (organic inner lining of tests), and conodonts. Such studies of organic-walled cysts and other microplankton have increased exponentially, beginning in the 1950s, and are now widely applied to oil exploration.

### *Where to begin and where to end?*

The starting point for the earliest twentieth-century descriptions and illustrations of Carboniferous spores in North America was two papers by Sellards (1902, 1903), who removed spores by maceration of fructifications embedded in Mazon Creek concretions (Upper Pennsylvanian) from Will County, Illinois. The earliest descriptions and illustrations of *dispersed* spores, which were obtained by maceration, were R. Thiessen's studies from 1907 to 1911 on coals from Alabama, Illinois, Indiana, Iowa, and Pennsylvania (White and Thiessen, 1913). Thiessen's work (see Chapter 11, this volume) continued in the 1930s with his assistants, G. C. Sprunk and Hugh O'Donnell, at the U.S. Bureau of Mines Central Experiment Station in Pittsburgh. Another starting point for paleopalynology in North America was the work of Bartlett (1929a, b), who described dispersed megaspores macerated from pebbles of coaly shales and coal from glacial gravels near Ann Arbor, Michigan, and discussed the megaspore genus *Triletes.* Bergquist (1939) demonstrated the probable location of the source of those spore-bearing coal and shale pebbles by their comparability to the Upper Pennsylvanian spore-coals at Williamston, Michigan.

A third starting point for paleopalynologic studies was the beginning of "pollen analysis," that is, the study of Pleistocene and Holocene peat deposits in North America. A number of persons working in this subdiscipline have often encountered recycled Devonian and Pennsylvanian palynomorphs in the residues. Some of those who first analyzed glacial-postglacial sample suites, including Leonard R. Wilson (see Chapter 20, this volume), were among the pioneers in North America to analyze coal samples in the 1930s. Arnold (1961, p. 245) stated:

> The important contributions to Pleistocene and Post-Pleistocene vegetational history that resulted from studies of pollen in the peat bogs of western Europe during the 1920's led to the belief that valuable information on older floras might be gained through studies of spores preserved in coal and shale. During the decade that began with 1930 several investigators began to examine fossil spores. The result was the beginning of a new phase of paleobotanical research and the development of a considerable body of literature.

An ending to this historical review of Carboniferous palynology is even more difficult to select. We have chosen the mid-1960s because of the accelerated growth of the science from its introduction into more than a dozen oil-exploration laboratories throughout the 1950s and the reciprocal growth of teaching centers in response to job availability and expanded petroleum exploration programs in the 1960s. The introduction of computers in the early 1960s helped to generate more reliable palynological data for the oil industry. Perhaps the biggest factor in selecting the mid-1960s as a cutoff date was the exponential growth and utilization of marine palynomorphs, including dinoflagellates and acritarchs, for stratigraphic and paleoecologic problem-solving during the 1960s. The energetic activity of W. R. Evitt and G. Norris, two of the pioneer advocates of marine palynology in America, and the burgeoning production of papers and catalogs from European specialists—A. Eisenack, G. Deflandre, C. Downie, W.A.S. Sargeant, G. L. Williams, and R. J. Davey—greatly enhanced palynological contributions to oil exploration. We have not included that area of paleopalynology in our historical review.

Several milestones in the form of conferences, publications, and visiting lecturers marked the progress from 1957 to 1967. The first was the 1957 publication of the first two volumes of the *Catalog of Fossil Spores and Pollen* at Pennsylvania State University. Also, Andrews's textbook, *Studies in Paleobotany* (1961), included a chapter on palynology by C. J. Felix. This was an appropriate recognition of the place in paleobotany of this subdiscipline in a classical textbook of paleobotany. The first International Palynological Conference (Tucson, 1962) and the 1962 symposium on "Palynology in Oil Exploration," which was sponsored by the Society of Economic Paleontologists and Mineralogists (SEPM) and the American Association of Petroleum Geologists (AAPG), were additional milestones. The first four volumes of Synopsis der Gattungen der Sporae dispersae (Potonié, 1956–1966) was a major summation of spore genera along with other summary volumes begun in the 1950s. The introduction of acritarchs as a group to divide the Paleozoic "hystrichospheres" from the dinoflagellates (Evitt, 1963) and the publication of the proceedings of the 1962 symposium (Cross, 1964) further enhanced the field. In 1964, the AAPG sponsored a four-month and 40-city "distinguished lecture" tour (Gries, 1991, p. 32) for A. T. Cross to visit and lecture to geological societies, petroleum companies, and universities in Canada and the United States on the potential and limitations of palynology—specifically, palynological techniques; laboratory operations; biostratigraphical, paleoenvironmental, and paleogeographic information; research directions; and the initiation and requirements for teaching programs and research directions. In 1966 the Second International Palynological Conference was held at Utrecht, The Netherlands, with hundreds attending. Many papers of the proceedings were published in 1967 as the first five volumes in the *Review of Palaeobotany and Palynology.* The publication of the "Bibliography of Palaeopalynology, 1836–1966," as volume 8 of the *Review of Palaeobotany and Palynology* (Manten, 1969) was a landmark volume of inestimable value in the preparation of this chapter.

## EARLIEST HISTORICAL THREADS—NINETEENTH CENTURY

### *Hildreth, Morton, and the Zanesville megaspores of 1836*

"Few places in the world, perhaps, afford plant fossils in such abundance and perfection as the mines about Zanesville. Many of the plates in the splendid work, 'Histoire des Végétaux Fossile' by Adolphe Brongniart [1828–1838], were figured from specimens furnished by the late Ebenezer Granger, Esquire, or from drawings sent by W. A. Adams, Esquire, all of which were procured near Zanesville. . . ." (Foster, 1838). One type of fossil is shown in Figure 1. This enthusiastic report of the Putnam Hill locality (Fig. 2) and others in the vicinity of Zanesville, Ohio, described one of the earliest sites in North America where fossil plants were recognized and collected for scientific study. The first detailed analysis of the geology and paleobiology of this classic site is in Samuel P. Hildreth's report (1836) on its geology and paleontology. Illustrations of the fossils were prepared as lithographs or woodcuts by Morton (1836). Two of the three species of fossil plants from Zanesville listed in Adolphe Brongniart's *Histoire* (1828–1838) came from this locality: *Neuropteris grangeri* Brongniart and *Pseudope-*

1
2
View of "Putnam's Hill" and the upper Bridge at Zanesville, on the Muskingum River.

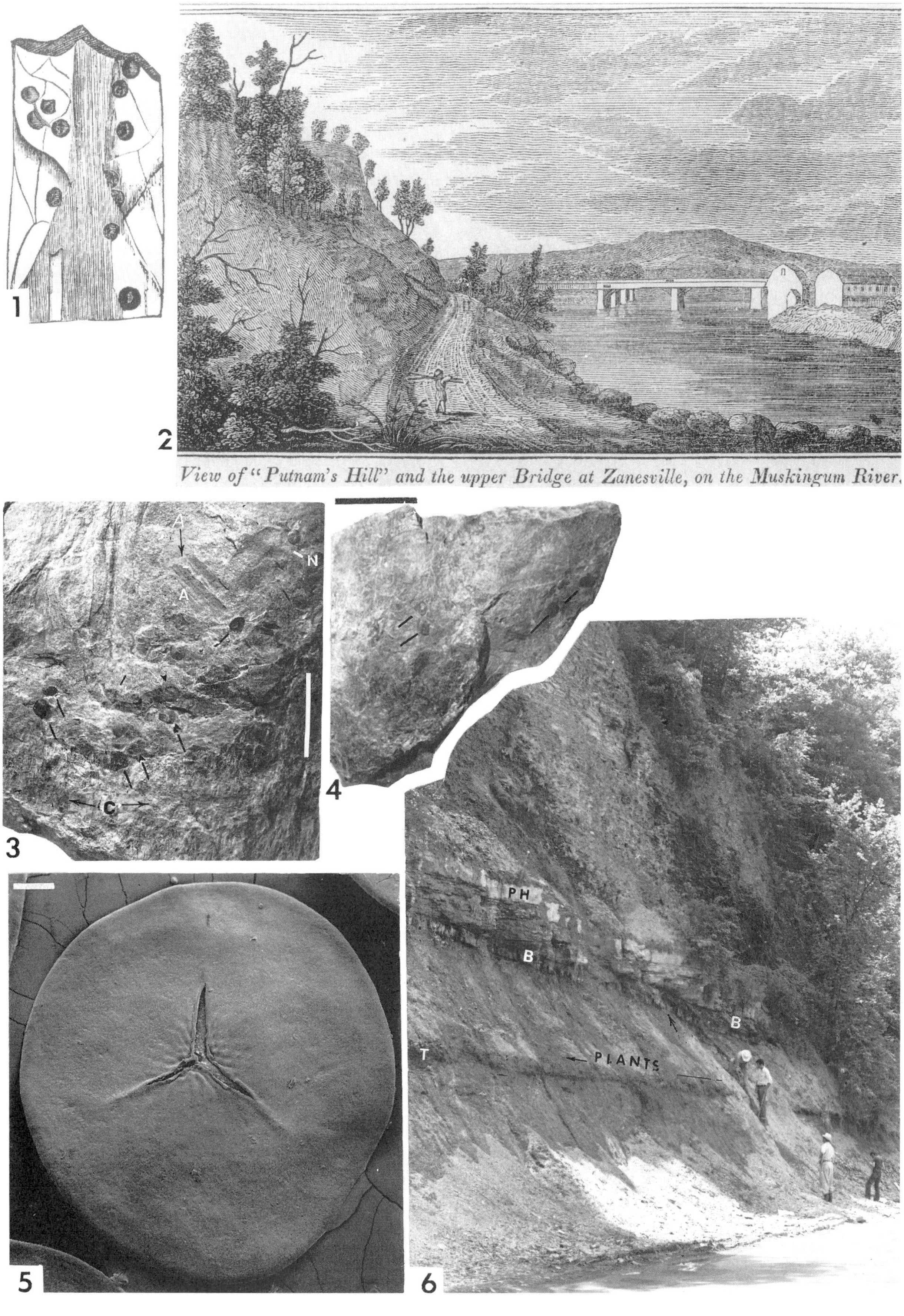

Figure 1. Original woodcut by Morton (1836, pl. 30, fig. 10) illustrating several circular bodies appressed together with macroscopic plant remains on a shale layer. These were interpreted as monocot seeds by Hildreth (1836, p. 35) preceding Morton's illustrations in the same volume (spores shown enlarged about 50%). (Reproduced from Morton, 1836, v. 29, pl. 30, fig. 10, by permission of the American Journal of Science.)

Figure 2. Morton's original woodcut of the Putnam Hill locality looking north up the Muskingum River toward the center of the City of Zanesville, Ohio, as it may have looked in the 1830s. The fossil locality is at the crest of the hill along the roadway on the west (left) bank. (Reproduced from Morton, 1836, v. 29, pl. 36, by permission of the American Journal of Science.)

Figure 3. Large chip of roof shale from just above Tionesta coal showing numerous specimens of *Triletes reinschii;* also *Annularia stellata* (2 leaves at A); *Neuropteris grangeri* (N); *Calamites* (C) all along lower edge. Bar scale = 10 mm.

Figure 4. Same as Figure 3. Bar scale = 10 mm.

Figure 5. *Triletes reinschii* (Ibrahim) Schopf, 1938. From shale above Tionesta coal (collection no. 5-21-77 III 1-2, A. T. Cross, Putnam Hill, Zanesville, Ohio, 1.95 mm diam.; SEM photo courtesy of George Massey, Amoco Production Co. Research Center, Tulsa, Oklahoma). Bar scale = 200 μm.

Figure 6. Exposure along Putnam Hill in 1976 at Hildreth's (1836) collection site. T—Tionesta coal bed of Pottsville Formation; PLANTS—Plant-bearing shale containing many megafossils; B—Brookville coal bed of Pottsville Formation; PH—Putnam Hill Limestone Member of the Allegheny Formation, upper bench.

*copteris (Pecopteris) sillimannii* Brongniart. The third species, *Calamites dubius* Artis, is also present at this Putnam Hill site, but the specimen illustrated by Brongniart may have come from a few meters lower, near the river.

Samuel G. Morton's woodcuts included an illustration of several megaspores (1836, pl. 30, figure 10) that were interpreted by Hildreth (1836, p. 35) as "seeds of some monocotyledonous plant, which are thickly scattered in patches amidst the fossil leaves in the shale. . . ." Morton's figure, reproduced here as Figure 1, is now recognized as the megaspore *Triletes reinschii* (Fig. 5). These spores are very abundant in the shale ("plant bed") just above the Tionesta coal bed (T) in Figure 6 (at about the level of the feet of the upper person in the photograph). Figures 3 and 4 show several megaspores each on the split surfaces along with *Annularia stellata* (A), *Neuropteris* (N), and *Calamites* (C). The large, smooth-walled megaspore has at least 30 small spores adhering to its upper surface in the clay and silt deposit that formed the shale matrix. Under higher magnification (SEM micrographs) at least three types of small spores (microspores or homospores)—*Lophotriletes, Granulatisporites,* and *Lycospora*—can be identified on the megaspore illustrated here (Fig. 5). Although first interpreted as seeds, they are the first figured fossil spores in North America (Fig. 1).

## *Dawson and classic European contributions*

Little is known of the historical aspects of palynology in North America between the early 1800s and the beginning of the twentieth century. An occasional European discovery or advance could have come to attention, but apparently the only one noticing was J. William Dawson of Nova Scotia and McGill University. In the late 1800s, he not only carried out his own investigations at several sites in eastern Canada but also sought personal approbation of his work abroad. In trying to decipher the origin of coal, he visited Thomas H. Huxley in England in the late 1860s. Huxley showed him coal thin sections containing large numbers of spores. Though Dawson considered these to be of little consequence in coal formation, when he returned he was prompted to look at his own coal sections from Horton Bluff, Nova Scotia, and even prepared some slides of Upper Devonian shales from Kettle Point, Ontario. In all of them he found spores, some in great profusion. In Dawson's (1871) publication on spore cases in coals, representative spores of Carboniferous vascular plants were figured. He also figured again the sporocarps of "Sporangites," now recognized as cysts of a planktonic prasinophycean alga, under the name *Tasmanites.*

Henry T. M. Witham "of Lartington" is generally credited with publishing the first illustration of spores (Witham, 1833, pl. XI, figs. 4, 5) in two sections of a cannel coal from Lancashire, England. He was successful in interpreting many of the accompanying quality illustrations of fossil wood. However, with wood anatomy on his mind, he was hesitant to attribute the figures in the two illustrations to anything except some type of wood structure. The full circles, tiny circles, small discs, and crescents (actually spores sliced through at different levels) were presumed to be a cross section of wood, but it was with reluctant obliquity that he inferred the structures to be possibly tubelike fibers of a monocotyledonous tissue. The section does illustrate spores, possibly microspores 25 to 35 μm in diameter with a possible reticulate megaspore, 120 μm in diameter, and a conspicuous *Pila* colony—a boghead alga typical of and similar to that found in Scottish torbanite (boghead coal).

John H. Balfour, professor of medicine and botany at the University of Edinburgh in 1840–1841, was one of Dawson's professors. Balfour was particularly interested in plant structures in coal. In a paper read before the Royal Society of Edinburgh in 1854, Balfour (1857) presented clear evidence of tissues of *Sigillaria* with cell anatomy in the Fordel coal of Fife, Scotland, to reject the popular concept that only conifer wood contributed to woody tissues preserved in coals. In Balfour's discourse, he was disposed to ask John Queckett, a well-known proponent of the concept, to reexamine the hollow circular structures that he used as evidence for cross sections of coniferous tubes and consider the possibility that they might be spores or spore cases, or even hollow spaces. Another professor at Edinburgh, John H. Bennett, had examined the Torbane Hill boghead cannel and recognized and figured several thin sections in color (Bennett, 1857, pl. I, figs. 1 to 9), five of which had small spores sliced parallel to the bedding and four perpendicularly. One very large megaspore with thick wall and part of the triradiate aperture showing appeared in his figure 1 and a small megaspore in figure 7. Abundant colonies of *Pila* are evident in all figures except his figure 3. In the illustrations for Balfour's paper, delineated by another scientist, R. K. Greville (Balfour, 1857, pl. II, figs. 12–18), several millimeter-size spores partially embedded in coal or freed from the matrix are shown on color plates! These are certainly the best figures of spores up to 1857.

Balfour (1857, p. 191) stated:

> Besides Sigillarias and Stigmarias, we also detected in the Fordel coal peculiar rounded organisms, which have the appearance of seeds (Plate II, figs. 12–13). Dr. Fleming informs me, that similar bodies have been observed by him in coal, and that he exhibited them to Witham [H. Witham, 1833] about twenty years ago. They have also been seen by Dr. Fleming at Lochgelly and Arniston parrot [a type of cannel], and in the coal at Boghead; and from having observed them in cherry, splint, and cannel coals, he is disposed to consider them . . . a common feature. . . . I have seen them at Miller-hill . . . as well as in coal from Fife. . . . The nearest approach to them is the Lycopodites, figured by Mr. Morris [1840], in the Appendix to Mr. Prestwich's paper on the Geology of Coal-Brook Dale [1840]. They appear to be certainly allied to the fructification of the Lycopodiaceae of the present day, . . . I therefore consider [them] to be the sporangia or spore cases of some plant allied to Lycopodium, . . . They have a rounded form, . . . [the] color is dark brown, . . . and they seem to be formed by two valves enclosing a cavity . . . often filled with black carbonaceous matter. . . . Under the microscope, the valves often present a reticulated appearance.

The contribution of Franz Schulze (1855) to developing the maceration technique for releasing liptinitic structures and resins from the coal matrix as well as for reducing the opacity of higher-rank coal particulates dispersed in sediments was of inestimable value to the development of palynology. Thiessen and Sellards both utilized this technique in their preparations. Reinsch (1884) also developed a technique for releasing spores from a coaly matrix or from high-rank coals. His photomicrographs of spores were the first to be published, but some of his interpretations were unsupportable. One of the more clearly defined Carboniferous spores of his *Subdivisio III* was described and named *Reinschospora* in Schopf et al. (1944). The classic work of Bennie and Kidston (1886) on Carboniferous spores of Scotland was the first major publication of a palynoflora.

### Introduction and spread of pollen analysis technology

A review of early North American pollen analysis research is appropriate here, and the significance will be defined more fully in a following section on the *Pollen Analysis Circular.* The basic techniques and procedures of collection, preparation, microscopical analysis, data handling, and statistical procedures have fundamental similarities to pre-Quaternary studies. The goals also have much in common. A number of early American workers in this field have had dual interests in both pre-Quaternary and Holocene sedimentary deposits for resolving mutual problems of interpreting paleoclimates and discerning paleoecological information, such as environmental controls and succession and floristic composition, and for analyzing factors involved in distribution and preservation of palynomorphs in different kinds of sedimentary accumulations. Finally, some workers whose training was principally in pollen analysis became early pre-Quaternary researchers.

Pollen analysis of Pleistocene and Holocene deposits of northern Europe spread very rapidly following Von Post's (1916) introduction of statistical pollen analysis to problems of sequential peat or sediment accumulations (see Godwin, 1968). Gunnar Erdtman, Von Post's student, made an exploratory tour of the British Isles in the mid-1920s to check out possible sites for research; he spoke at several institutions. He was an enthusiastic, eloquent, and forceful advocate of the potentials of application of statistical pollen analysis for interpretation of Wisconsin (late Pleistocene) and Holocene climates and history. He also toured the eastern United States with the same inspiring message. That did much to awaken northern Europe and eastern North America to the potentials of this fledgling science. During Erdtman's visits in England he had included a stop at Leeds University, where W. H. Burrell fell under his spell. Burrell immediately commenced investigating the upland peats of the Yorkshire Pennines. Shortly afterward, L. R. Wilson joined Burrell at Leeds for a year of foreign study. Wilson was an undergraduate from the University of Wisconsin and learned of this intriguing research tool as he worked with Burrell as an assistant in the museum. Wilson became inspired by the possibilities of applying this new research of pollen analysis to unraveling the maze of vegetation patterns and resolving the Pleistocene and Holocene history of the landscape back home in Wisconsin.

Perhaps as a result of Erdtman's visits, pollen analysis papers began to appear in North America (Auer, 1927a, b, 1930; Fuller, 1927). Stanley A. Cain (1939, p. 626) wrote: "The method was quickly realized. . . . Erdtman['s] bibliographies of the literature on pollen statistics . . . up to 1927 [Erdtman, 1927], consist of about 500 entries. During the next three years (1927–1929) about 275 additional papers appeared, while 1930–1931 saw the publication of 250 more, and at the present [1939] well over 1,000 papers have been published." This flurry of activity is reflected in the time taken and space covered for the pollen-analysis fever to reach across the Atlantic to North America. In rapid succession, publications appeared for most of the states around the Great Lakes and southeastern Canada: Fuller's 1927 introductory paper for the Lake Michigan vicinity; Väniö Auer's on peat bogs in southeastern Canada; Bowman's pollen analysis of a bog near the Matamek River in Quebec (1931); Draper (1929) and Sears

(1930a, b), the earliest studies of bogs in the Erie basin; L. R. Wilson's original study of the Two Creeks forest bed of Wisconsin (1932); Houdek's analyses of two bogs in Indiana (1933); Potzger's earliest paper from northern Michigan (1932); and Houdek's report on two bogs in southwestern Michigan (1935). Ira T. Wilson and Potzger reported on pollen records from lakes in Anoka County, Minnesota (1943). Lewis and Cocke's classic study (1929) resulted in a pollen profile of the Dismal Swamp of Virginia. This was the earliest research to be published on pollen analysis east of the Appalachians. Many papers appeared in quick succession as the studies spread westward across areas of glaciated terrane of the western Great Lakes and upper Midwest to the Pacific. These included Voss's pioneering works (1931, 1933, 1937) on interglacial and postglacial deposits in Illinois and Lane's (1941) paper on interglacial peats of Iowa. Henry P. Hansen moved his center of teaching and research activity from Wisconsin (1937) to Oregon, where the first of his many papers on bogs of the western United States and the Pacific Northwest area appeared in 1938. Hansen (1942, 1943) also published early papers on pollen of lake sediments of the Willamette Valley in Oregon and on peat profiles from the Lower Klamath Lake in California. Alaska was finally reached by Calvin J. Heusser (1952), although some secondary references to pollen analysis had been included in important studies on forests of Alaska by Cooper (1923, 1937, 1942) and Griggs (1934).

Centers of teaching and research developed at several institutions, generally around a single worker. The earliest were at the University of Chicago (G. D. Fuller) and University of Wisconsin (N. C. Fassett). Oberlin College (P. B. Sears) and Butler University (J. E. Potzger and R. C. Friesner) were both active centers by 1930. Coe College (L. R. Wilson) and University of Minnesota (W. S. Cooper and C. O. Rosendahl) programs were activated for pollen analysis in the early 1930s. Oregon State University (Hansen) and the University of Tennessee (S. A. Cain) were participating by the late 1930s.

## EARLIEST TWENTIETH-CENTURY STUDIES

### In situ *spores: Sellards and Gresley, 1901–1903*

A decade before the first major North American publication on dispersed spores in coals (Thiessen, 1913), Elias H. Sellards (see Chapter 12, this volume) and W. S. Gresley recognized and illustrated fossil spores and pollen extracted from Illinois Mazon Creek concretions and an Iowa coal ball. Sellards (1902) described spore-bearing fructifications and spores of *Crossotheca, Myriotheca,* and some marattiaceous fern spores (on *Pecopteris* foliage) from compressions embedded in Mazon Creek concretions. His Ph.D. dissertation at Yale University in 1903 was a continuation of those studies on a new "spore-bearing" organ, *Codonotheca caduca* (Sellards, 1903). He concluded that this was a homosporous fructification containing only one size "spore" of a large *Monoletes*-type. He indicated some curiosity about this lack of smaller spores in a plentiful record. A new group of seed plants, the Lyginopteridales, was proposed by Oliver and Scott (1904). That led to the substantiation of the Cycadofilicinean group (Pteridospermophyta), proposed by Potonié (1899). Sellards's (1907) paper on *Codonotheca* and its relation to the Cycadofilicales was like a sigh of relief in his quest for an explanation.

At about the same time as Sellards was studying fructifications in Mazon Creek ironstone concretions, W. S. Gresley described (1901) and "illustrated beautifully preserved reproductive structures of *Cordaites* (*Cardiocarpus* and *Cordaianthus*)" (Phillips et al., 1973, p. 11), which showed the *Florinites*-type spore, from "pyrite concretions" in the coal bed at What Cheer, Iowa. Gresley did not call these "coal balls" and perhaps did not know of European coal-ball studies (see Chapter 27, this volume). David White (see Chapter 10, this volume), who helped and advised Gresley with these concretions, did not recognize them as coal balls either, perhaps because he did not realize they were highly pyritized carbonate concretions.

### *Dispersed spores: Thiessen and the U.S. Bureau of Mines*

R. Thiessen, the true North American pioneer of fundamental studies of plant constituents of coal, was the first North American in the twentieth century to include research on dispersed spores. His extensive use of thin sections of coal led to the interpretation of the organic constituents of coals, mainly from North America. His research was a milestone in coal petrology (see Chapter 11, this volume). He is less well known for a considerable number of macerations of diverse coals by the Schulze technique, which was used to free spores, cuticles, and other dispersed tissues from their embedding matrices during 1907 to 1911 (White and Thiessen, 1913, p. 3–4). He noted that "the best conception of the exines may be gained by the maceration process" (Thiessen, 1913, p. 263). Thiessen included remarks on the size, morphology, and ornamentation of several different spore forms and noted that some were microspores; megaspores were also prominent, and often more conspicuous, although not as numerous as the smaller spores (less than 100 μm in diameter).

Figures 7 through 15, 17, and 19 through 29 depict some of the spores photographed from maceration residues by Thiessen (1913, 1920). Eight spore photomicrographs from the spore-rich Shelbyville coal from Shelby County, Illinois, are reproduced here as Figures 7, 8, 19, 20, 21, 25, 27, and 29. Many of the spores were referred to plant groups such as lycopods or ferns. Only one was referred to a particular plant. Thiessen (1913, p. 267) identified one of them, the dispersed megaspore, illustrated here as Figure 8. It was described by Thiessen in this way: "in this coal still another spore exine of rather large size and triangular shape is present. It has two borders, the outer one having fine radiation; the inner, denser part is always circular. . . . These are of the same structure and ap-

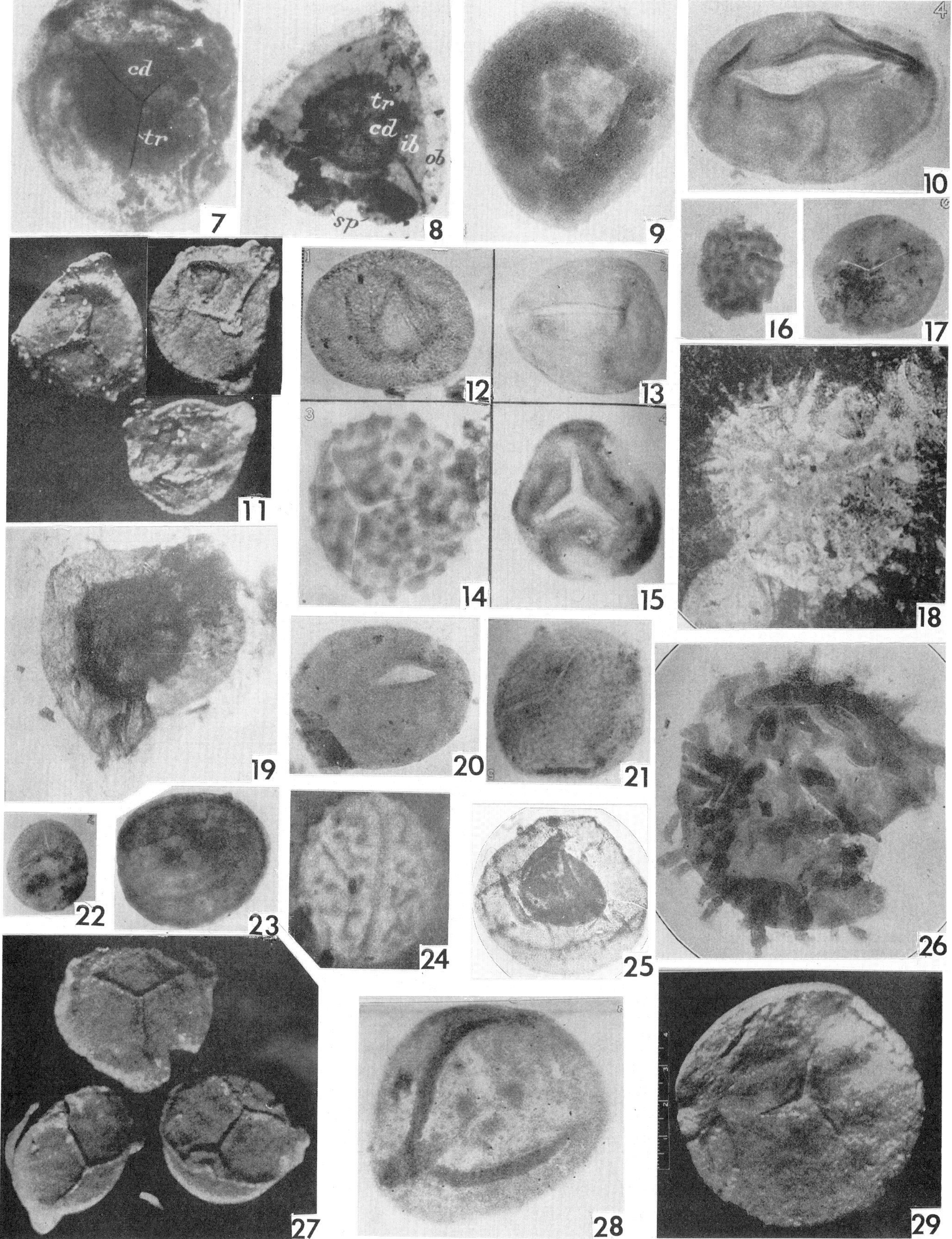
cd
tr
tr
cd
ib
ob
sp
7
8
9
10
11
12
13
14
15
16
17
18
19
20
21
22
23
24
25
26
27
28
29

Figure 7. *Endosporites globiformis* (Ibrahim) Schopf et al., 1944, 81 µm diam., Shelbyville coal, upper Bond Fm., upper Missourian; Shelbyville, Shelby County, Illinois. (Reproduced from Thiessen, 1913, pl. 53-A.)

Figure 8. *Spencerisporites gracilis* (Zerndt) Winslow, 1959, megaspore, 400 µm; same collection data as Figure 7. cd = central darker part (central body); ib = inner border (bladder); ob—outer border (= flange); tr—tetrasporic mark (= trilete suture); sp—two microspores (= *Endosporites*) (same locality and stratigraphic data as Fig. 7). (Reproduced from Thiessen, 1913, p. 377, pl. 53-C.) (See Leisman, 1962a, b.)

Figure 9. *Densosporites sphaerotriangularis* Kosanke, 1950, 54×52 µm; Black Creek coal, Black Creek coal Group, Pottsville Formation, near Sipsey, Walker County, Alabama. (Reproduced from Thiessen, 1920, p. 267, pl. 141-B.)

Figure 10. *Laevigatosporites ovalis* Kosanke, 1950, 55×37 µm; Pittsburgh coal, Monongahela Fm., Allegheny County, Pennsylvania. (Reproduced from Thiessen, 1920, p. 194, pl. 82-4.)

Figure 11. *Valvisporites (Triletes) auritus* (Zerndt) Potonié and Kremp, 1956, megaspore, varies 1.4–1.1 mm; Buxton, Monroe County, Iowa, age not defined. (Reproduced from Thiessen, 1920, p. 254, pl. 130.)

Figure 12. *Florinites mediapudens* (Loose) Potonié and Kremp, 1956, 50×36 µm (location and stratigraphy same as Fig. 11). (Reproduced from Thiessen, 1920, p. 260, pl. 136-1.)

Figure 13. *Laevigatosporites ovalis* Kosanke, 1950, 46×42 µm (all data same as Fig. 12, p. 260, pl. 136-2).

Figure 14. *Raistrickia* cf. *R. grovensis* Schopf, Wilson, and Bentall, 1944, 51 µm diam. (All data same as Fig. 12, pl. 136-3.)

Figure 15. *Deltoidospora* cf. *D. levis* (Kosanke) Ravn, 1986, 45 µm (all data same as Fig. 12, pl. 136-4).

Figure 16. *Thymospora thiessenii* (Kosanke) Wilson and Venkatachala, 1963, 18×21 µm; Pittsburgh coal, Monongahela Fm., near Pittsburgh, Allegheny County, Pennsylvania. (Reproduced from Thiessen, 1920, p. 194, pl. 82-1.)

Figure 17. *Cyclogranisporites obliquus* (Kosanke) Upshaw and Hedlund, 1967, 28 µm long dimension; Pittsburgh coal, near Pittsburgh, Allegheny County, Pennsylvania. (Reproduced from Thiessen, 1920, p. 194, pl. 82-3.)

Figure 18. *Raistrickia* cf. *R. aculeolata* Wilson and Kosanke, 1944 (possibly *R. crocea*) (specimen in thin section); body 54 µm diam., spiny processes up to 8 µm; Herrin (No. 6) coal, Carbondale Fm., Desmoinesian Series, Zeigler Mine, Franklin County, Illinois. (Reproduced from Thiessen, 1920, p. 229, pl. 111A.)

Figure 19. *Spencerisporites gracilis* (Zerndt) Winslow, 1959, megaspore, 2.8×3.1 mm overall, outer flange 350 to 400 µm wide, bladder 370–700 µm, body 130 µm; Shelbyville coal, Bond Fm., upper Missourian Series, near Shelbyville, Shelby County, Illinois. (Reproduced from Thiessen, 1920, p. 237, pl. 117–1.)

Figure 20. *Laevigatosporites* cf. *L. latus* Kosanke, 1950, 44×33 µm; Shelbyville coal, (same data as Fig. 19, p. 237, pl. 117-3).

Figure 21. *Triletes* cf. *T. globosus* Arnold var. (B), 1950 (see Winslow, 1959, p. 43–44), megaspore, 220 µm diam., Hart-Williams Mine, Herrin (No. 6) coal (locality as Fig. 18). (Reproduced from Thiessen, 1920, p. 230, pl. 112-5.)

Figure 22. *Cyclogranisporites obliquus* (Kosanke) Upshaw and Hedlund, 1967, 25×22 µm; Herrin (No. 6) coal, Royalton Mine, Franklin County, Illinois (age data as Fig. 18). (Reproduced from Thiessen, 1920, p. 230, pl. 112-2.)

Figure 23. *Schopfipollenites* sp., prepollen, 225×190 µm; Herrin (No. 6) coal, Hart-Williams Mine (same data as Fig. 7). (Reproduced from Thiessen, 1920, p. 230, pl. 112-4.)

Figure 24. *Thymospora thiessenii* (Kosanke) Wilson & Venkatachala (from thin section), 18×15.5 µm; Pittsburgh coal (same data as Fig. 16). (Reproduced from Thiessen, 1920, p. 194, pl. 82-2.)

Figure 25. *Endosporites globiformis* (Ibrahim) Schopf, Wilson, and Bentall, 1944, 90 µm; Shelbyville coal (same data as Fig. 19). Reproduced from Thiessen, 1920, p. 233, pl. 113-B.)

Figure 26. *Raistrickia crocea* Kosanke, 1950, body 64 µm, longest processes 15 µm; Herrin (No. 6) coal (same data as Fig. 18), Hart-Williams Mine, near Benton, Illinois. (Reproduced from Thiessen, 1920, p. 229, pl. 111-B.)

Figure 27. *Valvisporites (Triletes) auritus* (Zerndt) Potonié and Kremp, 1956, megaspores; 1.00–1.15 mm; Shelbyville, Illinois (same data as Fig. 19). (Reproduced from Thiessen, 1920, p. 239, pl. 119A.)

Figure 28. *Crassispora kosankei* (Potonié & Kremp) Bhardwaj, emend Smith and Butterworth, 1967, 63×57 µm; Shelbyville coal (same data as Fig. 19). (Reproduced from Thiessen, 1920, p. 237, pl. 117-2.)

Figure 29. *Triletes glabratus* Zerndt, 1930, megaspore, varies 1.85–1.95 mm; Shelbyville, Illinois (same data as Fig. 19). (Reproduced from Thiessen, 1920, p. 238, pl. 118A.). Figures 7 through 29 are reproduced here by courtesy of the U.S. Bureau of Mines.

pearance as those referred to in textbooks as the spores of *Spencerites*." The description of that megaspore is within the circumscription of *Spencerisporites gracilis* (Zerndt) Winslow. That and several others differ in vernacular from today's descriptions but not in reasonable descriptive clarity.

There have also been illustrations of *in situ* spores in Thiessen's thin sections. Two examples of these are reproduced here (Figs. 18 and 24). Figure 24, *Thymospora thiessenii,* is a reproduction of his photograph of a spore in a thin coal slice ground parallel with the bedding plane (Thiessen, 1920, p. 194). It compares quite favorably with Figure 16, which was photographed from a specimen isolated by maceration from the Pittsburgh coal bed near Pittsburgh. This is the "Pittsburgh spore" of Thiessen. The other spore, *Raistrickia* cf. *R. aculeolata* (Fig. 18), is remarkably well shown considering the coarsely spiny surface and the presence of some translucent mineral matrix embedding and masking it. Thousands of spores included in Thiessen's photomicrographs clearly depict spores in edge view cut perpendicular to the bedding planes of the coal. Altogether, Thiessen (1913, 1920) described spores isolated from the embedding matrix (coal, carbonaceous shale, and gray shales) from several states.

Thiessen (1920) recognized that some beds of coal contain one to several types of spore-exines that are predominant in or characteristic of that bed, or both. Further, he noted (p. 72),

> . . . even a casual examination of the photograph . . . from the different localities from the Pittsburgh bed . . . reveal[s] a striking similarity in the predominating mass of spore-exines. . . . In comparing the Black Creek coal from the Sipsey Mine with that from the Jagger bed, near Carbon Hill, in Alabama, the type of exine in each is strikingly different, but the predominating mass of exines in each coal are all of the same kind. . . . Other spore-exines are characteristic of a certain bed, but are not the predominant spore, as the beaked megaspore in the Buxton Coal [Iowa (Fig. 11)] or the heavily-spined exine from the Hart-Williams . . . and Ziegler mines [both from the Herrin No. 6 Coal, Illinois [Figs. 18, 26)]. . . . Although not enough coals from the Illinois and Indiana coals have been examined, the observations so far indicate that those coals may be grouped with respect to the predominating or the characteristic type of spores they contain. *This subject may find a wide application in determining the different beds in the Illinois coal fields, and it deserves careful development*. . . . [emphasis added]

This may have been the first time that identification of coal beds for correlation or stratigraphic placement on the basis of palynomorphs was postulated, certainly in North America.

In 1919, a cooperative program was generated between the U.S. Bureau of Mines, Carnegie Institute of Technology, and industrial users of coal in the Pittsburgh region of the northern Appalachian coalfields. Thiessen proposed a broad investigation of identification and correlation of coal beds using spore analysis (Thiessen and Staud, 1923). Thiessen and Staud included a review of spore distribution in coals of the Appalachian region, the maceration procedure for freeing the spores for analysis, and the general characteristics of the spore content of the commercial Upper Pennsylvanian Monongahela Formation coals (Pittsburgh, Redstone, and Sewickly) as observed in thin section.

Thiessen and F. E. Wilson (1924) published an account of the utilization of the coal thin-section method for the correlation of coal beds of the Allegheny (Middle Pennsylvanian) and Monongahela Formations of western Pennsylvania and Ohio. They identified eight coal beds by their spore content. The spores were not formally named. For example, they reported that the "Pittsburgh spore" (Figs. 16, 18) characterized the Pittsburgh (No. 8) coal bed and that the "pitted spore" *(Thymospora pseudothiessenii)* is characteristic of the Upper Freeport coal bed. Clearly, Thiessen and his associates were the first in North America to demonstrate the value of palynomorphs in the correlation of coals.

Sporangia containing spores were also recognized in all the coals examined by White and Thiessen (1913) and Thiessen (1920). Some of the sporangia were dispersed (called "middletonite" in the sense of Johnston [1838, p. 162–163]); others were isolated pieces of cones (Thiessen, 1920, pl. 134, from Buxton, Iowa) or whole cones (Thiessen, 1920, pl. 14, Benton, Illinois).

## COMING TOGETHER AND GETTING ORGANIZED IN NORTH AMERICA

### *The* Pollen Analysis Circular—*A unifying force*

The *Pollen Analysis Circular* (Sears, 1943) first appeared in May 1943 as a four-page mimeographed note containing a directory of 49 persons who were on the mailing list of P. B. Sears of Oberlin College, Ohio. This included several well-known botanists and several geologists or paleobotanists as well as those doing *pollen analysis.* The purpose of the *Circular* was to enable persons of similar interests, though widely dispersed by their varied military and civilian roles in World War II, to keep in touch with activities and publications. Thirty-six papers, dated 1939 to 1942, were listed, including one on dispersed coal microspores (Wilson and Coe, 1940) from Iowa. Several correspondents suggested continuation of the *Circular,* which included brief statements on work in progress, an announcement of Erdtman's new book (1943), and a note that Wilson would assemble and distribute the second and third *Circulars.* The third *Circular* (September 1943) included inquiries by Kirk Bryan and L. R. Wilson about the feasibility of forming a palynological organization. The 23 new publications listed included papers of interest to "peat workers." In the fourth *Circular* (November 1943), 16 of the 21 new references were on *pre-Quaternary* coals and oil shales that ranged in age from Devonian to Eocene, included were Wodehouse's classic papers (1932, 1933) on Tertiary microfossils. Several correspondents discussed the pros and cons of forming a more formal organization.

The growth of interest in the *Circular,* scope of topics, and the number of North American and foreign participants reached a high point by the eighth *Circular* (October 1944). Several letters indicated a desire for a bibliography, which Theodor Just offered to assemble as editor of the *American Midland Naturalist.* Ernst Antevs and Kirk Bryan endorsed the idea of forming an "informal organization." Antevs asked if the name *Pollen Analysis Circular* was adequate to encompass the diverse interests (including modern pollen morphology, anatomy, and chemistry; bog and lake analyses; aeroallergens; pollination; plant identification; coal studies; sediment and rock analyses; and statistical procedures) and if the name *Pollen Circular* would not be more appropriate. The eighth *Circular* included a key to Paleozoic genera of dispersed spores and pollen by L. R. Wilson. Thirty-two new papers were listed. A new branch of this budding science, the area of marine application, was introduced by Stanley A. Cain in a lengthy review of the papers on the geology and biology of North Atlantic deep-sea cores by Wilmot H. Bradley and others. Encouragement was sent by Henry N. Andrews, Jr., for the Paleobotanical Section of the Botanical Society of America and by William W. Rubey for the Division of Geology and Geography of the National Research Council. But, most important, in this circular was one letter, dated July 15, 1944, suggesting possible names for this new science. H. A. Hyde (Wales) and D. A. Williams (England) suggested: "*Palynology* (from Greek "paluno", to strew or sprinkle; cf. (pale), fine meal; cognate with Latin *pollen,* flour, dust): the study of pollen and other spores and their dispersal, and applications thereof."

The ninth *Circular* (January 1945) had a new name, *Pollen and Spore Circular,* and the eleventh *Circular* contained a reprint of a paragraph from *Nature* that gave notice of the *Pollen Analysis Circular.* "Palynology" was referred to in that article in *Nature* (March 3, 1945, v. 155, p. 264) as the term "for the study of pollen and other spores and their dispersal, and applications thereof."

Later *Circulars* became less frequent, but an effort was made to encompass a rapidly growing and diversifying field. The mailing list was over 200 for the seventeenth *Circular* (May 1949), which was sent to scientists in 22 countries. The issues were thicker and had fewer personal notes. There were summaries of programs at various meetings, and the directories showed it was truly international in scope. The eighteenth *Circular* (January 1954) was the final edition. However, by that time an arrangement had been completed with Brooks F. Ellis of the American Museum of Natural History to incorporate the information in the *Circular* into the journal *Micropaleontology,* which was published quarterly. Reports by volunteer writers from different regions or countries were published as submitted. However, this medium lacked the personal touch of the *Circular,* and the closeness of the group disappeared.

### *Umbrella organizations*

The Paleobotanical Section of the Botanical Society of America (PSBSA), which was founded in 1936, was the first organization to encourage this disparate group of scientists to participate in meetings with theme programs and field trips (Fig. 30, 31). For example, at the Dallas meeting in 1941, nine papers on the program were reports on plant microfossil research. The Paleobotanical Section occasionally functioned as an independent part of the Botanical Society by holding joint meetings with other organizations and conducting independent field trips. One such meeting, which was arranged by J. M. Schopf, was held in connection with the Paleontological Society meeting at Pittsburgh in 1959 and included a symposium on "Stratigraphic and ecologic interpretation of plant microfossils in petroleum geology."

Traverse (1960, p. 2–3) summarized succinctly several of these activities following the Ninth International Botanical Congress at Montreal.

> The trend in paleobotany toward more and more concern with plant microfossils. . . . [was] . . . reflected in the Section's program. . . . At the East Lansing meeting in 1955 and at [the] Storrs meeting in 1956, the Section made a memorable contribution to fossil pollen and spore systematics by special programs on the nomenclature and classification of plant microfossils. At the Palo Alto meeting in 1957 and at the Bloomington meeting in 1958, the Section played host to and helped organize the Fourth and Fifth National Pollen Conference[s], respectively. The 1958 Pollen Conference included papers on an especially wide range of palynological subjects, from Paleozoic to modern. . . . . [These programs] . . . demonstrated the increasing emphasis on plant microfossils in the laboratories of the various major oil companies. . . .

In 1948, a dedicated group of coal geologists and palynologists, guided by Gilbert H. Cady, Jack A. Simon, and Robert M. Kosanke of the Illinois State Geological Survey (see Simon et al., 1988), received unofficial sponsorship by the Society of Economic Geologists (SEM) for special meetings, symposia, and field trips. This background of unofficial meetings made it possible for Cady, Simon, Schopf, and Kosanke to obtain approval from the Council of the Geological Society of America (GSA) for the formation of the Coal Group (later Coal Division) beginning in 1955. This has been a very successful relationship. The two umbrella organizations, PSBSA and GSA, have sponsored meetings, symposia, special publications, and field trips annually at various places on the continent from Mexico to Canada (see Fig. 30).

SEPM also sponsored special symposia including paleopalynology, especially "Spores, pollen, and other microfossils in oil exploration" (New York, March 1955) and "Palynology in oil exploration" (San Francisco, March 1962) (Cross, 1964). Papers from several of these symposia have been published. The first North American palynological society, The American Association of Stratigraphic Palynologists (AASP), was not organized until 1968.

Figure 30. Schopf (left front with hat), Wilson (right front leaning on oar), Spackman (distant background behind Wilson's right shoulder), leader of field trip, and several colleagues knee deep in the Okefenokee Swamp. All three were involved in modern and ancient research and teaching of palynology, paleobotany, and coal geology during their careers. As an example of the broad base of study they promoted, they were examining modern coastal swamps and inland swamps and mires on this field trip of the Coal Geology Division of The Geological Society of America to the Okefenokee Swamp of Georgia and the Everglades, Keys, and coastal margins of southern Florida in November 1974. (Photograph by A. T. Cross.)

Figure 31. William C. Darrah (left) and James M. Schopf (right) in conversation; Ellis Yochelson (center) and James Barlow (behind Schopf) listening; William H. Smith (back to Darrah) and Garner Wilde (upper center) examining samples; and L. R. Wilson (in plaid shirt) collecting samples (as usual!); on the Dunkard Symposium field trip at a mid-Conemaugh Formation (Upper Pennsylvanian) section, in West Virginia, September 27, 1972. (Photograph by A. T. Cross.)

The first international conference on palynology (Tucson, 1962), which was organized by G. O. W. Kremp, included nearly 100 participants from several countries. The Second International Conference (Utrecht, The Netherlands, 1966) included a much larger and even more diversified group.

## EARLY STABILIZING RESEARCH CONTRIBUTIONS

### *Illinois State Geological Survey—Schopf and Kosanke*

Two epochal papers published by the Illinois State Geological Survey, the first by Schopf, Wilson, and Bentall (1944) and the second by Kosanke (1950), are classic contributions (see Figs. 32 and 33). These two papers stabilized the North American pre-Quaternary paleopalynology effort by their organization, critical review of earlier works, serious consideration of systematics and taxonomy, strict adherence to application of the International Rules of Botanical Nomenclature (now called the International "Code"), careful application of stratigraphic information, and extensive reference to western European and Russian literature as well as to significant American publications.

The Schopf et al. (1944) "Synopsis" came at a time when extensive diversity of taxonomic treatment was widespread in the works of authors from different countries and backgrounds. They proposed guiding principles, including the requirement that: (1) the International Rules of Botanical Nomenclature should be adhered to, (2) species classified within a genus must possess significant characteristics in common, (3) a conservative approach with regard to synonomy should be followed, (4) the specific epithet selected need not include the character or history of the taxon but should be adequate for ease of subsequent reference, and (5) specific identification must be based on positive morphological or anatomical features that can be compared critically. Following the discussion of the guiding principles, the authors critically described 23 major genera, including 22 from the Carboniferous. Six genera—*Punctatisporites, Granulatisporites, Reticulatisporites, Laevigatosporites, Monoletes,* and *Densosporites*—were described in detail. All but *Monoletes* are considered valid taxa today, although some of the original circumscriptions have been modified and new genera have been established on excluded forms. Five new genera—*Calamospora, Florinites, Lycospora, Raistrickia,* and *Reinschospora*—were proposed and described. A seminal review of *Tasmanites* Newton 1875, the sporelike cysts of a prasinophycean alga found widely dispersed in Devonian-Mississippian black shales and in unconsolidated sediments of the Great Lakes region, was the twenty-third genus included in their work. These microfossils were originally described under the designation "Sporangites" (Dawson, 1863). This 1944 landmark publication provided a stable basis for many subsequent palynological investigations here and abroad.

Figure 32. James M. Schopf, pioneer paleobotanist, palynologist, coal geologist with the Illinois State Geological Survey and U.S. Geological Survey. (Reprinted with permission from *Biostratigraphy of Fossil Plants* [Frontispiece], D. L. Dilcher and T. N. Taylor, eds., Copyright, 1980, by Dowden, Hutchinson and Ross, Inc., Stroudsburg, Pennsylvania.

Kosanke's (1950) treatise on the Pennsylvanian spores from 47 counties in Illinois included descriptions of 100 new species of the 130 spores differentiated. He summarized the generic characteristics of 19 genera, including five new genera: *Cadiospora, Illinites, Schopfites, Schulzospora,* and *Wilsonites.* His pioneering report was based on samples from more than 50 named coal beds from the Caseyville, Tradewater, Carbondale, and McLeansboro Formations. This basin-wide report demonstrated in detail the value of palynomorphs in the correlation of coal beds of the Illinois basin. A preliminary summary had appeared in Kosanke (1947). He depicted the stratigraphic occurrence of all 19 genera and 130 species described or identified in the Illinois basin (Kosanke, 1950). The extinction of *Lycospora* was placed near the Middle Pennsylvanian–Upper Pennsylvanian boundary. This was the first basinwide study of Pennsylvanian palynomorphs in the United States.

After Kosanke joined the U.S. Geological Survey in 1963, he was succeeded at the Illinois Survey in 1963 by one of his students, Russel A. Peppers. Peppers's (1964, 1970) memoirs on Pennsylvanian coals expanded on the palynology of Middle and Upper Pennsylvanian coals of Illinois.

### *Coe College and L. R. Wilson*

Several studies of Pennsylvanian palynomorphs were conducted on Iowa coal beds by L. R. Wilson (see Fig. 34)

Figure 33. Robert M. Kosanke, palynologist and coal geologist for U.S. Geological Survey and author of first comprehensive coal palynoflora of a major basin in North America; shown on the Everglades field trip, Geological Society of America Coal Geology Division, November 1964 (photograph courtesy of Jack A. Simon, retired director, Illinois Geological Survey).

and his students at Coe College during the decade beginning in 1936. The earliest report (Wilson and Brokaw, 1937) demonstrated the sequential changes in the spore floras of nine successive interval samples of a Desmoinesian coal that was collected near What Cheer, Iowa. Six unnamed principal spores were shown to vary in relative percentages in the coal bed. This was one of the earliest demonstrations of the potential of using palynomorphs to identify shifts of floral dominance (plant succession) driven by environmental changes (both climatic and edaphic) during peat formation. Wilson's extensive research on pollen analysis of peat swamps stimulated research in this aspect of coal geology.

Elizabeth A. Coe, an Honors thesis student of Wilson, published eleven new species and erected four new genera of Pennsylvanian spores (Wilson and Coe, 1940). Three of these new genera—*Cirratriradites, Endosporites,* and *Triquitrites*—are valid today. The genotype of a fourth new generic epithet, *Phaseolites,* was transferred by Schopf et al. (1944) to *Laevigatosporites.*

In 1940, another Honors student, Robert M. Kosanke (see Wilson and Kosanke, 1944), described three species belonging to three important and widespread new genera of upper Paleozoic spores: *Calamospora, Florinites,* and *Raistrickia.* Wilson and Kosanke's new species were *C. straminea, F. elegans,* and *R. aculeolata.* They also introduced two other new species assigned to the genus *Punctatisporites* (Ibrahim), then newly emended by Schopf et al. (1944). The other two new species described and illustrated were assigned to *Triquitrites* and *Cirratriradites.*

The final Honors thesis on Pennsylvanian spores directed by Wilson at Coe College, which was completed in 1946, was "Small spores of the Mystic coal (Desmoinesian) of Iowa" by Mart P. Schemel. His thesis was published with some revisions and expanded while Schemel was studying at West Virginia University on his Ph.D. program under Cross. He erected an important new genus of spores *Vesicaspora* and described *V. wilsonii* (Schemel, 1951). This genus was emended by Wilson and Venkatachala (1963) after restudying the original Iowa material and the examination of extensive new collections from several coals of the Desmoinesian Series and the Dawson coal of the Missourian Series of Oklahoma. It is noteworthy that Schemel, on the basis of palynology, correlated the Mystic coal of Iowa with the Herrin (No. 6) coal bed of Illinois.

Wilson (1943) published *Elaterites triferens* as a new Desmoinesian genus and species, the first fossil record of elater-bearing spores. It is generally very rare as a dispersed spore, probably because the elaters usually become detached from the *Calamospora*-type spore body, making it difficult to identify.

## REGIONAL PALYNOLOGICAL RESEARCH AND TEACHING

### *The Maritimes Region—Eastern Canada*

Grace Brush Somers published the two earliest palynological reports on Carboniferous strata in Canada (1952, 1953). She described the palynomorphic content of the Lower Jubilee and the Phalen coal beds of the Sydney coalfield, Nova Scotia. *Laevigatosporites* is the dominant palynomorph in both coal beds.

Wilson (1954) presented a preliminary report on the palynology of several coals of the more than 1,500-m-thick Pennsylvanian section along the classic Joggins sea cliffs of Nova Scotia. Wilson found that *Lycospora,* cf. *L. pellucida,* was the most abundant taxon, attaining a level of 60% or more in several assemblages. This indicated that arborescent lycopsids probably dominated the peat-forming swamps. Hamilton (1962), using spore analysis, updated the correlation of Pennsylvanian strata of New Brunswick.

Hacquebard (1957) (see Fig. 35) made the first study of palynomorphs in a Mississippian coal of Nova Scotia. He did

Figure 34. Leonard R. Wilson (circa 1971–1972). Teacher, early researcher in pollen analysis in United States and in coal and petroleum palynology. (Photograph courtesy of L. R. Wilson, from files of the School of Geology and Geophysics, University of Oklahoma, Norman.)

Figure 35. Peter A. Hacquebard, a pioneer in Canadian coal petrology and palynology, who has contributed significantly to the palynology of Carboniferous coal beds of Canada. (Photograph reproduced from A. R. Cameron, 1991, Peter Hacquebard, pioneer in Canadian coal petrology, International Journal of Coal Geology, v. 19, Special Issue, p. 3, by permission of Elsevier Science Publishers, B. V., Amsterdam, The Netherlands.)

the pioneering palynological work on the Horton Group (Middle Devonian to lower Carboniferous), which is sparsely distributed but widespread in Nova Scotia and New Brunswick. Hacquebard reported that *Punctatisporites* dominated the assemblage. Twenty genera were identified, including four new genera: *Leiozonotriletes, Lepidozonotriletes, Spinozonotriletes,* and *Vallatisporites.* Twenty-three species were described as new. One of the new species is *Punctatisporites irrasus,* which accounts for 31% of the assemblage. Hacquebard et al. (1960) presented a graphic summary of the stratigraphic significance of miospores (small spores) of Namurian A to Westphalian D ages in the Atlantic Maritime provinces at the Fourth International Congress on Carboniferous Stratigraphy and Geology. Comparative ranges for six regions in Euramerica were plotted, and megafloral correlations were given. Five spore divisions, four zones, and four subzones were delineated, and these approximate six zones based on plant megafossils by W. A. Bell. The Namurian A correlated best with the Midcontinent Chester Group of Late Mississippian age.

A major contribution to the palynology of the Carboniferous of Nova Scotia by Hacquebard (1961) was a study of the upper Carboniferous and Permian of the Pictou Group at Mabou and two subsurface samples of a comparable age from Prince Edward Island. Numerous spore-frequency diagrams, range-correlation charts, and illustrations were included. His study was amplified by Hacquebard and Donaldson (1964), who reported the palynological analysis of about 35 coal seams of the upper Carboniferous Cumberland Group. The coals analyzed were from the classic Joggins section and the Springfield coalfield. They represented 1,530 m of the 3,600-m-thick section. Histograms indicate that the lower 663-m interval is dominated by *Lycospora,* whereas in the upper 867-m interval *Lycospora* is highly variable in occurrence.

Geoffrey Playford (1964) identified 53 species of palynomorphs belonging to 28 genera from 13 localities in the Horton Group of Atlantic Maritime Canada. Twenty-one new species were described and illustrated. The abundant species were *Punctatisporites debilis, Lycospora torulosa, Vallatisporites vallatus, V. verrucosus, Granulatisporites crenulatus, Raistrickia clavata,* and *Pustulatisporites pretiosus.* Playford concluded that the palynomorphs have an affinity to Upper Devonian palynomorphs from the Russian Platform.

A major contribution to Canadian palynology was "Carboniferous and Permian Spores of Canada" (Barss, 1967), which included more than 900 original photomicrographs. Barss included 33 plates with 774 photomicrographs of Tournaisian, Viséan, Namurian, Westphalian, Stephanian, and younger Permian-age pollen and spores from the Yukon and

Northwest Territories and diverse localities in eastern New Brunswick, Prince Edward Island, and Nova Scotia. In an earlier paper, Barss et al. (1963) outlined the stratigraphic ranges of some of these spores in the Atlantic Maritime provinces and set up three miospore zones, corresponding to W. A. Bell's *Lonchopteris* and the *Linopteris obliqua* megafloral zones for Westphalian C and his *Ptychocarpus unitus* megafloral zone for the Westphalian D.

### *Northern Appalachian basin*

A very early paper for this region was Kosanke's (1943) report on the spores of the Pittsburgh and Pomeroy coal beds of southeastern Ohio. This study confirmed Thiessen's recognition of the common fern spore, *Laevigatosporites* (now *Thymospora*) *thiessenii* that is very abundant in and characteristic of the Pittsburgh coal. Kosanke named two new species belonging to the genera *Triquitrites* and *Punctatisporites,* and assigned another new species to *Punctatisporites* (reassigned to *Cyclobaculisporites* by Bhardwaj [1957, p. 91]). Kosanke was able to distinguish between the Pittsburgh (No. 8) coal and the closely overlying Pomeroy (No. 8a = Redstone) coal on the basis of the spore assemblages.

Cross (1947) reported on the Pennsylvanian megaspores extracted from 18 coal beds in the Kanawha, Allegheny, Conemaugh, and Monongahela Formations of West Virginia, eastern Kentucky, and eastern Ohio. He illustrated 185 specimens representing 60 species and varieties of megaspores. He also illustrated 15 types of palynomorphs representing nine genera of smaller spores—*Granulatisporites, Dictyotriletes, Densosporites, Laevigatosporites, Latosporites, Endosporites, Cirratriradites, Lycospora,* and *Calamospora*—from three beds of the Kanawha Formation. This work was carried out at the University of Cincinnati in 1941 and 1942 and was continued at the U.S. Bureau of Mines Central Experiment Station, Pittsburgh, in 1943 and 1944 while Cross was a National Research Council Fellow.

Cross and Schemel (1952) published on the palynomorph content of some Pennsylvanian and Lower Permian (Dunkard) coals and carbonaceous shales from West Virginia and surrounding states. These samples were from the Kanawha, Allegheny, Conemaugh, and Monongahela Formations, and the Dunkard Group. Abundance data for palynomorphs were also given for the Winifrede and Pittsburgh coal beds (Kanawha and Monongahela Formations, Middle and Upper Pennsylvanian, respectively). The authors described the importance of segment (increment) samples of coal beds to determine important changes in palynomorphs that might reveal paleoecological shifts. This was illustrated by palynomorphs of the Pittsburgh coal bed from Blaine Hill, Ohio.

Cross (1953), in a comprehensive paper on the geology of the Pittsburgh coal bed that is distributed over about 13,000 km$^2$ in four states, demonstrated the importance of sampling the coal in increments or by layers. *Thymospora* (*Laevigatosporites* of that report) *thiessenii,* the "Pittsburgh spore" of Thiessen, was dominant for bench or whole seam samples of the entire coal bed at every locality, but in lithologic benches (layers) or increment samples it ranged from less than 1% in a layer just above a lower main parting ("bearing-in-bench" of miners) to 95% relative abundance in certain layers of both the middle bench ("breast coal") and the roof coal.

During Cross's eight-year tenure (see Fig. 36) at the West Virginia Geological and Economic Survey and West Virginia University, he directed several theses, including five on the palynology of coals and coal-bearing Upper Pennsylvanian rocks. George H. Denton sampled lowermost Allegheny coals from 27 localities in the upper Ohio River Valley area, including 11 sites of test wells from which cuttings (rock chips) were taken in West Virginia, Pennsylvania, and Ohio. His thesis research (1957) demonstrated effectively that on the basis of palynology the Brookville, Clarion, and Lower Kittanning coal beds could be distinguished from each other throughout the region and from coals above and below.

Schemel's doctoral dissertation (1957) was based on several hundred increment samples taken from lower Allegheny coal beds from 31 localities in four states in the Appalachian basin. Significant differences were demonstrated between palynofloral assemblages of the southwestern part of the basin and those of the northern part, although long-standing miscorrelation by field geologists could not be ruled out. The older Pottsvillean-type floras apparently lingered on in the southwestern region after the typical Alleghenian floras had become established farther north. Cyclical changes in dominance of certain spores occurred upward through each coal bed.

Figure 36. Leonard R. Wilson (center), who first introduced Aureal T. Cross to geology and paleontology at Coe College, 1936, and his wife Marian and Cross (left) at the Wilsons' home in Norman, Oklahoma (photograph taken by Huang Wei, Michigan State University, June 15, 1992).

Ralph J. Gray's Master's thesis (1951) demonstrated the depositional history of the Waynesburg coal bed (Upper Pennsylvanian or Lower Permian) and the correlation of the main benches (coal split by partings) on the basis of selected pollen and spores. Four main benches accumulated with clay-shale and silty-shale partings, indicating three main interruptions and, in this case, some northward migration (following progradation?) of depositional sites of peat accumulation. The coal benches thin to the northwest. Palynologic correlation demonstrated that a thin southwestward extension of the bottom bench of the northern part of the field *overlies* the bottom bench in the southwestern area of the field. The rich palynoflora is dominated by ferns and seed ferns, with a lesser number of coniferalan saccate forms.

James C. Warman completed his M.S. thesis research (1952) on the palynology and petrography of the Washington coal bed (Washington Formation, Dunkard Group), near the Pennsylvanian-Permian boundary. This is the youngest widespread coal bed in the northern Appalachian basin. All the spores found were in common with those in seams down to the middle part of the Monongahela Formation below.

Also at West Virginia University, Horace R. Collins's Master's thesis (1959), which was conducted under the direction of William H. Gillespie, demonstrated the shifts in palynofloras from one lithologic layer to the next above. Four localities studied showed close correlation of the palynofloras (at the generic level) for each of five lithologic types: underclay, coal, coal roof shale, Ames Limestone Member of the Conemaugh Formation, and marine shale at the top. The spore content was markedly characteristic of each lithologic unit and contrasted sharply with palynomorphs of overlying and underlying beds.

John A. Clendening was employed by the West Virginia Geological Survey during the late 1950s and the 1960s while working on his M.S. and Ph.D. programs at West Virginia University. His Master's thesis (1962) was done under the direction of W. H. Gillespie, one of Cross's earliest doctoral students. It dealt with small spores of the middle Dunkard coals that are interbedded with freshwater limestones and fluvial red shales and sandstone in the controversial Pennsylvanian-Permian transition zone. In his research, the Stephanian B-C boundary was placed between the Nineveh and Windy Gap coals, the two youngest coals in the Dunkard basin. He considered none of the Dunkard sections to be younger than Virgilian, a view based on his investigation of comparable Upper Pennsylvanian sections in the Midcontinent. Clendening also wrote several papers describing small spores of the Redstone coal (1965), *Schopfipollenites varius* sp nov. and three other species of Stephanian age (Clendening, 1966). Clendening and Gillespie (1964) reviewed the spores characteristic of the Pittsburgh coal bed. Gillespie collaborated with Ira S. Latimer and Clendening to write a revision of a 1960 handbook on West Virginia plant fossils (Gillespie et al., 1966); several plates of fossil spores and pollen were included.

Daniel Habib's (1966) dissertation at Pennsylvania State University, under the direction of W. Spackman, described the palynology of the Lower Kittanning coal from western Pennsylvania. Coal beds were sampled from base to top in increments of 25-mm at 76-mm levels. Six excellent plates illustrate the palynomorphs. One hundred forty species (33 new) belonging to 59 genera (including two new genera, *Spackmanites,* and *Palaeospora*) were identified. Four of five "assemblages" were identified in the typical seam profile where marine or restricted marine facies overlie the coal; from bottom to top they were (1) *Lycospora-Guthoerlisporites,* (2) *Thymospora pseudothiessenii,* (3) *Punctatisporites obliquus,* and (4) *Densosporites oblatus.* Habib noted that assemblage 3 would terminate the succession when a freshwater facies directly overlies the coal bed. Edaphic factors were recognized by shifts in the palynoflora. For example, periods of high water levels in the coal swamp might be indicated by pulses of *Florinites pellucidus* in the *Thymospora pseudothiessenii* assemblage. The salinity factor was also interpreted as controlling some floristic responses, for example, the parent plants of *Lycospora* (i.e., the arborescent lycopods).

Norman O. Frederiksen, another Spackman student, did his Master's thesis (1961) on the palynology of the Brookville and Lower Clarion coal beds of western Pennsylvania. Another Master's thesis from Pennsylvania State University is the palynological investigation by P. K. H. Groth (1966) of shale in the Middle Pennsylvanian Columbiana shale Member of the Allegheny Formation of western Pennsylvania. Of several other graduate students in palynology, paleobotany, and coal-forming plants prior to 1968, Spackman and G. O. W. Kremp directed theses of Gilbert Brenner, Grace Somers Brush, Arthur Cohen, Russell Dutcher, Jacob Gerhard, Linda W. Groth, Walter Riegel, Wallace Riffelmacher, Ronald Stingelin, and Edward Stanley.

The palynology program at Pennsylvania State University commenced with the appointment of W. Spackman (see Fig. 37) to the Department of Geosciences, College of Earth and Mineral Sciences, in the early 1950s. He very rapidly developed a strong coal research program, a detour from his botanical agenda at Harvard University. Spackman, like Alfred Traverse, who came to "Penn State" in 1967, had done a thesis on the fossils and petrography of the Brandon Lignite (Miocene age) of Vermont under the tutelage of Elso S. Barghoorn. Spackman's knowledge of plants enabled him to demonstrate to the coal and steel industries that microscopic analysis of coal could lead to a great improvement in the utilization of coal for steelmaking and for coal by-product utilization. To this successful and dynamic coal petrology program, with a cadre of graduate students and some financial support from the coal and steel industries, he brought Gerhard Kremp from Germany. A new palynology program burst onto the scene. Kremp's dynamic energy and enthusiasm helped to launch the "Catalog of Fossil Spores and Pollen" (Kremp et al., 1957). Students gravitated to this center of academic and

Figure 37. William Spackman, teacher, coal geologist, palynologist. A, lecturing from a barge to participants of the Geological Society of America annual field trip on the southwestern Florida Everglades, 1964. B, Spackman and his wife, Virginia, at the First International Palynology Conference, Tucson, Arizona, 1962 (photographs by A. T. Cross).

applied coal geology and palynology, and two great energy industries benefited. When Traverse arrived on the scene the palynology program blossomed.

Papers by Winslow (1959) and Hoffmeister et al. (1955) provided a fundamental base for the palynology of the Upper Mississippian rocks in the eastern United States.

John H. Hoskins received his Ph.D. from the University of Chicago in 1924,where he studied under the direction of A. C. Noé, C. J. Chamberlain, J. M. Coulter, and A. C. Land. After a year of studying with D. H. Scott and A. C. Seward in England, Hoskins later embarked on a lifetime career of teaching paleobotany and plant morphology at the University of Cincinnati. Beginning in the late 1920s, he had a trickle of graduate students doing research on Carboniferous palynology and paleobotany for over 30 years, including R. M. Kosanke, A. T. Cross, and M. L. Abbott. Kosanke and Cross developed graduate programs divided between botany and geology.

Hoskins and Abbott (1956) described and illustrated a heterosporous cone, *Selaginellites crassicinctus,* from a Kansas coal ball. It contained *Cirratriradites annulatus* microspores and *Triangulatisporites (Triletes) triangulatus* megaspores. Leisman's (1961) observations on another Kansas specimen amplified their descriptions.

Maxine L. Abbott, in her Ph.D. dissertation (1963) described several compressed lycopod fructifications from the Upper Freeport (No. 7) coal (Allegheny Formation) of Ohio. One homosporous cone of *Lepidostrobus princeps* Lesq. (reassigned by her to a new genus *Lepidostrobopsis*) contained *Lycospora*-type microspores. A megasporangiate cone, *Lepidostrobus* (= *Lepidostrobopsis,* Abbott) *missouriensis* D. White, contained the auriculate megaspores, *Triletes* (= *Valvisisporites*) *auritus.* Two *Lepidocarpon*-type cones were described, both with *Cystosporites* megaspores. One type with *C. varius* spores was *Lepidocarpon* (= *Lepidocarpopsis* Abbott) *oblongifolius* Lesq. The other was a new species of *Cystosporites* found in *Lepidostrobophyllum* (= *Lepidocarpopsis* n. gen.) *lanceolatum* Lindley and Hutton. Two microsporangiate cones of a *Lepidocarpon*-type, both with *Lycospora*-type spores, and one new megasporangiate species of *Sigillariostrobus, S. leiosporous,* bearing *Triletes glabratus* spores, were also described by her. Schopf directed the research of several graduate students at Ohio State University. One of the studies, by Marcia Winslow, was published later by the U.S. Geological Survey (Winslow, 1962). In it she described spores of Devonian and early Mississippian shales of Ohio.

### Southern Appalachian region

Thiessen (1920) produced the earliest paleopalynology studies of southern Appalachian coals. Willard Berry (1937) reported a palynoflora of 14 new species, including a new genus, *Densosporites,* and three new species from the Pennington Shales (Upper Mississippian and Lower Pennsylvanian) of Tennessee. He applied binomial nomenclature following Ibrahim (1933). However, these were all poorly illustrated with extremely rudimentary sketches, so that only one species, *Densosporites covensis,* can be recognized today with certainty. Wilson (1959) redescribed and reillustrated that species from Berry's original specimens by photographs from a fragment of the holotype slide (after the slide had been crushed in the mail when sent by Berry to Wilson). However, the genus was later emended by Butterworth, Jansonius, Smith, and Staplin (1964), and redescribed as *Densosporites* (Berry), and *D. covensis* Berry 1937 was retained as the type species. The stratigraphic position of the Pennington Shale has been controversial, and possibly it is of different ages at different localities. Berry's samples came from a coal above the Bangor Limestone (Upper Mississippian) and is of Chesterian (early Namurian) age.

Fred Cropp conducted his Master's and doctoral researches at the University of Illinois and the Illinois Survey, under the direction of R. M. Kosanke. His Master's thesis (1960) is entitled "Pennsylvanian spore floras from the Warrior basin of Mississippi and Alabama." Precise correlation of coal beds between wells was not possible, mainly because the total Pennsylvanian section thins rapidly and coal beds feather out westward. However, the palynomorphs were determined to be Lower Pennsylvanian in all the coals he studied from the well samples of Mississippi. The stratigraphic ranges of several genera were extended farther downward than had been previously recognized in either the Illinois or Appalachian basins, more nearly comparable to ranges then recently defined for similar species in Europe and Asia by Potonié and Kremp (1954).

Cropp's Ph.D. dissertation, which was based on 15 coals from Tennessee, was completed in 1958 under Kosanke's guidance (Cropp, 1963). This stratigraphically oriented research was the first study published on coals from Tennessee since Berry's (1937). Bentall (1941) published an abstract on the palynological correlation of the Battle Creek and Angel coals of Tennessee, but none of the detailed supporting data were published except the description of *Reinschospora bellitas* Bentall sp. nov., in Schopf et al. (1944). However, the genotype was arbitrarily considered to be a later synonym of the species *R. (Alatisporites) speciosus* (Loose) Schopf et al., 1944, by Potonié and Kremp (1954). *Reinschospora* has been widely publicized since a sketch of it, from a drawing of *R. speciosa* in Potonié and Kremp (1954, p. 139), became the trademark for the *Catalog of Fossil Spores and Pollen.*

### *Illinois basin—Illinois*

Winslow (1959, p. 10) stated that the study of spores from coal macerations in Illinois began in 1931, under the supervision of G. H. Cady, with an investigation of plant remains in maceration residues from column samples of the Herrin (No. 6) coal. The reports of McCabe (1932) and Henbest (1933, 1936) emanated from this program.

Louis C. McCabe was the first of a long list of geologists and botanists from Illinois to publish studies of plant structures, including spores separated from coals by modified Schulze maceration techniques. Such research was conducted under the aegis of the Illinois State Geological Survey, often in collaboration with the faculty of the University of Illinois. McCabe (1932) macerated samples from 13 coal beds ranging from the Babylon coal of the Middle Pennsylvanian Abbott Formation to the Chapel (No. 8) coal of the Middle and Upper Pennsylvanian Modesto Formation. He made only a preliminary survey on plant structures, and a more detailed study was directed to palynomorphs in the Pope Creek Coal Member (Abbott Formation) and Springfield (No. 5) Coal Member (Middle Pennsylvanian Carbondale Formation). McCabe observed that some microspores and megaspores found in the Pope Creek coal, near the top of the Abbott, did not extend into younger layers in the Middle Pennsylvanian Spoon Formation and Carbondale strata.

Henbest (1933) described spores, cuticles, and woody detritus from macerated samples from Illinois. His interest in this had been encouraged by David White at the time when Henbest had been White's assistant during field studies in the Illinois basin. Henbest also published a short paper on some fossil lycopod and fern spores from Illinois coals (Henbest, 1936).

J. M. Schopf (1936a, b, 1938) reported on the spores from the Herrin (No. 6) coal. These publications primarily treated megaspores classified at that time under the genus *Triletes.* Three new genera—*Cystosporites, Parasporites,* and *Monoletes* (now *Schopfipollenites*)—and spores of ferns and calamites were described. Schopf (1938) illustrated a fern spore that is without question the species now known as *Thymospora pseudothiessenii.* The specific epithet calls attention to its similarity to *Thymospora thiessenii,* except for the larger size and coarser ornamentation.

In the late 1930s, Arnold Brokaw, under the direction of J. M. Schopf and H. R. Wanless, analyzed palynomorphs of the Springfield (No. 5) coal bed of Illinois for his Master's research (1942) at the University of Illinois. He concluded that the spore floras of the Springfield and Herrin coals could be distinguished from each other on the basis of their different spore contents. A. L. Eddings described the small spores of the Trivoli (No. 8) coal bed of Illinois in his Master's thesis (1947), which also was guided by Wanless and Kosanke. Eddings demonstrated that the No. 8 coal was distinctive from the No. 6 coal and used this as a basis for stratigraphic studies.

Several studies by W. C. Darrah included megaspores of *Macrostachya* (1936); *Codonotheca,* with excellent clusters of *Schopfipollenites (Monoletes)* from attenuated pairs of synangia (1937); *Crossotheca sagittata* (Lesquereux) Sellards, with microspores similar to those shown by Sellards from Mazon Creek material but incorrectly interpreted by Sellards (Darrah, 1937); *Selaginella* (Darrah, 1938b), which showed remarkable megagametophytes with well-preserved cell structure, nuclei, and mitotic figures (p. 127); and *Oligocarpia* (Darrah, 1938a) with abundant sori on pecopterid foliage containing masses of microspores. Darrah's textbook (1939, fig. 7) shows a megaspore of *Selaginella* from Illinois removed by the peel technique.

Schopf (1940, p. 39–45) revised the description of *Lepidocarpon corticosum* (Lesquereux) Schopf from a permineralized specimen in an ironstone nodule from Mazon Creek. Arnold (1938) described a new species of *Lepidostrobus* with a single functional megaspore on each sporophyll and three aborted spores toward the distal end of the megaspore. Chaloner (1958), in his review of the lycopod cone *Polysporia newberryi,* determined that *Lepidophyllum truncatum* Lesq. 1879, also collected at Mazon Creek, produced *Valvisisporites (Triletes) auritus* megaspores and *Endosporites* microspores. Arnold (1958) also described a permineralized *Calamostachys* cone with spores and gametophytes. Kosanke (1955) described a new genus and species of calamitean fructification, *Mazostachys pendulata,* from an ironstone concretion in the Francis Creek shale (Member of the Carbondale Formation) as part of his doctoral dissertation in 1952, directed by Wilson Stewart.

Hoskins (1923, 1926) and Fredda D. Reed (1926) were the first paleobotany students of Adolph C. Noé (see Chapter 14, this volume). They were also the first paleobotanists to describe plants from North American coal balls following their "recognition" in North America by Noé (1923). Plants had been previously described from permineralized "concretions" in coal seams by several investigators, including Gresley (1901) and Coulter and Land (1911, 1921). Hoskins and Reed conducted their Ph.D. researches on stems and fructifications from the Desmoinesian Danville (No. 7) coal of Illinois. *Scolecopteris minor* (Hoskins, 1926) was described on the basis of its fernlike sporangial clusters (sori) and spores attached to a *Pecopteris*-type frond. Hoskins assigned it, correctly, to the Marattiales. Reed (1926) described a *Botryopteris* fructification bearing both micro- and megasporangia and megaspores containing gametophytic tissue and three small stems, all from a 5-cm diameter coal ball! Roy Graham, one of Noé's last graduate students, completed his doctoral dissertation in 1933 and published three papers on coal-ball floras before his untimely death in 1939 in a rock fall in a mine in British Columbia. He described two new marattiaceous fructifications (*Scolecopteris latifolia* and *Cyathotrachus bulbaceus*) and *Telangium pygmaeum,* a medullosan fructification, in 1934, and a *Calamostachys* cone with spores in 1935.

Following these early studies on spore-bearing fructifications from coal balls, other students of Noé and several students guided by J. M. Schopf expanded on these studies in the late 1930s and early 1940s. Wilson N. Stewart, who founded the paleobotany program at the University of Illinois in 1947, had been introduced to coal-ball research by Schopf. Stewart and some of his students and some of their students, in the period 1947 to 1965, took up these studies and published a number of papers dealing with male fructifications and spores from Illinois and Indiana coal balls. An excellent review of this extensive body of research is to be found in Phillips et al. (1973, p. 14–15, 26–40; see also Chapter 27, this volume). Some of these contributions include illustrations and information on *in situ* spores and pollen of all the major groups of Carboniferous plants, many from Illinois coal balls. Andrews's reviews (1951, 1974) on selective evaluations of contributions to North American upper Carboniferous coal-ball floras provide more critical background information than we can include here. The clarity, charm of writing, and scientific insight make these two accounts very valuable. Another excellent reference is Phillips (1979), which is a compilation of paleobotanical studies carried out in the University of Illinois.

One of the earliest major monographs on coal-ball plants in the United States that included consideration of spores was Schopf's study of *Mazocarpon oedipternum* and its sigillarian relationships (1941). One of the most important results of the study was the recognition that one whole group of trilete spores, the "Aphanozonati," represents the sigillarians. Schopf (1941, p. 7) stated: "Recognition of the botanical affinity of the Aphanozonati greatly increases the usefulness of these spores in coal studies and in stratigraphic paleobotany." Schopf (1941) presented the first stratigraphic chart for correlation of the Mississippian and Pennsylvanian rocks of United States with the western European Carboniferous strata that included all known coal-ball occurrences.

Schopf's research at Illinois culminated in the Schopf et al. (1944) treatise, which was not published until shortly after he had moved to the U.S. Bureau of Mines Central Experiment Station at Pittsburgh in 1943. Kosanke joined the Illinois State Geological Survey in 1942, after receiving his Master's degree at the University of Cincinnati under J. H. Hoskins. Kosanke was appointed to the coal section by the eminent coal geologist, G. H. Cady, who had also brought Schopf to the Survey to participate in the research on all the branches of paleobotany and coal geology. When Schopf left Illinois in 1943, Kosanke took over Schopf's role in the Illinois Survey. Kosanke was appointed to the faculty of the University of Illinois in 1958 and directed the palynological research of several students until he joined the U.S. Geological Survey in Denver in 1963.

One of Kosanke's students was R. A. Peppers. As a Master's student in the Geology Department at Illinois, Peppers had contacted Kosanke concerning a problem with his research under William Merrill on the Cretaceous stratigraphy of the Powder River basin, Wyoming. Kosanke provided Peppers with a palynological aspect for his research that whetted Peppers's interest in palynology. Peppers's research for his Ph.D. dissertation in 1961, on the spores of Upper Pennsylvanian cyclothems in the Illinois basin, was published in 1964. When Kosanke left the Illinois Survey in 1963, Peppers took over Kosanke's work in the Coal Section and continued his extensive palynological studies to include the lower part of the Pennsylvanian rocks of Illinois, that is, the Spoon and Carbondale Formations (Peppers, 1970).

Hoffmeister et al. (1955) produced the first significant work on a lower Carboniferous palynoflora in the eastern United States. It was preceded only by Berry's (1937) report on the Pennington Shale of Tennessee and Schemel's (1950) excellent study of Upper Mississippian strata of Utah. Chesterian palynomorphs were sampled from cores and outcrop samples in Illinois and Kentucky. An objective was to establish criteria for differentiating Upper Mississippian from Lower Pennsylvanian deposits. Three new genera—*Auroraspora, Convolutispora,* and *Grandispora*—and 37 new species were proposed.

Winslow (1959) published a comprehensive study of the megaspores found in Mississippian and Pennsylvanian rocks of the Illinois basin. She was able to demonstrate stratigraphic usefulness of Carboniferous megaspores in the Illinois basin based on range limitations. She also described a new genus with one species, *Renisporites confossus.*

### *Illinois basin—Indiana*

Palynology of the Pennsylvanian coalfields of Indiana was first carried out by G. K. Guennel at the Indiana Geological Survey. Guennel's earliest work (1952) was a study of the palynomorphs occurring in nine Middle Pennsylvanian coal beds from 39 localities in Indiana. Guennel (1952, p. 10) coined the term "miospore" to represent "all fossil spores and spore-like bodies smaller than 0.20 mm, including homospores, true microspores, small megaspores, pollen grains, and pre-pollen, [which] are arbitrarily called miospores." The term "miospore" is widely used today by paleopalynologists.

Guennel's samples, taken from the Indiana III to VII coal-bed interval, represent nine coal zones. *Laevigatosporites* and *Lycospora* dominated the spores in this interval, followed by *Calamospora, Punctatisporites, Endosporites,* and *Granulatisporites.* The palynomorphic percentage relationships for those nine zones served as a correlation standard.

Another outstanding contribution by Guennel (1958) was his analysis of the miospores of the Pottsville-age coals of Indiana. Samples collected from 85 localities represent the entire range of the Mansfield (Lower and Middle Pennsylvanian) and Brazil (Middle Pennsylvanian) Formations and the base of the Staunton Formation (Middle Pennsylvanian). Twelve new miospore species belonging to eight genera were described. Guennel compared the palynology of the Indiana section with results of Kosanke's study (1950) of a comparable section in Illinois and identified striking palynomorphic similarities of the more widespread coal beds.

Guennel (1954) demonstrated that *Triletes* (now *Triangulatisporites*) *triangulatus* (Zerndt) Potonié and Kremp is the same as *T. gymnozonatus* when stripped of its reticulate perisporium. Schopf (1938) had noted the similarity but had maintained the two as separate species. Interesting accounts (Guennel and Neavel, 1961; Neavel and Guennel, 1960) of the true nature and source of a palynomorph of sporelike appearance, *Torispora securis* (Balme) Alpern, Doubinger and Horst, 1965, were based on material from the Indiana "paper coal." Guennel and Neavel recognized that these odd, acorn-shaped microfossils were the individual, dispersed entities of the outer layer or wall spores of a *Bicoloria* sporangium, which had formed the enclosing sac that contained other types of spores. Those two papers were taken from Guennel's Ph.D. dissertation that was prepared under the direction of James E. Canright of Indiana University.

Guennel and Neavel (*in* Zangerl and Richardson, 1963) also used palynologic evaluation and sedimentary petrology to interpret the depositional environments of the IIIa coal zone of Indiana. This study centered on the Mecca Quarry Shale Member of the Middle Pennsylvanian Linton Formation at the type locality near Mecca in Parke County, Indiana, a classic paleoecological locality (Zangerl and Richardson, 1963).

Occasional fructifications found in Indiana coal balls were found to contain identifiable spores. Reed (1939) and Benninghoff (1942) first described these plants bearing spores from the Springfield (No. 5) Coal Member (see Phillips et al., 1973, p. 26). Benninghoff (1942) reported both megaspores and microspores of lycopods, and spores from *Calamostachys.* One early record from near Lynnville was the report in the Ph.D. dissertation of Mamay (1950) at Washington University, St. Louis. The spores from *Scolecopteris minor* var. *parvifolia* Mamay are similar to *Punctatisporites trifidus* Felix and Burbridge 1967.

Joseph M. Wood, one of Canright's earliest students, studied sigillarian fructifications from compression material collected in Lower Pennsylvanian strata of Indiana. He had "worked up the Mazon Creek collection at the University of Michigan which included Arnold's fossils" (Phillips et al., 1973, p. 89). Wood's Ph.D. study on the Stanley Cemetery flora was completed in 1960 (see Wood, 1963). That flora was collected from the shales above the Lower Block coal (Lower Pennsylvanian) from surface mines in Indiana. The plants were contained mainly in ironstone concretions similar to those of Mazon Creek and, like those, they included many spore-bearing, compressed fructifications (cones and fertile fronds).

### *Michigan basin*

Bartlett (1929a), University of Michigan, used a modified maceration method to extract spores from weathered Pennsylvanian coal pebbles found in excavations of glacial gravels on the campus of the University of Michigan. Such pebbles, now in the University Museum, had been collected as early as 1874. Lycopsid megaspores, cuticles, and microsporangial material were obtained from these macerations. He assigned three species to the genus *Triletes,* which was originally proposed by Reinsch (1884) and subsequently validated by Bennie and Kidston (1886). Bartlett described the three new species—*T. rotatus, T. mamillarius,* and *T. superbus*—and illustrated them with exquisite photomicrographs. In the major reclassification of the diverse spores included in the genus *Triletes,* as emended by Schopf (1938), Potonié and Kremp (1954) reassigned *T. rotatus* as the genotype of *Rotatisporites* and *T. superbus* as the genotype of *Superbisporites. T. mamillarius* Bartlett was reassigned to *Tuberculatisporites* by Potonié and Kremp (1955–1956) as *T. mamillarius* (Bartlett) Potonié and Kremp (1955).

Chester A. Arnold (1961) reviewed and amplified Bartlett's paper (1929a). He noted "that the paper was an impressive demonstration of what might be accomplished with maceration techniques in isolating spores for study, and it establishes a precedent for naming spores in accordance with standard nomenclatorial practice."

An interesting deposit known as "Michigan spore-coal" was discovered in a small strip mine near Williamston (Bergquist, 1939). Two beds of this spore-coal overlie the main bituminous coal. The lower of the two spore-coal beds, 7.5 cm thick, is made up almost solidly of compressed megaspores mostly in loose contact (Bergquist, 1939). The spores are mainly the same as those reported earlier by Bartlett (1929a) from coal pebbles in glacial drift. George C. Sprunk, Reinhardt Thiessen's coworker, identified the spores as comparable to those in the Middle Pennsylvanian Elkhorn coal of Kentucky and the No. 2 Gas and Alma coal beds of West Virginia.

Arnold (1944) described and illustrated the megaspores and microspores of what appeared to be a sphenopsid cone from Grand Ledge, Michigan, as *Bowmanites delectus* with two spore sizes. The spores are similar to *Calamospora* when the perisporium is disregarded or lost. However, Arnold (1949, p. 220–221 and pl. 28) reassigned these cones to *Discinites* of the Noeggerathiales and reillustrated the two sizes of spores. He also illustrated a megaspore of *Lagenoisporites (Lagenicula) rugosus* from a *Lepidostrobus* cone and whittleseyinian sporocarps of *Aulacotheca campbelli* bearing the prepollen. *Schopfipollenites (Monoletes)* was also discussed and illustrated.

Arnold (1950) described the megaspores from coal and shale samples from 12 localities in the Michigan basin. Six of the samples were from Grand Ledge, and one of the samples was from the Williamston spore-coal locality. One of the species, *Triletes globosus* Arnold, was found by Winslow (1959) to be restricted to a portion of the upper part of the Caseyville Formation (Lower and Middle Pennsylvanian) of Illinois. Winslow reported that Arnold's varieties A, B, and C of *T. globosus* had useful restricted ranges.

William G. Chaloner, while studying with Arnold as a postdoctoral fellow, analyzed megaspores from Mississippian strata of three states. Michigan and Pennsylvania samples were of Mississippian age (Chaloner, 1954). The megaspores from his Michigan sample (Chaloner, 1954) were *Triletes angulatus* Zerndt, *T. subpilosus* forma *major* Dijkstra, and *Cystosporites giganteus* (Zerndt) Schopf. This was then the oldest stratigraphic record of *Cystosporites.* Three new species—*Triletes indianensis, T. echinoides,* and *T. aristatus*—were proposed from the Chesterian Beaver Bend Limestone of Indiana.

In 1961, A. T. Cross joined the Michigan State University faculties of geology and botany and initiated a broad program of studies in palynology, paleobotany, and coal geology. More than 30 doctoral dissertations have been completed under this research program. Only one of the eight Ph.D. degrees awarded during the 1960s dealt with Carboniferous palynology. Tidwell's dissertation, completed in 1966, was the elucidation of a flora of 59 species of plant megafossils of transitional Late Mississippian–Early Pennsylvanian age from the Manning Canyon Shale of Utah (Chesterian to Atokan). Repeated attempts failed to obtain palynomorphs of sufficient number and quality of preservation to be of significance in the study (Tidwell, 1968).

A long-range study of the palynology of Pennsylvanian and Michigan "Red Beds" of the Michigan basin was undertaken by Cross and his students in 1963. One early discovery resulted in excluding the "Red Beds" stratigraphic unit of the Michigan basin from the Pennsylvanian strata at the top of the Paleozoic section where it had been placed. Cross's preliminary study of samples from several well-cuttings in 1962 demonstrated usable pollen and spores of a later age. B. L. Shaffer's dissertation, "Palynology of the Michigan 'Red Beds,'" which was completed in 1969, demonstrated that the red beds are Jurassic (see Cross, 1975a, b).

Arnold (1949) illustrated and discussed microspores and megaspores of *Discinites delectus,* and *Calamospora,* and *Schopfipollenites* from *Aulacotheca campbellii* fructifications from the Michigan basin. Several oral reports on the Grand Ledge palynoflora by Cross and his students were presented during the 1960s.

### *Western Interior (north)—Missouri and Iowa*

The earliest publication on dispersed fossil spores in Missouri is that of Bailey (1936), whose Ph.D. dissertation was completed in 1934 at the University of Missouri under the direction of E. B. Branson and M. G. Mehl. This research was on the micropaleontology of the Middle Pennsylvanian Cherokee Shale and "Henrietta Formation" (abandoned) of Missouri. Bailey's main goal was to demonstrate the value of using combined assemblages of microfossils for biostratigraphic interpretations. Bailey briefly reviewed *Triletes* Reinsch and discussed the need for a classification scheme. He also reviewed three types of *Triletes* megaspores found in Missouri: the Laevigati-, Apiculati-, and Zonales-types.

Several unpublished Master's theses on palynology were completed by students of Athel G. Unklesbay and J. M. Wood at the University of Missouri, Columbia. These were D. L. Reinertson (1953), D. H. Jones (1957), R. G. Todd (1957), R. E. Schmieg (1959), and M. D. Mumma (1960). The direct supervision of the latter four well-illustrated theses was by

C. E. Upshaw, who in 1959 was completing his Ph.D. dissertation, "Palynology of the Frontier Formation (Cretaceous), northwestern Wind River basin, Wyoming" (see Upshaw, 1964). An unpublished Ph.D. dissertation, by Sujoy Gupta (1965), dealt with microfossils, including palynomorphs, in the Grassy Creek and Saverton Shales (transitional Devonian-Mississippian) of Missouri. His dissertation was also supervised by Wood. B. L. Shaffer's Master's thesis, which was directed by James W. Valentine and Raymond E. Peck in 1961, dealt with Permian salt palynomorphs from Kansas (Shaffer, 1964). W. F. von Almen wrote his M.S. thesis on the palynology of selected coals of Missouri in 1959, and R. N. Weiser completed his M.S. thesis in 1960 on the palynology of the Lexington and Mystic coals, both under the direction of A. G. Unklesbay, Valentine, and Peck.

Iowa coal balls have contributed to our knowledge of parent plant sources for a number of spores and pollen. Hoskins and Cross (1943) illustrated and discussed the spores of a new species of *Bowmanites* from a coal ball from the Atlas Mine. They illustrated a delicate perisporium or external membrane that extended in irregular spinelike protuberances, or conical spikes, giving the usually trilete spore a pseudoreticulate or spinous appearance that disappeared during the Schulze maceration technique. Hoskins and Cross (1941) described in detail a fragment of an intact *Lepidocarpon* cone from an Iowa coal. This unusual specimen contained many tetrads of large aborted megaspores. Hoskins and Cross (1946a, b), in describing a large *Pachytesta* seed in a coal ball from the Desmoinesian Series of Mahaska County, Iowa, illustrated several specimens of the pollen *Florinites antiquus.* About 50 such foreign *Florinites* pollen grains and fern spores were found in the micropyle and pollen chamber.

Edwin D. Levittan and E. S. Barghoorn (1948) described a very small sphenopsid cone, *Sphenostrobus thompsonii.* The sporangia, borne singly on each of 16 sporophylls per whorl, bear *Vestispora* spores. S. H. Mamay, as a part of his doctoral dissertation (1950) under H. N. Andrews, Jr., at Washington University, St. Louis, described spores from several Pennsylvanian-age fructifications in coal balls from the Shuler Mine of Iowa. This was the same mine from which most of W. C. Darrah's Iowa coal balls were collected, mainly by F. O. Thompson (see Chapter 22, this volume). Several other paleobotanists have studied some of the fructifications of plants found in coal balls from the Shuler Mine. Mamay (1954a) described a new genus of small sphenopsid cone that bore rather large spores with two sizes of reticules on the reticulation network, which indicates that they belong to *Vestispora clara.*

Another sphenopsid cone, *Kallostachys scottii,* from the Shuler Mine coal balls (Brush and Barghoorn, 1955) contained spores with an extremely thin exine with a fine to coarse reticulum, possibly a *Calamospora.* Brush and Barghoorn (1964) extracted *Punctatisporites* from a *Bowmanites* cone; *Calamospora* from cones of *Calamostachys germanica* and from *Macrostachya;* and *Reticulatisporites,* of the Schopf et al. (1944) interpretation, from *Sphenostrobus thompsonii,* all apparently from Iowa coal balls. They were able to free the fern spores, *Granulatisporites* and *Laevigatosporites,* respectively, from compressions of *Asterotheca* and *Ptychocarpus.* They also extracted *Florinites* pollen from cordaitaleans (Brush and Barghoorn, 1962) and *Monoletes (Schopfipollenites)* from the medullosan fructification *Whittleseya.*

Upshaw and Creath (1965) published an account of a unique occurrence of Pennsylvanian miospores extracted from a cave-filling in Devonian limestone near Jefferson City, Missouri. Sixty-nine spore taxa, including four new species, were differentiated. Lycopsid spores were dominant, and the age was interpreted as Desmoinesian. Similar occurrences of spores and other dispersed fossils in fluvial or other sedimentary deposits enclosed in Devonian limestones and caves were reported from Iowa (Wilson and Cross, 1939) and from the vicinity of Iowa City (Kosanke, 1964).

H. N. Andrews, Jr. (see Chapter 21, this volume), as a graduate student at Washington University, St. Louis, was encouraged by Robert E. Woodson to study paleobotany in England in 1937. His Ph.D. dissertation, under H. Hamshaw Thomas at Cambridge University, was on the anatomy of pteridosperms from coal balls. In 1940 he returned to Washington University, where for 25 years his teaching and research set an enviable standard for paleobotanical science. During most of that time, his research was mainly directed toward Carboniferous coal-ball studies (Andrews, 1951, 1961). Several of his students (E. Pannell, R. W. Baxter, S. H. Mamay, C. J. Felix, W. H. Murdy, and T. L. Phillips) made major contributions to morphological and anatomical research on fructifications from coal balls in Iowa, Kansas, Illinois, and Indiana.

Andrews and Pannell (1942) described cellular preservation within a *Lepidocarpon* megaspore. R. W. Baxter, while at the University of Kansas, produced several papers on fructifications. These included the pteridosperm *Microspermopteris,* from the Illinois basin (1949); the sphenopsid *Peltastrobus,* from the notable Wasson Mine in the Springfield coal (No. 5) of Indiana (1950); and several other sphenopsid and calamitean cones from Kansas (1955, 1962): *Palaeostachya* (1963), *Calamocarpon* (1964), and *Litostrobus* (1967).

Mamay (1950) compared fern fructifications and spores of coenopterid and marattiaceous ferns from localities in Iowa, Illinois, and Kansas. He made an excellent tabular comparison of morphological features of *Scolecopteris,* including the spores. Several other papers by Mamay on plants bearing fertile fructifications are those on *Sclerocelyphus* (1954b), *Acrangiophyllum* (1955), *Biscalitheca* (1957), and *Bowmanites* (1959).

Felix (1954) wrote an extensive review of American lycopod fructifications, the topic of his dissertation under Andrews. One new species, *Lepidostrobus diversus,* was based on several cones with either megaspores or microspores in some cones and both in others. The megaspores are a *Lagen-*

*oisporites rugosus* type, and the microspores belong to *Lycospora.* A *Botryopteris globosus* fructification with its spores was described by Murdy and Andrews (1957).

Phillips completed his Ph.D. dissertation (1961) on American species of *Botryopteris.* He also contributed extensively to descriptions of *in situ* spores of various plants (e.g., *Anachoropteris,* Phillips and Andrews, 1965).

### *Western Interior (south) and southwestern region, Kansas, Oklahoma, Texas*

The earliest major paleopalynologic study for the Midcontinent strata was that of Wilson and Hoffmeister (1956). They reported 13 genera, including one new genus, *Vestispora,* and nine species from exposures of the Croweburg coal bed (Pennsylvanian) of Oklahoma. They correlated the Croweburg palynoflora with that of the Colchester No. 2 coal bed of Illinois.

J. L. Morgan wrote his M.S. thesis (1955) under the direction of C. C. Branson, on spores of the McAlester coal of eastern Oklahoma. Hoffmeister gave technical guidance during the study. This thesis gives a description of one new species and shows several spore histograms.

During his first 10 years at the University of Oklahoma, L. R. Wilson directed the research of about 20 graduate students and research fellows. Unpublished Master's theses were completed in the years specified on Mississippian or Pennsylvanian coals or coal-bearing strata by the following: R. T. Clarke, in 1960, on the Secor coal; M. J. Higgins, in 1960, on an unnamed coal near Porter, Oklahoma; P. N. Davis, in 1961, on the Rowe coal; J. H. Ruffin, in 1962, on the Tebo coal; V. D. Wiggins, in 1962, on the Mississippian Goddard Formation; T. A. Bond in 1963, on the Weir-Pittsburg coal of Oklahoma and Kansas; K. M. Bordeau, in 1964, on the Drywood coal; E. D. Dolly, in 1965, on the Bevier coal; L. L. Urban, in 1965, on the Drywood and Blue Jacket coals; W. A. Edwards, in 1966, on the Morrowan Francis Formation near Ada; and R. B. Sanders, in 1967, on Desmoinesian and Missourian strata of the Elk City area. Wilson directed the Ph.D. dissertations of five graduate students on upper Paleozoic palynology: L. B. Gibson, in 1961, on the Iron Post coal; J. B. Urban, in 1962, on the Mineral coal of Oklahoma and Kansas; J. E. Dempsey, in 1964, on the Lower McAlester and Upper McAlester coals; R. W. Harris, Jr., in 1966, on the Sand Branch Member of the Upper Mississippian Caney Shale, and Bill E. Morgan, in 1967, on the Permian El Reno Group.

During the 1960s, Wilson was the author or coauthor, with his students or postdoctoral fellows, of 13 short papers or "notes" (in *the Oklahoma Geology Notes*) on the morphology, taxonomy, or revision of fossil spores and pollen. These publications include four "notes" on chitinozoans or acritarchs and six "notes" on palynological principles and techniques.

Wilson and Hoffmeister (1958) made a summary review of palynofloras of the 10 coal beds of the Senora Formation (Desmoinesian) of Oklahoma and Kansas. The stratigraphic ranges of 19 genera were compared, and 86 species of fossil spores were differentiated. These provided the basis of distinguishing and correlating coals; for example, the Henryetta coal on the basin margin was correlated with the Croweburg coal on the platform.

Felix and Burbridge (1967) studied 23 surface samples from the Springer Formation (Upper Mississippian and Lower Pennsylvanian) of Oklahoma. Their study also included subsurface samples of the overlying Morrow Formation (Lower and Middle Pennsylvanian). Ranges of 48 key palynomorphs for the Goddard, Springer, and Morrow Formations were determined. Four taxa were restricted to the Goddard, eight were restricted to the Springer, and ten originated in the Morrow. Many new species and seven new genera—*Costatascyclus, Cystoptychus, Tantillus, Trochospora, Hadrohercos, Nexuosisporites,* and *Scutulum*—were described in this study. Felix and Burbridge explained the intermingling of Mississippian and Pennsylvanian species as due to interdigitation of tongues of sediment in an oscillating transgressive sequence.

Felix and Paden (1964) described an unusual Morrowan spore, *Trinidulus diamphidios,* found in oil-well cores in the Anadarko basin of Oklahoma and Texas. Its unique trilete proximal surface with three spores of the tetrad in deep depressions has led to its recognition as an index fossil in the Lower Pennsylvanian of West Virginia, Kentucky, Iowa, and elsewhere.

A preliminary study by Wilson (1962) provided our first knowledge on the spectacular palynomorphic change that occurred in the Pennsylvanian and Permian Midcontinent region, Oklahoma, and Texas. The palynoflora of the Flowerpot Shale (Lower Permian) is dominated by saccate conifers: *Lueckisporites* (68%), *Strotersporites* (16%), *Vittatina* (2.5%) *Potonieisporites* (1.9%), and *Alisporites* (1.5%). Several new genera were introduced in this paper, including *Tririctus. Lunulasporites, Hoffmeisterites, Strotersporites, Rhizomaspora, Mucrosaccus, Hamiapollenites, Clavatasporites,* and *Trochosporites.*

Four papers amplified the early knowledge of Upper Pennsylvanian–Permian pollen and spores of the Midcontinent region. K. Jizba, for her Ph.D. dissertation that was completed in 1960 at the University of Illinois under the direction of Kosanke, studied 20 samples: six from the Virgilian of Kansas; 12 from Upper Permian rocks of Kansas, Oklahoma, and Texas; and two from Permian strata of Texas. Saccate pollen dominated the flora, including two new genera, *Complexisporites* and *Striatosaccites* (Jizba, 1962). Tschudy and Kosanke (1966) also reported on the vesiculate pollen from the Lower Permian of Texas in samples from an oil well in Callahan County. They demonstrated a distinctive break in the pollen flora at the top of the Pennsylvanian (Virgilian) by a definitive increase in diversity of saccate and striate palynomorphs.

Two other important researches should be mentioned here, though they were basically proprietary. Wilson and

Webster (1946, 1949) produced five volumes for the Carter Oil Company in which they detailed the palynomorphic correlation of strata in two wells in Texas. These are discussed further in the section on Applied Palynology that follows. A vast body of literature, data bases, and other fundamental information on Carboniferous palynology for the Midcontinent was built up during the 1950s and 1960s and is incorporated into proprietary records of oil companies.

In 1950, with the arrival of Robert W. Baxter at the University of Kansas, a new program was set up and directed toward the fructifications in Kansas, Illinois, Indiana, and Iowa coal balls. His reports include, in addition to those previously mentioned, a *Litostrobus* showing spores with preserved nuclei (Baxter and Leisman 1967) and a calamitean cone with *Elaterites triferens* spores so well preserved that the elaters opened when the spores were freed from the matrix (see Phillips et al., 1973, p. 35)! Among the postdoctoral and graduate students who studied with Baxter were A. A. Cridland and J. E. Morris.

In the early 1960s, Gilbert A. Leisman developed a research program at the Kansas State Teachers' College, Emporia (now Kansas State University). Leisman and his students published several important papers on Kansas coal-ball fructifications: a remarkable *Lepidocarpon* cone with its megaspores (Leisman and Spohn, 1962); another lycopod, *Selaginellites crassicinctus* (Leisman, 1961), with *Cirratriradites annulatus* microspores and *Triletes triangulatus* megaspores; sphenopsid cones of *Peltastrobus* from Kansas and Indiana (Leisman and Graves, 1964); and *Spencerites* megasporangia and their eye-catching *Spencerisporites* megaspores (Leisman, 1962a, b; Leisman and Stidd, 1967).

### *Western North America and Arctic islands*

The earliest contribution on Carboniferous palynology from the Rocky Mountains and western North America was the report on Upper Mississippian palynomorphs from Utah (Schemel, 1950). This report contained ten genera, including two new genera (*Rotaspora,* and *Tripartites*), and seven new species. Schemel demonstrated a distinctive stratigraphic character for the Upper Mississippian in western North America and compared it with the early Carboniferous palynofloras of the British Isles, Poland, and Russia.

Hacquebard and Barss (1957) described the palynomorphs of the first Carboniferous coal discovered in western Canada, a Mississippian coal from the Northwest Territories. Five new genera (*Cincturasporites, Labiadensites, Monilospora, Perianthospora,* and *Tendosporites*) and 24 new species were described. The palynomorphic assemblage suggested a close affinity to the lower Carboniferous (Viséan) of northern Russia. Barss (1967) illustrated 42 genera of Viséan spores from the Northwest Territories and Arctic islands.

Staplin (1960) illustrated 199 palynomorphs from the Chesterian Series of Alberta, Canada. Nine new genera—*Camptozonotriletes, Costaspora, Endoculeospora, Leioaletes, Retialetes, Retispora, Veliferaspora, Waltzispora,* and *Zonaletes*—were proposed.

Staplin and Jansonius (1964) analyzed densospore types with equatorial zona. Four new genera were proposed—*Asperispora, Clivosispora, Radiizonates,* and *Tumulispora*—and three were emended: *Cingulizonates, Cristatisporites,* and *Densosporites.*

D. C. McGregor's Ph.D. dissertation at McMasters University, which was supervised by Norman W. Radforth with assistance from Glenn E. Rouse, centered on the description of an unusual Devonian-Mississippian transitional palynoflora from a coal bed in the Canadian Arctic Archipelago (McGregor, 1960). This interesting palynoflora, which was found at several places across the High Arctic, was extracted from a coal bed at Steven's Head. *Biharisporites* and two new megaspore genera, *Hystricosporites* and *Circumsporites,* were found in abundance. *Lycospora* was dominant among the 24 new species of palynomorphs identified.

The earliest center of palynological research in western Canada developed within the laboratory of Imperial Oil, Ltd., of Canada. F. L. Staplin, Jan Jansonius, and Stanley A. J. Pocock contributed widely to various Paleozoic and Mesozoic palynological researches.

## APPLIED PALYNOLOGY AND PETROLEUM EXPLORATION

There are many excellent reviews of the early history of the application of paleopalynology to geology and exploration problems (Wilson, 1944, 1946) and Kuyl et al. (1955), Woods (1955), Hoffmeister (1959, 1960), Gutjahr (1960), Hopping (1967), and Traverse (1974, 1988). Hopping (1967, p. 24) noted: "Another great hydrocarbon industry pre-dates the oil industry in its use of palynology for correlation and age determination. The high quality of the coal palynologists' published work on the Late Paleozoic has been of invaluable assistance to the oil industry. . . ."

Palynology was first applied to the coal industry in North America by the U.S. Bureau of Mines, beginning with the research of R. Thiessen on dispersed organic remains in coals. Several studies (e.g., Thiessen 1913, 1920; Thiessen and Staud, 1923; Thiessen and Wilson, 1924) included analyses and interpretations of dispersed spores in coal. The first directed studies for petroleum palynology were requested in 1936, by a field paleontologist in Mexico, of the worldwide Royal Dutch/Shell petroleum operations (Hopping, 1967, p. 25–26). Rapidly other requests came in from around the world (Trinidad, Venezuela, Indonesia, British Borneo, and Colombia). Samples from Mexico, South America, and southeast Asia were analyzed, and early reports by Grimsdale (1937), Potonié (1938), Koch (1939), and Florschütz (1939) were instrumental in establishing a palynological section within the company. The first palynologists joined the Shell group in 1946.

The earliest publication of palynomorph illustrations from crude oil in North America (Sanders, 1937) were of Jurassic to Pliocene age from several widely dispersed wells in Mexico, mostly drilled by subsidiary divisions of Royal Dutch/Shell. Sanders differentiated and illustrated, by photomicrographs or camera lucida drawings, an assortment of spores, pollen, algal cysts (acritarchs, dinoflagellates, etc.), fungal spores and hyphae, trichomes and hairs, cuticle fragments, tracheids and small wood fragments, diatoms, radiolarians, and unidentifiable organic detritus.

William S. Hoffmeister, Carter Oil Company (see Fig. 38), received a report from L. R. Wilson in 1944 indicating some of the potentials of utilizing pollen and spores in oil exploration (Hoffmeister, 1959, p. 247). Hoffmeister engaged Wilson to demonstrate some of these potentials for an exploration site in Texas. Wilson and Webster (1946, 1949) completed an extensive investigation of palynomorphs occurring in two wells in Texas. The results of this proprietary study were compiled in five volumes containing about 9,600 photomicrographs. This pioneering study set the stage for subsequent investigations and a few copies were distributed through the American Museum of Natural History.

Two other early 1940s applications of palynology to oil exploration in the Americas should also be noted. Mrs. J. Wyatt Durham, employed by Tropical Oil Company in Colombia, used fossil spores for stratigraphic interpretations of Devonian to Tertiary strata (Hoffmeister, 1959, 1960). In the late 1940s, R. H. Tschudy also used palynology to resolve biostratigraphic problems and to correlate nonmarine to marine facies for the Creole Oil Company in Venezuela.

Figure 38. Gunnar Erdtman (left) and William Hoffmeister (right); each contributed significantly to palynology. Erdtman (1897–1973), Swedish palynologist and author of "Pollen Analysis"; Hoffmeister (1901–1980), a paleontologist-palynologist for Exxon Oil Company who promoted the use of palynology for biostratigraphy and petroleum exploration (reprinted from Alfred Traverse, *Palynology* [Boston, London, Sydney, and Wellington: Unwin Hyman, 1988] by permission of Dr. A. Traverse and the publisher).

Following these early applications of palynology to oil exploration, most of the major oil companies established palynologic research groups. Traverse (1988, p. 16), noted that when he joined Shell Development Company in 1955 as a palynologist, "Shell's palynological operations were worldwide from Nigeria to Canada." In the early 1950s Hoffmeister established a research group at Carter Oil Company at Tulsa. Among Hoffmeister's staff, W. R. Evitt, J. W. Funkhouser, F. L. Staplin, and L. E. Stover became better known later for their nonproprietary publications. A. T. Cross established a major palynology research group at Pan American Petroleum Corporation Research Center in Tulsa in 1957. He engaged C. E. Upshaw, D. W. Engelhardt, E. A. Stanley, C. Head, and W. B. Creath for pollen/spore analysis; H. Y. Ling and K. W. Klement for marine microplankton; A. Shaw for paleontologic data handling and analysis systems; R. W. Tschudy and R. M. Kosanke as consultants; and K. Brill to make special field and type-locality collections. Other palynology laboratories were established during the 1950s or slightly earlier at a number of major petroleum companies, including Atlantic-Richfield (ARCO), California Company, Chevron, Humble (Houston), Imperial (Calgary), Marathon, Mobil, Phillips, Sinclair, and Texaco. Under a joint program of Jersey Production Research Company and Pan American Petroleum Corporation Research Center, R. Potonié (Germany), G. Erdtman (Sweden), L. J. Grambast (France), and several others came to the United States for lectures and conferences.

In 1954, a stir was created among palynologists when it became known that Hoffmeister and his company had been granted a patent on one aspect of using palynomorphs as a new tool for prospecting for petroleum (Hoffmeister, 1954). However, the Jersey Production Research Company, an affiliate of Standard Oil Company of New Jersey, soon dedicated the patented principal to the public domain.

Large volumes of mostly proprietary palynological information accumulated at all major oil companies in North America. By the mid-1960s, this information had reached a level of maturity parallel with conventional micropaleontology. The utility of fossil spores and pollen began to be supplemented by a growing volume of information on marine microplankton that in the 1960s gradually became a more widely used base of information than terrestrial palynomorphs for analyzing marine deposits.

A symposium, "Palynology in oil exploration," was convened in San Francisco in 1962 by SEPM. The proceedings volume (Cross, 1964) included almost all the papers presented, which were directed toward basic principles. Noteworthy contributions were Tschudy's (1964) "Palynology and time-stratigraphic determinations" and Schopf's (1964) "Practical problems and principles in study of plant microfossils."

Early applications of palynology to petroleum exploration included studies by Kuyl et al. (1955) and Muller (1959) on the Orinoco delta and offshore shelves and bays; Koreneva's (1957, 1971) studies of pollen-spore distribution in the Sea of Okhotsk and the Mediterranean Sea; Cross et al.'s (1966) demonstration of palynomorph distribution in the Gulf of California and interactions with physical processes; Traverse and Ginsburg's research (1966) on factors accounting for palynomorph distribution in the surface sediments blanketing the Great Bahama Banks; the Rossignol (1962) study of marine Pleistocene sediments in Israel; and Wilson's (1964) cautionary paper on recycling and faulty techniques. Palynology had come of age!

## GENERAL REFERENCES, JOURNALS, BIBLIOGRAPHIES, CATALOGS

### *Pertinent foreign literature available by early 1940s*

Western European literature had a definite influence on the early palynological studies of J. M. Schopf, L. R. Wilson, C. A. Arnold, R. M. Kosanke, A. T. Cross, and others in North America. This literature included a paper by Raistrick and Simpson (1933) and several papers by Raistrick (1934, 1935, 1937, 1938, 1939) on microspores in coals of Northumberland, England, and their use in correlation of coal seams. Elizabeth Knox of Edinburgh depicted the spores of Pteridophyta and Bryophyta and compared them with microspores of Carboniferous coals of Fife, Scotland (Knox, 1938, 1939). All these papers were available to American palynologists before World War II. Knox's (1942) paper on microspores of coals of Fife also became available. She completed studies on several Scottish coalfields, mainly describing spores from coals and other rocks; her major treatise was on spores of Recent *Lycopodium, Phylloglossum, Selaginella* and *Isoetes* (1950). Papers by Paget (1936) correlating coal seams of North Staffordshire by microspore analysis and by Millott (1939) on microspores from the coal seams of North Staffordshire were also used.

Very few Russian papers were available until after World War II. Copies of such important works as those of Naumova (1938) and of Luber and Waltz (1938), on classification and stratigraphical value of Carboniferous spores, were obtained by Schopf in the early 1940s. However, the Luber and Waltz atlas (1941) on Paleozoic plant microfossils of the U.S.S.R. was not available until the late 1940s. Papers on spores of the Lower Gondwana glacial tillites of Australia and near contemporaneous shales of India (Virkki 1937, 1939, 1945) were also available.

Papers on pollen and spores of German Tertiary coals (Potonié and Gelletich, 1933; Potonié and Venitz, 1934; Wicher, 1934a, b) were of some value in classification and stratigraphy, but their scheme of nomenclature using abbreviations and symbols made them difficult to use. Potonié et al. (1932) and Loose (1932, 1934), made the earliest palynologic study of Carboniferous coal beds from the Ruhr region. A copy of Ibrahim's dissertation (1933) was also available in the mid-1930s. Hartung (1933), in a classic paper, presented excellent descriptions of calamitean spores from Carboniferous strata and discussed their morphology. J. Zerndt, Poland, published a pioneering series of papers on megaspores of Saxony (1932), Poland (1934), Bohemia (1937a), and the Saar (1940). His first paper on *Triletes giganteus* (1930) set a standard for his other papers. Zerndt's atlas (1934, 1937b) and Dijkstra's (1946) classic monograph on Carboniferous megaspores of The Netherlands were fundamental studies. Sahabi's (1936) paper on spores from French coals was also available to early American workers. Florin (1936, 1937) described the pollen grain structure of *Cordaites* and pteridosperms in classical detail.

### *General references*

Spores from several fossil plants were illustrated in Dukinfield H. Scott's *Studies in Fossil Botany* (1900, 1908, 1909, 1920, 1923) and in Seward's *Fossil Plants* (1898, 1910, 1917, 1919). The first textbook illustrations of fossil pollen were in Wodehouse's (1935) classic *Pollen Grains, Their Structure, Identification, and Significance in Science and Medicine.* This book still has considerable value today for pollen morphology. Darrah's textbook (1939) illustrated a number of spores using the peel and maceration techniques for a variety of Carboniferous plants. Erdtman's (1943) *An Introduction to Pollen Analysis* was a textbook of great value to the fundamental principles, techniques, morphology, and anatomy of pollen. Knut Faegri and Johannes Iverson published a very useful book, *Textbook on Pollen Analysis* (1950).

Gerhard O. W. Kremp's *Morphologic Encyclopedia of Palynology* (1965) dealt with the morphology of both modern and fossil spores and contains a useful compilation of the diverse terminology applied thereto. *Aspects of Palynology* (1969), by Robert H. Tschudy and Richard A. Scott, was the first English textbook on paleopalynology. Pokrovskaya's (1966) *Paleopalynologia,* a three-volume Russian work, illustrated an extensive array of fossil spores.

### *Journals and bibliographies*

Journals dedicated to palynology were not published in North America before 1968, but three major European journals partially filled that need. *Grana Palynologica* (now *Grana*), founded by Erdtman in 1954, included only a few paleopalynologic contributions. In 1959, *Pollen et Spores* was introduced from the Muséum National d'Histoire Naturelle, Paris, under the direction of M. van Campo. This journal included a greater variety of papers, paleopalynologic articles, and the annual bibliography of palynology, which was an extremely important service. The appearance of this bibliography resulted in termination, in 1957, of Erdtman's "Literature on Palynology," published irregularly in 19 issues under "Literature on Pollen Statistics" and other titles beginning in

1927 (Erdtman, 1927). The *Review of Palaeobotany and Palynology* was introduced in 1967. The first five volumes included some of the papers presented at the Second International Palynological Conference (Utrecht, The Netherlands, 1966). The journal *Micropaleontology* also included an occasional paper on paleopalynology, but more North American contributions appeared in the British journal, *Palaeontology*.

Another very significant contribution is the "Bibliography of Palaeopalynology 1836–1966" (Manten, 1969). That special issue of the *Review of Palaeobotany and Palynology* contains 12,557 paleopalynological publications cited up to 1966. This gigantic one-volume compilation is without parallel in palynology and contains cross-indexes by age and geographic areas. This special issue reflects the tremendous growth of palynology into the 1960s.

### *Catalogs*

The *Catalog of Fossil Spores and Pollen* (Kremp et al., 1957) was introduced by founding editors G. O. W. Kremp and W. Spackman at Pennsylvania State University. It began with two volumes and had reached 26 volumes by 1967, when A. Traverse became editor. H. T. Ames was associate editor over most of this period. The main purpose was to collect original descriptions of all taxa in one series of volumes for the convenience of systematists without proper access to some of the publications. The volumes included original descriptions and illustrations of several thousand taxa grouped in sequence as published in original papers, with some editorial comment but no revisions. Computerized catalogs did not become available until the 1970s.

## SUMMARY

Paleopalynology as a science in North America began with the publication of two definitive works: Schopf et al.'s (1944) "Annotated Synopsis of Paleozoic Fossil Spores" and Kosanke's (1950) "Pennsylvanian Spores of Illinois and Their Use in Correlation." These two major contributions brought together several crude threads, spun in earlier twentieth-century research pursuits on the origin, role, and significance of dispersed plants and their dissociated parts and geminules in sedimentary deposits.

Thiessen's pioneering efforts (1910–1920s) to determine the botanical basis of coal included determinations of spores and fructifications in coal and resulted in the first illustrations of dispersed Paleozoic spores in North America. Thiessen's utilization of palynomorphs for stratigraphic distribution represents the foundation of palynostratigraphy. Sellard's (1902, 1903) papers began a long thread of information on Carboniferous spores *in situ* in plant compressions that developed into a broad fabric of similar studies by the early 1950s. A source of pollen and spore information first came into focus through Noé's (1923) recognition of Pennsylvanian coal balls in North America and the first papers on fructifications bearing spores in coal balls by Noé's students, Hoskins and Reed, in the 1920s. Pollen analysis, based on Quaternary and Holocene peat studies by P. B. Sears, L. R. Wilson, S. A. Cain, J. E. Potzger, and their colleagues, was another conspicuous area of endeavor in the 1920s and 1930s that resulted in intercommunication and exchange of ideas and finally organization of a disparate group of early workers, some of whom took up the study of Paleozoic spores.

Teaching and research centers for Carboniferous palynology—initially introduced at the Illinois State Geological Survey by G. H. Cady, J. M. Schopf, and R. M. Kosanke and at Coe College in Iowa by L. R. Wilson in the early 1930s—spread to various universities in the late 1940s to early 1960s. These included University of Michigan, University of Cincinnati, Harvard University, Washington University (St. Louis), West Virginia University, Pennsylvania State University, Indiana University, University of Missouri, University of Oklahoma, University of Kansas, Michigan State University, and Emporia State University. The state geological surveys of Oklahoma, Indiana, West Virginia, and Ohio sponsored some research both academic and applied, as did the Geological Survey of Canada, the U.S. Geological Survey, and the Nova Scotia Research Foundation.

Brief reviews have been presented here for the principal and unusual publications and early unpublished theses conducted in each of the following regions: the Maritime basins of Nova Scotia and New Brunswick, the northern and southern Appalachians, the Midcontinent Illinois and Michigan basins, the Western Interior basin, the Arkoma and Ouachita basins and the southwestern coal regions of Texas, the western United States and intermontane fields of the Canadian Rockies, and the Arctic islands.

Palynology was introduced as an exploration tool to the petroleum industry in the early 1940s and was operational in most of the major oil company laboratories and some exploration offices by 1960. Application of marine microplankton for analyses of marine sequences had begun its very rapid growth to supplant spore/pollen paleopalynology for the biostratigraphy of such sequences by the late 1960s.

Few journals were dedicated exclusively to paleopalynology, and textbooks or general references were not available until Tschudy and Scott (1969). A few special publications such as *Palynology in Oil Exploration* (Cross, 1964) and Smith and Butterworth (1967) on spores in coal seams of Great Britain indicated the surge in applied paleopalynology by the end of the 1960s. *The Catalog of Fossil Spores and Pollen* (Kremp et al., 1957) was in the twentieth volume by the mid-1960s. The blossoming of palynology is indicated by the "Bibliography of palaeopalynology 1836–1966" (Manten, 1969), which includes over 12,500 publications worldwide. More than 100 authors who have published papers on Carboniferous paleopalynology of North America through 1969 are listed in this indispensable bibliographic source.

## ACKNOWLEDGMENTS

The authors gratefully acknowledge the assistance of R. A. Peppers of the Illinois State Geological Survey in identification and in updating the nomenclature in early reports by other early palynological workers of the United States. We also acknowledge, with thanks, Stephen R. Jacobson, Tom L. Phillips, L. R. and Marian Wilson for various types of assistance; Aleen Cross for extensive assistance in the manuscript preparation; and reviewers Paul C. Lyons, Alfred Traverse, Eleanora I. Robbins, and R. H. Wagner for their suggestions.

## REFERENCES CITED

Abbott, M. L., 1963, Lycopod fructifications from the Upper Freeport (No. 7) coal in southeastern Ohio: Palaeontographica, Abt. B, v. 112, p. 93–118, pls. 25–32.

Andrews, H. N., Jr., 1951, American coal-ball floras: Botanical Review, v. 17, p. 431–469.

Andrews, H. N., Jr., 1961, Studies in paleobotany: New York, John Wiley & Sons, 487 p.

Andrews, H. N., Jr., 1974, Paleobotany—1947–1972: Missouri Botanical Garden Annals, v. 61, p. 179–202.

Andrews, H. N., Jr., and Pannell, E., 1942, Contributions to our knowledge of American Carboniferous floras. Part 2: *Lepidocarpon:* Missouri Botanical Garden Annals, v. 29, p. 19–34.

Arnold, C. A., 1938, Note on a lepidophyte strobilus containing large spores, from Braidwood, Illinois: American Midland Naturalist, v. 20, p. 709–712.

Arnold, C. A., 1944, A heterosporous species of *Bowmanites* from the Michigan coal basin: American Journal of Botany, v. 31, p. 466–469.

Arnold, C. A., 1949, Fossil flora of the Michigan coal basin: University of Michigan Museum of Paleontology Contributions, v. 7, p. 131–269, 34 pl.

Arnold, C. A., 1950, Megaspores from the Michigan coal basin: University of Michigan Museum of Paleontology Contributions, v. 8, p. 59–111.

Arnold, C. A., 1958, Petrified cones of the genus *Calamostachys* from the Carboniferous of Illinois: University of Michigan Museum of Paleontology Contributions, v. 14, p. 149–165.

Arnold, C. A., 1961, Reexamination of *Triletes superbus, T. rotatus,* and *T. mamillarius* of Bartlett: Brittonia, v. 13, p. 245–252.

Auer, V., 1927a, Appendix: Botany of the interglacial peat beds in the Moose River Basin, *in* McLearn, F. H. I., The Mesozoic and Pleistocene deposits of the Lower Missinaibi, Opazatika, and Mattagami Rivers, Ontario: Ottawa, Canadian Department of Mines, Geological Survey, Summary Report 1926, Part C, Appendix, p. 45–47.

Auer, V., 1927b, Stratigraphical and morphological investigations of peat bogs of southeastern Canada: Communications Institute Forestry Fennicae, v. 12, 32 p.

Auer, V., 1930, Peat bogs in southeastern Canada: Canadian Geological Survey Memoir 162, 32 p.

Bailey, W. F., 1936, Micropaleontology and stratigraphy of the lower Pennsylvanian of central Missouri: Journal of Paleontology, v. 9, p. 453–502.

Balfour, J. H., 1857, On certain vegetable organisms found in coal from Fordel: Royal Society of Edinburgh Transactions, v. 21, p. 187–193.

Barss, M. S., 1967, Illustrations of Canadian fossils: Carboniferous and Permian spores of Canada: Ottawa, Department of Energy, Mines, and Resources, Geological Survey of Canada, Paper 67-11, 94 p., 38 pls.

Barss, M. S., Hacquebard, P. A., and Howie, R. D., 1963, Palynology and stratigraphy of some Upper Pennsylvanian and Permian rocks of the Maritime Provinces: Ottawa, Department of Mines and Technical Surveys, Geological Survey of Canada, Paper 63-3, 13 p.

Bartlett, H. H., 1929a, Fossils of the Carboniferous coal pebbles of the glacial drift at Ann Arbor: Michigan Academy of Science, Arts, and Letters Papers, v. 9, p. 11–28, 23 pls.

Bartlett, H. H., 1929b, The genus *Triletes* Reinsch: Michigan Academy of Science, Arts, and Letters Papers, v. 9, p. 29–38.

Baxter, R. W., 1949, Some pteridosperm stems and fructifications with particular reference to the Medullosae: Missouri Botanical Garden Annals, v. 36, p. 287–352.

Baxter, R. W., 1950, *Peltastrobus reedae:* A new sphenopsid cone from the Pennsylvanian of Indiana: Botanical Gazette, v. 113, p. 174–182.

Baxter, R. W., 1955, *Palaeostachya andrewsii,* a new species of calamitean cone from the American Carboniferous: American Journal of Botany, v. 42, p. 342–351.

Baxter, R. W., 1962, A *Palaeostachya* cone from southeast Kansas: Kansas University Science Bulletin, v. 43, p. 75–81.

Baxter, R. W., 1963, *Calamocarpon insignis,* a new genus of heterosporous, petrified calamitean cones from the American Carboniferous: American Journal of Botany, v. 50, p. 469–476.

Baxter, R. W., 1964, The megagametophyte and microsporangia of *Calamocarpon insignis:* Phytomorphology, v. 14, p. 481–487.

Baxter, R. W., 1967, A revision of the sphenopsid organ genus, *Litostrobus:* Kansas University Science Bulletin, v. 47, p. 1–23.

Baxter, R. W., and Leisman, G. A., 1967, A Pennsylvanian calamitean cone with *Elaterites triferens* spores: American Journal of Botany, v. 54, p. 748–754.

Bennett, J. H., 1857, An investigation into the structure of the Torbanehill mineral, and of various kinds of coal: Royal Society of Edinburgh Transactions, v. 21, part I, paper X, p. 173–185, pl. II, figs. 1–4 (color).

Bennie, J., and Kidston, R., 1886, On the occurrence of spores in the Carboniferous Formation of Scotland: Royal Physical Society of Edinburgh Proceedings, v. 9, p. 82–117.

Benninghoff, W. S., 1942, Preliminary report on a coal ball flora from Indiana: Indiana Academy of Science Proceedings, v. 52, p. 62–68.

Bentall, R., 1941, Application of spore studies to Pennsylvanian stratigraphic problems [abs.]: American Journal of Botany, v. 28, supplement, p. 7s–8s.

Bergquist, S. G., 1939, The occurrence of spore coal in the Williamston basin, Michigan: Journal of Sedimentary Petrology, v. 9, p. 14–19.

Berry, W., 1937, Spores from the Pennington coal, Rhea County, Tennessee: American Midland Naturalist, v. 18, p. 155–160.

Bhardwaj, D. C., 1957, The palynological investigations of the Saar coals. Part I: Morphography of sporae dispersae: Palaeontographica, Abt. B, v. 101, p. 73–125, pls. 22–31.

Bowman, P. W., 1931, Study of a peat bog near the Matamek River, Quebec, Canada, by the method of pollen analysis: Ecology, v. 12, p. 694–708.

Brokaw, A. L., 1942, Spores of coal No. 5 (Springfield-Harrisburg) in Illinois [M.S. thesis]: Urbana, University of Illinois, 28 p.

Brongniart, A., 1828–1838, Histoire des végétaux fossiles: Paris, G. Dufour and Ed. d'Ocagne, t. I (1828–1836), p. 1–488, pls. I–CLXVI; t. II (1837–1838), p. 1–72, pls. I–XXX.

Brush, G. S. (Somers), and Barghoorn, E. S., 1955, *Kallostachys scottii:* A new genus of sphenopsid cones from the Carboniferous: Phytomorphology, v. 5, p. 346–356.

Brush, G. S. (Somers), and Barghoorn, E. S., 1962, Identification and structure of cordaitean pollen: Journal of Paleontology, v. 36, p. 1357–1360.

Brush, G. S. (Somers), and Barghoorn, E. S., 1964, The natural relationships of some Carboniferous microspores: Journal of Paleontology, v. 38, p. 325–330.

Butterworth, M. A., Jansonius, J., Smith, A. H. V., and Staplin, F. L., 1964, *Densosporites* (Berry) Potonié and Kremp, and related genera: Report of C.I.M.P. Working Group No. 2: Congrès International de Stratigraphie et de Géologie du Carbonifère, 5e, Paris, 1963: Compte Rendu, t. III, p. 1049–1057.

Cain, S. A., 1939, Pollen analysis as a paleo-ecological research method: Botanical Review, v. 5, p. 627–654.

Chaloner, W. G., 1954, Mississippian megaspores from Michigan and adjacent states: University of Michigan Museum of Paleontology Contributions, v. 12, p. 23–35, 2 pls.

Chaloner, W. G., 1958, *Palysporia mirabilis* Newberry, a fossil lycopod cone: Journal of Paleontology, v. 32, p. 199–209, pls. 31, 32.

Clendening, J. A., 1962, Small spores applicable to stratigraphic correlation in the Dunkard basin of West Virginia and Pennsylvania: West Virginia Academy of Science Proceedings, v. 34, p. 133–142.

Clendening, J. A., 1965, Characteristic small spores of the Redstone coal in West Virginia: West Virginia Academy of Science Proceedings, v. 37, p. 183–189.

Clendening, J. A., 1966, *Schopfipollenites* in the Washington Formation: West Virginia Academy of Science Proceedings, v. 38, p. 169–176.

Clendening, J. A., and Gillespie, W. H., 1964, Characteristic small spores of the Pittsburgh coal in West Virginia and Pennsylvania: Journal of Paleontology, v. 35, p. 141–150.

Collins, H. R., 1959, Small spore assemblages of the Harlem coal and associated strata in the Morgantown, West Virginia area [M.S. thesis]: Morgantown, West Virginia University, 66 p.

Cooper, W. S., 1923, The interglacial forests of Glacier Bay, Alaska: Ecology, v. 4, p. 93–128.

Cooper, W. S., 1937, The problem of Glacier Bay, Alaska: A study of glacier variations: Geographical Review, v. 27, p. 37–62.

Cooper, W. S., 1942, Vegetation of the Prince Williams Sound region, Alaska; with a brief excursion into the post-Pleistocene climatic history: Ecological Monographs, v. 12, p. 1–22.

Coulter, J. M., and Land, W. J. G., 1911, An American *Lepidostrobus:* Botanical Gazette, v. 51, p. 449–453.

Coulter, J. M., and Land, W. J. G., 1921, A homosporous American *Lepidostrobus:* Botanical Gazette, v. 72, p. 106–108.

Cropp, F. W., 1960, Pennsylvanian spore floras from the Warrior basin, Mississippi and Alabama: Journal of Paleontology, v. 34, p. 359–367.

Cropp, F. W., 1963, Pennsylvanian spore succession in Tennessee: Journal of Paleontology, v. 37, p. 900–916.

Cross, A. T., 1947, Spore floras of the Pennsylvanian of West Virginia and Kentucky, *in* Wanless, H. R., ed., Symposium on Pennsylvanian problems: Journal of Geology, v. 55, pt. 2, p. 285–308.

Cross, A. T., 1953, The geology of the Pittsburgh coal: Stratigraphy, petrology, origin and composition, and geological interpretation of mining problems: Proceedings, Second Conference on the Origin and Constitution of Coal, Crystal Cliffs, Nova Scotia, June 1952: Halifax, Nova Scotia Department of Mines and Nova Scotia Research Foundation, p. 32–99.

Cross, A. T., ed., 1964, Palynology in oil exploration—A symposium, San Francisco, California, March 26–27, 1962: Society of Economic Paleontologists and Mineralogists Special Publication 11, 200 p.

Cross, A. T., 1975a, A comparison of the age of the Jurassic "Red Beds" of Michigan, with Jurassic rocks of North America [abs.]: Ann Arbor, Michigan Academy of Science, Arts, and Letters, Abstracts, p. 7.

Cross, A. T., 1975b, Jurassic plants of Michigan Basin [abs.]: American Association of Petroleum Geologists Annual Meeting Abstracts, p. 14–15.

Cross, A. T., and Schemel, M. P., 1952, Representative microfossil floras of some Appalachian coals: Congrès pour l'avancement des études de Stratigraphie et de Géologie du Carbonifère, 3e, Heerlen, 1951: Compte Rendu, t. I, p. 123–130.

Cross, A. T., Thompson, G. G., and Zaitzeff, J. B., 1966, Source and distribution of palynomorphs in bottom sediments, southern part of Gulf of California: Marine Geology, v. 4, p. 467–524.

Darrah, W. C., 1936, A new *Macrostachya* from the Carboniferous of Illinois: Harvard University Botanical Museum Leaflets, v. 4, p. 52–63.

Darrah, W. C., 1937, *Codonotheca* and *Crossotheca:* Polleniferous structures of pteridosperms: Harvard University Botanical Museum Leaflets, v. 4, p. 153–172.

Darrah, W. C., 1938a, A new fossil gleicheniaceous fern from Illinois: Harvard University Botanical Museum Leaflets, v. 5, p. 145–160.

Darrah, W. C., 1938b, A remarkable fossil *Selaginella* with preserved female gametophyte: Harvard University Botanical Museum Leaflets, v. 6, p. 118–136.

Darrah, W. C., 1939, Textbook of paleobotany: New York and London, D. Appleton-Century, 441 p.

Dawson, J. W., 1863, Synopsis of the flora of the Carboniferous period in Nova Scotia: Canadian Naturalist, new series, v. 8, p. 431–457.

Dawson, J. W., 1871, On spore cases in coals: American Journal of Science, 3d ser., v. 1, p. 256–263.

Denton, G. H., 1957, Plant microfossils in lower Allegheny coal beds of northeastern Ohio, western Pennsylvania, and northern West Virginia [abs.]: Geological Society of America Bulletin, v. 68, pt. 2, p. 1715–1716.

Dijkstra, S. J., 1946, Eine monographische Bearbeitung der karbonischen Megasporen: Mededeelingen Geologische Stichting, Ser. C-III-1, 101 p., 16 pls.

Draper, P., 1929, A comparison of pollen spectra of old and young bogs in the Erie basin: Oklahoma Academy of Science Proceedings, v. 9, p. 50–53.

Eddings, A. L., 1947, Correlation of the Trivoli (No. 8) coal bed in Illinois by plant microfossils [M.S. thesis]: Urbana, University of Illinois, 1947, 26 p.

Erdtman, G., 1927, Literature on pollen-statistics published before 1927: Geologiska Föreningens i Stockholm Förhandlingar, v. 49, p. 196–211.

Erdtman, G., 1943, An introduction to pollen analysis: Waltham, Massachusetts, Chronica Botanica Company, 240 p. (second edition, Waltham, Massachusetts, Chronica Botanica, 1954, 239 p.

Evitt, W. R., 1963, A discussion and proposal concerning dinoflagellates, hystrichospheres and acritarchs: U.S. National Academy of Science, Proceedings, v. 49, p. 158–164, 298–302.

Faegri, K., and Iversen, J., 1950, Textbook of pollen analysis: Copenhagen, Munksgaard Publ., 168 p. (third edition, 1979, New York: Hafner Publishing).

Felix, C. J., 1954, Some American arborescent lycopod fructifications: Missouri Botanical Garden Annals, v. 41, p. 351–394.

Felix, C. J., and Burbridge, P. P., 1967, Palynology of the Springer Formation of southern Oklahoma, U.S.A.: Palaeontology, v. 10, p. 349–425.

Felix, C. J., and Paden, P., 1964, A new lower Pennsylvanian spore genus: Micropaleontology, v. 10, p. 330–332.

Florin, R., 1936, On the structure of pollen grains in the Cordaitales: Svensk Botanisk Tidskrift, v. 30, p. 624–651.

Florin, R., 1937, On the morphology of the pollen grains of some Palaeozoic pteridosperms: Svensk Botanisk Tidskrift, v. 31, p. 305–338.

Florschütz, F., 1939, Rapport omtrent de resultaten van een microbotanisch onderzoek van gesteentemonsters uit Venezuela, Colombia en Z. O. Borneo: Amsterdam, Netherlands, Royal Dutch/Shell (unpublished report).

Foster, J. W., 1838, Second annual report of the Geological Survey of Ohio: Columbus, Ohio Geological Survey, 101 p.

Frederiksen, N. O., 1961, Sporomorphae of the Brookville and lower Clarion seams, near Brookville, Pennsylvania [M.S. thesis]: University Park, Pennsylvania State University, 273 p.

Fuller, G. D., 1927, Pollen analysis and postglacial vegetation: Botanical Gazette, v. 83, p. 323–325.

Gillespie, W. H., Latimer, I. S., Jr., and Clendening, J. A., 1966, Plant fossils of West Virginia (revised edition): Morgantown, West Virginia Geological and Economic Survey, Educational Series, 131 p. (3d revision, 1978, Educational Ser. 3A).

Godwin, H., 1968, The development of Quaternary palynology in the British Isles: Review of Palaeobotany and Palynology, v. 6, p. 9–20.

Graham, R., 1934, Pennsylvanian flora of Illinois as revealed in coal balls. I: Botanical Gazette, v. 95, p. 453–476.

Graham, R., 1935, Pennsylvanian flora of Illinois as revealed in coal balls. II: Botanical Gazette, v. 97, p. 156–168.

Gray, R. J., 1951, Plant microfossils and general stratigraphy of the Waynesburg coal [M.S. thesis]: Morgantown, West Virginia University, 129 p.

Gresley, W. S., 1901, Possible new coal plants in coal. Part 3: American Geologist, v. 27, p. 6–14.

Gries, R., 1991, AAPG celebrates 50 years 1941–1991 of distinguished lecturers: Tulsa, Oklahoma, American Association of Petroleum Geologists, Distinguished Lecture Committee, 48 p.

Griggs, R. F., 1934, The edge of the forest in Alaska and the reasons for its position: Ecology, v. 15, p. 80–96.

Grimsdale, T. F., 1937, Note upon correlation of lignitic series by means of fossil pollen: Trinidad, Royal Dutch/Shell (unpublished report).

Groth, P. K. H., 1966, Palynological delineation of environments in the Columbiana Shale (Pennsylvanian) of western Pennsylvania [M.S. thesis]: University Park, Pennsylvania State University, 192 p.

Guennel, G. K., 1952, Fossil spores of the Alleghenian coals in Indiana: Indiana Geological Survey Report of Progress 4, p. 1–40.

Guennel, G. K., 1954, An interesting megaspore species found in Indiana Block coal: Indianapolis, Indiana, Butler University Botanical Studies, v. 11, papers 8–17, p. 169–177.

Guennel, G. K., 1958, Miospore analysis of the Pottsville coals of Indiana: Indiana Geological Survey Bulletin 13, 101 p., pls. 1–6.

Guennel, G. K., and Neavel, R. C., 1961, *Torispora securis* Balme: Spore or sporangial wall cell?: Micropaleontology, v. 7, p. 207–212.

Gupta, S., 1965, Palynology of the Grassy Creek and Saverton shales of Missouri [Ph.D. thesis]: Columbia, University of Missouri, Botany Department, 247 p., 12 pls.

Gutjahr, C. C. M., 1960, Palynology and its application in petroleum exploration: Gulf Coast Association of Geological Societies Transactions, v. 10, p. 175–187.

Habib, D., 1966, Distribution of spore and pollen assemblages in the Lower Kittanning coal of western Pennsylvania: Palaeontology, v. 9, p. 629–666.

Hacquebard, P. A., 1957, Plant spores in coal from the Horton Group (Mississippian) of Nova Scotia: Micropaleontology, v. 3, p. 301–324.

Hacquebard, P. A., 1961, Palynological studies of some upper and lower Carboniferous strata in Nova Scotia: Proceedings, Third Conference on the Origin and Constitution of Coal, Crystal Cliffs, Nova Scotia, June 1956: Halifax, Nova Scotia Department of Mines and Nova Scotia Research Foundation, p. 227–256.

Hacquebard, P. A., and Barss, S., 1957, A Carboniferous spore assemblage, in coal from the South Nahanni River area, Northwest Territories: Canada Department of Mines and Technical Surveys, Geological Survey of Canada Bulletin 40, p. 1–63.

Hacquebard, P. A., and Donaldson, J. R., 1964, Stratigraphy and palynology of the upper Carboniferous Coal Measures in the Cumberland basin of Nova Scotia, Canada: Congrès International de Stratigraphie et de Géologie du Carbonifère, 5e, Paris, 9–12 septembre 1963, t. III, p. 1157–1169.

Hacquebard, P. A., Barss, S., and Donaldson, J. R., 1960, Distribution and stratigraphic significance of small spore genera in the upper Carboniferous of the Maritime Provinces of Canada: Congrès pour l'avancement des études de Stratigraphie et de Géologie du Carbonifère, 4e, Heerlen, 1958: Compte Rendu, t. I, p. 237–245.

Hamilton, J. B., 1962, Correlation of the Pennsylvanian rocks in the western part of the central Pennsylvanian basin of New Brunswick by means of fossil spores [M.S. thesis]: Fredericton, University of New Brunswick.

Hansen, H. P., 1937, Pollen analysis of two Wisconsin bogs of different age: Ecology, v. 18, p. 136–148.

Hansen, H. P., 1938, Postglacial forest succession and climate in the Puget Sound region: Ecology, v. 19, p. 528–542.

Hansen, H. P., 1942, A pollen study of lake sediments in lower Willamette Valley of western Oregon: Torrey Botanical Club Bulletin, v. 69, p. 262–280.

Hansen, H. P., 1943, A pollen study of peat profiles from Lower Klamath Lake of Oregon and California, *in* Cressman, L. S., ed., Archeological researches in the northern Great Basin: Carnegie Institute of Washington Publication 538, p. 103–118.

Hartung, W., 1933, Die Sporenverhältnisse der Calamariaceen: Preussische Geologische Landesanstalt, Institut für Paläobotanik und Petrographie der Brennsteine, Arbeiten, Band 3, Heft 1, p. 95–149, Tafn 8–11.

Henbest, O. J., 1933, Plant residues of Coal No. 6: Illinois State Academy of Science Transactions, 1932, v. 25, p. 147–149.

Henbest, O. J., 1936, Size and ornamentation of some modern and fossil lycopod spores: Illinois State Academy of Science Transactions, 1935, v. 28, p. 91–92.

Heusser, C. J., 1952, Pollen profiles from southeastern Alaska: Ecological Monograph, v. 22, p. 331–352.

Hildreth, S. P., 1836, Observations on the bituminous coal deposits of the valley of the Ohio, and accompanying rock strata; with notices on the fossil organic remains and relics of vegetable and animal bodies: American Journal of Science, v. 29, p. 1–148, and frontispiece.

Hoffmeister, W. S. [Assignor to Standard Oil Development Company], 1954, Microfossil prospecting for petroleum. Application August 10, 1951, serial no. 241,387 (Cl 23-230), 7 claims. Patented August 10, 1954, U.S. Patent Office no. 2,686,108, Washington, D.C., 4 p., Columns 1–8. [Dedicated to the public subsequent to July 5, 1955, Official Gazette, August 16, 1955.]

Hoffmeister, W. S., 1959, Palynology's first ten years as an aid to finding oil: Oil and Gas Journal, v. 57, p. 246–248, 250.

Hoffmeister, W. S., 1960, Palynology has important role in oil exploration: World Oil, April, 1960, 4 p. (Reprint, Gulf Publishing Co.).

Hoffmeister, W. S., Staplin, F. L., and Malloy, R. E., 1955, Mississippian plant spores from the Hardinsburg Formation of Illinois and Kentucky: Journal of Paleontology, v. 29, p. 372–399.

Hopping, C. A., 1967, Palynology and the oil industry: Review of Palaeobotany and Palynology, v. 2, p. 23–48.

Hoskins, J. H., 1923, A Paleozoic angiosperm from an American coal ball: Botanical Gazette, v. 75, p. 390–399.

Hoskins, J. H., 1926, Structure of Pennsylvanian plants from Illinois. I: Botanical Gazette, v. 82, p. 427–437, pls. 23–24.

Hoskins, J. H., and Abbott, M. L., 1956, *Selaginellites crassicinctus,* a new species from the Desmoinesian Series of Kansas: American Journal of Botany, v. 43, p. 36–46.

Hoskins, J. H., and Cross, A. T., 1941, A consideration of the structure of *Lepidocarpon* Scott based on a new strobilus from Iowa: American Midland Naturalist, v. 25, p. 523–547.

Hoskins, J. H., and Cross, A. T., 1943, Monograph of the Paleozoic cone genus *Bowmanites* (Sphenophyllales): American Midland Naturalist, v. 30, p. 113–163.

Hoskins, J. H., and Cross, A. T., 1946a, Studies in the Trigonocarpales. Part I: *Pachytesta vera,* a new species from the Des Moines Series of Iowa: American Midland Naturalist, v. 36, p. 207–250.

Hoskins, J. H., and Cross, A. T., 1946b, Studies in the Trigonocarpales. Part II: Taxonomic problems and a revision of the genus *Pachytesta:* American Midland Naturalist, v. 36, p. 331–361.

Houdek, P. K., 1933, Pollen statistics from two Indiana bogs: Indiana Academy of Science Proceedings, 1932, v. 42, p. 73–77.

Houdek, P. K., 1935, Pollen statistics for two bogs in southwestern Michigan: Michigan Academy of Science Papers for 1934, v. 20, p. 49–56.

Ibrahim, A. C., 1933, Sporenformen des Aegir-Horizonts des Ruhr-Reviers [Dissertation]: Würzburg, privately published by Konrad Triltsch, Universität Berlin, 47 p.

Jizba, K. M., 1962, Late Paleozoic bisaccate pollen from the United States mid-continent area: Journal of Paleontology, v. 36, p. 871–887.

Johnston, J. F. W., 1838, On the composition of certain mineral substances of organic origin: The London and Edinburgh Philosophical Magazine and Journal of Science, v. 12, p. 162–163.

Jones, D. H., 1957, Palynology of the Bevier coal of Missouri [M.S. thesis]: Columbia, University of Missouri, 148 p., 6 pls.

Knox, E. M., 1938, The spores of Pteridophyta, with observations on microspores in coals of Carboniferous age: Botanical Society of Edinburgh Transactions, Proceedings, v. 32, p. 438–466.

Knox, E. M., 1939, The spores of Bryophyta compared with those of Carboniferous age: Botanical Society of Edinburgh Transactions, Proceedings, v. 32, p. 477–487.

Knox, E. M., 1942, The microspores in some coals of the Productive Coal Measures in Fife: Institute of Mining Engineering (London) Transactions, v. 101, p. 98–112.

Knox, E. M., 1950, The spores of *Lycopodium, Phylloglossum, Selaginella,* and *Isoetes:* Botanical Society of Edinburgh Transactions, Proceedings, v. 35, p. 211–357.

Koch, R. E., 1939, Pollen grain research: Amsterdam, Netherlands, Royal Dutch/Shell, (unpublished).

Koreneva, E. V., 1957, Spore-pollen analysis of bottom sediments from the Sea of Okhotsk [in Russian]: Trudy Institut Okeanologyia, Akademia Nauk, S.S.S.R., v. 22, p. 221–251.

Koreneva, E. V., 1971, Spores and pollen in Mediterranean bottom sediments, *in* Funnel, B. M., and Riedel, W. R., eds., The micropaleontology of oceans: Cambridge, Cambridge University Press, p. 361–371.

Kosanke, R. M., 1943, The characteristic plant microfossils of the Pittsburgh and Pomeroy coals of Ohio: American Midland Naturalist, v. 29, p. 119–132.

Kosanke, R. M., 1947, Plant microfossils in the correlation of coal beds: Journal of Geology, v. 55, p. 280–284.

Kosanke, R. M., 1950, Pennsylvanian spores of Illinois and their use in correlation: Illinois State Geological Survey Bulletin 74, 128 p.

Kosanke, R. M., 1955, *Mazostachys*—A new calamite fructification: Illinois State Geological Survey Report of Investigations 180, 37 p.

Kosanke, R. M., 1964, Applied Paleozoic palynology, *in* Cross, A. T., ed., Palynology in oil exploration—A symposium, San Francisco, California, March 26–27, 1962: Society of Economic Paleontologists and Mineralogists Special Publication 11, p. 75–89.

Kremp, G. O. W., 1965, Morphologic encyclopedia of palynology: Tucson, University of Arizona Press, 186 p., + 76 unnumbered p. with 38 pls. (Revised printing 1968, 263 p.).

Kremp, G. O. W., and Spackman, W., eds., Kremp, G. O. W., Ames, H. T., and Grebe, H., compilers, 1957, Catalog of fossil spores and pollen: State College, Pennsylvania State University, v. 1, p. i–vii, 1–182, and index, 2 p. (unnumbered); and v. A, 2 p. (unnumbered); v. 2, p. i–ii, 1–171, and index, 7 p. (unnumbered).

Kuyl, O. S., Muller, J., and Waterbolk, H. T., 1955, The application of palynology to oil geology with reference to western Venezuela: Geologie en Mijnbouw, v. 17, p. 49–76.

Lane, G. H., 1941, Pollen analysis of interglacial peats of Iowa: Iowa Geological Survey, Annual Report, v. 37, p. 237–262.

Leisman, G. A., 1961, Further observations on the structure of *Selaginellites crassicinctus:* American Journal of Botany, v. 48, p. 224–229.

Leisman, G. A., 1962a, A *Spencerites* sporangium and associated spores from Kansas: Micropaleontology, v. 8, p. 396–402.

Leisman, G. A., 1962b, *Spencerites moorei,* comb. nov., from southeastern Kansas: American Midland Naturalist, v. 68, p. 347–356.

Leisman, G. A., and Graves, C., 1964, The structure of the fossil sphenopsid cone, *Peltastrobus reedae:* American Midland Naturalist, v. 72, p. 426–437.

Leisman, G. A., and Spohn, P. A., 1962, The structure of a *Lepidocarpon* strobilus from southeastern Kansas: Palaeontographica, Abt. B, Band 111, p. 113–125.

Leisman, G. A., and Stidd, B. M., 1967, Further occurrences of *Spencerites* from the middle Pennsylvanian of Kansas and Illinois: American Journal of Botany, v. 54, p. 316–323.

Levittan, E. D., and Barghoorn, E. S., 1948, *Sphenostrobus thompsonii:* A new genus of the Sphenophyllales: American Journal of Botany, v. 35, p. 350–358.

Lewis, I. F., and Cocke, E. C., 1929, Pollen analysis of the Dismal Swamp: Journal of the Elisha Mitchell Scientific Society, v. 45, p. 37–58.

Loose, F., 1932, Beschreibung von Sporenformen aus Flöz Bismarck, *in* Potonié, R., Ibrahim, A. C., and Loose, F., Sporenformen aus den Flözen Ägir und Bismarck des Ruhrgebietes: Stuttgart Neues Jahrbuch Mineralogie, Geologie und Paläontologie, Beilage-Band 67, Abt. B, p. 449–454, 7 Tafn.

Loose, F., 1934, Sporenformen aus dem Flöz Bismarck des Ruhrgebietes: Preussische Geologische Landesanstalt, Institut für Paläobotanik und Petrographie der Brennsteine, Arbeiten, Band 4, p. 127–164, 1 pl.

Luber (Lyuber), A. A., and Waltz (Valtz), I. E., 1938, Classification and stratigraphical value of spores of some Carboniferous coal deposits in the U.S.S.R. [in Russian]: Central Geological Prospecting Institute of Moscow Transactions, v. 105, 45 p., 10 pls.

Luber (Lyuber), A. A., and Waltz (Val'c), I. E., 1941, Atlas of microspores and of pollen grains of the Palaeozoic of the U.S.S.R. (Krishtofovich, A. N., ed.) [in Russian]: Moscow and Leningrad, All-union Scientific Research Institute Geologie (VSEGEI) Transactions, v. 139, p. 1–107.

Mamay, S. H., 1950, Some American Carboniferous fern fructifications: Missouri Botanical Garden Annals, v. 37, p. 409–476.

Mamay, S. H., 1954a, A new sphenopsid cone from Iowa: Annals of Botany, new series, v. 18, p. 229–239, pls. XIII–XIV.

Mamay, S. H., 1954b, Two new plant genera of Pennsylvanian age from Kansas coal balls: U.S. Geological Survey Professional Paper 254-D, p. 81–95.

Mamay, S. H., 1955, *Acrangiophyllum,* a new genus of Pennsylvanian Pteropsida based on fertile foliage: American Journal of Botany, v. 42, p. 177–183.

Mamay, S. H., 1957, *Biscalitheca,* a new genus of Pennsylvanian coenopterids, based on its fructification: American Journal of Botany, v. 44, p. 229–239.

Mamay, S. H., 1959, A new *Bowmanites* fructification from the Pennsylvanian of Kansas: American Journal of Botany, v. 46, p. 530–536.

Manten, A. A., ed., 1969, Bibliography of palaeopalynology 1836–1966: Review of Palaeobotany and Palynology, v. 8, p. 3–572.

McCabe, L. C., 1932, Some plant structures of coal: Illinois State Academy of Science Transactions, 1931, v. 24, p. 321–326.

McGregor, D. C., 1960, Devonian spores from Melville Island, Canadian Arctic Archipelago: Palaeontology, v. 3, p. 26–44, pl. 1.

Millott, J. O'N., 1939, The microspores in the coal seams of North Staffordshire. Part 1: The Millstone Grit–Ten Foot coals: London, Institute of Mining Engineering Transactions, v. 96, p. 317–353.

Morgan, J. L., 1955, Spores of McAlester coal: Oklahoma Geological Survey Circular 36, 54 p.

Morris, 1840. [see Prestwich, J.]

Morton, S. G., 1836, Being a notice and description of the organic remains embraced in the preceding paper [Hildreth, 1836]: American Journal of Science, v. 29, p. 149–154, pls. 1–36.

Muller, J., 1959, Palynology of Recent Orinoco delta and shelf sediments: Reports of the Orinoco Shelf Expedition: Micropaleontology, v. 5, p. 1–32, 1 pl., figs. 1–23.

Mumma, M. D., 1960, Palynology of selected coals of north-central Missouri [M.S. thesis]: Columbia, University of Missouri, 127 p., 12 pl.

Murdy, W. H., and Andrews, H. N., Jr., 1957, A study of *Botryopteris globosa* Darrah: Torrey Botanical Club Bulletin, v. 84, p. 252–257.

Naumova, S. N., 1938, Microspores from the coals of the Moscow Basin [in Russian]: Trudy vses. nauchno. issled. Inst. min. Syr'ya, v. 119, p. 21–32.

Neavel, R. C., and Guennel, G. K., 1960, Indiana paper coal: Composition and deposition: Journal of Sedimentary Petrology, v. 30, p. 241–248.

Noé, A. C., 1923, Coal balls: Science, v. 57, p. 385.

Oliver, F. W., and Scott, D. H., 1904, On the structure of the Paleozoic seed *Lagenostoma lomaxi,* with a statement of the evidence upon which it is referred to *Lyginodendron:* Philosophical Transactions Royal Society of London, v. 197B, p. 193–247.

Paget, R. F., 1936, The correlation of coal seams by microspore analysis: The seams of Warwickshire: London, Institute of Mining Engineering Transactions, v. 92, p. 59–88.

Peppers, R. A., 1964, Spores in strata of Late Pennsylvanian cyclothems in the Illinois basin: Illinois State Geological Survey Bulletin 90, 89 p.

Peppers, R. A., 1970, Correlation and palynology of coals in the Carbondale and Spoon Formations (Pennsylvanian) of the northeastern part of the Illinois basin: Illinois State Geological Survey Bulletin 93, 173 p.

Phillips, T. L., 1961, American species of *Botryopteris* from the Pennsylvanian [Ph.D. thesis]: St. Louis, Missouri, Washington University, 77 p.

Phillips, T. L., 1979, Paleobotanical studies and collection—With a bibliography of the research contributions of staff and students, 1932–1979; Prepared on occasion of the International Congress of Carboniferous Stratigraphy and Geology, 9th, Washington, D.C., and Champaign-Urbana, Illinois, May 17–26, 1979: Urbana, University of Illinois and Illinois State Geological Survey, p. i–v, 1–69.

Phillips, T. L., Pfefferkorn, H. W., and Peppers, R. A., 1973, Development of paleobotany in the Illinois basin: Illinois State Geological Survey Circular 480, 86 p.

Playford, G., 1964, Miospores of the Mississippian Horton Group, eastern Canada: Geological Survey of Canada Bulletin 107, p. 1–47.

Pokrovskaya, I. M., ed., 1966, Paleopalynologia [in Russian]: Leningrad, Nedra, Trudy VSEGEI, v. 1, 352 p.; v. 2, 446 p.; v. 3, 368 p.

Post, L. von, 1916, Om Skogsträdpollen i sydsvenska torfmosselagerföljder: Geologiska Föreningens i Stockholm Förhandlingar, v. 38, p. 384–390.

Potonié, H., 1899, Lehrbuch der Pflanzenpalaeontologie mit besonderer Rücksicht auf die Bedürfnisse des Geologen, Lief 4: Berlin, Ferd. Dümmler, 402 p., Tafn. I–III.

Potonié, R.,1938, Bericht über die mikropalaeontologische Untersuchung von Proben aus dem südöstlichen Mexiko: Mexico, Royal Dutch/Shell (unpublished report).

Potonié, R., 1956–1966, Synopsis der Gattungen der Sporae dispersae 1, Geologisches Jahrbuch, Beihefte, v. 23 (1956), p. 3–103; 2, v. 31 (1958), p. 3–114; 3, v. 39 (1960), p. 3–189; 4, v. 72 (1966), p. 3–244 + 15 pls.

Potonié, R., and Gelletich, J., 1933, Über Pteridophyten-Sporen einer eozänen Braunkohle aus Dorog in Ungarn: Sitzungsberichte Gesellschaft Naturforschende Freunde, Berlin, 1932, p. 517–528.

Potonié, R., and Kremp, G., 1954, Die Gattungen der paläozoischen Sporae dispersae und ihre Stratigraphie: Geologisches Jahrbuch, Beiheft 69, p. 111–194, Tafn. 20.

Potonié, R., and Kremp, G., 1955–1956, Die Sporae dispersae des Ruhrkarbons, ihre Morphographie und Stratigraphie mit Ausblicken auf Arten anderer Gebiete und Zeitabschnitte. 1: Palaeontographica, Abt. B, Band 98 (1–3), 1955, Teil I, p. 1–136; Band 99 (4–6), 1956, Teil II, p. 85–191; Band 100 (4–6), 1956, Teil III, p. 65–121.

Potonié, R., and Venitz, H., 1934, Zur Mikrobotanik des miozänen Humodils der niederrheinischen Bucht: Institut Paläobotanik und Petrographie der Brennsteine, Band 5, p. 5–58, Tafn. 1–6.

Potonié, R., Ibrahim, A. C., and Loose, F., 1932, Sporenformen aus den Flözen Ägir und Bismarck des Ruhrgebietes: Arbeiten Neues Jahrbuch Mineralogie Geologie und Paläontologie, Beilage-Band 67, Abt. B, p. 438–454.

Potzger, J. E., 1932, Succession of forests as indicated by fossil pollen from a northern Michigan bog: Science, v. 75, p. 366.

Prestwich, J., 1840, On the geology of Coalbrook Dale: Geological Society of London Transactions, ser. 2, v. 5, pt. 3, p. 413–495.

Raistrick, A., 1934, The correlation of coal seams by microspore content. Part I: The seams of Northumberland: London, Institute of Mining Engineering Transactions, v. 88, p. 142–153.

Raistrick, A., 1935, The microspore analysis of coal: Naturalist, no. 942, p. 145–150.

Raistrick, A., 1937, The microspores of coal and their use in correlation: Congrès pour l'avancement des études de Stratigraphie carbonifère, 2e, Heerlen, 1935: Compte Rendu, v. 2, p. 909–917.

Raistrick, A., 1938, The microspore content of some Lower Carboniferous coals: Leeds Geological Association Transactions, v. 5, p. 221–226.

Raistrick, A., 1939, The correlation of coal seams by microspore content. Part II: The Trencherbone Seam, Lancashire and the Busty Seams, Durham: London, Institute of Mining Engineering Transactions, v. 97, p. 425–431.

Raistrick, A., and Simpson, J., 1933, The microspores of some Northumberland coals, and their use in the correlation of coal seams: London, Institute of Mining Engineering Transactions, v. 85, p. 225–235.

Reed, F. D., 1926, Flora of an Illinois coal ball: Botanical Gazette, v. 81, p. 460–468, pl. XXXVIII.

Reed, F. D., 1939, Structure of some Carboniferous seeds from American coal fields: Botanical Gazette, v. 100, p. 769–787.

Reinertsen, D. L., 1953, Small spores from the Summit coal of Boone County, Missouri [M.S. thesis]: Columbia, University of Missouri, 55 p., 3 pls.

Reinsch, P. F., 1884, Micro-Palaeophytologia Formationis Carboniferae, Volume 1, Continens Trileteas et Stelideas: Erlangae, Germania, Theo Krische, p. vii–80, pls. 1–66.

Rossignol, M., 1962, Analyse pollinique de sédiments marins quaternaires en Israel. 2: Sédiments pleistocènes: Pollen et Spores, v. 4, p. 121–148.

Sahabi, Y., 1936, Recherches sur les spores des houilles françaises: Lille, Douriez-Bataille, 62 p.

Sanders, J. McC., 1937, The microscopical examination of crude petroleum: Journal of the Institution of Petroleum Technologists (London), v. 23, p. 525–573, + 17 plates without pagination.

Schemel, M. P., 1950, Carboniferous plant spores from Daggett County, Utah: Journal of Paleontology, v. 24, p. 232–244.

Schemel, M. P., 1951, Small spores of the Mystic coal of Iowa: American Midland Naturalist, v. 46, p. 743–750.

Schemel, M. P., 1957, Small spore assemblages of mid-Pennsylvanian coals of West Virginia and adjacent areas [Ph.D. thesis]: Morgantown, West Virginia University, 222 p., 7 figs.

Schmieg, R. E., 1959, Palynology of the Cabaniss coals in Henry County, Missouri [M.S. thesis]: Columbia, University of Missouri, 157 p., 7 pls.

Schopf, J. M., 1936a, The paleobotanical significance of plant structure in coal: Illinois State Academy of Science Transactions, 1935, v. 28, p. 106–110.

Schopf, J. M., 1936b, Spores characteristic of Illinois Coal No. 6: Illinois State Academy of Science Transactions, 1935, v. 28, p. 173–176.

Schopf, J. M., 1938, Spores of the Herrin (No. 6) coal bed in Illinois: Illinois State Geological Survey Report of Investigations 50, 73 p.

Schopf, J. M., 1940, *Lepidocarpon* Scott (1900), *in* Janssen, R. E., ed., Some fossil plant types of Illinois: Springfield, Illinois State Museum Scientific Papers, v. 1, p. 39–45.

Schopf, J. M., 1941, Contributions to Pennsylvanian paleobotany, *Mazocarpon oedipternum* sp. nov., and sigillarian relationships: Illinois State Geological Survey Report of Investigations 75, p. 1–53.

Schopf, J. M., 1964, Practical problems and principles in study of plant microfossils, *in* Cross, A. T., ed., Palynology in oil exploration—A symposium, San Francisco, California, March 26–27, 1962: Society of Economic Paleontologists and Mineralogists Special Publication 11, p. 29–57.

Schopf, J. M., Wilson, L. R., and Bentall, R., 1944, An annotated synopsis of Paleozoic fossil spores and the definition of generic groups: Illinois State Geological Survey Report of Investigations 91, 72 p.

Schulze, F., 1855, Über das Vorkommen wollerhaltener Zellulose in Braunkohle und Steinkohle: Berlin, Bericht Verhandlingoen Preussische Koeniglich Akademie Wissenschaften, p. 676–678.

Scott, D. H., 1900, Studies in fossil botany: London, Adam and Charles Black (second edition, 2 parts, 1908, 1909, 683 p.; third edition, 1920, Pteridophyta: v. 1, 434 p.; 1923, Spermophyta, v. 2, 446 p. (Reprinted 1962, New York, Hafner Publishing.)

Sears, P. B., 1930a, A record of post-glacial climate in northern Ohio: Ohio Journal of Science, v. 30, p. 205–217.

Sears, P. B., 1930b, Common fossil pollen of the Erie basin: Botanical Gazette, v. 89, p. 95–106.

Sears, P. B., 1943, Pollen analysis circular number 1: Oberlin, Ohio, Oberlin College, Department of Botany, 4 p. (Mimeographed).

Sellards, E. H., 1902, On the fertile fronds of *Crossotheca* and *Myriotheca*, and on the spores of other Carboniferous ferns from Mazon Creek, Illinois: American Journal of Science, ser. 4, v. 14, p. 195–202.

Sellards, E. H., 1903, *Codonotheca*, a new type of spore-bearing organ from the Coal Measures: American Journal of Science, v. 16, p. 87–95.

Sellards, E. H., 1907, Notes on the spore-bearing organ *Codonotheca* and its relationship with the Cycadofilicales: New Phytologist, v. 6, p. 175–178.

Seward, A. C., 1898–1919, Fossil plants, 4 volumes: Cambridge, Cambridge

University Press, v. 1, 1898, 452 p.; v. 2, 1910, 624 p.; v. 3, 1917, 656 p.; v. 4, 1919, 543 p. (Reprinted 1963, New York and London, Hafner Publishing)

Shaffer, B. L., 1964, Stratigraphic and paleoecologic significance of plant microfossils in Permian evaporites of Kansas, *in* Cross, A. T., ed., Palynology in oil exploration—A symposium, San Francisco, California, March 26–27, 1962: Society of Economic Paleontologists and Mineralogists Special Publication 11, p. 97–115.

Simon, J. A., Kosanke, R. M., and Cobb, J. C., 1988, Origin and history of the Coal Division of the Geological Society of America: Urbana, Illinois, presented in the Poster Display of the Division for the GSA Centennial Celebration, 1988, 10 p. (Privately printed.)

Smith, A. H. V., and Butterworth, M. A., 1967, Miospores in the coal seams of Great Britain: The Palaeontological Association, London, Special Papers in Palaeontology 1, 324 p.

Somers, G. [Brush, G. Somers], 1952, A preliminary study of the fossil spore content of the Lower Jubilee seam of the Sydney coalfield, Nova Scotia: Halifax, Nova Scotia Research Foundation, 30 p.

Somers, G., 1953, A preliminary study of the spores from the Phalen Seam in the New Waterford District, Sydney Coalfield, Nova Scotia: Proceedings, Second Conference on the Origin and Constitution of Coal, 1952, Crystal Cliffs, Nova Scotia: Halifax, Nova Scotia Department of Mines, and Nova Scotia Research Foundation, p. 219–247.

Staplin, F. L., 1960, Upper Mississippian plant spores from the Golata Formation, Alberta, Canada: Palaeontographica, Abt. B, Band 107, p. 1–40, Tafn. 1–8.

Staplin, F. L., and Jansonius, J., 1964, Elucidation of some Palaeozoic densospores: Palaeontographica, Abt. B, Band 114, p. 95–117, Tafn. 18–21.

Thiessen, R., 1913, Microscopic study of coal, *in* White, D. W., and Thiessen, R., The origin of coal: U.S. Department of Interior, Bureau of Mines Bulletin 38, p. 187–390, pls. 18–54.

Thiessen, R., 1920, Structure in Paleozoic bituminous coals: U.S. Bureau of Mines Bulletin 117, 296 p., 160 pls.

Thiessen, R., and Staud, J. N., 1923, Correlation of coal beds in the Monongahela Formation of Ohio, Pennsylvania, and West Virginia: Coal Mining Investigations: Pittsburgh, Carnegie Institute of Technology Bulletin 9, 64 p.

Thiessen, R., and Wilson, F. E., 1924, Correlation of coal beds of the Allegheny Formation of western Pennsylvania and eastern Ohio: Coal Mining Investigations: Pittsburgh, Carnegie Institute of Technology Bulletin 10, 56 p.

Tidwell, W. D., 1968, Flora of the Manning Canyon Shale. Part I: A lowermost Pennsylvanian flora from the Manning Canyon Shale, Utah, and its stratigraphic significance: Provo, Utah, Brigham Young University Geology Studies, v. 14 (December), p. 3–66.

Todd, R. G., 1957, Spore analysis of the "Alvis" coal of western Missouri [M.S. thesis]: Columbia, University of Missouri, 141 p., 6 pls.

Traverse, A., 1960, P.S.B.S.A.—The oldest organization of paleobotanists in the world: Plant Science Bulletin, Botanical Society of America, v. 6, p. 2–4.

Traverse, A., 1974, Paleopalynology, 1947–1972: Missouri Botanical Garden Annals, v. 61, p. 203–236.

Traverse, A., 1988, Paleopalynology: Boston, Unwin Hyman, 600 p.

Traverse, A., and Ginsburg, R. N., 1966, Palynology of the surface sediments of Great Bahama Banks, as related to water movement and sedimentation: Marine Geology, v. 4, p. 417–459.

Tschudy, R. H., 1961, Palynomorphs as indicators of facies environments in Upper Cretaceous and Lower Tertiary strata, Colorado and Wyoming *in* Symposium on Late Cretaceous Rocks of Wyoming: Wyoming Geological Association Guidebook, 1961, 16th Annual Field Conference, Green River, Washakie, Wind River and Powder River basins, p. 53–59, 1 pl.

Tschudy, R. H., 1964, Palynology and time-stratigraphic determinations, *in* Cross, A. T., ed., Palynology in oil exploration—A symposium, San Francisco, California, March 26–27, 1962: Society of Economic Paleontologists and Mineralogists Special Publication 11, p. 18–28.

Tschudy, R. H., and Kosanke, R. M., 1966, Early Permian vesiculate pollen from Texas, U.S.A.: The Palaeobotanist, v. 15, p. 59–71.

Tschudy, R. H., and Scott, R. A., 1969, Aspects of palynology: New York, Wiley-Interscience, 510 p.

Upshaw, C. F., 1964, Palynological zonation of the Upper Cretaceous Frontier Formation near Dubois, Wyoming, *in* Cross, A. T., ed., Palynology in oil exploration—A symposium, San Francisco, California, March 26–27, 1962: Society of Economic Paleontologists and Mineralogists Special Publication 11, p. 153–168.

Upshaw, C. F., and Creath, W. B., 1965, Pennsylvanian miospores from a cave deposit in Devonian limestone, Callaway County, Missouri: Micropaleontology, v. 11, p. 431–448.

Virkki, C., 1937, On the occurrence of winged spores in the Lower Gondwana rocks of India and Australia: Indian Academy of Science Proceedings, v. 6, sec. B, p. 428–431.

Virkki, C., 1939, On the occurrence of similar spores in a Lower Gondwana glacial tillite from Australia and in Lower Gondwana shales in India: Indian Academy of Science Proceedings, v. 9, sec. B, p. 7–12.

Virkki, C., 1946, Spores from the Lower Gondwana of India and Australia: National Academy of Science of India Proceedings, 1945, v. 15, p. 93–176.

Voss, J., 1931, Preliminary report on the paleoecology of a Wisconsin and an Illinois bog: Illinois State Academy of Science Transactions, v. 24, p. 130–137.

Voss, J., 1933, Pleistocene forests of central Illinois: Botanical Gazette, v. 94, p. 808–814.

Voss, J., 1937, Comparative study of bogs on Carey and Tazewell drift in Illinois: Ecology, v. 18, p. 119–135.

Warman, J. C., 1952, Some paleobotanic and petrographic aspects of the Washington coal [M.S. thesis]: Morgantown, West Virginia University, 79 p.

White, D., and Thiessen, R., 1913, The origin of coal: Washington, D.C., U.S. Department of the Interior, Bureau of Mines Bulletin 38, 390 p., 54 pls.

Wicher, C. A., 1934a, Zur Mikrobotanik der Kohlen und ihrer Verwandten. Sporenformen der Flammkohle des Ruhrgebietes: Berlin, Preussische Geologische Landesanstalt, Institut Paläobotanik und Petrographie der Brennsteine, Arbeiten, Heft 4, p. 165–212.

Wicher, C. A., 1934b, Über Abortiverscheinungen bei fossilen Sporen und ihre phylogenetische Bedeutung: Berlin, Preussische Geologische Landesanstalt, Institut Paläobotanik und Petrographie der Brennsteine, Arbeiten Heft 5, p. 87–96.

Wilson, I. T., and Potzger, J. E., 1943, Pollen records from lakes in Anoka County, Minnesota: A study on methods of sampling: Ecology, v. 24, p. 282–292.

Wilson, L. R., 1932, Two Creeks forest bed, Manitowac County, Wisconsin: Wisconsin Academy of Science, Arts and Letters Transactions, v. 27, p. 31–46.

Wilson, L. R., 1943, Elater-bearing spores from the Pennsylvanian strata of Iowa: American Midland Naturalist, v. 30, p. 518–523.

Wilson, L. R., 1944, Spores and pollen as microfossils: Botanical Review, v. 10, p. 499–523.

Wilson, L. R., 1946, The correlation of sedimentary rocks by fossil spores and pollen: Journal of Sedimentary Petrology, v. 16, p. 119–120.

Wilson, L. R., 1953, The plant microfossils of the Joggins Section: A progress report: Proceedings, Second Conference on Origin and Constitution of Coal, Crystal Cliffs, Nova Scotia, June 1952: Halifax, Nova Scotia Department of Mines and Nova Scotia Research Foundation, p. 208–218.

Wilson, L. R., 1959, Genotype of *Densosporites* Berry, 1937: Oklahoma Geology Notes, v. 19, p. 47–50.

Wilson, L. R., 1962, Permian plant microfossils from the Flowerpot Formation, Greer County, Oklahoma: Oklahoma Geological Survey Circular 49, 50 p.

Wilson, L. R., 1964, Recycling, stratigraphic leakage, and faulty techniques in palynology: Grana Palynologica, v. 5, p. 425–435.

Wilson, L. R., and Brokaw, A. L., 1937, Plant microfossils of an Iowa coal deposit: Iowa Academy of Science Proceedings, v. 44, p. 127–130.

Wilson, L. R., and Coe, E. A., 1940, Descriptions of some unassigned plant microfossils from the Des Moines Series of Iowa: American Midland Naturalist, v. 23, p. 182–186.

Wilson, L. R., and Cross, A. T., 1939, Fossil plants of a Des Moines sandstone cave deposit near Robins, Linn County, Iowa: Iowa Academy of Science Proceedings, v. 46, p. 225–226.

Wilson, L. R., and Hoffmeister, W. S., 1956, Plant microfossils of the Croweburg coal: Oklahoma Geological Survey Circular 32, 57 p.

Wilson, L. R., and Hoffmeister, W. S., 1958, Plant microfossils in the Cabaniss coals of Oklahoma and Kansas: Oklahoma Geology Notes, v. 18, p. 27–30.

Wilson, L. R., and Kosanke, R. M., 1944, Seven new species of unassigned plant microfossils from the Des Moines Series of Iowa: Iowa Academy of Science Proceedings, v. 51, p. 329–333.

Wilson, L. R., and Venkatachala, B. S., 1963, A morphologic study and emendation of *Vesicaspora* Schemel, 1951: Oklahoma Geology Notes, v. 23, p. 142–149.

Wilson, L. R., and Webster, R. M., 1946, Fossil spores and pollen, Bender No. 1 well, Montgomery County, Texas: Tulsa, Oklahoma, Carter Oil Company, Research Division, 150 p. (Proprietary but some copies distributed to American Museum of Natural History and several independent researchers.)

Wilson, L. R., and Webster, R. M., 1949, Fossil spores and pollen, Bender No. 1 and Griffin No. 5 wells in Montgomery County, Texas: Tulsa, Oklahoma, Carter Oil Company, Research Division, v. 1, 103 p.; v. 2, 294 p.; v. 3, 152 p.; v. 4, 192 p. (Proprietary but some copies distributed to American Museum of Natural History and several independent researchers.)

Winslow, M. R., 1959, Upper Mississippian and Pennsylvanian megaspores and other plant microfossils from Illinois: Illinois State Geological Survey Bulletin 86, 135 p.

Winslow, M. R., 1962, Plant spores and other microfossils from the Upper Devonian and Lower Mississippian rocks of Ohio: U.S. Geological Survey Professional Paper 364, p. 1–93.

Witham, H. T. M., 1833, The internal structure of fossil vegetables found in the Carboniferous and oolitic deposits of Great Britain: Edinburgh, Adam and Charles Black; London, Longman, Rees, Orme, Brown, Green, and Longman, 84 p.

Wodehouse, R. P., 1932, Tertiary pollen. I: Pollen of the living representatives of the Green River flora: Torrey Botanical Club Bulletin, v. 59, p. 313–340.

Wodehouse, R. P., 1933, Tertiary pollen. II: The oil shales of the Eocene Green River Formation: Torrey Botanical Club Bulletin, v. 60, p. 479–524.

Wodehouse, R. P., 1935, Pollen grains, their structure, identification, and significance in science and medicine: New York, McGraw-Hill Book Company, 574 p. (Reprinted, New York, Hafner Publishing, 1959.)

Wood, J. M., 1963, The Stanley Cemetery flora (Early Pennsylvanian) of Greene County, Indiana: Indiana Department of Conservation Geological Survey Bulletin 29, 73 p., 12 pls.

Woods, R. D., 1955, Spores and pollen—A new stratigraphic tool for the oil industry: Micropaleontology, v. 1, p. 368–375.

Zangerl, R., and Richardson, E. S., 1963, The paleoecological history of two Pennsylvanian black shales: Fieldiana, Geology, v. 4, 352 p.

Zerndt, J., 1930, *Triletes giganteus* n. sp., eine riesige Megaspore aus dem Karbon: Bulletin de l'Académie Polonaise des Sciences et des Lettres, sér. B, p. 71–79.

Zerndt, J., 1932, Megasporen aus dem Zwickauer und Lugau-Ölsnitzer Karbon: Jahrbuch für das Berg-und Huttenwesen in Sachsen, Jahrgang 1932, p. A9–A16.

Zerndt, J., 1934, Les mégaspores du bassin houiller polonais, Première Partie: Académie Polonaise des Sciences et des Lettres, 1934, Comité des Publications silésiennes, Travaux Géologiques, no. 1, 56 p.

Zerndt, J., 1937a, Megasporen aus dem Westfal und Stefan in Böhmen: Bulletin de l'Académie Polonaise des Sciences et des Lettres, sér. A, p. 583–599.

Zerndt, J., 1937b, Les mégaspores du bassin houiller polonais, Deuxième Partie: Académie Polonaise des Sciences et des Lettres, 1937, Comité des Publications Silésiennes, Travaux Géologiques, no. 3, 78 p.

Zerndt, J., 1940, Megasporen des Saarkarbons: Palaeontographica, Abt. B, Band 84, p. 133–150.

Manuscript Accepted by the Society December 28, 1994

Printed in U.S.A.

## ACKNOWLEDGMENTS ADDED IN PROOF: THE STEPHANIAN OF NORTH AMERICA (p. 293–314)

Elsie Darrah Morey is thanked for the use of correspondence from the W. C. Darrah Collection and for the photographs of W. C. Darrah, *Lescuropteris morrei* (Lesq.), and the "Hall of Discovery" photograph from the W. C. Darrah Collection. Jane Eggleston and Chris Wnuk, U.S. Geological Survey, graciously provided the photograph of *Sphenophyllum oblongifolium* from the Anthracite region of Pennsylvania. W. H. Gillespie generously provided a photograph of *Danaeites emersoni.* A. T. Cross, Michigan State University, graciously provided the photographs of W. J. Jongmans, H. Bode, W. Gothan, and P. Corsin. We are grateful to J. P. Laveine, C. Wnuk, and E. L. Zodrow for their very helpful reviews and suggestions.

# *Name Index*

[**Bold** page numbers indicate photographs]

## C



## D

## E



## F

## G



## H

## M



## N

## T



## U



## V



## W

## Y



## Z

# Taxa Index

[**Bold** page numbers indicate photographs]

## M



## N

**T**



**U**



**V**



**W**



**X**

**Z**